Social-Behavioral Modeling for Complex Systems

Stevens Institute Series on Complex Systems and Enterprises

Series Editor: William B. Rouse

Social-Behavioral Modeling for Complex Systems

Edited by
Paul K. Davis

Angela O'Mahony

Jonathan Pfautz

Registered Office
John Wiley & Sons, Inc., 111 River Street, Hoboken, NJ 07030, USA

Editorial Office
111 River Street, Hoboken, NJ 07030, USA

For details of our global editorial offices, customer services, and more information about Wiley products visit us at www.wiley.com.

Wiley also publishes its books in a variety of electronic formats and by print-on-demand. Some content that appears in standard print versions of this book may not be available in other formats.

Library of Congress Cataloging-in-Publication Data

Names: Davis, Paul K., 1943- editor. | O'Mahony, Angela, editor. | Pfautz,
    Jonathan, editor.
Title: Social-behavioral modeling for complex systems / edited by Dr. Paul
    K. Davis, Dr. Angela O'Mahony, Dr. Jonathan Pfautz.
Description: Hoboken, NJ, USA : Wiley, [2019] | Includes bibliographical
    references and index. |
Identifiers: LCCN 2018046976 (print) | LCCN 2018047389 (ebook) | ISBN
    9781119484981 (Adobe PDF) | ISBN 9781119484974 (ePub) | ISBN 9781119484967
    (hardcover)
Subjects: LCSH: Social psychology–Data processing. | Collective
    behavior–Simulation methods. | System analysis.
Classification: LCC HA29 (ebook) | LCC HA29 .S6736 2019 (print) | DDC
    302/.011–dc23
LC record available at https://lccn.loc.gov/2018046976

Cover Design: Wiley
Cover Image: © agsandrew/Shutterstock

Set in 10/12pt WarnockPro by SPi Global, Chennai, India

Printed in the United States of America

V10008699_031119

Contents

Foreword

Trying to understand human behavior has probably been a human passion since the first cavewoman tried to figure out why her mate could never remember to wipe his feet before coming inside. We're still at it, working with ever more sophisticated approaches and for increasingly important outcomes.

My own experience in this area started out when I was working at NASA to help design an autopilot for the Space Shuttle vehicle. It worked great when there was no *astronaut in the loop*, but when they started wrangling the controls themselves, the astronauts were not exactly enamored with my design and the way it behaved. Dismay would be a word that might apply here, but I went back to school to figure out why and soon became enmeshed in trying to understand human self-motion perception and control, by bringing together established theory, controlled experimentation, and computational modeling. It was an eye-opener on how complicated even the simplest of human behaviors could be, as well as the beginning of a long foray into developing and using computational models in this arena.

Of course, as I later discovered, there's a long history of this, perhaps going as far back as the 1880s with Ernst Mach and his pioneering work in visual and vestibular psychophysics and certainly at least the 1940s to Norbert Weiner with his introduction of cybernetics and the mathematical modeling of both humans and machines. Since then, many other disciplines have contributed and elaborated on this idea, from the basic sciences of neurophysiology and cognitive science to the more applied efforts in human systems engineering and robotics. Moving into the domain of computational representation has forced many of us to sharpen our approach to describing behaviors, formalizing our theories, and validating them against *real* data. As one of my mentors once told me (and I paraphrase a bit here): "The rubber hits the road when you start hacking code."

This movement by the research community has been documented in a number of efforts. In 1998, the National Research Council (NRC) published a review of potential models that might be usefully embedded in existing military simulations to provide greater realism by including the *human element* (National Research Council 1998). The report concluded that there was no single framework or architecture that could *meet all the simulation needs of the services*, but it did provide an extensive review of computational behavioral models, in-depth discussions of different functional areas (e.g. attention, memory, learning, etc.), and considerations for small unit representation (i.e. groups of individuals). A follow-on NRC study in 2008 provided a somewhat broader review, covering models of not only individuals but also organizations and societies (National Research Council 2008). This study also discussed different categories of models – both formal and informal – and common challenges across the community (e.g. interoperability, inconsistent frameworks, verification, etc.). And, like almost all NRC reports of this ilk, there were a number of recommendations proposed, in this case, covering areas from basic theory development to data collection methods and tools for model building.

Although many insights from these and other studies remain relevant, much has happened in the last decade, in terms of new basic research results, new applications afforded by the acceleration of technology (particularly in sensing, networking, computation, and memory), and, not least, a resurgence of a general interest in natural and artificial cognition, with the recent reemergence of artificial intelligence and machine learning. For example, on the basic research side, a revolution in neuroimaging methods is linking the underpinning of human thinking – across individual and societal levels. On the applied side, masses of data on human behavior are now being collected to describe and predict activity in a huge variety of applications spanning everyday consumer devices, socio-commercial networks, and population monitoring systems installed by local and national governments, to name a few. Online populations can now support *crowd-sourced* and *A/B* experiments that drive how corporations interact with their customers and governments with their citizens.

A reexamination of the issues addressed by the earlier studies is clearly called for, in light of what's happened over the last decade. This volume does just that and is a particularly welcome addition to the research community. It is structured to address issues of science, modeling, and relationships among theory, modeling, empirical research, and computational social science. It candidly emphasizes past shortcomings in these relationships and current progress that's been made in improving those relationships. One can sense a good deal of excitement among the contributors – both established researchers in their chosen fields and fresh PhDs – because so much is happening on so many fronts, including theory development, data collection, and computational methods of inquiry. It's also a delight to see chapters bringing to bear new insights from the study of nonhuman social systems, neuropsychology, psychology, and anthropology, among other disciplines. And, having spent much of my career concerned about real-world problems needing insights from the behavioral sciences, I was pleased to see several chapters addressing the frontier problem of how to develop and use social-behavioral modeling to assist human decision-making regarding profound social issues, some complicated by equally profound issues of privacy and ethics.

This book represents a carefully curated set of contributions that aim to inspire the current and next generation of researchers – and to encourage how the act of challenging current conceptual boundaries is needed to advance science. Social-behavioral modeling will continue to be beset by uncertainties because the social-behavioral phenomena occur in complex adaptive systems into which we have imperfect and sometimes contradictory insight. Nonetheless, such modeling – if undertaken appropriately – can help humans who must plan, operate, and adapt in their complex worlds. One of the editors' themes is most welcome: complexity need not be paralyzing, especially if we take a multidisciplinary use-centric approach to working on real-world problems.

All in all, this volume is a welcome contribution that will be valuable to diverse audiences in schoolhouses, research laboratories, and the workplace. It is a collection, not a textbook or single-author monograph, and it conveys an excellent sense of the current state of the art and the exciting opportunities that are now being exploited. The editors deserve credit for bringing about, organizing, and personally contributing heavily to this volume that merits a prominent place in the libraries of human behavior researchers, as well as those interested in helping solve some of the larger socio-technical issues facing us. I heartily recommend this volume to all of you.

22 July 2018

Greg L. Zacharias
Weston, Massachusetts

References

National Research Council (1998). *Modeling Human and Organizational Behavior: Application to Military Simulations.* Washington, DC: The National Academies Press https://doi.org/10.17226/6173.

National Research Council (2008). *Behavioral Modeling and Simulation: From Individuals to Societies.* Washington, DC: The National Academies Press https://doi.org/10.17226/12169.

List of Contributors

Tarek Abdelzaher
Computer Science Department
University of Illinois at
Urbana–Champaign
Champaign
IL 61801
USA

Scott Appling
Georgia Tech Research Institute
Atlanta
GA 30318
USA

Rebecca Balebako
RAND Corporation
Santa Monica
CA 90401
USA

Christopher L. Barrett
Biocomplexity Institute and Initiative
University of Virginia
Charlottesville
VA 22904
USA

Danielle S. Bassett
Department of Bioengineering
University of Pennsylvania
Philadelphia
PA 19104
USA

and

Department of Electrical & Systems
Engineering
University of Pennsylvania
Philadelphia
PA 19104
USA

and

Department of Neurology
University of Pennsylvania
Philadelphia
PA 19104
USA

and

Department of Physics & Astronomy
University of Pennsylvania
Philadelphia
PA 19104
USA

Rahmatollah Beheshti
School of Public Health
Johns Hopkins University
Baltimore
MD 21218
USA

Leslie M. Blaha
Visual Analytics
Pacific Northwest National
Laboratory
Richland
WA 99354
USA

David Blumstein
Charles River Analytics
Cambridge
MA 02138
USA

Bethany Bracken
Charles River Analytics
Cambridge
MA 02138
USA

Matthew E. Brashears
Department of Sociology
University of South Carolina
Columbia
SC 29208
USA

Erica Briscoe
Georgia Tech Research Institute
Atlanta
GA 30318
USA

Kathleen M. Carley
Institute of Software Research
School of Computer Science and
Engineering and Public Policy
Carnegie Institute of Technology
Carnegie Mellon University
Pittsburgh
PA 15213
USA

Steven R. Corman
Hugh Downs School of Human
Communication
Arizona State University
Tempe, AZ
USA

Gene Cowherd
Department of Anthropology
University of South Florida
Tampa
FL 33620
USA

Paul K. Davis
Pardee RAND Graduate School
Santa Monica
CA 90407
USA
CA 90401
USA

Andrea de Silva
Department of Epidemiology and
Preventive Medicine
Alfred Hospital
Monash University
Clayton
VIC 3800
Australia

Laura Epifanovskaya
Sandia National Laboratories
California
Livermore
CA 94551
USA

Joshua M. Epstein
Department of Epidemiology
Agent-Based Modeling Laboratory
New York University
New York
NY 10003
USA

Leonard Eusebi
Charles River Analytics
Cambridge
MA 02138
USA

Emily B. Falk
Annenberg School for
Communication
University of Pennsylvania
Philadelphia
PA 19104
USA

and

Department of Psychology
University of Pennsylvania
Philadelphia
PA 19104
USA

and

Marketing Department
Wharton School
University of Pennsylvania
Philadelphia
PA 19104
USA

Michael Gabbay
Applied Physics Laboratory
University of Washington
Seattle
WA 98105
USA

Ivan Garibay
Department of Industrial
Engineering and Management
Systems
College of Engineering and
Computer Science
University of Central Florida
Orlando
FL 32816
USA

Traci K. Gillig
Annenberg School for
Communication and Journalism
University of Southern California
Los Angeles
CA 90007
USA

Prasanna Giridhar
Computer Science Department
University of Illinois at
Urbana–Champaign
Champaign
IL 61801
USA

Christopher G. Glazner
Modeling, Simulation,
Experimentation, and Analytics
The MITRE Corporation
McLean
VA 22103
USA

Emily Saldanha
Data Sciences and Analytics Group
National Security Directorate
Pacific Northwest National
Laboratory
Richland
WA 99354
USA

Mark Greaves
Data Sciences and Analytics Group
National Security Directorate
Pacific Northwest National
Laboratory
Richland
WA 99354
USA

Marco Gribaudo
Department of Computer Science
Polytechnic University of Milan
Milan
Italy

Sean Guarino
Charles River Analytics
Cambridge
MA 02138
USA

Chathika Gunaratne
Institute for Simulation and Training
University of Central Florida
Orlando
FL 32816
USA

Mirsad Hadzikadic
Department of Software and
Information Systems
Data Science Initiative
University of North Carolina
Charlotte
NC 28223
USA

Nathan Hodas
Data Sciences and Analytics Group
National Security Directorate
Pacific Northwest National
Laboratory
Richland
WA 99354
USA

Mauro Iacono
Department of Mathematics and
Physics
Università degli Studi della
Campania "Luigi Vanvitelli"
Caserta
Italy

Michael Jenkins
Charles River Analytics
Cambridge
MA 02138
USA

David C. Jeong
Annenberg School for
Communication and Journalism
University of Southern California
Los Angeles
CA 90007
USA

and

CESAR Lab (Cognitive Embodied
Social Agents Research)
College of Computer and
Information Science
Northeastern University
Boston
MA 02115
USA

Mubbasir Kapadia
Department of Computer Science
Rutgers University
New Brunswick
NJ
USA

Melvin Konner
Department of Anthropology and
Neuroscience and Behavioral Biology
Emory University
Atlanta
GA 30322
USA

Bart Kosko
Department of Electrical Engineering
and School of Law
University of Southern California
Los Angeles
CA 90007
USA

Kiran Lakkaraju
Sandia National Laboratories
Albuquerque
NM 87185
USA

Daniel Lende
Department of Anthropology
University of South Florida
Tampa
FL 33620
USA

Josh Letchford
Sandia National Laboratories
California
Livermore
CA 94551
USA

Alexander H. Levis
Department of Electrical and
Computer Engineering
George Mason University
Fairfax
VA 22030
USA

Corey Lofdahl
Systems & Technology Research
Woburn
MA 01801
USA

Christian Madsbjerg
3ReD Associates
New York
NY 10004
USA

Achla Marathe
Biocomplexity Institute and Initiative
University of Virginia
Charlottesville
VA 22904
USA

Madhav V. Marathe
Biocomplexity Institute and Initiative
University of Virginia
Charlottesville
VA 22904
USA

Luke J. Matthews
Behavioral and Policy Sciences
RAND Corporation
Boston
MA 02116
USA

and

Pardee RAND Graduate School
Santa Monica
CA 90401
USA

Rod McClure
Faculty of Medicine and Health
School of Rural Health
University of New England
Armidale
NSW 2351
Australia

Laura McNamara
Sandia National Laboratories
Albuquerque
NM 87123
USA

Lynn C. Miller
Department of Communication and
Psychology
University of Southern California
Los Angeles
CA 90007
USA

and

Annenberg School for
Communication and Journalism
University of Southern California
Los Angeles
CA
USA

Kent C. Myers
Net Assessments
Office of the Director of National
Intelligence
Washington
DC 20511
USA

Benjamin Nyblade
Empirical Research Group
University of California Los Angeles
School of Law
Los Angeles
CA 90095
USA

Angela O'Mahony
Pardee RAND Graduate School
Santa Monica
CA 90407
USA
CA 90401
USA

Mark G. Orr
Biocomplexity Institute & Initiative
University of Virginia
Charlottesville
VA 22904
USA

Osonde Osoba
RAND Corporation and Pardee
RAND Graduate School
Santa Monica
CA 90401
USA

Christopher Paul
RAND Corporation
Pittsburgh
PA 15213
USA

Theodore P. Pavlic
School of Computing, Informatics,
and Decision Systems Engineering
and the School of Sustainability
Arizona State University
Tempe
AZ 85287
USA

Glenn Pierce
School of Criminology and Criminal
Justice
Northeastern University
Boston
MA 02115
USA

Jonathan Pfautz
Information Innovation Office (I20)
Defense Advanced Research Projects
Agency
Arlington
VA 22203
USA

William Rand
Department of Marketing
Poole College of Management
North Carolina State University
Raleigh
NC 27695
USA

Stephen J. Read
Department of Psychology
University of Southern California
Los Angeles
CA 90007
USA

Scott Neal Reilly
Charles River Analytics
Cambridge
MA 02138
USA

Jason Reinhardt
Sandia National Laboratories
California
Livermore
CA 94551
USA

William B. Rouse
School of Systems and Enterprises
Stevens Institute of Technology
Center for Complex Systems and
Enterprises
Hoboken
NJ 07030
USA

Scott W. Ruston
Global Security Initiative
Arizona State University
Tempe, AZ
USA

Arun V. Sathanur
Physical and Computational Sciences
Directorate
Pacific Northwest National
Laboratory
Seattle
WA 98109
USA

Davide Schaumann
Department of Computer Science
Rutgers University
New Brunswick
NJ
USA

Steve Scheinert
Department of Industrial
Engineering and Management
Systems
University of Central Florida
Orlando
FL 32816
USA

Katharine Sieck
Business Intelligence and Market
Analysis
RAND Corporation and Pardee
RAND Graduate School
Santa Monica
CA 90401
USA

Amy Sliva
Charles River Analytics
Cambridge
MA 02138
USA

Mallory Stites
Sandia National Laboratories
Albuquerque
NM 87185
USA

Gita Sukthankar
Department of Computer Science
University of Central Florida
Orlando
FL 32816
USA

Samarth Swarup
Biocomplexity Institute and Initiative
University of Virginia
Charlottesville
VA 22904
USA

Jason Thompson
Transport, Health and Urban Design
Research Hub
Melbourne School of Design
University of Melbourne
Parkville
VIC 3010
Australia

Andreas Tolk
Modeling, Simulation,
Experimentation, and Analytics
The MITRE Corporation
Hampton
VA 23666
USA

Steven H. Tompson
Human Sciences Campaign
U.S. Army Research Laboratory
Adelphi
MD 20783
USA

and

Department of Bioengineering
University of Pennsylvania
Philadelphia
PA 19104
USA

Hanghang Tong
School of Computing, Informatics
and Decision Systems Engineering
(CIDSE)
Arizona State University
Los Angeles, CA
USA

Raffaele Vardavas
RAND Corporation and Pardee
RAND Graduate School
Santa Monica
CA 90401
USA

Jean M. Vettel
Human Sciences Campaign
U.S. Army Research Laboratory
Adelphi
MD 20783
USA

and

Department of Bioengineering
University of Pennsylvania
Philadelphia
PA 19104
USA

and

Department of Psychological and
Brain Sciences
University of California
Santa Barbara
93106 USA

Svitlana Volkova
Data Sciences and Analytics Group
National Security Directorate
Pacific Northwest National
Laboratory
Richland
WA 99354
USA

Liyuan Wang
Annenberg School for
Communication and Journalism
University of Southern California
Los Angeles
CA 90007
USA

Jon Whetzel
Sandia National Laboratories
Albuquerque
NM 87185
USA

Joseph Whitmeyer
Department of Sociology
University of North Carolina
Charlotte
NC 28223
USA

Levent Yilmaz
Department of Computer Science
Auburn University
Auburn
AL 36849
USA

Niloofar Yousefi
Department of Industrial
Engineering and Management
Systems
College of Engineering and
Computer Science
University of Central Florida
Orlando
FL 32816
USA

About the Editors

Paul K. Davis is a senior principal researcher at RAND and a professor of policy analysis in the Pardee RAND Graduate School. He holds a BS in chemistry from the University of Michigan and a PhD in chemical physics from the Massachusetts Institute of Technology. After several years at the Institute for Defense Analyses focused primarily on the physics, chemistry, and interpretation of rocket observables, he joined the US government to work on strategic nuclear defense programs and related arms control. He became a senior executive in the Office of the Secretary of Defense and led studies relating to both regional and global military strategy and to program development. Subsequently, Dr. Davis joined the RAND Corporation. His research has involved strategic planning; resource allocation and decision aiding; advanced modeling, simulation, gaming, and analysis under uncertainty; deterrence theory; heterogeneous information fusion; and integrative work using social sciences to inform national strategies in defense and social policy. Dr. Davis teaches policy analysis and modeling of complex problems. He has served on numerous national panels and journals' editorial boards.

Angela O'Mahony is a senior political scientist at the RAND Corporation and a professor at the Pardee RAND Graduate School. Her research has focused on how international political, economic, and military ties affect policy-making. Some of the topics she has examined are the effectiveness of US security cooperation and defense posture, the implications of international political and economic scrutiny on governments' decision-making, the causes and consequences of transnational political behavior, public support for terrorism, and the role of social media in public policy analysis. From 2003 to 2011, O'Mahony was an assistant professor at the University of British Columbia. She received her PhD in political science from the University of California, San Diego.

xlii | *About the Editors*

Jonathan Pfautz is a program manager at DARPA and previously led cross-disciplinary research and guided system development and deployment at Charles River Analytics. His efforts spanned research in social science, neuroscience, cognitive science, human factors engineering, and applied artificial intelligence. Dr. Pfautz holds a doctor of philosophy degree in computer science from the University of Cambridge. He also holds degrees from the Massachusetts Institute of Technology: a master of engineering degree in computer science and electrical engineering, a bachelor of science degree in brain and cognitive sciences, and a bachelor of science degree in computer science and engineering. Dr. Pfautz has published more than 60 peer-reviewed conference and journal publications and 5 book chapters. He holds five patents.

About the Companion Website

This book is accompanied by a companion website:

www.wiley.com/go/Davis_Social-Behavioralmodeling

The website includes:

- Supplementary material to the eighth chapter.

Part I

Introduction and Agenda

1

Understanding and Improving the Human Condition: A Vision of the Future for Social-Behavioral Modeling

Jonathan Pfautz[1], Paul K. Davis[2], and Angela O'Mahony[2]

[1] Information Innovation Office (I2O), Defense Advanced Research Projects Agency, Arlington, VA 22203-2114, USA
[2] RAND Corporation and Pardee RAND Graduate School, Santa Monica, CA 90407-2138, USA

Technology is transforming the human condition at an ever-increasing pace. New technologies emerge and dramatically change our daily lives in months rather than years. Yet, key aspects of the human condition – our consciousness, personalities and emotions, beliefs and attitudes, perceptions, decisions and behaviors, and social relationships – have long resisted description in terms of scientific, falsifiable *laws* like those found in the natural sciences. Past advances in our knowledge of the human condition have had valuable impacts,[1] but much more is possible. New technologies are providing extraordinary opportunity for gaining deeper understanding and, significantly, for using that understanding to help realize the immense positive potential of the humankind.

In the information age our understanding of the human condition is deepening with new ways to observe, experiment, and understand behavior. These range from, say, identifying financial and spatiotemporal data that correlate with individual well-being to drawing on the narratives of social media and other communications to infer population-wide beliefs, norms, and biases. An unprecedented volume of data is available, an astonishing proportion of which describes human activity and can help us explore the factors that drive behavior. Statistical correlations from such data are

---

1 Examples include epidemiological modeling and an understanding of collective behavior that has helped make it possible to blunt, contain, and then defeat epidemic outbreaks; the destigmatizing of mental health problems; the use of game theory in developing nuclear deterrence theory; and applications of *nudge theory* in behavioral economics, as in instituting *opt-in-as-a-default* policies resulting in more workers availing themselves of economic benefits in pension plans.

Social-Behavioral Modeling for Complex Systems, First Edition.
Edited by Paul K. Davis, Angela O'Mahony, and Jonathan Pfautz.
© 2019 John Wiley & Sons, Inc. Published 2019 by John Wiley & Sons, Inc.
Companion website: www.wiley.com/go/Davis_Social-Behavioralmodeling

already helping to inform our understanding of human behavior. New experimentation platforms have the potential to support both theory-informed and data-driven analysis to discover and test the mechanisms that underlie human behavior. For example, millions of users of a social website or millions of players of online games can be exposed to different carefully controlled situations – within seconds – across regional and cultural boundaries. Such technologies enable heretofore impossible forms and scales of experimentation. At the same time, these new capabilities raise important issues of how to perform such experimentation, how to correctly interpret the results, and, critically, how to ensure the highest ethical standards.

Such advances mean that theory development and testing in the social-behavioral sciences are poised for revolutionary changes. Behavioral theories, whether based on observation, in situ experiments, or laboratory experiments can now be revisited with new technology-enabled instruments. Applying these new instruments requires confronting issues of reproducibility, generalizability, and falsifiability. Doing so will help catalyze new standards for scientific meaning in the social-behavioral sciences. The massive scale of some such studies will require complex experimental designs, but these could also enable substantially automated methods that can address many problems of reproducibility and generalizability.

Similarly, representation of knowledge about the human condition is poised for revolution. Using mathematics and computation to formally describe human behavior is not new (Luce et al. 1963), but new and large-scale data collection methods require us to reconsider how to best represent, verify, and validate knowledge in the social and behavioral sciences. New approaches are needed to capture the complex, multiresolution, and multifaceted nature of the human condition as studied with different observational and experimental instruments. Capturing this knowledge will require new thinking about mathematical and computational formalisms and methods, as well as attention to such engineering hurdles as achieving computational tractability.

Advances in knowledge representation will also motivate advances in social and behavioral science. The need for accuracy and precision in describing the current understanding of the human condition will require models with structured descriptions of data sources, data interpretations, and related assumptions. These will support calibration, testing, and integration and, critically, identifying gaps in current theory and instruments. Mathematical, computational, and structured qualitative can models provide a comprehensive epistemology for the social and behavioral sciences – describing not only what is known but also the certainty and generalizability of that knowledge. This will allow comparing different and even conflicting sources of knowledge and resulting theories, as well as identifying future research needs.

The combination of computational models with new technology-enabled instruments for studying human behavior should result in a tightly coupled and partly automated ecosystem that spans data collection, data analysis, theory development, experimentation, model instantiation, and model validation. Such an ecosystem, if constructed to maximize the soundness of the science, would radically transform how we pursue our understanding of the human condition.

Revolutions in the scientific process of creating and encoding knowledge about human behavior will allow applications that aid human decision-making. Computational models provide the means to readily apply (and democratize nonexpert access to) knowledge of human behavior. However, access to an ever-expanding body of knowledge about the human condition must be appropriately managed, especially as techniques for reflecting and combining the inherent uncertainties in the growing knowledge base are developed. Also, we must understand how to use these increasingly accurate models and how to quantify and share information transparently, including information about uncertainty, so as actually to assist human decisionmaking rather than increase confusion. Accurately representing the growing bounds of our knowledge about the human condition is essential for ensuring ethical application of the knowledge and maximizing its benefit for society (Muller 2018).

Challenges

We are at the beginning of an era in which sound computational models of human behavior and its causes can be constructed. Such models have vast potential to positively affect the human condition. Yet, despite this heady promise, a great deal of creativity and innovation will be needed to surmount the considerable challenges. Even describing these challenges remains a subject for scholarly debate; therefore, the list of challenges below should be considered representative rather than comprehensive (see also Chapter 2, Davis and O'Mahony 2018).

Challenge One: The Complexity of Human Issues

First and foremost, it is for good reasons that the social and behavioral sciences have not progressed in the same way as the natural sciences. The human condition is inherently complex with overlapping multiresolution features across multiple dimensions (e.g. from short to long time frames, from individual behavior to group activities and governance, and from individual neuronal physiology to brain-region activation, psychophysical action, cognitive task performance, and intelligence and consciousness). Further, the factors driving behavior change at different rates as a function of the physical world, technology, and social interactions, resulting in emergent, population-wide change. Such complexities contribute to what have been called *wicked problems* (Churchman 1967; Rittel and Noble 1989). Recognizing the implications of these complexities in studying the human condition is a critical first step.

Challenge Two: Fragmentation

The social and behavioral sciences are fragmented within and across such constituent disciplines as psychology, sociology, political science, and economics. Each discipline has made many contributions to our knowledge and methods, but few attempts have been made to integrate that knowledge. Fragmentation occurs as the result of (i) studying separately different aspects of the human condition (e.g. trust in information, mental health, community formation), (ii) studying at different levels of detail (e.g. psychomotor systems, individual cognition, economic systems, political systems), (iii) using diverse methods (e.g. qualitative observation, structured interviewing, surveys, controlled laboratory experiments, in situ experiments, analysis of existing data sets, natural experiments, computational simulation), and (iv) applying differing concepts of and standards for meaningful *scientific rigor* as evidenced by variations of acceptance criteria across peer-reviewed publications (e.g. acceptance criteria for studies on just-noticeable differences in psychomotor responses necessarily differ from those for studies on delayed gratification in infants). Some prominent researchers have argued for standards within their disciplines (King et al. 1994) but different standards are clearly needed for different conceptual and empirical approaches (Brady and Collier 2010).

A fifth important manifestation of fragmentation is the awkward and overly narrow relationship between theoretical and empirical inquiry (Davis et al. 2018) (see also Chapter 2). In the domain of social-behavioral work, the theory and models are overly narrow – focusing on some particular variables and ignoring others. They are also not adequately related to empirical data, in part because such data has been hard to come by. Most data analysis is insufficiently informed by suitable theory: empirical comparisons are often made to particular very narrow theories, but not to theories that *put the pieces together*. Relationships among theory, empirical, and computational work are not yet well drawn. Let us comment briefly on some of these matters.

Empirical Observation

Observational approaches view behavior in naturalistic and uncontrolled settings (i.e. with no managed interventions). They include traditional trained-observer approaches (e.g. manual coding of behaviors according to some schema, ethnographic observation) as well as newer approaches applying statistical and other methods to the masses of data on human activity now available. These approaches identify factors meriting additional experimental study and confounding factors to be managed in such experiments. One complicating issue is that analytic results can vary significantly not just across disciplines but also from study to study as uncontrolled factors influence results. Technology continues to change the realm of the possible in such work, but the few attempts to create standard practices have quickly fallen behind the state of the technologically possible.

Another issue is that observational data are necessarily interpreted by the researcher. That is, the meaning of the observed behavior is inferred, often without standards on how such meaning is derived (e.g. the notion of *economic health* could be based on the price of electricity, the number of transactions at restaurants, or stock market fluctuations). Another issue is that the opportunity to study a phenomenon may be transient and impossible to re-create because of the real-world circumstances that led to the phenomenon or because of limitations in access to reliable data (e.g. regulatory limits on access to financial data). We must recognize, then, that the ability to derive and substantiate causal factors from observation alone is likely to be highly limited, although more recent work on finding *natural experiments* (also called quasi-experiments) is sometimes quite valuable. Observational approaches rely on the space of what has happened or is happening and do not permit creating new situations. Analytic results from such approaches are ripe for misinterpretation by nonexperts (and, sometimes, experts) who generalize outcomes without understanding the limits of the observations and the quality of inferences made from those observations. Unfortunately, the social and behavioral sciences also lack consistent standards for quantifying the certainty around such analytic results (e.g. while data scientists may report standard deviations, they may fail to report the span of the underlying data and/or noise-correction methods used).

Empirical Experiments

Experimental approaches hold conditions constant except for those variables that are systematically varied (sometimes referred to as *managing an intervention*). Such experimentation is a pillar of traditional science. Experimental approaches lie along a spectrum of controlled to semi-controlled – that is, the degree to which potentially confounding factors can be controlled, independent variables manipulated, and dependent variables measured. For example, a typical laboratory experiment will differ in these dimensions from a randomized controlled trial conducted in situ. Experiments necessarily introduce some artificiality (e.g. participants in a study may change their behaviors simply because they are aware of being in a study). Managing this artificiality is difficult and relies on the researcher's expertise and his or her discipline's accepted practices.

Artificiality in terms of the participant population is a related problem. The social and behavioral science community has recognized, for example, a broad bias toward Western, educated populations from industrialized, rich, developed countries (Henrich et al. 2010). A further difficulty in experimental approaches is that even a carefully designed experiment may fail to accommodate some unknown confounding variable. Worse, should such a confounding factor be found after the original experiment was conducted, it is unlikely that that additional data on that factor was collected in the original experiment, meaning that the overall validity of that experiment may be dramatically reduced. The discovery of new variables therefore has the ability to undermine the value of past empirical work. This leads to the problem of reproducibility, such as the degree to which empirical studies adequately document all aspects of an experiment so that they may be independently tested and potentially unknown confounding factors can be exposed. These problems have been well documented in a number of recent studies (Nosek and Open Science Forum 2015). As with observational approaches, experimental data is also subject to a degree of interpretation, although standards for statistical analysis are generally clear and accepted. Regardless, the potential for unknown confounds means that while statistical confidence intervals describe one form of uncertainty, other forms are difficult to quantify.

Generative Simulation

As discussed above and in Chapter 2, we see computational social science as one of the three pillars of modern science (along with theory and empiricism). One crucial aspect of computational social science is its generative ability. Generative simulation supports study of (i) heretofore unencountered and/or hard-to-anticipate situations, (ii) situations that are hard to study or measure because of access restrictions, and (iii) situations for which real-world experiments could be considered overly artificial (e.g. a population's response to a deadly pandemic or disaster) (Waldrop 2018). Significantly, the data from such generative models can be studied by both observation and experimentation. That is, the output of simulated data can be observed in an uncontrolled way or after systematic changes to some variables.

Such simulations range in complexity. The most complex simulations generate masses of data and require advanced analytic capabilities to explain the emergent behavior of the simulated system. Simpler simulations (e.g. some game theory-based simulations) also have the potential to inform understanding of individual and social behavior. In both simple and complex simulations, describing the propagation of uncertainty remains a significant issue. These approaches (even if rough) supplement empirical observational and experimental methods.

Unification

A major challenge, then, is how to unify across theoretical, observational, experimental and generative approaches (and variant sub approaches within each). Any such unification requires understanding the qualities of each approach. This is akin to challenges in the natural sciences where conclusions draw on theory (and sometimes competing theories), experimental data obtained with different instruments and methods, and computations based on the best physical models available. Although no single set of standards is in the offing for such unification, there is need for a common understanding of sound ways to unify knowledge.

Challenge Three: Representations

Creating mathematical or computational representations of complex human behavior is itself a challenge. While applied mathematicians and computer scientists offer many representational formalisms (e.g. differential equations, hidden Markov models, Bayesian networks, theorem approaches, system dynamics), deciding which to use in describing a particular social-behavioral phenomenon requires careful consideration. For example, when and where do behavior processes follow different probability distributions? A core issue is coping with the multiple types of uncertainty across different data sources and the under-specification of social-behavioral theories. That is, not only is the selection of particular computational or mathematical formalisms deserving of scrutiny but also their ability to represent the phenomena of interest and, critically, the many associated uncertainties and assumptions. For example, what computational representations can best integrate laboratory – experiment data on a specific population with the statistical analysis of correlations across populations? Selecting and integrating multiple formalisms, to include dealing with the propagation of uncertainty, and developing mappings across them, will be an active area for future research.

Challenge Four: Applications of Social-Behavioral Modeling

The final and most consequential challenge is translating knowledge of the human condition into the decisions and actions of individual citizens, of corporate or government leadership, or, increasingly, automated systems. Communicating current knowledge of the human condition and its uncertainties is critical. Society does some of this today, as when engineers use knowledge of human behavior to design roads and place warning signs appropriately. However, modern social-behavioral knowledge is also being applied in domains that raise profound ethical and moral considerations that have not been adequately explored. At the root of such concerns is yet another type of problem – the need to ask how the consumer of knowledge about human behavior can realistically use that knowledge. Fortunately, some aspects of this challenge are already active areas of research, such as work on communicating uncertainty, on creating explanations of complex analytic methods (e.g. explaining machine learning techniques), and on how to help decision-makers to assess the trustworthiness of data and analyses of data. So also, research is underway on how to allow humans to better understand privacy issues and how to protect themselves (see Chapter 3, Balebako et al. 2018). In any case, given that improved and accessible understanding of the human condition can be used for both good and bad purposes, more research is needed to understand the issues and develop protections.

About This Book

Interest in social-behavioral modeling for complex systems has waxed and waned, but is now very much ascending. This book came about because we recognized that a large community was very interested and deeply involved, as illustrated by a large number of studies and conferences, many of them related to data and data analytics and others being broader in scope.[2] Identifying and articulating common questions, challenges, and opportunities across all of these activities has become a priority.

With that in mind, we began the book project and solicited chapters from prominent and diverse researchers to represent a broad perspective on recent developments and the future of research on modeling of the human condition. Their chapters offer a great deal of wisdom and numerous suggestions based on their experience and visions. Many of the chapters posit pathways for future work in specific areas, each of which could revolutionize our understanding of the human condition. Although even this sizable volume is addressing only a sampling of the myriad efforts underway, it draws on work in at least the following areas: anthropology, cognitive psychology, complex adaptive systems, computer science, data science, economics, electrical engineering, health research, law, medicine, modeling and simulation, neuroscience, physics, policy analysis, political science, social psychology, and sociology. Interestingly, many of the

---

2 A few of many examples include the ongoing "Decadal Survey of the Social and Behavioral Sciences and Applications to National Security" by the National Academy of Sciences, DARPA-sponsored RAND workshops on privacy and ethics (Balebako et al. 2017), and priority challenges for social and behavioral modeling (Davis et al. 2018) and various professional conferences by, e.g. the Society for Computational Social Sciences and Social Computing, Behavioral-Cultural Modeling & Prediction and Behavior Representation in Modeling (SBP-BRIMS). Chapter 2 of this volume also cites numerous foundational documents from earlier work, including an important National Academy study from a decade ago (Zacharias et al. 2008).

contributors no longer identify exclusively with a single academic discipline but rather engage problems that cross disciplines. Many are also concerned with application to real-world problems.

With all humility, then, we hope the book will be an inspirational crosscutting milestone volume that helps set the agenda for the next decade of research and development. We hope that it will prove stimulating to a number of audiences: (i) scientists and modelers; (ii) organizations that support basic and applied research; (iii) practitioners who apply social-behavioral science and computational modeling; and (iv) decision-makers responsible for addressing individual and societal-level problems. All of the chapters reflect the ultimate goal of this book: to reflect on the opportunities for technology to help transform the social and behavioral sciences while – at the same time – honoring our duty to create scientifically sound knowledge and assure that it is responsibly used to benefit mankind. We (the editors) see researchers as having heavy and nontrivial responsibilities for creating models with a solid basis in reality and for ensuring that the models and related analytic tools are used responsibly to benefit humanity.

Roadmap for the Book

The book is divided into distinct parts (see Table 1.1). Part I sets the stage. Chapter 2 is an overview of social-behavioral modeling for complex systems. It structures the subject area in terms of three pillars: theory, empirical observation and experimentation, and computational observation and experimentation. These exist in an ecology that includes infrastructure, governance, and technology. Chapter 2 also establishes goals of *usefulness* of social-behavioral modeling and identifies priorities. Chapter 3 recognizes the fundamental importance of seeing opportunities and challenges in a framework that recognizes ethical and privacy issues from the very start. Since modeling is an attempt to represent and exploit knowledge, it needs foundations in science. Part II, Social-Behavioral Science Foundations, is a set of nine chapters about frontier aspects of social-behavioral science and its implications for modeling. Part III is a set of nine chapters discussing how social-behavioral modeling can and should be informed by underlying theory and modern ubiquitous data, not only traditional data from social science research, but new forms reflecting social media, recreational games, and laboratory experiments. One of the themes is the need to better relate theory and data: theory should be informed by data, and both data collection and data analysis should be informed by theory; but neither should be unduly constrained by the other, and competition of inquiry should be encouraged. Part IV is about modeling itself, describing innovations and challenges in doing so. Some of the 12 chapters describe new methods; some address frontiers in crossing the boundaries of model types and formalisms and of spatial,

Table 1.1 A view of the book's composition.

Part	Chapter	Authors	Title (abbreviated)	Overview	Theory and its modeling	Social science data	Theory-data linkages	Validation	Modeling theory and technology	Models for decision aiding	Model-based decision aiding
I	1	Pfautz et al.	Understanding and Improving the Human Condition								
	2	Davis and O'Mahony	Improving Social-Behavioral Modeling								
	3	Balebako et al.	Ethical and Privacy Issues in Social-Behavioral Research								
II	4	Nyblade and O'Mahony	Building on Social Science								
	5	Brashears	How Big and How Certain? Defining Levels of Analysis								
	6	Corman et al.	Generative Narrative Models of Conflict and Its Resolution								
	7	Read and Miller	A Neural Network Model of Everyday Decision-Making in Social Behavior								
	8	Matthews	Dealing with Culture as Inherited Information								
	9	Cowherd and Lende	Social Media, Global Connections, and Information Environments								
	10	Thompson et al.	Using Neuroimaging to Predict Behavior								
	11	Pavlic	Social Models from Nonhuman System								
	12	Sieck et al. (panel)	Moving Social-Behavioral Modeling Forward: Insights from Social Scientists								
III	13	Gabbay	Integrating Computational Modeling and Experiments								
	14	Sliva et al.	Combining Data-Driven and Theory-Driven Models for Causality Analysis								
	15	Rand	Theory-Interpretable, Data-Driven Agent-Based Modeling								
	16	Miller et al.	Bringing the *Real World* into the Lab: Transformative Designs								
	17	Lakkaraju et al.	Online Games for Studying Human Behavior								
	18	Guarino et al.	Using Sociocultural Data from Online Gaming and Game Communities								
	19	Osoba and Davis	An Artificial Intelligence/Machine Learning Perspective on Social Simulation								
	20	Giridhar and Abdelzaher	Social Media Signal Processing								
	21	Grace et al.	Evaluation and Validation Approaches for Simulation of Social Behavior								
IV	22	Garibay et al.	The Agent-Based Model Canvas: A Modeling Lingua Franca								
	23	Tolk and Glazner	Representing Socio-behavioral Understanding with Models								
	24	Yilmaz	Toward Self-Aware Models as Cognitive Adaptive Instruments								
	25	Osoba and Kosko	Beyond DAGs: Causal Modeling with Feedback Using Fuzzy Cognitive Maps								
	26	Swarup et al.	Simulation Analytics for Social and Behavioral Modeling								
	27	Sukthankar and Beheshti	Using Agent-Based Models to Understand Health-Related Social Norms								
	28	Hadzikadic and Whitmeyer	Lessons from a Project on Agent-Based Modeling								
	29	Scaumann and Kapadia	Modeling Social and Spatial Behavior in Built Environments								
	30	Orr	Multi-scale Resolution of Human Social Systems								
	31	Gribaudo et al.	Multi-formalism Modeling of Complex Social-Behavioral Systems								
	32	Carley	Social-Behavioral Simulation: Key Challenges								
	33	O'Mahony et al.	Panel Discussion: Moving Social-Behavioral Modeling Forward								
V	34	Rouse	Human-Centered Design of Model-Based Decision Support								
	35	Thompson et al.	Understanding Policy and Management Interventions in Health Care								
	36	Lofdahl	Modeling Information and Gray Zone Operations								
	37	Paul	Modeling Narrative in Human Understanding								
	38	Sieck	Aligning Behavior with Desired Outcomes: Lessons from the Marketing World								
	39	Myers	Future Social Science That Matters for Statecraft								
	40	Davis	Lessons on Decision Aiding for Social-Behavioral Modeling								

temporal, and other scales of description. Part V addresses how models and modeling can and should relate to decision-making and those who must make those decisions. It is more applied in character, drawing on examples from international security, health research, and marketing, among others.

References

Balebako, R., O'Mahony, A., Davis, P.K., and Osoba, O.A. (2017). Lessons from a workshop on ethical and privacy issues in social-behavioral research. *PR-2867*.

Brady, H.E. and Collier, D. (2010). *Rethinking Social Inquiry: Diverse Tools, Shared Standards*. Rowman & Littlefield Publisher.

Churchman, C. (1967). Wicked problems. *Management Science* 14 (4): B-141.

Davis, P.K. and O'Mahony, A. (2018). Improving social-behavioral modeling. In: *Social-Behavioral Modeling for Complex Systems* (ed. P.K. Davis, A. O'Mahony and J. Pfautz). Hoboken, NJ: Wiley.

Davis, P.K., O'Mahony, A., Gulden, T. et al. (2018). *Priority Challenges for Social-Behavioral Research and Its Modeling*. Santa Monica, CA: RAND Corporation.

Henrich, J., Heine, S.J., and Norenzavia, A. (2010). Beyond weird: towards a broad-based behavioral science. *Behavioral and Brain Sciences* 33 (2–3): 111–135.

King, G., Keohane, R.O., and Verba, S. (1994). *Designing Social Inquiry: Scientific Inference in Qualitative Research*. Princeton, NJ: Princeton University Press. Retrieved from http://www.loc.gov/catdir/toc/prin031/93039283.html http://www.loc.gov/catdir/description/prin021/93039283.html.

Luce, R.D., Bush, R.R., and Galanger, E. (eds.) (1963). *Handbook of Mathematical Psychology, Volumes I and II*. New York, NY: Wiley.

Muller, J. (2018). *The Tyranny of Metrics*. Princeton, NJ: Princeton University Press.

Nosek, B. and Open Science Forum (2015). Estimating the reproducibility of psychological science. *Science* 349 (6251): aac4716.

Rittel, H.W.J. and Noble, D.E. (1989). Issue-based information systems for design. Working Paper. Berkeley, CA: Institute of Urban and Regional Development, University of California, Berkeley.

Waldrop, M. (2018). What if a nuke goes off in Washington, D.C.? Simulations of artificial societies help planners cope with the unthinkable. *Science* https://doi.org/10.1126/science.aat8553.

Zacharias, G.L., MacMillan, J., and Van Hemel, S.B. (eds.) (2008). *Behavioral Modeling and Simulation: From Individuals to Societies*. Washington, DC: National Academies Press.

2

Improving Social-Behavioral Modeling

Paul K. Davis and Angela O'Mahony

RAND Corporation and Pardee RAND Graduate School, Santa Monica, CA 90401, USA

Aspirations

Social-behavioral modeling (SBM) has many valuable functions, but an aspiration should be that such modeling help policy-makers understand classes of social-behavioral (SB) phenomena with national significance, anticipate how those phenomena may unfold, and estimate effects of potential events or government interventions. It should inform policy-making on such security and social-policy issues as radicalization for terrorism, weakening of democracy by foreign influence campaigns, improving prospects for stability after international interventions, managing population behaviors after natural disasters, and dealing with the opioid crisis and epidemics. Today's SBM is contributing far less to the study of such matters than it could. Major advances are needed, but in what? We offer suggestions distilled from a longer study (Davis et al. 2018).

Before proceeding, it may be useful to give two examples of what may in the future be possible.

Vignette 1

Today (March 1, 2035), the US government released its report on home-grown terrorism with good news: the number of youths seeking to join violent extremist groups such as the descendants of ISIS and Al-Qaeda has dropped precipitously. This is apparently due to programs helping to inoculate impressionable youth against the attractions and propaganda of such organizations. The programs involve schools, neighborhood and religious organizations, public service announcements, commercial recreational online games, and individual-level interventions. The programs were initially

Social-Behavioral Modeling for Complex Systems, First Edition.
Edited by Paul K. Davis, Angela O'Mahony, and Jonathan Pfautz.
© 2019 John Wiley & Sons, Inc. Published 2019 by John Wiley & Sons, Inc.
Companion website: www.wiley.com/go/Davis_Social-Behavioralmodeling

informed and continue to be updated by the efforts of a dedicated virtual social and behavioral science laboratory that employs state-of-the-art models representing the best social science and a flow of data unavailable in earlier years. Early in the 2020s, the simulation laboratory tested alternative portfolios of intervention activities to find a package that would be likely to succeed with low risks of such bad side effects as acrimonious community rejection. Many ideas thought initially to be promising were discarded because the simulations credibly suggested a significant probability of counterproductive effects. As the simulations anticipated, successful implementation has also required rapid adaptation as social science is improved, social patterns and communications change, external events occur, and data indicates what works and does not. Overall, models have been less about predicting the future than about helping to chart a roughly right course and prepare for adaptations (sometimes in minutes or days, rather than months).

Vignette 2

It is 2035. Students in the nation's best business-administration program now take a labor-management negotiations course using a simulation laboratory in which computer agents represent labor and management actors at top, middle, and individual levels and also community governments with authority for considering incentives and constraints. Sometimes, human teams play these roles, interchangeably with agents. Students often have a background in business economics and game theory, in which case they imagine initially that negotiations will be just a matter of finding a mutually agreeable optimum agreement. They discover, however, that the opposing teams are distrustful because of opposed views and a long history of troubled relationships. Further, they seem erratic in their demands (due, in some cases, to factional internal disputes, as when workers and their labor negotiators go back and forth, or when a corporation's board and the company managers disagree among themselves about objectives and responsibility to the community). Neither side understands the other well and, at the outset, neither respects the others' values. Exogenous events also occur, as when recessions strike, the president announces a new trade policy, or the Federal Reserve changes interest rates. Negotiations stop and start, sometimes because of temper tantrums or events or sometimes because of seemingly impossible impasses. Sometimes, a given simulation leads to bargains emerging; sometimes, it does not. Students learn about how to build on success and how to change losing games. They learn how to adapt their own value structures or how to understand the values of others. They practice the art of finding possible acceptable outcomes and then working backward to infer necessary or helpful ingredients in their approach. The models employed *stress* the student negotiators in ways that reflect empirically confirmed human behaviors. Students come out with no illusions about the simulations' ability to predict results, but they emerge better able to negotiate and with more mature concepts of victory, compromise, and failure.

To make such vignettes real, many challenges must be overcome.

Classes of Challenge

Social sciences, the allegedly soft sciences, are in fact notoriously hard (Diamond 1987), primarily because they deal with "inexact science and technology dealing with complex problems in the real world" (Simon 1996, p. 304). Social sciences differ from the physical sciences in seeking to understand the behavior of *people*, the nature of explanations and evidence, and the methods that predominate. We discuss the challenges to SBM in two groups: inherent challenges and challenges relating to specific issues that are difficult to address with current mainstream practices.

Inherent Challenges

Four inherent challenges relate to individual cognition, social systems as complex adaptive systems (CAS), the dynamic and storytelling character of people, and dealing with wicked problems. All are especially relevant to the future of SBM.

Individual Cognition and Behavior

Social science in the mid- to late twentieth century leaned heavily on the economist's notion of a rational actor. Although very useful, it has fatal flaws. Herbert Simon's Nobel Prize celebrated his discussion of human *bounded rationality*, the fact that human decision-making is limited by finite and imperfect information, cognitive limitations, and time. As a result, humans employ shortcuts to find strategies that are good enough for the purposes at hand (i.e. that *satisfice*) (Simon 1978). Simon discussed the matter as follows:

> … the first consequence of the principle of bounded rationality is that the intended rationality of an actor requires him to construct a simplified model of the real situation in order to deal with it. He behaves rationally with respect to this model, and such behavior is not even approximately optimal with respect to the real world. To predict his behavior we must understand the way in which this simplified model is constructed, and its construction will certainly be related to his psychological properties as a perceiving, thinking, and learning animal.
>
> (Simon 1957, p. 198)

The items to which Simon alluded in the last sentence have been exhaustively studied by psychologists for a half century. Early work revealed systematic cognitive biases in human reasoning, relative to that of the idealized economic actor (Kahneman and Tversky 1982). Subsequent research noted that – despite these biases – so-called naturalistic or intuitive human behavior also has great virtues (Klein 1998; Gigerenzer and Selten 2002). Kahneman discusses the virtues and shortcomings of cognitive biases, and of intuitive thinking, distinguishing between what he calls fast and slow thinking, modes for both of which are built in biologically and have survival advantages (Kahneman 2011). The emerging holistic view is an example of science at its best, a remarkable *synthesis*.

This departure of human cognition and behavior from that of the idealized rational actor has profound implications, as illustrated by the rise of behavioral economics (Thaler 1994; Sunstein 2014), versions of deterrence theory that recognize the role of perceptions and biases (Jervis et al. 1985; National Research Council 2014), modeling of adversaries (Davis 2018), and many other contexts where cultural factors are important (Faucher 2018). Agent-based modeling (ABM) represents such matters, but, so far, it has typically not incorporated such matters well (see, however, Read and Miller 2019; Sukthankar and Behesti 2019). As discussed recently by Robert Axtell, a pioneer in ABM, numerous assumptions need to be abandoned if ABM is to reach its potential (Axtell et al. 2016; O'Mahony and Davis 2019). Interestingly, a different tack will be to represent behavior as emanating more from primal considerations than reasoning, as suggested by behavior of nonhuman species (Pavlic 2019).

Social Systems as Complex Adaptive Systems (CAS)

If an individual's reasoning is complicated, that of humans as a whole is even more so because humans are social: their values, beliefs, and behaviors depend on interactions and learning. Indeed, social systems are CAS, the behavior of which is famously troublesome with even small changes in initial conditions or events over time sometimes having outsized effects on the system's dynamics (Holland and Mimnaugh 1996; Page 2010).

This said, it is not hopeless to study CAS with an eye toward predictions and a degree of influence. A CAS is uncontrollable in only certain of its states. Key issues are recognizing which states of the system are subject to a degree of influence and – for those that are – what can be achieved and with what risks. For CAS that are not in controllable states, it is sometimes possible to identify the alternative states that may emerge. Confronting the CAS challenge has many implications for SBM, but these include the need to do better in representing variable-structure systems and emergent phenomena.

The Dynamic and Storytelling Character of People and Social Systems

A related matter is that people and societies depend on storytelling or narratives to make ostensible sense of the world (Corman 2012; Corman et al. 2019; Cowherd and Lende 2019; Paul 2019). The narratives can change markedly and sometimes unpredictably. Narratives are not simple objects, attributes, or processes (the building blocks of usual computational models). How should they be represented in modeling? Will prospective models be able to *generate* these features, i.e. to have them be emergent rather than baked in? A number of mechanisms for doing so may exist. One is defining *alternative* agents with different characters that embody the different narratives. These might coexist (as when humans act as if they have conflicting personas) or might be activated or inactivated in the course of events. Some early experiments along these lines occurred with Cold War modeling of Red and Blue agents in automated war gaming (Davis 2018). Future approaches will need to allow for narratives to emerge that have not previously been anticipated.

Wicked Problems

A fourth inherent challenge has a different character, although aspects relate to the CAS phenomena of social systems and the emergence of narratives. The success of SBM will depend on its value in aiding human decisions. Much formal education and modeling focuses on finding optimum solutions to well-defined problems. Policy-makers, however, are often confronted with so-called wicked problems that, by definition, have no preordained solution or even a priori objectives to be achieved (Rosenhead and Mingers 2002). Such problems are central to much policy analysis and strategic planning. In favorable cases of wicked problems, people come in the course of interacting to agree on some actions and perhaps some objectives. If so, these agreements emerge rather than being predetermined or implied. The socialization process is crucial because it creates new or adjusted perceptions and values. To put the matter differently, the notion that individuals have fixed utility functions is often quite wrong. Through interactions and other experiences, people learn, change values, and sometimes find ways to compromise. Doing so is crucial for humans going beyond the solitary, poor, nasty, brutish, and short lives that Thomas Hobbes discussed in the seventeenth century.

How can SBM help with wicked problems? One imperative is confronting the realities of CAS, as mentioned above. We also believe that it will be necessary for SBM to move away from the common ideal of *closed* simulation to embrace man–machine interaction, human gaming, and man–machine laboratories where people can gain experiential knowledge (Rouse 2015, 2019) and interact creatively, competitively, and with corresponding emotions. Hybrid approaches are also possible, as illustrated by recreational games that include automated opponents or a 1980s project in analytic war gaming that allowed for substitution of human teams or artificial intelligence models when conducting a game or a game-structured simulation (Davis 2018).

Selected Specific Issues and the Need for Changed Practices

Having discussed inherent challenges, we now touch on some specific matters, starting with background on current-day practices. We then discuss the need for a rebalancing of the portfolio of methods and models used in social science and its modeling, confronting uncertainty, relating theory to experimental information, finding synthesis opportunities, and rethinking the concept of model validity.

Background on Fragmentation of SB Theories

Today's SB theories exist in bewilderingly large numbers[1] but are typically narrow and fragmented and often conflicting. Part of obtaining an advanced degree in social science is learning to navigate through the corresponding thickets. Professors strive to help such navigation, but the current situation is nonetheless troubled.

The Nature of *Theory*

One manifestation of the fragmentation is in the view of *theory*. In contrast to the treatment of theory in physical sciences, in which theory unifies, broadens, and explains, social science theories tend to be simple conceptual models that work in only very limited contexts rather than more generally (i.e. context dominates). Social scientists tend to be empirically driven and are comfortable with a diversity of theories. As a contrast, imagine that physicists still talked separately about Boyle's and Charles' gas laws, perihelion theory, Bohr's theory, wave theory, friction theory, and so on (an observation motivated by a recent

---

1 For example, browsing the Web for *social psychology theories* yields an illustrative list with attribution theory, cognitive dissonance theory, drive theory, elaboration likelihood model, motivation crowding theory, observational learning theory, positioning theory, schemata theory, self-perception theory, self-verification theory, social comparison theory, social exchange theory, social penetration theory, socioemotional selectivity theory, terror management theory, and triangular theory of love.

chapter (Hadzikadic and Whitmeyer 2019)). Physics, however, has been studied for millennia, and many of its fragments have been integrated.

In physics, perhaps because of the field's relative maturity, *theory* connotes something offering generality, coherence, and explanatory power – e.g. classical mechanics (Newton), classical electromagnetics (Maxwell), quantum mechanics (Schrödinger, Heisenberg, Dirac), and relativity (Einstein). Although mysteries persist, much is understood about how these theories relate to each other. These exemplify the understanding of scientific theory expressed by the US National Academy (National Academy of Sciences 1999, p. 2):

> … theories are the end points of science. They are understandings that develop from extensive observation, experimentation, and creative reflection. They incorporate a large body of scientific facts, laws, tested hypotheses, and logical inferences.

The authors might better have said that theories are at least for now end points, since even good theories are sometimes extended, corrected, or otherwise enriched over time. The National Academy's point, however, was correct.

Similarities and Differences

In reviewing the claimed differences between the social and physical sciences, we have benefited from the more recent philosophy of science literature (Cartwright 1999; Haack 2011), which has come far since the discussions of Popper, Kuhn, Lakatos, and Feyerabend that we studied earlier in our professional careers. We conclude that the inherent differences between the physical and social sciences should not be exaggerated. Although the social sciences are dominated by different methods than the physical sciences, the primary differences probably lie in the nature of explanations and evidence. Social science "tries to understand people's behavior by coming up with explanatory hypotheses about their beliefs, goals, etc." (Haack 2011, p. 166). And, although fragmented in many respects, it is possible – from a remove – to provide considerable unity in the concepts that dominate social science (Nyblade et al. 2019).

We also pondered the ubiquitous claim that context matters more in social science than in the physical sciences. We argue that that is misleading. *The maturity of the physical sciences consists in significant degree in their having identified and defined the factors that constitute context.* Physical theories specialize for varied contexts (e.g. dilute gases vs. dense gases; fluids at ordinary temperatures vs. temperatures where superfluidity appears). Indeed, the notion that well-known physical theories are general is misleading because, in fact, they pertain to idealized worlds. Their value in the real world depends on substantial modifications accounting for real circumstances (Cartwright 1983).

Emergence is often mentioned as something unique to social CAS.[2] It is worth noting, however, that physics and chemistry have analogues to CAS behavior, albeit without sentient agents. Think of phase transitions, such as precipitation from liquids or the transition of a solid to a condition of superconductivity. Or think of self-organization in nonequilibrium systems, which was studied by theoretical chemists at much the same that CAS theory began to be appreciated, sometimes using similar mathematics (Nicolis and Prigogine 1977).

Despite these cautions about exaggerating inherent differences, differences certainly exist today in practice between the physical and social sciences. Table 2.1 shows some comparisons, drawn starkly for the sake of exposition. Social science modeling tends to be empirically driven with statistical methods for finding correlations that fit data with as few variables as possible. To social scientists, to say that a model *explains the data* usually means only that a regression formula fits the data with acceptably small residuals. Physical science models (especially system models) tend to be concepts driven, to emphasize causal relationships, to include more variables, and to regard a model as explanatory only if it provides a satisfying and coherent description of cause–effect relationships. The aspirations of social science theories are relatively modest, the number of theories is large, and the number of inconsistencies across theories is large as well. Although they are sometimes important in informing policy choices (e.g. when health trials suggest that certain types of exercise may prolong health into old age), correlational theories have known shortcomings (Pearl and Mackenzie 2018). Good system theories are better because causation is more explicit.

Reality is more complicated than Table 2.1 would have it. Great advances have occurred in social science. For example, while quantitative statistical analysis was once regarded by many scholars as more scientific than qualitative research even when merely correlational, today's graduate students learn about the strengths and weaknesses of both quantitative and qualitative approaches. The latter is often insightful about causality. Further, econometricians now emphasize causal relationships and have developed quasi-experimental methods to infer them from *natural experiments* (Angrist and Pischke 2009). Some qualitative research is now well structured and highly regarded, as with comparative case studies (George and Bennett 2005). Hybrid approaches that use both case studies and quantitative analysis well are especially useful (Sambanis 2004).

Nonetheless, on balance, most social science products are a poor match for the needs of those seeking to advance SB science and apply it problems using SB

---

2 Interestingly, disagreements exist on the definition of *emergence* and beliefs about to what extent simulation can generate emerge, as discussed in a *panel chapter* in this volume (O'Mahony and Davis 2019).

Table 2.1 Comparisons.

Attribute	Mainstream social science modeling	Physical science modeling (especially system modeling)
Character	Data and hypothesis driven	Concepts driven
Methods	Quantitative: statistical models	Causal models (in mathematics or computational form)
	Qualitative: textual arguments, perhaps with tables and notional diagrams	
Number of variables	Few, with parsimony in mind[a]	Few, but as many as needed[b]
Aspiration	Sound correlations, for context	Causal explanation and sometimes prediction of a relatively general nature
Explanation	Good fits to data (small residuals)	Causal explanation that fits data adequately and is otherwise coherent and persuasive[c]
Aspiration on generality	Modest	High
Number of theories	Many, fragmented and inconsistent	A few, which are coherent and related
Relevance to policy choices	Weak on explanation and informing decisions about the future	Potentially good (for valid models)
Treatment of human behavior	Yes (that is the point)	No

a) Parsimony in social science: use the simplest theory *that can explain the empirical evidence.*
b) Ascribed to Einstein: theory should be "as simple as possible, *but no simpler.*"
c) Statistical analysts may say that the variables of a regression *cause* the results for a given context after testing for certain other explanations, but they are defining *cause* differently than would a physicist.

system models. The modeling community is not receiving what it needs from the science community (McNamara et al. 2011). At the same time, many in the modeling community do not try very hard to get into the scientific subtleties because of being more passionate about the programming than the science and because it is so difficult to translate empirically driven and statistically interpreted social science into causal system models. A sore point with social scientists is that computerized SBM often claim to be based on a named social science theory, but actually represent the body of work with nothing more than some name-dropping and a parameter.

Rebalancing the Portfolio of Models and Methods

We conclude that achieving the potential of SBM requires a rebalancing of social science to increase the relative emphasis on theory that is causal, multivariate, systemic, sometimes nonlinear, coherent, and general (although explicitly context sensitive).

Causal description is a core element of most science and a rather fundamental aspect of language. Causal models are *crucial* for addressing many issues. As Judea Pearl has emphasized,

> we can't really deal with issues such as influence, effect, confounding, holding constant, intervention, and explanation without causal models.
> (Pearl 2009)

The issues that he identifies are crucial in policy-making (Pearl and Mackenzie 2018).

The requirement for multivariate and systemic modeling reflects the fact that many important SB phenomena are due to the interaction of numerous factors. Further, system models are often nonlinear. The ubiquity of nonlinearity is illustrated by physical systems such as aircraft with a number of critical components (*each* component must work or the system fails), the product form of Cobb–Douglas production functions in economics, the nonlinear nature of many system dynamics models in diverse business or public-policy areas (Sterman 2000), or the nonlinear qualitative theory needed to understand the rise of and public support for terrorism (Davis and O'Mahony 2017). We have already discussed the need for more models that are unifying, coherent, and relatively more general than the existing set of fragmented models.

As noted earlier, the corresponding modeling should also allow for the very structure of systems to change with new entities coming into existence and others disappearing and with their value structures changing as well. Overall, we need richer system models in which all of these phenomena can *emerge* naturally, as occurs with CAS. Achieving such models and learning how to use them analytically is a profound multifaceted challenge.

What We Are Not Saying. None of this means that we doubt the value of much existing social science theory for traditional purposes or that we harbor notions of some grand overall theory of everything. Such a theory does not yet exist even in the more aged field of physics, and it is implausible in the human domain (Cartwright 1999; Konner 2003). We do see the need, however, to move in the direction of more unified and general theories.

Confronting Uncertainty

Another fundamental challenge for SBM is routinely building into models the capability to deal with uncertainty. Doing so is crucial for CAS for which reliable prediction is fraught. It is difficult, however, especially when dealing with *deep uncertainty*, i.e. the uncertainty that obtains when we do not know (or agree on) how actions relate to consequences, probability distributions for inputs to the models, which consequences to consider, or their relative importance.[3] Fortunately, much progress has been made on dealing with uncertainty in other fields, and the methods can be brought over into SB work (Davis 2019).

Combination, Synthesis, and Integration

If fragmentation is a problem in SB science and its modeling, then the solution would seem to be models that unify. But what does that mean? The available approaches include combining models, synthesizing new models, and developing families of models.

A triumph of computer science has been to develop methods for combining models in what are sometimes called multimodels, to include doing so in model federations. These methods assure interoperability and can assist in organizing meaningful composition (Levis 2016; Gribaudo et al. 2019). Model federations were highlighted in an earlier review of SBM (Zacharias et al. 2008); they have been pursued within the Department of Defense (DoD) since the 1990s (Kött and Citrenbaum 2010), notably for training and exercising of military forces distributed across the country and globe. The subject is mature enough to have professional societies and texts (Tolk 2012).

We are much less sanguine about using such federations to aid higher-level decision-making. Good analysis requires that the analyst team understand and control its models. After all, the analysts are *responsible* for the soundness of their conclusions or suggestions. Using federations of models for analysis within a tightly knit group is possible, but often requires considerable code checking and adaptation, not simple stapling together of models.

The second problem is that while the model federation approach has great allure for organizations seeking to improve efficiencies, such organizations routinely desire standardization. In the DoD, for example, this led in recent decades to a standardization of large combat models, scenarios for investigation, and databases to feed the models. This standardization had notable value for establishing shared baselines of knowledge, but profound problems when supporting decision-makers planning military forces and capabilities for a highly uncertain world. Uncertainty analysis was severely circumscribed, as was ability quickly to respond to decision-maker *what-if* questions. Because of such concerns, a large DoD modeling group was disbanded with prejudice (Davis 2016). Given the current state of SB science and its modeling, the last thing needed is standardization on allegedly authoritative models and databases. Instead, what is needed is vigorous competition of ideas and rapid evolution.

---

3 Definitions vary somewhat. See the website of the society for decision-making under deep uncertainty (www.deepuncertainty.org).

Another profound problem with the federation approach, especially as it is usually pursued, is that it is often neither analytically sensible nor pragmatically desirable to be working with large, complex assemblages, especially high-resolution assemblages, when conducting analysis. Instead, analysis requires simpler models, albeit ones that deal with uncertainty.

Families of Multiresolution, Multiperspective Models

In some fields, it has long been customary to develop families of models that are seldom used at the same time but that have known relationships and that can be used to mutually inform each other and, ideally, can be used together with all available data for self-consistent calibration across the model family. The members of the family may differ in resolution, formalism, and other matters. Figure 2.1 illustrates this schematically for DoD modeling, which has different models at the engineering, engagement, mission, and campaign levels (Mullen 2013). The number of low-level (detailed) models is large (more than a 1000), whereas only a few campaign-level models exist, the purpose of which is to provide an integrative higher-level view. For details, one must go deeper into the pyramid. Even military modeling has SB aspects. The relationship between mission and campaign levels, for example, depends on the complex behavior of vast numbers of individuals at different levels of organization. This is affected by factors such as morale, physical state, discipline, and doctrine.

Examples of model families are also possible from physics. We mentioned earlier the unity provided by its major theories. Much is known about how those theories relate to each other and when they should be used together or separately. For example, the macroscopic laws of thermodynamics used by engineers have understood foundations in quantum statistical analysis. Newton's laws are fully valid in certain understood limiting cases; in some other cases, quantum corrections are needed; and in still other cases (e.g. on

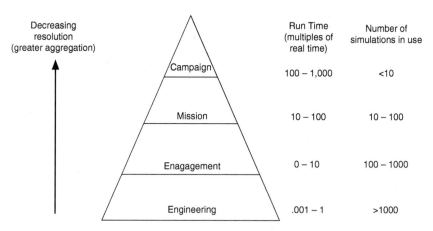

Figure 2.1 A pyramid of military simulations. Source: adapted from Mullen (2013).

atomic scales), the phenomena of interest may need to be addressed with quantum mechanics alone. Similarly, cases exist in which relativity theory must be invoked to correct what might naively be expected to follow classic mechanics. This is the case when exploiting navigation satellites (Ashby 2002). Many examples could also be drawn from systems engineering, to include allowing for and anticipating emergent phenomena (Rainey and Jamshidi 2018).

Nothing analogous to a mature family-of-models approach yet exists in social and behavioral research. This suggests questions for SBM, such as the following:

- Can ABM be used experimentally to better inform macroscopic modeling such as might be accomplished with system dynamics?
- Can conclusions from system dynamics research and empirical work inform how most productively to go about ABM research and where regularities might be expected?
- How should *modules* be conceived given that it is often important to consider both issues of resolution or scale and issues of constancy as discussed elsewhere in this volume (Brashears 2019)?
- When does it make sense to directly couple different types of SBM, rather than have one level of analysis inform another? How is that *informing* to be accomplished? The influences should run both ways (Orr 2019).

Composability

As mentioned above, achieving interoperability among models has been a triumph of computer science. Achieving *meaningful* composition is different (Davis and Anderson 2003; National Research Council 2006; Zeigler and Hammond 2007; Tolk et al. 2013; Taylor et al. 2015). At a simple level, suppose that model B requires an input X, which model A is said to provide as an output. A and B can then be connected and *run*. Suppose, however, that model A's variable X has the same label as the input to B, but actually means something different. Or, more subtle, suppose that model A calculates X using assumptions that seemed to the model builder reasonable, but were actually valid only for the context of his immediate work, with model B requiring an estimate of X that comes from a different concept of how X comes about (e.g. is a decision based on rational actor theory or heuristics and biases?). In several of the latter cases, A and B can be combined and will run (i.e. they interoperate), but the composition makes no sense. Successful composition depends on the modelers understanding both context and the models to be composed. Obviously, it will be easier to the extent that the separate models are transparent and comprehensible (i.e. subject to inspection and understandable as a whole, respectively).

Model composition demands a deeper understanding than is sometimes realized by those doing the composing. Further, pressures may exist to do a quick composition to *demonstrate* the idea without claiming validity. Strong pressures often exist to standardize, thereby suppressing important variations. All of this is a recipe for failure when modeling complex SB phenomena.

To summarize, it is necessary to better understand how to achieve meaningful model compositions, but the goal should not be *blessed federations* and databases, but the ability to compose models for a given purpose and context, while allowing for uncertainty, disagreement, and competition. That is, a composition should be *fit for purpose.* This should be much easier with community-available libraries of well-vetted modules (including competitive modules) and sufficiently rich parameterization visible at the interface. How should such modules be developed, peer reviewed, and shared? We conclude that two general admonitions apply:

- No general solution exists for assuring valid composition. This situation will not change with new technology. Subject-area expertise will often be necessary.
- It is more reasonable to seek relatively general processes and procedures by which to validate a composition for a given context and to identify methods and tools to help.[4]

Connecting Theory with Evidence

If theory and both empirical and computational experimentation are the pillars of science, it is hardly controversial to say that they should be well connected. Theory should inform observation, and experimentation should affect theory iteratively and constructively. However, SB research is out of kilter in this respect: the linkages are often not very strong. Improving the linkages may prove controversial because of deep-seated differences in philosophy between those who are data driven and those who are theory driven, but examples of linkage seeking occur in recent work in this volume (Gabbay 2019; Rand 2019; Sliva et al. 2019). Another chapter discusses a structured way to do such way (Garibay et al. 2019).

Rethinking Model Validity

The Five Dimensions of Model Validity

Given the difficulties in measuring faithfully the many variables important to human and social behavior, and given the difficulty and sometimes the impossibility of predicting CAS behavior, it is necessary to redefine *validity*

---

4 The process will often involve ontologies to assure consistent vocabularies and consistent meanings of variables, mappings from one set of variable names to another, and other devices by which computer science can help (Gribaudo et al. 2019). A useful formalism is fuzzy cognitive maps (FCMs), which allow straightforward combining of FCMs from multiple experts (Osoba and Kosko 2017).

of theories and models. Much discussion has already occurred, and it is time to draw conclusions and move on rather than debate the matters endlessly. Fortunately, it is possible to proceed by merely extending the classic concept. As discussed earlier (Davis et al. 2018, pp. 25–30), we suggest the following redefinition of *validation*:

> Validation: The process of determining the degree to which a model and its associated data are an accurate representation of the real world from the perspective of the intended uses of the model (Department of Defense 2009). Validation should be assessed separately with respect to (1) description, (2) causal explanation, (3) postdiction, (4) exploratory analysis, and (5) prediction.

For the five criteria, we have the following in mind:

- *Description*: Identify salient structure, variables, and processes.
- *Causal explanation*: Identify causal variables and processes and describe reasons for behavior in corresponding terms.
- *Postdiction*: Explain past behavior quantitatively with inferences from causal theory.
- *Exploratory analysis*: Identify and parameterize causal variables and processes; estimate approximate system behavior as a function of those parameters and variations of model structure (coarse prediction).
- *Prediction*: Predict system behavior accurately and even precisely (as assessed with empirical information).

The criterion of description is straightforward, but causality is more difficult. The concept of causality is deep, ambiguous, and debated. Whether two variables have a causal relationship, and what the direction of that relationship is, depends on the model and context, including assumptions about what is *normal* (Halpern 2016). Effects often have multiple causes, and it seldom makes sense to ask whether B is caused by A, but rather to ask whether A is *one* of the causal influences on B. Feedback effects are also common in systems. As a result, it is sometimes more appropriate to think in terms of *balances* and about different regions of the system's state space. Such thinking is familiar in ecology where the populations of predators and prey rise and fall over time.

Postdiction (sometimes called *retrodiction*) can have the negative connotation of after-the-fact rationalization, but the connotation here is positive, that of explaining previously observed system behavior using theory. Physicist Steven Weinberg uses the example of Einstein's explanation of the observed anomaly in Mercury's orbit. Weinberg argues that explanatory and postdictive analyses are as important and persuasive in science as predictions (Weinberg 1994, p. 96ff). Explaining Mercury's behavior was in part persuasive to physicists because of the theory's elegance, even though the agreement of

Einstein's predictions with empirical data was equivocal for years. Geological science, of course, also depends on postdiction. A more recent example might be that economists can persuasively explain the economic crash of 2007–2008 by looking at information about the state of the economy (and central-bank policies) before the crash – information that was available in principle beforehand, but that was not adequately appreciated at the time. In retrospect, the crash was avoidable (Financial Crisis Inquiry Commission 2011).

Exploratory analysis and coarse prediction refer to studying the model's behavior over the uncertain space of inputs (also called scenario space), perhaps to identify regions with *good* and *bad* characteristics. The purpose is not to predict what will happen, but to understand what *may* happen, and to estimate the circumstances under which various behaviors are most likely. But how might one judge validity for such purposes? The primary observation here is that exploratory analysis (analysis across the case space implied by the values of all model parameters) is not useful if the model used is structurally incorrect, as with omitting key variables or significantly misrepresenting their effects and interactions. A military example would be to use a model that ignores the influence of airpower on ground-force maneuver. In the social-policy domain, an example would be to evaluate various urban-planning options without considering the value that people ascribe to community, parks, and safety. To conclude that a model is valid for exploratory analysis of a particular problem area, then, is nontrivial and often depends on qualitative social science as well as more usual methods of evaluating models. Sometimes, a model will be good enough to help disentangle conflicting influences and to help motivate a rough understanding of events (Lofdahl 2019). Sometimes, it will be adequate for recognizing high- and low-risk regions of a system's state space (Davis and O'Mahony 2013). Sometimes, the model is deemed good enough to give decision-makers valid insights about measures to consider and avoid and about how to monitor and adapt as empirical information comes in (Rouse 2015 and Thompson et al. 2019).

Figure 2.2 illustrates how a model might be characterized in this five-dimensional framework. The strength of a given model along a given dimension of validity is measured on a 0–5 scale, and the result is a *spider plot* (sometimes called *radar plot*). The notional model characterized in Figure 2.1 is said to be descriptive, to have a good sense of the causal variables and processes, and to be good for postdiction and exploratory analysis, but to be poor for prediction. Why the latter? Perhaps the values of the key causal variables are not known currently – i.e. they are knowable in principle, but not currently (a circumstance common in strategic planning and science). This is why so much social science is expressed in contingent terms. That *wishy-washiness* may make decision-makers unhappy, but predicting the details of future states of the world is often not in the cards. In such cases, *good* theory is contingent, not narrowly predictive.

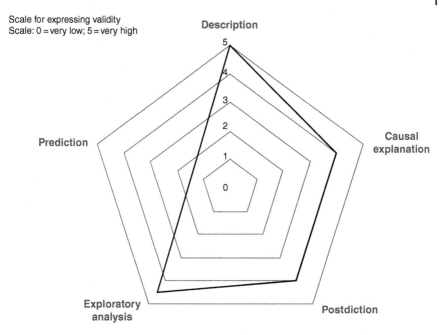

Figure 2.2 Spider chart of model validity by five criteria (adapted from Davis, et al. 2018).

Other models would have very different spider-diagram characteristics. For example, some models are predictive but not descriptive with respect to causal variables (as when macroeconomic models predicting next year's rise in gross domestic product have no obvious relationship to microeconomic mechanisms).

Assessing a Model's Validity in a Context
As has long been understood, validity can only be judged for a purpose and context (Grace et al. 2019). For example, is the model being asked whether a policy intervention will have a positive effect or what the magnitude of that effect will be? Is the question being asked when the state of the system is near equilibrium and the intervention effects would be marginal and captured by elasticities, or is it being asked when the system is *on the edge of chaos*?

A related question is whether the system has *stationary behavior* or whether its structure or basic relationships are changing (a point mentioned in a chapter in this volume by Kathleen Carley that also discusses validation, including incremental validation; Carley 2019). Weather models were long considered to be statistically valid, as in characterizing once-in-a-century storms. They are no longer valid because the frequency of high-intensity storms is increasing, presumably due to climate change. As a second example, classic two-party American political models are no longer valid because so many individuals regard themselves as independent. Further, polarization has increased with the disappearance of *moderates.*

Some General Criteria for Validation

Theorists and modelers have been struggling with the issue of validation for many years. A good deal of de facto consensus exists at the practical level. Some methods apply to each of the five dimensions discussed above. Our own summary is that judgments about the validity of theories and models are and should be based on:

1. *Falsifiability* (if a model cannot be falsified, it fails as meaningful science).
2. *Roots* in deeper theory regarded as valid.
3. *Logic.*
4. *Confirmational evidence*, especially model successes in cases chosen to attempt falsification, but also in broad patterns of success.
5. *Elegance*: Theoretical physicists are famous or notorious for their attention to matters such as coherence and beauty. Others look for consistency and power of description and explanation.
6. *Due diligence*: Since judgment is involved, a key is whether due diligence has occurred in considering all of the foregoing.

Our list is pragmatic. We do not attempt to define necessary and sufficient conditions for *validity*, which would arguably be to chase a red herring because validity is ultimately about whether we have adequate confidence in the model for a particular purpose.

Table 2.2 summarizes the way we recommend discussion of model validity in the SB regime (and, actually, more generally).

Strategy for Moving Ahead

An earlier report (Davis et al. 2018) provided more details on our recommendations on how to move ahead in SB research. What follows is a brief summary. Figure 2.3 from that report suggests a way ahead with three pillars of progress in social science: theory and modeling, empirical observation, and the newer pillar of computational experimentation. Activity occurs in a larger ecology of enabling activities. The first step in moving forward may be defining a few difficult national challenge problems for multiyear efforts forcing interdisciplinary work and providing the concrete context that motivates problem-solving.

Each such problem would have sub-challenges: (i) tightening links among theory, modeling, and experimentation; (ii) seeking better and more unifying theories while retaining alternative perspectives and narratives; (iii) challenging experimenters to find new theory-informed ways to obtain relevant

Table 2.2 A syntax for discussing model validity.

Dimensions	Criteria
Description, explanation, postdiction, exploration, prediction	Falsifiability and falsification; roots in grounded theory; logic; conformational evidence; elegance, as with coherence and power; due diligence in evaluations

Illustrative syntax

Example 1. For the purpose of comparing these options for this purpose, we conclude – on the basis of independent studies and workshops to debate issues and find holes – that the model is certainly falsifiable, has not been falsified, is logical and motivated by broader knowledge, has significant empirical evidence to support it, and seems to characterize our knowledge as well as possible at this point in time. It seems good for description and explanation, but it (and our uncertainties about its input values and certain details of structure) is not adequate for either postdiction or prediction. It is as valid as anything we have for exploring the problem across the range of uncertainties to decide which of the options available would be most flexible, adaptive, and robust. No special weight should be given to the model's outputs for *best estimate* values of its inputs. The model should be reevaluated over time as we gain further information.

Example 2 (Different model, different purpose). We have six months of massive relevant data on the system in question. After calibration with test-set data, the model has proven highly accurate for predicting results for the remaining empirical data. We conclude that so long as the system is relatively stable (and the existing data does not suggest any ongoing changes), the model should be a good basis for predicting future behavior (and for exploration). The model is the algorithmic result of applying machine intelligence methods. At this point in time, such methods do not generate models with explanatory power.

information and analyze data; (iv) improving methods for using models to inform decisions; (v) challenging theorists and technologists to provide new methods and tools; and (vi) nurturing the overall ecology. We discuss a few of these, briefly, in the following paragraphs.

Tightening the Theory–Modeling–Experimentation Research Cycle

A major issue is how to improve interactions among SB scientists on the one hand and *modelers* on the other. It is odd, in a sense, that this should be necessary. Why are the scientists and modelers not the same people, as they often are in, e.g. physics, engineering, and economics? A related issue is the need to improve the degree to which theories and models can be comprehended, reproduced, peer reviewed, debated, and iterated within the scientific community.

Sometimes idealizations suggest directions. Figure 2.4 shows an idealized way to relate the real and model worlds. The imagery is that a real system

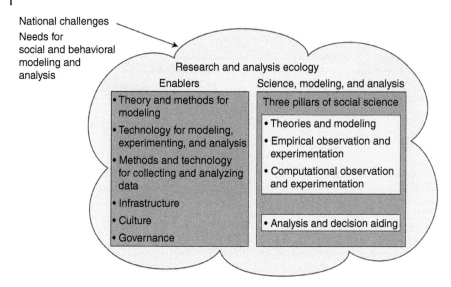

Figure 2.3 The ecology to respond to national challenges. Source: Davis et al 2018. Reproduced with permission of the RAND Corporation.

exists (the social system of interest, item 1). Real-world observation and experimentation (item 2) help in in forming hypotheses about the system's elements, relationships, and processes. Because theory and modeling are simplifications, we must have particular objectives in mind when asking about the real world or how to model it. We may construct one or more rough-cut system theories in our heads to describe the relevant reality (item 3). Often, alternative notions about the system exist, reflecting different hypotheses, perspectives, or both. This is symbolized by the stacking of icons.

Moving rightward, we construct coherent conceptual models of the system (item 4) – including aspects that are important but not observable directly. The conceptual models may be combinations of essays, listings of objects and attributes, or such qualitative devices as influence, stock-and-flow, or network diagrams. The next step, when feasible, is to develop a formal model (item 5), i.e. one that specifies all the information needed for computation of model consequences. A specified model must have tight definitions, equations, algorithms, or tables.

In this idealized image, the formal model is independent of programming language or is expressed in a high-level language that is easily comprehended by non-programmers (i.e. in a week rather than a month). The intent is to lay bare the essence of the model without the mind-muddling complications of most computer code and to permit replication, peer review, debate, and iteration

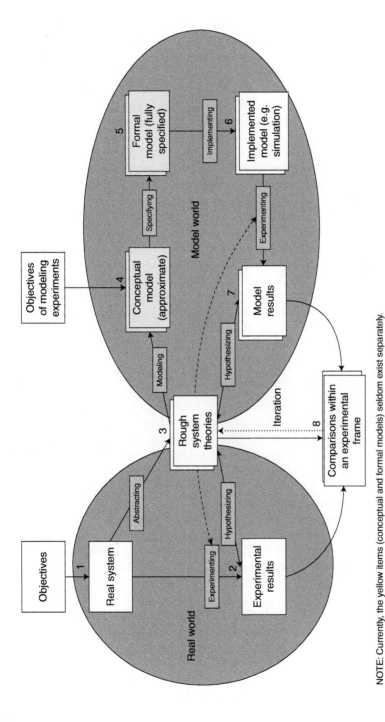

NOTE: Currently, the yellow items (conceptual and formal models) seldom exist separately.

Figure 2.4 An idealized system view of theory, modeling, and experimentation. Currently the conceptual and formal models at the top right seldom exist separately. Source: Davis et al. (2018). Reproduced with permission of RAND Corporation.

among scientists. After those occur, a formal model can (moving downward) be implemented in other programming languages as convenient (item 6). Moving leftward, the implemented model can then be used with an experimental design to generate model results across the cases implied by different values of model inputs. Results of these exploratory computational experiments (item 7) may falsify, enrich, or support earlier beliefs. For example, they may suggest that the system will show troublesome behavior in certain circumstances not previously considered. Such coarse computational predictions should be compared to experimental results from the real system (item 8). To do so sensibly requires defining the relevant *experimental frame,* i.e. specifying the conditions under which the real and model systems are to be observed and how model results are being used (Zeigler 1998). A model can be considered valid for a particular application in a particular context if the result of using the model's outputs is believed – using all available knowledge – to be adequately close to the result of using the real system's behavior for that experimental frame. The cycle continues (dotted arrow). Overall, Figure 2.4 shows a virtuous research cycle with a holistic view in which theory, models, empirical inquiry, and computational inquiry are all part of a closely networked system. Today's actual research process in SB work is quite different. If the idealization is attractive, what can be done to move toward it? There is need for new ideas, debate, and example setting.

Improving Theory and Related Modeling

Tables 2.3 and 2.4 list some priorities for SB research, with some of the items having been discussed earlier in this chapter. For the purposes of SB system modeling, the primary focus of this chapter and the larger volume, Table 2.3 urges an approach to theory and modeling that is closer to the science, puts a premium on multivariate causal models, translates *context* into concrete variables and cases, and confronts uncertainty seriously. Such modeling should include multiresolution, multiperspective families of models, should synthesize across current simplistic theories, and should allow translations among different views of the system. Models should be designed to promote mainstay features of good science, including reproducibility, comprehensibility, peer review, and iteration. Table 2.3 suggests going about this by organizing for both collaborative and competitive interdisciplinary work.

Table 2.4 lists a similar summary relating to priorities for both computational and empirical experimentation. It recommends moving away from statistical work using data of convenience toward seeking to find the data most relevant to the underlying phenomena, even though doing so is notoriously difficult and will require more in-depth case studies and fieldwork, as well as new methods. Exploiting opportunities and shortening cycles are straightforward admonitions, but the next one is not. In comparing both computational and empirical

Table 2.3 Priorities for improving theory and modeling.

Item	Comment
Move modeling farther into the science	Example: an anthropologist professor and graduate student do the modeling directly or with exceptionally tight teaming across departments
Put a premium on causal models including all important variables, whether or not soft, and allowing for system characteristics to be emergent	This will require variable-structure simulation
Define context by defining what characterizes it, so as to be better able to define general theories that specialize by context	
Build in uncertainty analysis from the outset	Include both structural and parametric deep uncertainties
Use portfolios of methods	Example: simulation, equilibrium models, static models, qualitative models, and human-in-loop simulation or gaming
Seek multiresolution, multiperspective families of models	Cut across levels of detail (e.g. individual, local, regional, and global, or short and longer time scales); respect different perspectives, including culture-sensitive views
Synthesize across theories where feasible	Example: unifying theory may suggest behavior will be due to any of several pathways depending on circumstances and history
Translate across theories	This should include mappings among equivalent formalisms and highlighting of incommensurate theories (e.g. individual-centric and culture-centric models)
Design for good practice	Design for reproducibility, comprehensibility, peer review, and iteration
Organize for collaborative and competitive interdisciplinary work	Problem-focused inquiry will not honor disciplinary boundaries

data to *theory*, the emphasis should be on testing and enriching theories that are more unified and comprehensive, rather than testing theories that we know a priori only capture narrow aspects of the phenomenon. This will require having some more unified theories to test, as well as new data-driven methods for finding such theories with computational methods. The purpose should not be to find which of various simplistic theories happens to fit particular data best, but rather to help test or establish more unified synthesizing theories. Another purpose, especially in the modern era of ubiquitous SB data of some

Table 2.4 Improving computational and empirical experimentation.

Item	Comment
Seek the right data even if inconvenient. Focus on problem (e.g. a national challenge)	Address difficult-to-measure and sometimes latent qualitative variables, preferring representative albeit uncertain data rather than more precise but inadequate proxies
Exploit opportunities	Better exploit modern and emerging data sources, which are sometimes massive, sometimes sparse, and sometimes riddled with correlations and biases
Shorten the cycle	Greatly increase the speed with which needed data can be obtained and processed
Give nontrivial theory a chance, paying less attention to testing of obviously simplistic theories	Use integrative multivariate causal theory to inform specifications used in statistical analyses while not imposing that theory. Use data to test and refine theory
Simultaneously encourage competition between theory-informed and data-driven approaches	For some purposes, data-driven work may yield better predictive capability. For other purposes, theory-informed work may improve better explanation and better ability to anticipate consequences of major changes
Use exploratory analysis routinely	Analyze computational data with the methods of exploratory analysis, looking at outcome landscapes and phase diagrams rather than point outcomes

types, should be to encourage competition between theory-informed and more data-driven approaches. Data-driven approaches will sometimes be superior (especially as artificial intelligence methods improve with respect to generating explanatory models) and sometimes faster to develop and faster to adapt to changing circumstances. In other cases, theory-informed modeling will prove superior, especially for explaining, exploring, more deeply understanding phenomena, and for anticipating changes. Let the competition proceed.

Attending to the priorities of Tables 2.3 and 2.4 requires methods and technology, some of which do not yet exist. Some priority questions for providers of modeling theory, methods, and tools are: *how* do we do the functions called for above, such as routine multidimensional deep-uncertainty analysis, variable-structure simulation, and theory-informed empirical and computational experiments? Some other examples are: *how* do we make the best use of modern data sources, infer causal relations from observational data, accomplish heterogeneous fusion, and otherwise manage knowledge across variations of scale, formalism, and character? So also, there are profound issues involved in finding ways to build conceptual and formal models that are comprehensible to subject-area scientists and that are subject to reproducibility, peer review, debate, sharing, and iteration. Validation has always been a challenge but is even more problem when accommodating analysis under multidimensional deep uncertainty and improving comprehensibility and the ability to check composition validity.

Finally, a pressing challenge is improving methods for aiding decision-makers, which requires substantive analysis and effective communication. Great strides have been made over the last 25 years related to model-based analysis for complex problems, and some of the corresponding lessons need to be assimilated in conceiving and nurturing research on SBM modeling (Davis 2019).

Social-Behavioral Laboratories

One mechanism for proceeding is to have, for each national challenge, a virtual social-behavioral modeling laboratory (SBML). This might be an organized program akin to the Genome Project of years past, but we use the term laboratory to convey a sense of purposeful and organized scientific inquiry to *crack* a particular challenge (Davis et al. 2018).

A given SBML (Figure 2.5) would exist for 5–10 years and would enable interdisciplinary sharing and synergism. An SBML approach would not seek a monolithic standardized model federation with approved structure and databases. Instead, the approach would be dynamic and iterative with routine competition, iteration, and evolution. Meaningful model compositions would be constructed for specific purposes. The SBML activities would include both simulation modeling (generating system behavior over time) and other forms of qualitative and quantitative modeling, to include participative modeling and such other forms of human interaction as gaming.[5] Related conferences would focus on the national challenge, the state of related social science, the degree to which modeling represents that science, the products of empirical work and computational experimentation, and how to characterize knowledge and inform decisions. Comparing lessons from multiple national challenges would reveal further generalizations.

A successful SBML approach would foster a new *epistemic culture* in SB research: those participating would be building knowledge in a different and more multifaceted way than is customary. The result might reflect not just scientific knowledge and craft, but also what Aristotle called phronesis (practical wisdom and thoughtfulness, reflecting an understanding of ethics and situational subtleties). The word may be Greek, but the ideas endure.

---

5 Some chapters in this volume describe interactive methods that use agents (Miller et al. 2019), gaming (Lakkaraju et al. 2019), and simulation analytics (Swarup et al. 2019) in laboratory approaches. Others describe use on online gaming as data (Guarino et al. 2019; Lakkaraju et al. 2019).

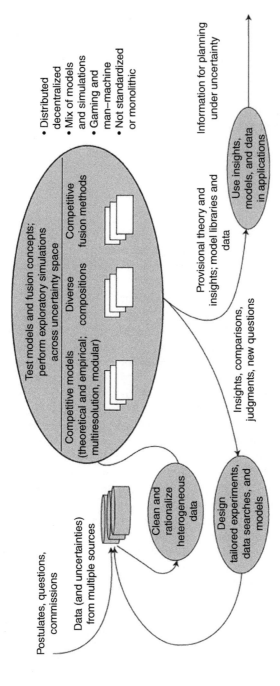

Figure 2.5 An SBML for a particular national challenge. Source: Davis et al. (2018). Reproduced with permission of RAND Corporation.

Conclusions

It may be good to end this chapter by indicating how we suggest addressing the overarching complications mentioned at the outset:

Human cognition: ABM should move smartly in the direction of better representing the human and social behaviors very different from those of pure rational actors. Doing so would represent cognitive biases, the fast and slow modes of thinking, and social analogues (i.e. we believe organizations and even societies have analogous characteristics). This should be in addition to modeling that represents some behavior as having more primal basis (Pavlic 2019).

CAS: We recommend an emphasis on beyond-what-if simulation and analysis that asks questions such as "Under what circumstances will the optional intervention probably be useful and relatively free from side effects, rather than making things potentially worse?" This can mean major changes in the form of analysis based on simulation (Davis 2019). We also recommend that simulation be supplemented with interactive activities, to include human gaming and collaborative exercises. Some of the most profound applications of modeling and simulation occur when they are used interactively and collaboratively so that relevant decision-makers and stakeholders understand and buy into the modeled representation of issues and have a substantial body of common knowledge. Such circumstances are more likely to generate ways ahead that are broadly acceptable and moving developments *in the right direction*. Decision-makers will draw upon insights from modeling and simulation (e.g. understanding the potential adverse consequences of actions and the need to plan for adaptation), but will not need to believe the detailed predictions of models. The modeling itself may recede to the background, rather than being touted as the basis for decision, but the exploratory experience with model-generated scenarios may have a profound and beneficial effect (as, historically, human gaming has helped educate policy-makers).

Narratives: Modelers need to develop effective methods for representing narratives in a combination of inputs and outputs, in part because narratives can sometimes be seen as important causal factors and in part because communicating about the system and the implications of system interventions will be more effective when it connects with competing narratives. Because competing narratives are a major source of tension and conflict, SBM should also seek to falsify seriously incorrect narratives in convincing ways while giving visibility to alternative competing but legitimate narratives. Analysis should sometimes seek options that can be attractive under alternative narratives. It should sometimes seek options that could influence the prevalent narrative or options that would be effective in a societal context with a particular narrative. All of this may require alternative agents and alternative depictions of system processes. It should also be possible for research-level SBM to demonstrate realistic emergence of new or altered narratives in populations, whether by conceiving them in advance and allowing the simulation to choose the more appropriate ones in a situation or by allowing intelligent agents to recognize and report on new narratives.

Wicked problems: Dealing with wicked problems must confront differences of values and perceptions, which is difficult to do in modeling without using interactive methods (including gaming) to bring out the relevant tensions. In some cases, the resulting insights can then be mapped into agents with the expectation of simulation in which resolutions (or failures) will emerge. In other cases, as in the paragraph above, it will prove more valuable to think in terms of interactive settings, including games, with modeling and simulation support.

Acknowledgments

This chapter derives from research accomplished for the Defense Advanced Research Projects Agency (DARPA) as in a previous report (Davis et al. 2018).

References

Angrist, J.D. and Pischke, J.-S. (2009). *Mostly Harmless Econometrics: An Empiricist's Companion*. Princeton, NJ: Princeton University Press.

Ashby, N. (2002). Relativity and the global positioning system. *Physics Today* 55 (5): 41–47.

Axtell, R., Carella, R., and Casstevens, R. et al. (2016). *Agentization: Relaxing Simplistic Assumptions with Agent Computing*. Fairfax, VA: George Mason University.

Brashears, M.E. (2019, this volume). How big and how certain? a new approach to defining levels of analysis for modeling social science topics. In: *Social-Behavioral Modeling for Complex Systems* (ed. P.K. Davis, A. O'Mahony and J. Pfautz). Hoboken, NJ: Wiley.

Carley, K.M. (2019, this volume). Social-behavioral simulation: key challenges. In: *Social-Behavioral Modeling for Complex Systems* (ed. P.K. Davis, A. O'Mahony and J. Pfautz). Hoboken, NJ: Wiley.

Cartwright, N. (1983). *How the Laws of Physics Lie*. Oxford: Oxford University Press.

Cartwright, N. (1999). *The Dappled World: A Study of the Boundaries of Science*. Cambridge: Cambridge University Press.

Corman, S.R.N. (2012). Understanding sociocultural systems through a narrative lens. In: *A Sociocultural Systems Primer for the Military Thinker: Multidisciplinary Perspectives and Approaches* (ed. L.L. Brooks, B. Strong, M. Zbylut and L. Roan), 71–86. Leavenworth, KS: U.S. Army Research Institute.

Corman, S.R.N., Suston, S.W., and Tong, H. (2019, this volume). Toward generative narrative models of the course and resolution of conflict. In: *Social-Behavioral Modeling for Complex Systems* (ed. P.K. Davis, A. O'Mahony and J. Pfautz). Hoboken, NJ: Wiley.

Cowherd, G. and Lende, D.H. (2019, this volume). Social media. Global connections, and information environments: building complex understandings of multi-actor interactions. In: *Social-Behavioral Modeling for Complex Systems* (ed. P.K. Davis, A. O'Mahony and J. Pfautz). Hoboken, NJ: Wiley.

Davis, P. (2016). *Capabilities for Joint Analysis in the Department of Defense: Rethinking Support for Strategic Analysis*. Santa Monica, CA: RAND Corporation.

Davis, P.K. (2018). Simple culture-informed models of the adversary. In: *Advances in Culturally-Aware Intelligent Systems and in Cross-Cultural Psychological Studies* (ed. C. Faucher), 235–264. Cham: Springer International Publishing.

Davis, P.K. (2019, this volume). Lessons on decision aiding for social-behavioral modeling. In: *Social-Behavioral Modeling for Complex Systems* (ed. P.K. Davis, A. O'Mahony and J. Pfautz). Hoboken, NJ: Wiley.

Davis, P.K. and Anderson, R.H. (2003). *Improving the Composability of Department of Defense Models and Simulations*. Santa Monica, CA: RAND Corporation.

Davis, P.K. and O'Mahony, A. (2013). A Computational Model of Public Support for Insurgency and Terrorism: A Prototype for More General Social-Science Modeling. TR-1220, Santa Monica, CA: RAND Corporation.

Davis, P.K. and O'Mahony, A. (2017). Representing qualitative social science in computational models to aid reasoning under uncertainty: national security examples. *Journal of Defense Modeling and Simulation* 14 (1): 1–22.

Davis, P.K., O'Mahony, A., Gulden, T. et al. (2018). *Priority Challenges for Social and Behavioral Research and its Modeling*. Santa Monica, CA: RAND Corporation.

Department of Defense (2009). DoD Modeling and Simulation (M&S) Verification, Validation, and Accreditation (VV&A). Number 5000.61.

Diamond, J. (1987). Soft sciences are often harder than hard sciences. *Discover* 8: 34–39.

Faucher, C. (ed.) (2018). *Advances in Culturally-Aware Intelligent Systems and in Cross-Cultural Psychological Studies*. Cham: Springer International.

Financial Crisis Inquiry Commission (2011). Financial Crisis Inquiry Report.

Gabbay, M. (2019, this volume). Integrating experimental and computational approaches to social influence. In: *Social-Behavioral Modeling for Complex Systems* (ed. P.K. Davis, A. O'Mahony and J. Pfautz). Hoboken, NJ: Wiley.

Garibay, I., Gunaratne, C., Yousefi, N., and Scheitert, S. (2019, this volume). The agent-based model canvas: a modeling lingua franca for computational social science. In: *Social-Behavioral Modeling for Complex Systems* (ed. P.K. Davis, A. O'Mahony and J. Pfautz). Hoboken, NJ: Wiley.

George, A.L. and Bennett, A. (2005). *Case Studies and Theory Development in the Social Sciences (BCSIA Studies in International Security)*. Cambridge, MA: MIT Press.

Gigerenzer, G. and Selten, R. (2002). *Bounded Rationality: The Adaptive Toolbox*. Cambridge, MA: MIT Press.

Grace, E., Blaha, L.M., Sathanur, A.V. et al. (2019, this volume). Evaluation and validation approaches for simulation of social behavior: challenges and opportunities. In: *Social-Behavioral Modeling for Complex Systems* (ed. P.K. Davis, A. O'Mahony and J. Pfautz). Hoboken, NJ: Wiley.

Gribaudo, M., Iacono, M., and Levis, A.H. (2019, this volume). Multi-formalism modeling of complex social-behavioral systems. In: *Social-Behavioral Modeling for Complex Systems* (ed. P.K. Davis, A. O'Mahony and J. Pfautz). Hoboken, NJ: Wiley.

Guarino, S., Eusebi, L., Bracken, B., and Jenkins, M. (2019, this volume). Using sociocultural data from online gaming and game communities. In: *Social-Behavioral Modeling for Complex Systems* (ed. P.K. Davis, A. O'Mahony and J. Pfautz). Hoboken, NJ: Wiley.

Haack, S. (2011). *Defending Science – within Reason: Between Scientism and Cynicism*. Amherst, NY: Prometheus Books.

Hadzikadic, M. and Whitmeyer, J. (2019, this volume). Lessons and musings for experience with agent-based modeling. In: *Social-behavioral Modeling for Complex Systems* (ed. P.K. Davis, A. O'Mahony and J. Pfautz). Hoboken, NJ: Wiley.

Halpern, J.Y. (2016). *Actual Causality*. Cambridge, MA: The MIT Press.

Holland, J.H. and Mimnaugh, H. (1996). *Hidden Order: How Adaptation Builds Complexity*. New York: Perseus Publishing. ISBN: 0201442302.

Jervis, R., Lebow, R.N., and Stein, J.G. (1985). *Psychology and Deterrence*. Baltimore, MD: Johns Hopkins University Press.

Kahneman, D. (2011). *Thinking, Fast and Slow*, 1e. New York: Farrar, Straus and Giroux.

Kahneman, D. and Tversky, A. (1982). Intuitive prediction: biases and corrective procedures. In: *Judgment Under Uncertainty: Heuristics and Biases* (ed. D. Kahneman, P. Slovic and A. Tversky), 414–421. New York, NY: Cambridge University Press.

Klein, G. (1998). *Sources of Power: How People Make Decisions*. Cambridge, MA: MIT Press.

Konner, M. (2003). *The Tangled Wing: Biological Constraints on the Human Spirit.* New York, NY: Henry Holt and Company.

Kött, A. and Citrenbaum, G. (eds.) (2010). *Estimating Impact: A Handbook of Computational Methods and Models for Anticipating Economic, Social, Political and Security Effects in International Interventions.* New York: Springer.

Lakkaraju, K., Epifanovskaya, L., States, M. et al. (2019, this volume). Online games for studying behavior. In: *Social-Behavioral Modeling for Complex Systems* (ed. P.K. Davis, A. O'Mahony and J. Pfautz). Hoboken, NJ: Wiley.

Levis, A.L. (2016). Multi-formalism modeling of human organization. In: *Seminal Contributions to Modeling and Simulation. Basel* (ed. K. Al-Begain and A. Bargiela), 23–46. Switzerland: Springer International.

Lofdahl, C. (2019, this volume). Modeling information and gray zone operations. In: *Social-Behavioral Modeling for Complex Systems* (ed. P.K. Davis, A. O'Mahony and J. Pfautz). Hoboken, NJ: Wiley.

McNamara, L.A., Trucano, T.G., and Gieseler, C. (2011). *Challenges in Computational Social Modeling and Simulation for National Security Decision-Making.* Albuquerque, NM: Sandia National Laboratories.

Miller, L.C., Wang, L., Jeong, D.C., and Gillig, T.K. (2019, this volume). Bringing the "real world" into the experimental lab: technology-enabling transformative designs. In: *Social-Behavioral Modeling for Complex Systems* (ed. P.K. Davis, A. O'Mahony and J. Pfautz). Hoboken, NJ: Wiley.

Mullen, F.E. (2013). *Dynamic Multilevel Modeling Framework Phase I – Feasibility.* Washington, DC: Modeling & Simulation Coordination Office (M&SCO) of the U.S. Department of Defense.

National Academy of Sciences, E (1999). *Science and Creationism: A View from the National Academy of Sciences*, 2e.

National Research Council (2006). *Defense Modeling, Simulation, and Analysis: Meeting the Challenge.* Washington, DC: National Academies Press.

National Research Council (2014). *U.S. Air Force Strategic Deterrence Analytic Capabilities: An Assessment of Methods, Tools, and Approaches for the 21st Century Security Environment* (December 20, Trans.). Washington, DC: National Academies Press.

Nicolis, G. and Prigogine, I. (1977). *Self-Organization in Nonequilibrium Systems: From Dissipative Structures to Order Through Fluctuations.* Wiley. ISBN: 0471024015.

Nyblade, B., O'Mahony, A., and Sieck, K. (2019, this volume). State of social and behavioral science theories. In: *Social-Behavioral Modeling for Complex Systems* (ed. P.K. Davis, A. O'Mahony and J. Pfautz). Hoboken, NJ: Wiley.

O'Mahony, A. and Davis, P.K. (2019, this volume). Panel discussion: moving social-behavioral modeling forward. In: *Social-Behavioral Modeling for Complex Systems* (ed. P.K. Davis, A. O'Mahony and J. Pfautz). Hoboken, NJ: Wiley.

Orr, M.G. (2019, this volume). Multi-scale resolution of human social systems: a synergistic paradigm for simulating minds and society. In: *Social-Behavioral Modeling for Complex Systems* (ed. P.K. Davis, A. O'Mahony and J. Pfautz). Hoboken, NJ: Wiley.

Osoba, O. and Kosko, B. (2017). Fuzzy knowledge fusion for causal modeling. *Journal of Defense Modeling and Simulation* 14 (1): 17–32.

Page, S.E. (2010). *Diversity and Complexity (Primers in Complex Systems)*, 1e. Princeton, NJ: Princeton University Press.

Paul, C. (2019, this volume). Homo narratus (the storytelling species) the challenge (and importance of modeling narrative in human understanding). In: *Social-Behavioral Modeling for Complex Systems* (ed. P.K. Davis, A. O'Mahony and J. Pfautz). Hoboken, NJ: Wiley.

Pavlic, T.P. (2019, this volume). Social models from non-human systems. In: *Social-Behavioral Modeling for Complex Systems* (ed. P. Davis, A. O'Mahony and J. Pfautz). Hoboken, NJ: Wiley.

Pearl, J. (2009). *Causality: Models, Reasoning, and Inference*. Cambridge, MA: Cambridge University Press.

Pearl, J. and Mackenzie, D. (2018). *The Book of Why: The New Science of Cause and Effect*, 1e. Basic Books.

Rainey, L.B. and Jamshidi, M. (Eds.). (2018). *Engineering Emergence: a Modeling and Simulation Approach*. Boca Raton, FL: CRC Press, Taylor & Francis Group.

Rand, W. (2019, this volume). Theory-interpretable, data-driven agent-based modeling. In: *Social-Behavioral Modeling for Complex Systems* (ed. P.K. Davis, A. O'Mahony and J. Pfautz). Hoboken, NJ: Wiely.

Read, S.J. and Miller, L.C. (2019, this volume). A neural network model of motivated decision-making in everyday social behavior. In: *Social-Behavioral Modeling for Complex Systems* (ed. P.K. Davis, A. O'Mahony and J. Pfautz). Hoboken, NJ: Wiley.

Rosenhead, J. and Mingers, J. (eds.) (2002). *Rational Analysis for a Problematic World Revisited: Problem Structuring Methods for Complexity, Uncertainty and Conflict*, 2e. New York: Wiley.

Rouse, W.B. (2015). *Modeling and Visualization of Complex Systems and Enterprises*. Hoboken, NJ: Wiley.

Rouse, W.B. (2019, this volume). Human-centered design of model-based decision support for policy and investment decisions. In: *Social-Behavioral Modeling for Complex Systems* (ed. P.K. Davis, A. O'Mahony and J. Pfautz). Hoboken, NJ: Wiley.

Sambanis, N. (2004). Using case studies to expand economic models of civil war. *Perspectives on Politics* 2 (02): 259–279.

Simon, H.A. (1957). A behavioral model of rational choice. In: *Models of Man: Social and Rational: Mathematical Essays on Rational Human Behavior in a Social Setting* (ed. H.A. Simon), 241–260. New York: Wiley.

Simon, H.A. (1978). Nobel Prize Lecture: Rational Decision-Making in Business Organizations (February 3, Trans.). NobelPrize.org (accessed 31 August 2018).

Simon, H.A. (1996). *The Sciences of the Artificial*, 3e. Cambridge, MA: The MIT Press.

Sliva, A., Really, S.N., Blumstein, D., and Peirce, G. (2019, this volume). Combining data-driven and theory-driven models for causality analysis in sociocultural systems. In: *Social-Behavioral Modeling for Complex Systems* (ed. P.K. Davis, A. O'Mahony and J. Pfautz). Hoboken, NJ: Wiley.

Sterman, J.D. (2000). *Business Dynamics: Systems Thinking and Modeling for a Complex World*. Boston, MA: McGraw-Hill. ISBN: 007238915X.

Sukthankar, G. and Beheshti, R. (2019, this volume). Using agent-based models to understand health-related social norms. In: *Social-Behavioral Modeling for Complex Systems* (ed. P.K. Davis, A. O'Mahony and J. Pfautz). Hoboken, NJ: Wiely.

Sunstein, C.R. (2014). *Why Nudge?: The Politics of Libertarian Paternalism (The Storrs Lectures Series)*. New Haven, CT, Yale University Press.

Swarup, S., Marathe, A., Marathe, M.V., and Barrett, C.L. (2019, this volume). Simulation analytics for social and behavioral modeling. In: *Social-Behavioral Modeling for Complex systems* (ed. P.K. Davis, A. O'Mahony and J. Pfautz). Hoboken, NJ: Wiley.

Taylor, S.J.E., Kahn, A., Morse, K.L. et al. (2015). Grand challenges for modeling and simulation: simulation everywhere – from cyberinfrastructure to clouds to citizens. *Simulation* 91: 648–655.

Thaler, R.H. (1994). *Quasi Rational Economics*. New York: Russell Sage Foundation.

Thompson, J., McClure, R., and DeSilva, A. (2019, this volume). A complex systems approach for understanding the effect of policy and management interventions on health system performance. In: *Social-Behavioral Modeling for Complex Systems* (ed. P.K. Davis, A. O'Mahony and J. Pfautz). Hoboken, NJ: Wiley.

Tolk, A. (ed.) (2012). *Engineering Principles of Combat Modeling and Distributed Simulation*. Hoboken, NJ: Wiley.

Tolk, A., Diallo, S., Padilla, J.J., and Herencia-Zapana, H. (2013). Reference modeling in support of M&S – foundations and applications. *Journal of Simulation* 7: 69–82.

Weinberg, S. (1994). *Dreams of a Final Theory: The Scientist's Search for the Ultimate Laws of Nature*. New York: Vintage.

Zacharias, G.L., MacMillan, J., and Van Hemel, S.B. (eds.) (2008). *Behavioral Modeling and Simulation: From Individuals to Societies*. Washington, DC: National Academies Press.

Zeigler, B.P. (1998). A framework for modeling and simulation. In: *Applied Modeling and Simulation: An Integrated Approach to Development and Operations* (ed. D.S. Cloud and L.B. Rainey), 67–103. New York: McGraw Hill.

Zeigler, B.P. and Hammond, P.E. (2007). *Modeling & Simulation-Based Data Engineering: Introducing Pragmatics Into Ontologies for Information Exchange*. New York: Academic Press.

3

Ethical and Privacy Issues in Social-Behavioral Research

Rebecca Balebako[1], Angela O'Mahony[2], Paul K. Davis[2], and Osonde Osoba[2]

[1] *RAND Corporation, Santa Monica, CA 90401, USA*
[2] *RAND Corporation and Pardee RAND Graduate School, Santa Monica, CA 90407, USA*

In today's environment, personal data collection and exploitation is pervasive. Government, industry, researchers, and society grapple with dilemmas about how to make use of data to provide benefits to individuals and society while safeguarding individuals' privacy and following ethical procedures. Another puzzle for public policy is how and when the government should collect or use data (and protect them), particularly when those data are already being collected by industry, researchers, criminal networks, and even foreign adversaries. It is difficult to develop guidelines on such matters, and it is also likely that what makes sense will change over time, suggesting the need for adaptive policies. However, as we contemplate increasing the effectiveness of social-behavioral modeling and populating these models with data drawn from our everyday lives, how to balance privacy and efficacy is of paramount importance. The social-behavioral modeling community needs to be informed by a deep appreciation of current privacy and ethical issues and be prepared for new challenges that will emerge in the future.

This chapter is the result of a small workshop held at the RAND Corporation in December 2016 to understand what scholars and practitioners working in the field of data privacy considered the most troublesome tensions, where strict boundaries must be erected, and how tensions might be mitigated or resolved, in part with technology. Based on previous research and the workshop discussions, we identified six important topics to consider for reducing problems related to privacy and harms caused by misuse of personal data, ranging from how individuals are notified of data use to how to deter harms arising from data use. The topics were how to:

- Assure meaningful user notice and choice.
- Provide usable and accurate access control.

Social-Behavioral Modeling for Complex Systems, First Edition.
Edited by Paul K. Davis, Angela O'Mahony, and Jonathan Pfautz.
© 2019 John Wiley & Sons, Inc. Published 2019 by John Wiley & Sons, Inc.
Companion website: www.wiley.com/go/Davis_Social-Behavioralmodeling

- Anonymize data.
- Validate and audit data algorithms to avoid inadvertent harms.
- Challenge and redress data harms.
- Deter abuse.

Looking across these six issues, we had three main observations:

- Developing solutions to prevent harms caused by misuse of personal data is difficult, complicated, and beset by tensions (no simple divide exists between *good guys* and *bad guys*).
- The challenges of and potential solutions to data harms are interrelated. For example, providing mechanisms for redress are intertwined with increasing transparency about how data are used.
- Single-approach solutions are unlikely to be effective. Decision-makers should adopt a portfolio approach that looks for mixes of incentives, education, market solutions and related technologies, government regulations, and standards. The either-or fallacy should be avoided: it is evident that in most of the problem areas, a mixed approach is necessary for success.

What follows identifies priority topics for attention and action and summarizes considerations. In each case, much more in-depth work will be necessary in moving toward effective and feasible solution mixes that deal appropriately with dilemmas and trade-offs. We include possible solution elements that may or may not turn out to be feasible and attractive. Some related technologies may arise naturally from market forces; others are likely not to arise without government sponsorship and/or coordination.[1]

Improved Notice and Choice

Diagnosis

An approach with long precedents in legal affairs is for vendors to provide individuals with long-form notices about what will and will not be done with their data (Ware 1973; Cranor 2012; Solove and Hartzog 2014). Long privacy policies such as those posted on websites or in smartphone apps have been typically relied to provide consumers notice about company's data protection and privacy practices. Individuals who use products with long-form notices are presumed to have read and agree with the data practices, simply because the notices exist, regardless of the placement or clarity of the notice. However, long privacy policies, especially with legalese and hedging terms, often do not serve

---

1 The importance of these issues was dramatized in 2018 by the discovery that a private company, Cambridge Analytica, had improperly harvested Facebook data on 87 million users, including political beliefs, interests, and friends' information. In some cases, this even included the full content of private messages (Frenkel et al. 2018). The issue arose in connection with investigation of Russian attempts to influence the 2016 US election. Attempts to identify remedies will require dealing with all of the issues discussed in this chapter.

the implied purpose of informing the user as they are vague or ambiguous (Reidenberg et al. 2015). To understand and compare privacy policies would require users to spend significant time and effort – perhaps 81–253 hours each year just to skim privacy policies from websites that they use (McDonald and Cranor 2008). As for granting vendors permission to use data, users cannot realistically understand, internalize, and appreciate the consequences of data sharing practices that may be invisible and untraceable by the user (Acquisti and Grossklags 2005).

Although notice and choice has long been an important part of privacy protections in the US legal system – including being a central element of the Federal Trade Commission's (FTC) protection of consumer privacy in the United States (Solove and Hartzog 2014) – more recent and upcoming regulations, including the European Union's General Data Protection Regulation (GDPR), have recognized the failure of such forms of notice and are pushing for more user-friendly notices.

When parties such as researchers or governments use data that are nominally publicly available (such as open social network profiles or content), notice and choice is particularly problematic. Generally, these data are considered public since they have already been shared with minimal controls on who can access it. In this case, researchers may wish to use such data without providing additional notice to the individual, assuming the long-form privacy policy of the social network is sufficient. In these cases, the data subject, e.g. the person who wrote the tweet, Facebook post, or about whom the public record concerns, is not given notice that the data will be repurposed for research, health, or financial decisions. However, reusing data ignores the context in which people shared data (Hartzog 2016). Data subjects may not be aware that social network data have been used to make inferences about behavior that is often considered private, such as health information (Inkster et al. 2016). People who have logged on to a service using an account name and password may have some privacy expectations for their data and may not fully understand the implications of public data sharing, including the availability and power of data reuse (Acquisti and Grossklags 2005).

We see prospects for the existing model of long-form privacy policies to mitigate data harms as dismal. More recent and upcoming regulations, including the EU's GDPR, have recognized the failure of such forms of notice and are pushing for more user-friendly notices. Furthermore, technological changes are making possible more nuanced and personalized approaches.

Prescriptions

Many possibilities exist. These include:

- Strongly protective software defaults that provide security protection both against malicious, illegal attacks and opportunistic privacy invasions by third parties.

- Standardized easy-to-understand user agreements prohibited from including harmful permissions as a default or in order to use basic features of the service and that provide transparent implications for optional user choices that would increase risks.
- Context-specific user authorization. This would require services to reask users for permission to use their data as new contexts arise. This may be particularly useful for technologies in which no single default setting meets the majority of users' needs.
- Tools that make it easy to review and adjust permissions and preferences across programs and service.
- Tools that alert users at the time as to how their data may be used and the potential or likely implications (including harms) and that offer users the opportunity to tighten or loosen their agreements.
- A user privacy-authorization clearinghouse – a centralized (virtual) location that maintains user privacy preferences so that users can receive notice and make choice at a central point (by analogy, a few examples of information clearinghouses are law school applications, international financial transactions, and international police databases).
- An adjudication process to review changes to user agreements. User agreements are a contract in which currently only one side can easily modify the contract. Service providers can (and do) update user agreements frequently. Users' only options are to accept the change or opt out of service. An adjudication process, potentially under the authority of an administrative court judge, would provide both parties to user agreements, the user and the provider, opportunity to control the contract.

Usable and Accurate Access Control

Diagnosis

It is important that the right people have access to individuals' data as needed (e.g. a medical specialist to whom they have been referred). The general problem, however, is that too much information is available to the wrong people at some point, particularly after sharing, sales, and interceptions of information occur over time. To make things worse, information may be available indefinitely. This can lead to harms related to long-ago events, or news, fake or otherwise that never goes away, and so on (Mayer-Schönberger 2011). However, the ability to control access to data can be difficult not just for end users, but also for more sophisticated system administrators, who may need to decide who can get access to data and for how long, what the exceptions might be, and how blacklists and whitelists interplay (Kim et al. 2010; Mazurek et al.

2010). So while access control can be difficult to implement and use, access control might be necessary tool for protecting the privacy of the data subjects (Byun and Li 2008; Kagal and Abelson 2010; Raykova et al. 2012).

Prescriptions

The suggestions that arise about this class of problems include mechanisms allowing for the *right to be forgotten* and managing data access permissions. Many have to do with attaching metadata or tags to individuals' data. While these approaches can be applied relatively easily to their initial uses, enforcing them when data are used in other contexts and updating them to capture changes in individuals' preferences are more difficult. For example, Snapchat photos disappear quickly – unless someone took a photo of the screen before they disappear. Twitter allows individuals to expunge their tweets, but cannot control access to tweets that have already been downloaded. Also, even nominally protected and encapsulated data may be vulnerable to hacking. These speculative prescriptions raise questions about the extent to which, even in principle, data can be or should be erased. Possibilities (all with shortcomings) include:

- *Data expiration dates*: Develop mechanisms for expunging data included in databases on an announced timetable. A key challenge for this prescription is attaching the expiration date to the information so that it is applied even if the data are copied.
- *Data access permissions for end users*: Develop mechanisms to specify what types of users and uses have rights to access information.
- *Data access permissions for system administrators*: Develop sophisticated mechanisms to understand and build complicated access control policies for institutions and groups.
- *Data tracking*: One of the more speculative prescriptions discussed in the workshop was the possibility of an individual-specific digital ID that would be included as metadata for every piece of information recorded about that individual. Users would be able to track what individual data were collected and how they were used.

Anonymization

Diagnosis

One of the more unpleasant realities is that much current discussion about anonymization is based on myth. In reality, identities can often be backed out despite alleged anonymization (Sweeney 2000; Golle 2006; Narayanan and

Shmatikov 2010; Ohm 2010; de Montjoye et al. 2013; Caliskan-Islam et al. 2015). Another unpleasant reality is that with ubiquitous cameras, tracking, facial recognition, and the like, the potential for individual-level inference is growing extremely fast. Although stronger methods of anonymization of data are available (Sweeney 2002; Dwork and Naor 2008; Davis and Osoba 2016), such approaches raise serious problems because any such method involves a trade-off between preserving information because it is useful and reducing information to protect privacy. The availability of flexible and relatively cheap computing resources makes it easier to implement de-anonymization algorithms at scale. Which information is to be preserved depends on the particular activity and context, whether research, design of medical protocols, or, say, terrorist detection. Moreover, emerging analytic techniques are leveraging the availability of massive amounts of information rather than focusing on limiting the amount of information necessary. There are not yet widely agreed-upon standards to determine what the right trade-off between anonymization and individual data is for accurate analysis of the data.

Prescriptions

Prescriptions range from guidelines for how to use data effectively while preserving individuals' anonymity to strategies that individuals can take to remove observation of their information. In particular, a level of data *anonymization* could arise naturally if some people adopt multiple identities and go out of their way to defeat tracking and recognition methods. It is conceivable that some new methods and tools for re-identification and inference could defeat countermeasures (e.g. disguises, multiple identities) but would be practically feasible only by motivated and highly resourced agencies. Possibilities include:

- *Curated data for research and analysis*: Although researchers often are constitutionally uncomfortable with anything less than *all* the available data, not all data are equally important. Speculatively, on-demand data anonymization tools could enable more flexible and adaptive data access. Researchers might initially access limited, potentially synthetic data sets for analysis. If it turned out that crucial information had been excluded, an iteration could be made that would provide that crucial information while still cloaking other information.
- *Digital identity makeovers*: Workshop participants predicted a burgeoning business in providing individuals with new digital identities. This could include services to expunge and/or obfuscate past information about them and to create the basis for new digital identities.

Avoiding Harms by Validating Algorithms and Auditing Use

Diagnosis

Numerous privacy harms arise because inferences made by algorithms are *wrong*, as when innocents are placed on no-fly risks or medical procedures are recommended that, for the specific person in question, are far too risky even if they are good procedures on average. Many of these are referred to informally as being the result of *bias* in data analytics. Yet algorithms and other data-driven systems have proven very useful at handling hard problems even if they are fallible.

Discussion of such matters is complicated by the sloppiness of language. In particular, we need to distinguish among types of *bias*:

- Analysis is *wrong*. Someone may be categorized incorrectly because of algorithms and data characterizing an over-aggregated population, an irrelevant small population, biases inherited from previous human decisions, or a lack of feedback and updates to a system when people or society change. Even when analysis is seemingly competent, it may be biased because the *base rates* used (the baseline frequency against which observed events are compared) are inappropriate. This problem is rampant because the data for highly differentiated base rates often do not exist.
- Results of analysis and inference are unacceptable to society, even if statistically reasonable. Societies prohibit certain kinds of inferences because they have such deleterious effects on individual rights. Examples of this often arise in connection with racial or other kinds of profiling. Algorithms may be accused of *bias*, but the issue is not so much the technical competence of the algorithm as a matter of balancing objectives and values.

It should be recognized that the *bias* problems are real and serious (Gandy 2010). People do suffer harm by being denied insurance (or charged enormous rates), by not getting a fair shake when applying for jobs, or when applying for passports or visas and being improperly deemed high risks (Bollier and Firestone 2010; Gangadharan et al. 2014; Office of the President 2016).

Prescriptions

Prescriptions for mitigating harms that can arise from using algorithms for decision-making involve greater transparency, validation, and auditing of algorithms and their uses. Possibilities include the following:

- *Create transparent algorithms*: Increasing transparency will not be easy because machine learning algorithms are often opaque and there is no clear-cut theory for reviewing them. Further, the data on which they depend may have complex and questionable provenance or relevance. There are many opportunities for scientific and technological innovation here, to include tests for bias (of both kinds).
- *Validate algorithms*: Validating algorithms is also challenging. As anyone familiar with the concept of validation knows, validity must typically be judged with respect to a particular context and purpose. This means that good tools must accommodate particulars of context and purpose when needed: generalizations are not sufficient.
- *Audit algorithms and their use*: Even if the science underlying an algorithm is excellent, its application practices may be sloppy or at least inconsistent. Thus, a class of prescriptions involve what we call *auditing*: does an organization using data, perhaps for social-behavioral research, routinely conduct audits of process and routinely test particular results to assure that everything is being done according to best practices?

Challenge and Redress

Diagnosis

Often, people (or organizations) are unaware that their data are being used in a way that may cause them harm. Much less are they aware of what inferences have been drawn from their data (e.g. a higher propensity for cancer, for being a terrorist, or for not repaying debts). If they become aware, they often lack any mechanism for challenging the results (e.g. pointing out errors and getting off a no-fly list) or seeking redress.

Prescriptions

Developing effective mechanisms to allow for challenge and redress to occur will need to build on the foundations elucidated above for greater transparency of process, methods, data, inference, the uses to which data is being put, and the potential harms thereof. In addition, we see the need for fresh thinking and mechanisms for challenge and appeal (a kind of due process) and, important, mechanisms for redress (e.g. apologies and generous compensation). While these issues were touched upon in the workshop, few concrete suggestions were forwarded. This may reflect the fact that mechanisms for challenge and redress are likely to involve policy choices rather than technical solutions and that challenge and redress are tied so closely with notice and choice.

Deterrence of Abuse

Diagnosis

The incentives for abuse of individuals' data are many and strong. Within industry, the incentives include professional advancement, responsibility, salary, and – for the companies involved – profit and market share. Strong disincentives exist for raising doubts, or *pushing back*. Within government, the incentives include catching criminals, detecting terrorists, and coercing suspects or prisoners into providing confessions or information. Consider that in novels and movies, it is seen as normal and virtuous for the hero to cut corners as necessary to catch the villain. For scientists, the incentives include being able to answer a vexing question, discovering causal relationships, or discovering methods for, e.g. early detection of cancer. Some of the incentives are *good*; some are *bad*; all are very human.

Prescriptions

As in other domains, deterring problems requires both carrots and sticks – increasing benefits and self-esteem for doing the *right* thing and increasing costs for doing the *wrong* thing. It is necessary both to include active and passive defenses (e.g. better security, better practices) and to reduce the likelihood that people with access to data will abuse their access by using it in illegal or otherwise harmful ways. One aspect of deterrence is to reduce the attractiveness of attempting to do what would constitute abuse. Another is to catch and punish those who do so. For researchers, guidelines for ethical research exist and continue to evolve. Possibilities include the following:

- Increase awareness of and sensitivity to potential data harms. Serious data use problems occur in part because those involved are insufficiently aware of the dangers, harms, and their own personal options and responsibilities.
- Develop training programs on how to prevent data harms. Organizations need to provide training in recognizing and deterring data harms. This would be akin to current training on cybersecurity, discrimination, and human-research-subjects protection.
- Increase options for tracing data across systems to understand which decisions were made based on information.
- Update ethical principles for research in line with emerging changes in data and analysis. The Department of Health and Human Services has updated the *Common Rule* specifying US federal policy for the protection of human subjects. However, these changes do not fully reflect changes in risks to human subjects stemming from data and analytic innovations.

- Develop enforcement mechanisms. These can take many forms:
  - Mechanisms to challenge and demand redress for data abuses.
  - Mechanisms and protection for whistleblowers.
  - Laws and regulations that punish abusers, including officials, by fines and incarceration.
  - Research-grant criteria in which grantees are held accountable for data protection. Precedents for this exist. Research houses have become acutely aware of the need to protect health data in part because of some well-known instances in which large research contracts were denied because of past errors.
  - *Feedback into the system*: Algorithms may make decisions that turn out to be incorrect; these data points need to be fed back into the system so that the decisions can be revised. Naturally, there is a concern that individuals who were incorrectly labeled by a system must remain tracked by that same system in order for mistakes to be corrected. Thus, allowing for such feedback may only be appropriate in specific circumstances.

And Finally *Thinking Bigger* About What Is Possible

Developing solutions to prevent data harms will be necessary for maintaining societal trust in and support for publicly funded research in social-behavioral modeling. As a community, we need to ensure that our research is conducted ethically if it is to remain a valuable and effective contribution to social welfare. These solutions will need support, particularly from government research agencies. While many of the investments needed to accomplish the prescriptions discussed in this chapter are likely to focus on technological approaches, given the deep entanglement between technology and society, we believe social-behavioral approaches should be emphasized as well.

Looking across all of the prescriptions, two technologically focused research areas appeared to us to be particularly strong candidates for government investments in data privacy – they are real challenges, they are unlikely to be adequately addressed in the absence of investment, and they hold the promise of significant advances to deter data harms. Other challenges, such as advancing data anonymization techniques, already have strong and ongoing research streams or rest more squarely in the realm of regulations than research, such as developing adjudication procedures for redress and contract negotiations.

One promising research area is to build a context-specific data-use authorization framework to take context seriously. Many of the prescriptions across topics touched upon managing access to data based on individuals' contextualized preferences. Developing such a framework raises important questions about how a context is defined, recognized, and enforced. It touches upon technological solutions involving tagging data, developing data clearinghouses, and designing contextualized access procedures that are responsive to data users' and producers' needs.

A second promising research challenge area is to develop a framework for algorithm accountability. As the magnitude of data and the power of big data analytic techniques increase, data algorithms are playing an increasingly important role in decision-making. However, no commonly accepted process exists for evaluating algorithms. Developing such a framework raises important questions about technologies (how can algorithms be made transparent, what is required to validate algorithms) and responsibilities (who can be trusted with oversight and enforcement authorities, and for what purpose).

We are struck by the degree to which discussion of ethical procedures and privacy protections is dominated by acceptance of the inexorability of trends and behaviors that infringe in myriad ways on individual rights. There is need, in our view, for deeper systemic studies to understand how, if it chooses, society could bring about major changes. Such studies would require deep looks at, e.g. incentive structures for business, government, and individuals and how they could be changed. Can market mechanisms be brought to bear by changing some incentives over which society – through its government and educational programs – has some influence?

Our current project has only touched upon privacy issues as one part of a much larger effort, but we hope that some of the ideas we have discussed will prove useful in stimulating subsequent result-oriented work. The time is past for wringing hands; the problems of privacy are well known (e.g., Executive Office of the President 2014). The challenge is what to do about them.

References

Acquisti, A. and Grossklags, J. (2005). Privacy and rationality in individual decision making. *IEEE Security and Privacy* 3 (1): 26–33.

Bollier, D. and Firestone, C.M. (2010). *The promise and Peril of Big Data*. Washington, DC: Aspen Institute, Communications and Society Program.

Byun, J.-W. and Li, N. (2008). Purpose based access control for privacy protection in relational database systems. *The VLDB Journal* 17 (4): 603–619.

Caliskan-Islam, A., Harang, R., Liu, A. et al. (2015). De-anonymizing programmers via code stylometry. In: *24th USENIX Security Symposium (USENIX Security 15), 2015*, 255–270.

Cranor, L.F. (2012). The economics of privacy: necessary but not sufficient: standardized mechanisms for privacy notice and choice. *Journal on Telecommunications and High Technology Law* 10 (2): 273–445.

Davis, J.S. and Osoba, O. (2016). *Privacy Preservation in the Age of Big Data*. Santa Monica, CA: RAND.

Dwork, C. and Naor, M. (2008). On the difficulties of disclosure prevention in statistical databases or the case for differential privacy. *Journal of Privacy and Confidentiality* 2 (1): 8.

Executive Office of the President (2014). *Big Data: Seizing Opportunities, Preserving Values*. Washington, DC: White House.

Frenkel, S., Rosenberg, M., and Confessore, N. (2018). Facebook data collected by quiz app included private messages. Technology. *New York Times*, online. (accessed 30 April 2018).

Gandy, O.H. Jr., (2010). Engaging rational discrimination: exploring reasons for placing regulatory constraints on decision support systems. *Ethics and Information Technology* 12 (1): 29–42.

Gangadharan, S.P., Eubanks, V., and Barocas, S. (2014). *Data and Discrimination: Collected Essays*. Open Technology Institute.

Golle, P. (2006). Revisiting the uniqueness of simple demographics in the US population. In: *Proceedings of the 5th ACM Workshop on Privacy in Electronic Society*, 77–80.

Hartzog, W. (2016). There Is No Such Thing as "Public" Data: And it's not OK for researchers to scrape information from websites like OkCupid. Slate.com, May 16, 2016.

Inkster, B., Stillwell, D., Kosinski, M., and Jones, P. (2016). A decade into Facebook: where is psychiatry in the digital age? *The Lancet Psychiatry* 3 (11): 1087–1090.

Kagal, L. and Abelson, H. (2010). Access control is an inadequate framework for privacy protection. In: *W3C Privacy Workshop*, 1–6.

Kim, T.H.-J., Bauer, L., Newsome, J. et al. (2010). Challenges in access right assignment for secure home networks. 5th USENIX Workshop on Hot Topics in Security.

Mayer-Schönberger, V. (2011). Delete: the Virtue of Forgetting in the Digital Age. Princeton, NJ: Princeton University Press.

Mazurek, M.L., Arsenault, J.P., Bresee, J. et al. (2010). Access control for home data sharing: Attitudes, needs and practices. In: *Proceedings of the SIGCHI Conference on Human Factors in Computing Systems, 2010*, 645–654.

McDonald, A.M. and Cranor, L.F. (2008). The cost of reading privacy policies. *I/S: A Journal of Law and Policy for the Information Society* 4 (3): 540–540.

de Montjoye, Y.-A., Hidalgo, C.A., Verleysen, M., and Blondel, V.D. (2013). Unique in the crowd: the privacy bounds of human mobility. *Scientific Reports* 3 (1): 1376. https://doi.org/10.1038/srep01376.

Narayanan, A. and Shmatikov, V. (2010). Myths and fallacies of "Personally Identifiable Information". *Communications of the ACM* 54 (6): 24–26.

Office of the President (2016). Big Data: A Report on Algorithmic Systems, Opportunity, and Civil Rights https://www.whitehouse.gov/sites/default/files/microsites/ostp/2016_0504_data_discrimination.pdf (accessed 28 August 2018).

Ohm, P. (2010). Broken promises of privacy: responding to the surprising failure of anonymization. *UCLA Law Review* 57: 1701.

Raykova, M., Zhao, H., and Bellovin, S.M. (2012). Privacy enhanced access control for outsourced data sharing. In: *International Conference on Financial Cryptography and Data Security, 2012*, 223–238.

Reidenberg, J.R., Breaux, T., Cranor, L.F. et al. (2015). Disagreeable privacy policies: Mismatches between meaning and users' understanding. *Berkeley Technology Law Journal* 30: 39.

Solove, D.J. and Hartzog, W. (2014). *The FTC and the New Common Law of Privacy*. Columbia Law Review.

Sweeney, L. (2000) Uniqueness of Simple Demographics in the US Population. Technical Report LIDAP-WP4. Carnegie Mellon University.

Sweeney, L. (2002). k-anonymity: A model for protecting privacy. *International Journal of Uncertainty, Fuzziness and Knowledge-Based Systems* 10 (05): 557–570.

Ware, W.H. (1973). *Records, Computers and the Rights of Citizens*, P-5077. Santa Monica, CA: RAND Corp.

Part II

Foundations of Social-Behavioral Science

4

Building on Social Science: Theoretic Foundations for Modelers

Benjamin Nyblade[1], Angela O'Mahony[2], and Katharine Sieck[3]

[1] *Empirical Research Group, University of California Los Angeles School of Law, Los Angeles, CA 90095, USA*
[2] *RAND Corporation and Pardee RAND Graduate School, Santa Monica, CA 90401, USA*
[3] *Business Intelligence and Market Analysis, Pardee RAND Graduate School, RAND Corporation, Santa Monica, CA 90407, USA*

Background

The social sciences encompass a wide range of disciplinary and theoretical approaches. While researchers in these fields may not always be conscious of the philosophy of science underpinnings of their work, the vast bulk of social and behavioral researchers are drawn to a *middle ground* when it comes to beliefs about the generalizability and universality of social and behavioral theories. Most researchers reject both the claim that generalizable social scientific theories are impossible or unhelpful and the claim that theories must be grounded in a universal *covering law* model of the scientific enterprise. Social and behavioral researchers tend to work with theories that they do not claim to be universal, but that they instead suggest should be generalizable to those contexts or situations in which the researchers expect the causal mechanism to operate similarly.

The fact that social and behavioral researchers are thus working with contextual mid-level theories makes reviewing the state of the social science theories challenging, and for modelers who seek to speak to researchers across the social sciences, the diversity of social scientific theorizing can be daunting. In this chapter we highlight key common building blocks used in theories across the social sciences, focusing on three broad classes of theory: atomistic micro-level theories that can explain individual behavior without reference to society or social dynamics, social micro-level theories that explain individual human behavior focusing on how social factors influence individual behavior, and theories of collective or group behavior and interactions. While the range

Social-Behavioral Modeling for Complex Systems, First Edition.
Edited by Paul K. Davis, Angela O'Mahony, and Jonathan Pfautz.
© 2019 John Wiley & Sons, Inc. Published 2019 by John Wiley & Sons, Inc.
Companion website: www.wiley.com/go/Davis_Social-Behavioralmodeling

of theories used in the social sciences are diverse, they generate a common set of challenges for data collection and model building; they also generate opportunities for modelers who are willing to read across social scientific disciplines and build on a diversity of insights and approaches.

Atomistic Theories of Individual Behavior

The Belief–Desire Model

While "no man is an island," many social scientific and behavioral theories set out to explain individual behavior without reference to interaction with other individuals or society. Not only are these theories among the most important and widely used in the social and behavioral sciences, but also these atomistic theories include some of the key building blocks used in social scientific theories of broader social interaction.

Perhaps the most common building block in the social sciences is the belief–desire model of human behavior. In this approach, individual actors have desires and beliefs about how they can achieve their desires, and their actions reflect their decisions as to how best achieve their desires in light of their beliefs. Belief–desire models and their variants go under many guises. Most prominently, belief–desire psychology underpins the *rational actor* model. These are models of rational actors in the Humean sense of instrumental rationality, in which reason is a "slave of the passions."

The belief–desire model is also perhaps the most common *folk psychology* of human behavior – the informal models by which people commonly explain the behavior of those around them. The simplest answer one can give to "Why did she do X?" for many people is "She did X because she wanted to do X." Instrumental theories of intentional human behavior typically go at least one step further, suggesting that one can answer the same question by saying, "She did X because she wanted Y and believed doing X would help her achieve Y."

Among academic disciplines, this simplistic but commonsensical approach to explaining human behavior is perhaps most dominant, explicit, and formalized in economics. In most economic research, individual actors are modeled as utility maximizers.[1] Formally, Della Vigna (2009) and Rabin (2002) lay out the canonical approach, which can be traced back at least as far as von Neumann and Morgenstern (1944) if not farther as follows. Individual i at time

---

1 There are long-standing and wide-ranging alternatives that have had some influence in economics, including work that has resulted in multiple Nobel Prizes (Simon, Kahneman); however even scholars long associated with these lines of work recognize that it continues to have only a niche role within the discipline (e.g. Camerer 2003). We discuss some of the key strands of behavioral economics later in this section.

$t = 0$ maximizes expected utility subject to a probability distribution $p(s)$ of the states of the world ($s \in S$):

$$\max_{x_i^t \in X_i} \sum_{t=0}^{\infty} \delta^t \sum_{s_t \in S_t} p(s_t) U(x_i^t \mid s_t).$$

The utility function $U(x|s)$ is defined over payoffs x at time t to player i, and future utility is discounted at a (consistent) discount factor δ. This equation is merely a formal restatement of the same idea expressed above: an individual actor will make choices at a given time that she believes will maximize her discounted future utility given her understanding of the world. Thus the framework requires understanding two key types of variables: actors' beliefs and desires (utility). The assumed cognitive process – whether conscious or unconscious – that maps beliefs and desires into choices is that actors maximize their expected utility in light of their beliefs about the state of world and the consequences of their choices.

We discuss below each of these components of desire–belief psychology, which underpin many formal and informal causal theories in the social and behavioral sciences, but before that let us briefly note four major critiques of the general belief–desire approach. The first two push on the framework at theoretical and philosophical perspectives, while the third and fourth highlight systematic efforts from within the social and behavioral sciences to modify or supplant the framework.

First, when this model is taken in isolation, utility maximization has been critiqued as trivial or *empty* (e.g. Barry 1988). As the framework, in and of itself, gives little elucidation as to what beliefs or desires people hold, how they come to hold such beliefs and desires, and how those beliefs and desires may change, people have suggested there is inadequate substance to the framework to be of value in understanding human behavior. Indeed, as we will see below, this basic framework is frequently combined with other *atomistic* theories of individual human behavior, as well as social theories of individual human behavior and interaction. To its critics, the fact that the belief–desire framework is promiscuous – frequently combining with a wide range of theories of actors' preferences and desires – makes it less compelling. To its defenders this flexibility of the framework is a strength.

Second, the key pieces of this framework – desires, beliefs, and cognition – are for the most part unobservable, which critics suggest combines with the flexibility of the framework to allow almost any behavior to be explained post hoc (e.g. Green and Shapiro 1996). Thus critics have suggested that the framework cannot be falsified: when a defender is confronted with behavior that appears at first glance to contradict the framework, she can invariably rationalize the apparently irrational behavior by suggesting that beliefs or desires have been improperly understood. Rather than rationalism or belief–desire psychology being flawed, to resolute defenders of the framework, it is always the particular understanding of beliefs or desires that must be flawed or a misunderstanding of the appropriate scope for the framework.[2] The challenges of falsification of a theory or theoretic perspective in light of this sort of challenge are not unique to the social and behavioral sciences and have been extensively discussed and debated by philosophers of science in other contexts, but the fact that this critique may apply beyond this instance does not negate its relevance.

A third line of critique, one that is typically more grounded in the empirical developments of the social and behavioral sciences in recent decades, particularly in cognitive psychology, is that even when human behavior is driven by actors' beliefs and desires, human cognition is more complex than the *utility maximizer* formalization of actors making the best choice to achieve their desires in light of their beliefs. In the standard belief–desire framework, thinking is costless, and decision-making optimal (in light of the information actors have).[3] *Rational* actors never make a mistake on a math test or have bad breath (cf. Cox 2004). However, decades of research point to instances in which people appear to *satisfice* rather than *optimize* (e.g. Simon 1972) and systematically fall prey to a large number of cognitive biases in processing information that are inconsistent with (at least prototypical versions of) rationalist theories (e.g. Kahneman 2011). For the most part, most defenders of rationalist theories, or of belief–desire psychology more generally, consider many of these critiques to be friendly. Just as all maps are both simplifications and *wrong*, defenders of the framework suggest that models based on belief–desire psychology are simplified maps of human behavior that are useful in many cases and that the systematic critiques mentioned are cases in which having another (more detailed) map to guide our understanding is useful.

A fourth major line of critique of belief–desire psychology grounded in alternative understandings of cognition is the suggestion that much of human action is driven by subconscious choice. If people's beliefs and desire may in fact result from their mind rationalizing their subconscious choices rather than the choices being driven by (prior) beliefs and desires, then the basic belief–desire psychology framework has reversed cause and effect. Newer

---

2 As a matter of pure logic, utility maximization as an explanation certainly can be empirically falsified (for example, one could seek to identify situations in which outcomes vary despite utilities and beliefs remaining constant). However, such falsification rests upon accurate measurement of utility and beliefs – a challenging endeavor.

3 This is not to say the decisions are correct or optimal, but rather that the decision-making model envisioned suggests that decisions made are optimal in light of the information available at hand as to how the options are believed to affect actors' utility. Search (information acquisition) need not be costless even in the basic framework, and of course, it is common in rationalist models that lack of information can lead to suboptimal decision-making. But without a more complex cognitive model, the actual process of thinking through the choice between options entails no cognitive burden.

work on cognitive processing similarly hints that even complex actions can be executed before a full decision-making process may come to conscious awareness (e.g. Bechara et al. 1997; Strack and Deutsch 2004; Custers and Aarts 2010). While slightly different than the rationalization critique, it similarly challenges the basic premise that actions are a response to deliberative decision-making and reverses the order of cognitive processing.

This critique is deeper than a generic concern about the beliefs and preferences being driven by (inter)action rather than the converse. It is not a concern that beliefs and preferences at time $t + 1$ are influenced by what occurred at time t; rather it is that beliefs and preferences at time t in fact are rationalizations of (effectively) simultaneous choices at time t. Fundamentally, this critique suggests that people (and researchers) may be subject to false consciousness with regard to what drives human behavior: our beliefs and desires may adapt to conform to our behavior rather than our behavior being driven by (causally prior) beliefs and desires. This critique is compatible with a range of alternative explanations of the subconscious decision-making that occur.

Given these extensive and long-standing critiques of the rationalist framework and simplistic versions of belief–desire psychology, it strikes many as odd that reliance on utility maximization as a framework for understanding human behavior remains so widespread. Rather than provide a full-throated defense of the framework, at this point we should highlight two basic arguments that most social scientists rely upon for not discarding the framework. The first is a pragmatic Friedman-esque argument. Many social scientists believe that, at least in the context they study, humans behave "as if" they are instrumentally rational. The second argument is that even social scientists who do not rely on rationalist assumptions rarely argue that we should ignore beliefs and desires in explaining social behavior. This suggests that even if social scientists wish to adopt alternative approaches still must consider the same crucial questions: "What do people want?" and "What do people believe?" These are the questions we turn to in the next sections.

Desires

To the extent that individual actors have agency and they are consciously making choices to pursue that which they desire, an understanding of what people want becomes a central concern for modeling human behavior. Indeed, better understanding of the question of what people want (and why) is central in classical philosophy, in a wide range of religious traditions, and in the contemporary social and behavioral sciences. The list of approaches to understanding desires is sufficiently extensive that we focus here on theoretic approaches that need not be based on social interaction, turning to the social bases of human action in later sections.

Theories of human desires typically suggest that people want many things and face scarcity in achieving those desires. So a fully developed theory of human desires also must become a theory of the hierarchy or prioritization of those desires – thus the emphasis in rationalist theories on developing theories of preferences. Theories of preferences are definitionally ordered (preferences entail, at a minimum a preference of X over Y), but frequently theories require cardinal utility functions (knowing how much actors prefer X to Y) in order to explain the choices people make.

The most common family of theories of human desires is purely self-regarding theories, most of which in turn focus on materialist interests (although overreliance on materialistic notions of utility has long been critiqued as well; see Mill 1863). There is nothing intrinsic to rationalist theorizing that requires pairing it with materialist theories of human desires, although, as previously discussed, some critics have argued that rationalist theories are *empty* without being paired with such an individualistic theory of human desire. However, although materialist theories need not be rationalist in either form or substance, as an empirical matter materialist rationalist theorizing is quite common in the social and behavioral sciences.

Given the potential importance of nonmaterial factors in understanding actors' utility, why do many social scientific theories rest on assumptions that human behavior is primarily driven by people's desire to improve their individual material well-being? A normative argument about the nature of social scientific theorizing might point to the strong norms among researchers concerning the validity of such theories (more so in some social sciences than in others) or to a Friedman-esque argument that such theories have proven empirically useful in many cases, which by itself may be a pragmatic reason to rely on such theories. However, there are theoretic reasons that researchers point to in justifying materialist understandings of human desire as a basis for social and behavioral theories. Some have pointed to psychological theories about fundamental human needs, such as Maslow's hierarchy of needs, as justification for a focus on materialistic concerns. At the base of such models of human needs are material concerns – air, water, food, shelter, etc. – that scholars can use to justify, suggesting that the material wherewithal to ensure such needs are met is the most central and fundamental factor driving human behavior. In a similar vein, cultural materialism (cf. Harris 1968) posits that basic survival needs are a primary driver of both individual behavior and collective action, resulting in many of the social institutions that define human life.

However, while a materialistic focus may be common, it is certainly not the only basis of understanding human desires. Some scholars focus more on human biology and physiology as a basis to understanding human behavior. One version of this approach is in scholarship that focuses on the biological and neurological basis of human behavior. To do something of a disservice to the more nuanced versions of scholarship in this line of thinking, one approach is to understand humans as hedonists, at least in the sense that their behavior is driven by what rewards the pleasure centers of their brain and a desire to avoid what causes pain. This approach to understanding human desires suggests that understanding the science of pleasure and pain is crucial for understanding human behavior. While it is beyond the scope of this chapter to review the extensive literature in psychology and neuroscience that seeks to ground our understanding of human behavior in a biological foundation, interested readers should look to the discussions in Read and Miller (2019) and Thompson et al. (2019).

Another approach to understanding human desire rests on evolutionary biology. This work views human behavior through a broader historical lens and posits that behaviors that enabled the successful propagation of the species must *fit* into the contexts that defined our evolutionary past or the environment of evolutionary adaptedness (cf. Bowlby 1969). In some areas, theories of human behavior based on evolutionary biology may overlap with either physiology of pleasure/pain or psychology of needs approaches, but evolutionary biology approaches explicitly place not just desire for survival of an individual front and center, but desire for successful procreation and the success of those who will carry on one's genes (cf. Dawkins 1976). These biological theories of alternative desires may (but need not necessarily) be combined with the idea that human beliefs and desires may in fact be rationalizations of subconscious processes that make decisions before conscious thought even comes into play.

One aspect of an evolutionary biology approach to understanding human desires is that it typically places certain very specific other-regarding preferences (i.e. a desire for furthering the well-being of one's genetic successors) at a more *fundamental* level than the other two nonsocial approaches to understanding human preferences discussed above. We see this reflected in developments in studying the neurobiology of altruism and social connection that extends the "short and brutish" model of our evolutionary past to include theories beyond selfishness, pleasure, and pain (e.g. de Waal 2008).

One clear consequence of building on any of these approaches to understanding human desires is the recognition that context matters. In particular, each approach can be taken to suggest that when the most fundamental human needs or wants are at stake, humans are more likely to act in their narrower (often material) self-interest, whereas when people feel assured as to their current and future material well-being, they may place greater emphasis on nonmaterial desires.

Beliefs

Even when social scientists have a clear understanding of people's desires to the extent that people are making conscious decisions to pursue their desires, researchers need to have an understanding of what people believe about how to achieve those desires. The theories that people hold – their beliefs as to how to achieve those desires – thus are a second fundamental building block in social and behavioral theories of conscious human choice. While beliefs may not play as central a role in theories of human behavior that do not rest on conscious decision-making, relatively few theories of human behavior rest solely on subconscious or unconscious decision-making.

How do people develop beliefs about how the world works and the (likely) consequences of choices they face? There is a vast literature across a wide range of social and behavioral sciences on this topic, but for the most part, it is fair to say that most theories of beliefs focus on how they are *learned*. Beliefs about the consequences of one's actions may be learned through direct experience (typically this is modeled as learning from trial and error), beliefs may come about from *reasoning* (an internal cognitive process through which people extrapolate from preexisting beliefs to develop new beliefs), or they may result from persuasion by or learning from others (direct modeling/instruction or feedback about the appropriateness of any particular course of action).

One common approach to modeling how beliefs change is to treat people as Bayesians who update their beliefs about the state of the world and the consequences of their actions following Bayes' law. When they become aware of new information, they use this to update their beliefs in an optimal manner based on the strengths of their prior beliefs and their beliefs about the quality of the new information. A vast literature explores the extent to which people fail to be perfect Bayesians because of having a wide range of cognitive biases and engaging in *motivated reasoning*, such that new evidence about the world is not incorporated into people's beliefs in a neutral fashion or in certain cases may not be incorporated into people's beliefs in a neutral fashion or in certain cases may not be incorporated at all. As most of the more complex models of change in beliefs rest on richer understandings of social interaction, we return to this topic in our later section focused on interactions.

Cognition

A fully specified set of desires and beliefs takes social and behavioral scientists of a rationalist persuasion much of the way toward developing a fully specified theory of human decision-making, but in and of themselves, they are limited. These beliefs and desires must be processed by actors as they determine how to act and what choices to make. In these sets of theories, beliefs and desires are *mapped* to choices through a cognitive process.

The default cognitive assumption in most belief–desire models of human behavior is utility maximization. Desires have cardinal weights, allowing for a fully specified utility function, actors rely on their beliefs about the consequences of choices to calculate the relative costs and benefits of their options, and they choose options that maximize their expected utility.

In the simplest version of the utility maximization approach to decision-making, the act of making a choice is simple. There is no cognitive burden, nor any other sort of challenge to coming to a decision. And there are no mistakes or decisions that are not optimal, at least from the perspective of the actor at the moment of the decision.

While it is possible that actors may learn that past choices may have been wrong – at least in the sense that they may develop a belief that making a different choice in the past would have resulted in higher utility – decisions are always optimal in the sense that the choices they make maximize expected utility in light of the beliefs actors hold at the time the choice is made. While an actor may be indifferent between options they face, or the calculations may involve many factors, that does not make them hard or costly in any sense that is built into the model.

This simplistic version of cognition involved in canonical utility maximization theory has long been criticized. One of the most influential early lines of critique came from the decidedly cross-disciplinary work of Herbert Simon (1955, 1956, 1957, 1972), whose influence spanned psychology, economics, political science, and organization theory. Simon was an empirically grounded versatile social science theorist whose work pointed to the fact that information is costly and decisions are not necessarily easy. In an informal sense, Simon suggested that decisions may involve so many options with unclear payoffs that trying to maximize utility is unrealistic and actors may prefer to *satisfice* – latch on to a satisfactory alternative rather than *search* interminably for that which is optimal. More recent research, indeed, finds more direct evidence that decisions may, in fact, be costly. In multiple studies, researchers document *decision fatigue* (cf. Vohs et al. 2008) and track impaired quality of decisions as people make multiple choices through a day. The subsequent literature on heuristics (e.g. Tversky and Kahneman 1974) and additional approaches to bounded rationality (e.g. Gigerenzer and Selten 2002) can trace their roots to the influence of Simon as well.

For modeling purposes, there are multiple approaches to incorporating satisficing into decision-making, at both the individual and organizational levels. At the broadest level, these approaches generally assume decision-making occurs under incomplete information and model a cost to searching for information and/or evaluating options. Thus, actors develop a threshold of acceptability that determines whether they should stop searching/evaluating once they have identified a satisfactory choice. There is an extensive literature in psychology that highlights substantial individual variation in the extent to which people are satisficers or maximizers. Schwartz et al. (2002) is a particularly prominent study that finds that satisficers generally report higher levels of happiness, self-esteem, and life satisfaction and lower levels of depression.

Decision-making under uncertainty also induces people to rely on heuristics to cut down on information search and decision-making costs (Tversky and

Kahneman 1974; Gigerenzer and Todd 1999). Heuristics are *rules of thumb* or *information shortcuts* that may not be optimal, but cut down on the information search and cognitive burden of decision-making under uncertainty. For example, while an individuals' and politicians' policy preferences may not match perfectly with a particular political party, those individuals and politicians may rely on party labels or endorsements and vote the party line (or engage in straight ticket voting) rather than investigate every option they must consider in detail (Lupia 1995). While the reliance on heuristics may be advantageous for individuals and at times be in the collective interest, it can induce systematic biases in collective decision-making (Kuklinski and Quirk 2000; Lau and Redlawsk 2001).

There are three additional dimensions of cognition that are particularly well studied that researchers have suggested systematically influence decision-making in ways that may be worthwhile for modelers to consider as well: asymmetries in valuation of losses and gains (e.g. the prospect theory of Kahneman and Tversky (1979)), inconsistencies in valuation of present costs and benefits relative to future costs and benefits,[4] and biases in processing low probabilities and large numbers (which may have multiple neurological bases, e.g. Doya (2008)).

While most of the cognitive models discussed above posit systematic directions in which decision-making departs from optimization, other models of decision-making have been developed in which choices are intendedly optimal but people make errors. Unlike the decisions that are suboptimal post hoc, actors at the time a decision is made know it is a mistake, but they *pressed the wrong button*. In some instances, theories may be agnostic as to the cause of errors (e.g. the trembling hand equilibrium in Selten (1975)), although in others there may be specific theories as to the nature of the errors (see Kőszegi 2014, Section 6 for a review of modeling approaches to systematic mistakes in an economic context).

Alternative Atomistic Theories of Individual Behavior

While the bulk of individualistic theories of human desires rest on some version of a belief–desire framework, a range of psychological theories can be understood (in certain contexts) as having an atomistic basis and may be used as either complementary or alternatives to a belief–desire perspective. These tend to be theories that highlight that people differ along a range of cognitive,

---

4 The experimental and observational evidence for a "present bias" appears fairly compelling, but the preferred functional form to incorporate both present bias and future discounting continues to be extensively debated (e.g. the extensive debate over hyperbolic discounting; see also Halevy 2008, Benhabib et al. 2010).

emotional, and personality dimensions and that these differences may affect the desires and beliefs people hold and the ways in which they make conscious and subconscious decisions.

For example, personality theories may suggest that people's desires and beliefs depend (in part) on innate personality types – extraverts will be more likely to desire social interaction, and pessimists may be more risk-averse. Theories of how emotion affects human behavior may tie in to how our emotions drive what we desire as well. People who are happy may desire different things than those who are sad, and those who are angry may engage in a different cognitive process than those who are calm.

These sorts of emotional and personality-based theories of human desires are less frequently used in general social scientific models of human behavior (outside of disciplinary subareas), but they are increasingly popular in the business world where they are used in employee screening tools, in performance review matrices, and for segmentation analyses. However, to the extent that there are calls for richer models of human behavior in social scientific modeling, this is one direction that modelers could pursue.

Social Theories of Individual Behavior

Norms

The dominant approach to understanding individual human behavior in economics is the individualistic belief–desire framework discussed above, which has been characterized as laying out a theory of humans as *homo economicus*, which can be contrasted to *homo sociologicus*, whose behavior cannot be understood without reference to the social forces that have shaped her over time (e.g. Elster 1989). Whereas homo economicus is *pulled* by her desires, homo sociologicus is *pushed* by the norms she has internalized.

In thinking about norms, it is worthwhile to lay out three distinct (but related) usages of the concept. One way of using the term norm is primarily descriptive and empirical: one observes norms by looking at what behavior is *normal* (the *frequency* or *descriptive* side of norms). A second use of the term norm is a social expectation: norms are how society expects individuals to behave (the *social expectation* aspect of norms). A third use of the term norm is that of a moral prescriptive: norms are how one should behave (the *moral obligation* aspect of norms). While most social and behavioral science researchers would no longer suggest that that which is most frequently observed is necessarily what is socially expected or morally imperative, the fact that the term can reflect each of these three usages reflects a historical belief among many early social scientists that things that are *abnormal* are socially and morally problematic (e.g. Galton).

Descriptive Norms

How might individual human behavior be driven by frequency aspect of norms? The most natural way descriptive norms may drive human behavior is due to mimicry. Mimicry can be conscious or unconscious. We may adopt the behavior and mannerisms to the people around us even without a conscious intention to do so ("monkey see, monkey do"). We may also consciously choose to attempt to mirror the behavior around us, and this choice may be *strategic* (driven by a belief that doing so will help us achieve our desires, as per the framework discussed earlier) or normative (driven by either norms as social expectations or moral imperatives as discussed in the subsequent section). Moreover, from a developmental perspective, behaviors that were once consciously learned and practiced may subsequently become second nature to us and require little thought or effort. For example, children learn appropriate personal space through direct teaching, but as they mature, these lessons become *just part of the way we do things.*

Extensive experimental research has demonstrated unconscious mimicry in a variety of circumstances. For example, when people are exposed to emotional facial expressions, they spontaneously mimic those expressions faster than would be possible through conscious thought (Dimberg et al. 2000), and a substantial literature examines the neurological basis of such mirroring behavior (e.g. Rizzolatti et al. 2001). Indeed, perceiving the pain of others may stimulate the same areas of the brain that are activated when someone experiences pain directly (Jackson et al. 2005). Thus we may be *hardwired* to mimic and even empathize with others (at least in some circumstances). Work on mirror neurons (Gallese and Goldman 1998) provides a unique perspective on how critical social engagement skills may be hardwired for our species survival.

Not all mimicry is unconscious. It is often a conscious choice to mimic the behavior of others; indeed it has long been suggested as one of the wisest rules of thumb to guide people's behavior. While the idiom "When in Rome, do as the Romans do" has been attributed to both St. Augustine and St. Ambrose, similar aphorisms are found in around the world.[5] Choices to mimic others may in some cases be based on conscious cost–benefit calculation but may also be driven by normative consideration of social expectations and/or moral beliefs.

The fact that certain normative behavior may be internalized and exhibited even when it appears to conflict with cost–benefit calculation buttresses the emphasis many social scientists give on social norms in their theory building.

---

5 In Japanese the equivalent aphorism to "when in Rome…" is "when in the city…" (郷に入れば郷に従え). In some languages such as Czech and Russian, the more common phrase suggests that if "you wish to live with wolves, you must howl like them" (Chceš-li s vlky žíti, musíš s nimi výti).

To take a relatively innocuous example, many drivers stop at stop signs even when no one is around, and they have no reason to believe anyone is observing their behavior. That social norms can drive human behavior even when they are not in a social situation is key to the argument that norms are central to understanding human behavior.

Norms as Social Expectation

While the frequency dimension of norms is the most directly observable facet of norms, most social scientific theories of norms focus on them as the embodiment of social expectations. And while the *is*, the *expected*, and the *ought* frequently go together, this need not always be the case. It is possible that most norms may be like those regarding theft: norms against stealing things from other people in your own community are typically strong, with almost all people abiding by the norm, expecting others to do so, and believing that it is the ethical thing to do. However, there can be cases in which people perceive social expectations that in fact are at odds with what is the behavioral norm (in the frequency sense of the term). For example, adolescents routinely overestimate the amount of sexual activity engaged in by their peers, conflating social expectation norms (engage in intimate behavior) with the actual behavioral norms (Black et al. 2013).

Such conflicts may arise for a variety of reasons. On the one hand, we must remember that societies and cultures are dynamic, and norms can and do change over time, especially in response to new technologies or major shifts in other contextual factors (e.g. trade partners, social institutions). Alternatively, people may hold to norms that govern one situation when they transition into a new context and either miss or forget that a new set of expectations governs interactions (Goffman 1959). Such *frame violations* (Shore 1996) are a key way in which norms are learned and discussed.

And, of course, social expectations and frequency aspects of norms need not match up with moral prescriptions. Some social norms may be local conventions that are not tied to an ethical or moral perspective (e.g. the side of the street on which to drive), while other social norms may conflict with the moral or ethical beliefs of many or even most members of a society. For example, in many European countries, the majority of people believe that there is nothing immoral per se about public nudity, but as a matter of norms, public nudity is far more circumscribed than would be suggested by the ethical beliefs people hold.

This example of public nudity also highlights that the frequency, expectation, and ethical components of norms may change differentially over time. These norms dynamics tend to come about through social interaction and as such are discussed primarily in a latter section when we focus on social interaction as a driver of belief dynamics.

Norms as Moral and Ethical Obligations

To the extent that social expectations overlap with what an individual perceives to be morally and ethically correct, the distinction between norms as social expectations and norms as a moral and ethical code may be less important. However, as individuals' moral and ethical code may not match social expectations, understanding how people behave when social expectations and individuals' personal normative (moral and ethical) commitments conflict can become quite important.

Under what circumstances might we see conflict between social expectations and personal ethical codes? Perhaps the most frequent normative conflict is combined with social expectations: when individuals are members of multiple social groups or communities that have different norms. People are often members of distinct social groups concurrently, thus leading to perennial conflicts such as when people are torn between the norms of behavior among their friends and their families. But people also may transition from one group to another, as in major life cycle transitions (e.g. completing school and entering the workforce) or migration. When social expectations of groups differ, people may become more uncertain about the moral and ethical component of norms.

Even when social expectations are clear and consistent to a person, it is possible that her ethical and moral views develop in such a way that she develops an individual moral code that is distinguishable from social norms in the communities she is part of. This is because culture and social norms are not a rigid imprint on people, but embed themselves through the more unique circumstances of each person's life. In this case, the individual may feel more like Cassandra than the prototypical person torn between groups; however to the extent that development of a moral code always involves the creation of an (in part) imagined community of ethical people to which the individual strives to be part of, it is not clear that this needs to be theorized in a qualitatively different fashion than other instances in which norms come into conflict. These points are elaborated in another chapter in this volume (Cowherd and Lende 2019).

An additional point is that cultures build in space for conflict in social and moral norms as a way to create a dynamic tension that enables the society to grow and evolve over time (Turner 1995 [1966]; Nuckolls 1998). If we recall that moral dilemmas are perceived as tensions between competing goods or competing evils (never simply between good and evil), this dialectic may become particularly salient in moments of social, economic, and cultural transition. For example, the increased salience of a *pro-choice/pro-life* debate was not simply about changing moral or ethical norms, but came about as a by-product of medical advances in fertility control and infant survival, economic advances in opportunities for women, and legal advances in women's rights. These external factors drove the rise of a social and moral debate in which both sides lay claim to representing core American values.

The Relationship between Normative and Rationalist Explanations of Behavior

In contrasting normative and rationalist explanations of behavior, Olsen and March (2008) cast the key difference as being one of competing *logics*. The belief–desire framework discussed earlier sees human behavior as driven by a logic of consequences: actions are instrumental, as actors strive to achieve their desires by choosing actions that they believe will best help them achieve their ultimate ends. Normative approaches see human behavior as driven by a logic of appropriateness. People do what they perceive to be most appropriate in a situation given the norms they have previously been exposed to and have internalized.

It is not difficult to understand how either approach might be seen as encompassing the other. Behavior that appears consistent with an instrumental rationalist logic may be driven by social norms suggesting that – to an actor that in this situation – cost–benefit calculation is normatively appropriate. Thus instrumental rational action could be circumscribed to situations in which norms permit such behavior. Alternatively, it may be that people's behavior that appears to be driven by norms in fact results from a rational choice being made by individuals. Individuals may rationally believe that norm abidance maximizes their individual utility (in particular instances or more generally).

However, it is important to note that even if it is the case that one of these approaches does truly encompass the other, this may not have as much practical effect on applications of social and behavioral theorizing as one might expect. The typical efforts to encompass one theory in the other (in either direction) result in a situation in which social and behavioral researchers still need to understand the particular context in which we are seeking to understand human behavior, and these resolutions will typically suggest that in some contexts, rational theorizing will be more efficient (even if rational action rests on a normative logic of appropriateness) and in others normative theories will be more efficient (even if normative action rests on a rational logic of consequences).

What is likely to be more important than a fundamental resolution of these competing approaches is a clearer understanding of the situations in which human behavior is driven more by instrumental logic or by a logic of appropriateness. Perhaps equally as important, social scientists need to build better models of how these drivers of human behavior may interact. Indeed, it is clear that in many cases people seek to take advantage of or alter social norms in an instrumental fashion and that people may also respond normatively in the face of strategic self-interested action by others. There are a range of modeling strategies that seek to account for the interaction of rational self-interest and social norms, an excellent example of which includes the models of smoking cessation efforts developed in Chapter 27 of this volume (Sukthankar and Behesthti 2019).

Theories of Interaction

From Individual Behavior to Social Interaction

Humans are deeply social animals. This is fundamental and has biological roots. Humans have one of the longest phases of dependency among all animals, and we cannot survive without intensive investment by others (Konner 2010). This includes more than the provision of food and basic safety – infants reared without adequate social and emotional interaction *fail to thrive* and die (Spitz 1952). Social scientific theories often recognize not just that social factors influence individual behavior, as discussed in the previous section, but that human behavior in groups is distinct from how they act when people are not around.

While there is an extraordinarily diverse set of models of social interaction, we highlight four key areas of the development of social and behavioral theories of social interaction. First, we consider two lines of social scientific theorizing and research that highlight that even apparently straightforward attempts to aggregate individual behavior may be challenging: first we consider collective action and social dilemmas, and then turn to models of bargaining. These theories demonstrate how relatively minimal assumptions about how social interaction affects individuals' beliefs and identities result in group behavior that would not be predicted based on simply aggregating individuals' preferences. We then turn to how social interaction can dynamically affect the drivers of human behavior, considering how social interaction affects beliefs and then how social interaction affects identity and culture. These theories demonstrate that understanding individuals' beliefs and identities in the absence of the social interactions that shape them can profoundly misunderstand the social roots of human behavior.

We suggest that modelers face three key challenges in moving from the determinants of individual behavior to modeling social interaction. First, modelers need to understand the micro–macro distinction, and their models inevitably must make assumptions about how individual and collective actors may (or may not) differ. For many applications, this will require modelers to work at multiple levels of analysis (for a more detailed consideration of the factors involved in making levels of analysis choices, see the discussion in Brashears (2019)). Second, modelers must consider the possibility of recursive dynamics: they need to consider whether and how to incorporate the dynamics by which social interaction influences the key drivers of human behavior previously discussed, not just how those drivers affect interaction. Third, modelers need to determine the scope of social interaction to include. Each of these challenges is apparent in the four types of interactive models explored below.

Social Dilemmas and Collective Decision-Making with Common Interests

A social dilemma, broadly speaking, is a situation in which people acting in what appears to be their individual best interest make themselves worse off collectively. One of the major changes in the social and behavioral sciences in the second half of the twentieth century, compared with the social theory that predates that period, is the recognition that even when people share similar – or even identical – preferences, beliefs, and norms, social dilemmas are a common challenge. While most social scientists today take for granted that social dilemmas are a central concern of the social sciences, this under-standing is a fairly new development in the long scale of social theory and theories of human behavior. To take a prominent example, whereas the bulk of Marx's work (and that of most other nineteenth-century social theorists) simply assumes that similarly situated members of a group will work together in their collective interest, many of the most influential works of social science in the twentieth century call into question the premise that collective interests result in collective action.[6]

Counterexamples certainly exist: it is not hard to point to prominent examples of social dilemmas in earlier social theory and philosophy – Rousseau's discussion of the stag hunt and Hobbes' explication of the challenges of cooperation in a (hypothetical) state of nature are examples. However, in these examples the dilemmas are discussed as isolated in a pre-society (pre-social contract) setting. Perhaps the most distinctive break in the social sciences in the twentieth century was the emphasis on how these sorts of dilemmas are far more pervasive.

Given that social dilemmas are now widely recognized as a central concern and that models treating them are well established, we briefly highlight two families of social dilemmas: collective action and instability of majority decision-making.

---

6 Olson's *Logic of Collective Action*, Axelrod's *Evolution of Cooperation*, Coleman's *Foundation of Social Theory*, Putnam's *Bowling Alone*, and Ostrom's *Governing the Commons* all have this as a central theme (despite their differences) and are among the 50 most cited books in the social sciences (http://blogs.lse.ac.uk/impactofsocialsciences/2016/05/12/what-are-the-most-cited-publications-in-the-social-sciences-according-to-google-scholar/). This is also a central theme in Hardin's "Tragedy of the Commons" article, which, like these books, also has been cited over 30 000 times according to Google Scholar.

The most common social dilemma that social scientists tend to emphasize is that of collective action. Self-interested actors may have desires that are best achieved through collective action, but even if faced with the choice to join in collective action, individuals may have incentives to free ride (Olson 1965). This logic of collective action has been relied upon to suggest that policies with concentrated costs and diffuse benefits will face more intense opposition than policies with diffused costs and concentrated benefits (e.g. diffuse environmental benefits vs. concentrated job losses). And while the classical portrayal of collective action problems has been that the dilemma arises due to rationalist calculation of narrow self-interest, researchers increasingly have recognized the role of social norms in both creating and mitigating collective action problems (e.g. Ostrom (2009) on the range of local strategies used to manage common pool resources).

A second, perhaps less widely appreciated, sort of social dilemma is the instability of collective choice. Drawing on classic work by Condorcet and building on the seminal work of Arrow (1950), which highlighted the theoretical challenges in moving from individual ordinal preferences over multiple choices to a coherent understanding of social welfare, social choice theorists starting in the 1970s came to emphasize that when preferences over particular goods or policies are multidimensional, there may be no stable choice that garners the support of a majority (e.g. McKelvey 1979). Even when individual preferences are stable, collective choice may be unstable, as a majority could rationally cycle among multiple options.

A simple example illustrates this counterintuitive concept. Suppose that three people must agree over what flavor of ice cream to buy in the supermarket. Anna prefers Vanilla to Chocolate to Strawberry. Bill prefers Chocolate to Strawberry to Vanilla. Cathy prefers Strawberry to Vanilla to Chocolate. A majority of the group (Anna and Bill) prefers Chocolate to Strawberry. A majority (Anna and Cathy) prefers Vanilla to Chocolate. If collective preferences were transitive the way rationalist theories typically assume individual preferences are, we should thus expect that the group would also prefer Vanilla to Strawberry – and yet a majority (Bill and Cathy) in fact prefer Strawberry to Vanilla. Thus whatever ice cream is chosen, a majority would support replacing it with a different flavor.[7]

While easily established theoretically, the extent to which such *majority cycles* are a major empirical challenge in understanding collective decision-making has been debated (e.g. Kurrild-Klitgaard 2001). However the fact that aggregation of individual interests may be unstable has played an

---

7 Unfortunately, Neapolitan ice cream is not a solution to all of life's problems.

important role in theorizing about the development of social and political institutions and organizations, as structure may help induce stable (equilibrium) outcomes of collective choice even when aggregation of individual preferences is insufficient to do so (e.g. Shepsle 1979).

Fundamentally, both lines of theorizing point in a similar direction for modeling social behavior: even in the simplest instances, the behavior of collective actors frequently cannot be seen as a simple aggregation of the behavior of individual actors. The collective interest of a group may not be defined, and even when it is defined, it is not always the case that collective actors will pursue that interest.

Bargaining over Conflicting Interests

Social dilemmas are most starkly presented when individuals have clearly identified collective interests and may arise even when actors have identical beliefs and desires and operate under common norms. However, in other cases we seek to understand the behavior of actors with competing interests. A wide range of models and theoretic approaches would fall under this umbrella, but consider a simple two-actor situation in which interests are in conflict: a fixed-sum bargaining situation in which whatever one actor garners comes directly at a cost to the other.

Informally, analysts refer to the relative bargaining power of the two actors to predict outcome. In the literature on negotiation, bargaining power is typically seen as deriving from the "best alternative to a negotiated agreement" (BATNA). In practice, this bargaining power may derive from a variety of different means: for example, if only one party can easily leave the negotiation and garner similar gains by bargaining with another actor, or the bargaining occurs in the shadow of credible threat of coercion by one party, these unsurprisingly give the actor greater leverage. However, bargaining power is not necessarily solely a function of material power or outside options (e.g. Scott 1985). Much of the bargaining power of *weak* actors may derive from their greater patience and commitment. Rubinstein (1982) developed the canonical bargaining model formalizing how an actor's willingness to outwait or outlast another leads to the more impatient actor being willing to make greater concessions.

Bargaining models, both formal and informal, have been applied to a wide variety of social contexts. For example, Fearon (1995) relies on bargaining models to assess explanations for war, suggesting that only a few rationalist explanations account for the fact that as war is costly, there should be incentives ex ante to strike bargains to avoid wars. From Fearon's perspective, war is an example of bargaining failure; war should be driven by broader factors that inhibit bargaining. These include incentives to misrepresent private information about capabilities or resolve, a lack of credible commitment mechanisms, and the indivisibility of issues over which there may be conflict.[8]

Fearon (1995) is explicitly focused on rationalist explanations for war and treats war as a discrete choice. However, if going to war is not simply a discrete choice, but a series of sequential steps in a bargaining process, as in Slantchev (2003), war is one of many possible equilibrium outcomes, even in a full information model in which actors may credibly commit. Furthermore, as Fearon (1995) recognizes, if the choice by states (which are collective actors) is not about maximizing some version of their collective good, but the interests of individual decision-makers, it may be that war is in the interests of leaders even when it is not in the interest of the state as a whole. Thus, models that define the scope of the decision or the nature of actors differently might result in very different results.

In addition, a sizable literature in a number of disciplines has suggested that social and cultural norms, including norms of fairness, influence human behavior in bargaining contexts, suggesting that the assumptions of the bargaining used by Fearon may not be universal. For example, Gintis (2009) suggests norms of inequity aversion and (strong) reciprocity are crucial to incorporate in social scientific models of bargaining and interaction, and the work of Henrich (2004) and his work with various colleagues provide strong evidence of systematic cross-cultural variation in bargaining behavior that typical rationalist models fail to capture. Henrich et al. (2010) go even further in noting the strong biases and limitations inherent in the bulk of social scientific research that has both built theory and tested empirics based on Western, educated, industrialized, rich, and democratic (*WEIRD*) populations. Henrich et al. (2010) note that there is a growing recognition in the social and behavioral sciences that many of the cognitive and motivational processes that the scholarly literature has previously assumed to be universal may be in fact be more context bound than has been appreciated.

Social Interaction and the Dynamics of Beliefs

While the drivers of individual human behavior – desires, beliefs, cognition, norms, emotions, and personality – are frequently taken as fixed, essentially every theoretic approach suggests that they can be influenced or even determined by social interaction. Dynamic models that incorporate change in these characteristics as a consequence of social interaction thus are particularly important in improving the match between models and social scientific theory.

---

8 The indivisibility of particular issues should only undermine bargaining when divisible side payments of sufficient value are unavailable to compensate for the indivisible issues.

While these all can be modeled as being influenced by social interaction, we focus in this section on the well-studied social influence on beliefs.

Suggesting that beliefs are (at least partially) determined by social interaction is commonsensical. Most people currently believe the Earth is round because of social interaction (what they have learned from others, what others believe, etc.) rather than because they have clear personal experience or have done experiments or carefully reasoned this out from evidence they themselves have experienced. If your neighbors do not believe in unidentified flying objects (UFOs), anthropogenic climate change, or the health benefits of eating vegetables, you are less likely to believe in them.

The relevant question among social and behavioral scientists is not whether social interaction influences beliefs, but rather by what mechanisms it does so, and whether this dynamic is important to include in a given model. In this section we briefly explore four basic mechanisms through which social interaction might influence beliefs: mimicry, Bayesian updating, motivated reasoning, and social persuasion.

Perhaps the simplest mechanism is mimicry. As discussed earlier, the basic logic is that people's beliefs will come to mimic those of the people around them over time. The cognitive theory underpinning mimicry may vary: it may be conscious or unconscious, driven by adaptation to social expectations, rational self-interested calculations, etc. But from a modeling perspective, changes in an individual's beliefs can be simply modeled as a function of (probabilistic) adoption of the beliefs of people they interact with (see Osoba and Davis (2019), in this volume, for a more sophisticated example regarding meme propagation).

Another common approach to modeling changes in beliefs is to assume that actors are intendedly rational when it comes to assessing the evidence for their beliefs and that they *update* those beliefs according to Bayes' rule (adjusting their prior beliefs in accordance with the relative strength of new evidence they may come across as a result of new experiences and new interactions). For example, the contact hypothesis (aka intergroup contact theory; e.g. Allport, 1979 [1954]), which suggests that intergroup prejudices can be reduced through positive social interaction between members of different groups and can be interpreted as reflecting individuals updating their beliefs about others in a Bayesian fashion (e.g. Lupia et al. 2015).

However, *neutral* Bayesian models of information updating have been extensively critiqued, with the bulk of the psychological and cognitive literature suggesting that people exhibit a wide range of biases in updating their beliefs. Perhaps most extensively studied is various forms of motivated reasoning (Kunda 1990), including confirmation bias (Nickerson 1998). People tend to seek out information that confirms their prior beliefs, contributing to selection bias in the information an actor receives. Furthermore, when faced with evidence that conflicts with their beliefs, they tend to *counterargue*, while uncritically accepting evidence that reinforces their prior beliefs (Taber and

Lodge 2006). Motivated reasoning and biased updating have consequences ranging from political polarization in an era of greater media diversity (Stroud 2010) to people tending to be overly optimistic in estimating of how long tasks will take (Buehler et al. 1997).[9]

Reviewing the psychology literature on social influence and persuasion, Wood (2000) suggests that changes in attitudes and beliefs occur from three primary mechanisms, including two normative mechanisms: (i) ensuring beliefs about the self are coherent and favorable, (ii) ensuring good relations with others (given the rewards and punishments they can provide), and one informational mechanism: (iii) a desire to hold beliefs that conform to reality. Thus persuasion and social influence may occur along multiple tracks, with, for example, those seeking to influence others appealing to various aspects of identity and social ties, as well as tying their appeals to empirical evidence.

These multiple tracks of influence and persuasion are well understood in business and marketing. Sales pitches for cars are almost never solely based on providing information that a neutral Bayesian will use to update her beliefs about whether to buy a car, and if so, which to purchase. Rather, advertisements highlight how the vehicle would reinforce a positive self-perception (e.g. certain SUV advertisements reinforcing a rugged outdoor sports self-image) and enhance the positive perception of a person by others (e.g. the number of car commercials in which the purchase of a new vehicle apparently leads to greater adoration by a devoted spouse and the awe of coworkers). The goal is to sell more cars by influencing beliefs about how certain cars are associated with personal identity, self-worth, and social esteem.

While the literature on how beliefs change in response to social interaction is well established, the consequences for modelers will vary based on the social phenomena modelers seek to understand. While beliefs are clearly influenced by social factors, beliefs can be resistant to change due to motivated bias and additional reasons for *status quo bias* discussed in Samuelson and Zeckhauser (1988). Thus, one of the most important decisions a modeler must consider is whether including the possibility of dynamic change in beliefs in a particular model will be valuable.

Social Interaction and the Dynamics of Identity and Culture

To many social scientists, identity is more than a set of beliefs: identity is intrinsically a social phenomenon, i.e. something constituted primarily through social interaction. Crucially, an individual's identity is understood through how that person situates themselves vis-à-vis the various groups to

---

9 This, of course, is not something that would ever apply to the writing of this chapter (or any other project any of the authors have ever been involved in).

which the person belongs. Understanding and modeling identity thus becomes a challenge of understanding a person's relations with a wide range of groups.

At the crux of this challenge lies the idea of boundaries: When, where, and how are social groups defined? Where does one group *end* and another *begin*? What is entailed in joining or being a member of the community? What are the dynamics that produce new communities or mobilize existing ones? How do the boundaries between groups flex and shift? What drives those changes?

These questions – long a concern of anthropologists and sociologists – become increasingly complex and urgent in our hyperconnected world. We must recognize that, however defined, boundaries are fluid, flexible, and overlapping. When effects of American policies and media permeate daily life in countries across the globe, just as Japanese anime and Al-Jazeera percolate through US communities, researchers must question the value of traditional *bounded* communities from an analytical standpoint. Can we identify something distinctive as *American* vs. *Korean* among cohorts of young people whose lives are parallel on many fronts?

One way to address these questions is to assume that identities are neither equal nor neutral, but are experienced and *managed* by people in accordance with hierarchies, relations of power and prestige, and ability to control the presentation of self (Goffman 1959) in different circumstances. How people represent *race* in the U.S. Census is one relevant example of how even allegedly *fixed* categories shift over time (Harris and Sim 2002). To address these questions, researchers must understand why people tie allegiance to some models of identity over others in situations of competing demands. Research on cognitive and cultural *motivated models* (Holland and Quinn 1987; D'Andrade and Strauss 1992) and *acts of meaning* (Bruner 1990) is part of more recent line of work aimed at establishing broader theories about how humans negotiate and create *self* amid competing social categories.

How might these approaches be useful for modeling? Understanding how people prioritize and align to different identities in different situations is important for modeling dynamic change. For example, we may hypothesize that a particular identity will become the grounding/salient driver of behavior when it is the marker of difference for those in a minority role: a woman physicist will be more cognizant of her status as a woman when at a professional conference dominated by men, whereas she may be more cognizant of her status as a professional physicist when among a group of at-home mothers. Alternatively, we can hypothesize that people seek out points of connection to others when they are in the minority category; the dominant behavioral expectations will govern behavior. Both hypotheses, while essentially positing opposite dynamics, could be born out through research and data and then modeled into a simulation that does not assume all identities are equally interchangeable.

The focus on the idea that identity drives behavior – that is to say, the ideas, attitudes, and beliefs that people hold about themselves and others shape and direct the way they behave – may be appropriate in many contexts[10] and is extremely common. Failure to abide by a code of conduct for one's chosen role/identity will often inspire social groups to take efforts to bring individuals back in line, enforcing and reinforcing the power of the social expectations side of norms.[11]

However, identity may similarly change in response to new behaviors. Lightfoot's (1997) work on the way risk-taking behaviors shape adolescent social group dynamics, and thus each teens' perceptions of herself, is a nice example of identity responding to behavior: when teens refused to cross new risk lines with their peers, they shifted identity from one social network to another. We see similar findings in Csordas' (1997) study of charismatic Catholics and Masquelier's (2001) work on bori possession cults in Niger, in which unexplained behaviors drive the search for a new identity that better aligns and explains the behaviors. Luhrmann's work on schizophrenia and prayer (2012) is a fascinating example of how people align new behaviors to specific social categories and thus transform something that is a *disorder* into an acceptable path.

This line of research may in part help explain the common *not me* response in the wake of everything from gang rapes to soccer mobs to raves or even marathons. Specifically, in post hoc reflections on such experiences, participants often cite that their behavior was atypical and unusual given their past history. Such responses are challenging for rationalist models of behavior because people behave in ways that they themselves describe as nonsensical and suggest an important area in which more common models of social interactions come up short.

Even in areas like the interactive evolution of group identity and cultural information, which have long been central in sociology and cultural anthropology, there remains a great deal of room for the development of more sophisticated models. These topics are explored further in Cowherd and Lende (2019) and Matthews (2019) later in this volume.

From Theory to Data and Data to Models

Even the most cursory overview of social scientific theories should make clear at least two interrelated challenges for the collection of data for

---

10 At a fundamental level, this makes intuitive sense as we look at people who change their behavior to align with their identity. For example, religious renunciates explicitly align their actions to conform with a particular chosen identity, as do military personnel, the newly married, and new parents. A more developmental take on this dynamic include studies of gender and racial bias among young children in which they self-limit their actions based on perceptions of what is acceptable for kids "like me" (Maccoby 1988; Witt 1997). In every case, the role/status assigned to a person is expected to drive behavior.

11 The nail that sticks out will be hammered back in (in Japanese, 出る杭は打たれる).

social-behavioral research. First, almost all of the crucial moving pieces of theories of human behavior are mental states that are unobserved and often intrinsically unobservable. Working with concepts that can only be measured through imperfect proxies is not unique to the social and behavioral sciences – it is common throughout the sciences. Second, the extent to which theory and measurement of data in the social sciences are context dependent is all too frequently overlooked or misunderstood. The combination of unobserved mental states and often unclear context dependency means that social science researchers typically are working with data for their key concepts that are weak proxies and may have limited external validity.

To take a simple example, consider how we can learn people's desires (one of the most frequent mental states social scientists wish to understand). One can derive some ideas from theory, but there often exist competing theories, and the theories are rarely specific enough to generate either universal or context-specific cardinal values for the weights people place on their various desires.

So, if people's desires matter, how do we overcome this challenge? Multiple approaches exist, but they most frequently boil down to first-person explication, third-party interpretation, or revealed preferences – that is to say, our empirical evidence about what a social actor wants is based on what that person says, on what another person says, or on behavior that the analyst believes accurately reveals what the person wants.

To take a prosaic example, how does a parent know what his 10-year-old child wants for dinner? Well, the child might announce as the parent picks her up from school, "I want sushi for dinner," and most people would take that as fairly clear evidence that was what the child wanted. And if she does not announce it, a parent (or a researcher) could ask her. Of course, it is clear that these speech acts may not be true revelation of desires, but may intentionally[12] or unintentionally[13] fail to accurately capture what the child wants. And often researchers are interested in the desires of people with whom they cannot interact for whom there is no detailed record of stated preferences.[14]

Overreliance on first-person explication of desires can also bias the subjects we study. A 10-year-old child may be very vocal about her desires, but when she was 10 months old, her parents could not necessarily expect her to tell them what she wanted for dinner – at least not in words. At times like this

---

12 This may be strategic misrepresentation of desires. For example, the child may not actually want sushi but knows that getting her parent to say no makes the parent feel guilty and thus be more likely to indulge her in other ways.

13 This may be false consciousness: i.e. the child may believe she wants sushi for dinner, but she in fact does not truly want sushi for dinner.

14 Although perhaps with the move to much richer big data more may be possible than was in the past. Do kids these days post what they want for dinner on social media?

we often resort to third-party interpretation (i.e. expert opinion). New parents introducing foods to a 10-month-old might ask other parents or a pediatrician what foods their child would like. Social scientists frequently ask third parties or experts their informed opinion about the mental states of subjects of interest, and this entails many of the same challenges as to reliance on first-party explication: the measurements are filtered through the a third-party perspective – a perspective that is, at best, *differently* subjective than a first-party explication of mental states. And concerns about strategic misrepresentation and false consciousness still arise.

How do experts develop beliefs about the mental states of human subjects? The same way social scientists do: through a rough mix of extrapolation from theory, projection of their own mental states to others, and drawing on a lifetime of inferring mental states from human behavior. Revealed preferences are perhaps the most commonly relied-upon source of information regarding preferences, but they too face comparable challenges to the other approaches. While we might infer that a 10-month-old child does not like the latest blended mush being offered because she pushes it away or onto the floor, we typically wish to make inferences about mental states for something beyond the immediate context. Is the 10-month-old refusing pumpkin–squash–blueberry for dinner making only a simple comparison (eating the bland blend vs. no food)? Or is a 10-month-old sufficiently cognitively developed that the fact that she had the much tastier apple–banana–spinach mush for lunch and liked it consciously or subconsciously influences this decision? One might imagine that pushing away the bland blend could increase the probability of getting more of the food that was previously a hit. The behavior may not be a true preference revelation that pumpkin–squash–blueberry is less desirable than no food, but in fact a strategic gamble that pushing it away increases the likelihood of receiving apple–banana–squash.

Social scientific disciplines have emphasized different data sources to a different degree. In economics, the emphasis in most subfields has been primarily on how the researcher can interpret choices and actions in terms of revealed preferences. In some branches of psychology, researchers have typically adopted the role of expert and assess for themselves what humans want, need, or desire in a particular context. In contrast, anthropology has long held that people are the experts on their lives, and the work of the ethnographer is not to reinterpret reports about choices, beliefs, and so on, but to provide the contextual framework within which such reports make sense. Ultimately the sources for data regarding the mental states that form the fundamental building blocks of most social scientific theorizing are often both crude and imprecise proxies for the underlying concepts, and the external validity of the measurements is highly context dependent.

Given the inherent challenges in measuring mental states discussed above, much of the empirical work in the social sciences is behaviorist in the sense that it treats mental states as a *black box* that links variables that are more readily observable. While a theory linking civil war to relative economic deprivation will likely ultimately involve the mental states of members of different groups in a society, a researcher might simply black-box that relationship in their analyses and test for a relationship between (observed) relative economic deprivation and (observed) civil war. This *behaviorist* approach has the advantage of not relying on data on unobservables.

However, to the extent that key links in a causal chain are black-boxed, we have difficulty distinguishing between theories that give similar predictions for the observed outcomes but may have different theoretical or policy implications. For example, an observed association between relative economic deprivation and civil war might be driven by beliefs held by members of marginalized socioeconomic groups that civil war is superior to the status quo in terms of extracting economic benefits for their group relative to other groups (a materialist group conflict model). Or perhaps the empirical pattern results from beliefs about violation of certain societal norms (a normative social contract model). While the empirical pattern might appear the same, the distinct mental models underpinning the observed correlation could very well suggest different policy prescriptions for helping resolve or prevent civil wars and could have different implications for models predicting civil war. When our research fails to get at the actual mechanism(s) that underpin empirical relationships we seek to understand, the value of our research is ultimately much more limited.

The rise of new sources of data, and the scale of data we can now analyze, may suggest a greater opportunity to draw on empirical evidence to parameterize models. While we believe this to be the case, the rise of big data has not obviated the intrinsic challenges entailed in properly collecting and understanding social scientific data. Fundamentally, the challenge of inferring unobserved mental states from imperfect proxies remains, and new sources and greater quantities of data do not change this. Modelers must choose between relying on imperfect measures and relying on untested assumptions about key components of the social and behavioral dynamics that they explore.

While detailed discussion of the challenges of going from data to models is beyond the scope of this chapter, the fundamental importance of understanding the *data-generating process* that underlies measures is difficult to underestimate. Modelers need to be particularly concerned with the validity of the data they use to parameterize any model across three dimensions: concept validity, measurement validity, and external validity.

In thinking about concept validity, modelers need to consider the correspondence between what the concept they are modeling and the measure they are relying on. For example, empirical indicators of mental states can be affected by multiple unobserved factors. Indicators of aggressive anger may also partially be measuring fear. Attempts to measure beliefs through interpretation of speech patterns may be valid for only some actors (e.g. not infants) and may be culturally specific.

In thinking about measurement validity, modelers need to be attentive to both the reliability and the possibility of bias in their measures. Data collected from small samples or from particularly noisy processes may exhibit greater random variation, and the assessment of the validity of data that purports to measure unobserved mental states is particularly challenging, because there is no objective standard by which to validate the measure. For example, relying on speech as a true representation of a mental state without accounting for how speakers may tailor their speech to different audiences runs substantial risks of propagating measurement bias into a model.

Finally, modelers must consider concerns of external validity. In some ways, this is the most crucial concern, given the emphasis in the social sciences on the context-dependent nature of both empirical and theoretical research. Modelers who wish to draw on empirical data, like all other researchers, must grapple with the challenge of generalizing from measures taken in a particular time and place to other contexts. While visible behaviors in discussion among friends and family might candidly reveal anger or angst, the same person's emotions might be suppressed or expressed differently when interrogated by authorities or enemies. The expression of political attitudes may be quite different during wartime, election campaigns, and economic booms (to highlight just three distinct contexts).

Ultimately, two of the key challenges of theorizing in the social sciences – that most theories rest on factors that are unobserved and potentially unobservable and that most theorizing is *mid level* and bounded – present serious challenges to researchers who simply might wish for the data to speak for itself. Modelers thus need to take care to understand any data they wish to incorporate in parameterizing or empirically assessing their models. While these problems may seem basic and perhaps even trivial to some readers, in truth they are enduring challenges intrinsic to social science research and modeling efforts. For more detailed consideration and applied examples, the discussions in both Carley (2019) and Grace et al. (2019) in this volume consider the challenges of validating models using empirical data in more detail.

Building Models Based on Social Scientific Theories

Social scientific research builds on a wide range of theoretic building blocks. This chapter has discussed many, including beliefs, desires, cognitive processes, norms, emotions, personality, and identity. Even so, the discussion in this chapter still gives short shrift to several important topics considered in detail in other chapters of this volume, including culture and the neurological bases for human behavior.[15] However, we do not believe it is the diversity of theoretic perspectives in the social sciences that makes social scientific modeling most challenging.

Building on social science theory is challenging because the foundations are invisible and the reliability of the conceptual building blocks may vary in different contexts. While the former point may be frustrating for developing and testing general social science theory (just as the limits of physical observation hamper our ability to develop a more unified set of physical laws), it is the latter point that is most crucial for most modelers and social scientists. This is because most social scientists and modelers are more like builders than theoretical physicists.

Building a sturdy home does not require knowledge of subatomic physics. A builder needs to know which building materials go together well and are suited to the local environment. Houses built in different climates often rely on different techniques and materials, and when builders move, their preferred materials are often less well adapted to their new conditions. And yet, borrowing building materials, techniques, and styles developed in other environments is one of the most powerful forces for innovation and advancement in architecture and construction.

In the social sciences we see a similar dynamic. It is at the interstices of disciplinary boundaries where researchers draw on new types of building blocks and often push fields forward. For example, in recent decades in economics, behavioral economics, political economy, and neuroeconomics all have become active and innovative subfields. We should not expect all such efforts at cross-disciplinary construction to be successful – efforts to incorporate distinct building styles or materials often lead to spectacular architectural failures. But over time these efforts are crucial in driving innovation in the social sciences.

The other major source of innovation in the social sciences (and in architecture) is the development of new technology. Over the past century, the rise of experimental social science, developments in sampling theory, and advances in computing all have dramatically altered social scientific and behavioral research. We have tools to observe, measure, and analyze social behavior today that social scientists a century ago did not even dream of. Opportunities to leverage new technology are likely to continue to drive innovation in social science in the future.

Modelers are particularly well suited to identify opportunities to leverage technological innovations and interdisciplinary perspectives in the social

---

15 We commend to readers the discussion in other chapters of this volume of culture (Matthews 2019), motivational systems (Read and Miller 2019), and neuropsychological bases of behavior (Thompson et al. 2019). A panel discussion chapter highlights additional important issues that modelers should consider based on theory building and testing in the social sciences (Sieck et al. 2019).

sciences. Most social scientists are firmly ensconced in their discipline; their research relies on well-established building blocks that are perceived to be appropriate and (hopefully) well suited to the disciplinary terrain. In keeping up with advances in modeling, modelers are inevitably exposed to cross-disciplinary perspectives and technological developments that may not be as visible to those who are primarily based on a single social scientific discipline.

The fragmented nature of social scientific theorizing can be frustrating. And yet the diversity in conceptual building blocks generates opportunities for creative and innovative modeling that draws on these building blocks to improve our understanding of social and behavioral dynamics.

Acknowledgments

This chapter builds on many conversations that the authors have had, both with each other and with colleagues and students over the years. In particular, this chapter draws on many years of teaching social science theory to political science graduate students at the University of British Columbia, conversations with members of the Canadian Institute for Advanced Research's interdisciplinary Institutions, Organizations and Growth Program, and wide-ranging conversations that took place at a workshop focused on the frontiers of social science modeling held at RAND in Santa Monica in April 2017.

References

Allport, G.W. (1979 [1954]). *The Nature of Prejudice*. Basic Books.

Arrow, K.J. (1950). A difficulty in the concept of social welfare. *Journal of Political Economy* 58 (4): 328–346.

Barry, B. (1988). *Sociologists, Economists, and Democracy*. University of Chicago Press.

Bechara, A., Damasio, H., Tranel, D., and Damasio, A.R. (1997). Deciding advantageously before knowing the advantageous strategy. *Science* 275 (5403): 1293–1295.

Benhabib, J., Bisin, A., and Schotter, A. (2010). Present-bias, quasi-hyperbolic discounting, and fixed costs. *Games and Economic Behavior* 69 (2): 205–223.

Black, S.E., Devereux, P.J., and Salvanes, K.G. (2013). Under pressure? The effect of peers on outcomes of young adults. *Journal of Labor Economics* 31 (1): 119–153.

Bowlby, J. (1969). *Attachment and Loss: Attachment*. Basic Books.

Brashears, M.E. (2019, this volume). How big and how certain? A new approach to defining levels of analysis for modeling social science topics.

In: *Social-Behavioral Modeling for Complex Systems* (ed. P.K. Davis, A. O'Mahony and J. Pfautz). Hoboken, NJ: Wiley.

Bruner, J. (1990). *Acts of Meaning*. Cambridge: Harvard University Press.

Buehler, R., Griffin, D., and MacDonald, H. (1997). The role of motivated reasoning in optimistic time predictions. *Personality and Social Psychology Bulletin* 23 (3): 238–247.

Camerer, C. (2003). *Behavioral Game Theory: Experiments in Strategic Interaction*. Princeton University Press.

Carley, K.M. (2019, this volume). Social-behavioral simulation: key challenges. In: *Social-Behavioral Modeling for Complex Systems* (ed. P.K. Davis, A. O'Mahony and J. Pfautz). Hoboken, NJ: Wiley.

Cowherd, G. and Lende, D.H. (2019, this volume). Social media, global connections, and information environments: building complex understandings of multi-actor interactions. In: *Social-Behavioral Modeling for Complex Systems* (ed. P.K. Davis, A. O'Mahony and J. Pfautz). Hoboken, NJ: Wiley.

Cox, G.W. (2004). Lies, damned lies and rational choice analyses. In: *Problems and Methods in the Study of Politics* (ed. I. Shapiro, R.M. Smith and T.E. Masoud), 167–185. Cambridge University Press.

Csordas, T. (1997). *The Sacred Self: A Cultural Phenomenology of Charismatic Healing*. Berkeley, CA: University of California Press.

Custers, R. and Aarts, H. (2010). The unconscious will: how the pursuit of goals operates outside of conscious awareness. *Science* 329 (5987): 47–50.

D'Andrade, R. and Strauss, C. (eds.) (1992). *Human Motives & Cultural Models*. Cambridge: Cambridge University Press.

Dawkins, R. (1976). *The Selfish Gene*. Oxford University Press.

Della Vigna, S. (2009). Psychology and economics: evidence from the field. *Journal of Economic Literature* 47 (2): 315–372.

de Waal, F.B. (2008). Putting the altruism back into altruism: the evolution of empathy. *Annual Review of Psychology* 59: 279–300.

Dimberg, U., Thunberg, M., and Elmehed, K. (2000). Unconscious facial reactions to emotional facial expressions. *Psychological Science* 11 (1): 86–89.

Doya, K. (2008). Modulators of decision making. *Nature: Neuroscience* 11: 410–416.

Elster, J. (1989). Social Norms and Economic Theory. *Journal of Economic Perspectives* 3 (4): 99–117.

Fearon, J.D. (1995). Rationalist explanations for war. *International Organization* 49 (03): 379–414.

Gallese, V. and Goldman, A. (1998). Mirror neurons and the simulation theory of mind-reading. *Trends in Cognitive Sciences* 2 (12): 493–501.

Gigerenzer, G. and Selten, R. (2002). *Bounded Rationality: The Adaptive Toolbox*. MIT Press.

Gigerenzer, G. and Todd, P. (1999). *Simple Heuristics That Make Us Smart*. Oxford: Oxford University Press.

Gintis, H. (2009). *The Bounds of Reason: Game Theory and the Unification of the Behavioral Sciences*. Princeton University Press.

Goffman, E. (1959). *The Presentation of Self in Everyday Life*. New York: Doubleday Books.

Grace, E., Blaha, L.M., Sathanur, A.V. et al. (2019, this volume). Evaluation and validation approaches for simulation of social behavior: challenges and opportunities. In: *Social-Behavioral Modeling for Complex Systems* (ed. P.K. Davis, A. O'Mahony and J. Pfautz). Hoboken, NJ: Wiley.

Green, D. and Shapiro, I. (1996). *Pathologies of Rational Choice Theory: A Critique of Applications in Political Science*. Yale University Press.

Halevy, Y. (2008). Strotz meets allais: diminishing impatience and the certainty effect. *American Economic Review* 98 (3): 1145–1162.

Harris, M. (1968). *The Rise of Anthropological Theory: A History of Theories of Culture*. AltaMira Press.

Harris, D.R. and Sim, J.J. (2002). Who is multiracial? Assessing the complexity of lived race. *American Sociological Review* 67 (4): 614–627.

Henrich, J.P. (2004). *Foundations of Human Sociality: Economic Experiments and Ethnographic Evidence from Fifteen Small-Scale Societies*. Oxford University Press.

Henrich, J., Heine, S.J., and Norenzayan, A. (2010). Most people are not WEIRD. *Nature* 466 (7302): 29–29.

Holland, D. and Quinn, N. (eds.) (1987). *Cultural Models in Language & Thought*. Cambridge: Cambridge University Press.

Jackson, P.L., Meltzoff, A.N., and Decety, J. (2005). How do we perceive the pain of others? A window into the neural processes involved in empathy. *Neuroimage* 24 (3): 771–779.

Kahneman, D. (2011). *Thinking, Fast and Slow*. Macmillan.

Kahneman, D. and Tversky, A. (1979). Prospect theory: an analysis of decision under risk. *Econometrica* 47 (2): 263–292.

Konner, M. (2010). *Evolution of Childhood: Relationships, Emotion, Mind*. Cambridge, MA: Harvard University Press.

Kőszegi, B. (2014). Behavioral contract theory. *Journal of Economic Literature* 52 (4): 1075–1118.

Kuklinski, J.H. and Quirk, P.J. (2000). Reconsidering the rational public: cognition, heuristics and mass opinion. In: *Elements of Reason: Cognition, Choice, and the Bounds of Rationality* (ed. Lupia, M.D. McCubbins and S.L. Popkin), 153–182. Cambridge University Press.

Kunda, Z. (1990). The case for motivated reasoning. *Psychological Bulletin* 108 (3): 480.

Kurrild-Klitgaard, P. (2001). An empirical example of the Condorcet paradox of voting in a large electorate. *Public Choice* 107 (1–2): 135–145.

Lau, R.R. and Redlawsk, D.P. (2001). Advantages and disadvantages of cognitive heuristics in political decisionmaking. *American Journal of Political Science* 45 (4): 951–971.

Lightfoot, C. (1997). *The Culture of Adolescent Risk-Taking*. New York: Guilford Press.

Luhrmann, T. (2012). *When God Talks Back: Understanding the American Evangelical Relationship with God*. New York: Vintage Books.

Lupia, A. (1995). Shortcuts versus encyclopedias: information and voting behavior in California insurance reform elections. *American Political Science Review* 88 (1): 63–76.

Lupia, A., Casey, L.S., Karl, K.L. et al. (2015). What does it take to reduce racial prejudice in individual-level candidate evaluations? A formal theoretic perspective. *Political Science Research and Methods* 3 (01): 1–20.

Maccoby, E.E. (1988). Gender as a social category. *Developmental Psychology* 24 (6): 755–765.

Masquelier, A. (2001). *Prayer Has Spoiled Everything: Possession, Power & Identity in an Islamic Town in Niger*. Durham: Duke University Press.

Matthews, L.J. (2019, this volume). Dealing with culture as inherited information. In: *Social-Behavioral Modeling for Complex Systems* (ed. P.K. Davis, A. O'Mahony and J. Pfautz). Hoboken, NJ: Wiley.

McKelvey, R.D. (1979). General conditions for global intransitivities in formal voting models. *Econometrica* 1085–1112.

Mill, J.S. (1863). *Utilitarianism*. London: Parker, Son, and Bourn.

Nickerson, R.S. (1998). Confirmation bias: a ubiquitous phenomenon in many guises. *Review of General Psychology* 2 (2): 175.

Nuckolls, C.W. (1998). *Culture: A Problem That Cannot Be Solved*. Madison, WI: University of Wisconsin Press.

Olsen, J.P. and March, J.G. (2008). The logic of appropriateness. In: *The Oxford Handbook of Public Policy* (ed. R.E. Goodin, M. Moran and M. Rein), 689–708.

Olson, M. (1965). *The Logic of Collective Action; Public Goods and the Theory of Groups*. Cambridge: Harvard University Press.

Osoba, O. and Davis, P.K. (2019, this volume). An artificial intelligence/machine learning perspective on social simulation: new data and new challenges. In: *Social-Behavioral Modeling for Complex Systems* (ed. P.K. Davis, A. O'Mahony and J. Pfautz). Hoboken, NJ: Wiley.

Ostrom, E. (2009). Collective action theory. In: *The Oxford Handbook of Comparative Politics* (ed. C. Boix and S.C. Stokes). Oxford: Oxford University Press.

Rabin, M. (2002). A perspective on psychology and economics. *European Economic Review* 46 (4): 657–685.

Read, S.J. and Miller, L.C. (2019, this volume). A neural network model of motivated decision-making in everyday social behavior. In: *Social-Behavioral Modeling for Complex Systems* (ed. P.K. Davis, A. O'Mahony and J. Pfautz). Hoboken, NJ: Wiley.

Rizzolatti, G., Fogassi, L., and Gallese, V. (2001). Neurophysiological mechanisms underlying the understanding and imitation of action. *Nature Reviews Neuroscience* 2 (9): 661–670.

Rubinstein, A. (1982). Perfect equilibrium in a bargaining model. *Econometrica* 50 (1): 97–109.

Samuelson, W. and Zeckhauser, R. (1988). Status quo bias in decision making. *Journal of Risk and Uncertainty* 1 (1): 7–59.

Schwartz, B., Ward, A., Monterosso, J. et al. (2002). Maximizing versus satisficing: happiness is a matter of choice. *Journal of Personality and Social Psychology* 83 (5): 1178–1197.

Scott, J.C. (1985). *Weapons of the Weak: Everyday Forms of Peasant Resistance.* New Haven, CT: Yale University Press.

Selten, R. (1975). Reexamination of the perfectness concept for equilibrium points in extensive games. *International Journal of Game Theory* 4 (1): 25–55.

Shepsle, K.A. (1979). Institutional arrangements and equilibrium in multidimensional voting models. *American Journal of Political Science* 27–59.

Shore, B. (1996). *Culture in Mind: Cognition, Culture and the Problem of Meaning.* Oxford: Oxford University Press.

Sieck, K., Brashears, M., Madsbjerg, C., and McNamara, L.A. (2019, this volume). Moving social-behavioral modeling forward: insights from social scientists. In: *Social-Behavioral Modeling for Complex Systems* (ed. P.K. Davis, A. O'Mahony and J. Pfautz). Hoboken, NJ: Wiley.

Simon, H.A. (1955). A behavioral model of rational choice theory. *Quarterly Journal of Economics* 59: 99–118.

Simon, H.A. (1956). Rational choice and the structure of the environment. *Psychological Review* 63: 129–138.

Simon, H.A. (1957). *Models of Man, Social and Rational: Mathematical Essays on Rational Human Behavior.* New York: Wiley.

Simon, H.A. (1972). Theories of bounded rationality. *Decision and Organization* 1 (1): 161–176.

Slantchev, B.L. (2003). The power to hurt: costly conflict with completely informed states. *American Political Science Review* 97 (01): 123–133.

Spitz, R. (1952). Psychogenic diseases of infancy. *The Psychoanalytic Study of the Child* 6 (1): 255–275.

Strack, F. and Deutsch, R. (2004). Reflective and impulsive determinants of social behavior. *Personality and Social Psychology Review* 8 (3): 220–247.

Stroud, N.J. (2010). Polarization and partisan selective exposure. *Journal of Communication* 60 (3): 556–576.

Sukthankar, G. and Behesthti, R. (2019, this volume). Using agent-based models to understand health-related social norms. In: *Social-Behavioral Modeling for Complex Systems* (ed. P.K. Davis, A. O'Mahony and J. Pfautz). Hoboken, NJ: Wiley.

Taber, C.S. and Lodge, M. (2006). Motivated skepticism in the evaluation of political beliefs. *American Journal of Political Science* 50 (3): 755–769.

Thompson, S.H., Falk, E.B., Bassett, D.S., and Vettel, J.M. (2019, this volume). Using neuroimaging to predict behavior: an overview with a focus on the moderating role of sociocultural context. In: *Social-Behavioral Modeling for Complex Systems* (ed. P.K. Davis, A. O'Mahony and J. Pfautz). Hoboken, NJ: Wiley.

Turner, V. (1995 [1966]). *The Ritual Process: Structure and Anti-Structure*. New York: Aldine Press.

Tversky, A. and Kahneman, D. (1974). Judgment under uncertainty: heuristics and biases. *Science* 185: 1124–1131.

Vohs, K.D., Baumeister, R.F., Schemichel, B.J. et al. (2008). Making choices impairs subsequent self-control: a limited-resource account of decision making, self regulation, and active initiative. *Journal of Personality and Social Psychology* 94 (5): 883–898.

Von Neumann, J. and Morgenstern, O. (1944). *Theory of Games and Economic Behavior*. Princeton, NJ: Princeton University Press.

Witt, S.D. (1997). Parental influence on children's socialization to gender roles. *Adolescence* 32 (126): 253–259.

Wood, W. (2000). Attitude change: persuasion and social influence. *Annual Review of Psychology* 51 (1): 539–570.

5

How Big and How Certain? A New Approach to Defining Levels of Analysis for Modeling Social Science Topics

Matthew E. Brashears

Department of Sociology, University of South Carolina, Columbia, SC 29208, USA

Introduction

When modeling social processes, what level of analysis is most appropriate? Should we stand back to a macro-level and treat nation-states as unitary entities? Should we stand at the micro-level and attempt to model the knowledge, motivations, and choices of millions of simulated actors in the hope that they aggregate into consistent results? Or should we try to locate ourselves somewhere in between, in the meso-level, where predictions are easier and data requirements are lighter? The answer is critical to producing useful results, but it is often not obvious which level of analysis offers the most promise. This problem is further exacerbated by the fact that each descriptor, micro-, macro-, and meso-level, applies to a range of scales rather than just one, depriving these terms of analytic precision. Here, I argue that the appropriate level of analysis is often difficult to identify because our concept of levels is inadequate. The appropriate level of analysis depends not solely on the size of the subject to be modeled but also on the constancy of the system itself, or the degree to which its behavior varies between fully determined and complex. By explicitly considering the interaction between the size and the constancy of the system, we improve our ability to build rigorous social science models.

These challenges are worth confronting because the construction of accurate predictive social science models is of primary importance. The key existential threats facing society in the twenty-first century, including global climate change and transnational terrorism, reflect the unintentional by-products of our highly developed physical technology. The legacy of the industrial revolution and medical advances permitting rapid population and consumption growth can literally be felt in rising global temperatures. Likewise, improvements in chemical, biological, and nuclear technology make greater levels

Social-Behavioral Modeling for Complex Systems, First Edition.
Edited by Paul K. Davis, Angela O'Mahony, and Jonathan Pfautz.

of power accessible to smaller groups, ultimately placing weapons of mass destruction within the grasp of hostile sub-state actors. These threats can be reduced in part through better technology (e.g. improved radiation sensors), but to a large extent their solutions require social change (e.g. decreased carbon footprints, demobilization of hostile forces, verification of arms agreements). As such, the critical frontier for science in the twenty-first century is firmly in the realm of the social.

In the remainder of this chapter, I begin by reviewing how levels of analysis are usually regarded in the social sciences. I draw attention to how this conception is inadequate in many instances and especially how it creates difficulties when attempting to produce reliable models. I argue for a revised perspective on levels of analysis that attends to both scale and constancy. While I cannot claim to have solved the problems of multi-level modeling in social science, this chapter will hopefully spark fruitful consideration of how new conceptions of levels are necessary in order to make progress.

Traditional Conceptions of Levels of Analysis

What level of a phenomenon should a scientist focus on? All scientific disciplines grapple with issues of scale. For example, the behavior of steel beams in a bridge is partly dependent on the behavior of individual molecules in the beams' structures. In turn, the behavior of these molecules depends upon their constituent atoms, subatomic particles, and perhaps vibrating strings. Likewise, the behavior of human collectivities, including nation-states, cultures, social movements, and mobs, is at least partly determined by the behavior of smaller constituent elements (e.g. individuals). International alliances are shaped by the behaviors of countries, which are shaped by the behavior of their civil societies, which are shaped by smaller communities, which are made up of individuals, who possess specific thoughts, feelings, desires, and beliefs. In short, almost every phenomenon in nearly any science rests upon a set of nested smaller elements and in many cases supports a further set of nested larger elements. On what level should one focus? For many social sciences, this issue of scale is what is most often referred to as *level of analysis*: the decision of how large a social entity one will attempt to describe, explain, and predict and with how much detail.

In social science disciplines, levels of analysis are most often referred to as macro-, meso-, or micro-level. The macro-level encompasses large social formations (e.g. nation-states, cities) and studies them as, essentially, coherent entities. At this level, the beliefs and ideas of individual people are typically viewed as irrelevant or as outcomes of larger-scale processes. In contrast, the micro-level tends to be focused on individuals or small groups (e.g. dyads, triads) and is often concerned with individual beliefs, behaviors, and ideas. Finally, the meso-level often emerges as an attempt to understand the connections between the macro- and micro-levels or to study intermediate-level processes. By way of an analogy to biology, the macro-level is akin to the study of overall physiology, the micro-level is akin to the study of cellular activity, and the meso-level is centered on individual organ function. Certain social science disciplines are sometimes thought of as having an affinity for particular levels, such as psychology with the micro-level or political science with the macro-level, but this affinity is more apparent than real. Relevant processes (e.g. voter behavior) often happen off of the supposedly native level of analysis for a discipline. Likewise, certain disciplines, such as sociology and economics, consciously span all levels (e.g. microeconomics, which deals with individual economic decisions, and macroeconomics, which grapples with the function of entire economies).

The centrality of the level of analysis is often far greater than the attention it garners. This is somewhat surprising in that the level of analysis fundamentally determines the units that will be studied, as well as the types of measurements that can be used, and by extension the kinds of conclusions that will be reached, thus making it one of the most consequential decisions in the research process. Moreover, questions of proper level of analysis are often confused with other related issues. This can be observed in the study of social networks, where there has been much discussion of the *boundary specification problem* (Laumann et al. 1992). The boundary specification problem refers to the difficulty in determining which nodes should be regarded as within, or a part of, a particular network.[1] For example, when conducting a network analysis of a corporate department, who should be regarded as a member? Only permanent employees, or do temporary workers also count? What about suppliers or customers? Typically, the boundary of a network is determined using either a realist or a nominalist approach. In the realist approach,[2] researchers define the network based on the views of network members (i.e. the network consists of those individuals who are consensually agreed upon by most other members). In the nominalist approach, the network contains those nodes that fit a set of criteria specified by the researcher (e.g. all those who worked in Department A more than 20 h/wk). These two strategies have obvious advantages and disadvantages but also inadvertently drive the analysis toward a particular level of analysis. In that the realist approach relies on the views of network members, it increases the relevance of their individual beliefs and ideas, pushing the analysis as a whole toward the micro-level. Likewise, by emphasizing analyst-set factors, the nominalist approach tends to favor a more meso- or macro-level perspective. It

---

1 There is a parallel problem in deciding which types of edges (e.g. friendship, material exchange, authority) should be measured. This issue is not usually viewed as quite as significant, however, because node-based sampling is dominant in network research.

2 Readers from political science should be especially cautious as the term, as used here, differs from how it is often used in that discipline. However, the difficulties imposed by the cross-disciplinary proliferation of terminology are beyond the domain of this chapter.

should be noted that this issue is not primarily one of scale; network studies of social media that include thousands or millions of nodes are still forced to adopt one of these perspectives with the resultant impacts on their level of analysis (see also Lazer and Radford 2017). Thus, decisions that are often not viewed in terms of level of analysis nonetheless partially decide it. By extension, these analytic decisions that shape or constrain the level of analysis may play an outsized role in determining the conclusions that will be drawn from the research, without the researcher necessarily being aware of it.

The result of this confusion is that earlier choices of *what* is to be studied are often partially made by logically subsequent choices of *how* to study it. For example, a significant body of research shows that individual perceptions of networks do not match actual contact patterns in a straightforward manner. Indeed, the disconnect between who individuals claim to have interacted with in a given period of time and who they can be observed to interact with over the same period is so large that researchers have questioned whether humans can be said to recall their interactions at all (e.g. Bernard and Killworth 1977; Bernard et al. 1979/1980). Reassuringly, subsequent work has shown that while humans are often poor at remembering specific interactions, they are quite good at remembering typical interactions and at identifying subgroups and subgroup members (e.g. Freeman and Romney 1987; Freeman et al. 1987; Freeman 1992). Yet, consider this finding in light of the boundary specification problem – a network defined by the perceptions of its members will necessarily be skewed toward their perceptions of typical patterns of contact and away from specific and concrete interactions. As a result, this simple decision of how to draw a boundary around a network ultimately determines not only which nodes are included but the meaningfulness of the resulting network. As such, it is necessary to make deliberate decisions about the level of analysis before related analytic decisions are made. Obviously, decisions can be revisited and altered later in the research process, but choices should be made deliberately by researchers and should be made early given the reflexivity of human research subjects. Likely, the failure to proceed in such a manner derives partly from the difficulty involved in identifying the proper level of analysis, an issue to which I now turn.

Incompleteness of Levels of Analysis

The difficulty with the traditional conception of levels of analysis is that in many instances it is quite simply insufficient to describe the actual level of inquiry. Moreover, this insufficiency makes effective modeling exceptionally difficult. This can be seen especially clearly in two papers deriving from the *structuralist* tradition in sociology (e.g. Blau 1977; Mayhew 1980, 1981), which emphasizes maximal parsimony and minimal attention to the beliefs and ideas of individuals. The first (Mayhew and Levinger 1976a) attempts to explain the level of interaction in human groups purely using their size. The possible number of undirected relations, L_{Max}, increases as $n(n-1)/2$ where n is group size; as a result linear increases in group size generate greater than linear increases in the potential number of relations. Since individuals have limited resources to devote to interaction (e.g. time), large systems must include large numbers of potential relations that are either null (i.e. totally unused) or contain close to zero investments of time and energy. Consequently, if residents of small towns seem friendlier than those of large cities, it is not necessarily a cultural distinction, but rather is likely a direct mathematical result of the interaction between system size and limited individual resources.[3]

The second paper (Mayhew and Levinger 1976b) attempts to explain the robust tendency toward oligarchic forms of government using human biological limitations. The authors argue that decisions must be made during, or following, an interaction sequence (e.g. series of speaking or influence events). However, while increases to the size of a group require longer interaction sequences in order for each member to have a say, human working memory has a more or less fixed capacity at approximately seven elements (Miller 1956; Baddeley 1986). This limitation means that longer sequences cannot be held in memory and thus decisions must be made based on smaller sequences involving interaction between a subset of group members. Moreover, since this limitation does not scale with system size, larger groups must be governed by smaller, more oligarchic subgroups.

The most obvious point that can be drawn from these two papers is that neither of them works strictly at a single level of analysis. The first paper uses a macro-level feature of a system, its size, partly in order to predict behavior at the micro-level, the richness of interaction. Likewise, the second paper uses individual-level limitations, in the form of human memory constraints, in an effort to explain a macro-level outcome, the centralization of power in large groups. In both cases these efforts are imperfect; the size of the system requires that many interactions become less rich, but says nothing about which interactions will suffer, while memory limits provide little traction in estimating how large an oligarchy will be or how it will be structured.

A second issue is that both explanations are fundamentally compromised by an attempt to bridge between levels without paying sufficient attention to processes occurring at either. While it is true that increases in system size must produce a decrease in interaction richness overall, the interactions that are preserved will not be randomly distributed. Instead, interactions are guided by homophily (Lazarsfeld and Merton 1954; McPherson et al. 2001), or the tendency for individuals to associate with similar others (e.g. "Birds of

---

3 Improvements in technology can make interaction more efficient, partially counteracting this tendency, but they also tend to increase the practical size of the system (e.g. cell phones make it less costly to interact but also dramatically increase the number of people who can be interacted with). As such, efficiency gains tend to be overwhelmed by system size increases.

a feather flock together"), and thus increases in system size are likely to cause fragmentation rather than a uniform decrease in interaction density. While intergroup interaction rates are partly governed by population proportions (Blau 1977), homophily tendencies are typically measured as departures from this baseline rate (Marsden 1987, 1988), and homophily varies by sex (Brashears 2008a) as well as over time (Smith et al. 2014; Brashears 2015). Doubtless forces beyond homophily would also play a role, but the broader point should be clear. System size ultimately provides very little insight into how decreases in interaction richness will be expressed and provides little predictive power for the inevitable fragmentation of the social system or quality of life therein. Moreover, not all characteristics on which individuals can be homophilous are socially meaningful, or salient, and which characteristics confer meaningful status is itself the outcome of a social process (Ridgeway 1991; Ridgeway and Balkwell 1997), leading to cross-cultural variation (Brashears 2008b). As a result, the macro-level interaction density is fundamentally impacted by micro-level conceptions of what is, or is not, relevant in particular circumstances. Similarly, while it is true that working memory is a limited resource, humans exhibit numerous strategies for circumventing these limitations (e.g. *compression heuristics* that allow more relations to be accommodated in the same memory space; Brashears 2013; Brashears and Brashears Forthcoming). By failing to take account of these strategies, the underlying connection between working memory and oligarchy is weakened to the point of irrelevance. Put another way, the limits to working memory are considerably more flexible than Mayhew and Levinger (1976b) realize and, as a result, allow quite a range of solutions to the problem of oligarchy. Additionally, working memory is paired with long-term memory as well as various artificial memory aids (e.g. written language), allowing decisions to be made using a wider range of views by trading off against the capabilities of each system. By sprawling awkwardly across levels of analysis, these papers are able to make useful statements in a broad sense (e.g. small towns do not seem more friendly than cities because the inhabitants are qualitatively different) but do so imprecisely, hampering efforts to construct models or empirical tests.

The third, and most significant, issue is that within and across levels, the nature of the predictions made is highly variable. System size influences the density of interaction in a deterministic way (more people necessarily leads to lower interaction density), but the allocation of those interactions is stochastic. Indeed, with the intervention of homophily, individuals within the system may fail to notice that overall interaction density is decreasing because their local density remains essentially unchanged (i.e. the system becomes more clustered). Similarly, while cognitive limitations restrict the scope of interaction that social groups can productively track, this argument is at best stochastic, in that many approaches to circumventing these limits are available and can be linked to the emergence of a particular oligarchy only through a vastly more involved model. As a result, it is unclear what level of evidence would be necessary to support or falsify the models' predictions and what time scale is needed for model outcomes to manifest. For example, is a precise mathematical agreement between system size and density of interaction required or only approximate? Without a more precise definition of what is being studied, accurately specifying the necessary outcomes becomes difficult verging on impossible.

What becomes clear in the above is that *the level of analysis, as typically conceived, only partly describes the phenomenon to be studied. Because this description is only partial, it is more likely that the analysis will yield vague, difficult-to-model predictions, at worst making hypothesis testing a matter of rhetoric rather than evidence.* This is not to say that the above papers are entirely irrelevant, but rather only that their impact is blunted and their utility for constructing predictive models is reduced. To produce the best research, it is necessary to think deliberately about the proper level of analysis. Moreover, it is necessary to employ a conception of level of analysis that accounts not only for the *scale* of the phenomenon, as the level of analysis is usually thought to specify, but also for its *constancy*, a concept to which I now turn.

Constancy as the Missing Piece

Useful empirical analysis and modeling requires not just an understanding of the level of scale that characterizes the phenomenon, but the tendency of the phenomenon to unfold in consistent, easily predictable ways. I refer to this as *constancy* in an effort to capture the variation in whether the underlying connection between inputs and outputs, or causes and effects, is reliable. While constancy is most safely viewed as a somewhat smoothly varying property, we can conceptualize three ideal types to capture how variations in constancy matter for modeling and analysis.

At the highest level of constancy, the phenomenon is deterministic. In this case, a given input will invariably yield a specific output, and thus the phenomenon behaves in a manner consistent with classical conceptions of physical law. Much as Newton's laws of motion deterministically explain many mechanical systems,[4] a phenomenon exhibiting maximal constancy will yield identical outcomes given identical inputs. The above theory linking system size to interaction (Mayhew and Levinger 1976a) is situated at this level of constancy, arguing that system size will, by itself, always yield consistent decreases in interaction density. For obvious reasons, many scientists, including social scientists,

---

4 It goes without saying that Newton's laws are only approximations that have been supplanted by newer conceptions (e.g. relativity), but they remain valid at the scales and energies humans typically operate at and thus remain a useful exemplar.

have long hoped to uncover a set of social theories that would deterministically explain social behavior (i.e. maximize constancy). The collective inability to derive such high constancy rules has generally been blamed on the large number of factors at work in determining social outcomes. Yet, whether this account is correct or not, the lack of deterministic theory has led social science to focus on probabilistic models instead.[5]

At intermediate levels of constancy, the phenomenon can be described as stochastic. In this case, inputs are linked to outputs via a fundamentally probability-based mechanism that does not guarantee the same effect for equal sets of causes. Yet, while the precise outcome at any given time is unknown, the likelihood of discrete outcomes is known, and the distribution thereof is essentially stationary. For example, when rolling a pair of six-sided dice, one is ignorant of the specific combination of values that will appear but can characterize accurately the likelihood of particular sums.[6] Social phenomena at this level of constancy behave like any other probability distribution in which the general nature of the outcome can be characterized with high accuracy. Much of modern social science has become rooted at this level of constancy, eschewing the possibility of deterministic laws in favor of a fundamentally stochastic view of social organization. For example, Lizardo (2006) used log-linear models (i.e. more advanced forms of chi-squared tests for predicting allocation of observations to cells in a cross-classification table) and instrumental variables regression to establish that cultural consumption patterns shape the development of network ties. In short, widely accessible forms of culture provide a foundation for association with many others and so tend to support the formation of superficial relationships, while narrowly available culture tends to support deeper, more intimate connections. These causal connections are meaningful, and theoretically important, but are entirely stochastic; the presence of common culture makes relationships more likely, but does not guarantee them. Access to common culture will increase the number of relationships possessed, but to a variable extent. Moreover, there is no theoretical reason to think that this stochasticity is due to imprecision in measurement; even if association would be facilitated by common culture, the actual formation of a tie should always be partly a matter of chance. Thus, for example, you may find yourself chatting about a recent television program with someone you meet on the subway, but this does not inevitably produce a new durable relationship.

---

5 Some argue that social scientists have given up on a grand theory entirely, preferring to focus on smaller domain-specific theories (Swedberg 2017).

6 Of course, the specific outcome might be predictable given sufficiently precise measurements of a host of variables (e.g. weight distributions in the dice, initial orientation, initial vector, exact state of the air, etc.), but even if it were possible to measure all factors sufficiently accurately, in practice it is unlikely to be practical or useful to do so. Thus, a specifically stochastic model is useful and practical.

Finally, at the lowest level of constancy, the phenomenon is fundamentally complex. Complex systems "…consist of situated, adaptive, diverse individuals whose interactions produce higher-order structures (self-organization) and functionalities (emergence)" (Page 2015, p. 22). Such systems include a variety of entities engaged in interdependent interaction, and the constituent entities are capable of learning and adaptation. As a result, a single stationary probability distribution is insufficient to link causes to effects. While some attention has been devoted to the possibility that many social systems are fundamentally complex (e.g. Ceteno et al. 2015), most social science research remains wedded to the notion that causes combine in some broadly linear fashion to yield stochastic effects (but see also Davis 2011; Ragin 2014). Exceptions to this rule, however, demonstrate its promise. For example, Centola and Macy (2007) used a simulation model to show that for social contagions (e.g. new behaviors or technologies that spread via social networks) that are *complex* and require multiple contacts to diffuse (e.g. the benefits of adoption rise as more others have already adopted), long-range bridging ties do *not* facilitate rapid spread. Instead, redundant, densely tied network regions are the most critical drivers of adoption. For example, accepting a new medical claim (e.g. that vaccines cause autism) promulgated on social media may be more likely when many others to whom one is connected espouse this claim in quick succession. Thus, this belief will spread more rapidly in denser portions of the network. But critically, later work (Barash et al. 2012) shows that this is only the case early in a diffusion process and that long-range bridging ties (i.e. network connections that link otherwise disconnected clusters together) play a more significant role in diffusing complex contagions later on. Thus, the relevance of specific types of ties is not well described by a stationary probability distribution and depends instead on how much progress the contagion has already made. Complex contagions initially benefit from dense network regions but not from long-range connections while later come to benefit from bridges as well. Similarly, Shi et al. (2017) used an agent-based model to study how organization-level pressure for members to devote time to an organization and to limit their social ties to other organization members impacted organizational growth and longevity. The model additionally explored how these factors changed in their effects depending on the degree of competition an organization experienced from other organizations, as well as the strategies adopted by those competitors. Their results show that positive growth is attainable with the right strategy combinations, that effective combinations differ based on competition, and that the range of such effective combinations shifts dramatically based on competitor strategy. Indeed, the simplest summary of the ideal competitive strategy is "be slightly less demanding than one's competition," and so the ideal approach cannot be identified at all without knowing something of the other entities and their natures. As a result, selecting an optimum approach to recruiting and retaining members depends heavily on an ongoing sequence of decisions made by rivals and *cannot be modeled as a single stationary probability distribution.*

It may appear at first glance that constancy is an attribute of the analysis rather than of the thing being analyzed and thus should not be integrated with the notion of analytic level. This position is understandable, but fundamentally incorrect. Put simply, the goal of most sciences is not, strictly speaking, to describe and assess *things*, but rather to describe and understand *processes*. Since any process is a chain of inputs and outputs, or of causes and effects, binding entities in interaction, then the constancy with which those inputs and outputs are connected fundamentally determines the nature of the thing under study. In other words, the scale of a process is clearly relevant to the level of analysis, but since the goal of analysis is to link causes to effects, so too is the constancy of those processes. Moreover, it is clear that the degree of constancy to a phenomenon is not simply a function of the quality of our explanatory models. While Newtonian mechanics operates in a satisfactorily deterministic way within its domain of applicability, quantum mechanics appears to be fundamentally stochastic. Likewise, while the gas laws are pleasingly deterministic, they are built on a foundation of stochastic behavior at the particle level and are a factor in shaping global weather, which is generally viewed as a complex system.[7] While analyses can approach a phenomenon as though it were stochastic when it really is not, producing what may be a useful approximation, processes themselves possess some underlying degree of constancy, and the most accurate and useful models will be rooted in this degree. As such, the almost reflexive tendency of most social science to adopt a *stochastic orientation to all problems is a clear shortcoming.* Likewise, social scientists often default to a multiple regression-based framework (see McPherson 2004), thereby assuming linearity, certain underlying distributions to the data, and a set of static processes. When all you have is a hammer, everything starts to look like a nail, and when all you have is regression, everything will start to look like a stochastic linear process. Such problems become worse with overly strict interpretations of Occam's razor or models of human behavior that are excessively rigid (e.g. rational actor assumptions).

Constancy is related to, but distinct from, the concept of predictability and should not be confused with it. In short, two systems may both be predictable,

---

7 Astute readers may note that in contrast to the earlier definition, the weather system does not contain entities capable of adaptation or learning. This is reasonable, but only partly true: the behavior of physical systems can exhibit sudden nonlinear responses (e.g. the *clathrate gun*), and the presence of living organisms has a powerful aggregate effect (e.g. the *oxygen catastrophe*). Thus, even if learning is an incorrect term, the system still exhibits behavior that is a reasonable facsimile.

in that inputs (causes) can be linked reliably to outputs (effects), but may vary in the nature of the connection (e.g. deterministic vs. stochastic). While a totally unpredictable outcome by definition has a minimal level of constancy (i.e. causes and effects are related to each other with zero consistency), processes across a wide range of constancy levels can all be viewed as *predictable*. Likewise, constancy is also distinct from the richness or complexity of the predictors (e.g. causes) used to infer an outcome (e.g. effect); few or many variables may be necessary in order to model a process that is fundamentally deterministic, stochastic, or complex. In this sense, constancy is similar in logic to the *Big O notation* used in computer science to express the impact of increasing input size on run time or space requirements for program execution. Much as larger inputs can be connected to run time in a sublinear, linear, or superlinear fashion, more and less complex sets of causes can be connected to effects in a deterministic, stochastic, or complex manner. In all cases, there is a connection between input size and run time, but the nature of this connection can vary in characteristic ways. Likewise, a process may have a large and complicated phase diagram while nonetheless remaining fundamentally deterministic, and we should be cautious not to confuse modeling complexity with process constancy.

Constancy should be understood as distinct from the scale of a phenomenon, which usually dominates our thinking in regard to levels of analysis. Returning to an earlier example, physical processes occurring at a very small (quantum) scale appear to be fundamentally stochastic, but likewise the sum obtained from a throw of two dice is also stochastic. Despite the extreme scale differences, the constancy is equivalent. Likewise, the overall behavior of national economies may be deterministic based on several factors (that may themselves be difficult to predict or control), but the success or failure of individual corporations under these conditions may be stochastic, while the fate of economic sectors may be complex. At any given scale processes can exhibit deterministic, stochastic, or complex levels of constancy, and failure to take account of these differences can compromise the resulting models.

Putting It Together

Ultimately, levels of analysis are most valuable when understood in terms of both scale and constancy. Micro processes can be essentially deterministic, stochastic, or complex. For example, work using power dependence theory has shown that particular structures of exchange relations (e.g. economic transactions, choices among prospective dating partners) reliably produce specific degrees of advantage regardless of the intent of the actors (e.g. Emerson 1962).

Moreover, the availability of coercion produces only minor deviations from the predicted advantages (Molm 1990). Thus, individual exchange advantage is best modeled at the micro-deterministic level. At the other end of the micro spectrum, Simpson et al. (2014) show that the cooperation and defection behavior of individuals, even given stable individual preferences, depends on the relationship they maintain with their exchange partner (e.g. friends vs. strangers). Moreover, this appears to be the case in both experimental and natural settings, indicating that realized behavior is determined by an interaction between consistent individual preferences and particular prior sequences of outcomes that gave rise to the current relationship. This work therefore operates primarily at the micro-complex level of analysis.

Alternatively, while Shi et al.'s (2017) model includes a stochastic component to individual agent behavior, it is built from the ground up as a complex system. The order of actions profoundly influences the future development of the system, and the model connects organization-level and individual-level behavior in a reciprocal fashion. Similarly, Brashears and Gladstone (2016) show that the accuracy of information diffusion depends on both the length of diffusion chains and the success or failure of individual efforts to repair errors in the contagion. Earlier, unpredictable failures in repairs produce dramatic changes in the success of future diffusion. These papers both operate at the meso-complex level of analysis.

Work by McPherson (1983) and Mark (1998) shows how organizations and musical styles compete for members much as biological organisms compete for food. Organizations require many resources to survive, but the one resource that they indisputably require is the time and attention of human members. Since individual time is limited, there exists a finite time budget that organizations rely on for their survival. Organizations that compete for the same types of individuals are competing over the same limited pool of time much as organisms sharing a biological niche compete for food and shelter resources. The same logic applies to cultural elements, such as musical styles (e.g. harpsichord vs. country music) and languages (e.g. Latin vs. English), such that an element that claims no time from living humans ceases to exist as a living entity. This logic allows recruitment into groups, familiarity with cultural artifacts, and even the future demographic makeup of a group (McPherson and Ranger-Moore 1991) to be determined by intergroup or inter-style competition. Moreover, these predictions are heavily deterministic based on the degree of competition for needed resources. This tradition therefore operates largely at the meso-deterministic level.

Ultimately, a wide range of combinations of scale and constancy are possible and have been explored to some degree in the existing literature. These are illustrated in Table 5.1 with examples drawn from the discussion above.

Table 5.1 Levels of analysis as scale by constancy combinations with published examples.

	Deterministic	Stochastic	Complex
Macro	Mayhew and Levinger (1976a)	Blau (1977)	Centola and Macy (2007) Barash et al. (2012)
Meso	McPherson (1983) Mark (1998) McPherson and Ranger-Moore (1991)	Mayhew and Levinger (1976b)	Shi et al. (2017) Brashears and Gladstone (2016)
Micro	Emerson (1962) Molm (1990)	Lizardo (2006)	Simpson et al. (2014)

Implications for Modeling

One of the critical advantages to be had from a revised conception of levels of analysis is improvements in modeling. However, making such improvements requires that researchers pay careful attention to a number of issues stemming directly from the level of analysis.

Concretely, this expanded concept of levels of analysis will require changes in the selection of independent and dependent variables, as well as greater attention to the time scale of the phenomenon under study. First, selection of independent and dependent variables must ensure that they are matched not only in terms of the scale of the unit of analysis but also that the proposed linkage between them is at a homogeneous level of constancy. An independent variable-to-dependent variable connection that requires many transitions between deterministic, stochastic, and complex levels of constancy will be difficult to substantiate and model in a useful way. Likewise, unit homogeneity will become more important in analyses as increasing heterogeneity also implies variation in constancy. This will inevitably drive social scientific theorization toward a focus on smaller mechanisms and will discourage modelers from excessively general efforts, but *encourage the development of a more mechanism-focused approach to understanding social life.*

Second, social scientists often devote relatively little attention to how long a process requires to play out, not least as a result of data limitations, or rely on simulation models that are inherently difficult to connect to clock time. Yet, attention to constancy will require a greater awareness of how long a mechanism should require to operate, and processes that occur on very different time scales may not belong to the same level of constancy. *Logically, processes occurring at different levels of constancy should not be treated as occurring at the same level of analysis.*

The above changes to the research process will be challenging but should ultimately contribute to the development of effective multi-scale models. While this new approach encourages defining mechanisms in smaller terms using both scale and constancy, it should result in a modular set of models. Each small model, capturing the behavior of a mechanism operating at a particular scale and constancy, can more easily be linked together to create larger, more comprehensive systems. Moreover, the increased focus on the temporal dimension to mechanisms will make these connections more organic. The connection of micro events directly to macro outcomes in research will be somewhat de-emphasized, leading to greater attention to intermediate-level linkages that help smaller processes aggregate into larger outcomes. Ideally, we will ultimately see a library of mechanisms that are more naturally compatible with each other for the construction of full-scale multi-level models.

The increasing availability of automatically logged and time-stamped interaction data (e.g. email, social media interaction, etc.) offers new opportunities to bring time and constancy into focus. The large data sets available from harvesting social media allow issues of temporal ordering to be clarified, while the larger data sizes make it more straightforward to distinguish among constancy levels. While not without risks (Lazer et al. 2014), the rise of *big data* is promising.

Previously, modeling has typically chosen a level of analysis and then viewed approach (e.g. formal deterministic vs. agent-based simulation) as a methodological issue of lesser importance. However, the key insight that I wish to emphasize is that both of these elements are integral to defining the phenomenon that is to be modeled. If the phenomenon acts as a stochastic system, then a stochastic approach should be used. In contrast, if it is essentially deterministic, then it should be modeled as such. The goal is to fit the modeling strategy to the comprehensive level of the phenomenon to be studied, consisting of both scale and constancy. This is especially important given that the same social system may behave in both deterministic and complex ways, depending on time or other factors. It will be necessary to understand how these levels interface in order to produce accurate predictive models.

Adopting this approach will require active participation from both social scientists and modelers. Social scientists have the theoretical and substantive knowledge necessary to understand a phenomenon but often are ill equipped to construct formal or computational models. Modelers possess this expertise but lack an understanding of the phenomenon that they hope to model. Thus, successful efforts will be rooted in collaboration, but it cannot be *collaboration as usual*. For social scientists, more attention will need to be directed to the expected behavior of the process under study. This will be difficult in some cases as many social science disciplines currently carve up their field of interest purely in terms of scale (e.g. microeconomics) or based on substantive interest (e.g. sociology of religion) rather than based on the underlying process. However, forcing attention to these issues should facilitate improved integration across subareas, as disparate substantive areas often share common processes of interest (e.g. recruitment is of interest to researchers studying religion, formal organizations, and social movements, among others). Moreover, many theories are at least somewhat level agnostic (e.g. ecological theory) or have distinct predictions and proposed mechanisms operating at particular levels. Thus it will often be possible to adopt this more refined conception of levels of analysis without abandoning the existing corpus of theory or substantive knowledge.

For modelers, this shift will require a degree of humility. Rather than modeling a social process based on the skills the modeler already has (e.g. "I am an agent-based modeler"), it will be necessary to think seriously about the concordance between modeling method and underlying process. This will require considerable knowledge of the theory and substantive findings present in a given area because scale, which is relatively easy to identify, provides little guidance as to constancy. Prior assumptions that scale is closely linked to constancy collapse under only light scrutiny and should not be relied upon. Modelers should be especially cautious when exploring social topics because the identification of something as being, or not being, a problem to be solved is itself often the outcome of a social process. For example, technical limitations and bureaucratic inertia led to a systematic failure to properly account for the impact of mass fire following a nuclear strike, leading to significant underestimates of the damage likely to result from an exchange of strategic nuclear weapons (Eden 2004). Similarly, an incremental process of accommodation led NASA engineers and managers to ignore growing signs of aberrant behavior, ultimately resulting in the loss of the space shuttle *Challenger* (Vaughan 1996). In both of these examples, the conception of the problem was, itself, the driver of significant failures and oversights. Rapidly identifying a system to be modeled without input from social science experts runs a high risk of baking a large number of biases and problematic assumptions in from the start.

The needed changes for both social scientists and modelers will be more easily achieved if they work together in active collaboration throughout the research process. Critically, this means that teams of modelers who add a social scientist as, essentially, window dressing or teams of social scientists who add a modeler as a mere programmer are to be avoided. Instead, relatively equal teams will be needed that feature frank discussions between both types of researchers. Teams will also need to avoid past tendencies to use collaborations to add depth to specific areas of a study rather than to build connections between different areas (Leahey 2016). Ideally, these efforts will lead over time to an increase in the role of modeling within social science, as well as a group of modelers who have greater familiarity with social science.

Conclusions

In this chapter I discussed the limitations of existing concepts of levels of analysis and argue for a new, expanded approach. Typical concepts of levels of analysis are limited to questions of scale and therefore ignore the degree of constancy exhibited by the processes of interest. As a result, it is often highly uncertain what is actually being studied. This uncertainly hinders efforts to construct accurate models and therefore damages our chances of accumulating useful general models of social behavior. Moreover, decisions impacting the level of analysis are often inadvertently made by certain analytic decisions, potentially compromising the ability of research to speak to its intended topics.

Moving past these challenges will require several changes. Social scientists and modelers must collaborate in a genuine two-way manner. While it is hypothetically attractive to simply plug a social scientist or modeler into a team to increase its attractiveness to funders, this makes poor use of resources and contributes nothing to improving the state of the art. Both types of researchers can add to the process if given genuine opportunities. Likewise, we must give up the notion that improvements in social science will necessarily lead to deterministic *laws* of social behavior. Some determinism is surely present, but many social systems are complex and therefore inherently defy such efforts. If the underlying phenomenon is complex, then we must theorize it, and model it, as such, and make use of the resulting predictions in a way that is consistent with the contingent and uncertain nature of complex systems.

The challenges are obvious, but the severity of existential challenges facing our society, as well as the fundamentally social nature of their solutions, makes the benefits clear. Only by more accurately grasping the nature of the processes we hope to understand can we learn to model and control them.

Acknowledgments

This paper was partially supported by grants from the Defense Threat Reduction Agency (HDTRA-1-10-1-0043) and the Army Research Institute (W911NF-16-1-0543). Special thanks to Paul Davis, Jeff Elmore, Michael Gabbay, and Angel O'Mahoney for their feedback on earlier drafts.

References

Baddeley, A.D. (1986). *Working Memory*. Oxford: Clarendon Press.
Barash, V., Cameron, C., and Macy, M. (2012). Critical phenomena in complex contagions. *Social Networks* 34: 451–461.

Bernard, H.R. and Killworth, P.D. (1977). Informant accuracy in social network data II. *Human Communication Research* 4: 3–18.

Bernard, H.R., Killworth, P.D., and Sailer, L. (1979/1980). Informant accuracy in social network data IV: a comparison of clique-level structure in behavioral and cognitive network data. *Social Networks* 2: 191–218.

Blau, P.M. (1977). A macrosociological theory of social structure. *American Journal of Sociology* 83: 26–54.

Brashears, M.E. (2008a). Gender and homophily: differences in male and female association in Blau space. *Social Science Research* 37 (2): 400–415.

Brashears, M.E. (2008b). Sex, society, and association: a cross-national examination of status construction theory. *Social Psychology Quarterly* 71 (1): 72–85.

Brashears, M.E. (2013). Humans use compression heuristics to improve the recall of social networks. *Nature Scientific Reports* 3: 1513.

Brashears, M.E. (2015). A longitudinal analysis of gendered association patterns: homophily and social distance in the general social survey. *Journal of Social Structure* 16: 3.

Brashears, M.E. and Brashears, L.A. (Forthcoming). Compression heuristics, social networks, and the evolution of human intelligence. *Frontiers in Cognitive Psychology*.

Brashears, M.E. and Gladstone, E. (2016). Error correction mechanisms in social networks can reduce accuracy and encourage innovation. *Social Networks* 44: 22–35.

Centola, D. and Macy, M. (2007). Complex contagions and the weakness of long ties. *American Journal of Sociology* 113: 702–734.

Ceteno, M.A., Nag, M., Patterson, T.S. et al. (2015). The emergence of global systemic risk. *Annual Review of Sociology* 41: 65–85.

Davis, P.K. (2011). *Dilemmas of Intervention: Social Science for Stabilization and Reconstruction*. Santa Monica, CA: RAND Corporation.

Eden, L. (2004). *Whole World On Fire: Organizations, Knowledge, and Nuclear Weapons Devastation*. Ithaca, NY: Cornell University Press.

Emerson, R.M. (1962). Power-dependence relations. *American Sociological Review* 27: 31–41.

Freeman, L.C. (1992). Filling in the blanks: a theory of cognitive categories and the structure of social affiliation. *Social Psychology Quarterly* 55: 118–127.

Freeman, L.C. and Kimball Romney, A. (1987). Words, deeds and social structure: a preliminary study of the reliability of informants. *Human Organization* 46: 330–334.

Freeman, L.C., Kimball Romney, A., and Freeman, S.C. (1987). Cognitive structure and informant accuracy. *American Anthropologist* 89: 310–325.

Laumann, E.O., Marsden, P.V., and Prensky, D. (1992). The boundary specification problem in network analysis. In: *Research Methods in Social Network Analysis* (ed. L.C. Freeman, D.R. White and A.K. Romney). New Brunswick, NJ: Transaction Publishers.

Lazarsfeld, P.F. and Merton, R.K. (1954). Friendship as a social process: a substantive and methodological analysis. *Freedom and Control in Modern Society* 18: 18–66.

Lazer, D. and Radford, J. (2017). Data ex machina: introduction to big data. *Annual Review of Sociology* 43: 19–39.

Lazer, D., Kennedy, R., King, G., and Vespignani, A. (2014). The parable of google flu: traps in big data analysis. *Science* 343: 1203–1205.

Leahey, E. (2016). From sole investigator to team scientist: trends in the practice and study of research collaboration. *Annual Review of Sociology* 42: 81–100.

Lizardo, O. (2006). How cultural tastes shape personal networks. *American Sociological Review* 71: 778–807.

Mark, N. (1998). Birds of a feather sing together. *Social Forces* 77: 453–485.

Marsden, P.V. (1987). Core discussion networks of americans. *American Sociological Review* 52: 122–131.

Marsden, P.V. (1988). Homogeneity in confiding relations. *Social Networks* 10: 57–76.

Mayhew, B.H. (1980). Structuralism versus individualism: part 1, shadowboxing in the dark. *Social Forces* 59: 335–375.

Mayhew, B.H. (1981). Structuralism versus individualism: part II, ideological and other obfuscations. *Social Forces* 59: 627–648.

Mayhew, B.H. and Levinger, R.L. (1976a). Size and density of interaction in human aggregates. *American Journal of Sociology* 82: 86–110.

Mayhew, B.H. and Levinger, R.L. (1976b). On the emergence of oligarchy in human interaction. *American Journal of Sociology* 81: 1017–1049.

McPherson, M. (1983). An ecology of affiliation. *American Sociological Review* 48: 519–532.

McPherson, M. (2004). A Blau space primer: prolegomenon to an ecology of affiliation. *Industrial and Corporate Change* 13: 263–280.

McPherson, J.M. and Ranger-Moore, J.R. (1991). Evolution on a dancing landscape: organizations and networks in a dynamic Blau space. *Social Forces* 70: 19–42.

McPherson, J.M., Smith-Lovin, L., and Cook, J.M. (2001). Birds of a feather: homophily in social networks. *Annual Review of Sociology* 27: 415–444.

Miller, G.A. (1956). The magical number seven, plus or minus two: some limits on our capacity for processing information. *Psychological Review* 63: 81.

Molm, L.D. (1990). Structure, action, and outcomes: the dynamics of power in social exchange. *American Sociological Review* 55: 427–447.

Page, S.E. (2015). What sociologists should know about complexity. *Annual Review of Sociology* 41: 21–41.

Ragin, C. (2014). *The Comparative Method: Moving Beyond Qualitative and Quantitative Strategies*. Oakland, CA: University of California Press.

Ridgeway, C. (1991). The social construction of status value: gender and other nominal characteristics. *Social Forces* 70: 367–386.

Ridgeway, C.L. and Balkwell, J.W. (1997). Group processes and the diffusion of status beliefs. *Social Psychology Quarterly* 60: 14–31.

Shi, Y., Dokshin, F.A., Genkin, M., and Brashears, M.E. (2017). A member saved is a member earned? The recruitment-retention trade-off and organizational strategies for membership growth. *American Sociological Review* 82: 407–434.

Simpson, B., Brashears, M.E., Gladstone, E., and Harrell, A. (2014). Birds of different feathers cooperate together: no evidence for altruism homophily in networks. *Sociological Science* 1: 542–564.

Smith, J.A., McPherson, M., and Smith-Lovin, L. (2014). Social distance in the United States: sex, race, religion, age, and education homophily among confidants, 1985 to 2004. *American Sociological Review* 79: 432–456.

Swedberg, R. (2017). Theorizing in sociological research: a new perspective, a new departure? *Annual Review of Sociology* 43: 189–206.

Vaughan, D. (1996). *The Challenger Launch Decision: Risky Technology, Culture, and Deviance at NASA*. Chicago, IL: University of Chicago Press.

6

Toward Generative Narrative Models of the Course and Resolution of Conflict

Steven R. Corman[1], Scott W. Ruston[2], and Hanghang Tong[3]

[1] Hugh Downs School of Human Communication, Arizona State University, Tempe, AZ, USA
[2] Global Security Initiative, Arizona State University, Tempe, AZ, USA
[3] School of Computing, Informatics, and Decision Systems Engineering (CIDSE), Arizona State University, Los Angeles, CA, USA

All cultures tell stories – or, more accurately, no cultures have yet been discovered that do not tell stories. In common usage, the word *story* implies a rather limited episode, for instance, grandpa telling a story about his childhood or that big battle in World War II. But in fact, most stories depend on larger narratives for meaning and context. Stories of the American Revolution take on their meaning of democratic empowerment only in the context of stories about British imperial rule in the pre-revolution colonies. Contemporary stories like those surrounding the US *Tea Party* movement are meaningful, in part because of the context of stories of the American Revolution. Understanding the narrative of Al-Qaeda requires knowledge of stories about the Soviet invasion of Afghanistan, as well as the US response to Iraq's invasion of Kuwait in 1990 and subsequent stationing of troops in Saudi Arabia (Wright 2006). Thus, stories rarely stand alone, but instead form *narrative systems* that both constrain and potentiate possibilities for meaning and are a key tool for understanding the world.

The systematic and universal nature of narrative creates the potential for a new approach to social science modeling, yet the way narratives are conceptualized, and the way models are currently built, prevents this potential from being realized. In this chapter, we outline limitations of current conceptualizations of narrative, propose a framework for modeling of generative narrative systems that overcomes these limitations, illustrate its application to the simple case of a well-known fairy tale, and describe the potential for applying the framework to social science modeling of real conflict events.

Social-Behavioral Modeling for Complex Systems, First Edition.
Edited by Paul K. Davis, Angela O'Mahony, and Jonathan Pfautz.
© 2019 John Wiley & Sons, Inc. Published 2019 by John Wiley & Sons, Inc.
Companion website: www.wiley.com/go/Davis_Social-Behavioralmodeling

Limitations of Current Conceptualizations of Narrative

Over the last 15 years, interest in narrative as a concept for supporting national defense has grown considerably. Islamist extremists' heavy use of narrative to influence contested populations to support their cause has led researchers to consider the role of narrative in countering Islamist ideologies (Quiggin 2009; Halverson et al. 2011), in sharpening the public messaging of the United States in its military information operations and public diplomacy efforts (Sloggett and Sloggett 2009; Paruchabutr 2012), in illuminating complex information environments where rumors and disinformation proliferate (Bernardi et al. 2012), and in explaining extremism in general (Corman 2016). Important as these efforts may be, they miss a potentially more important aspect of narrative – that it is a logic of action that may be useful for organizing or predicting the behavior of adversaries and the course of events in conflict and competition situations (whether in security, business, or other arenas).

For example, one reason DOD efforts to *win the battle of the narrative* have yielded limited results is the DOD's reliance on a definition of narrative rooted in the spoken or written word. *Joint Publication 3-0 Joint Operations* defines *narrative* as "a short story used to underpin operations and to provide greater understanding and context to an operation or situation" (US Department of Defense 2017, p. III-18). This outdated definition equates narrative with reactive efforts of public affairs officers when a mission or strike has gone awry or the five-paragraph description of a commander's operational design found elsewhere in US DOD joint doctrine (e.g. US Department of Defense 2011, p. IV-44).

Contemporary conceptualizations treat narrative as a cognitive process of comprehension, a consensus that spans the fields of psychology (Mandler 1984; Bruner 1991), sociology (Cortazzi 2001), communication (Fisher 1984), literature (Jahn 1997; Herman 2003), and media studies (Branigan 1992; Ryan 2006). What unites these fields' approaches to narrative is a movement away from an antiquated and strictly textual (whether oral, graphical, or written) definition of narrative, and a recognition that the phenomenon of narrative is better and more usefully understood, as Branigan describes, "as a distinctive strategy for organizing data about the world, for making sense and significance" (1992, p. 18). Recognizing that narrative is more than a set of spoken or written words (or images), but rather is also a means of organizing information – notably events, actors, and desires, and an underlying logic of action – unlocks the power of narrative as a means to understand beliefs, attitudes, and behaviors. This more nuanced and powerful definition of narrative may be unusual or unsettling to those more familiar with the dictionary definition provided by US DOD doctrine as noted above. However, leveraging the insights and advances of the humanities and social sciences over the past 30 years offers opportunities for modeling narrative systems that can illuminate probable target audience reactions and decision consequences.

Branigan, taking a cognitive (rather than strictly textual) approach to narrative, argues that we can think of narrative as "a mechanism that systematically *tests* certain combinations and transformations of a set of basic elements and propositions about events…not simply to enumerate causes but to discover the causal efficacy of an element" (1992, p. 9, emphasis original). Data elements (what we describe below as locations, events, actions, and participants (LEAP) are perceived and then slotted into functional roles, which support an evaluation of whether the combination makes sense. Whether the combination makes sense depends on the plausibility of the arrangement of data elements and the congruency with past comprehension episodes. In other words, narratives embody a unique form of rationality composed of *coherence* and *fidelity* (Fisher 1984) that helps decide what should happen next in a constellation of narrative elements. It is distinct from the instrumental rationality typically used to explain behavior in conflict situations (Eriksson 2011; Wittek et al. 2013). Nissen (2013) uses the idea of a comprehension process with its own form of rationality to argue for narrative as a basis for making command intent clear in military operations.

Unfortunately, narrative as it is traditionally applied is also primarily descriptive in nature and does not support its use in discovery and forecasting. For example, the *narrative arc* (Abbott 2001; Halverson et al. 2011) is a heuristic used to organize the elements and flow of a story. It specifies a conflict (or other deficiency) that creates desire, motivating characters in the story. The desire gives rise to a set of LEAP factors that are connected over time and lead to a resolution of the desire. In fictional or historical stories, the resolution is known, and the arc furnishes a narrative-rational explanation of how the characters arrived at the resolution. In most cases this is how the arc is applied, as an analytical framework for post facto explanation of a known resolution.

Though the narrative arc is a useful tool for abstract analysis of simple stories, it oversimplifies how narrative operates in the real world. First, in most real-world stories – especially those of interest to the defense community – the resolution lies in the future, is not known, and is multiplex. The problem is not explaining how a resolution came about, but which of many potential resolutions might come to be. Such an analysis would provide insight into participant actions/events that would *constitute progress* toward any such potential resolution. Accommodating for the linkage between any potential resolution and emotionality, sacred values, and connection to related narrative systems (or other criteria) can offer analysts and modelers a path toward assessing the attractiveness or potential power of a particular narrative trajectory among the range of plausible trajectories.

Second, as we argue above, real-world narratives are part of a system (Halverson et al. 2011). They may influence each other and may share common elements. A good example of this in a present-day context is the conflict in Syria. The narratives expressing, shaping, and driving the conflict in Syria overlap and exert influence on adjacent narratives in complicated ways that would benefit from new tools of thinking, modeling, analysis, and visualization. These narratives are not exclusively the words used to describe events; rather, they are the glue that individuals and communities use to understand the consequences of the events and agents in the world.

Using narratives, then, is a kind of cognitive strategy and also a potential mechanism for influence. Figure 6.1 offers a simplified schematic of the narrative landscape of Syria vis-à-vis the civil war, humanitarian crisis, and linkage to terrorism. The columns represent the functional roles of a macro-level narrative structure (the traditional narrative arc, discussed above), and the arrows indicate dependencies between the elements of different narrative trajectories related to the Syria conflict. For example, the Arab Spring is an event complicating Assad's trajectory of maintaining power (first row), and it is also an event representing a threat to the Ummah (conflict, third row). Actions in the global war on terrorism (GWOT) trajectory, such as air strikes (event, fourth row), enhance the perception of the threat to the Ummah that motivates ISIS's narrative trajectory (conflict and desire, third row). Meanwhile the projected resolution of the ISIS sub-narrative, a caliphate, enhances the terrorism conflict that motivates the GWOT trajectory (conflict, fourth row). In this, there are feedback loops that potentially escalate the conflict.

Narratives	Conflict	Desire	Sample LEAP 1	Sample LEAP 2	Sample LEAP 3	Resolution
Assad	Revolution	Regime survival	Arab Spring	Russia ally	Gassing of rebels	Military victory; negotiated settlement
Humanitarian crisis	Civil war	Protect civilians	Establish refugee camps	Assad gases rebels	Cease fire	Negotiated settlement; emigration program
ISIS	Threat to Ummah	Protect Islam	Seize Raqqa	Inspire terror	Defend Mosul	Aspiration: caliphate
US (GWOT)	Terrorism	Protect US interest	Stabilize Iraq	Strike ISIS leaders	Assist Iraqi Army	Defeat ISIS; deter terrorism
US global security	Globalization	Protect US interests	Assert influence Middle East	Influence global institutions	Advance US values	Stable, prosperous world

Figure 6.1 An illustration of the narrative complexity of the Syria conflict.

In the real world, then, there is not one conflict but many. Each conflict instills desires in the actors it motivates, and there may be multiple – even conflicting – desires. Each desire is connected to a set of LEAP elements in complex ways, and LEAP elements may be connected to each other. For example, events may take place in typical locations (e.g. protests in public squares or near government buildings), and participants may favor certain kinds of action or only have access to certain locations. Conversely, actions may require a certain kind of participant. The LEAP elements are connected to multiple possible resolutions, and configurations of them might make one resolution more likely than another. Over time, resolutions may themselves lead to escalation or de-escalation of conflicts or creation of new ones.

A Generative Modeling Framework

Thus, a real-world narrative system can be conceptualized as a time-based directed graph, as shown in Figure 6.2. In such a graph, the vertices may be assigned weights. For example, one desire may be stronger than another, or some participant may be more powerful or consequential than another. The edges may also have weights indicating probability or strength of connection,

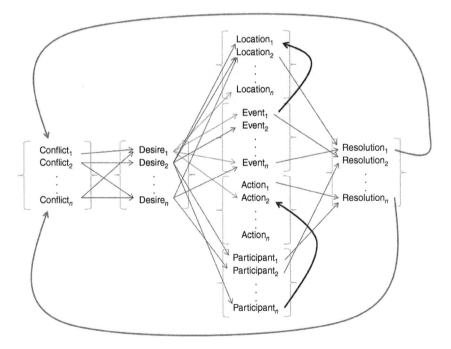

Figure 6.2 A notional generative network.

here conceptualized by number of co-occurrences of pairs of connections. Other approaches could include subjective assignments of narrative importance. For instance, there may be a higher probability of action$_x$ given desire$_y$. Or there may be a higher probability of location$_x$ given participant$_y$ or a weaker connection between event$_x$ and resolution$_y$.[1]

Application to a Simple Narrative

To partially illustrate the application of the framework just described, we apply it to the simple narrative system of a well-known fairy tale, *The Three Little Pigs*. Doing so capitalizes on advantages of story familiarity and a limited scope and helps us identify some of the challenges involved in applying our framework to real-world examples. The text of the story we used comes from Ashliman's Folktexts (1996–2017). Of course, real-world conflicts that are not at all fanciful have narrative bases as well (Halverson et al. 2011; Bernardi et al. 2012; Corman 2016), but a short story with a limited scope makes illustration of the generative narrative concept more tractable.

We began by trying to identify the LEAP factors in the story paragraph by paragraph, looking for cases where actions connect locations, events, and participants. For example, one of the early events in the story is when the first little pig gets straw from the man with straw to build his house. We coded this as indicated in Table 6.1.

However, this reveals one possible problem with the LEAP factors identified above: *straw* is not an agent and therefore cannot take actions on its own. We decided to add a fourth type of arc factor called *things* and continued identifying locations, events, actions, participants, and things (LEAPT) factors (shown in appendix "Locations, Events, Actions, Participants, and Things in the *Three Little Pigs*") and their connections in the rest of the story. We expect that *things* will help contextualize relationships between participants, actions, and events when the narrative system is abstracted into a graph, but

Table 6.1 LEAP coding example.

Participant	Action	Participant
Pig 1	Meet	Straw vendor
Pig 1	Obtain	Straw

---

1 This scheme bears some resemblance to SRI's SEAS system (Lowrance 2007), but that system differs from the perspective described here in that it is intended to support reasoning based on classical rationality and relies on hierarchies represented by directed acyclic graphs.

whether they warrant status as a separate element or a subset of the participant element requires further research. Finlayson, for example, classifies animate and non-animate objects as *entities* that either act or are acted upon within the context of key narrative actions serving plot functions (2011, p. 48). This approach works for constrained systems, such as the folktales Finlayson uses as his data source. As our approach is applied to a less constrained narrative system such as a corpus of news articles from a region, the greater fidelity afforded by an additional subclass of LEAP(T) element may be necessary to distill narratives out of an unruly corpus.

It is this latter big data context that in part also differentiates our approach from the field of computational interactive narrative, which pursues the related challenge of generating believable narrative experiences responsive to user input in a virtual environment. Like our approach, these emphasize entities and events organized in a structure. Some approaches distinguish between characters and objects (e.g. Sánchez and Lumbreras 1997; Murray 2005), while most focus primarily on characters as entities (Mateas and Stern 2002; Hartmann et al. 2005), owing to their more constrained systems and focus on plot actions. The degree to which specific objects aid in correctly categorizing different event–participant interactions is an open research question for our approach in unconstrained contexts.

A second issue we encountered in coding the story had to do with locations. Typically, words used in the story to describe locations are also used to describe things. For example, the wolf visits the first little pig's house, then huffs and puffs, and blows the house down. The wolf's actions transform the thing, but the location still exists. Thus, we described the wolf going to the house with edges representing both the thing and the location, as shown in Table 6.2.

A third issue encountered in coding the story had to do with events, specifically their scope. Technically, each subject/action/object triple could be considered as an event. This approach would create one event for each action taken. However, the conceptual category we recognize and label as an *event* typically encompasses numerous actions – in other words, a collection of actions that make something worthy of being called an event. For example, a terrorist attack involves planning, obtaining weapons, traveling to the target, and executing the attack. Also included in the *terrorist event* are the designation of the attack as a terrorism incident and numerous actions involving victims. If any of these

Table 6.2 Dual coding of things and locations.

Subject	Action	Object
Wolf	Visit	Straw house (thing)
Wolf	Visit	Straw house (location)

Table 6.3 Nodes in Figure 6.3 with highest current flow betweenness.

Locations		Actions		Participants		Things	
Merry Garden	0.47	Visit/travel to	8.18	Wolf	11.54	Fair	2.46
Mr. Smith home field	0.47	Obtain	7.24	Pig 3	10.15	Apple tree	2.36
Shanklin	0.47	Identify	3.53	Pig 1	3.76	Apples	1.89
Stick house	0.00	Meet	3.33	Pig 2	3.76	Brick house	1.55
Wolf home	0.00	Descend	3.26	Old sow	0.00	Straw house	1.52
Straw house	0.00	Ask about	3.20	Straw vendor	0.00	Stick house	1.52
Brick house	0.00	Build	2.93	Stick vendor	0.00	Churn	1.36
		Seek	2.50	Brick vendor	0.00	Turnip field	0.90
		Kill	2.12			Fright	0.75
		Reveal	1.91			Pot of water	0.51

elements were missing, the collection would not be an event we would call a *terrorist attack* – rather they would be some collection of unconnected actions. Thus, we treated events as collections of subject/action/object triples that execute a function within the narrative structure of *The Three Little Pigs*.[2] A complete list of the triples and their associated events for the story are shown in appendix "Edges in the *Three Little Pigs* Graph".

The edges listed in appendix "Edges in the *Three Little Pigs* Graph" can be represented as a weighted graph[3] as shown in Figure 6.3, where edge thickness represents co-occurrence and node shape represents the type of LEAPT factor: triangles are locations, squares are actions, circles are participants, and hexagons are things. Nodes are sized to reflect degree centrality, essentially how much they are connected to other nodes and thus *involved* in the story. Table 6.3 shows current flow betweenness (Brandes and Fleischer 2005; a measure of information value) of the highest-value nodes of each type.[4]

From these *off-the-shelf* network measures (Table 6.1) and visualization (Figure 6.3), we can discern several structural details about the narrative

---

2 Note how a story can relate an *event*, which is a further illustration of the principle of narratives as *systems of stories*.

3 Social network analysis uses similar representations to describe relationships between people, but graph analysis uses a more general concept that describes any kind of relationships between entities.

4 In network analysis, betweenness is the proportion of shortest paths through the network on which a given node lies. Current flow betweenness is similar but calculates the influence of a node considering all the paths in a network, not just the shortest ones. It is more appropriate in this case because it represents the importance of a node in the entire narrative system rather than just the most efficient subsystem.

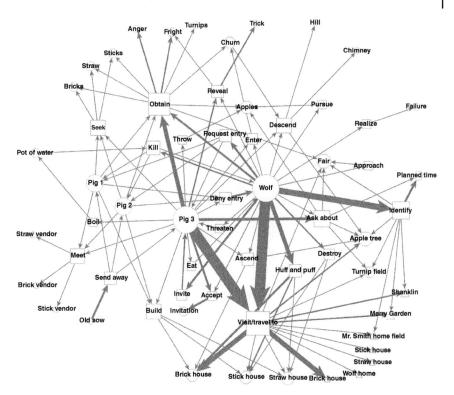

Figure 6.3 Narrative network of the *Three Little Pigs* story.

without reading the text. The wolf and pig 3 are the central participants, whereas the other pigs are less central and the other participants are peripheral. Most of the story focuses on going to places and obtaining things. Betweenness calculations (Table 6.1) reveal the most important locations are Merry Garden, Mr. Smith's home field, and Shanklin – all locations are where we see extended descriptions of pig 3's efforts to escape the wolf, leading to other actions. In Figure 6.3, we can also see that all the houses were huffed and puffed at (the brick house twice as much) and that the straw and brick houses were destroyed, whereas the brick house was not. In constrained narrative systems, such as this folktale and other *single text* examples, betweenness calculations may be subject to distortion and mis-proportion due to authorial intent (such as subplots or *red herrings*); however, our ultimate goal is unconstrained narrative systems discernable within large bodies of data, such as 1000s of news articles and blog posts from a community, in which *artistic license* should wash out as a factor influencing the network.

Real-World Applications

In the *Three Little Pigs*, all the LEAPT factors are known because the graph is derived from a complete narrative. However, in real-world contexts, especially defense-relevant ones, information is frequently missing. An important question is whether the information in the graph is *generative* in the sense that we can infer unknown information from known information. To answer this question, we demonstrate that two techniques can be used to infer at least some missing edge weights and missing relationships in the graph in Figure 6.3.

Regarding the first of these, suppose we knew the basic relationships in Figure 6.3, but did not know the edge weights. By the notation of *line graphs*, we can turn each edge of a narrative system into a node of its line graph, with connections between these being created by nodes in the original graph. Thus, the edge-states inference problem becomes the problem of inferring the node states on this line graph. We created the line graph of Figure 6.3 and computed current flow betweenness centrality as a measure of the importance of the nodes (which, again, represent the links in the original Figure 6.3 graph). We then used these values to replace the link weights in the original graph. The result is shown in Figure 6.4.

The correlation between the original weights (co-occurrence counts) and the estimated values from the line graph is $r = 0.583$ ($p < 0.001$, $R^2 = 0.34$). On the one hand, though we regard this as promising performance, it would clearly be better to have more than 34% accuracy in predicting edge weights in a practical context. On the other hand, we can see that many of the same qualitative conclusions about the story drawn above would hold true. The most important edges still involve pig 3 and the wolf and their actions of visiting/traveling to places and obtaining things.

A second kind of missing information in the narrative graph in Figure 6.3 could be missing links between the nodes. From graph analysis perspective, this is a classic example of the *link prediction problem*, where we want to infer the unknown/missing links based on the known/observed links between nodes. To accomplish this, we applied random-walk-based methods (Tong et al. 2006; Zhu 2006). Specifically, we created a one-mode projection of the graph in Figure 6.3, representing the locations, events, participants, and things as connected by actions. For every edge (i, j), we deleted it and then computed the ranking value $R(i, j)$ using a random walk with restart (RWR) (Tong et al. 2006). It summarizes the weighted summation of all the possible paths from $node_i$ to $node_j$, i.e. the more short paths from $node_i$ to $node_j$, the higher the ranking value $R(i, j)$ is. The nodes with rank 1 for $R(*, j)$ are shown in Table 6.4.

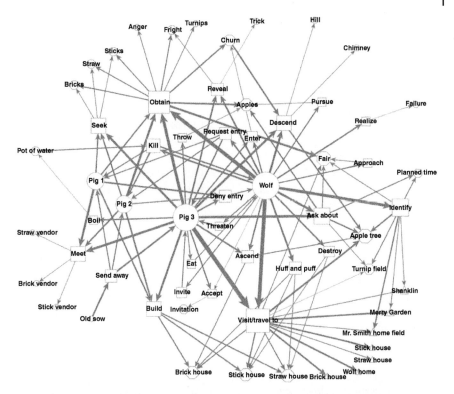

Figure 6.4 Original graph with link weights replaced by link graph betweenness values.

Many of the connections between either pig 3 or the wolf and other nodes can be inferred based on this result, because one or more indirect paths exist between these node pairs, which in turn indicates a strong proximity/association even without the direct link between these node pairs. For example, due to the existence of an indirect path from pig 3, through wolf, to brick house, RWR would infer a strong association between pig 3 and wolf, thus suggesting the existence of a possible link between them. We cannot infer *which* verb should connect the elements but that there should be some action-based connection. This includes several connections that are key to the story. For example, pig 3 did something with a brick house, pig 3 did something with apples, and the wolf had an action related to the pot of water. This demonstrates that analysis of the graph has the ability to fill in missing information or at the least point to relationships that should be further investigated.

Table 6.4 Links recovered using random walk with restart.

Edge	Triplet	Weight	Rank in R(i, *)	Rank in R(*, j)
(11, 32)	P4: little pig 3 > ask about/visit > T17: fair	0.190 864 600 326	9	1
(11, 3)	P4: little pig 3 > visit > L3: brick house	0.190 864 600 326	11	1
(11, 4)	P4: little pig 3 > visit > L4: Mr. Smith's home field	0.190 864 600 326	11	1
(11, 5)	P4: little pig 3 > visit > L5: Merry garden	0.190 864 600 326	11	1
(11, 6)	P4: little pig 3 > visit > L6: Shanklin	0.190 864 600 326	11	1
(11, 21)	P4: little pig 3 > build/visit > T6: brick house	0.190 864 600 326	10	1
(11, 23)	P4: little pig 3 > ask about/visit > T8: turnip field	0.190 864 600 326	12	1
(11, 25)	P4: little pig 3 > ask about > T10: planned time	0.190 864 600 326	10	1
(11, 26)	P4: little pig 3 > ask about/visit/ascend/descend > T11: apple tree	0.190 864 600 326	9	1
(11, 30)	P4: little pig 3 > obtain/throw > T15: apples	0.190 864 600 326	10	1
(11, 31)	P4: little pig 3 > obtain > T16: freight	0.190 864 600 326	10	1
(15, 3)	P8: wolf > visit > L3: brick house	0.146 818 923 328	10	1
(15, 4)	P8: wolf > identify > L4: Mr. Smith's home field	0.146 818 923 328	10	1
(15, 5)	P8: wolf > identify/visit > L5: Merry garden	0.146 818 923 328	9	1
(15, 6)	P8: wolf > identify/visit > L6: Shanklin	0.146 818 923 328	9	1
(15, 17)	P8: wolf > visit/attack/destroy > T2: straw house	0.146 818 923 328	12	1
(15, 19)	P8: wolf > visit/attack/destroy > T4: stick house	0.146 818 923 328	9	1
(15, 21)	P8: wolf > visit/attack/ascend > T6: brick house	0.146 818 923 328	11	1
(15, 23)	P8: wolf > identify > T8: turnip field	0.146 818 923 328	11	1
(15, 25)	P8: wolf > identify > T10: planned time	0.146 818 923 328	11	1
(15, 26)	P8: wolf > identify/visit > T11: apple tree	0.146 818 923 328	11	1
(15, 30)	P8: wolf > ask about/pursue > T15: apples	0.146 818 923 328	10	1
(15, 31)	P8: wolf > obtain/reveal > T16: freight	0.146 818 923 328	10	1
(15, 32)	P8: wolf > identify/approach > T17: fair	0.146 818 923 328	10	1
(15, 35)	P8: wolf > enter > T20: pot of water	0.128 022 759 602	12	1

Challenges and Future Research

Analysis Challenges

As noted above, the line graph weights in the *Three Little Pigs* graph explained about 34% of the variance in the actual edge counts. While this demonstrates the potential of the method, in practical contexts, higher performance is desirable. Even with this method, a potential bottleneck for large-scale application of this technique lies in computation, as the size of a line graph could be significantly larger than its original graph (i.e. the input generative system). However, the low-rank structure of the line graph, with the help of the source/target incident matrices, might be the key to addressing the computational issue, as indicated in our prior work (Kang et al. 2011). Possible strategies for improving performance of the method include the heterogeneous information network model (Sun and Han 2012) and factor graph inference methods (McCallum et al. 2009).

Our application of the RWR method to infer missing links shows promise, but at this point can only show that some action should relate two nodes, not which action. Theoretically, application of narrative logic principles could further narrow the possibilities of what type of relationship exists between two nodes. Stories that recur within narrative systems often share common structures and thus common node relationships. For example, in the original Boston Tea Party story, the revolutionaries protest a distant government making taxation decrees without representation. In 2010, political activists perceived administration efforts to impose taxation mandates related to funding healthcare initiatives as an equivalent governmental decree without representation. Thus, the edge connecting 2010 political activists and US government would be *oppose* just as it is in the original Boston Tea Party narrative network.

Other methods have the potential to improve the performance of this technique. First, in terms of data modeling, there are different types of nodes in generative narrative systems (e.g. conflict, desire, locations, resolutions, etc.). This suggests that the heterogeneous information network model (Sun and Han 2012), which encodes both topological and attribute information of nodes and links, might be a better choice than the plain unipartite graph representation. Second, in terms of inference algorithms, we note that the fundamental assumption behind random-walk-based methods is homophily, which might oversimplify the complicated relationship of a generative narrative system.[5] Factor graph inference methods (McCallum et al. 2009) might be a more effective way to infer node states in the presence of such a heterophily effect by introducing a compatibility matrix between different node types,

---

5 For example, two opposite desires might originate from the same conflict and/or partially share the same participants and locations.

which in turn allows random walks to transition between heterogeneous node types. There are other missing information problems we have not addressed in the example. For example, a node's state/weight could be missing from the information used to create a graph. Possible methods for addressing this include graph-based semi-supervised learning (Zhu 2006).

At a more macro scale, the abstract model of Figure 6.2 suggests that certain conflicts may be more likely than others to lead to a given resolution. For instance, in Syria, is the conflict between the Kurds and the Turks more likely to lead to a restoration of Assad's power than the conflict between the United States and Russia? Such a question poses an interesting theoretical challenge: the conventional answer would take the form of a historical explanatory narrative (after the conflict resolves). Generating plausible narrative trajectories might illuminate what potential answers are possible when combining the communication and narrative theory models discussed above. The cognitive narratology approach stipulates that narratives are mental processes of data acquisition and organization via culturally provided templates. The narrative validity approach stipulates criteria for acceptance (coherence and fidelity). These point to potential avenues for eliciting potential narrative trajectories from a large corpus of data.

Computationally, this becomes a problem of predicting the most likely pathway in each narrative network. As a starting point for solving this problem, the *connection subgraph* (Tong and Faloutsos 2006) is a natural way to infer a small connection subgraph from one or more desired nodes (the source nodes) to one or more resolution nodes (the target nodes). The detected connection subgraph contains most important intercorrelated paths from the source node(s) to the destination node(s). Here a potential risk lies in the *outlying/noisy querying nodes*. If the input query contains one or more irrelevant desire/resolution nodes or multiple desire/resolution nodes form multiple independent groups, the standard connection subgraph tools might introduce false connections between the irrelevant desire and resolution. We hypothesize that an effective solution to address this issue is graph summarization (Akoglu et al. 2013): given a set of desire and resolution nodes on a narrative system, the aim would be to simultaneously find (i) the grouping/clustering structure of desire/resolution nodes and (ii) the optimal connection/pathway within each group.

Scale Challenges

The second research problem is the macro-level scope and scale of the model. On one hand, if we take seriously the idea that narratives are part of a system, then potentially all narratives that could interact with a generative narrative model are relevant to its operation. For instance, in the Syria example given above, Turkey and its opposition to Kurdish groups is not included. Neither is Israel's recent attack on the Golan Heights. Are these elements of the Syrian narrative too critical to ignore? On the other hand, as the model scales up, it becomes less tractable, and per Bonini's paradox (Bonini 1963; Starbuck 1976), there is a point of diminishing returns where everything becomes connected and the model becomes no more understandable than the reality it seeks to explain. Adding Israel into the example Syria narrative could entail the entire *al Nakba* narrative (Halverson et al. 2011, chapter 12) and others related to it, making the graph impossibly complicated. The challenge would be to locate a *sweet spot* where it is possible to specify the model at a just large enough scale to encompass significant influences on the course of a conflict without making it overly complex.

A second and related problem is scaling of the elements shown in Figure 6.1. The Syria conflict could be framed in terms of the Arab Spring revolution, or of Assad's desire to stay in power, or of various groups vying for control of the country, or of a proxy war between the United States, Russia, Iran, and the Arab states. Desires for Assad could be defined in sweeping terms like maintaining sovereignty and power or in strategic terms like control of a given part of the country. Locations could be large areas like countries or provinces or very small ones like cities, towns, and villages. Events can be unitized at different time scales, participants could scale from organizations to groups to individuals, and so on. The challenge would be defining a tractable way of defining the elements at similar and conformable scales.

Sensitivity Challenge

A final problem is determining the sensitivity of a generative narrative model to missing information. The sensitivity of networks to missing information in descriptive applications is well known (Holland and Leinhardt 1973). In a complex network such as what we are describing, the problem could be worse: missing elements could completely change the conclusion of the model, or it may have less effect owing to the ability of other known elements to compensate for missing ones. The challenge would be to test a prototype model to determine the effect of removing elements on the output of the model.

Conclusion

From an information science and communication perspective, this approach has the potential to create a paradigm shift of narrative analysis, from simple description to discovery and forecasting. We envision that the proposed generative narrative network offers a potentially ideal way to model a holistic/comprehensive picture of narratives (including multiple intercorrelated conflicts, temporal dynamics, and the feedback effect of resolution to conflict) that has largely been ignored in the existing models. It would also open the door to the vast machinery of graph mining tools and algorithms, which would in turn allow us to understand/discover, forecast, and even influence narratives in a desired way.

Generative narrative models also have broader research applications. Agent-based modeling (ABM) has assumed an increasingly important place in the study of social systems. Originally, ABMs operated on the basis of component agents making decisions about their behavior based on rules, which processed information about their own states and the states of other agents to which they were connected (Kotseruba et al. 2016). Since the late 1990s, so-called cognitive ABMs have become increasingly popular because "the use of cognitive architectures leads to a more detailed (and deeper) computational model of social phenomena" (Sun 2007, p. 36).

Cognitive ABMs incorporate elements of agency typical of human behavior, including memories and motivations, which link goals with actions and actions with outcomes. In other words, they imbue agents with rationality. Additionally, they imbue agents with decision-making processes influenced by reward and punishment, which is also known as reinforcement learning. Q-learning, a scheme of reinforcement learning, for example, is used in Clarion (Sun 2007), and many architectures of cognitive ABMs use some form of reinforcement to implement associative learning (Kotseruba et al. 2016). Advances in deep learning represent a complementary track for cognitive ABMs, enabling, for example, the learning of temporal and language sequences using long short-term memory units (Greff et al. 2017) and the generation and discrimination of novel patterns using generative adversarial network (Goodfellow et al. 2014), all of which enrich cognitive ABMs beyond associative learning. Such techniques have been applied in (as yet imperfect) generation of narratives from photo streams (Wang et al. 2018).

Existing cognitive ABMs are based on instrumental rationality (see Lehman et al. (2006)) for a simple example applied to baseball pitching). But, as we have argued, narrative rationality is an alternative and possibly more realistic alternative, based on motivations (desires) that create collections of interrelated LEAPT factors, which themselves lead to outcomes (resolutions). Thus, generative narrative networks constitute a potential new avenue for describing the cognitive architecture of computational agents.

From an applied perspective, a generative narrative model would offer the means to explore the potential impact of exogenous influences on a narrative landscape – a testbed for potential defense response options (whether informational such as a psychological operations [PSYOP] series or kinetic such as a drone strike) and their effect on prevailing narratives. Often, narratives provide an explanation of *why* events resulted in a particular conclusion. Attention to the key components of the narrative reveals motivations and key causal relationships. But humans understand themselves in terms of narratives (Polkinghorne 1988), and narrative logic shapes our understanding of the world (Branigan 1992; Fisher 1984). If we could forecast the narrative trajectory of a given situation, given the principles of narrative fidelity, we should be able to equally forecast the likely outcomes or at least bound the range of options likely available to an actor/agent/adversary.

Increasingly the *cognitive terrain* of populations, communities, and leadership elites is a central concern, whether in terms of business performance, political advocacy or security, and military applications. As the first Battle of Fallujah in March 2004 illustrates, tactical battlefield victories are hollow if dominance of the cognitive terrain is not equally achieved.[6] The relevance of the paradigm proposed here, and research that supports it, is a better understanding of how meaning and perception shape actions and responses over time. Decisions by leaders, communities, and populations take place in the context of the conflicts, motivations, and goals that frame the situation – they are embedded in narrative landscapes.

Thus, a generative narrative analysis capability could enhance not only military intelligence preparations of adversary courses of action but also business and marketing strategies, public health outreach campaigns, and innumerable other situations. Such generative narrative analysis could also provide insight into how potential US government, business, or nongovernmental organization actions might alter a community's narrative landscape and thus those critical public perceptions and meanings. Along with shaping campaign plans at the strategic level, a generative narrative analysis paradigm would contribute to decision-making at a tactical level.

Acknowledgment

The authors thank Alex Yahiya for helpful comments on a draft of this manuscript.

Locations, Events, Actions, Participants, and Things in the Three Little Pigs

Locations	Events	Actions	Participants	Things
L1 = straw house	E1 = quest begins	A1 = send away	P1 = old sow	T1 = straw
L2 = stick house	E2 = house created	A2 = seeks	P2 = little pig 1	T2 = straw house
L3 = brick house	E3 = house destroyed	A3 = obtain	P3 = little pig 2	T3 = sticks

---

6 Conventional wisdom holds that the US retaliation against insurgent killing of four US contractors temporarily rooted out insurgent elements but turned community allegiance toward the insurgents, forcing subsequent US withdrawal from the city in April and ceding it to insurgent control.

Locations	Events	Actions	Participants	Things
L4 = Mr. Smith's home field	E4 = pig killed	A4 = build	P4 = little pig 3	T4 = stick house
L5 = Merry garden	E5 = wolf attempts to trick pig	A5 = meet	P5 = straw vendor	T5 = bricks
L6 = Shanklin	E6 = pig tricks wolf	A6 = huff & puff A7 = request entry	P6 = stick vendor	T6 = brick house
L7 = wolf's home	E7 = confrontation	A8 = deny entry	P7 = brick vendor	T7 = failure
		A9 = kill	P8 = wolf	T8 = turnip field
		A10 = visit/ travel to		T9 = invitation
		A11 = destroy		T10 = planned time
		A12 = realize		T11 = apple tree
		A13 = identify		T12 = turnips
		A14 = ask about		T13 = trick
		A15 = invite		T14 = anger
		A16 = accept		T15 = apples
		A17 = find		T16 = freight
		A18 = reveal		T17 = fair
		A19 = ascend		T18 = churn
		A20 = descend		T19 = hill
		A21 = throw		T20 = pot of water
		A22 = pursue		T21 = chimney
		A23 = enter		
		A24 = approach		
		A25 = threaten		
		A26 = boil		
		A27 = eat		

Edges in the Three Little Pigs Graph

Sequences	Description	Edges	Event
1	Sow sends pigs off	P1 > A1 > P2,P3,P4	E1
2	Pig 1 seeks materials	P2 > A2 > T1	E2
3	Pig 1 meets straw vendor	P2 > A5 > P5	
4	Pig 1 obtains straw	P2 > A3 > T1	
5	Pig 1 builds house	P2 > A4 > T2	
6	Wolf visits house	P8 > A10 > T2	E3
6	…at location	P8 > A10 > L1	
7	Wolf says let me in	P8 > A7 > P2	
8	Pig 1 refuses wolf	P2 > A8 > P8	
9	Wolf attacks house	P8 > A6 > T2	
10	Wolf destroys house	P8 > A11 > T2	
11	Wolf eats Pig 1	P8 > A9 > P2	E4
12	Pig 2 seeks materials	P3 > A2 > T3	E5
13	Pig 2 meets stick vendor	P3 > A5 > P6	
14	Pig 2 obtains sticks	P3 > A3 > T3	
15	Pig 2 builds house	P3 > A4 > T4	
16	Wolf visits house	P8 > A10 > T4	E6
16	…at location	P8 > A10 > L2	
17	Wolf says let me in	P8 > A7 > P3	
18	Pig 2 refuses wolf	P3 > A8 > P8	
19	Wolf attacks house	P8 > A6 > T4	
20	Wolf attacks house again	P8 > A6 > T4	
21	Wolf destroys house	P8 > A11 > T4	
22	Wolf eats Pig 2	P8 > A9 > P3	E7
23	Pig 3 seeks materials	P4 > A2 > T5	E8
24	Pig 2 meets stick vendor	P4 > A5 > P7	
25	Pig 2 obtains bricks	P4 > A3 > T5	
26	Pig 2 builds house	P4 > A4 > T6	
27	Wolf visits house	P8 > A10 > T6	E9
27	…at location	P8 > A10 > L3	
28	Wolf says let me in	P8 > A7 > P4	
29	Pig 2 refuses wolf	P4 > A8 > P8	

Sequences	Description	Edges	Event
30	Wolf attacks house	P8 > A6 > T6	
31	Wolf attacks house again	P8 > A6 > T6	
32	Wolf attacks house again	P8 > A6 > T6	
33	Wolf discovers he can't blow house down	P8 > A12 > T7	
34	Wolf identifies turnip field	P8 > A13 > T8	E10
35	Pig 3 asks where	P4 > A14 > T8	
36	Wolf says where	P8 > A13 > L4	
37	Wolf invites pig to meet	P8 > A15 > P4	
38	Pig 3 accepts	P4 > A16 > T9	
39	Pig 3 asks what time	P4 > A14 > T10	
40	Wolf says 6:00	P8 > A13 > T10	
41	Pig 3 visits turnip field	P4 > A10 > T8	E11
41	…at location	P4 > A10 > L4	
42	Pig gets turnips	P4 > A3 > T12	
43	Pig returns home	P4 > A10 > T6	
43	…at Location	P4 > A10 > L3	
44	Wolf goes to pig's house	P8 > A10 > T6	
44		P8 > A10 > L3	
45	Pig reveals deception	P4 > A18 > T13	
46	Wolf gets pissed	P8 > A3 > T14	
47	Wolf identifies apple tree	P8 > A13 > T11	E12
48	Pig asks where	P4 > A14 > T11	
49	Wolf says where	P8 > A13 > L5	
50	Wolf invites pig	P8 > A15 > T9	
51	Pig 3 accepts	P4 > A16 > T9	
52	Wolf says meet at 5	P8 > A13 > T10	
53	Pig visits apple tree	P4 > A10 > T11	E13
53	At location	P4 > A10 > L5	
54	Pig climbs tree	P4 > A19 > T11	
55	Pig gets apples	P4 > A3 > T15	
56	Wolf gets there	P8 > A10 >T11	
56	At location	P8 > A10 > L5	
57	Pig gets scared	P4 > A3 > T17	
58	Wolf asks if apples are good	P8 > A14 > T15	
59	Pig throws apple	P4 > A21 > T15	

Sequences	Description	Edges	Event
60	Wolf chases apple	P8 > A22 > T15	
61	Pig climbs down	P4 > A20 > T11	
62	Pig returns home	P4 > A10 > T6	
62	...at Location	P4 > A10 > L3	
63	Wolf visits house	P8 > A10 > T6	E14
63	...at Location	P8 > A10 > L3	
64	Wolf notes fair	P8 > A13 > T17	
65	Pig asks where	P4 > A14 > T17	
66	Wolf says where	P8 > A13 > L6	
67	Wolf invites pig	P8 > A15 > P4	
68	Pig 3 accepts	P4 > A16 > T9	
69	Wolf says meet at 3	P8 > A13 > T10	
70	Pig visits fair	P4 > A10 > T17	E15
70	At location	P4 > A10 > L6	
71	Pig buys butter churn	P4 > A3 > T18	
72	Wolf approaches fair	P8 > A24 > T17	
72	At location	P8 > A10 > L6	
73	Pig gets scared	P4 > A3 > T16	
74	Pig gets in churn	P4 > A23 > T18	
75	Churn rolls down hill	T18 > A20 > T19	
76	Wolf gets scared	P8 > A3 > T16	
77	Wolf runs home	P8 > A10 > L7	
78	Pig returns home	P4 > A10 > T6	
78	...at Location	P4 > A10 > L3	
79	Wolf visits house	P8 > A10 > T6	E16
79	...at Location	P8 > A10 > L3	
80	Wolf says churn scared him	P8 > A18 > T16	
81	Pig says it was him	P4 > A18 > T13	
82	Wolf gets pissed	P8 > A3 > T14	
83	Wolf tells pig he will eat him	P8 > A25 > P4	
84	Wolf climbs house	P8 > A19 > T6	
85	Pig boils water	P4 > A26 > T20	
86	Wolf climbs down chimney	P8 > A20 > T21	
87	Wolf falls into pot	P8 > A23 > T20	
88	Pot of water boils wolf	T20 > A9 > P8	
89	Pig eats wolf	P4 > A27 > P8	E17

References

Abbott, H.P. (2001). *The Cambridge Introduction to Narrative*, 2e. Cambridge: Cambridge University Press.

Akoglu, L., Chau, D., Faloutsos, F. et al. (2013). Mining connection pathways for marked nodes in large graphs. In: *Proceedings of the SIAM international conference on data mining*, 37–45. Society for Industrial and Applied Mathematics.

Ashliman, D.L. (1996–2017). Folklore and mythology electronic texts. Retrieved from http://www.pitt.edu/~dash/folktexts.html (accessed 29 August 2018).

Bernardi, D.L., Cheong, P.H., Lundry, C., and Ruston, S.W. (2012). *Narrative Landmines: Rumors, Islamist Extremism and the Struggle for Strategic Influence*. New Brunswick, NJ: Rutgers University Press.

Bonini, C.P. (1963). *Simulation of Information and Decision Systems in the Firm*. Englewood Cliffs, NJ: Prentice-Hall.

Brandes, U. and Fleischer, D. (2005). Centrality measures based on current flow. In: *STACS 2005*, Lecture Notes in Computer Science (ed. V. Diekert and B. Durand), 533–544. Heidelberg, Germany: Springer-Verlag.

Branigan, E. (1992). *Narrative Comprehension and Film*. New York: Routledge.

Bruner, J. (1991). The narrative construction of reality. *Critical Inquiry* 18: 1–21.

Corman, S.R. (2016). The narrative rationality of violent extremism. *Social Science Quarterly* 97 (1): 9–18.

Cortazzi, M. (2001). Narrative analysis in ethnography. In: *Handbook of Ethnography* (ed. P. Atkinson, A. Coffey, S. Delamont, et al.), 384–394. Thousand Oaks, CA: Sage.

Eriksson, L. (2011). *Rational Choice Theory: Potential and Limits*. New York: Palgrave Macmillan.

Finlayson, M.M.A. (2011). Learning narrative structure from annotated folktales. Doctoral dissertation. Massachusetts Institute of Technology.

Fisher, W.R. (1984). Narration as a human communication paradigm: the case of public moral argument. *Communication Monographs* 51: 1–22.

Goodfellow, I., Pouget-Abadie, J., Mirza, M. et al. (2014). Generative adversarial nets. In: *Advances in Neural Information Processing Systems* (ed. Z. Ghahramani, M. Welling, C. Cortes, et al.), 2672–2680. La Jolla, CA: Neural Information Processing Systems Foundation.

Greff, K., Srivastava, R.K., Koutník, J. et al. (2017). LSTM: a search space odyssey. *IEEE Transactions on Neural Networks and Learning Systems* 28 (10): 2222–2232.

Halverson, J.R., Goodall, H.L., and Corman, S.R. (2011). *Master Narratives of Islamist Extremism*. New York: Palgrave-Macmillan.

Hartmann, K., Hartmann, S., and Feustel, M. (2005). Motif definition and classification to structure non-linear plots and to control the narrative flow in interactive dramas. In: *Proceedings of the 3rd International Conference on Virtual Storytelling* (ed. G. Sobsol), 158–167. Berlin: Springer.

Herman, D. (ed.) (2003). *Narrative Theory and the Cognitive Sciences (No. 158).* Stanford, CA: CSLI.

Holland, P.W. and Leinhardt, S. (1973). Structural implications of measurement error in sociometry. *Journal of Mathematical Sociology* 3: 85–111.

Jahn, M. (1997). Frames, preferences, and the reading of third-person narratives: toward a cognitive narratology. *Poetics Today* 18: 441–468.

Kang, U., Papadimitriou, S., Sun, J., and Tong, H. (2011). Centralities in large networks: algorithms and observations. In: *Proceedings of the SIAM International Conference on Data Mining*, 119–130. Society for Industrial and Applied Mathematics.

Kotseruba, I., Gonzalez, O.J.A., and Tsotsos, J.K. (2016). A review of 40 years of cognitive architecture research: focus on perception, attention, learning and applications. arXiv preprint. Retrieved from https://arxiv.org/pdf/1610.08602 .pdf (accessed 29 August 2018).

Lehman, J.F., Laird, P., and Rosenbloom, P. (2006). A gentile introduction to SOAR, an architecture for human cognition: 2006 update. Retrieved from http://ai.eecs.umich.edu/soar/sitemaker/docs/misc/GentleIntroduction-2006 .pdf (accessed 29 August 2018).

Lowrance, J.D. (2007). Graphical manipulation of evidence in structured arguments. *Oxford Journal of Law, Probability and Risk* 6 (1–4): 225–240.

Mandler, J. (1984). *Stories, Scripts, and Scenes: Aspects of a Schema Theory.* Hillsdale, NJ: Lawrence Erlbaum Associates.

Mateas, M. and Stern, A. (2002). Towards integrating plot and character for interactive drama. In: *Socially Intelligent Agents* (ed. A. Bond, B. Edmonds and L. Canamero), 221–228. Boston, MA: Springer.

McCallum, A., Schultz, K., and Singh, S. (2009). Factorie: probabilistic programming via imperatively defined factor graphs. In: *Advances in Neural Information Processing Systems* (ed. M. Mozer, M. Jordan and T. Petsche), 1249–1257. Cambridge, MA: MIT Press.

Murray, J. (2005). Did it make you cry? Creating dramatic agency in immersive environments. In: *Proceedings of the 3rd International Conference on Virtual Storytelling* (ed. G. Sobsol), 83–94. Berlin: Springer.

Nissen, T. (2013). Narrative led operations. *Militært Tidsskrift* 141 (4): 67–77.

Paruchabutr, G. (2012). *Understanding and Communicating Through Narratives.* Monograph, School of Advanced Military Studies. Retrieved from http://www .dtic.mil/cgi (accessed 29 August 2018).

Polkinghorne, D.E. (1988). *Narrative Knowing and the Human Sciences.* Albany, NY: SUNY Press.

Quiggin, T. (2009). Understanding al-Qaeda's ideology for counter-narrative work. Perspectives on Terrorism, 3 (2). Retrieved from http://www.terrorismanalysts .com/pt/index.php/pot/article/download/67/138 (accessed 29 August 2018).

Ryan, M.-L. (2006). *Avatars of Story*. Minneapolis, MN: University of Minnesota Press.

Sánchez, J. and Lumbreras, M. (1997). HyperHistoires: narration interactive dans des mondes virtuels. In: *Hypertextes et Hypermédias*, vol. 1, 2–3. Paris: Editorial Hermes.

Sloggett, D. and Sloggett, C. (2009). Reframing the narrative of the global war on terrorism. *Journal of Information Warfare* 8 (2).

Starbuck, W.H. (1976). Organizations and their environments. In: *Handbook of Industrial and Organizational Psychology* (ed. M.D. Dunnette), 1069–1123. Chicago, IL: Rand.

Sun, R. (2007). Cognitive social simulation incorporating cognitive architectures. *IEEE Intelligent Systems* 22 (5): 33–39.

Sun, Y. and Han, J. (2012). *Mining Heterogeneous Information Networks: Principles and Methodologies*. Williston, VT: Morgan & Claypool Publishers.

Tong, H. and Faloutsos, C. (2006). Center-piece subgraphs: problem definition and fast solutions. Proceedings of the 12th ACM SIGKDD International Conference on Knowledge Discovery and Data Mining (pp. 404-413). New York: ACM.

Tong, H., Faloutsos, C., and Pan, J. (2006). Fast random walk with restart and its applications. In: *Proceedings of the 6th International Conference on Data Mining*, 613–622. New York: ACM.

U.S. Department of Defense (2011). *Joint Publication 5-0 Joint Operation Planning*. Washington, DC: United States Department of Defense.

U.S. Department of Defense (2017). *Joint Publication 3-0 Joint Operations*. Washington, DC: United States Department of Defense.

Wang, J., Fu, J., Tang, J. et al. (2018). Show, reward and tell: automatic generation of narrative paragraph from photo stream by adversarial training. Retrieved from https://www.microsoft.com/en-us/research/publication/show-reward-tell-automatic-generation-narrative-paragraph/ (accessed 29 August 2018).

Wittek, R., Snijders, T., and Nee, V. (2013). Introduction: rational choice social research. In: *The Handbook of Rational Choice Social Research* (ed. R. Wittek, T. Snijders and V. Nee), 1–32. Stanford, CA: Stanford University Press.

Wright, L. (2006). *The Looming Tower*. New York: Alfred Knopf.

Zhu, X. (2006). Semi-Supervised Learning Literature Survey. Computer Science, University of Wisconsin-Madison. Available from http://pages.cs.wisc.edu/~jerryzhu/pub/ssl_survey.pdf (accessed 29 August 2018).

7

A Neural Network Model of Motivated Decision-Making in Everyday Social Behavior

Stephen J. Read[1] and Lynn C. Miller[2,3]

[1] Department of Psychology, University of Southern California, Los Angeles, CA 90089, USA
[2] Annenberg School for Communication and Journalism, University of Southern California, Los Angeles, CA, USA
[3] Department of Communication and Psychology, University of Southern California, Los Angeles, CA 90007, USA

Introduction

Our aim is to model everyday motivated decision-making over time as grounded in the motivational dynamics of the individual. These motivational dynamics arise from organized motivational systems within the individual interacting with the social and physical environment, as well as with the internal environment of the individual. A particular focus of our model is on capturing individual differences in motivation and decision-making and the underlying neurobiological mechanisms that are responsible for those individual differences. The model we discuss is implemented as a neural network model.

One benefit of our model is that it provides a framework that allows researchers to create a relatively lightweight set of psychological mechanisms for an agent that can quickly and efficiently generate a wide range of plausible behaviors. At the same time, because the model is firmly grounded in a wide array of research and theory, it provides a psychologically and neurobiologically plausible account of these mechanisms.

Much, if not most, of the work on models of decision-making focuses on how people make single decisions. In contrast, our focus is on the dynamics of decision-making as people move through situations and across time as they move through their daily lives. We are interested in how motivation and decisions change over time as a function of changes in the situation and the individual's internal state, which are influenced by external forces as well as the individual's own actions.

Most models of decision-making are variants of subjective expected utility models, where the individual is treated as having a stable utility function for

Social-Behavioral Modeling for Complex Systems, First Edition.
Edited by Paul K. Davis, Angela O'Mahony, and Jonathan Pfautz.
© 2019 John Wiley & Sons, Inc. Published 2019 by John Wiley & Sons, Inc.
Companion website: www.wiley.com/go/Davis_Social-Behavioralmodeling

the current circumstances and where the focus is on the abstract values of the attributes of the potential choices and the probability that these values will be attained. In contrast, we model decision-making in terms of the dynamics of underlying motivational systems interacting with a changing internal and external environment. Instead of modeling decisions in terms of abstract values, we model them in terms of the competing influence of multiple concrete motives within broad motivational systems. Further, our model assumes that the values of choices will vary over time and situation as a function of the internal and external environment. For example, the value of food depends on how hungry we are, and the value of social contact depends on how lonely we are. Our characterization of these motivational systems is grounded in what is currently known about the neurobiological bases of the relevant motivational/reward processing systems.

Overview

Our neural network model of motivated decision-making and personality (Read et al. 2010, 2017a,b, in press) argues that decisions and behavior in everyday life arise from the interactions between structured motivational systems and the motive affordances of situations. By motive affordances we mean that a key aspect of the representation of the situations that people find themselves in is a representation of the possibilities that the situation affords (or provides) for the pursuit of different motives. For example, a cafeteria affords the pursuit of food, whereas a classroom affords the pursuit of achievement (although both situations also provide opportunities for social interaction). We argue that situations activate representations of motives that can be pursued in that situation and that different situations have different distributions of affordances (afford the pursuit of different motives). Because the distributions of affordances vary across situations, as situations change, an individual's relevant motives change. In addition, based on learning and possible neurobiological differences, individuals may differ in the motive affordances of situations. For example, one individual may see a fraternity party as affording social affiliation and mating, whereas another individual may see the same party as affording social rejection or even physical harm.

More specifically, we propose that everyday motivated decision-making is a joint function of the individual's current motivational state and the motive affordances of the situations in which the individual finds herself. Further, the individual's current motivational state is a joint function of (i) the individual's *baseline or chronic motivations*, which typically vary across individuals, (ii) the current motivationally relevant bodily or *interoceptive states* (e.g. information regarding current internal states, such a sense of hunger, exhaustion, or loneliness) that provide information as to whether various needs (e.g. for food, water, companionship) are currently met, and (iii) the current *motive affordances in the situation*. For example, an individual may have a moderately high typical or *baseline* need for people, may be lonely at the particular time, and may see that the situation offers the opportunity for social contact.

Interoceptive state, the motive affordances of situations, and the strength of competing motives all vary over time and situations. Further, the individual's choices and behaviors can then modify his/her current interoceptive state as well as the current situation.

As a result of all these factors, everyday decision-making, the sequence of choices that an individual makes as they move through daily life, is a highly dynamic process: situations and bodily states change as a function of the individual's own actions and the actions of other people and the environment. As a result, the individual's motivational state and the situation to which they are responding are in a continual state of flux. Thus, choices and behavior will vary dynamically over the course of a day.

Constraint Satisfaction Processing

It is useful to think of this kind of decision-making process as a *constraint satisfaction* process that operates to make choices as a joint function of a number of different factors, such as the environment, the individual's current bodily states, chronic and momentary motives, and their previous behaviors. This process would typically be highly nonlinear and dependent on the order of information received, the order of behavior, and the feedback relationships. Such constraint satisfaction processes, such as occur in neural networks, are typically only locally optimal. There is no guarantee that a global maxima will be reached.

Our approach is influenced by Atkinson and Birch's (1970) dynamics of action (DOA) model and Revelle and Condon's (2015) reparameterization of the DOA, called the cues, tendencies, action (CTA) model. Revelle and Condon suggest that their reparameterized DOA can be modeled as a neural network model.

In other work (e.g. Read et al. 2010; Read et al. 2017a) we have argued that personality (individual differences in patterns of behavior over situations) arises from individual differences in characteristics of motivational systems and individual differences in the motivational affordances of situations. We address that issue to a limited extent in the current chapter, but the above papers provide more detailed accounts.

Theoretical Background

In the following section, we describe the major components of our model and how they are tied together. We first describe the structure and nature of

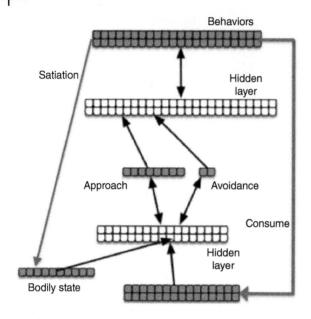

Figure 7.1 Diagram of neural network implementation of the motivated decision-making model.

the motivational systems that are central to our model. We then describe our conceptualization of situations and how the affordances of situations interact with the motivational systems. Following this, we discuss the central concept of wanting and how people's wanting for things is a multiplicative function of the affordances of the situation and the current interoceptive or bodily state. See Figure 7.1 for the overall structure of the model and the flow of influence among elements. After describing the conceptual structure of the model, we describe how it is implemented as a neural network.

Motivational Systems

At the broadest level in our model, based on considerable neurobiological and psychological work, there are two broad motivational systems. There is an Approach or behavioral approach system (BAS) (e.g. Gray 1987; Depue and Collins 1999; Gray and McNaughton 2000; Clark and Watson 2008) that governs sensitivity to cues signaling reward, as well as cues signaling the omission of punishment. And there is an Avoidance or behavioral inhibition system (BIS) (e.g. Gray 1987; Gray and McNaughton 2000; Clark and Watson 2008) that governs sensitivity to cues of punishment or threat. Gray (1987) originally termed this system the BIS, but he subsequently renamed it the fight, flight, freeze system (FFFS) (Gray and McNaughton 2000), although it plays the same general role as the original BIS.

Individual differences in the Approach system have been argued to map onto the broad personality trait of Extraversion (Depue and Collins 1999; Gray and McNaughton 2000) and to depend heavily on individual differences in tonic (or chronic) dopamine levels, especially within the ventral striatum (nucleus accumbens). And individual differences in the Avoidance system have been argued to map onto the broad personality trait of Neuroticism (Gray and McNaughton 2000; Corr et al. 2013; DeYoung and Allen in press), which is characterized by such things as social anxiety and fearfulness. Although there seems to be broad consensus that dopaminergic systems play a key role in Extraversion, there is much less consensus about the neurobiological bases of Neuroticism.

Within each of these two broad motivational systems are nested more specific, task-based motives that have evolved to manage various everyday life tasks (Kenrick and Trost 1997; Bugental 2000; Chulef et al. 2001; Talevich et al. 2017). Among the more specific motives within the Approach system are such motives as mating, nurturance of young, affiliation and bonding with peers, establishing dominance hierarchies, insuring attachment to caregivers, acquiring resources, and more physical motives dealing with such things as hunger, thirst, and sleep. Important motives within the Avoidance system are avoiding social rejection and avoiding physical harm. We should note that for highly social species, such as humans, being a member of a group is critical for long-term survival. Thus, avoiding social rejection plays an important role in avoiding physical harm.

Individuals obviously differ in the chronic importance of all of these motives. These individual differences in chronic motive importance are an important basis of personality. Although there are chronic individual differences in a variety of different motives, the proximal driver of choice will be the current levels of different motives (which are strongly, although not uniquely, influenced by chronic importance of motives).

Situations

A key part of our model is our conceptualization of situations. Although many areas of psychology (e.g. social psychology, personality, ecological psychology) view situations as fundamental to understanding human social behavior, there is surprisingly little agreement as to how to think about situations, and the suggestions that have been made for how to characterize situations are not particularly conducive to creating simulations of the interactions between people and situations. Years ago we argued (Read and Miller 1989; Miller and Read 1991) that situations can be conceptualized as motive or goal-based structures that consist of the motives afforded by the situation, physical attributes of the situation, and the typical roles and scripts that can be enacted in the situation (Argyle et al. 1981). Information about the current situational affordances then activates the potential motives that an individual might pursue in that situation. A situation that affords academic pursuits provides various affordances for achievement and will activate motives related to achievement. A situation that affords romantic pursuits provides very different affordances.

In addition to arguing for the validity of this conceptualization of situations, we also propose that this conceptualization is extremely useful for creating simulations of human social behavior. Many models of individual agents in AI and cognitive science focus on agents as pursuing particular goals and engaged in planning and goal-directed behavior. With this conceptualization of agents and our conceptualization of situations, it is straightforward to create a model of how an agent would interact with different situations. Recognition of the situation that an agent is in provides information about the relevant goals that can be pursued at the current time in the current situation, and this information can be easily integrated with the goal-directed systems in the agent to generate behavior.

Interoceptive or Bodily State

The strength of motivation or desire is not just a function of the chronic motive and the strength of situational cues. It is also a function of the current degree of satiation or need, which is conveyed by a signal about the current **interoceptive or bodily state** relevant to that cue (e.g. Bechara and Naqvi 2004; Tindell et al. 2006; Berridge 2007, 2012; Zhang et al. 2009; Berridge and O'Doherty 2013). Interoceptive state provides information about the current state of such things as hunger, thirst, loneliness, etc. (need states).

Wanting

Berridge (2007, 2012) has made a convincing case for the distinction between wanting something and liking it once we have attained it. This is the distinction between the desire for something and the degree to which it is perceived as rewarding, once it is attained. Wanting can be viewed as the *strength* of desire for something. According to Berridge wanting is a multiplicative function of the strength of the situational cue (its affordances) and the current interoceptive or bodily state. For example, wanting social interaction is not simply a function of how attractive other people are, but it is also a function of how lonely you are and how much you need social contact.

Wanting encapsulates two separate sources of variability in behavior over time: the **strength of the cue (or situational affordance)** and **interoceptive state**. As a result of the multiplicative relationship between cue strength and interoceptive state, each factor can play a gating role in controlling behavior. For example, if there is attractive food (strong environment cue strength), but we are not hungry (weak interoceptive signal), we will probably not eat, or if there are interesting people around, but we do not need social contact, then we will not approach the people to hang out.

Competition Among Motives

A final important aspect of our model is that motivation is competitive. By that we mean that the strength of motives or different *wantings* competes with each other for the control of behavior. Thus, the likelihood of behavior is a function of the strength of wanting one thing *compared with* the strength of wanting alternative things. For example, if we are moderately hungry with food around, but we are very thirsty with water around, we will drink. However, in the absence of thirst, the same amounts of hunger and food will lead to eating.

Motivation Changes Dynamically

The strength of motives or *wantings* changes dynamically, as a function of changes in the environment and changes in interoceptive state. As a result, behavior changes dynamically in response to changes in motivation. For example, in the example given above where we are moderately hungry with food around but very thirsty with water around, we will initially drink. However, as we drink, our interoceptive state indicating thirst will decrease, and therefore our wanting for water will decrease. At some point the wanting for food will then be greater than our wanting for water, and we will stop drinking water and start eating. Further, if there is a limited amount of water, removing water from the environment by drinking it will remove that affordance and therefore decrease the wanting for water. This will allow wanting for food to take control of behavior. Finally, if there was no water available initially, then a moderately hungry animal with food around would initially eat. However, if water became available, then the wanting for water would increase, and the animal might switch from eating to drinking.

Neural Network Implementation

Our model is implemented as a neural network model. This allows us to simulate and examine how our model captures an individual's motivational dynamics and changes in decisions and behavior over time. To create our neural network models, we use a software system called *emergent* (Aisa et al. 2008; O'Reilly et al. 2012). *emergent* provides a framework for creating all parts of a network and implements several different neural network architectures. We use a particular architecture called Leabra, which is designed to be biologically plausible. Leabra stands for local, error-driven and associative, biologically realistic algorithm.

Neural networks are constructed of nodes and the weighted links among them. Processing in a neural network proceeds by the passage of activation among nodes. Nodes are analogous to neurons or systems of neurons and represent features or concepts. Each node's level of activation represents its current importance and its readiness to fire and send activation to the other nodes to which it is connected.

Nodes receive activation from sending nodes, integrate the incoming activation from all the sending nodes, and then send activation to other nodes. The relationship between the incoming activation and the output of the node is captured by an activation function. This activation function may capture different forms of relationships. The most typical are linear relationships, S-shaped or sigmoidal relationships, and binary relationships (on or off).

Nodes are organized in layers that correspond to processing components within the model. In Leabra, nodes within a layer compete for activation, and one can manipulate the degree of competition between nodes within a layer, as well as manipulating various other parameters of all the nodes within that layer, such as gain on the activation function or the threshold for firing.

In the current model (see Figure 7.1), at the input level we have the Situational Features and the Interoceptive or Bodily State. We then have two layers that represent the two broad motivational systems: Approach and Avoidance. Following this is the Behavior layer. Typical neural networks also have what are called Hidden layers, intervening between the different processing layers. Hidden layers learn higher-order representations or conjunctions of lower-level features.

Nodes are connected by weighted links, where the strength and direction of the link represents the degree and direction of influence. These connection strengths correspond to synapses in neural systems. Links can be unidirectional or bidirectional. Given the massive bidirectional connectivity of the human brain, bidirectional connections are more plausible. Bidirectional connectivity allows for such things as modeling processing dynamics over time, top-down influences, parallel constraint satisfaction processing, and pattern completion from incomplete data.

The strength of connections can be modified by learning or experience, and the extent of learning from experience is a parameter of the links that can be modified. The simplest kind of learning is Hebbian learning (Hebb 1949), which enables learning the correlational structure of the environment. Here the weight between a sending and a receiving node increases if the two nodes are active at the same time. "Nodes that fire together, wire together."

A simple version of Hebbian learning is

$$\Delta w = \gamma * a * b$$

where the change in weight Δw is equal to the product of the activation of the two nodes, a and b, times a learning parameter γ.

A second form of learning is error-correcting (or task-based) learning, which seeks to minimize the sum squared error of prediction of the receiving node given the activation from the sending node. Here there is a teaching signal. When the receiving node fires, its activation is compared with the teaching signal, and the difference (the error of prediction) is used to modify the weight between the sending and receiving nodes. Error-correcting learning is also called the delta rule (Widrow and Hoff 1960) and looks like

$$\Delta w = \gamma * (t - o) * a$$

where the weight change, Δw, is equal to the difference between the predicted activation o and the teaching activation t, times the input activation a, multiplied by a learning parameter γ. Because delta-rule learning seeks to minimize the summed squared error, a network with a single output node is essentially learning a simple linear regression model.

With error-correcting learning you might do something like pronounce a word, and then a teacher tells you the correct way to pronounce the word. You can then calculate the difference between your output (how you pronounced the word) and how the teacher says you should have pronounced the word. A key point about error-correcting learning is that there is a clear target or correct answer.

A third kind of learning is reinforcement learning, where you are trying to maximize rewards or minimize punishment. Here you are simply told whether or not you got the reward or punishment. There is no information about what the correct response should have been.

Each form of learning is important under different conditions. Hebbian learning learns the correlational structure of the environment. Error-correcting or delta-rule learning learns from an explicit teaching signal that tells us what the correct answer is, and reinforcement learning allows us to learn from the rewarding or punishing effects of our actions. Although each is useful, they are most powerful when combined.

General Processing in the Network

Processing in the network proceeds as follows. Information about the affordances of the situation (external information) and interoceptive state concerning all the potential motives (internal state) are applied to the Situational Feature layer and the Interoceptive (Bodily) State layer, respectively. Activations from situational features and interoceptive states are then sent to the nodes in the Approach and Avoidance layers. The resulting activation of the Approach and Avoidance motive nodes is a multiplicative function of the corresponding Situational feature activation and the Interoceptive state activation, as well as individual differences in chronic motive activations, overall sensitivities or gains of the general Approach and Avoidance systems, and chronic dopamine level. Motives within each layer compete with each other, and the motives that *win* the competition then send activation to behaviors in the Behavior layer that can help to satisfy those motives. Since only one behavior can be enacted at a time, the behaviors within the layer compete with one another, and only the most highly activated behavior is enacted.

Finally, the individual's behavior can change her interoceptive state (Satiation) (e.g. eating food will reduce hunger, hanging out with friends will reduce loneliness) and can potentially change the situation or characteristics of the situation to which the individual is responding (Consume). These changes in external and internal state are an important source of dynamics.

Social behavior is highly dynamic and varies considerably over time, and our model captures that dynamic. A major source of those behavioral dynamics is motivational dynamics, changes in what we *want* to do and what we decide to do. In our model motivation varies over time due to changes in affordances (opportunities in the environment for motive or goal-related behavior) and due to changes in our current interoceptive state that result from our own behavior as well as from the passage of time. As a result, behavior will vary over time and situations.

Individual differences in the response of the motive systems can be modeled in terms of a number of different parameters. First, one can manipulate individual differences in the broad Approach and Avoidance tendencies. This can be captured by modifying the overall sensitivity or gains of the activation function for the nodes within each system. Second, individual differences in the importance of specific motives can be modeled in terms of individual differences in the baseline or chronic activation of the motive. Third, experience can shape the strength of the links between the situational features and the motives. As a result, different individuals may respond differently to the same situational features.

As we noted in the preceding, when creating our neural network models, we use a particular architecture called Leabra, which is implemented in a software system for creating neural network called *emergent* (Aisa et al. 2008). *emergent* provides a framework for creating all parts of a network and implements several different neural network architectures. As in most neural network architectures, Leabra includes learning. The learning rules in Leabra allow the network to model the strength of associations between nodes as a function of their frequency of co-occurrence, as well as allowing the network to use prediction error to tune the strength of relevant connections. The basic forms of learning in Leabra are similar to the forms we outlined above, although they differ in some of their details.

The activation function in Leabra has an S-shaped form and has several key components (see Eq. (7.1)). g_e is the sum of the excitatory activation into the node, whereas g_e^θ is the threshold for firing and is a function of the inhibitory activation into the node. Thus, one key aspect of this activation function is that the level of firing of the node is a function of the balance between the excitatory and inhibitory activation into the node. Further, γ is a gain parameter that

Figure 7.2 Graph of output of Leabra activation function. Source: Adopted from O'Reilly et al. (2012).

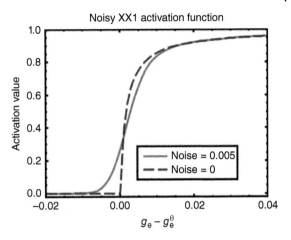

controls how steep the slope of the activation function is. Higher γ results in a steeper slope. With the gain parameter one can control whether the output function is relatively linear or is progressively more S shaped. The result of this activation function can be seen with the dashed line in Figure 7.2, where there is a sharp discontinuity at the threshold for firing:

$$y = \frac{1}{\left(1 + \frac{1}{\gamma[g_e - g_e^\theta]}\right)} \tag{7.1}$$

However, the transition is more gradual when the results of the equation are convolved with random noise as can be seen in the solid line in Figure 7.2. Without random noise added to the activation function, there is a sharp threshold for firing; with noise the slope is more graded. The basic idea is that neurons are inherently noisy so that even when the activation of the node is below threshold, it will still occasionally fire.

The network in Figure 7.3 provides a simplified version of the environmental affordances and potential behaviors of a college student. In this example, there are five Environmental features that identify five different situations with different situational affordances: Friend, Library, Food, Social Situation (Soc-Sit), and Danger. There are also five corresponding Interoceptive States: Affiliation (nAff), Achievement (nAch), Hunger, Social Anxiety (SocAnx), and Fear. Activation from the Environment and the Interoceptive State simultaneously flow to the motivational systems where they are multiplicatively combined. In the Approach system, there are three motives: Affiliation (AFF), Achievement (ACH), and Eat (Hunger). In the Avoid system there are two: Avoid Social Rejection (REJ) and Avoid Physical Harm (HRM). The two layers can have different gains: this makes it possible to model the two systems as having different sensitivities to their inputs. For example, if the Avoid layer has a higher gain than the Approach layer, this captures the notion that people are more sensitive

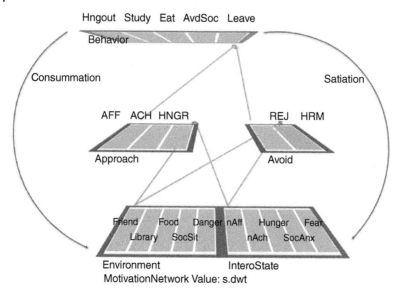

Figure 7.3 Simple model of a college student.

to negative events or losses. In addition, the different motives in a layer can differ in importance, represented by differences in their baseline activation. Thus, the activation of each motive will be a function of the input activation (a multiplicative function of Environment and Interoceptive State), the sensitivity or gain of the system, and the baseline activation of each motive.

Motives in each layer compete for activation and then send activation to the Behavior Layer, where behaviors that are relevant to each of the motives then compete for activation. The winning behavior is enacted. Potential behaviors are Hangout (Hngout), Study, Eat, Avoid Socializing (AvdSoc), and Leave. Enacted behaviors can then feedback and modify both the Environment and the Interoceptive State. For example, eating can reduce the amount of food available, and hanging out with friends for a while can reduce the Interoceptive State associated with Affiliation (loneliness). In addition, there can be changes in the Environment caused by external factors (e.g. a friend walks in) and changes in Interoceptive state due to things such as the passage of time (e.g. getting hungry). These changes to the Environment and the Interoceptive state are then fed back to the system on the next time step and may result in different behavior.

It is also important to bear in mind that there are competitive dynamics at both the motivational level and the behavioral level. This means that the ultimate activation of a motive or a behavior is dependent on its own activation *compared with* the relative degree of activation of competing motives and behaviors.

A run through of one example of a college student moving through an ordinary day is shown in Figure 7.4. The bottom graph shows one possible set

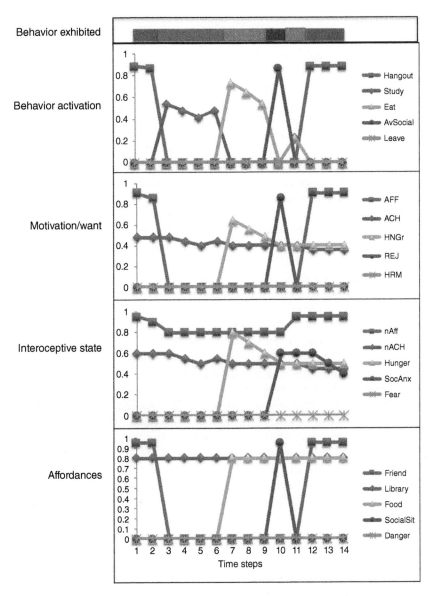

Figure 7.4 Graph of affordances, interoceptive state, motivation, and behavior over 14 time steps.

of Affordances during one day. A friend is available at time 1 and 2, but then the friend leaves. The student is in the library from time step 1 through 14. At time step 7, food is available and remains available. At time step 10 a social situation appears that is relevant to this student's social anxiety. At time step 12 the friend comes back. Thus, during the course of the day, some of the affordances available to the student will change. Notice that different affordances have somewhat different strengths, representing differences in the cue strength of the different situational cues to the Affordances.

The second graph from the bottom shows the student's Interoceptive State across the 14 time steps. Note that various Interoceptive States change over time both as a function of the student's behavior as well as simply due to the passage of time (e.g. Hunger). Here the student starts out with a high need for Affiliation and a somewhat weaker Achievement need, both of which decrease somewhat over time as a function of their behavior. Then at time step 7, the student's Hunger jumps and then slowly decreases because they Eat, as shown in the Behavior Activation graph. Then at time step 10 the student's Social Anxiety jumps and then declines slightly.

When we look at the motivation/want graph, we can see that the motivations of the student change considerably over time. For time steps 1 and 2, the motivation for Affiliation is considerably stronger than the motivation for Achievement, and consequently the Behavior Activation for Hngout is the strongest. Then at time step 3, when the friend has left, the motivation for achievement is the strongest motivation, resulting in Study. Then at time step 7 when Food and Hunger occur, the Hunger motivation becomes the strongest, and the student Eats. As the student eats the Interoceptive State of Hunger decreases, and the Hunger motivation also drops. However, for several time steps there are not any competing motivations, and the student continues to eat. Then at time step 10 when the threatening Social Situation occurs and the Interoceptive State for Social Anxiety rises, the resulting motivation to Avoid Social Rejection spikes and is larger than any other motivation, resulting in the Behavior Activation for Avoiding the Social Situation. Note that when the student Avoids the Social Situation, the corresponding Affordance disappears, and the resulting motivation drops to 0. When that motivation drops to 0, the Hunger motivation is momentarily the strongest, resulting in a brief spike for Eat. Finally, at time step 12, the student's friend returns, and this new Affordance together with the continuing Interoceptive State of need for Affiliation results in a very strong motivation to Hngout and the student HangsOut for the last time steps.

This example shows how the dynamics of motivation and behavior can be modeled in terms of the dynamics of changes in the Affordances in the Environment and the individual's Interoceptive States. Further, although we have not demonstrated it here, various kinds of individual differences can be captured in the model. One can vary such things as the importance or baseline activations of individual motives, which will influence their activation strength, and the gains of the Approach and Avoidance layer, which will influence the sensitivity of the network to rewards and punishments. Further, one can model differences in learning from experience, for example, some people will see social situations as opportunities for Affiliation, whereas others will see the same situation as affording the possibility of Social Rejection. Together these different aspects of the model allow one to model a wide range of factors that influence the dynamics of everyday motivated decision-making.

Conclusion

This neural network provides a model of the dynamics of motivated decision-making and behavior over time and changing situations. In this model decisions and behaviors are the result of interactions of situations with structured motivational systems and of the interactions within these motivational systems. The behaviors that result from these dynamics can then lead to changes in both the external environment and the internal environment (interoceptive state) that can then lead to changes in decisions and behavior.

A number of different factors influence the dynamics of decision-making in this model. First, there are individual differences in motivational states, represented by the overall sensitivities (gains) of the two motivational systems and the importance (baseline activations) of the individual motives. Second, different situations provide different affordances, different opportunities for goal pursuit, and potential attainment. Third, there are individual differences in experience that lead to differences in the perceptions of the likely affordances of different situations (e.g. is a party an opportunity for affiliation or an opportunity for social rejection?). Fourth, interoceptive states multiplicatively combine with situational affordances to provide the inputs to the motivational systems. These interoceptive states vary over time due to a variety of factors. Fifth, there are competitive dynamics among motives and among behaviors that ensure that the enactment of a behavior is not simply a function of its *absolute* strength, but rather is the result of the *relative* strength of the behavior and its relevant motives compared with alternatives.

In addition to its usefulness as a psychological and neurobiological model, this network can provide a computationally *lightweight* framework that could be used to simulate the motivational processes and behaviors of agents in broad social simulations in a psychologically and neurobiologically plausible way. Moreover, because of the highly interactive dynamics of this model, a relatively modest model can generate a high degree of complexity and variability.

Although the model is currently implemented within the Leabra neural network architecture within the *emergent* framework, it would be relatively straightforward to take something like the student model presented here and implement it as a relatively simple program. Both the environment and the interoceptive state inputs can be represented as vectors, in this case each with five elements. These inputs can then be multiplied and input into a standard Leabra activation function for each of the nodes in the approach and avoid layers, where the nodes would compete with each other. The resulting output activations would then go to the behavior layer where one could either implement competition in the layer or perhaps use something like a Softmax activation function (which gives similar results to using a competitive layer) to choose the behavior. The behavior output could then be run through several simple equations to model the effects of consummation on environmental features and satiation on interoceptive states. This would modify the inputs to the model on the next time step. The behaviors of other agents and forces could also modify the environmental inputs to the agent. Finally, a variety of different parameters of the model, such as differences in chronic motive importance, gain or sensitivity to inputs, and relative strength of approach and avoidance sensitivity, could be modified to investigate the behavior of the agent both alone and when interacting with other agents.

This model is based on a broad base of knowledge from the fields of personality, developmental psychology, evolutionary psychology, and the neurobiology of motivation and reward. It provides a model of the dynamics of everyday choice and decision-making that is firmly grounded in theory and research. As we gain new findings and theory develops, we can modify the model to capture these new insights.

References

Aisa, B., Mingus, B., and O'Reilly, R. (2008). The emergent neural modeling system. *Neural Networks* 21 (8): 1146–1152. https://doi.org/10.1016/j.neunet.2008.06.016.

Argyle, M., Furnham, A., and Graham, J.A. (1981). *Social Situations*. Cambridge [Eng.]. New York: Cambridge University Press.

Atkinson, J.W. and Birch, D. (1970). *The Dynamics of Action*. Oxford: Wiley.

Bechara, A. and Naqvi, N. (2004). Listening to your heart: interoceptive awareness as a gateway to feeling. *Nature Neuroscience* 7 (2): 102–103.

Berridge, K.C. (2007). The debate over dopamine's role in reward: the case for incentive salience. *Psychopharmacology* 191 (3): 391–431. https://doi.org/10.1007/s00213-006-0578-x.

Berridge, K.C. (2012). From prediction error to incentive salience: mesolimbic computation of reward motivation. *European Journal of Neuroscience* 35 (7): 1124–1143. https://doi.org/10.1111/j.1460-9568.2012.07990.x.

Berridge, K.C. and O'Doherty, J.P. (2013). From experienced utility to decision utility. In: *Neuroeconomics, Second Edition. Decision Making and the Brain* (ed. P.W. Glimcher and E. Fehr), 325–341. Academic Press.

Bugental, D.B. (2000). Acquisition of the algorithms of social life: a domain-based approach. *Psychological Bulletin* 126 (2): 187–219. https://doi.org/10.1037/0033-2909.126.2.187.

Chulef, A.S., Read, S.J., and Walsh, D.A. (2001). A hierarchical taxonomy of human goals. *Motivation and Emotion* 25 (3): 191–232. https://doi.org/10.1023/A:1012225223418.

Clark, L.A. and Watson, D. (2008). Temperament: an organizing paradigm for trait psychology. In: *Handbook of Personality: Theory and Research*, 3e (ed. O.P. John, R.W. Robins and L.A. Pervin), 265–286. New York: Guilford Press.

Corr, P.J., DeYoung, C.G., and McNaughton, N. (2013). Motivation and personality: a neuropsychological perspective. *Social and Personality Psychology Compass* 7 (3): 158–175. https://doi.org/10.1111/spc3.12016.

Depue, R.A. and Collins, P.F. (1999). Neurobiology of the structure of personality: dopamine, facilitation of incentive motivation, and extraversion. *Behavioral and Brain Sciences* 22 (03): 491–517. https://doi.org/10.1017/S0140525X99002046.

DeYoung, C.G. and Allen, T.A. (in press). Personality neuroscience: a developmental perspective. In: *The Handbook of Personality Development* (ed. D.P. McAdams, R.L. Shiner and J.L. Tackett). New York: Guilford Press.

Gray, J.A. (1987). *The Psychology of Fear and Stress*, 2e. New York, NY: Cambridge.

Gray, J.A. and McNaughton, N. (2000). *The Neuropsychology of Anxiety: An Enquiry into the Functions of the Septo-Hippocampal System*, 2e. New York, NY: Oxford University Press.

Hebb, D.O. (1949). *The Organization of Behavior*. New York: Wiley.

Kenrick, D.T. and Trost, M.R. (1997). Evolutionary approaches to relationships. In: *Handbook of Personal Relationships: Theory, Research, and Interventions* (ed. S. Duck), 151–177. Chichester: Wiley.

Miller, L.C. and Read, S.J. (1991). On the coherence of mental models of persons and relationships: a knowledge structure approach. In: *Cognition in Close Relationships* (ed. G.J.O. Fletcher and F. Fincham), 69–99. Hillsdale, NJ: Erlbaum.

O'Reilly, R.C., Munakata, Y., Frank, M.J. et al. (2012). *Computational Cognitive Neuroscience*, 1e. Retrieved from http://ccnbook.colorado.edu.

Read, S.J. and Miller, L.C. (1989). Inter-personalism: toward a goal-based theory of persons in relationships. In: *Goal Concepts in Personality and Social Psychology* (ed. L.A. Pervin), 413–472. Hillsdale, NJ: Erlbaum.

Read, S.J., Monroe, B.M., Brownstein, A.L. et al. (2010). A neural network model of the structure and dynamics of human personality. *Psychological Review* 117 (1): 61–92. https://doi.org/10.1037/a0018131.

Read, S.J., Droutman, V., and Miller, L.C. (2017a). Virtual personalities: a neural network model of the structure and dynamics of personality. In: *Computational Social Psychology* (ed. R.R. Vallacher, S.J. Read and A. Nowak), 15–37. New York, NY: Routledge.

Read, S.J., Smith, B., Droutman, V., and Miller, L.C. (2017b). Virtual personalities: using computational modeling to understand within-person variability. *Journal of Research in Personality* 69: 237–249. http://dx.doi.org/10.1016/j.jrp.2016.10.005.

Read, S.J., Brown, A.D., Wang, P., and Miller, L.C. (in press). Neural networks and virtual personalities: capturing the structure and dynamics of personality. In: *The Handbook of Personality Dynamics and Processes* (ed. J.F. Rauthmann). Elsevier.

Revelle, W. and Condon, D.M. (2015). A model for personality at three levels. *Journal of Research in Personality* 56: 70–81. https://doi.org/10.1016/j.jrp.2014.12.006.

Talevich, J.R., Read, S.J., Walsh, D.A. et al. (2017). Toward a comprehensive taxonomy of human motives. *PLoS One* 12 (2): e0172279. https://doi.org/10.1371/journal.pone.0172279.

Tindell, A.J., Smith, K.S., Pecina, S. et al. (2006). Ventral pallidum firing codes hedonic reward: when a bad taste turns good. *Journal of Neurophysiology* 96: Retrieved from http://www.ncbi.nlm.nih.gov/pubmed/16885520.

Widrow, B. and Hoff, M.E.J. (1960). Adaptive switching circuits. In: *IRE WESCON Convention Record*, vol. 4, 96–104.

Zhang, J., Berridge, K.C., Tindell, A.J. et al. (2009). A neural computational model of incentive salience. *PLoS Computational Biology* 5 (7): e1000437. https://doi.org/10.1371/journal.pcbi.1000437.

8

Dealing with Culture as Inherited Information

Luke J. Matthews

Behavioral and Policy Sciences, RAND Corporation, Boston, MA 02116, USA

Galton's Problem as a Core Feature of Cultural Theory

Improving models for social behavior eventually will require dealing with culture theoretically and empirically and in the statistical models that link theory and evidence. Developing ways to meaningfully include culture could increase the validity and credibility of social-behavioral modeling. Culture is theorized to be inherited information that influences beliefs and behaviors and that is acquired through social learning (Cavalli-Sforza and Feldman 1981; Boyd and Richerson 1985; Durham 1991; Laland and Kendal 2003; Richerson and Boyd 2005; Dean et al. 2014; Henrich 2015; Mesoudi et al. 2016). This conception of culture is accepted by nearly all researchers working today on culture who employ a generally scientific approach, by which I mean striving to operationalize measurements of culture and to test hypotheses about how it operates. Yes, most scientific cultural researchers also do a great deal of qualitative research that may be solely descriptive rather than oriented toward testing hypotheses, but they do not see this solely descriptive work as an end point for the scientific study of culture.

The idea that culture is socially inherited information in fact characterizes even Edward B. Tylor's first definition of culture: "Culture, or civilization, taken in its wide ethnographic sense, is that complex whole which includes knowledge, belief, art, morals, law, custom, and any other capabilities and habits acquired by man as a member of society." Although Tylor's definition often is panned by critics as the sum of all things people do, a close reading in context of Tylor's book reveals it is not. First, he specifies that culture is only that which is acquired by man, i.e. it does not include that which is inborn by genetic inheritance. Tylor was famously skeptical that there were any behavioral differences among individuals due to genetics – he coined

Social-Behavioral Modeling for Complex Systems, First Edition.
Edited by Paul K. Davis, Angela O'Mahony, and Jonathan Pfautz.
Companion website: www.wiley.com/go/Davis_Social-Behavioralmodeling

the phrase "psychic unity of mankind" – but regardless of whether genetic differences in fact affect behavior, Tylor's definition specifically excludes them from being considered "culture." Tylor also excludes behavioral differences that arise solely from individualistic rational calculation, for culture is "acquired by man as a member of society." Tylor employs this definition in his study of the evolution of religion from animism through monotheism and scientism, and this application makes clear he intends that the cultural aspects of religion are those learned from others within one's society (Langness 1987).

Theorizing culture as inherited information presents complications for utility maximization or other mechanistic interactions that would cause different traits to be correlated with each other. These analytic complications were realized early on when in 1889, Edward Tylor gave a presentation to the Royal Anthropological Institute that detailed his correlational analysis of rules for marriage and property inheritance across hundreds of societies. During the presentation, Francis Galton questioned Tylor as to how he could account for the fact that various groups of people were differentially related to one another by varying degrees of shared inherited beliefs and behaviors (e.g. all the Scandinavian countries share a more common heritage with each other than they do with other European countries, Pacific Northwest Native Americans share much cultural heritage, etc.). Galton argued that the fact cultures were interrelated in this way meant the data were statistically nonindependent and therefore inappropriate for correlation analysis that assumed independence of data points (Naroll 1961).

Thus, the issue of statistical nonindependence of societies in cross-cultural analysis came to be known as "Galton's problem." Unfortunately, Galton had no solution for Tylor. Among other factors, Galton's problem led some subsequent anthropologists to write off the entire practice of cross-cultural analysis. For example, Franz Boas, who had a tremendous influence upon twentieth-century anthropology, considered Galton's problem to be fatal to the cross-cultural research enterprise (Naroll 1961). It should be noted that nonindependence is not the same as unmeasured variables that are omitted from an analysis (termed "unmeasured confounding" in statistics). An unmeasured confounding variable itself has a consistent relationship with the outcome of interest. In Tylor's analysis of inheritance and marriage rules, perhaps an unmeasured confounding of inheritance would be a society's overall wealth. We might propose that being richer or poorer is consistently associated with some types of inheritance and not others. When an important variable is omitted from an analysis, it actually can bias the analysis to find an association of two variables when no association exists, because they are both in fact determined by the unmeasured variable.

Unmeasured pathways of cultural or genetic inheritance are not consistently correlated with the outcome, however, and do not produce the same kind of bias as can an omitted variable (Figures 8.1 and 8.2). Galton was not arguing that Scandinavian societies would have higher or lower outcomes consistently

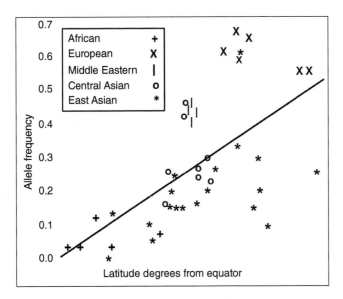

Figure 8.1 Galton's problem in a regression of cross-cultural genetic data. Source: Redrawn from Coop et al. 2010.

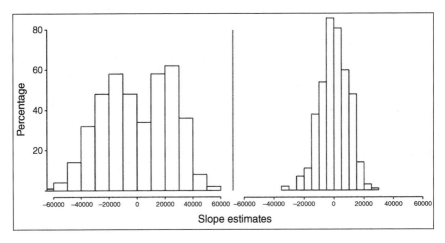

Figure 8.2 Inflated variance of slope estimates without and with correcting for nonindependence. Left panel does not make the correction; the right panel does. Source: Redrawn from Matthews and Butler (2011).

for any trait, but that for any trait they will tend to deviate in the same way from what we would otherwise expect for them. In a regression context, this shows up not like an omitted variable that biases the estimated regression line, but as deviations from a regression line that are associated with the cultural or genetic ancestry of the data points.

Although Tylor and Galton debated the statistical effects of inheritance specifically in the context of culture (i.e. information acquired through social learning), virtually the same statistical issues arise if traits are inherited genetically (Nunn 2011). Some adjustments are needed to model genetic versus cultural inheritance processes, but they share many similarities. The underlying similarity of biology and culture as inheritance processes was noted early on by Darwin, who said the diversification of biological species from shared origins was analogous to the diversification of human languages. In fact, the first fundamental advances in the algorithms for reconstructing evolutionary trees – branching diagrams of lineages linked by inheritance – were made by German historical linguists in the late 1800s and were incorporated into evolutionary biology only in the 1960s. The reconstruction of evolutionary trees and analyses of patterns on them are termed "phylogenetics" in the evolutionary anthropology and biology literatures. This specific term is used to distinguish the study of branching relationships of organisms (phylogenetics) from other aspects of evolutionary or natural history such as adaptation by natural selection.

For example, the data depicted in Figure 8.1 show that instead of populations from different regions being scattered at random across the regression line, populations that share more common inheritance tend to be either above or below the estimated line. For example, Middle Eastern populations are all above the line, and East Asian populations are all below the line. This pattern of clustered deviations from the line results in slope estimates for neutral genes that are randomly too high or too low. Since neutral genes are not expressed in body cells as proteins, they generally are not expected to be acted on by natural selection and so should not be related to ecological contexts. They will be affected by inheritance, however, because just as with expressed genes, they are inherited by offspring from their parents (Rohlf 2006; Matthews and Butler 2011; MacLean et al. 2012).

Figure 8.2 shows the effects of such nonindependence across many regressions of neutral genes all against migration distance from our species' origin in Africa. Note that even before correcting for nonindependence, the estimated slopes are centered on 0 and they are not biased to be high or low as would occur if inheritance were an unmeasured confounding variable. This inflation of slope estimates around 0, however, still results in many false findings for statistical significance in which either a negative or positive association between the gene frequency and migration distance is found (Matthews and Butler 2011).

Another potentially helpful way to understand cultural nonindependence is its similarity to spatial nonindependence, which occurs when data points close together in space tend to exhibit similar values. Kissling and Carl (2008) articulated an important distinction about processes that cause spatial nonindependence, which they term "spatial autocorrelation:"

> Two types of spatial autocorrelation might be distinguished depending on whether endogenous or exogenous processes generate the spatial structure … In the case of endogenous processes, the spatial pattern is generated by factors that are an inherent property of the variable itself, for instance, distance-related biotic processes such as reproduction, dispersal, speciation, extinction or geographical range extension. On the other hand, spatial autocorrelation can be induced by exogenous processes that are independent of the variable of interest. These are most likely spatially structured environmental factors such as geomorphological processes, wind, energy input or climatic constraints…

This important distinction of endogenous and exogenous causes of nonindependence can be applied to cultural nonindependence as well. Galton's original observation specified endogenous nonindependence, which is a concern only if one takes seriously that culture is socially learned information that therefore has its own endogenous properties of replication, diffusion, social inheritance through time, etc. In principle, there is no set of other variables that can properly correct for endogenous nonindependence, as it is not caused by other variables but instead arises from the inherent properties of social inheritance across individuals and down through time.

How to Correct for Treelike Inheritance of Traits Across Groups

Early Attempts to Correct Galton's Problem

Researchers have developed several distinct approaches to attempt solutions to Galton's problem (Naroll 1961). An early approach to correct for cultural nonindependence is to reduce the societies held in an analysis to a set of distantly related ones. This is most commonly implemented through the standard cross-cultural sample (SCCS), which was a set of 186 societies devised by Murdock to be representative of global cultural variation but all distantly related to each other (Murdock and White 1969; Mace et al. 1994). The 186 societies are pulled from the fuller set of 1250 societies in the Murdock Ethnographic Atlas, which has a set of societies closely related to the sample in the Human Relations Area Files (HRAF – http://hraf.yale.edu/).

The SCCS approach, although relatively common in the literature, suffers from two crippling flaws. First, it is ineffective for most characteristics that can be measured on a continuous scale, such as rates of domestic violence or rates of religious participation, because it fails to eliminate the data point nonindependence within the reduced sample. Since much cultural evolution has played out

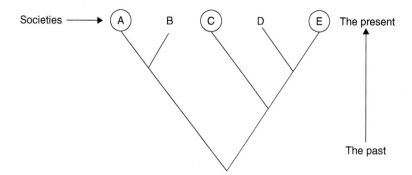

Figure 8.3 Example of the standard cross-cultural sample approach to dealing with Galton's problem. Only the circled distantly related societies (A, C, E) are selected for the sample. Unfortunately, Galton's problem still exists because societies C and E are more closely related to one another than either is to A, and a new statistical problem has been created because information about the traits of interest from societies B and D has been lost entirely.

through branching processes of descent with modification, continuous characteristics will never fully lose the nonindependence due to more ancient shared inheritances (Mace et al. 1994, Figure 8.3).

In the case of binary or multistate characteristics, the SCCS may effectively reduce or eliminate the nonindependence because these characteristics, having a finite number of states, can become saturated with changes over time that eliminate the nonindependent patterns (Smith and Smith 1996). This phenomenon is well known in genetic data as "rate saturation" and occurs because each position in a DNA sequence can take on only one of four possible states.

Second, although the SCCS may correct for nonindependence for multistate characteristics, it does so at the cost of throwing out most of the data. Out of thousands of distinct languages and cultures, the SCCS retains only 186! That simply is not an opportunity for future growth of cultural modeling and analysis, which should be striving to have better measurements and more sophisticated models, as should any branch of science.

Another approach to deal with Galton's problem can be to measure a great many various characteristics of societies to include in statistical tests as control variables. The World Values Survey (WVS) is likely the most comprehensive set of such cultural variables for modern nation-states (Welzel 2013; World Values Survey Association 2016), while the HRAF is the most commonly used for samples of traditional societies (http://hraf.yale.edu/). The recently launched D-PLACE website provides many of the traditional society data from literature sources related to the HRAF (Kirby et al. 2016). All three are tremendous resources for the social science community. Their ability to correct for Galton's problem is limited, however, principally because these databases all capture population phenotypes that likely arise from a mix of cultural, genetic, and environmental factors. Galton's problem theorizes culture not as the sum of all things people do, which is the frequent caricature of Tylor's original definition (Tylor 1871), but rather specifically as the social learned behaviors that people inherit from previous or contemporaneous groups. Culture is not a directly observed phenomenon; it is an inferred mechanistic process of inheritance via social learning. Considering the WVS specifically, one of the main axes of variation across its many variables is "survival versus individualist" values (Welzel 2013), which correspond roughly to Hofstede's individualism–collectivism cultural axis (Hofstede 1984, 2001). Although these authors always discuss this axis as cultural in nature, it also corresponds to well-characterized genetic differences between populations in novelty-seeking genes that likely arose as a selective response from the out-of-Africa migration (Matthews and Butler 2011) and likely still impact economic behaviors and development today (Gören 2017). By measuring population pheno-types, approaches like WVS are providing a valuable service, but something additional is needed to disentangle the specifically cultural component of Galton's problem.

More Recent Attempts to Correct Galton's Problem

More recent approaches to deal with Galton's problem focus on measuring the pathways through which the social and/or genetic inheritances took place for any given set of data and then correct the statistical model accordingly. One simple method, commonly employed in health research, is to use individuals' self-reported race or ethnicity. Race is included as a control variable in regres-sion models for the outcome of interest. Many have pointed out, however, that self-reported race is likely to be a flawed correction for ancestry, since race is largely a recently constructed social identity rather than an accurate reflec-tion of cultural or genetic inheritance. For example, Irish, Italian, and European Jewish people have quite distinct cultural and genetic inheritances but are all "Whites" in terms of American racial categories.

A step further is to use actual ancestry information to construct scientifically more informed buckets of individuals in place of traditional American race. The ancestry category is then included as a control variable in regression mod-els. This approach is informed by actual information about cultural or genetic inheritance pathways but reduces a complex web of relationships to a simple categorical assignment.

Several methods attempt to make use of the full set of inheritance pathways. One is to construct a set of continuous variables, usually through some form of principal component analysis or multidimensional scaling, that capture most of the variation in the original cultural or genetic relationships. These variables are then included as independent predictors in a regression model.

Another approach is to transform the matrix of cultural or genetic distances between data points into a pairwise matrix of expected correlations between them. Thus, you end up with a matrix that specifies the expected correlation for any trait between data points 1 and 2, data points 1 and 3, data points 2 and 3, etc. There are two basic implementations for this method, one that allows a flexible network structure and one that constrains the network to be a treelike model of bifurcating relationships. The matrix then transforms the regression residuals in an autoregression model (the vertical deviations from the regression line as shown in Figure 8.1). By transforming these residuals, the autoregression is intended to fully correct for Galton's problem but does not compete with the traditional independent variables because it does not "explain" any portion of the variance.

I tested these approaches to correcting Galton's problem with a set of 1000 characteristics that I evolved independently along the branches of the language tree of Indo-European countries (Figure 8.4). Each characteristic evolved from a starting trait value that then had amounts randomly added or subtracted down the branches of the tree. This simulation process, termed Brownian motion in the literature, draws the random deviations from the same distribution, but each tip shares the same deviation amounts in direct proportion to their level of shared ancestry in the tree. This results in closely related groups having more similar values for an individual trait because they had more common inheritance of the traits value. Importantly, all the traits evolved completely independently of each other – none of them were correlated, and no parts of the tree were enforced to have consistently high or low trait values. I then tested 500 random and nonoverlapping pairings from among the 1000 variables and applied the above methods to test the rate at which they inferred a statistically significant correlation using a 0.05 significance level (Table 8.1).

The results show that only the phylogenetic autoregression model produced the correct 0.05 proportion of false findings for statistical significance. A phylogenetic autoregression specifies the expected similarity in trait values between every pair of data points based on their quantity of shared ancestry on the tree. Thus, it differs from techniques that use ancestry groups (random effects, aka "clustered error" models, aka "mixed hierarchical" models) as these do not specify on a continuous scale the expected similarities among all data points, but instead simply specify that data points within the same group are expected to be somewhat more similar to each other than they are to data points in other groups.

The simple linear regression produced false findings for significance in over half the cases, and applying standard American racial categories performed equally poorly. Besides the phylogenetic autoregression, the other ancestry-informed techniques cut the false positive rate almost in half, but the rates still far exceed anything acceptable. Across all methods, the estimated slopes were centered on zero, even when a false inference of significance was

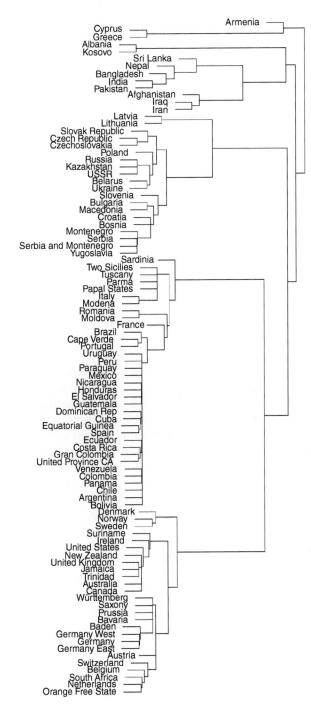

Figure 8.4 Tree of Indo-European languages based on Bayesian phylogenetic analysis of cognate basic words (see Matthews et al. 2016 for methods).

Table 8.1 False positive rates for simulations of treelike inheritance on the Indo-European linguistic tree.

Statistical model	False positive proportion (should be = 0.05)
Ancestry uninformed	
Simple linear regression	0.64
Race random effects	0.62
Ancestry informed	
Ancestry random effects	0.34
Ancestry principal components	0.35
Network autoregression	0.45
Phylogenetic autoregression	0.056

made (data supplement). Thus, the nonindependence, as expected, did not bias the slope estimates but made them arbitrarily overly negative or overly positive.

Having the correct nonindependence network or tree obviously is critical for correctly correcting for Galton's problem in an empirical analysis. Most often researchers have used some form of linguistic phylogeny that reflects the historical separation times of languages inferred from statistical analyses of cognate word lists (Matthews et al. 2016). Cognate words are useful because (i) they are reasonably available, (ii) they are definitely inherited culturally rather than genetically or as a response to local environmental conditions, and (iii) a substantial number of studies have shown many cultural traits are co-inherited with language (Collard et al. 2006; Jordan et al. 2009; Currie et al. 2010; Mathew and Perreault 2015). So as long as researchers can be confident they are applying a tree that is relevant to the traits being studied, there are no downsides to applying phylogenetic approaches (Rohlf 2006). Correcting for nonindependence in this way does not insert control variables into the regression that soak up variation and thereby reduce statistical power. The only problem arises when an inappropriate phylogenetic model is applied, meaning one that does not reflect the actual cultural inheritance process. Thus, some *a priori* understanding on the cultural mechanics is needed on the part of the researcher.

Indeed, nonlinguistic modes of inheritance should receive more attention in future research. For example, it is highly plausible that some cultural traits are co-inherited more with religious lineages than linguistic ones (Matthews 2012; Matthews et al. 2013). Additionally, some of the observed nonindependence of population data points might arise from genetic rather than cultural inheritances. Obtaining genetic divergences estimated from neutral genes is quite achievable for most populations, but in general, genetic divergences have not often been used in autoregression models to resolve Galton's problem (see Matthews and Butler 2011 for an exception).

Example Applications

Several applied studies have been published that incorporated statistical models for nonindependence. Atkinson et al. (2016) examined whether more eco-friendly cultural beliefs or technologies were more important to the rate of deforestation on Austronesian islands. They used phylogenetic autoregression modeling to correct for data point nonindependence that arose from both cultural beliefs and technologies being socially learned phenomenon inherited from prior generations. Their final analysis showed that technological capacity was much more determinative of deforestation than were cultural beliefs, which only mitigated environmental impacts to a small degree.

Matthews et al. (2016) recently tested whether linguistic divergence pre-dicted political and economic behaviors of modern nation-states. This prediction was based on the observation that language divergence predicts many cultural traits in traditional societies. One mechanism that may play out in the modern context is that countries influence each other regarding their politics and economies, and culturally more similar countries may be of greater influence than are more culturally distant ones, either because countries want to differentiate from culturally similar countries or because they more readily adopt innovations from them. Consistent with this hypothesis, the study found that changes in the autocratic versus democratic character of a country, and decisions to default on national debt, frequently correlated either positively or negatively across ties of language divergence over the past 100 years. Related work has shown that patterns of economic development appear substantially predicted by population phylogenies (Spolaore and Wacziarg 2009, 2011, 2013).

Dealing with Nonindependence in Less Treelike Network Structures

Related issues of nonindependence occur when we study how culture diffuses along social network ties that are less treelike than those just considered. Often such networks are studied as relationships among individuals, but they can characterize relationships among groups as well. Whatever entities are linked together by connections are called "nodes" in network analysis terminology. Whether the nodes of the network are individuals or groups, however, the very process of cultural diffusion means that nodes that share traits they acquired from each other are nonindependent. Network structural measurements like the number of ties a node has (termed network degree)

Figure 8.5 Nonindependence in a simple social network. Network metrics are fundamentally nonindependent because they are relational. Node A has a network degree of 3, B and D each have degree 2, and C has degree 1. There are only four ties in the network, but the sum of degrees is 8 (3 node A + 2 node B + 1 node C + 2 node D) because each tie to A is counted twice and the tie between B and D is counted twice – once in calculating the degree for B and once for calculating the degree for D.

also are nonindependent, since each tie is connected to two nodes and thus contributes to both of their network metrics (Figure 8.5).

Network autoregression models can be used to account for these issues in network data (Leenders 2002; O'Malley and Marsden 2008). However, compared with the ancient cultural lineages like linguistic distances, the interpretation of correlations across network ties is somewhat more complex (Shalizi and Thomas 2011). This is because the network ties themselves can form and dissolve in addition to the cultural traits diffusing on them. In contrast, England and Germany cannot change that they are both Germanic-speaking countries, while France is not.

The fact that network ties can change means any correlation we find in traits on networks could arise from cultural processes (i.e. social learning) or from a tendency for individuals with similar traits to form connections with each other. This latter phenomenon would require no social learning per se and is well documented and known as homophily in social network science (McPherson et al. 2001; Wimmer and Lewis 2010). Individuals exhibit homophily based on sex, race, class, and a number of other attributes (McPherson et al. 2001). Most traits probably arise from a mix of homophily and social influence.

Determining Which Network Is Most Important for a Cultural Trait

Regardless of the process for any particular case, applied modelers often need to know which of several potential networks are most involved in the cultural diffusion or homophily dynamics. Even if distinguishing homophily and influence with observational data is fundamentally confounded (Shalizi and Thomas 2011), knowing the most important network for a certain set of traits is a first step necessary for most applied and theoretical analyses. Which networks should be considered among the "potentials" usually will be highly context and data set specific, but within that context-specific set, we should have statistical methods that reliably discriminate their relative importances.

For example, in a study of violent religious extremists, Matthews et al. (2013) assessed whether letter correspondences among leaders or historical congregational connections were most important to groups sharing violent ideologies in common. Knowing the answer to such a question clearly can guide whether policy interventions should target leaders or congregations regardless of whether the generative mechanism for the correlation is homophily or influence.

Here again, autoregression is one approach that could be applied, but previous studies have shown that at least the most common implementation (lnam in R) suffers from poor statistical performance (Leenders 2002; Mizruchi and Neuman 2008; Neuman and Mizruchi 2010; Matthews et al. 2016; Dittrich et al. 2017; Karimov and Matthews 2017). Another long-standing technique relies on using permutations to assess the statistical significance of trait correlations across network ties. The classical version of this is the Mantel permutation of the dependent variable, which is known to have statistical failures when the network independent variables are correlated with each other (Dekker et al. 2007). More recently developed permutation approaches permute the regression residuals rather than the dependent variable and by doing so exhibit much more robust statistical properties (Dekker et al. 2007).

Another approach to deal with nonindependence in network data is to use random effects that effectively allow for some individuals to have high or low values in the dependent variable just based on their identity. The idea here is that network data conform in a sense to a kind of repeated measure – all of my network connections have a repeated identity in common – myself. This approach was developed independently by de Nooy (2011) and O'Malley and Christakis (2011) in the context of modeling the formation and dissolution of ties in networks. Matthews et al. (2013) proposed this method might be equally useful for dependent variables that center on the similarity of a diffusing trait on a network rather than the formation of the network ties themselves.

More recently Karimov and Matthews (2017) published a reasonably complete set of agent-based diffusion simulations to systematically evaluate the statistical performance of autoregression (implemented through the *lnam* function in the R package *sna*), the permutation approach of Dekker et al. (2007) (implemented through the *netlm* function in R package *sna*), and the random effects method (implemented through the *lmer* function in R package *lme4*). We found that the random effects method provided the highest statistical power and nearly always produced statistically correct false positive rates. The permutation method had close to the power of random effects and usually acceptable false positive rates except in some conditions with treelike (i.e. bifurcating) network structures. The autoregression method demonstrated low power and consistently elevated false positive rates. Our final recommendation was that the random effects methods appears to be the most robust current implementation for assessing which of several plausible networks are most related to the dynamics of cultural trait diffusion (Karimov and Matthews 2017; Matthews et al. 2018).

Correcting for Network Nonindependence When Testing Trait–Trait Correlations

Having determined which network appears more important for a cultural trait of interest, we then face a highly similar problem as in phylogenetic systems if we want to test correlations of this trait with other traits. Because one or more of the traits may have diffused through the network, the data are statistically nonindependent. Although cultural diffusion on networks usually is not conceptualized as cross-generational inheritance, it is still an inheritance process in that each node is acquiring trait values from other nodes over some amount of time lag.

As before, I conducted a set of 1000 simulated continuous traits, but this time the traits diffused across the network of connections across countries implied by the Indo-European language tree (Figure 8.6). To simulate a rapid diffusion process, in each simulation the trait values were drawn at random from a normal distribution. Then each node adopted 20% of their connections' average deviation from their own trait value. The 20% adoption was weighted by the values of the network ties, which were the years of separation between countries due to language divergence or colonial independence. I then evaluated the same five statistical models on their ability to produce statistically correct false positive rates (Table 8.2).

Compared with the phylogenetic simulations (Table 8.1), more of the statistical methods performed well and produced acceptable false positive rates for the network diffusion simulations. Once again, however, simple linear regression and regression with the American racial categories produced unacceptably high false positive rates. Of the methods that used ancestry information, only the lnam R function that implements a network autoregression model failed to produce the expected 0.05 false positive rate. Since the math underlying this function was intended to correct for exactly this type of simulated process (Leenders 2002; O'Malley and Marsden 2008), it suggests there may be some fundamental errors in how lnam currently is implemented into code. As with the phylogenetic simulations and as expected from theory, the slope estimates were centered on zero such that the false significance findings were equally for negative or positive associations between the uncorrelated variables (data supplement).

Example Applications

Network models have been applied in numerous studies of cultural diffusion. Studies on the diffusion of health behaviors have been particularly prominent. Among the most prominent would be the findings that obesity spread across

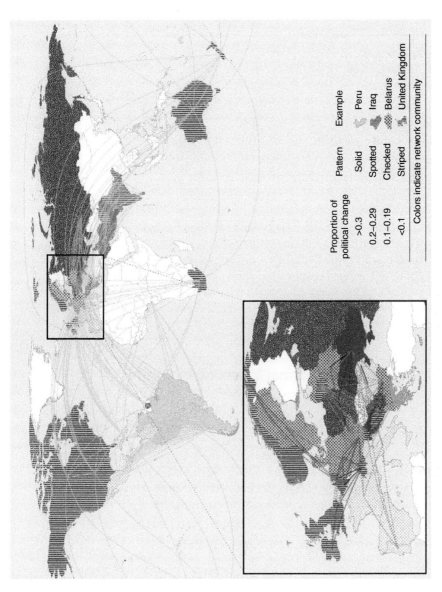

Figure 8.6 Linguistic network for Indo-European countries based on Bayesian phylogenetic analysis of cognate basic words. Source: Redrawn from Matthews et al. (2016).

Table 8.2 False positive rates for simulations of diffusion of trait variation Indo-European linguistic network.

Statistical model	False positive proportion (should be = 0.05)
Ancestry uninformed	
Simple linear regression	0.54
Race random effects	0.49
Ancestry informed	
Ancestry random effects	0.058
Ancestry principal components	0.044
Network autoregression	0.14
Phylogenetic autoregression	0.050

network ties in the Framingham Heart Study (Christakis and Fowler 2007). This study was critiqued on a number of statistical issues, a prominent critique being that Christakis and Fowler had failed to correct for data point nonindependence (Lyons 2011). Subsequent research has shown that the original obesity findings from Framingham are likely correct (VanderWeele 2011), but a good deal of time, effort, and controversy might have been saved if issues like network nonindependence had been dealt with up front.

Future Directions for Formal Modeling of Culture

Social scientists outside of scientific anthropology, the latter being admittedly a small social clique, may be surprised to learn that a reasonably comprehensive set of formal methods have been devised for dealing with cultural data at the levels of variation across groups, variation across relationships within groups of varying size, and variation among individuals within small and even tightly interrelated groups. While the methods are comprehensive, there are still important methodological developments to undertake.

Improved Network Autoregression Implementations

At least in contemporary settings after the effects of colonialism and modernity, strictly bifurcating phylogenetic models *may* be inappropriate for cross-cultural analyses, although in my network simulations above the phylogenetic model actually performed well. Network autoregression techniques are conceptually nearly the same as phylogenetic models but allow for greater flexibility in the structure of the expected correlations among the data points. The most commonly used implementation of the network autoregression, however, has been shown to exhibit statistical biases in multiple studies by different research groups (Mizruchi and Neuman 2008; Neuman and Mizruchi 2010; Matthews et al. 2016; Dittrich et al. 2017; Karimov and Matthews 2017). The community needs an improved implementation and preferably in R as it is the most commonly used coding language by computationally minded social scientists. Likely there is simply some implementation problem in lnam, as statistical theory would argue that this model should work as advertised. There could be unresolved statistical issues as well.

A Global Data Set of Expected Nonindependence to Solve Galton's Problem

While this chapter and this volume as a whole both focus on modeling, future development of formal cultural models will be seriously compromised if more and better data collection is not conducted. More than 128 years after Galton identified the problem of cultural nonindependence, anthropologists still have not produced a global assessment of expected nonindependence among cultural data points. Linguistic, religious, and genetic divergences would provide complementary views for alternative pathways by which traits could be inherited. Sufficient linguistic and genetic data are available to produce a global network of expected nonindependence at least for country-level data points. These linguistic and genetic data are, however, mostly published in anthropological data sets that are organized by ancient ethnic origins rather than by modern nation-state. This is because anthropologists have been interested primarily in reconstructing the genetic and linguistic histories of populations rather than applying knowledge of these histories to the modern period. For example, during genetic data collection, anthropologists have frequently excluded individuals for whom all four grandparents were not from the same small geographic area.

Constructing a global religious network presents greater challenges. Even though the historical divergences among major religions are reasonably well known, the percentage of adherents by country is more poorly recorded for non-Christian denominations than for Christian ones (Johnson and Grim 2008, 2013; Hackett et al. 2015). For example, data on even the basic Sunni–Shiite division within Islam are scarce for most countries, and data on adherents of the theological schools within these two branches are virtually nonexistent. The reasons that one of humanity's most ancient and enduring cultural systems (religion) is so systematically unmeasured are several: much of social science since Marx has had antipathy toward the subject, some governments like the United States do not collect religious affiliation in census data due to separation of church and state issues, many governments have inaccurate census records because the government persecutes minority religions, and counts reported by religious groups themselves are routinely inflated and inaccurate.

Better Collection of Behavioral Trait Variation Across Populations

Leaving aside cultural nonindependence, basic variation in cultural traits has similarly gone unmeasured. Articles with sophisticated statistical models rely routinely on decades-old binary measurements of continuous variables. Several modern efforts are ongoing that may provide foundations for more systematic data collection in the future. D-PLACE is one resource that improves the ability for researchers to systematically use existing data by linking HRAF data with ecological and linguistic resources in a modern database structure. However, the data themselves are often relatively coarse-grained variables that were coded 50+ years ago (Kirby et al. 2016).

The WVS is arguably the best example of an ongoing systematic data collection effort relevant to cultural research. The WVS samples individual-level data longitudinally and from about 80 contemporary countries (Inglehart et al. 2005; Welzel 2013). This is an impressive accomplishment but obviously leaves out historical data, data from many smaller countries, and WVS focuses on values and so excludes more ecological variables.

Two other newer efforts are *The Database of Religious History* and *Seshat: Global History Databank* (Turchin et al. 2015; Slingerland and Sullivan 2017). Both are trying to conduct new data collection by systematically coding primary and secondary source documents on various contemporaneous and historical world cultures and religious traditions. As the efforts are new, only the future will tell if they will be able to sustain the level of effort needed to build ongoing and robust data resources. Advantages of these efforts are that they are not constrained to modern populations or to national samples that tend to capture mostly majority viewpoints. Both are attempting to sample the historical record and to include religious or cultural traditions of minority groups.

While selecting samples of global variations can be perfectly valid for testing certain hypotheses, in my view greater global sampling would be a key contribution that *The Database of Religious History* or *Seshat* might provide. For example, *Seshat* is currently working on a World 30 sample that provides data on 30 representative locations across the globe over historical time. Because only group-level attributes are being collected, I think one advance something like *Seshat* can provide over WVS would be to vastly increase sample size beyond the ~80 modern nations included in WVS.

The question might be asked, "but how much sample is enough?" I agree that large samples are not always necessary. I myself have published statistical analyses with quite small sample sizes. Many questions in the social sciences are, in fact, only applicable to constrained subsets of global variation. Aspects of my own articles on Christian religious evolution are cases in point in that non-Christians do not have a view on matters like the trinity or the Eucharist (Matthews 2012; Matthews et al. 2013). That being said, reasonably comprehensive global-scale cultural databases potentially could facilitate global-scale analyses of broadly applicable questions and also provide easily accessible data subsets for more targeted analyses. At least for group-level measurements, the data size involved is not large by modern "big data" standards. There are about 7000 languages globally (Simons and Fennig 2018) and likely far fewer distinct religious denominations. When I worked in industry, I had access to datasets with records on tens of millions of individual Americans. Finding and justifying funding for social science is always a challenge, but we as a society should take a step back to question why we are willing to make the monetary investment required to aggregate records on tens of millions of individuals to sell more consumer goods, but we can't find the resources to study group-level characteristics of several thousand languages or religions that continue to shape many features of our modern world.

Acknowledgments

I would like first to thank my coauthors on *A Manual for Cultural Analysis*, David Kennedy and Ryan Brown, who both have influenced my views of these topics over the past couple years we spent writing the manual together. Thanks also to Andy Parker for his helpful critiques of some of the ideas in this chapter. Great thanks are due to my students from my Pardee RAND Graduate School Course *Cultural Measurement and Analysis for Policy Research*. Their enthusiasm and questioning underscored for me the importance of Galton's problem to the whole history of cultural analysis. This work was supported in part by NIH grant R34MH114696 and by the RAND Corporation.

References

Atkinson, Q.D., Coomber, T., Passmore, S. et al. (2016). Cultural and environmental predictors of pre-European deforestation on Pacific Islands. *PLoS ONE* 11 (5): e0156340.

Boyd, R. and Richerson, P.J. (1985). *Culture and the Evolutionary Process*. University of Chicago Press.

Cavalli-Sforza, L.L. and Feldman, M.W. (1981). *Cultural Transmission and Evolution: A Quantitative Approach*. Princeton University Press.

Christakis, N.A. and Fowler, J.H. (2007). The spread of obesity in a large social network over 32 years. *New England Journal of Medicine* 357 (4): 370–379.

Collard, M., Shennan, S.J., and Tehrani, J.J. (2006). Branching, blending, and the evolution of cultural similarities and differences among human populations. *Evolution and Human Behavior* 27 (3): 169–184.

Coop, G., Witonsky, D., Di Rienzo, A., and Pritchard, J.K. (2010). Using environmental correlations to identify loci underlying local adaptation. *Genetics* 185 (4): 1411–1423.

Currie, T.E., Greenhill, S.J., Gray, R.D. et al. (2010). Rise and fall of political complexity in island South-East Asia and the Pacific. *Nature* 467 (7317): 801–804.

De Nooy, W. (2011). Networks of action and events over time. A multilevel discrete-time event history model for longitudinal network data. *Social Networks* 33 (1): 31–40.

Dean, L.G., Vale, G.L., Laland, K.N. et al. (2014). Human cumulative culture: a comparative perspective. *Biological Reviews* 89 (2): 284–301.

Dekker, D., Krackhardt, D., and Snijders, T.A. (2007). Sensitivity of MRQAP tests to collinearity and autocorrelation conditions. *Psychometrika* 72 (4): 563–581.

Dittrich, D., Leenders, R.T.A., and Mulder, J. (2017). Bayesian estimation of the network autocorrelation model. *Social Networks* 48: 213–236.

DRH (2018). The database of religious history. Retrieved from https:// religiondatabase.org/landing/ (accessed 30 August 2018).

Durham, W.H. (1991). *Coevolution: Genes, Culture, and Human Diversity*. Stanford University Press.

Gören, E. (2017). The persistent effects of novelty-seeking traits on comparative economic development. *Journal of Development Economics* 126: 112–126.

Hackett, C., Connor, P., Stonawski, M. et al. (2015). *The Future of World Religions: Population Growth Projections, 2010–2050*. Washington, DC: Pew Research Center.

Henrich, J. (2015). *The Secret of Our Success: How Culture Is Driving Human Evolution, Domesticating Our Species, and Making Us Smarter*. Princeton University Press.

Hofstede, G. (1984). *Culture's Consequences: International Differences in Work-Related Values*, vol. 5, Sage.

Hofstede, G.H. (2001). *Culture's Consequences: Comparing Values, Behaviors, Institutions and Organizations Across Nations*. Sage.

Inglehart, R., Puranen, B., Pettersson, T. et al. (2005). *The World Values Survey*. The Institute for Comparative Survey Research.

Johnson, T.M. and Grim, B.J. (2008). *World Religion Database*. Leiden/Boston, MA: Brill.

Johnson, T.M. and Grim, B.J. (2013). *The World's Religions in Figures: An Introduction to International Religious Demography*. Wiley.

Jordan, F.M., Gray, R.D., Greenhill, S.J., and Mace, R. (2009). Matrilocal residence is ancestral in Austronesian societies. *Proceedings of the Royal Society of London B: Biological Sciences* 276: 1957–1964.

Karimov, R. and Matthews, L.J. (2017). A simulation assessment of methods to infer cultural transmission on dark networks. *The Journal of Defense Modeling and Simulation: Applications, Methodology, Technology* 14: 7–16.

Kirby, K.R., Gray, R.D., Greenhill, S.J. et al. (2016). D-PLACE: a global database of cultural, linguistic and environmental diversity. *PLoS ONE* 11 (7): e0158391.

Kissling, W.D. and Carl, G. (2008). Spatial autocorrelation and the selection of simultaneous autoregressive models. *Global Ecology and Biogeography* 17 (1): 59–71.

Laland, K.N. and Kendal, J.R. (2003). What the models say about social learning. In: *The Biology of Traditions: Models and Evidence* (ed. D. Fragaszy and S. Perry), 33–55. Cambridge University Press.

Langness, L.L. (1987). *The Study of Culture*. Chandler Sharp Publishers.

Leenders, R.T.A. (2002). Modeling social influence through network autocorrelation: constructing the weight matrix. *Social Networks* 24 (1): 21–47.

Lyons, R. (2011). The spread of evidence-poor medicine via flawed social-network analysis. *Statistics, Politics, and Policy* 2 (1): 1–26.

Mace, R., Pagel, M., Bowen, J.R. et al. (1994). The comparative method in anthropology [and comments and reply]. *Current Anthropology* 35 (5): 549–564.

MacLean, E.L., Matthews, L.J., Hare, B.A. et al. (2012). How does cognition evolve? Phylogenetic comparative psychology. *Animal Cognition* 15 (2): 223–238.

Mathew, S. and Perreault, C. (2015). Behavioural variation in 172 small-scale societies indicates that social learning is the main mode of human adaptation. *Proceedings of the Royal Society B: Biological Sciences* 282: https://doi.org/10.1098/rspb.2015.0061.

Matthews, L.J. (2012). The recognition signal hypothesis for the adaptive evolution of religion. *Human Nature* 23 (2): 218–249.

Matthews, L.J., Brown, R.A., and Kennedy, D.P. (2018). *A Manual for Cultural Analysis*. Retrieved from Santa Monica, CA.

Matthews, L.J. and Butler, P.M. (2011). Novelty-seeking DRD4 polymorphisms are associated with human migration distance out-of-Africa after controlling for neutral population gene structure. *American Journal of Physical Anthropology* 145 (3): 382–389.

Matthews, L.J., Edmonds, J., Wildman, W.J., and Nunn, C.L. (2013). Cultural inheritance or cultural diffusion of religious violence? A quantitative case study of the radical reformation. *Religion, Brain and Behavior* 3 (1): 3–15.

Matthews, L.J., Passmore, S., Richard, P.M. et al. (2016). Shared cultural history as a predictor of political and economic changes among nation states. *PLoS ONE* 11 (4): e0152979.

McPherson, M., Smith-Lovin, L., and Cook, J.M. (2001). Birds of a feather: homophily in social networks. *Annual Review of Sociology* 27 (1): 415–444.

Mesoudi, A., Chang, L., Dall, S.R., and Thornton, A. (2016). The evolution of individual and cultural variation in social learning. *Trends in Ecology & Evolution* 31 (3): 215–225.

Mizruchi, M.S. and Neuman, E.J. (2008). The effect of density on the level of bias in the network autocorrelation model. *Social Networks* 30 (3): 190–200.

Murdock, G.P. and White, D.R. (1969). Standard cross-cultural sample. *Ethnology* 8 (4): 329–369.

Naroll, R. (1961). Two solutions to Galton's problem. *Philosophy of Science* 28 (1): 15–39.

Neuman, E.J. and Mizruchi, M.S. (2010). Structure and bias in the network autocorrelation model. *Social Networks* 32 (4): 290–300.

Nunn, C.L. (2011). *The comparative approach in evolutionary anthropology and biology*. University of Chicago Press.

O'Malley, A.J. and Marsden, P.V. (2008). The analysis of social networks. *Health Services and Outcomes Research Methodology* 8 (4): 222–269.

O'Malley, A.J. and Christakis, N.A. (2011). Longitudinal analysis of large social networks: estimating the effect of health traits on changes in friendship ties. *Statistics in Medicine* 30 (9): 950–964.

Richerson, P. and Boyd, R. (2005). *Not By Genes Alone: How Culture Transformed Human Evolution*. Chicago: University of Chicago Press.

Rohlf, F.J. (2006). A comment on phylogenetic correction. *Evolution* 60 (7): 1509–1515.

Shalizi, C.R. and Thomas, A.C. (2011). Homophily and contagion are generically confounded in observational social network studies. *Sociological Methods and Research* 40 (2): 211–239.

Simons, G.F. and Fennig, C.D. (2018). *Ethnologue: Languages of the World*, 21e. Dallas, TX: SIL International. Retrieved from www.ethnologue.com.

Slingerland, E. and Sullivan, B. (2017). Durkheim with data: The Database of Religious History. *Journal of the American Academy of Religion* 85 (2): 312–347.

Smith, J.M. and Smith, N. (1996). Synonymous nucleotide divergence: what is "saturation"? *Genetics* 142 (3): 1033–1036.

Spolaore, E. and Wacziarg, R. (2009). The diffusion of development. *The Quarterly Journal of Economics* 124 (2): 469–529.

Spolaore, E. and Wacziarg, R. (2011). Long-term barriers to the international diffusion of innovations. Paper presented at the NBER International Seminar on Macroeconomics 2011.

Spolaore, E. and Wacziarg, R. (2013). How deep are the roots of economic development? *Journal of Economic Literature* 51 (2): 325–369.

Turchin, P., Brennan, R., Currie, T.E. et al. (2015). Seshat: history the global databank. *Cliodynamics: The Journal of Quantitative History and Cultural Evolution* 6 (1): 77–107.

Tylor, E.B. (1871). *Primitive Culture: Researches into the Development of Mythology, Philosophy, Religion, Art, and Custom*, vol. 2. Murray.

VanderWeele, T.J. (2011). Sensitivity analysis for contagion effects in social networks. *Sociological Methods and Research* 40 (2): 240–255.

Welzel, C. (2013). *Freedom Rising*. Cambridge University Press.

Wimmer, A. and Lewis, K. (2010). Beyond and below racial homophily: ERG models of a friendship network documented on Facebook 1. *American Journal of Sociology* 116 (2): 583–642.

World Values Survey Association (2016). World values survey wave 6 2010–2014 official aggregate v. 20150418. Aggregate File Producer: Asep/JDS, Madrid Spain.

9

Social Media, Global Connections, and Information Environments: Building Complex Understandings of Multi-Actor Interactions

Gene Cowherd and Daniel Lende

Department of Anthropology, University of South Florida, Tampa, FL 33620, USA

A New Setting of Hyperconnectivity

Russian spies. Fake news. Bots. Trolls. Social media. Information operations. With news stories and television reports rife with references to recent alleged operations from Russian-linked actors, it would be easy to conclude that we are living in an age of unprecedented activity of international information operations activity, complete with Cold War era intrigue and espionage. Yet historians would be quick to point out that despite this recent wave of hysteria, this latest activity is only the latest crest in the larger ebb and flow of a tide of international tensions. Intelligence professionals have similarly described a constant din of monitoring and activity that has been largely invisible to the general public.

So, what has changed? A new state of technological hyperconnectivity has provided governments and associated actors with the ability to connect directly with local populations in unprecedented ways. The unification of social media and the news into a signal platform has catalyzed the effectiveness of recent efforts to color political activity and discourse. Given the local impact, once surreptitious tactics in information operations are increasingly part of public discourse. Overall, changing technology has altered the complex interactions that have long happened between state actors and local populations and provided new and often contested abilities to shape how people think and act.

This chapter utilizes an anthropological framework to illustrate these changes and to outline how computational approaches need to embrace some of the complexities in these new types of human interactions. One of the chapter's main points is "ethnography meets modeling," in which our ability to produce better data and better analyses relies on grasping the relevant factors that play out on the ground via technological hyperconnectivity. Ethnographic research

Social-Behavioral Modeling for Complex Systems, First Edition.
Edited by Paul K. Davis, Angela O'Mahony, and Jonathan Pfautz.
© 2019 John Wiley & Sons, Inc. Published 2019 by John Wiley & Sons, Inc.
Companion website: www.wiley.com/go/Davis_Social-Behavioralmodeling

relies on thick description and the identification of relevant properties in social and linguistic interactions and thus can promote better efforts in computational models. Our approach here will be to consider both actors and the information environments we create and manipulate as essential to producing the types of complex understandings needed to model and analyze the enormous data that gets generated about our social interactions via social media.

We will use the example of hybrid warfare to analyze some of the changes that have happened to the strategies that actors employ and the ways that people understand and act on the information in their environment. Historically, diplomacy, political elections, and policy votes have all been significant targets of hybrid warfare operations, the goal being to subvert, disrupt, and influence diplomatic and political forces in the target nation. Today, professionally trained information operators can now interact directly, and en masse, with the American populace, even as the US forces rely on these same social media sources to guide their own actions. How can we better understand and model the changes in our information environment?

The Information Environment

The Department of Defense (DoD) defines the information environment in the Joint Publication 3-13 as "the aggregate of individuals, organizations, and systems that collect, process, disseminate, or act on information" (Joint Chiefs of Staff 2013). The DoD describes the information environment as three domains that interact with one other: the physical, the informational, and the cognitive domains. The physical domain refers to those geographical, maritime, airspace, and cyberspace components that contain the physical infrastructure supporting the reception, transmission, and storage of information. The information domain itself refers to the flow of data, text, images, and other forms of media that staffs can collect, analyze, and report. The cognitive domain contains the minds of targeted individuals who are affected by and act upon the information. These minds may be military or government civilian individuals, government contractors, or civilians in the general population. As this domain seeks to influence the complexities of human social life, it focuses primarily on the societal, cultural, religious, and historical contexts that influence the perceptions of individuals in the target nation (Department of the Army 2013).

Anthropology offers up another domain that shapes the information environment – the cultural domain. Rather than reduce societal, religious, and historical contexts to how they shape perceptions, it is important to recognize that culture structures the information domain on its own (and not solely through physical, informational, and cognitive dimensions). Differences in language, in what counts as important, and in the types of official and informal procedures used to interact with the flow of information are all dynamically shaped by cultural and societal processes. For example, current understandings of how Western audiences search for, consume, and share online information are largely derived from theoretical models based on secondary data sources and demographic statistics collected by third parties. The success of recent foreign IO efforts suggests the need for a disruptive approach to increase our understanding of how culture shapes individual actors' use of social media.

We propose that what has been missing is a "ground-truth" element to understanding online culture that social science research theories and methods can provide. Anthropological approaches work well with system theory, complexity theory, and computational approaches. Key to this is identifying interdependent parts and interrelating structures within complex conglomerations. One added advantage is observing the study population *in situ* rather than in the laboratory. Observing study participants as they interrelate and share information in a natural "in the wild" setting throughout their day provides significant insights to bolster both modeling efforts within the behavioral sciences and national security initiatives.

Social Media in the Information Environment

The Internet has transformed the way people produce, store, exchange, and consume information, creating a new niche within the information environment that both adapts to the information requirements of its users and shapes their perceptions by the information it provides. Algorithms employed by search engines learn from our information consumption patterns and tailor the flow of search returns to cater to our preferences. This ultimately has the effect of homogenizing the perspectives reflected in the search returns and mirrors back to us information that reinforces our worldview. The ability to connect people to ideas is at the heart of the impact of the information domain.

This feedback function is readily apparent and easily accessible on social media. In the last decade, social media has revolutionized the way we interact with others and share information. It has provided a convenient way to connect with friends and relatives and the flow of events in our lives irrespective of the geographical space separating them. The widening use of mobile technology has fueled a steady increase in the volume of information shared and the speed of which it can be consumed. In the United States alone, adults ages 18–35 have increased their use of mobile technology from an average of just over two hours per day in 2013 to just under four hours per day in 2018, with 70% of that time devoted to social networking (Statistica 2018).

Social media has also provided a unified platform for the consumption of news, political commentary, and the organization of grassroots movements into one consolidated space. In 2017, Pew research found that 67% of Americans surveyed reported that they got at least some of their news from social media sites, with 26% of those reporting that social media was their primary source of news (Shearer and Gottried 2017). Their research also shows a steady uptick in the number of people who report social media as their primary source of news from 15% in 2013 to 18% in 2016 at the height of the US presidential election to 26% in 2017. By far, Facebook and Twitter were the most popular social media platforms for news with 83% of those reporting using social media for news citing them as their preferred news aggregator.

With this growth in online consumption of information, social media has also introduced a layer of complexity into the information environment that influences users' perception of verity and prevarication of information. For example, social media amplifies the spread of false news. A study conducted by Vosoughi et al. (2018) showed that false news travels faster, reaches further, and lingers longer on social media than true news and that emotions like fear and anxiety, as well as the novelty of the news, are significant driving factors of this effect.

The researchers analyzed 126 000 stories tweeted by 3 million people more than 4.5 million times between 2006 and 2017. The stories were designated as true or false based on six independent fact-checking organizations. Their findings found that false tweets triggered a cascade of retweets, resulting in false information traveling faster and reaching further than true news. They also found that this effect was more pronounced in fake political news than in fake news about terrorism, natural disasters, or other tragedies. This same study highlighted the emotional impact of these false news stories, too. The researchers found that the false stories triggered the emotions of fear, disgust, and surprise, whereas the true stories inspired anticipation, sadness, joy, and trust.

Interestingly, Vosoughi and colleagues were able to control for the role that bots play in disseminating data. What they found was that the bots shared both true and false stories indiscriminately, and when controlled for in their analysis, they found that it was humans, not bots, that played a greater role in the dissemination of false information. Translated into anthropological terms, social media in the information environment works to promote and reinforce particular sociocultural views. No independent vetting of the information occurs, and the feedback happens through social groups that one views as familiar and trustworthy.

Anthropologists use the terms emic and etic to denote the insider's view and the outside anthropological assessment of the phenomena in question. The implication of social media in the information environment is that only emic views get promoted and reinforced. There is no independent evaluation but rather a multitude of competing views that run concurrently and often in isolated fashion inside our information environment.

Integrative Approaches to Understanding Human Behavior

The integration of cognitive science and anthropology offers great potential for advancing theoretical and methodological models for understanding human behavior. There are many endeavors that build from the cognitive sciences toward the social sciences and humanities. The general assumption is that modeling of the human mind will then solve the theoretical problems that the social sciences have long studied. There is much value in this sort of work, but an alternative approach is also possible. We can build from the social sciences and humanities toward the cognitive sciences. Indeed, we advocate for the utter importance of that, because together, each approach can reveal the biases and weaknesses of other side and often correct them or at the very least offer interesting data and insights.

Anthropologists have long advocated for the importance of ethnography and qualitative research to provide the sort of evidence-based approach to how people interpret and interact in specific settings. This type of work has shown value in understanding how pilots manage to navigate such complex machines as airline airplanes; only by observing and understanding how pilots actually interact with all the controls in the cockpit of an airplane can we come to understand how cognition and context interact (Hutchins 1995). Neuroanthropology represents one recent endeavor to integrative social science and cognitive science, with an emphasis on building from anthropology toward cognitive science (Lende and Downey 2012). An emphasis on ethnography (or understanding "brains in the wild" not just in the laboratory or clinic), a critical take on both anthropological and cognitive science approaches, and a focus on understanding the processes can account for the variation revealed by field-based studies, whether those processes are social, cognitive, or some interesting synergy – these are the hallmarks of how neuroanthropology tackles research topics.

At the same time, anthropologists have embraced agent-based modeling, system theory, and computational approaches as part of their research (Agar and Wilson 2002; Franz and Mathews 2011; Hofer 2013). This work has relied on ethnographic insights in conjunction with modeling to understand complex problems. However, this work has had difficulties in modeling complex agents that accurately represent how people actually behave. Neuroanthropology draws on the substantial body of evidence showing that human perceptions and behavior are dynamically and inextricably intertwined with the cultural environment (Lende and Downey 2012).

We now present an overview of three challenges for theoretical and computational approaches to understanding what is happening with the interaction of people and social media in our globalized information environments. These challenges are based on the combination of ethnographic assessment of what is happening presently and the use of cognitive and anthropological approaches to identify relevant features of these current challenges. Below we present an overview of the challenges before turning to an ethnographic modeling of each one.

Muddy the Waters

A primary tactic today is to use disinformation to (i) confirm an extreme position, (ii) create confusion and a sense of threat so people rely more on "fast" assessment and favor their particular social group against out groups, and (iii) generate counter-narratives so people "don't know what to think" and thus rely on one-sided positions that seem to offer clarity and certitude. Stir the pot, so people want security, stability, and sharedness. Given this analysis, a key question emerges: How can we turn from muddying waters to making meaning in the middle, bringing together ideas from cognitive science, anthropology, and communications?

Missing It

In both counterintelligence operations and in everyday life, many people do not really see how media shapes what they think and how they think. They also do not have many of the cognitive/cultural tools to critically assess what is happening. Focusing specifically on the national security side, operations are too often viewed through a technical lens rather than realizing that intelligence processes are also cultural. Intelligence analysis and the counterintelligence operations it informs are cultural products. There is a lack of embedded and applied research that can get at how analysis plays out in real-world settings and the types of vulnerabilities that offer up for other actors to exploit. Applied ethnography here offers a vital way to develop data and applied insights.

Wag the Dog

How and why do people follow along with what "makes sense?" Rather than taking cultural consensus as a given (as anthropologists often do), an important question today is how people get caught between attention-grabbing media and larger cultural narratives. Using interdisciplinary approaches, we need to ask, "Why do people "follow consensus"?" And what ways exist to insert the types of prediction errors, sensory feedback, and critical assessment that help people reevaluate their ongoing engagement? Otherwise, we continue to put ourselves in a position where other actors can manipulate our own institutions and values – they find the levers that let them exploit "what makes sense" to us.

The Ethnographic Examples

Below we present a series of linked ethnographic examples as a representational model. We move through three topics with three different actors, and in each case, aim to produce an idealized version derived from on-the-ground sources. We view this idealized approach as similar to modeling – identifying relevant features of the actor and the information environment and the interactions between each person and social media. We also aim to convey some of the complexity of the physical, informational, cognitive, and cultural domains of the information environment and thus highlight the challenges faced as we advance computational approaches and complex modeling.

The ethnographic examples come from a combination of research, professional experience, and the use of secondary sources. Prior research includes work on cybersecurity, how people interact with social media, and understandings of our rapidly changing information environment. One of the authors worked for nearly two years first as an intelligence analyst and later as an operational environment designer for multinational joint training exercises. Finally, primary sources reporting on social media, news reporting, political conflicts, and local actors' lives provided further detail. The stories and persons represented here constitute an amalgam of this information. As such, the characters and situations depicted here are abstract figures that typify the salient trends that have emerged through this research. Before presenting the ethnographic examples, however, let us set the stage with an example.

> It is true that there is nothing new under the sun. All knowledge is within our reach; we need only to scour the vast depths of the Internet itself.
> - Abraham Lincoln, 1765

Nothing about the above quote is true. None of it. We stole the idea from a social media trend where pictures of historical figures are accompanied by satirical text. This was not a one-off ad lib, however. We did put some effort into crafting lines that used a parlance and cadence that sounded literary, inspirational, and plausible enough to get the reader to go along with it right up to the very end for a little comedic payoff. In so doing, our goal was to blur the lines, to play with the reader's ability to make predictions about future events based on incoming information in the hopes of getting them to suspend disbelief just long enough to defy those predictions and, in the end, leave them with a lingering emotional experience. In short, our goal was to engage in a bit of theater.

We blur the lines here intentionally to illustrate the underlying point of this section – that increasingly, the spaces between theater and fact in our information sharing venues are blurring, and this is hampering our ability to vet information. This blurring creates opportunities for people to intentionally muddy the waters, producing confusion and turmoil for political ends. Let us now give three ethnographic examples illustrating the phenomena.

Muddying the Waters: The Case of Cassandra

Cassandra is a 58-year-old woman living in Central Florida's Gulf Coast in a small ranch house she shares with her boyfriend. Jonathan is a 62-year-old disabled veteran who works part-time as a handyman and fixes motorcycles on the side. They live in a small community about five miles from the beach. An American flag flies on their front porch, and another one is tacked above Jonathan's workbench in his garage.

Cassandra comes from three generations of farmers in rural New Jersey but decided in her teens that farming was not in her future. After graduating high school, Cassandra decided to leave New Jersey and work her way around the country until eventually coming to rest in Florida, where she has lived for over 30 years. Cassandra never had any children and never attended college. She has made her living working in the restaurant business as a waitress and tending bar.

Cassandra recently passed through a crucial juncture in her life. She had several health scares over the past year related to the same hereditary heart disease that claimed her father's life six years ago. After watching him deteriorate and facing her own health issues, Cassandra has begun to make changes in her life. She quit smoking and planted a vegetable garden in her backyard. She has started walking regularly and is active in her community. She takes great pleasure sharing her harvest with friends and those in need nearby.

Cassandra has also taken a renewed interest in the Christian faith of her childhood and now attends church regularly. She started a prayer group that she continues to lead. Cassandra has also become more involved in local politics by attending council meetings and recently established Facebook and Twitter groups aimed at organizing community support for important local issues. Through these efforts, Cassandra has gained prominence in her community, and that has given her a sense of purpose. She feels both connected and valued.

Before, she felt she faced a system that left her feeling powerless and alone. The health costs her father incurred in the final years of his life were staggering. Then she was left with the additional costs of his funeral after his death. She sees a system that is biased, broken – or worse, rigged – against people like her. People in her social circle have often had to make tough choices between receiving medical care and paying bills or buying groceries. Then other people told them their situation was the product of bad character, lack of education, and poor choices. In 2016 she sees an opportunity to help wipe the slate clean.

Cassandra starts "Patriots for America," a grassroots movement dedicated to organizing voting momentum to change the status quo in Washington. Her goal is to organize voters to oust establishment politicians and usher in a new era of representation. She begins to do online research to identify candidates who are Washington outsiders, with a proven track record of successes independent of the political arena, namely, business. She believes these people will right the sinking ship of the American political economic system:

> "America has always relied on hard-working people," reads one Facebook ad she sees. "If you remove jobs, you'll remove our country from the map."

That statement rings true for Cassandra. She is working full time as a waitress and still needs to find a way to pay for insurance and medication co-pays. If it were not for the extra income from her boyfriend's disability and odd job work, they would not be able to make ends meet. The text is accompanied by a picture of candidate Trump in a hard hat surrounded by Pennsylvania coal miners holding signs that read "Trump Digs Coal."

The ad comes from a group called "Being Patriotic," and it advertises an event called Miners for Trump to be held in Pennsylvania. Cassandra investigates their home page and sees that they have over 200 000 followers and have successfully organized several rallies in the Northeast. She clicks the "Follow" button, and before long she receives a message from a representative of the organization:

> My name is Ted and I represent a conservative patriot community named as "Being Patriotic" ... So we're gonna organize a flash mob across Florida to support Mr. Trump. We clearly understand that the elections winner will be predestined by purple states. And we must win Florida. We got a lot of volunteers in 25 locations and it's just the beginning. We're currently choosing venues for each location and recruiting more activists. This is why we ask you to spread this info and participate in the flash mob.

Cassandra feels the same spark of connection and purpose as when she delivers her vegetable care packages and attends city council meetings. She now knows she is a part of something larger than herself, something meaningful, a part of a movement that may bring about a more hopeful future. She replies to the message and begins a back-and-forth on Facebook messenger with the person known as Ted.

In the ensuing months, news that Being Patriotic and other online groups are Russian actors would reach Cassandra. She finds these reports revolting and divisive. Trump supporters like her are now under attack. Perhaps this signaled the beginnings of being persecuted for her beliefs.

After all, she had personally interacted with these people online, and though she had never met them in person, their causes resonate with her convictions. She knows that they are patriotic Americans, just like the other Americans who have shown up at the rallies she has helped organize.

With muddying the waters, a primary tactic used is to disseminate information that allows outside confirmation of an extreme position and promotes a sense of threat so people assess that information quickly based on their immediate social group. As we will show in the final vignette, these state actors will also generate counter-narratives so that people are confused about what to think and thus tend to rely on one-sided positions that seem to offer clarity and certitude. These divisive efforts can then be fleshed out in the real world to help form movements around more extreme ideas. The end goal is to generate division between people and reduce trust in the state.

Missing It: The Case of SSgt Michaels

Staff Sergeant (SSgt) Michaels is a 28-year-old US Marine deployed with his platoon to North Africa. He, along with 2000 service members from allied nations, is here to participate in "African Ghost," a massive joint training exercise that brings together conventional and special forces for the purpose of honing their craft in a coordinated mission simulation that takes place in a real-world environment. Exercises like these typically last about two to four weeks, take a year to plan, and have become standard in the past two decades as increasingly there has been a shift in focus away from unilateral action toward allied partnership when military action is used. Exercises like African Ghost and others provide a mechanism for military forces to work together and build versatility.

It is March of 2017. SSgt. Michaels is currently in the middle of a heated divorce with his wife of three years. He is worried that he may lose his 8-month-old daughter in the split, and during breaks in the action, he can usually be found in the corner of the sandy courtyard on base, chain-smoking American Spirit cigarettes while answering emails from his lawyer on his smartphone. He looks confident in his uniform and dark tinted aviator sunglasses when addressing his squad, but in the unguarded moments of the breaks, he chews his fingernails and rubs the bags under his eyes.

Michaels and his team are part of a Marine Intelligence unit trained to collect, integrate, and analyze information from military sources as well as publicly available sources such as news, blogs, and social media. This information can be used to gain an understanding of the enemy's location and plan of action and how the local population is perceiving and responding to the military presence. Michaels' team forms part of the command post operations, which is located in a room where a network of computers and servers project onto screens simulated satellite imagery, drone feeds, maps, and the notional locations of the units who are participating in this exercise.

In a separate room sits the White Cell – the people running the exercise simulations and those in charge of throwing curveballs like equipment failure, adverse weather conditions, ambush attacks, and other surprises to keep the command staff on their toes. In the back corner of the White Cell sits a team of intelligence operational environment designers. It is their job to ensure that the information that Michaels and his team interact with is replicated as close as possible to what they would see in a real-world operation. To accomplish this, the operational design team consists of writers, graphic artists, and media production specialists who are able to produce content, graphics, photographs, and videos that mirror the tone and style of various real-world news sites and blogs that respond in real time to the exercise events as they unfold. Their sole purpose is to create an information environment that mimics the real world so well that the training audience forgets they are looking at reports of a notional reality.

On the second day of the exercise, Major Youssef Mahmoud, one of the White Cell officers in charge of overseeing the Intelligence Division, is called aside and given a special task by his commanding officer to test the joint intel cell's readiness to detect and counter misinformation operations similar to those used in the 2016 US election. He is asked to simulate a "fake news hack." His objective is to create a false news scenario that the notional antagonist for the exercise, Amir Sham, has already been killed by coalition forces.

Mahmoud agrees, and the design team gets to work. Mahmoud orders that the stories they create work as a direct counter-narrative to the military intelligence (MI) reports that will come the next morning, which will not reflect a raid or any activity relating to Amir Sham.

The next day, the terrorist blog posts an entry saying that their illustrious leader has been martyred. Shortly after, a tweet appears containing a picture of the role player playing Amir Sham and holding an AK-47. It is juxtaposed against a picture of a different role player in a ditch, presumably dead. Mahmoud orders that the face and head in the second image be digitally blurred and that any other identifying features be obscured to help sell the ruse that the two pictures show the same individual. "Terrorist cell leader Amir Sham killed in overnight raid" reads the tweet, which also contains a link to the terrorist blog post.

The design team also publishes an article on the exercise's mainstream news site, such as Global News Network (GNN), that was patterned graphically and stylistically after the BBC. The article uses the blog post and tweet to report the story but does not supply any new or otherwise confirming information.

At the day's briefing, SSgt Michaels and his team report the ruse as fact. Their PowerPoint brief includes the picture of the role player and the blurred-out body in the ditch and a screenshot of the tweet and utilizes as confirmation sources the bad guy blog and the GNN article.

The training commanders are taken aback. Amir Sham's death effectively signals the end of an exercise that has taken a year to plan. The senior officers call for an explanation, and Major Mahmoud steps in to explain that the requested scenario was meant to test readiness against the emerging use of social media to spread disinformation in the intelligence environment.

SSgt Michaels, still standing at the front of the room, clinches his teeth and shifts his weight side to side. He craves a cigarette as Major Mahmoud walks the SSgt back through his own slides, pointing out that what looked like data triangulation was in fact circular reporting. The blog is a lie, the picture fabricated, and the GNN article parroted both without verifying against an outside source. Major Mahmoud points to the overall weakness in the information vetting process as something that needs to be addressed in future training scenarios.

Later that evening, SSgt. Michaels talks things over with his gunnery sergeant. He says he knew that open source intelligence (OSINT) story and MI were in conflict with one another. Then he says:

> The OSINT report was just easier to read. Everything was right there. MI comes across on a dot-matrix printer, it's in all caps, it's not interactive, it's really difficult to get through a ton of it, quickly, and it's not always as up to date as the OSINT. That OSINT looked legit though. It was sourced, it had graphics and pictures, it had time stamps.

Even among highly trained professionals, quality of production and visual presentation plays a role in deciding which information to trust. The example of SSgt. Michaels' experience is not just a fluke; it is substantiated with research showing that false news that conforms to users' preconceived notions of the truth is 70% more likely to be shared than true news and that the quality of the medium shapes our perceptions of the validity and importance of the information and its source. In this case, the colorful graphics, gritty picture, and interactivity provided an easier path to follow than the all-caps block text of the intelligence reporting format. Moreover, SSgt. Michaels was forced to choose between conflicting narratives, and resolving which one was accurate required critical assessment that is not easy to do in time-pressured settings like command operations.

Wag the Dog: The Case of Fedor the Troll

It is tempting to theorize a model that explains how salacious and visually compelling information will spread further, be more convincing, and have a longer online half-life than true information that is not packaged quite as neatly. But these models run the risk of casting the Cassandras and SSgt. Michaels of the world as victims of their psychobiology – fish that bits the hook after chasing a shiny object wriggling through murky waters. Such approaches discount some of the complexity in the physical and cognitive domains that influence perceptions. This is one area where ethnography can fill in the gaps by adding an empirical dimension of human experience that is crucial to understanding both the how and the why of behavior in specific settings.

Fedor is a 24-year-old Russian software designer looking for work. He has just completed an 18-month internship at VxTx, a small voice-over IP service started by Pyotr, a friend of his oldest brother. Pyotr ran the start-up out of a warehouse next to the train yard about an hour walk from the apartment building where Fedor grew up.

Fedor is the youngest of five children. He grew up in what amounted to a single-parent household in Ramenskoye, a town south of Moscow. His father, Aleksandr, drove a logging truck for a Russian milling company; the job took him away from the family for months at a time. His job was treacherous, hauling 38 tonnes of freshly hewn pine, spruce, and cedar from Siberia's forests in a 1970s Soviet era semitruck over an unpredictable terrain of frozen marsh, partially thawed rivers, and mud-slicked roads.

His father was extremely proud of his job. Russian forestry was a major supplier of wood worldwide, and the timber renowned for its strength and superior quality. Aleksandr approached his job with a proud smugness that came with obliging the Western world's need for what Russia had to offer. Timber represented the bounty of Russia and was a symbol of superior socialist might. He tried to instill this pride in his sons in three or four days a month he was home in between runs.

Fedor's father did not provide much of an income for the family. After the bills were paid, there was little left for anything extra. His mother worked at the local convenience store in the winter to help offset the cost of using the electric stove to heat water for washing dishes and bathing when the central heat was turned off or broken. But their poverty did allow the boys to qualify for federal tuition waivers for college. Fedor's brothers majored in business, while Fedor studied computer science. Fedor had dreams of joining ROSCOSMOS, the Russian space program, as a programmer designing software for the International Space Station. He was a bright student, but his grades fell short of qualifying him for a competitive graduate school tuition waiver.

At the start of his internship, Fedor is charged with writing code to help VxTx integrate cryptocurrency into their smartphone app. Cryptocurrency is quickly becoming a popular trend among Russia's youth as it offers them a means of circumventing the regulations of the central bank and allows users to obscure transactions for goods and services they would rather not make public. Cryptocurrency exists in a realm beyond restriction and confiscation in an online space governed by a set of negotiations that are neither capitalist nor socialist. This emerging frontier is where the wild of a burgeoning digital economy meets with the hard assets of the real world.

During the internship, Fedor begins to idolize Pyotr, founder and CEO of VxTx. Pyotr started VxTx – a Latin alphabet stylization for Vox transmission, with a nod to punk music roots – as a means of circumventing the government's practice of monitoring private phone conversations. VxTx already has over 100 000 subscribers, and cryptocurrency represents the last step in making these transactions truly and legally untraceable.

Pyotr dresses like a punk rocker – platinum blonde buzz cut, a dog chain necklace, faded black Sex Pistols T-shirt, tight black jeans, and Converse Chuck Taylors. Pyotr admires the US ideal of free speech and loves watching torrented files of American TV shows like *ER*, *Friends*, and *Lost*. However, Pyotr is also a fierce Russian patriot, given to frequent tirades on how the expanding capitalist system of the West encroaches on Russian affairs.

Influenced by Pyotr, Fedor begins to listen to punk music. He also decorates his laptop with a Dead Kennedys sticker, a ROSCOSMOS logo, and a CCCP sickle and hammer decal.

Pyotr finally notices. One day he comes over to Fedor. "Fedor, why are you here? For what we pay you, why do you volunteer to stay late to clean after we have gone home?"

Fedor freezes as he struggles for a plausible explanation. His eyes dart to the propaganda poster of Stalin tacked up on the wall behind Pyotr's desk and then to the Russian military Ushanka and bayonet on his shelf.

Pyotr tracks Fedor's gaze back to his office. "I think you have a desire to serve your country, is that it? To serve Mother Russia?" Pyotr starts walking to his office, and Fedor follows.

Pyotr moves methodically to his desk. "Don't think I haven't noticed you snooping in my office, my friend. Looking at my book shelves. You are trying to figure out 'who is this guy,' I think."

He opens a desk drawer. "Well, these will help you; at least they did me." He holds out two old paperbacks with frayed edges. "My father gave me Stalin's books when I left for University. I've gotten my use of them."

As Fedor leaves the office, Pyotr brays,

> Chtoby imet' bol'she, my dolzhny proizvodit' bol'she. Chtoby poluchit' bol'she, my dolzhny znat' bol'she.

> To have more, we must produce more. To produce more, we must know more.

Four months later, his internship over, Fedor moves to Moscow, resumé in hand. Every day he goes to the business district and heads to an upscale coffee shop with elegant furniture, free Wi-Fi, and a decent cup of Turkish coffee. He hopes opportunity might present itself. He places Pyotr's books at the edge of table and positions his laptop so people can see him polishing the code from his senior project, reading about cryptocurrencies, and surveying blogs about how Western capitalism undercuts Russia's return to world prominence.

> East Bay Ray was underappreciated in his time.

Fedor looks up from his laptop to see a balding, heavy set man in dark slacks, a polo shirt, and a sport jacket wearing mirrored sunglasses pulling up a chair to his table.

The heavy set man continues, "His work in the 1970s was a melodic deconstruction of the modal forms used by white American rock musicians as a critique of their cultural appropriation of black slave spirituals to sell records."

The man motions for coffee. Fedor sits stunned, trying to make sense of what this man just said.

"The Dead Kennedys." The man gestures to the sticker on Fedor's laptop. "East Bay Ray, the guitarist."

"Ah! Yes!" Fedor responds.

"On the other hand," the man continued, "Stalin's genius is well known."

Fedor again stumbles to respond.

The man laughs. "I'm sorry, I have put you on the spot. My name is Mishka."

He extends his hand to Fedor. "I'm a recruiter for a software company. We work mostly with social media and news aggregation."

He hands Fedor a business card. "I've seen you in here for the past few weeks. I wonder if we might talk."

Fedor sits up straight. He launches into his qualifications and work experience. He hands Mishka a resumé. Mishka places it facedown on the table and then asks, "So, tell me, what is your philosophy on cryptocurrency?"

Next, "Why the space program, was it national pride or professional interest?"

Finally, "How's your English?"

Fedor switches to heavily accented English. "No problem."

Fedor moves to Saint Petersburg and accepts a position at the Russian Internet Research Agency. He is paid roughly 50 000 Russian rubles a month, substantially more than the 32 000 per month offered elsewhere. At the start, all he has to do is become familiar with political issues in the United States, watch American pop TV, and lurk on the US social media sites.

By the time of the 2016 election, Fedor is part of Russia's Веб-бригады – web brigade –more colloquially known as Russia's troll army. He creates online personas on the US news sites and posts commentaries that match conservative narratives, and he creates divisive discussions on Facebook and Twitter using multiple fake personas. He gets well paid to do it. One of his favorite creations is a meme to suppress voting featuring a Left-leaning US musician telling people to "Stay home. Avoid the line. Vote from home." A pop musician, not a punk musician.

The Russian web brigade is made up of Russian citizens with backgrounds in computer programming, product marketing, social media networking, journalism, and graphic design that are part of Kremlin's propaganda machine. Their chief role is to influence the tide of thought of any populace – domestic or international – toward Kremlin's agenda.

The term "Internet troll" or "troll" has historically been used as Internet slang, but recent relevance has elevated its use, and it has become codified in the American lexicon. A troll is someone who intentionally engages in online discussion to sow discord, redirect, and inflame online discussion groups with the overall goal of provoking an emotional response. Trolls are effective precisely because their attacks provoke strong emotional responses that serve to cloud judgment of facts. Professional trolls such as Fedor are trained to elicit emotional reactions as a means to dilute social momentum on hot topics by using distraction, misdirection, and misinformation techniques.

Conclusion

The three vignettes illustrate a more complex reality than often considered in modeling exercises. Yet at the same time, certain commonalities emerge. The three vignettes show specific actors interacting with specific information environments. Looking at the actors, economic, emotional, political, and ideological factors motivate their actions and their interpretations of the information environment. On the environment side, there are the immediate mechanisms that provide content and feedback to actors. But there is also the context of the specific information environment and linkages among elements in that environment. The influences of context and connections are harder for actors to discern – that is one element of the interaction. The other feature of the interactions between actors and environments is the intersection of emotions and social media immediacy, as well as how feedback from the environment confirms rather than questions economic and political factors. Emotional immediacy in a confirming social context is a main feature for why actors do not see the larger contextual determinants of their behaviors. Rather than being agentive, they are passively shaped because they lack access to these features of their environments.

This summary points to a role where qualitative approaches of anthropology can come together with the computational approaches common to modeling and to assessing big data. Actors need a signal to assess where the social media feed they are getting fits into larger contextual determinants of the information environment. Here we propose the creation of a new approach that combines geographic information services with computational approaches while at the same time bringing in qualitative and critical insights that come from anthropology's approach to culture.

The method of metainformation systems (MIS) is an approach to capture, manage, analyze, and present the cultural data within a specifically defined information environment. MIS conceives of any information environment as a cultural product – something produced by a group of people that feeds back to shape the thoughts, emotions, language, and behaviors of that group. It is not enough to just look at the physical, computational, and cognitive features of an information environment. We also need a tool that can assess the cultural dimension of the information environment.

There are several key elements to doing this assessment. First, MIS recognizes that there is not one "truth" out there. Different information environments will function according to different ideas about what counts and what can be excluded from considerations. These larger framings, often presented as ideologies, shape consensus with the information environment and are rarely questioned. That actors and the cultural contexts of information environments are shaped by larger framings and ideologies is a central part of what makes actors within that environment open to manipulation by people who know how to exploit larger framings.

Second, on the anthropological side, there is a need to move toward assessing larger etic perspectives. Anthropology makes a distinction between emic and etic, where emic is the insider's perspective and etic is the outsider or researcher's perspective. However, research in metamodernism, competing ideologies, and science and technology studies all point to the recognition that these etic perspectives are in themselves ideologies that shape the interpretations of the researchers. For example, the worldview that comprises "science" is an etic perspective but one that is increasingly contested in many of today's information environments. MIS can work to assess that etic framing of "science" as it exists in different information environments. But to be able to do that, there is a need to go meta – to recognize that ideologies or framings that look truthful from the emic perspective (they make sense, are consistent, lead to consensus, and so forth) are a particular meta dimension of that specific information environment.

MIS relies on the integration of multiple methods. One key approach, which we call MIX, begins with assessing key features quantitatively and deriving a latent variable based on these cultural items. The latent variable is thus something that moves from emic (what actors perceive) to etic (an assessment of a latent, often invisible factor that accounts for the variation in the cultural items). But that latent variable is also assessed using a qualitative approach that uses a systematic approach to identifying features of a specific cultural theme (e.g. the latent cultural variable). Coding in qualitative research is basically an operationalization of a latent variable (the theme) that emerges from the data. One important innovation that the MIX tool offers is its ability to judge the latent cultural variable not just via item data (actors' emic views) but also via a qualitative assessment of that latent variable in relation to cultural dynamics that shape the information environment.

The main point for this chapter is that agent-based modeling needs a computational method to better consider features of the cultural environment within which human agents operate. With continued development, MIS offers a way to assess specific information environments better than existing techniques, which rely on calculations from surveys and scales. Pushing further, MIS integration can then produce assessments of how close or far from a particular meta-narrative individuals are and thus provide an error-feedback mechanism for people. By producing a latent variable measurement that is then adjusted to match overarching themes (MIX), it becomes possible to anticipate providing an assessment to individuals on whether a particular social media feed they have chosen matches up with larger cultural themes. For example, even if the messaging matches what Cassandra believes, does that messaging actually come from local actors? Does SSgt Michaels' particular intelligence assessment line up with military consensus or social media consensus? Are Fedor's actions consistent with punk values or with Russian state values?

What is clear is that anthropology and computational modeling need to come together to better assess the cultural domains that shape our everyday lives – not just our everyday lives but also the lives of people who come from different backgrounds and espouse different values but are nonetheless interested in learning more from their information environments. Applied modeling, using an integrative theoretical and methodological approach typified by MIS, is an important component to taking advantage of the emerging technologies, big data, and changing social interactions that typify our global world today.

References

Agar, M. and Wilson, D. (2002). Drugmart: Heroin epidemics as complex adaptive systems. *Complexity* 7 (5): 45–52. https://doi.org/10.1002/cplx.10040.

Department of the Army (2013). *Inform and influence activities (FM 3-13)*. Washington, DC: Department of the Army. Retrieved from https://usacac.army.mil/sites/default/files/misc/doctrine/CDG/cdg_resources/manuals/fm/fm3_13.pdf.

Franz, M. and Matthews, L.J. (2011). Social enhancement can create adaptive, arbitrary and maladaptive cultural traditions. *Proceedings of the Royal Society B: Biological Sciences* 277 (1698): 3363–3372.

Hofer, L.D. (2013). Unreal models of real behavior: the agent-based modeling experience. *Practicing Anthropology* 35 (1): 19–23.

Hutchins, E. (1995). *Cognition in the Wild*. Cambridge: MIT Press.

Joint Chiefs of Staff (2013). *Information Operations (JP 3-13)*. Joint Chiefs of Staff. Retrieved from http://www.jcs.mil/Portals/36/Documents/Doctrine/pubs/jp3_13.pdf.

Lende, D.H. and Downey, G. (2012). *The Encultured Brain: An Introduction to Neuroanthropology*. Cambridge: The MIT Press.

Shearer, E. and Gottried, J. (2017). *News Use Across Social Media Platforms 2017*. Pew Research Center http://www.journalism.org/2017/09/07/news-use-across-social-media-platforms-2017/.

Statista (2018). *Social Media Usage in the United States*. Statista. Retrieved from https://www.statista.com/study/40227/social-social-media-usage-in-the-united-states-statista-dossier/.

Vosoughi, S., Roy, D., and Aral, S. (2018). The spread of true and false news online. *Science 359* (6380): 1146–1151. https://doi.org/10.1126/science.aap9559.

10

Using Neuroimaging to Predict Behavior: An Overview with a Focus on the Moderating Role of Sociocultural Context

Steven H. Tompson[1,2], Emily B. Falk[3,4,5], Danielle S. Bassett[2,6,7,8], and Jean M. Vettel[1,2,9]

[1] Human Sciences Campaign, U.S. Army Research Laboratory, Adelphi, MD 20783, USA
[2] Department of Bioengineering, University of Pennsylvania, Philadelphia, PA 19104, USA
[3] Annenberg School for Communication, University of Pennsylvania, Philadelphia, PA 19104, USA
[4] Department of Psychology, University of Pennsylvania, Philadelphia, PA 19104, USA
[5] Marketing Department, Wharton School, University of Pennsylvania, Philadelphia, PA 19104, USA
[6] Department of Electrical & Systems Engineering, University of Pennsylvania, Philadelphia, PA 19104, USA
[7] Department of Neurology, University of Pennsylvania, Philadelphia, PA 19104, USA
[8] Department of Physics & Astronomy, University of Pennsylvania, Philadelphia, PA 19104, USA
[9] Department of Psychological and Brain Sciences, University of California, Santa Barbara, 93106, USA

Introduction

How do we predict how an individual will behave in a particular situation? Across several decades, social scientists have identified many self-report measures that account for individual variability in behavior, yet a large percentage of the variance remains unaccounted for by these introspective reports (Armitage and Conner 2001; O'Keefe 2018). Recent advances in analytic approaches and computational tools have provided new, complementary avenues to investigate this difficult question. Noninvasive neuroimaging approaches (e.g. functional magnetic resonance imaging, fMRI; functional near-infrared spectroscopy, fNIRS; and electroencephalography, EEG) measure brain activity while participants view stimuli and make decisions, providing a powerful tool to capture objective measurements of individual differences during task performance (Berkman and Falk 2013; Tompson et al. 2015). Analytic tools for quantifying patterns of activation within and between brain regions further advance the power of neuroimaging approaches to predict how individuals will behave (Kriegeskorte 2011; Bassett and Sporns 2017), how they will interact with one another, and how groups of individuals will make decisions. In short, the foundation of social neuroscience posits that measuring brain activity provides

Social-Behavioral Modeling for Complex Systems, First Edition.
Edited by Paul K. Davis, Angela O'Mahony, and Jonathan Pfautz.
© 2019 John Wiley & Sons, Inc. Published 2019 by John Wiley & Sons, Inc.
Companion website: www.wiley.com/go/Davis_Social-Behavioralmodeling

access to psychological processes and neural circuitry that may serve as the underlying mechanisms that explain individual differences in behavior.

In this review, we first discuss evidence that demonstrates the association between brain activation and individual decisions and behaviors. Studies within the domains of health and consumer behaviors have identified a consistent set of brain regions associated with individual decisions and behaviors. These brain regions have been implicated in processing information about the reward value of choice and behavioral options (valuation) as well as processing social information about the mental states of others (mentalizing). Second, we discuss evidence that these same brain regions are also linked to aggregate behavior for groups of individuals. Third, we discuss evidence that these brain regions are associated with how individuals will behave in social interactions. Finally, we discuss evidence that the association between brain activation and behavior is moderated by social factors including social network position, culture, and socioeconomic status. Throughout, we highlight recent advances that leverage multivariate and network analysis approaches that emphasize different components of brain activity patterns to understand the neural mechanisms underlying social behavior.

The Brain-as-Predictor Approach

The human brain is a massively interconnected network consisting of 86 billion neurons with trillions of connections between neurons (Azevedo et al. 2009). Human cognition requires coordinated communication across macroscopic brain systems composed of both gray matter (cell bodies) and white matter (axons; Bassett and Sporns 2017). The gray matter is typically divided into brain regions composed of large groups of adjacent neurons that have similar properties, and these regions demonstrate specialized information processing and knowledge representation. The white matter provides the structural connections between distant brain regions and is often described as the wiring in the brain (Vettel et al. 2017). Together, brain networks support cognition and human behavior by communicating information among brain regions for integrated processing and rely on the structural connections to enable efficient and rapid responses across distant brain regions (Passingham et al. 2002). Consequently, coordinated communication across the brain is fundamentally constrained by specialized processing in individual brain regions and patterns of interconnections reflected in functional connectivity of synchronized activity between regions.

The brain-as-predictor approach measures brain activation while individuals evaluate information about various behavioral options, and then uses that activation to predict subsequent behavioral outcomes, often over the course of weeks, months, or even years (see Cascio et al. 2015b; Falk and Scholz 2018; Knutson and Genevsky 2018 for a review). The majority of these studies use functional magnetic resonance imaging (fMRI) to measure brain activation, although other imaging modalities such as electroencephalography (EEG), functional near-infrared spectroscopy (fNIRS), magnetoencephalography (MEG), or positive emission tomography (PET) could also be used. To date, the brain-as-predictor approach has been applied to predict both individual behaviors and aggregate group behaviors in diverse domains, including health (Falk et al. 2015b; Cooper et al. 2017), consumer (Levy et al. 2011; Genevsky et al. 2017), and political behaviors (Rule et al. 2009). Importantly, in many of these studies, brain activation provides additional information about the likelihood of engaging in a particular behavioral outcome, beyond what is explained by self-reported intentions, preferences, and other questionnaire items (Venkatraman et al. 2015; Falk et al. 2015b; Genevsky et al. 2017).

In addition to improving our ability to predict behavior, the brain-as-predictor approach can also yield important insights into the psychological processes underlying these behaviors. Studies using fMRI to predict behaviors frequently implicate three sets of brain regions that are broadly involved in processing self-relevance, social relevance, and overall value of incoming information (see Figure 10.1). In particular, ventral medial prefrontal cortex

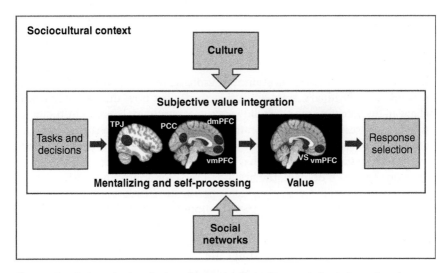

Figure 10.1 Brain activation. Brain activation implicated in processing information about the self, others' mental states, and reward value can be used to predict how people will behave and make decisions. The relationship between brain activation and behavior is in turn moderated by sociocultural factors, including culture and social networks. TPJ, dmPFC, and PCC are frequently implicated in thinking about the mental states of others (mentalizing). vmPFC and PCC are also implicated in thinking about the self. Ventral medial prefrontal cortex (vmPFC) and VS are frequently implicated in processing reward value. Specialized processing in these regions as well as communication of information between brain regions is thought to directly support behavior and decision-making, at least in part through integration of self and mentalizing processing into a subjective value signal.

(vmPFC) and posterior cingulate cortex (PCC) have been implicated in processing the relevance of information to the self (Denny et al. 2012; Martinelli et al. 2013), such as whether a word or product describes the self or is part of the individual's identity (Kelley et al. 2002; Kim and Johnson 2012). The dorsal medial prefrontal cortex (dmPFC) and temporoparietal junction (TPJ) have been implicated in social processing, including considering the mental states of others (i.e. mentalizing; Denny et al. 2012; Saxe and Kanwisher 2013). The vmPFC and ventral striatum (VS) have been implicated in integrating information from different sources to compute a signal of the subjective value of the information (Martinelli et al. 2013). It should be noted that these processes often overlap and that several of the brain regions listed are implicated across these functions as well. Indeed, researchers have argued that these brain regions work together to process information about the fit of the behavioral options to an individual's values, beliefs, and goals as well as to broader social norms, which are then integrated into a single value signal indicating the subjective value of the behavioral options being considered (e.g. whether to quit smoking, choose product A or B, donate to a charity, or share a news article; Tompson et al. 2015; Scholz et al. 2017; Knutson and Genevsky 2018).

Predicting Individual Behaviors

The majority of research using brain activation to predict behaviors has focused on using an individual's brain activation to predict how that individual will behave in the weeks or months following the experimental session. Although a diverse set of research has used brain activation to predict behavior, we highlight two domains that have successfully used brain activation to predict behavior outside the scanner: health behavior and consumer behavior.

Within the health domain, the brain-as-predictor approach measures brain activation, while an individual evaluates persuasive health messages and then tracks their behavior over the next week or month. A study by Falk et al. (2010) measured individuals' brain activation while they viewed health messages promoting the benefits of wearing sunscreen. They then tracked individuals' sunscreen usage the week after the study and compared it with their sunscreen usage the week prior to the study. Participants who recruited vmPFC more during message evaluation were more likely to use sunscreen afterward, even after controlling for sunscreen usage the week prior to the study (Falk et al. 2010; Vezich et al. 2017). Similarly, research in other health domains has found that individuals who exhibit greater mPFC activation during message exposure are more likely to engage in health behaviors endorsed by those messages, including increasing physical activity (Falk et al. 2015a) and reducing smoking (Chua et al. 2011; Falk et al. 2011; Wang et al. 2013; Cooper et al. 2015).

Why does vmPFC activation predict behavior change? Research suggests that vmPFC processes the relevance and value of the messages to the individual. In particular, the subregion of vmPFC involved in predicting behavior change overlaps with subregions of vmPFC known to be involved in both thinking about the self and processing the value of objects to the self (Cooper et al. 2015). Moreover, self-affirmation prior to being exposed to health messages promoting increased physical activity led to greater vmPFC activation (relative to presenting the messages without the self-affirmation) and greater subsequent behavior change (Falk et al. 2015a), providing experimental evidence that self-processing plays a key role linking vmPFC activation and behavior change (see Figure 10.2).

In addition to predicting health behaviors, brain activation can predict consumer behaviors. Complementary experimental designs on consumer

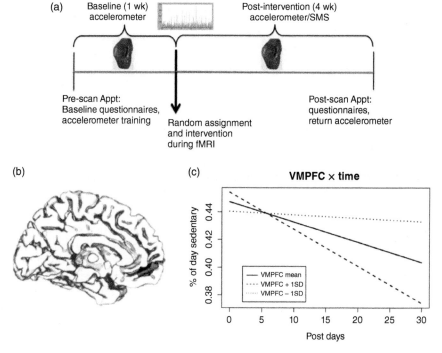

Figure 10.2 Physical activity before and after health messages. Falk and colleagues measured physical activity before and after participants were presented with health messages promoting exercise while their brain activity was measured in an MRI scanner (a). Participants exhibited greater activation in vmPFC when they were first given an opportunity to affirm positive attributes about the self (b), and individuals who recruited vmPFC more during message exposure were more likely to have fewer sedentary days in the month after the study (c). Source: Adapted from Falk et al. (2015a), by permission of Oxford University Press.

behaviors examined brain activation, while participants evaluated explicit appeals designed to persuade the participant to purchase a particular product (e.g. Genevsky et al. 2017). Consumer products that elicited greater vmPFC and VS activation were more likely to be chosen by individuals (Knutson et al. 2007; Levy et al. 2011). Interestingly, brain activation can predict consumer choices even when participants are not explicitly evaluating the choice options. Levy et al. (2011) had participants passively view consumer products in an MRI scanner without making any explicit judgments about the products and then had participants choose which products they wanted to own after the scan. The researchers could accurately predict which product a participant would choose based on activation in vmPFC and VS during the passive viewing task (Levy et al. 2011).

Across both domains, brain activity in vmPFC and VS reliably predicts individual differences in health change and consumer behaviors. These regions have been implicated in self-related and reward processing (Lieberman 2007; Adolphs 2009; Denny et al. 2012; Bartra et al. 2013; Schurz et al. 2014), suggesting that people are more likely to engage in behaviors or make choices that are high in self-relevance and subjective value. Importantly, in many cases brain activation predicts behaviors with an accuracy above and beyond those obtained from self-report measures of preferences or intentions (Falk et al. 2011; Genevsky et al. 2017). Objective measures of subjective value indexed by vmPFC and VS might therefore be providing novel insight into the value or relevance of the outcomes to the individual's salient values, beliefs, and goals.

Interpreting Associations Between Brain Activation and Behavior

How can understanding the relationship between brain activation and behavior improve our understanding of the psychological processes underlying health and consumer behaviors? In this review, we highlight three core neuroimaging analysis approaches that examine how the brain processes and represents the content of persuasive messages and how this processing predicts behavior. The first and most common univariate analysis compares brain activity between two conditions, where the only difference between them is the cognitive process of interest. This type of analysis underlies the research reviewed in the last section, identifying that brain activity in vmPFC and VS reliably predicts individual differences in health change and consumer behaviors. The second approach complements the univariate analysis by looking at multivariate patterns within a region. The core intuition is that the knowledge represented in a region may be distributed across the smaller units of brain tissue within a region (known as voxels in fMRI data). Whereas the conventional univariate analysis simply averages the functional activity across all subcomponents of an imaged brain region, a multivariate analysis assumes that understanding how a region gives rise to psychological processes of interest is coded by the distributed pattern in a region. Finally, our review highlights a third neuroimaging analysis approach that uses connectivity methods to estimate task-relevant network activity. Connectivity research posits that synchronized activity between regions demarcates the integration of information across regions, and thus the functional network dynamics capture the spatiotemporal processes necessary for the brain to enable behavior.

Multivariate pattern analysis (MVPA) and representational similarity analysis (RSA) are two analysis methods to quantify how patterns of activation across voxels within a brain region (or across the whole brain) relate to behavior. They use the relative similarity of neural activity between pairs of trials to make inferences about the content encoded in that region (Norman et al. 2006; Kriegeskorte 2011; Nili et al. 2014). For example, Pegors et al. (2017) used RSA to investigate how vmPFC represents information about persuasive messages. The multivariate patterns of activation within vmPFC successfully differentiated information about whether persuasive messages contained information about health, social, or risk consequences of smoking cigarettes (Pegors et al. 2017). Furthermore, individual differences in the representation of message content in vmPFC predicted whether individuals would reduce their smoking behavior after the study (Pegors et al. 2017).

Recently, researchers have also begun to employ connectivity-based approaches to predict behavior across a diverse set of domains (Garcia et al. 2017; Muraskin et al. 2017; Passaro et al. 2017; Brooks et al. 2018). Within the behavioral change literature, researchers have studied brain connectivity patterns while participants viewed persuasive health messages, and results demonstrated that greater connectivity within a network of brain regions associated with the processing of subjective value was linked to greater likelihood of engaging in the health behaviors being promoted in the messages (Cooper et al. 2018, 2017). These results suggest that integration of information about the subjective value of the messages to the individual is an important pathway through which persuasive messages lead to successful behavior change.

Collectively, all three of these neuroimaging analysis approaches suggest that the individuals are more likely to be persuaded by persuasive appeals, choose consumer products, and engage in health behaviors when they evaluate those options as more relevant and valuable.

Predicting Aggregate Out-of-Sample Group Outcomes

Brain activation in regions that predict behavior change at the individual level can also be used to predict aggregate behavior across groups of individuals whose brains are not scanned (Berns and Moore 2012; Falk et al. 2012, 2015a;

Genevsky and Knutson 2015). In a persuasive messaging task, brain activation in a small group of participants predicted the population-level success of a set of health messages (Falk et al. 2012, 2015b); however, the relevance of the message content moderated the relationship between brain activation and aggregate group response (Falk et al. 2015b). For instance, the relationship between the activation in vmPFC, dmPFC, PCC, and TPJ in a small group of participants who viewed anti-smoking messages and the percentage of individuals who clicked on an email link containing one of the ads (as part of a large-scale email campaign in the state of New York) was significantly stronger for ads that were smoking relevant than for ads that were compositionally similar but behaviorally irrelevant (Figure 10.3) (Falk et al. 2015b).

Additional research has shown that brain activation can also predict aggregate consumer behavior (Berns and Moore 2012; Genevsky and Knutson 2015; Venkatraman et al. 2015; Kühn et al. 2016; Genevsky et al. 2017). One study demonstrated that VS activation while a small sample of individuals listened to songs of relatively unknown artists predicted how popular those songs will be over the next three years, such that songs that elicited greater activation in VS sold more albums over the next three years (Berns and Moore 2012). Activation in vmPFC and VS also predicted crowdfunding outcomes (Genevsky and Knutson 2015; Genevsky et al. 2017). Images and descriptions for crowdfunding projects that elicited greater vmPFC and VS activation while a small group of participants evaluated the projects were more likely to receive enough investments to ultimately be funded (Genevsky and Knutson 2015; Genevsky et al. 2017). Additionally, ads for chocolate products that elicited greater activation in vmPFC and VS led to greater increases in chocolate sales in a supermarket where they were sold (Kühn et al. 2016).

While the above research focused on average activation in single brain regions, multivariate patterns can in some case better predict behavior than average activation within a single brain region or group of brain regions (Genevsky et al. 2017). Whereas models incorporating average brain activation in vmPFC and VS (as well as amygdala, insula, and inferior frontal gyrus) successfully predicted funding outcomes for projects on a crowdfunding website with 59–61% accuracy (significantly better than chance), whole-brain multivariate patterns were able to successfully predict funding outcomes with 65–67% accuracy (Genevsky et al. 2017).

Across domains, the studies described above show consistently that brain activation improves our ability to predict aggregate group behaviors. Brain activation predicts group-level popularity of music songs (Berns and Moore 2012) as well as the group-level success of persuasive health messages (Falk et al. 2012, 2015b), consumer product ads (Venkatraman et al. 2015; Kühn et al. 2016), and crowdfunding ads (Genevsky and Knutson 2015; Genevsky et al. 2017) even after controlling for self-report measures of behavioral intentions or preferences of the test sample in response to the messages. In many cases, brain

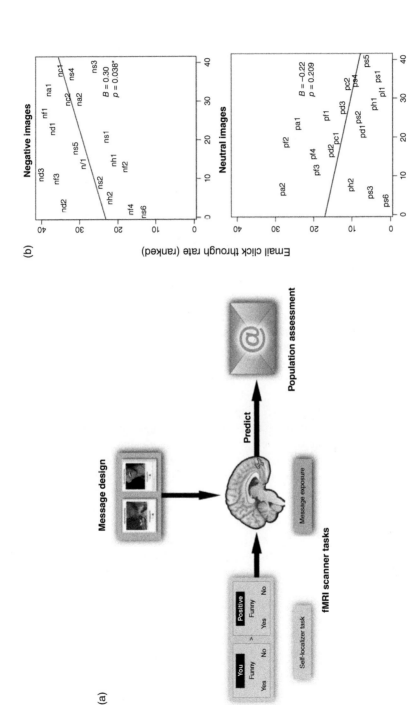

Figure 10.3 Brain activation. Brain activation in subregion of vmPFC identified in self-localizer task predicted which messages would elicit greater responses in a large-scale email campaign (a). Behavior-relevant messages that recruited vmPFC more were more likely to be clicked on, whereas behavior-irrelevant messages that recruited vmPFC were not any more likely to be clicked on (b). Source: Adapted from Falk et al. (2015a), by permission of Oxford University Press.

activation provides additional information about aggregate group behavior beyond self-report measures (Berns and Moore 2012; Genevsky and Knutson 2015; Venkatraman et al. 2015; Falk et al. 2015b; Genevsky et al. 2017; Scholz et al. 2017).

Predicting Social Interactions and Peer Influence

In addition to being influenced by mass media campaigns (as described above), peoples' behavior is also routinely influenced by social norms and interpersonal influence (Cialdini et al. 1991); to this end, a growing body of literature has explored brain processes associated with changing attitudes and behavior in response to peer influence (Klucharev et al. 2009; Cascio et al. 2015a; Wasylyshyn et al. 2018) and, on the other side of the coin, what motivates people to share information with others (Falk et al. 2013; Baek et al. 2017; Scholz et al. 2017).

Brain activity within the value system, as well as regions that help people understand the mental states of others (e.g. TPJ, dmPFC), has been implicated in conformity to peer judgments (Klucharev et al. 2009; Mason et al. 2009; Campbell-Meiklejohn et al. 2010; Zaki et al. 2011; Nook and Zaki 2015; Cascio et al. 2015a). A recent study found that individuals who had greater brain activity in VS and TPJ showed stronger susceptibility to conform to their peers' preferences (Cascio et al. 2015a). In this case, both value regions and mentalizing regions are associated with adapting in response to information about others' preferences, suggesting that people may be integrating these two types of information (the value of objects to the self and the value to others). More broadly, people who are more sensitive to social cues in general are also more susceptible to conforming to peer influences. Individuals who exhibited the greatest activation in mentalizing regions during the common social experience of exclusion were then more susceptible to peer influence in a risky driving task in a driving simulator a few weeks later (Falk et al. 2014); these results suggest that if one's brain is more sensitive to potential social threats, it may be adaptive to fit in by conforming to peer influences. Individuals who showed greater connectivity between mentalizing regions and the rest of the brain were also more susceptible to peer influence (Wasylyshyn et al. 2018). These results support the idea that filtering information through mentalizing systems is an important pathway through which conformity in social groups operates.

In addition to capturing how people respond to others, brain activation in mentalizing and value regions is associated with how people exert influence on others. One way of measuring this behavior is by looking at the brain regions and psychological processes that underlie individuals' choices to share novel information with others. Here, greater activation in vmPFC, VS, and TPJ predicted what ideas for television shows individuals were more likely to share

(Falk et al. 2013). In other work, greater activation in vmPFC, VS, PCC, dmPFC, and TPJ was associated with decisions to select and share news articles (Baek et al. 2017).

Brain activation can also provide additional information about aggregate group social interactions. Scholz et al. (2017) tested whether brain regions involved in self-processing (e.g. vmPFC and PCC), mentalizing (e.g. dmPFC and TPJ), and subjective value (e.g. vmPFC and VS) would be associated with how viral a *New York Times* article was, indexed by how often people shared the article. Activation in all three sets of brain regions was positively correlated with greater article virality. More specifically, the effects of self-processing and mentalizing on article virality were mediated through subjective value, suggesting that brain regions involved in self and social processing index the relevance of the information to the self and close others and that the relevance across domains is combined into a value index, which then determines whether people share the article (Scholz et al. 2017).

Collectively, these results indicate that brain activation is a reliable predictor of broader susceptibility to social influences on behavior, demonstrating an influence on behavior that extends beyond explicitly persuasive messages. Activation in brain regions linked to processing the value of the social behaviors to the self as well as considering the social value of the behavior to others is particularly relevant to these broader forms of social influence. As we will see below, however, the relationship between brain activation and social behavior is also often contingent on social context. Social network position, cultural background, and socioeconomic status (SES) all influence what values, beliefs, and goals are salient for an individual (Markus and Kitayama 1991; Visser and Mirabile 2004; Stephens et al. 2014), which in turn influences how they process behavioral options such as whether to quit smoking (Pegors et al. 2017), respond to peer influence (O'Donnell et al. 2017), or donate to a charity (Park et al. 2017).

Sociocultural Context

The majority of studies examining the link between brain activation and behavior focus on a direct relationship between brain activation and behavior, but to improve our ability to understand social behavior, it is also important to understand the heterogeneity in the relationship between brain and behavior (Tompson et al. 2015). There are many ways that sociocultural context could influence behavior as well as the relationship between brain activation and behavior; however, the two primary routes include (i) normative influence, where individuals engage in a behavior because of what other people care about, and (ii) individual values and beliefs, where individuals engage in a behavior because of their personal interests and concerns. A number of psychological theories argue that both individual attitudes and social norms influence behavioral intentions (Fishbein and Ajzen 1975; Cialdini et al. 1991). Extending this logic, one's position in their social network, cultural background, and SES will also likely influence what norms and beliefs surround an individual. As such, sociocultural context may reinforce and promote different types of normative beliefs about how people should act and what they should care about; these norms may in turn influence the types of goals, values, and beliefs that people hold (Markus and Kitayama 1991; Riemer et al. 2014). Moreover, sociocultural context should also influence how sensitive individuals are to these norms (Riemer et al. 2014; Stephens et al. 2014), which may be reflected in the brain regions recruited when making a choice or evaluating various behavioral options.

One social factor that likely influences how people evaluate behavioral options is social network composition. One study hypothesized the extent to which individuals have close friends who smoke should influence how people evaluate anti-smoking messages (Pegors et al. 2017). In particular, individuals with more smokers in their network might be exposed to more examples of negative impacts of smoking and conversations about the desire to stop smoking, which might influence how much the messages resonate with them (Pegors et al. 2017). Results confirmed that individuals with more smokers than nonsmokers in their social networks who also had stronger multivariate patterns representing message content in vmPFC were more likely to reduce smoking when exposed to anti-smoking messages (Pegors et al. 2017). This work shows that persuasive content can affect behavior differently based on social context and thus how it is interpreted and received.

In addition to social network composition, an individual's position in their network also influences how they respond to information about others' opinions. Some individuals in a network are more ideally positioned to encounter, adopt, and share new information (Burt et al. 2013). This experience sharing information between groups of individuals might relate to an individual's ability or motivation to take the perspective of others, which would in turn influence how they evaluate information about others' opinions. One study found that people who are in a more central position with greater potential opportunities to broker information between people in their social network exhibited greater activation in mentalizing brain regions when incorporating peers' preferences into a rating of a smart phone app (O'Donnell et al. 2017). Taken together, these studies suggest that social network properties influence how individuals process behavioral options.

Similarly, Schmälzle et al. (2017) examined functional connectivity during one common social experience, social exclusion, and found that individuals showed stronger connectivity between brain regions involved in mentalizing during exclusion compared with inclusion. Interestingly, they also found that this relationship was moderated by social network density, such that individuals with less dense friendship networks showed a stronger link between mentalizing network connectivity and rejection sensitivity (Schmälzle et al. 2017). It is possible that social network composition influences what strategies individuals use when interacting with others, which may in turn influence how they respond to social exclusion. Frequently interacting with people who are not connected with others in your social group may sensitize individuals to potentially excluding others and make them more likely to consider others' perspectives during social interactions (Figure 10.4).

Culture also influences brain activation, including the link between brain activation and behavior. Across various different social, cognitive, and affective tasks, people from Western cultures were more likely to show greater activation in self-processing or value regions including vmPFC, whereas people from Asian cultures were more likely to show greater activation in mentalizing regions, including dmPFC and TPJ (Han and Ma 2014). Results demonstrated that when making trait judgments either about the self or a friend, Chinese participants were more likely to recruit TPJ to make these judgments, whereas Danish participants were more likely to recruit vmPFC, and these cultural differences were mediated by differences in interdependence (Ma et al. 2012).

Cultural differences in normative beliefs also influence how people behave when asked to donate to charities. People from East Asian cultures value balanced emotions over high arousal, highly positive emotions (Tsai 2007) and, as a result, are more likely to donate to recipients whose emotional expression matches their cultural norms (Park et al. 2017). That is, East Asians trust and donate more money to charities represented by people with calm, balanced facial expressions, whereas European Americans trust and donate more money to charities represented by people with excited facial expressions (Park et al. 2017). The researchers further found that increased trust for people expressing culturally sanctioned emotions led to a stronger value signal but reduced mentalizing and ultimately greater likelihood of donating. Specifically, cultural differences in donations were linked to differences in brain activation in VS, PCC, and TPJ (Park et al. 2017). TPJ activation was negatively correlated with the fit between an individual's cultural beliefs about what emotions are valued, and individuals were more likely to donate to charities that elicited greater VS activation but weaker PCC and TPJ activations (Park et al. 2017).

Lastly, socioeconomic factors such as parents' educational status strongly influence how individuals perceive choice options (Stephens et al. 2007) and their motivation to influence vs. adjust to their environment (Savani et al. 2008). As described above, activation in social pain and mentalizing regions when individuals are excluded from a group was associated with how susceptible individuals were to peer influence in a driving simulation weeks later (Falk et al. 2014); however, this effect was moderated by SES: adolescents from lower SES backgrounds showed a stronger relationship between social pain regions and susceptibility to peer influence, whereas individuals from

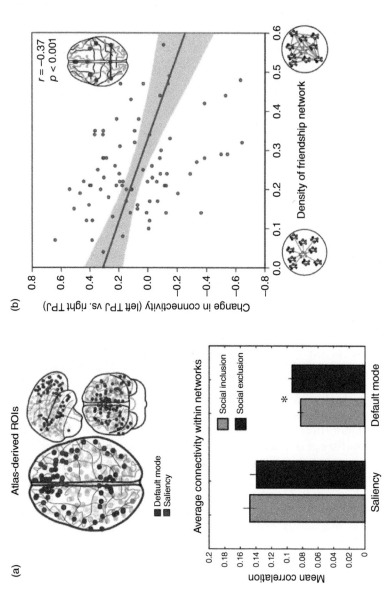

Figure 10.4 Brain networks and social networks. Recent work shows that network connectivity within parts of the default mode subnetwork is greater following social exclusion (a), and this effect is moderated by the density of an individual's social network (b). Source: Adapted with permission from Schmälzle et al. 2017.

high SES backgrounds were more susceptible to peer influence when they exhibited weaker activation in social pain and mentalizing regions (Cascio et al. 2017). This work suggests that how individuals respond to and manage negative affective reactions to exclusion differs as a function of SES, and activation in social pain regions might actually promote susceptibility to peer influence.

Taken together, these three lines of research on moderating roles of social networks, culture, and SES show that the relationship between brain activation and behavior is context dependent. In order to accurately predict behavior, it is therefore important to consider biological factors such as brain activation as well as sociocultural factors such social networks, culture, or SES. However, none of these studies have considered how multiple sociocultural factors might interact. Future research could examine whether social network composition might buffer against the effects of SES on neural responses to social exclusion or investigate if differences in cultural values modulate the relationship between social network position and behavior.

Future Directions

Across the research highlighted in this review, results demonstrate that incorporating measures of brain activation improves our ability to predict human behavior at both the individual and group levels. Furthermore, the specific brain regions and network connectivity patterns provide insight into the psychological processes underlying social behaviors such as persuasion, peer influence, and information sharing. Critically, these associations were identified between brain activation and real-world behaviors, indicating the utility of laboratory research to capture behavioral variability within our daily lives; however, there is still progress and advances to be made in the experimental paradigms employed to better capture the richness and complexity of real-world behaviors and social interactions.

The majority of the research reviewed here utilized fMRI where participants lie on their backs in a scanner with minimal head and body movement, responding to isolated stimuli presented on a single computer monitor. While this environment is designed for studying specific cognitive processes, without excessive noise overriding the physiological signal of interest or confounds of concurrent tasks, the laboratory may not fully capture how tasks are performed in the real world, where our bodies and eyes move freely while we perform multiple, concurrent tasks (Vettel et al. 2012). Even in laboratory settings, complex, naturalistic stimuli elicit different patterns of activation and interregional connectivity compared with more controlled, experimentally manipulated stimuli (Hasson et al. 2010). This complements additional studies that have also identified performance differences when tasks are embedded in naturalistic contexts (Kingstone et al. 2003; Gramann et al. 2011; Oie and McDowell 2011; Shackman et al. 2011).

Recent advances in portable neuroimaging technologies, such as several commercial EEG systems (Hairston et al. 2014; Ries et al. 2014), make it possible to measure brain activation while individuals are navigating complex environments (McDowell et al. 2013; Oliveira et al. 2016a,b; Melnik et al. 2017). Thus, we contend that the brain-as-predictor approach provides a productive framework to study real-world behaviors when combined with ongoing innovations in mobile neuroimaging and artifact rejection techniques (Lawhern et al. 2012; Oliveira et al. 2017). For example, in many studies on social exclusion and peer influence, participants never actually meet their interaction partners. Mobile neuroimaging makes it possible to measure brain activation while individuals are navigating complex social dynamics, which can provide further insight into how individuals behave in naturalistic social interactions. For instance, the degree to which students in a classroom have brain patterns that are in sync with one another predicts classroom engagement and social dynamics, suggesting that shared attention in group settings is a potentially important feature of successful teaching (Dikker et al. 2017). Ongoing work in our laboratories examines how dyadic communication is influenced by the real-world risk of driving along the interstate while a passenger communicates stories to the driver (Vettel et al. 2018). Likewise, tools from computational social science (e.g. social network analysis, computational linguistic methods, geolocation tracking) are rapidly making it more possible to integrate large amounts of information about an individual's specific social environment into models of brain–behavior relationships.

In addition to considering how new tools (such as mobile neuroimaging) might enable researchers to better model brain–behavior relationships, it is also important to consider how this knowledge might be applied to influence or inform how individuals behave. In other cases, this work might be helpful in understanding how to increase more personal resilience to deception or manipulation. The studies described above reveal a few key insights. First, neuroimaging research can help identify which messages or techniques are most likely to be effective in influencing behavior (e.g. Falk et al. 2015b). Second, neuroimaging research can help identify which individuals are more or less susceptible to influence (e.g. Wasylyshyn et al. 2018). However, the third, and most important, insight is that these studies indicate that brain connectivity is shaped by, and malleable to, environmental factors. Contextual factors such as social networks (e.g. Pegors et al. 2017) and short-term shift in mind-set (e.g. Falk et al. 2015a) can influence the brain–behavior relationship, including how people respond to persuasive appeals. Thus, while neuroimaging research can potentially identify which individuals will be more likely to change their behaviors, it is also possible to use neuroimaging to identify how and when persuasive appeals are more or less effective.

Conclusion

In this chapter, we have discussed evidence that brain activation, including multivariate patterns of activation within and connectivity between brain regions, is associated with individual and aggregate health behaviors (Falk et al. 2011, 2015a; Cooper et al. 2015), individual and aggregate consumer behaviors (Levy et al. 2011; Genevsky and Knutson 2015; Genevsky et al. 2017), responses to social exclusion (Schmälzle et al. 2017), responses to peer influence (Falk et al. 2013; Cascio et al. 2015a; Wasylyshyn et al. 2018), and information sharing (Baek et al. 2017; Scholz et al. 2017). In many cases, brain activation provides information that predicts behavior with an accuracy that is above and beyond that obtained from self-report measures of attitudes, preferences, or intentions (Venkatraman et al. 2015; Falk et al. 2015b; Genevsky et al. 2017).

Across these diverse domains, brain regions involved in social and reward processing are frequently associated with behavior, suggesting that brain activation might be providing insight into how people process information about the reward value of various options or outcomes as well as how they consider the perspective and mental states of others. In particular, vmPFC and VS are most often associated with behavior in domains where behaviors are primarily self-focused (e.g. health behavior change, consumer choices). By contrast, dmPFC and TPJ are more often implicated in domains where others' thoughts and opinions are relevant (e.g. information sharing, social influence, and exclusion).

Moreover, the association between brain activation and individual and group behavior is context dependent. Social factors including social network position, culture, and SES influence individuals' beliefs, values, and goals (Markus and Kitayama 1991; Visser and Mirabile 2004; Stephens et al. 2014), which in turn influence how they make decisions (Perry-Smith and Shalley 2003; Stephens et al. 2007; Riemer et al. 2014), interact with close and distant others (Xie et al. 1999; Wagner et al. 2012), and process persuasive messages (Uskul and Oyserman 2010; van Noort et al. 2012). These effects of social context are then reflected in the relationship between brain activation and behavior. The extent to which social and reward-related brain regions are linked to individual behavior is influenced by social network position (O'Donnell et al. 2017; Pegors et al. 2017), culture (Tompson et al. 2015; Park et al. 2017), and SES (Muscatell et al. 2012; Cascio et al. 2017).

Recent advances in multivariate approaches for analyzing brain data can also improve our ability to predict behavior and provide additional insight into the psychological processes mediating this effect. MVPA and RSA provide insight into how individual brain regions encode information about behavioral options (Kriegeskorte 2011), whereas network approaches provide insight into how brain regions work together to evaluate information about the behavioral options (Bassett and Sporns 2017).

In sum, neuroimaging research can advance understanding of how group-level dynamics emerge, including how public service announcements influence ad campaign success (Venkatraman et al. 2015; Falk et al. 2015a), how information spreads throughout a group (Baek et al. 2017; Scholz et al. 2017), and how information about group members' opinions influence individual behaviors (Berns et al. 2010; Tomlin et al. 2013; Cascio et al. 2015a). These methods provide novel insights into how individuals behave in a social world and will serve as useful tools for researchers aiming to understand and predict human behavior.

References

Adolphs, R. (2009). The social brain: neural basis of social knowledge. *Annual Review of Psychology* 60: 693–716. https://doi.org/10.1146/annurev.psych.60 .110707.163514.

Armitage, C. and Conner, M. (2001). Efficacy of the theory of planned behaviour: a meta-analytic review. *British Journal of Social Psychology* 40 (4): 471–499. https://doi.org/10.1348/014466601164939.

Azevedo, F.A.C., Carvalho, L.R.B., Grinberg, L.T. et al. (2009). Equal numbers of neuronal and nonneuronal cells make the human brain an isometrically scaled-up primate brain. *The Journal of Comparative Neurology* 513 (5): 532–541. https://doi.org/10.1002/cne.21974.

Baek, E.C., Scholz, C., O'Donnell, M.B., and Falk, E.B. (2017). The value of sharing information: a neural account of information transmission. *Psychological Science* 28 (7): 851–861. https://doi.org/10.1177/0956797617695073.

Bartra, O., McGuire, J.T., and Kable, J.W. (2013). The valuation system: a coordinate-based meta-analysis of BOLD fMRI experiments examining neural correlates of subjective value. *NeuroImage* 76: 412–427. https://doi.org/10 .1016/j.neuroimage.2013.02.063.

Bassett, D.S. and Sporns, O. (2017). Network neuroscience. *Nature Neuroscience* 20 (3): 353–364.

Berkman, E.T. and Falk, E.B. (2013). Beyond brain mapping: using neural measures to predict real-world outcomes. *Current Directions in Psychological Science* 22 (1): 45–50. https://doi.org/10.1177/0963721412469394.

Berns, G.S. and Moore, S.E. (2012). A neural predictor of cultural popularity. *Journal of Consumer Psychology* 22 (1): 154–160. https://doi.org/10.1016/j.jcps .2011.05.001.

Berns, G.S., Capra, C.M., Moore, S., and Noussair, C. (2010). Neural mechanisms of the influence of popularity on adolescent ratings of music. *NeuroImage* 49 (3): 2687–2696. https://doi.org/10.1016/j.neuroimage.2009.10.070.

Brooks, J.R., Passaro, A.D., Kerick, S.E. et al. (2018). Overlapping brain network and alpha power changes suggest visuospatial attention effects on driving performance. *Behavioral Neuroscience* 132 (1): 23–33. https://doi.org/10.1037/bne0000224.

Burt, R.S., Kilduff, M., and Tasselli, S. (2013). Social network analysis: foundations and frontiers on advantage. *Annual Review of Psychology* 64 (1): 527–547. https://doi.org/10.1146/annurev-psych-113011-143828.

Campbell-Meiklejohn, D.K., Bach, D.R., Roepstorff, A. et al. (2010). How the opinion of others affects our valuation of objects. *Current Biology* 20 (13): 1165–1170. https://doi.org/10.1016/j.cub.2010.04.055.

Cascio, C.N., O'Donnell, M.B., Bayer, J. et al. (2015a). Neural correlates of susceptibility to group opinions in online word-of-mouth recommendations. *Journal of Marketing Research* 52 (4): 559–575. https://doi.org/10.1509/jmr.13.0611.

Cascio, C.N., Scholz, C., and Falk, E.B. (2015b). Social influence and the brain: persuasion, susceptibility to influence and retransmission. *Current Opinion in Behavioral Sciences* 3 (3): 51–57. https://doi.org/10.1016/j.cobeha.2015.01.007.

Cascio, C.N., O'Donnell, M.B., Simons-Morton, B.G. et al. (2017). Cultural context moderates neural pathways to social influence. *Culture and Brain* 5 (1): 50–70. https://doi.org/10.1007/s40167-016-0046-3.

Chua, H.F., Ho, S.S., Jasinska, A.J. et al. (2011). Self-related neural response to tailored smoking-cessation messages predicts quitting. *Nature Neuroscience* 14 (4): 426–427. https://doi.org/10.1038/nn.2761.

Cialdini, R.B., Kallgren, C.A., and Reno, R.R. (1991). A focus theory of normative conduct: a theoretical refinement and reevaluation of the role of norms in human behavior. *Advances in Experimental Social Psychology* 24: 201–234.

Cooper, N., Tompson, S., O'Donnell, M.B., and Falk, E.B. (2015). Brain activity in self-and value-related regions in response to online antismoking messages predicts behavior change. *Journal of Media Psychology* 27 (3): 93–109. https://doi.org/10.1027/1864-1105/a000146.

Cooper, N., Bassett, D.S., and Falk, E.B. (2017). Coherent activity between brain regions that code for value is linked to the malleability of human behavior. *Scientific Reports* 7: 43250. https://doi.org/10.1038/srep43250.

Cooper, N., Tompson, S., O'Donnell, M.B. et al. (2018). Associations between coherent neural activity in the brain's value system during antismoking messages and reductions in smoking. *Health Psychology: Official Journal of the Division of Health Psychology, American Psychological Association* 37 (4): 375–384. https://doi.org/10.1037/hea0000574.

Denny, B.T., Kober, H., Wager, T.D., and Ochsner, K.N. (2012). A meta-analysis of functional neuroimaging studies of self- and other judgments reveals a spatial gradient for mentalizing in medial prefrontal cortex. *Journal of Cognitive Neuroscience* 24 (8): 1742–1752. https://doi.org/10.1162/jocn_a_00233.

Dikker, S., Wan, L., Davidesco, I. et al. (2017). Brain-to-brain synchrony tracks real-world dynamic group interactions in the classroom. *Current Biology* 27 (9): 1375–1380. https://doi.org/10.1016/j.cub.2017.04.002.

Falk, E.B. and Scholz, C. (2018). Persuasion, influence, and value: perspectives from communication and social neuroscience. *Annual Review of Psychology* 69 (1): 329–356. https://doi.org/10.1146/annurev-psych-122216-011821.

Falk, E.B., Berkman, E.T., Mann, T. et al. (2010). Predicting persuasion-induced behavior change from the brain. *The Journal of Neuroscience: The Official Journal of the Society for Neuroscience* 30 (25): 8421–8424. https://doi.org/10.1523/jneurosci.0063-10.2010.

Falk, E.B., Berkman, E.T., Whalen, D., and Lieberman, M.D. (2011). Neural activity during health messaging predicts reductions in smoking above and beyond self-report. *Health Psychology: Official Journal of the Division of Health Psychology, American Psychological Association* 30 (2): 177–185. https://doi.org/10.1037/a0022259.

Falk, E.B., Berkman, E.T., and Lieberman, M.D. (2012). From neural responses to population behavior. *Psychological Science* 23: 439–445. https://doi.org/10.1177/0956797611434964.

Falk, E.B., Morelli, S., Welborn, B.L. et al. (2013). Creating buzz: the neural correlates of effective message propagation. *Psychological Science* 24: 1234–1242. https://doi.org/10.1177/0956797612474670.

Falk, E.B., Cascio, C.N., O'Donnell, M.B. et al. (2014). Neural responses to exclusion predict susceptibility to social influence. *Journal of Adolescent Health* 54: S22–S31. https://doi.org/10.1016/j.jadohealth.2013.12.035.

Falk, E.B., O'Donnell, M.B., Cascio, C.N. et al. (2015a). Self-affirmation alters the brain's response to health messages and subsequent behavior change. *Proceedings of the National Academy of Sciences of the United States of America* 112 (7): 1977–1982. https://doi.org/10.1073/pnas.1500247112.

Falk, E.B., O'Donnell, M.B., Tompson, S. et al. (2015b). Functional brain imaging predicts public health campaign success. *Social Cognitive and Affective Neuroscience* 11 (2): 204–214. https://doi.org/10.1093/scan/nsv108.

Fishbein, M. and Ajzen, I. (1975). *Belief, Attitude, Intention, and Behavior: An Introduction to Theory and Research*. Reading, MA: Addison-Wesley.

Garcia, J.O., Brooks, J., Kerick, S. et al. (2017). Estimating direction in brain-behavior interactions: proactive and reactive brain states in driving. *NeuroImage* 150: 239–249. https://doi.org/10.1016/j.neuroimage.2017.02.057.

Genevsky, A. and Knutson, B. (2015). Neural affective mechanisms predict market-level microlending. *Psychological Science* 26 (9): 1411–1422. https://doi.org/10.1177/0956797615588467.

Genevsky, A., Yoon, C., and Knutson, B. (2017). When brain beats behavior: neuroforecasting crowdfunding outcomes. *The Journal of Neuroscience: The Official Journal of the Society for Neuroscience* 37 (36): 8625–8634. https://doi.org/10.1523/jneurosci.1633-16.2017.

Gramann, K., Gwin, J.T., Ferris, D.P. et al. (2011). Cognition in action: imaging brain/body dynamics in mobile humans. *Reviews in the Neurosciences* 22 (6): 593–608. https://doi.org/10.1515/rns.2011.047.

Hairston, W.D., Whitaker, K.W., Ries, A.J. et al. (2014). Usability of four commercially-oriented EEG systems. *Journal of Neural Engineering* 11 (4): 46018. https://doi.org/10.1088/1741-2560/11/4/046018.

Han, S. and Ma, Y. (2014). Cultural differences in human brain activity: a quantitative meta-analysis. *NeuroImage* 99: 293–300. https://doi.org/10.1016/j.neuroimage.2014.05.062.

Hasson, U., Malach, R., and Heeger, D.J. (2010). Reliability of cortical activity during natural stimulation. *Trends in Cognitive Sciences* 14 (1): 40–48. https://doi.org/10.1016/j.tics.2009.10.011.

Kelley, W.M., Macrae, C.N., Wyland, C.L. et al. (2002). Finding the self? An event-related fMRI study. *Journal of Cognitive Neuroscience* 14 (5): 785–794. https://doi.org/10.1162/08989290260138672.

Kim, K. and Johnson, M.K. (2012). Extended self: medial prefrontal activity during transient association of self and objects. *Social Cognitive and Affective Neuroscience* 7 (2): 199–207. https://doi.org/10.1093/scan/nsq096.

Kingstone, A., Smilek, D., Ristic, J. et al. (2003). Attention, researchers! It is time to take a look at the real world. *Current Directions in Psychological Science* 12 (5): 176–180. https://doi.org/10.1111/1467-8721.01255.

Klucharev, V., Hytönen, K., Rijpkema, M. et al. (2009). Reinforcement learning signal predicts social conformity. *Neuron* 61: 140–151. https://doi.org/10.1016/j.neuron.2008.11.027.

Knutson, B. and Genevsky, A. (2018). Neuroforecasting aggregate choice. *Current Directions in Psychological Science* 27 (2): 110–115. https://doi.org/10.1177/0963721417737877.

Knutson, B., Rick, S., Wimmer, G.E. et al. (2007). Neural predictors of purchases. *Neuron* 53 (1): 147–156. https://doi.org/10.1016/j.neuron.2006.11.010.

Kriegeskorte, N. (2011). Pattern-information analysis: from stimulus decoding to computational-model testing. *NeuroImage* 56 (2): 411–421. https://doi.org/10.1016/j.neuroimage.2011.01.061.

Kühn, S., Strelow, E., and Gallinat, J. (2016). Multiple "buy buttons" in the brain: forecasting chocolate sales at point-of-sale based on functional brain activation using fMRI. *NeuroImage* 136: 122–128. https://doi.org/10.1016/j.neuroimage.2016.05.021.

Lawhern, V., Hairston, W.D., McDowell, K. et al. (2012). Detection and classification of subject-generated artifacts in EEG signals using autoregressive models. *Journal of Neuroscience Methods* 208 (2): 181–189. https://doi.org/10.1016/j.jneumeth.2012.05.017.

Levy, I., Lazzaro, S.C., Rutledge, R.B., and Glimcher, P.W. (2011). Choice from non-choice: predicting consumer preferences from blood oxygenation level-dependent signals obtained during passive viewing. *Journal of Neuroscience* 31 (1): 118–125. https://doi.org/10.1523/jneurosci.3214-10.2011.

Lieberman, M.D. (2007). Social cognitive neuroscience: a review of core processes. *Annual Review of Psychology* 58: 259–289. https://doi.org/10.1146/annurev.psych.58.110405.085654.

Ma, Y., Bang, D., Wang, C. et al. (2012). Sociocultural patterning of neural activity during self-reflection. *Social Cognitive and Affective Neuroscience* 9 (1): 73–80. https://doi.org/10.1093/scan/nss103.

Markus, H.R. and Kitayama, S. (1991). Culture and the self: implications for cognition, emotion, and motivation. *Psychological Review* 98: 224–253.

Martinelli, P., Sperduti, M., and Piolino, P. (2013). Neural substrates of the self-memory system: new insights from a meta-analysis. *Human Brain Mapping* 34: 1515–1529. https://doi.org/10.1002/hbm.22008.

Mason, M.F., Dyer, R., and Norton, M.I. (2009). Neural mechanisms of social influence. *Organizational Behavior and Human Decision Processes* 110 (2): 152–159. https://doi.org/10.1016/j.obhdp.2009.04.001.

McDowell, K., Lin, C.-T., Oie, K.S. et al. (2013). Real-world neuroimaging technologies. *IEEE Access* 1: 131–149. https://doi.org/10.1109/access.2013 .2260791.

Melnik, A., Legkov, P., Izdebski, K. et al. (2017). Systems, subjects, sessions: to what extent do these factors influence EEG data? *Frontiers in Human Neuroscience* 11: 150. https://doi.org/10.3389/fnhum.2017.00150.

Muraskin, J., Sherwin, J., Lieberman, G. et al. (2017). Fusing multiple neuroimaging modalities to assess group differences in perception-action coupling. *Proceedings of the IEEE* 105 (1): 83–100. https://doi.org/10.1109/jproc .2016.2574702.

Muscatell, K.A., Morelli, S.A., Falk, E.B. et al. (2012). Social status modulates neural activity in the mentalizing network. *NeuroImage* 60: 1771–1777. https:// doi.org/10.1016/j.neuroimage.2012.01.080.

Nili, H., Wingfield, C., Walther, A. et al. (2014). A toolbox for representational similarity analysis. *PLoS Computational Biology* 10 (4): e1003553. https://doi .org/10.1371/journal.pcbi.1003553.

Nook, E.C. and Zaki, J. (2015). Social norms shift behavioral and neural responses to foods. *Journal of Cognitive Neuroscience* 27 (7): 1412–1426. https://doi.org/ 10.1162/jocn_a_00795.

van Noort, G., Antheunis, M.L., and van Reijmersdal, E.A. (2012). Social connections and the persuasiveness of viral campaigns in social network sites: persuasive intent as the underlying mechanism. *Journal of Marketing Communications* 18 (1): 39–53. https://doi.org/10.1080/13527266.2011.620764.

Norman, K.A., Polyn, S.M., Detre, G.J., and Haxby, J.V. (2006). Beyond mind-reading: multi-voxel pattern analysis of fMRI data. *Trends in Cognitive Sciences* 10 (9): 424–430. https://doi.org/10.1016/j.tics.2006.07.005.

O'Donnell, M.B., Bayer, J.B., Cascio, C.N., and Falk, E.B. (2017). Neural bases of recommendations differ according to social network structure. *Social Cognitive and Affective Neuroscience* 12 (1): nsw158. https://doi.org/10.1093/scan/ nsw158.

O'Keefe, D.J. (2018). Message pretesting using assessments of expected or perceived persuasiveness: evidence about diagnosticity of relative actual persuasiveness. *Journal of Communication* 68 (1): 120–142. https://doi.org/10.1093/joc/jqx009.

Oie, K. and McDowell, K. (2011). Neurocognitive engineering for systems' development. *Synesis: A Journal of Science, Technology, Ethics, and Policy* 2 (1): T26–T37.

Oliveira, A.S., Schlink, B.R., Hairston, W.D. et al. (2016a). Induction and separation of motion artifacts in EEG data using a mobile phantom head device. *Journal of Neural Engineering* 13 (3): 36014. https://doi.org/10.1088/1741-2560/13/3/036014.

Oliveira, A.S., Schlink, B.R., Hairston, W.D. et al. (2016b). Proposing metrics for benchmarking novel EEG technologies towards real-world measurements. *Frontiers in Human Neuroscience* 10: 188. https://doi.org/10.3389/fnhum.2016.00188.

Oliveira, A.S., Schlink, B.R., Hairston, W.D. et al. (2017). A channel rejection method for attenuating motion-related artifacts in EEG recordings during walking. *Frontiers in Neuroscience* 11: 225. https://doi.org/10.3389/fnins.2017.00225.

Park, B., Blevins, E., Knutson, B., and Tsai, J.L. (2017). Neurocultural evidence that ideal affect match promotes giving. *Social Cognitive and Affective Neuroscience* 12 (7): 1083–1096. https://doi.org/10.1093/scan/nsx047.

Passaro, A.D., Vettel, J.M., McDaniel, J. et al. (2017). A novel method linking neural connectivity to behavioral fluctuations: behavior-regressed connectivity. *Journal of Neuroscience Methods* 279: 60–71. https://doi.org/10.1016/j.jneumeth.2017.01.010.

Passingham, R.E., Stephan, K.E., and Kötter, R. (2002). The anatomical basis of functional localization in the cortex. *Nature Reviews Neuroscience* 3 (8): 606–616. https://doi.org/10.1038/nrn893.

Pegors, T.K., Tompson, S., O'Donnell, M.B., and Falk, E.B. (2017). Predicting behavior change from persuasive messages using neural representational similarity and social network analyses. *NeuroImage* 157: 118–128. https://doi.org/10.1016/j.neuroimage.2017.05.063.

Perry-Smith, J.E. and Shalley, C.E. (2003). The social side of creativity: a static and dynamic social network perspective. *Academy of Management Review* 28 (1): 89–106. https://doi.org/10.5465/amr.2003.8925236.

Riemer, H., Shavitt, S., Koo, M., and Markus, H.R. (2014). Preferences don't have to be personal: expanding attitude theorizing with a cross-cultural perspective. *Psychological Review* 121 (4): 619–648. https://doi.org/10.1037/a0037666.

Ries, A.J., Touryan, J., Vettel, J.M. et al. (2014). A comparison of electroencephalography signals acquired from conventional and mobile systems. *Journal of Neuroscience and Neuroengineering* 3 (1): 10–20. https://doi .org/10.1166/jnsne.2014.1092.

Rule, N.O., Freeman, J.B., Moran, J.M. et al. (2009). Voting behavior is reflected in amygdala response across cultures. *Social Cognitive and Affective Neuroscience* 5 (2–3): 349–355. https://doi.org/10.1093/scan/nsp046.

Savani, K., Markus, H.R., and Conner, A.L. (2008). Let your preference be your guide? Preferences and choices are more tightly linked for North Americans than for Indians. *Journal of Personality and Social Psychology* 95 (4): 861–876. https://doi.org/10.1037/a0011618.

Saxe, R. and Kanwisher, N. (2013). People thinking about thinking people: the role of the temporo-parietal junction in "theory of mind". In: *Social Neuroscience: Key Readings*, 171–182. Taylor & Francis Group. https://doi.org/10.4324/ 9780203496190.

Schmälzle, R., O'Donnell, M.B., Garcia, J.O. et al. (2017). Brain connectivity dynamics during social interaction reflect social network structure. *Proceedings of the National Academy of Sciences of the United States of America* 114 (20): 5153–5158. https://doi.org/10.1073/pnas.1616130114.

Scholz, C., Baek, E.C., O'Donnell, M.B. et al. (2017). A neural model of valuation and information virality. *Proceedings of the National Academy of Sciences of the United States of America* 114 (11): 2881–2886. https://doi.org/10.1073/pnas .1615259114.

Schurz, M., Radua, J., Aichhorn, M. et al. (2014). Fractionating theory of mind: a meta-analysis of functional brain imaging studies. *Neuroscience and Biobehavioral Reviews* https://doi.org/10.1016/j.neubiorev.2014.01.009.

Shackman, A.J., Maxwell, J.S., McMenamin, B.W. et al. (2011). Stress potentiates early and attenuates late stages of visual processing. *The Journal of Neuroscience: The Official Journal of the Society for Neuroscience* 31 (3): 1156–1161. https://doi.org/10.1523/jneurosci.3384-10.2011.

Stephens, N.M., Markus, H.R., and Townsend, S.S.M. (2007). Choice as an act of meaning: the case of social class. *Journal of Personality and Social Psychology* 93 (5): 814–830. https://doi.org/10.1037/0022-3514.93.5.814.

Stephens, N.M., Markus, H.R., and Phillips, L.T. (2014). Social class culture cycles: how three gateway contexts shape selves and fuel inequality. *Annual Review of Psychology* 65 (1): 611–634. https://doi.org/10.1146/annurev-psych-010213- 115143.

Tomlin, D., Nedic, A., Prentice, D.A. et al. (2013). The neural substrates of social influence on decision making. *PLoS ONE* 8 (1): e52630. https://doi.org/10.1371/ journal.pone.0052630.

Tompson, S., Lieberman, M.D., and Falk, E.B. (2015). Grounding the neuroscience of behavior change in the sociocultural context. *Current Opinion in Behavioral Sciences* 5: 58–63. https://doi.org/10.1016/j.cobeha.2015.07.004.

Tsai, J.L. (2007). Ideal affect: cultural causes and behavioral consequences. *Perspectives on Psychological Science* 2 (3): 242–259. https://doi.org/10.1111/j .1745-6916.2007.00043.x.

Uskul, A.K. and Oyserman, D. (2010). When message-frame fits salient cultural-frame, messages feel more persuasive. *Psychology & Health* 25 (3): 321–337. https://doi.org/10.1080/08870440902759156.

Venkatraman, V., Dimoka, A., Pavlou, P.A. et al. (2015). Predicting advertising success beyond traditional measures: new insights from neurophysiological methods and market response modeling. *Journal of Marketing Research* 52 (4): 436–452. https://doi.org/10.1509/jmr.13.0593.

Vettel, J.M., Lance, B., Manteuffel, C. et al. (2012). Mission-based scenario research: experimental design and analysis. Proceedings of the Ground Vehicle Systems Engineering and Technology Symposium.

Vettel, J.M., Cooper, N., Garcia, J.O. et al. (2017). White matter tractography and diffusion-weighted imaging. In: *eLS*, 1–9. Chichester: Wiley https://doi.org/10 .1002/9780470015902.a0027162.

Vettel, J.M., Lauharatanhirun, N., and Wasylyshyn, N. et al. (2018). Translating driving research from simulation to interstate driving with realistic traffic and passenger interactions. International Conference on Applied Human Factors and Ergonomics.

Vezich, I.S., Katzman, P.L., Ames, D.L. et al. (2017). Modulating the neural bases of persuasion: why/how, gain/loss, and users/non-users. *Social Cognitive and Affective Neuroscience* 12 (2): 283–297. https://doi.org/10.1093/scan/nsw113.

Visser, P.S. and Mirabile, R.R. (2004). Attitudes in the social context: the impact of social network composition on individual-level attitude strength. *Journal of Personality and Social Psychology* 87 (6): 779–795. https://doi.org/10.1037/ 0022-3514.87.6.779.

Wagner, J.A., Humphrey, S.E., Meyer, C.J., and Hollenbeck, J.R. (2012). Individualism-collectivism and team member performance: another look. *Journal of Organizational Behavior* 33 (7): 946–963. https://doi.org/10.1002/job .783.

Wang, A.-L., Ruparel, K., Loughead, J.W. et al. (2013). Content matters: neuroimaging investigation of brain and behavioral impact of televised anti-tobacco public service announcements. *The Journal of Neuroscience* 33 (17): 7420–7427. https://doi.org/10.1523/jneurosci.3840-12.2013.

Wasylyshyn, N., Falk, B.H., Garcia, J.O. et al. (2018). Global brain dynamics during social exclusion predict subsequent behavioral conformity. *Social Cognitive and Affective Neuroscience* 13 (2): 182–191. https://doi.org/10.1093/ scan/nsy007.

Xie, H., Cairns, R.B., and Cairns, B.D. (1999). Social networks and configurations in inner-city schools: aggression, popularity, and implications for students with

EBD. *Journal of Emotional and Behavioral Disorders* 7 (3): 147–155. https://doi
.org/10.1177/106342669900700303.

Zaki, J., Schirmer, J., and Mitchell, J.P. (2011). Social influence modulates the
neural computation of value. *Psychological Science: A Journal of the American
Psychological Society / APS* 22 (June): 894–900. https://doi.org/10.1177/
0956797611411057.

11

Social Models from Non-Human Systems

Theodore P. Pavlic

School of Computing, Informatics, and Decision Systems Engineering and the School of Sustainability, Arizona State University, Tempe, AZ 85287, USA

It is natural to eschew consideration of the nonhuman world when seeking understanding of the complex social-behavioral landscape of human phenomena. Humans build cities, form governments, make laws, write poetry, and make decisions that are evidently products of deliberate, long-term planning – all things that seem to be absent from most of the nonhuman world. Anthropologists sometimes look to nonhuman primates for insights into human social behavior because they are a closely related out-group that can be used as a reference for understanding the evolutionary trajectory of hominids. Despite the use of the idiom "the birds and the bees" as a euphemism for an important human social behavior, the use of birds and bees and other nonhuman organisms is almost entirely restricted to cases where there is high correspondence between physiological mechanisms under study. An ant colony seems to be a poor model for a human society because it lacks realistic detail and scaling, and so any analogy between ant colonies and human groups is largely thought to be "for the birds."

However, experienced quantitative modelers recognize that models live on a continuum – from analogous to metaphorical – and are defined by how they are used and not what they are made of. Box (1979) is well known for the aphorism that "all models are wrong but some are useful," which is meant to highlight both that no model can be a perfect representation of a system and that even the most imperfect representation of a system can be insightful. Furthermore, to Box (1976), the utility of a model cannot be evaluated *a priori* – "we cannot know that any statistical technique we develop is useful unless we use it." Physics has made great progress predicting the macroscopic behavior of ensembles of particles using mathematical and computational models that make microscopic assumptions about those

Social-Behavioral Modeling for Complex Systems, First Edition.
Edited by Paul K. Davis, Angela O'Mahony, and Jonathan Pfautz.
© 2019 John Wiley & Sons, Inc. Published 2019 by John Wiley & Sons, Inc.
Companion website: www.wiley.com/go/Davis_Social-Behavioralmodeling

particles that are almost certainly wrong. Models are a means to end; they are tools on which theories are built, and theories are tools used to deepen the shallows of explanatory abilities (Wilson and Keil 1998) through an iterative process of conceptual exploration and experimentation.

Thus, the differences between complex human societies and complex societies of ants and birds are important to recognize but should not be viewed as weaknesses offhand; these animal systems may still be mined, with caution, for transferable insights. In fact, just as a simple, smoothly varying differential equation may be an attractive model for a population of discrete individuals, the empirical tractability that might be gained by working with nonhuman systems could compensate for losses in realism. Furthermore, given that the wide variety of nonhuman social systems is accompanied by a wide range of differing social behaviors, methods and perspectives developed to understand nonhuman behavioral ecology in a unified evolutionary and developmental context may themselves be valuable when considering avenues for exploring complex human social behavior.

This chapter explores several motivational examples for how nonhuman systems can contribute to the study of complexity in human societies. The discussion is divided into three main parts: (i) emergent social patterns in self-interested individuals, (ii) the effect of competition between groups, and (iii) how information is used within tightly integrated groups. As the examples move the focus from low integration to high integration, the animal models move from vertebrate systems – primarily social foraging birds – to arthropod systems, social spiders and ants. This chapter then concludes with some general comments about the generalizability of studies with nonhuman social systems when used as a model for human social systems.

Emergent Patterns in Groups of Behaviorally Flexible Individuals

A particularly fruitful animal model for studying sociality in nonhuman systems is the nutmeg mannikin (Figure 11.1), a dull-colored bird common to both the pet trade and behavioral ecology (Tinbergen 1951; Moynihan and Hall 1955; Restall 1997). These birds form natural groups that can vary in size but can be as large as 100, similar to estimates of early Pre-Pottery Neolithic human community sizes (Kuijt 2000). Like humans, their behaviors have evolved against a background of living in potentially large groups within habitats that vary in resource availability. Because there is natural temporal variability in resources as well as group composition, selection has put pressure on these birds to flexibly and adaptively defend against as well as take advantage of their social context for survival. Consequently, as reviewed by

(a)   (b)

Figure 11.1 The nutmeg mannikin, *Lonchura punctulata*. (a) Immature nutmeg manikin. (b) Foraging group. The *Lonchura punctulata*, otherwise known as a nutmeg mannikin, spice finch, or scaly-brested/spotted munia. Juveniles, as in (a), have a brownish underside that will develop into more well-defined, highly contrasting scales. Nutmeg mannikins regularly forage in groups, as in (b), where each individual's tactic varies with the composition of the group. Source: The image in (a) is in the public domain, originally taken by Vicki Nunn. The image in (b) is by W. A. Djatmiko and is provided under a CC-BY-SA 3.0 (Creative Commons 2018c) license.

Giraldeau and Dubois (2008) and echoed in this section, nutmeg mannikins are well suited for laboratory studies of social behavior, especially in the case of social exploitation.

When nutmeg mannikins forage in view of each other, birds can be observed finding new food patches on their own as well as joining other birds that have recently found new patches. Early conceptual models of these scenarios took an information sharing perspective (Clark and Mangel 1984), whereby each group member could simultaneously search for new patches while monitoring the success of other members. The competing approach, first discussed by Barnard and Sibly (1981), assumes that individuals must balance their time between two mutually exclusive tactics – a producer tactic focused on the private search for new patches and a scrounger tactic focused on the social task of monitoring others. It is this latter perspective that has been successful in describing the group-level behaviors of nutmeg mannikins in social foraging tasks. Moreover, it also demonstrates how these individual-level trade-offs can lead to nontrivial patterns in large groups. Thus, the rest of this section focuses on the interplay of mathematics and controlled social foraging experiments in the service of building confidence in exploitative conceptual models of social behavior.

From Bird Motivations to Human Applications

When nutmeg mannikins forage in view of each other, the birds sort them-selves consistently into one of two tactics – *producer* and *scrounger* (Giraldeau and Dubois 2008). By observing groups of these birds in controlled settings in the laboratory, it is possible to study how each bird chooses a tactic. A common assay is to release a group of birds onto a large array of wells, some of which are empty and some of which contain seeds. Birds adopting the producer tactic will pitch their beaks down toward the array and will have a tight correlation between the time they spend searching and the number of seed-filled wells they find (Coolen et al. 2001). However, birds adopting the scrounger tactic keep their heads upright and rarely make new discoveries of seed-filled wells; instead, scrounger foraging time is tightly correlated with number of seed-filled wells joined after a producer discovery of the well (Coolen et al. 2001; Wu and Giraldeau 2005). Putting aside differences in efficiency as a producer, the benefit to joining a group as a scrounger must depend on the composition of producers and scroungers in the existing group. Despite the apparent benefits of being a parasitic scrounger, the tactic is a poor choice if a group consists of nearly all scroungers. Studying the behavior of birds making these decisions led to the development of a game-theoretic *producer–scrounger* (*PS*) *game* that is general enough to model a wide range of other exploitative social phenomena in animal behavior (Barnard and Sibly 1981; Vickery et al. 1991; Giraldeau and Beauchamp 1999; Giraldeau and Caraco 2000). In fact, the PS game has also been used to study human social behavior, including cultural learning (Kameda and Nakanishi 2002; Mesoudi 2008), the evolution of leadership (King et al. 2009), and the coevolution of technology with culture (Lehmann and Feldman 2009). This important conceptual framework came about through the study of the social foraging behavior of birds, and it can be used as a tool for translating insights from the study of bird social behavior to under-standing patterns, either existing or possible, in analogous human social scenarios.

Game-Theoretic Model of Frequency-Dependent Tactic Choice

In order to make predictions that could be tested in empirical studies of social foraging behavior in these birds, a simplified game-theoretic model was built that captured the salient features of how group size and foraging returns were modulated by tactic frequency. The basic 2-tactic PS game (Giraldeau and Caraco 2000) considers a group of G foragers, of which some fraction p are producers and $(1 - p)$ are scroungers. On average, each of the pG producers finds λ patches of food items per unit time, and each patch carries F units of food. The producer takes a finder's advantage of a food units and divides the remaining $F - a$ units among itself and the $(1 - p)G$ scroungers that join. Thus, the producer and scrounger intake rates, respectively, are

$$I_p = \lambda \left(\overbrace{a}^{\substack{\text{Private} \\ \text{Reward}}} + \overbrace{\frac{F - a}{1 + (1 - p)G}}^{\text{Social Reward}} \right) \text{ and } I_s = \overbrace{pG\lambda}^{\substack{\text{Group} \\ \text{Production} \\ \text{Rate}}} \overbrace{\left(\frac{F - a}{1 + (1 - p)G} \right)}^{\text{Social Reward}}$$

$$(11.1)$$

For the PS game, the equilibrium concept is the behaviorally stable strategy (BSS), which is an invasion-resistant specialization of a Nash equilibrium assumed to be attainable through behavioral and cognitive mechanisms (Harley 1981; Giraldeau and Dubois 2008). In principle, there is a stable equilibrium frequency (SEF) p^* of producers such that if $p = p^*$, then the two intake rates will be equivalent, which ensures that there is no benefit for any individual to unilaterally change her chosen tactic. Assuming that a is sufficiently small and G is sufficiently large, the SEF for the PS game has the expression

$$p^* = \overbrace{\frac{a}{F}}^{\substack{\text{Finder's} \\ \text{Fraction}}} + \overbrace{\frac{1}{G}}^{\substack{\text{Social} \\ \text{Fraction}}}$$

$$(11.2)$$

That is, the equilibrium fraction of producers reflects the fraction of each patch that is guaranteed to the finder. Although a large group size decreases the social fraction, the finder's advantage ensures that the BSS always includes some producers.

Mathematical Model as Behavioral Microscope on Carefully Prepared Birds

The role of the model described by Eqs. (11.1) and (11.2) is to generate predictions consistent with the hypothesis that individual birds switch between two mutually exclusive tactics and modulate their use of those tactics based on frequency-dependent rewards. Thus, the game-theoretic mathematical model is not being tested itself but is instead a logical tool used to test the more general idea that individual birds are sensitive to their individual reward rates in social scenarios and switch between distinct behaviors in accordance with those rates. The PS game is intentionally simplistic because it is intended to generate predictions that will be tested in tightly controlled laboratory environments for which the PS game is an appropriate tool for predicting social foraging behavior. In other words, the goal of the game-theoretic approach is not to build a detailed and accurate representation of a particular organism. Instead, the mathematical, game-theoretic approach acts like a microscope that reveals deep insights into phenomena that would otherwise be unobservable if not carefully prepared, secured to a glass slide, and well lit for observation. The mathematical model focuses on the salient features of the hypothesis, and the subsequent experiment carefully controls the model system so that important details are in focus through the lens of the mathematical model.

Consistent with the hypothesis that birds switch between two mutually exclusive search modes based on their individual intake rates, the PS game accurately predicts the equilibrium frequency of producers in controlled laboratory experiments with nutmeg mannikins. In particular, by observing groups of a fixed size G foraging on patches of a fixed size F, the finder's advantage a (or, more importantly, finder's share a/F) can be estimated, and the SEF p^* can be predicted. For example, Giraldeau and Livoreil (1998) show that experimentally manipulating the finder's share generated shifts in the frequency of scrounging that match shifts consistent with a PS game. Similar PS-consistent shifts in scrounging frequency were observed by Coolen (2002) for different group sizes. To move beyond these qualitative matches, Mottley and Giraldeau (2000) were able to design an apparatus that could precisely control the finder's share and even force birds constrained to certain parts of the apparatus to only use one of the two tactics. This apparatus could also change these precisely controlled intake rates in the middle of an experiment in order to test how quickly birds can adjust to the change. Social groups tested on the apparatus not only converged on the predicted SEF values, but they also could converge on a new SEF value in one or two days after the apparatus had been reconfigured for the new intake rates. Thus, the birds have some mechanism that allows them to respond to changes in intake rate and strategically adjust their individual tactics to maximize their individual returns subject to the social foraging context.

The precise form of Eq. (11.2) depends upon the simplistic intake rates expressed by Eq. (11.1) that are only justified in tightly controlled social for-aging experiments in the laboratory. However, the hypothesis being tested by those experiments is far more general. Having accumulated confidence in that hypothesis under tightly controlled settings, it is then possible to extrapolate to more general social scenarios without simplistic mathematical models. In other words, the simplistic mathematical model and tightly controlled laboratory setup were not meant to be direct surrogates for complex social behavior but rather simple tests to characterize the feasibility of an underlying conceptual model. The outcomes of those experiments are consistent with the hypothesis that birds are sensitive at an individual level to the cost of each tactic. Consequently, even though tactic-specific costs are not captured by Eq. (11.1), qualitative predictions about increases in the cost of one strategy over the other can be made and tested in experiments. For nutmeg mannikins, the cost of producing can be adjusted in the laboratory by weighing down lids that cap food patches (Giraldeau et al. 1994), and such a manipulation reduces the frequency of producers observed in groups, as would be expected. Similar results have been observed when the cost of producing or scrounging were altered in Carib grackles, another bird known to steal food from neighboring flockmates (Morand-Ferron et al. 2007). Thus, although a precise model predicting the SEF for the case of asymmetric costs was not developed, the intuition imported from the more simplistic case was valuable and helped to build more confidence in a general model of how stable patterns at the population level are driven by individuals making locally optimal decisions within their social context.

Synthesizing all of these observations together suggests that heterogeneity in producer and scrounger performance may lead to consistent tactical patterns in natural flocks. In particular, if each bird has a different intrinsic efficiency at producing and scrounging and can flexibly respond to its intake rate, then each bird may have a predictable choice of tactic use for each particular combination of flockmates it is placed with. This consistency may appear to be a kind of personality, but it is instead an adaptive response to the heterogeneity in the surrounding flockmates. To test this idea, Morand-Ferron et al. (2011) followed individual nutmeg mannikins within social foraging tasks and observed whether (i) there were persistent individual-specific strategies of tactic use and (ii) these strategies were a function of the group composition. In fact, individuals observed in socially foraging groups would adopt the same PS strategies in the same groups even after being separated for as much as six months; however, individuals would adopt new strategies when placed in novel groups. Thus, apparent idiosyncratic preference for certain tactics emerges naturally out of heterogeneity and sensitivity to local rewards and can be completely erased by even minor changes in group composition.

Transferable Insights from Behavioral Games to Human Groups

In human groups, such as faculty meetings, there may be anecdotal evidence that certain individuals always seem to emerge as a strong leader when in some groups but are almost invisible in other groups. To try to explain this phenomenon, complex models specially tuned to human uniqueness could be built. Those models might incorporate proxies for emotion, anxiety, memory, personality, or other highly cognitive processes assumed to play a role in all human decision-making. Psychologists and management scientists have developed statistical methods, such as structural equation models (Bagozzi and Yi 2012), for formally building systems of latent variables up from an introspective approach to human behavior. For example, the technology acceptance model (TAM) (e.g., Davis 1989; Davis et al. 1989) explains the adoption of new technologies in terms of latent variables like attitude and subjective norm and intention. Attempting to apply the same models to other animal systems would likely be ill-advised because it would be suspicious to assume that attitude, intention, and subjective norm had meaning outside of humans. However, despite these latent variables being meaningless in animal groups that at least superficially seem to produce similar patterns of behavior, they are often deemed necessary for the analysis of human behavior. In fact, computational simulation studies of human behavior even insert some mechanistic approximation of emotional dynamics without any understanding of the actual mechanisms that give rise to the personal experience of emotion (Popescu et al. 2013; Belhaj et al. 2014; Fan et al. 2018).

Although models derived from introspection may be viewed as superior to other models because they incorporate variables that are otherwise impossible to observe, the introspective gambit comes with much risk without much reward. Measurements of brain activity have shown that outcomes of human decisions are encoded in the brain as much as 10 seconds before a subject becomes aware of the decision (Soon et al. 2008), which may suggest that the introspective experience of a human is actually a side effect of more fundamental processes likely to be shared with other animals. Even if aspects of the conscious experience do play a significant role in explaining human behavior, little is known about the underlying mechanisms that give rise to phenomena such as emotion. Consider a computational model of human agents built using some numerical approximation of emotion and other relevant factors. Simulations of these numerical humans (n-humans) might reveal that individual assertion of leadership is a function of the surrounding group. However, the insights gained from such an n-human simulation will be inextricably linked to poor approximations of underlying mechanisms that are inferred to exist purely based on introspective experience of the modeler and not on objective operational descriptions. A more conservative approach is to approach the analysis of human behavior as in the analysis of the behavior of any other organism – based only on observations made by an objective, third-party observer.

Just as mathematical models can be used to give insight into the social behaviors of flocks of birds, birds themselves can be viewed as conceptual models for understanding social behavior in less tractable social groups, such as human social groups. Although there are significant differences between nutmeg mannikins and humans, groups of these birds searching a shared space for limited resources face social problems analogous to those seen in human groups. Consequently, studying how the birds naturally solve these problems provides perspective not only on the birds but also on the structure of the problem itself. The birds are just as arbitrary of a model system as an n-human simulation, but the bird behaviors have been shaped by similar selective pressures as human behaviors, and thus it is more reasonable to assume congruence between the bird behaviors and human society than the computational behaviors and human society. Properly using birds in this way takes great care, but good computational modeling should also be done carefully. Like a computational simulation, studying the birds in a quasi-natural laboratory environment allows for exquisite control over factors that are outside of the scope of the investigation. With great control allows for the application of mathematical methods that provide even more insight into the underlying processes. By iterating between laboratory experiments and mathematical modeling, more sophisticated conceptual models of social behavior can be built. In the case of the nutmeg mannikins, this iterative process led to a conceptual model that could explain the emergence of group-composition-dependent individual tactics. This perspective could also be used to understand the emergence of different roles in human groups where positive and negative rewards are frequency dependent. In fact, the PS game and related insights from social foraging theory have already been applied to human cases (e.g. Kameda and Nakanishi 2002; Mesoudi 2008; King et al. 2009; Lehmann and Feldman 2009), which demonstrates the generality of this approach.

Model Systems for Understanding Group Competition

The behaviorally stable strategies discussed for the nutmeg mannikins take place on short time scales and help to ensure that every individual receives the same direct benefits as all other individuals. It is not the case that birds in the scrounger role are cheating and receiving more direct benefits than birds in an altruistic producer role. No bird can receive higher direct benefits by choosing a different tactic, and all patterns that are idiosyncratic to the group are epiphenomena that emerge from inter-individual competition and have no group-level adaptive value on their own. Moreover, if one individual is added or replaces another, the previously consistent mixture of strategies may completely change.

In a human business organization, individuals may similarly self-sort into behavioral roles based on individual benefit, and those assortments may be similarly fragile when individuals are added, removed, or changed. However, the particular mixture of roles may not be optimal in terms of the organization's productivity output as a whole. In the case of a monopoly, suboptimal performance of the organization may be frustrating to its leadership but is unlikely to lead to the downfall of the company. However, when multiple companies compete with each other and consist of relatively immobile employees, the incentives of each employee are more tightly correlated with the output of their employer. Thus, tight competition among groups can shift an individual's strategic focus toward behavioral tactics that may appear to be altruistic but are actually entirely consistent with the self-interest of the individual. Such shifts in individual focus can occur at an evolutionary time scale in nonhuman animal social systems as well. A mathematical framework for this kind of nested tug-of-war was formalized by Reeve and Hölldobler (2007) to give a possible explanation for the evolution of high reproductive asymmetry in the so-called eusocial insects, a special group of social insects including all ants as well as social bees and wasps (Hölldobler and Wilson 1990, 2009). In their framework, the high levels of intergroup competition lead to an emergence of superorganismic groups that truly do have group-level properties that give rise to group-level performance. In this section, two related examples are presented – one from social spiders and the other from ants – that demonstrate how natural selection acting on group-level productivity can lead to different mixtures of individuals within those groups.

Social Spiders as Model Systems for Understanding Personality in Groups

A particularly striking natural example of emergent group-level properties comes from the socially polymorphic cobweb spider *Anelosimus studiosus* (Figure 11.2). These spiders live exclusively in large groups of related individuals. Unlike social insects, these spiders have no strict division of labor; each female could potentially lay eggs that develop to maturity. However, members of the colony do assist each other in prey capture, parental care, and web maintenance. In these spiders, each individual can be grouped into one of two behavioral types – *docile* and *aggressive* – that each correspond to a syndrome of consistent variations across a wide range of behavioral contexts

(a)   (b)

Figure 11.2 The social cobweb spider, *Anelosimus studiosus*. (a) Individual spider. (b) Large colony. The social cobweb spider *A. studiosus*, which is otherwise known as the comb-footed or tangle-web spider. Individual spiders, as in (a), form large colonies, as in (b), of related individuals that assist each other in prey capture, parental care, and web maintenance. Each colony contains a mix of two distinct behavioral phenotypes, docile and aggressive, and the particular docile-to-aggressive ratio for a colony not only has fitness effects on colony reproduction but also is heritable from mother colony to daughter colony. Source: Both images are provided under a CC-BY 2.0 (Creative Commons 2018a) license. The image in (a) is by Judy Gallagher. The image in (b) is by Sarah Zukoff.

(Pruitt et al. 2008; Pruitt and Riechert 2009b; Pruitt 2012). In other words, the behavioral type alone is a strong predictor of several behavioral traits simultaneously. These behavioral syndromes are analogous to personalities in humans – knowing that someone has an aggressive personality allows for good predictions of responses in a wide range of behavioral contexts. Likewise, spider personalities can be determined in laboratory assays by scoring spiders in a range of behavioral tests over different contexts (Pruitt et al. 2008). Furthermore, the docile–aggressive behavioral type of a spider is heritable – the deviation in offspring behavioral type from the population mean is 32% of the deviation of the parents' behavioral types from the population mean (Pruitt and Riechert 2009b). In other words, the offspring of two aggressive parents will be more likely to be aggressive itself than would otherwise be predicted by the natural variation in behavioral types. Natural colonies contain nontrivial mixtures of docile and aggressive individuals, and there is high variation across colonies in the within-group composition of different behavioral types. Additionally, spider groups of an artificially determined size and personality mixture can be generated and studied in the lab or transplanted back into a natural field habitat for observation (Pruitt and Riechert 2009a). Together, the tractability of determining a personality of an individual spider and the ability to create colonies of artificial size and personality composition make these spiders ideal model systems for studying the effect of different personality mixtures on group performance.

To say that groups are in competition means that there is some intrinsic group-level property that:

- Varies among groups.
- Is inherited, to some extent, from a parent group to offspring groups.
- Leads some groups to have more group progeny than other groups.

In fact, these are the same properties that are necessary for natural selection to act on an individual-level trait. Showing that a system exhibits these three properties at the level of a natural group has historically been very difficult until a carefully designed study by Pruitt and Goodnight (2014) on *A. studiosus* social spiders. In the study, spiders were collected from several different sites that could broadly be put into two groups – high and low resource availability – under the assumption that the different group conditions would put different selective pressures on colonies. The docile-to-aggressive ratio was measured in natural colonies from each site to determine the local mixture, and then colonies with artificial docile-to-aggressive mixtures were generated in the laboratory and transplanted either into the native site or into a foreign site that differed in resource availability. Pruitt and Goodnight (2014) then measured the reproductive output of each of the released groups (i.e. the number of offspring colonies produced) as well as the docile-to-aggressive composition of the released groups two generations after release. In summary:

- Experimental colonies configured with docile-to-aggressive ratios that greatly differed from their local mixture did not survive or produce any offspring colonies.
- Experimental colonies with docile-to-aggressive ratios that matched the local mixture produced 10 times as many offspring as those colonies configured with ratios different from but near the local mixture.
- After two generations, experimental colonies placed in foreign sites reverted to docile-to-aggressive ratios that were closer to their original site's local mixture.

The first two results imply that the local mixture is optimal for the environment – it provides higher fitness than other mixtures to the group. Alone, this result might mean that colonies are responding to local costs and benefits similar to the nutmeg mannikins described earlier. However, the third result shows that the docile-to-aggressive ratio is truly a group-level trait that is maintained by the group independent of the environment. Thus, the personality mixtures emerge from processes intrinsic to the group, and groups with different personality mixtures compete with each other to determine which groups proliferate over time. Exactly why certain personality mixtures are well suited for particular sites is not understood and a potentially interesting future research direction.

Ants as Model Systems for Understanding the Costs and Benefits of Specialization

The fact that certain behavioral mixtures of social spider groups were shown to be advantageous in certain habitats suggests that personality mixtures may correspond to certain divisions of labor. Aggressive spiders, for example, may be specialists on nest defense and prey capture, whereas docile spiders may be specialists on brood care. A colony needs both, but the right mixture may vary depending on the demands put on each colony. Whereas social spiders are a good model system to study the costs and benefits of different mixtures of specializations, other animal model systems are well suited to study the value of specialization itself. One such example studied by Jongepier and Foitzik (2016) is the system of *Temnothorax longispinosus* acorn ants and their social parasite *Temnothorax americanus*.

Temnothorax longispinosus is a small ant with colonies that reside in individual acorns in nature but can be readily transplanted into artificial laboratory nests. Because the colonies are so small and the tempo of the ants so slow, every individual within the nest can be observed simultaneously, which makes them ideal for studying division of labor. In some areas, these ants share their habitat with the social parasite *T. americanus*, which has lost the ability to raise workers that care for its own brood. Instead, newly mated *T. americanus* queens or

0.5 mm

(a)   (b)

Figure 11.3 *Temnothorax longispinosus* host and *Temnothorax americanus* social parasite. (a) *T. longispinosus* queen. (b) *T. americanus* queen with slave workers. The *T. longispinosus* acorn ant and its raiding, slave-making, social parasite *T. americanus*. A *T. longispinosus* queen, which is also the largest individual in a colony, is shown in (a). A colony of these small ants can live within a single acorn. In (b), the large ant is a queen of *T. americanus*. Workers of *T. americanus* will raid colonies of *T. longispinosus* (and others) and abscond with the brood (eggs, larvae, and pupae), which are brought back to the nest of the *T. americanus* colony where they will develop into slave workers of that colony. Alternatively, recently mated queens of *T. americanus* can invade a functioning colony of *T. longispinosus* and kill all workers, effectively stealing all brood as well as the cavity of the host *T. americanus* colony. Source: The image in (a) is by April Nobile from www.AntWeb.com and provided under a CC-BY 4.0 (Creative Commons 2018d) license. The image in (b) is by Gary Alpert and provided under a CC-BY-SA 3.0 license (Creative Commons 2018b).

workers from established *T. americanus* colonies seek out other ant colonies, including *T. longispinosus*, and raid those colonies in search of brood (i.e. eggs, larvae, and pupae) that can be enslaved to serve the reproductive interests of the *T. americanus* queen. These raids are potentially fatal to an appreciable number of ants from the host colony being raided. Queens of both of these ants are shown in Figure 11.3, where the *T. americanus* queen in Figure 11.3b is shown with worker slaves of yet another *Temnothorax* species.

At least since Smith (1776), specialization has been held up as an enormous benefit to functioning societies because it minimizes the costs of task switching. Nevertheless, real societies are rarely free of generalists and usually have levels of specialization far less than what is theoretically possible. One possible explanation for the relatively low level of specialization in society is that developmental constraints prevent the realization of maximally specialized mixtures. Alternatively, there may be costs to specialization, and the intermediate level of task specialization reflects a balance of its costs and benefits. It is this latter hypothesis that Jongepier and Foitzik (2016) tested using *T. longispinosus* as a model system.

In ant colonies, there are a wide variety of task types that must be completed by workers. During raiding events of *T. longispinosus* by *T. americanus*, there are two mutually exclusive tasks that should be of critical importance to survive the raid, namely:

- On encountering a raiding ant, a host ant can attempt to defend the colony and immobilize or kill the invader.
- On encountering an undefended brood item, a host ant can attempt to carry the brood item out of the nest out of reach of the raiders.

Similar to the personality test and artificial group generation described for social spiders, Jongepier and Foitzik (2016) developed a laboratory procedure to find ants that either specialized in one of these tasks or generalized and could complete both. They then created subsets of ant colonies with artificially defined compositions of specialists and generalists. By subjecting these colony fragments to raids, they could use the number of brood saved by the end of the raid as a measure of the success of that particular mixture of specialists and generalists. Their results showed clearly that colonies composed of generalists saved more brood than those composed of specialists. Effectively, colonies with generalists could continuously perform both tasks even after a large number of host ants were lost as casualties of the conflict. In specialist colonies, one task type could be completely lost during the raid. Moreover, to confirm that this result was not a laboratory artifact, they sampled the actual level of specialization in natural colonies in areas with and without *T. americanus* raiding ants present. Those colonies in areas without *T. americanus* showed high levels of specialization, whereas those colonies in areas with raiding ants had more generalists, which is consistent with their laboratory results. A colony with more specialists is less resilient to raids, but this resilience cost is modulated by how likely a colony is to be raided. Thus, although specialization is advantageous in some contexts, it may be disadvantageous in others.

Personality and Specialization: From Nonhuman to Human Groups

In the first section, the nutmeg mannikins were used as a model of how the introduction of a novel individual in a group could lead to widespread behavioral changes in all others in the group. Thus, any properties that are apparently idiosyncratic to the group are actually emergent properties from the interaction among the group members and their environment. However, for the social spiders from the second section, groups truly do have their own separate sets of properties, and those properties can have causal influence on which member compositions persist or are lost in the meta-population of groups. Whereas it was more parsimonious to discuss the individual-level costs and benefits in the case of nutmeg mannikins, it is more parsimonious to discuss group-level costs and benefits in social spiders. The social spiders provide a conceptual model for considering how personality mixture effects the performance of a group. Furthermore, the example of the social spiders demonstrates that some groups will reinforce certain personality mixtures over others even after being perturbed from their original mixture. Projecting onto human cases, organizations wishing to alter their personality mixture (e.g. to shift away from a workforce dominated by aggressive individuals) may need to do much more than simply adjust the mixture of personality types within the organization or else the original group composition will return, possibly at the detriment of the group performance. For example, the "cultural DNA" of the organization or other factors favoring the retention of aggressive individuals and encouraging the attrition of docile individuals may also need to be changed to ensure long-lasting changes in group dynamics.

Just as human organizations often have staff that are cross-trained on multiple tasks, ant groups like the ones described above often include mixtures of specialists and generalists. Cross-training is often viewed from the perspective of maximizing the utilization of human resources – harvesting idle time from the cross-trained group of generalists to better respond to sudden bursts of demand for one task type or another. However, the ant model shows that generalists also buffer against catastrophic events, and the generalist-to-specialist ratio is effectively a sliding scale of resilience. Having a single specialist for each task type means having a single point of failure; the system depends on every individual being present and working without error. If conditions exist that remove individuals from the workforce or introduce faults, highly specialized groups will not be able to recover. Furthermore, in the utopian, hyper-specialized vision of Smith (1776), a highly efficient society would also be very fragile. Consequently, natural ant colonies that are highly specialized coincide with fail-safe areas where raiding is very unlikely, and naturally occurring ant colonies with appreciable numbers of generalists are safe to fail and are thus resilient to periodic invasions or other dislocations of the workforce.

Both of these examples examined group-level performance in a sequence of steps starting with careful laboratory observations of individuals and building up to observations of real groups in nature. Each layer of investigation was informed by conceptual developments and tools developed in the layer before it. Furthermore, mathematical models linking individual performance to group performance like the game-theoretic approach of Reeve and Hölldobler (2007) provide a unifying perspective for comparing and contrasting the evolutionary pressures that ultimately shaped these complex societies. Colonies of cobweb spiders or acorn ants may seem unrelated to human groups, but they are sufficiently rich examples of complex sociality that they can reveal general principles that transcend taxonomic boundaries. Furthermore, in the sequence of steps leading up to understanding a complex human social phenomenon, natural models even as phylogenetically distant as arthropods can still have value at intermediate stages of the sequence. In fact, computer simulation models developed by sociophysicists to describe the behavior of human groups may be even more distant from real human groups than animal models that were at least shaped by the same evolutionary pressures. In reality, careful laboratory and field observations of animal social systems can be complementary to mathematical and computational methods meant to directly capture human-specific behaviors. Just as in the case of *T. longispinosus* ants, specializing on a single modeling method may increase the efficiency of the research community, but it will necessarily decrease its resilience.

Information Dynamics in Tightly Integrated Groups

Patterns of group composition are one of the most conspicuous features of social groups; however, they may also be one of the most superficial. A tightly integrated group has individuals that share information so as to coordinate collective actions. It may seem natural to use large, social vertebrates as models for coordination and information processing in human systems; however, most information processing work in the social vertebrate literature (e.g. Hare and Tomasello 2005; Greenberg et al. 2010; Plotnik et al. 2011; Sasaki and Biro 2017) attempts to measure the similarity between those systems and humans as opposed to discovering new insights for further human studies. Furthermore, those animal models often involve small groups – not the kind of large-scale information processing that might characterize teams, companies, cities, and societies. One exception is the study of collective motion, as in the study of fish schools that collectively escape the attacks of predators (Handegard et al. 2012). However, just as it was described for nutmeg mannikins, fish in these schools are not truly sharing information so much as exploiting available information from fish around them. Coordination, information integration, and deliberation is not well represented in studies of these vertebrate groups.

Alternatively, organismal biologists have studied collective decision-making in a wide range of social-insect systems without explicit reference to human decision-making (e.g. Camazine et al. 1999; Detrain et al. 1999; Franks et al. 2002; Kulahci et al. 2008). Because of strongly aligned interests among insects in a social-insect colony, natural selection has generated truly coordinated behaviors – where the actions of groups of individuals are mutually beneficial to all in the group – as opposed to purely exploitative behaviors between individuals, where the actions of individuals are pragmatic responses to the actions of others. For example, a worker ant is physiologically constrained so that she has little-to-no opportunity to produce her own offspring, and so it is in her best interests to join with her similarly constrained sisters to raise the brood of their mother, who does have the ability to produce offspring. Thus, the competition between two ant colonies is effectively the competition between queens; the workers are so tightly integrated into the queen's interests that they may be viewed as her extended phenotype. Workers in an ant colony coordinate in much the same way cells of a multicellular organism do, which is one reason for the use of the term superorganism to describe colonies of these so-called *eusocial* insects – that is, insects with extreme reproductive asymmetry among overlapping generations of workers that practice cooperative brood care (Hölldobler and Wilson 1990, 2009). So eusocial insects are macroscopic systems with integration rivaling that of the integration between microscopic components of a monolithic organism.

Furthermore, because similar collective decision-making challenges have been solved in different ways across different insect taxa, the social insects are a natural comparative laboratory for understanding how different information processing mechanisms are matched to different contexts. Thus, these social-insect systems can be useful models for understanding effective information processing in large groups under significant communication constraints. The wide range of information processing examples in social insects can serve as inspiration for the development of novel decision support systems to improve decision-making in large human socio-technological systems where tight integration and high group-level performance is desired. This section describes several examples of how different tactics for sharing information among individuals lead to important differences in collective decision-making in ants. For more details on the connections between crowd-sourced human computation and a wide range of social insects (including examples from ants, bees, and wasps), see the extensive review by Pavlic and Pratt (2013).

Linear and Nonlinear Recruitment Dynamics

Collective decision-making in ant colonies often requires each ant to recruit others to new opportunities she discovers, such as a new food source. As discussed above, ant colonies are eusocial, which implies that it is in each worker's best interests to increase the productivity of her mother queen. So the sharing of information is not altruistic, nor is it exploitative; it is mutually beneficial to all in the colony. Much of the earliest work on collective decision-making by ants studied recruitment during group foraging in the black garden ant, *Lasius niger* (Figure 11.4). When an *L. niger* worker finds a food source, she can recruit other workers to that food source by depositing a chemical pheromone trail between the food source and the home nest. Workers that would otherwise be waiting at the nest or searching randomly through the environment will preferentially follow that trail and, after finding the corresponding food source, can reinforce the trail with additional chemical deposition. Unreinforced, the chemical trail will evaporate, and foragers will return to a random search pattern. However, if the trail attracts enough recruits to reinforce it at a rate faster than the evaporation rate, it will persist and even grow in its attractiveness. In Figure 11.5, hypothetical relationships between the strength of the trail (i.e. how many ants are committed to the

Figure 11.4 Workers of the black garden ant, *Lasius niger*, tending to a herd of mealybugs. Source: This image is by Katja Schulz and is used here under a CC-BY 2.0 license (Creative Commons 2018a).

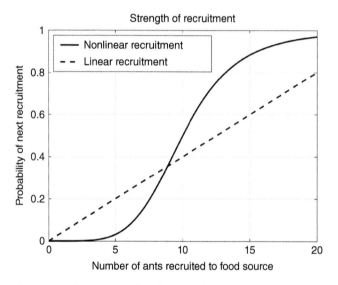

Figure 11.5 Comparison of nonlinear and linear recruitment. The dashed line is an example of linear recruitment, which represents that a single recruiter has the same ability to recruit another ant regardless of how many other ants are currently committed to a site. Thus, the attractiveness of a site rises linearly with the number of ants recruiting for it. The solid line shows an example of nonlinear recruitment, where an ant may be very unlikely to follow a recruiter when there are few ants committed to a site but increasingly more likely to follow a single recruiter when several recruiters are committed to the site.

corresponding site) and the probability that a new ant will follow the trail are shown. Mathematical models (Sumpter and Beekman 2003) suggest that if the shape of this curve has an accelerating portion, as in the solid line in Figure 11.5, the foraging ants will come to consensus on a single site. That is, so-called *nonlinear recruitment* will lead to random search being abandoned, and all ants will be concentrated on a single trail that is sustained by their continuous reinforcement. Furthermore, laboratory experiments with *L niger* confirm that their trail-laying dynamics do lead colonies to come to consensus on a single foraging site (Beckers et al. 1990, 1993).

Although the chemical communication between trail-laying ants like *L. niger* may seem foreign compared to modes of human communication, similar forms of emergent consensus building exist in human socio-technological systems. For example, humans engaged in online commerce may be given the opportunity to leave publicly accessible reviews on purchased products. The cumulative record of reviews for a product is not unlike the accumulation of pheromones for a particular food source. A product with a few reviews may not attract much more attention than a product with no reviews. However, once a product reaches some threshold level of reviews, it can dominate the attention of naïve consumers. Even the notion of ratings (e.g. 4 out of 5 stars) has an analog in trail-laying ants because ants can deposit different amounts of pheromone based on the perceived quality of the food source. So just as nonlinear recruitment is expected to drive trail-laying ants to consensus on a single option, the persistent review record on online commerce sites combined with human psychology can lead to an apparent consensus on a single product. Thus, studying the accuracy and flexibility of such recruitment systems in ants can also shed light on important scenarios for analogous human information aggregation systems.

Herd Behavior and Information Cascades in Ants

Continuing the analogy with human markets, the recruitment process described for *L. niger* is formally similar to sequential decision models developed by economists to understand herd behavior in human markets (Banerjee 1992; Bikhchandani et al. 1992; Welch 1992). In those economic models, the transition from private information to social information led to a reduction in system flexibility and a risk of an information cascade where consumers would ignore novel, potentially better products if those products arrived too late in the process. Despite the term herd behavior having obvious connections to nonhuman biology, early economists studying sequential models of human markets were apparently unaware of much of the work in social-insect studies. Bikhchandani et al. (1992) do review examples from the vertebrate literature and even note that it may be advantageous to study zoological examples because animals are less likely to engage in extended discourse and mass media campaigns that might affect individual decision making about alternative actions. In other words, Bikhchandani et al. (1992)

recognize that animal models can be used to control otherwise confounding effects so as to validate insights from theoretical models, and they discuss particular results where cascades apparently occur in vertebrate systems. However, they ignore or are unaware of results like those of de Groot (1980) where vertebrate social systems are able to avoid the deleterious effects of information cascades. If they had recognized that vertebrate social systems exist that are both prone to cascades and immune to their effects, it may have motivated more sophisticated economic models for understanding the complexities of human behavior.

Beyond the vertebrate literature, it is unfortunate that early economists studying herd behavior and information cascades were apparently unaware of work in the invertebrate literature on similar problems. Beckers et al. (1990) studied both *L. niger* and the pavement ant, *Tetramorium immigrans* (then called *Tetramorium caespitum*), in laboratory experiments where a low-quality feeder was introduced long before a high-quality feeder. They show that *L. niger* fixates on the low-quality feeder and is unable to break the social inertia when the high-quality feeder is introduced; however, the pavement ant is able to detect the later-introduced high-quality feeder and reallocate foragers to exploit both feeders simultaneously, with a greater proportion of ants allocated to the high-quality feeder, thus sacrificing consensus for flexibility. Beckers et al. (1990) also introduce a simple mathematical model that is able to reproduce both trajectories so long as *L. niger* is assumed to use purely nonlinear recruitment and the pavement ants are assumed to use a mixture of linear and nonlinear recruitments. One possible explanation for the putative linear portion of the pavement ant response is that pavement ants interact both with an accumulating pheromone trail and directly with each other. However, mechanisms of direct ant-to-ant interaction in pavement ants were not well understood. However, direct ant-to-ant recruitment is observed in several species of ants in the genus *Temnothorax*, and any use of pheromone trails in these ants is apparently limited to individual navigation as opposed to communication (Maschwitz et al. 1986; Möglich et al. 1974; Möglich 1979; Pratt et al. 2001; Franks and Richardson 2006; Franklin et al. 2011). Consequently, Shaffer et al. (2013) were able to show that *Temnothorax rugatulus* allocate foragers to feeders in proportion to quality and are able to track changes in quality, just as would be predicted in the mathematical model of Beckers et al. (1990) with purely linear recruitment. The distinction between linear and nonlinear recruitment has become a fundamental component of modern models of social-insect recruitment dynamics (Nicolis and Deneubourg 1999; Detrain and Deneubourg 2008; Sumpter 2010).

Although *L. niger* has historically been used as a typical model of recruitment by trail laying, more recent work has shown that many other trail-laying ants can respond to changes in feeder quality (Latty and Beekman 2013) similar to the pavement ants studied by Beckers et al. (1990). Furthermore, rather than assuming that these ants achieve their flexibility through linear recruitment mechanisms, Dussutour et al. (2009) introduced a modification of the classic (Beckers et al. 1990) model that uses entirely nonlinear recruitment combined with some rate of individual ant errors. That is, rather than assuming that there is some unknown ant-to-ant recruitment mechanism that cannot be easily observed, they assume that individual ants sometimes fail to choose paths of highest chemical concentration. The resulting failure rate makes it possible for ants to effectively reach consensus on one option while still being able to switch to a better option that is introduced later. Their model fits the recruitment dynamics of *Pheidole megacephala* in laboratory experiments where feeders are made to change positions during the experiment and ants are observed making errors at decision points in the laboratory apparatus. Thus, Dussutour et al. (2009) show both with a mathematical model and with evidence from controlled laboratory experiments with *P. megacephala* that deleterious information cascades can also be avoided if there is some small rate of infidelity to the current socially selected option. That is, it is possible for the occurrence of individual-level mistakes to be a mechanism that paradoxically reduces the likelihood of group-level mistakes – possibly at the cost of reducing the equilibrium level of group agreement or the convergence rate of a group decision.

From Ants to Human Decision Support Systems

In this section, connections have been made between sequential decision models in human systems and foraging recruitment dynamics in ants. Whereas the risk of information cascades is of critical importance in the human systems, the ant models show at least two mechanisms that can prevent deleterious information cascades, namely:

- Linear recruitment, which can be implemented by one individual only recruiting a finite number of others, generates social allocations in proportion to quality. Consensus on a single option is lost, but the social allocations can respond flexibly to changes in quality.
- Individual errors, which represent the propensity of an individual to choose an option that differs from the social choice, allow for nonlinear recruitment mechanisms that nonetheless are sensitive to changes in option quality. Groups effectively reach consensus on a single option, but enough error-prone individuals will monitor for changes in the distribution of quality and be able to steer the group consensus toward new, high-quality options.

For the case of markets, the latter case of individuals making social errors is essentially capturing the effect that Bass (1969) attributed to advertising in his classic model of innovation diffusion through markets. That is, discovery of a high-quality option is only possible by enticing individuals to stray from an existing option, which is far more likely when those new options are advertised. Once enough individuals adopt the newer, higher-quality option, word of mouth generates an autocatalytic growth in the adoption of that new product. However, the use of linear recruitment to prevent information cascades has not yet been investigated in human socio-technological systems. Returning to the online commerce example from earlier, the process of maintaining reviews for products could be adjusted to implement a linear recruitment model. For example, if a review recorded for a product automatically expired after a finite number of times it was accessed, then every consumer of a product would be limited to recruiting a finite number of new candidate consumers to that product. In other words, automatically deleting reviews would induce an approximation of linear recruitment. In the case of *T. rugatulus* ants, linear recruitment has the effect of allocating foragers to sites in proportion to their relative quality, and those allocations update automatically with changes in quality (Shaffer et al. 2013). The analogous effect for online commerce would be an automatic allocation of potential consumers in proportion to quality of competing products, and newly generated products would be guaranteed to be discovered and gain market share.

Additional Examples: Rationality and Memory

In simple foraging experiments like the ones described so far, each option varies in its sugar concentration, and colonies are assumed to prefer higher concentrations over lower ones. In crevice-dwelling species, such as European honeybees and ants in the genus *Temnothorax*, colonies must frequently choose a new nest site from a finite number of choices in the environment (Franks et al. 2002). Moreover, these nest sites can vary in a wide range of criteria of apparent interest to the insect decision-makers (Visscher 2007). Consequently, ant colonies such as *T. rugatulus* can be useful models for both understanding how individual-level decision making scales to group-level decision making and how natural systems solve multicriteria decision-making problems.

In a nest-site selection problem, the choice of a colony of ants is clear when the colony moves into one of several presented candidate nest cavities. For comparison, a similar test can be given to individual ants. In particular, individual *T. rugatulus* ants can be placed in controlled decision-making experiments by placing them in an arena with a group of brood items (e.g. eggs, larvae, or pupae). If multiple cavities are also placed in the arena, the individual will scrutinize the cavities and eventually transport brood into one of them, which indicates her choice. Thus, individual-level decision-making can be directly compared with group-level decision making. For example, Sasaki and Pratt (2012) presented a pair of nests – one that is constructed to be highly preferred over the other – to individual ants as well as to groups of ants. In that case, both the individual and the group selected the nest constructed to be higher quality. However, when four pairs of high- and low-quality nests were given, the ability of an individual ant to choose a high-quality nest from the eight candidates became no better than chance. Despite the difficulty observed for the individual ant, colonies continued to perform well in the 4-pair test; a high-quality nest was overwhelmingly preferred. These results indicate that cognitive overload does exist in individual ants but is somehow mitigated by the information aggregation process at the level of the colony. This observation suggests that carefully selected information aggregation mechanisms may be able to pool assessments in human groups so that the group can make good decisions among a group of alternatives so large that would overwhelm the cognitive capacities of a single individual. For example, rather than asking individuals to evaluate every option simultaneously, it may be more effective at the group level to have individuals nominally only evaluate one option and recruit others to those options based on their enthusiasm. Asking a customer to leave a product rating (e.g. 4 out of 5 stars) without any reference to competing alternatives may be viewed as a crude approximation of this kind of recruiting.

Even when choosing among a small number of alternatives, decisions can be complicated by multiple, competing criteria whose relative value may change over time and experience. In the case of *T. rugatulus* ants choosing among nests that vary in darkness and entrance size, the best nest will be the darkest and have the smallest entrance size. However, depending on the available nests, there may be a conflict between the objective of maximizing darkness and minimizing entrance size. Furthermore, if a colony migrates from one region to another, the importance of attending to one of the features (e.g. darkness) may vary (e.g. if all nests in the new area have the same darkness). To test whether ant colonies vary in their attentiveness to certain features based on their history, Sasaki and Pratt (2013) gave colonies of *T. rugatulus* the choice of two nests that varied in both darkness and entrance size. This allowed for a measurement of a baseline preference for the two nests. Colonies could then be subjected to an intermediate choice trial of two nests that varied in only one of the two features – darkness or entrance size – and then brought back to the original baseline choice context. Ants that were subjected to an intermediate treatment where both nests matched in darkness were found to more strongly prefer entrance size when returning to the baseline case. Similarly, ants that experienced the intermediate treatment where both nests matched in entrance size had less of a preference for entrance size when brought back to the baseline case. Thus, somewhere within the *T. rugatulus* ant colony, there is a social memory that stores the preferences of the colony and can be changed with experience to new habitats. Similarly, in a study of *Atta insularis* leaf-cutter ants, Nicolis et al. (2013) found that activity patterns at the nest entrance were best described by a dynamical system model with parameters indicating that the system is at the edge of chaos (Feigenbaum 1978; Langton 1990; Lewin 1999; Mora and Bialek 2011). In this region, the activity patterns of the ants naturally oscillate. However, when the oscillations vary greatly from the externally generated temperature oscillations, the internal oscillations of the ant colony can self-organize to entrain on the external temperature oscillations. Thus, like the *T. rugatulus* preferences for one criterion over another, the leaf-cutter ants can tune a colony-level operational parameter based on experience. These two examples – *T. rugatulus* and *A. insularis* – demonstrate that these ant systems can provide insights into how social memory emerges within a group, drives the actions of that group, and can be altered by the experience of the group over time.

Conclusions

Throughout this chapter, examples have been given that showed how behavioral ecologists combine mathematical modeling with laboratory assays and field work to test hypotheses of ever-increasing complexity. In every case, the particular research question of interest guides the choice of mathematical model as well as animal model. The kinds of research questions addressed in the first section on the emergent patterns in large groups of self-interested individuals are qualitatively different than the questions from the previous section on the behavior of tightly integrated groups of individuals with aligned interests. Consequently, the animal model systems for the research questions are similarly diverse. Many examples exist of human groups with high levels of individual competition as well as examples of human groups with aligned interests due to high levels of group competition. No animal model is meant to be a perfect match for a human society. Like a mathematical model, it is meant to be a surrogate for a specific case that would be difficult to study directly.

When combining mathematical modeling with the study of animal behavior, it is easy to make the mistake in assuming that animals must possess mathematical skills to be able to behave according to the mathematical model. As argued by Mottley and Giraldeau (2000), relatively simple learning heuristics can be used by a nutmeg mannikin to determine when to switch from the producer to scrounger tactic and back. Although predicting the SEF requires the behavioral analyst to have some familiarity with calculus, there is no reason to believe that the birds can do any level of mathematics. However, even under the assumption that some birds are expertly trained mathematicians and could conceivably deliberate on strategic foraging decisions, there is no reason to believe that large-scale demographic patterns in their tactic use are due to those sophisticated deliberative processes. Although anthropomorphism is rampant in popular explanations of nonhuman animal behavior, the suggestion that nonhuman behavior can provide a mental model for understanding human behavior is often met with skepticism because human cognition is held up as being singularly different. Even taking for granted that human cognition is so marvelous, there is no reason to believe that the higher cognition that humans are endowed with plays a significant role in the day-to-day patterns seen in human groups.

Furthermore, if a particular social behavior can only be explained by assuming individuals have advanced cognitive skills, a complex artificial computer simulation is arguably no better than an animal model for some questions.

Although there is no nonhuman system that will ever be a perfect match for a given human social system, there is a great diversity of nonhuman social systems from which to learn. As said by Box (1979), "all models are wrong but some are useful." The goal of modeling is not to find the most correct model but to find the most useful models. An ant is no more of a human than an equation, a statistic, or a line of code is. In the search for the most illuminating models, excluding insights from nonhuman systems is necessarily introducing more error in our understanding, not less.

Acknowledgments

The writing of this paper was supported by DARPA under the Bio-Inspired Swarming seedling project, contract FA8651-17-F-1013. Images in this chapter that were not already in the public domain were used either under the explicit permission of the image owner or according to a CC BY 2.0 (Creative Commons 2018a), CC BY 3.0 (Creative Commons 2018c), CC BY-SA 3.0 (Creative Commons 2018b), or CC BY 4.0 (Creative Commons 2018d) license.

References

Bagozzi, R.P. and Yi, Y. (2012). Specification, evaluation, and interpretation of structural equation models. *Journal of the Academy of Marketing Science* 40 (1): 8–34. https://doi.org/10.1007/s11747-011-0278-x.

Banerjee, A.V. (1992). A simple model of herd behavior. *The Quarterly Journal of Economics* 107 (3): 797–817. https://doi.org/10.2307/2118364.

Barnard, C.J. and Sibly, R.M. (1981). Producers and scroungers: a general model and its application to captive flocks of house sparrows. *Animal Behaviour* 29: 543–550. https://doi.org/10.1016/S0003-3472(81)80117-0.

Bass, F.M. (1969). A new product growth for model consumer durables. *Management Science* 15 (5): 215–227. https://doi.org/10.1287/mnsc.15.5.215.

Beckers, R., Deneubourg, J.-L., Goss, S., and Pasteels, J.M. (1990). Collective decision making through food recruitment. *Insectes Sociaux* 37 (3): 258–267. https://doi.org/10.1007/BF02224053.

Beckers, R., Deneubourg, J.-L., and Goss, S. (1993). Modulation of trail laying in the ant *Lasius niger* (Hymeoptera: Formicidae) and its role in the collective selection of a food source. *Journal of Insect Behavior* 6 (6): 751–759. https://doi .org/10.1007/BF01201674.

Belhaj, M., Kebair, F., and Ben Said, L. (2014). Agent-based modeling and simulation of the emotional and behavioral dynamics of human civilians during emergency situations. In: *Proceeding of the 12th German Conference on*

Multiagent System Technologies (MATES 2014), 266–281. Stuttgart, Germany: Springer International Publishing. https://doi.org/10.1007/978-3-319-11584-9_18.

Bikhchandani, S., Hirshleifer, D., and Welch, I. (1992). A theory of fads, fashion, custom, and cultural change as informational cascades. *Journal of Political Economy* 100 (5): 992–1026. https://doi.org/10.1086/261849.

Box, G.E.P. (1976). Science and statistics. *Journal of the American Statistical Association* 71 (356): 791–799. https://doi.org/10.1080/01621459.1976.10480949.

Box, G.E.P. (1979). Robustness in the strategy of scientific model building. In: *Robustness in Statistics* (ed. R.L. Launer and G.N. Wilkinson), 201–236. Academic Press.

Camazine, S., Visscher, P.K., Finley, J., and Vetter, R.S. (1999). House-hunting by honey bee swarms: collective decisions and individual behaviors. *Insectes Sociaux* 46 (4): 348–360. https://doi.org/10.1007/s000400050156.

Clark, C.W. and Mangel, M. (1984). Foraging and flocking strategies: information in an uncertain environment. *The American Naturalist* 123 (5): 626–641. https://doi.org/10.1086/284228.

Coolen, I. (2002). Increasing foraging group size increases scrounger use and reduces searching efficiency in nutmeg mannikins (*Lonchura punctulata*). *Behavioral Ecology and Sociobiology* 52 (3): 232–238. https://doi.org/10.1007/s00265-002-0500-4.

Coolen, I., Giraldeau, L.-A., and Lavoie, M. (2001). Head position as an inducator of producer and scrounger tactics in a ground-feeding bird. *Animal Behaviour* 61 (5): 895–903. https://doi.org/10.1006/anbe.2000.1678.

Creative Commons (2018a). Attribution 2.0 Generic (CC BY 2.0). Retrieved from http://creativecommons.Org/licenses/by/2.0/ (accessed 14 September 2018).

Creative Commons (2018b). Attribution 3.0 ShareAlike 3.0 Unported (CC BY-SA 3.0). Retrieved from http://creativecommons.org/licenses/by-sa/3.0/ (accessed 14 September 2018).

Creative Commons (2018c). Attribution 3.0 Unported (CC BY 3.0). Retrieved from http://creativecommons.Org/licenses/by/3.0/ (accessed 14 September 2018).

Creative Commons (2018d). Attribution 4.0 International (CC BY 4.0). Retrieved from http://creativecommons.org/licenses/by/4.0/ (accessed 14 September 2018).

Davis, F.D. (1989). Perceived usefulness, perceived ease of use, and user acceptance of information technology. *MIS Quarterly* 13 (3): 319. https://doi.org/10.2307/249008.

Davis, F.D., Bagozzi, R.P., and Warshaw, P.R. (1989). User acceptance of computer technology: a comparison of two theoretical models. *Management Science* 35 (8): 982–1003. https://doi.org/10.1287/mnsc.35.8.982.

de Groot, P. (1980). Information transfer in a socially roosting weaver bird *(Quelea que- lea:* Ploceinae): an experimental study. *Animal Behaviour* 28 (4): 1249–1254. https://doi.org/10.1016/s0003-3472(80)80113-8.

Detrain, C. and Deneubourg, J.-L. (2008). Collective decision-making and foraging patterns in ants and honeybees. In: *Advances in Insect Physiology,* vol. 35 (ed. S.J. Simpson), 123–173). Academic Press. https://doi.org/10.1016/S0065-2806(08)00002-7.

Detrain, C., Deneubourg, J.-L., and Pasteels, J.M. (eds) (1999). *Information Processing in Social Insects.* Basel, Switzerland: Springer Birkhauser Basel. https://doi.org/:10.1007/978-3-0348-8739-7.

Dussutour, A., Beekman, M., Nicolis, S.C., and Meyer, B. (2009). Noise improves collective decision-making by ants in dynamic environments. *Proceedings of the Royal Society B* 276 (1677): 4353–4361. https://doi.org/10.1098/rspb.2009.1235.

Fan, R., Xu, K., and Zhao, J. (2018). An agent-based model for emotion contagion and competition in online social media. *Physica A: Statistical Mechanics and its Applications* 495: 245–259. https://doi.org/10.1016/j.physa.2017.12.086.

Feigenbaum, M.J. (1978). Quantitative universality for a class of nonlinear transformations. *Journal of Statistical Physics* 19 (1): 25–52. https://doi.org/10.1007/bf01020332.

Franklin, E.L., Richardson, T.O., Sendova-Franks, A.B. et al. (2011). Blinkered teaching: tandem running by visually impaired ants. *Behavioral Ecology and Sociobiology* 65 (4): 569–579. https://doi.org/10.1007/s00265-010-1057-2.

Franks, N.R. and Richardson, T. (2006). Teaching in tandem-running ants. *Nature* 439: 153. https://doi.org/10.1038/439153a.

Franks, N.R., Pratt, S.C., Mallon, E.B. et al. (2002). Information flow, opinion polling and collective intelligence in house-hunting social insects. *Philosophical Transactions of the Royal Society B: Biological Sciences* 357 (1427): 1567–1583. https://doi.org/10.1098/rstb.2002.1066.

Giraldeau, L.-A. and Beauchamp, G. (1999). Food exploitation: searching for the optimal joining policy. *Trends in Ecology & Evolution* 14 (3): 102–106. https://doi.org/10.1016/S0169-5347(98)01542-0.

Giraldeau, L.-A. and Caraco, T. (2000). *Social Foraging Theory.* Princeton, NJ: Princeton University Press.

Giraldeau, L.-A. and Dubois, F. (2008). Social foraging and the study of exploitative behavior. In: *Advances in the Study of Behavior,* vol. 38 (ed. H.J. Brockmann, T.J. Roper, M. Naguib et al.), 59–104. Elsevier. https://doi.org/10.1016/s0065-3454(08)00002-8.

Giraldeau, L.-A. and Livoreil, B. (1998). Game theory and social foraging. In: *Game Theory and Animal Behavior* (ed. L.A. Dugatkin and H.K. Reeve), 16–37. New York: Oxford University Press.

Giraldeau, L.-A., Soos, C., and Beauchamp, G. (1994). A test of the producer-scrounger foraging game in captive flocks of spice finches *Loncbura punctulata. Behavioral Ecology and Sociobiology* 34 (4): 251–256. https://doi.org/10.1007/bf00183475.

Greenberg, J.R., Hamann, K., Warneken, F., and Tomasello, M. (2010). Chimpanzee helping in collaborative and noncollaborative contexts. *Animal Behaviour* 80 (5): 873–880. https://doi.org/10.1016/j.anbehav.2010.08.008.

Handegard, N.O., Boswell, K.M., Ioannou, C.C. et al. (2012). The dynamics of coordinated group hunting and collective information transfer among schooling prey. *Current Biology* 22 (13): 1213–1217. https://doi.org/10.1016/j.cub.2012.04.050.

Hare, B. and Tomasello, M. (2005). Human-like social skills in dogs? *Trends in Cognitive Sciences* 9 (9): 439–444. https://doi.org/10.1016/j.tics.2005.07.003.

Harley, C.B. (1981). Learning the evolutionarily stable strategy. *Journal of Theoretical Biology* 89 (4): 611–633. https://doi.org/10.1016/0022-5193(81)90032-1.

Hölldobler, B. and Wilson, E.O. (1990). *The Ants*. Harvard University Press.

Hölldobler, B. and Wilson, E.O. (2009). *The Superorganism: The Beauty, Elegance, and Strangeness of Insect Societies*. W. W. Norton & Company.

Jongepier, E. and Foitzik, S. (2016). Fitness costs of worker specialization for ant societies. *Proceedings of the Royal Society B: Biological Sciences* 283 (1822): 2015–2572. https://doi.org/10.1098/rspb.2015.2572.

Kameda, T. and Nakanishi, D. (2002). Cost-benefit analysis of social/cultural learning in a non- stationary uncertain environment. *Evolution and Human Behavior* 23 (5): 373–393. https://doi.org/10.1016/s 1090-5138(02)00101-0.

King, A.J., Johnson, D.D.P., and Vugt, M.V. (2009). The origins and evolution of leadership. *Current Biology* 19 (19): R911–R916. https://doi.org/10.1016/j.cub.2009.07.027.

Kuijt, I. (2000). People and space in early agricultural villages: exploring daily lives, community size, and architecture in the late Pre-Pottery Neolithic. *Journal of Anthropological Archaeology* 19 (1): 75–102. https://doi.org/10.1006/jaar.1999.0352.

Kulahci, I.G., Dornhaus, A., and Papaj, D.R. (2008). Multimodal signals enhance decision making in foraging bumble-bees. *Proceedings of the Royal Society B: Biological Sciences* 275 (1636): 797–802. https://doi.org/10.1098/rspb.2007.1176.

Langton, C.G. (1990). Computation at the edge of chaos: phase transitions and emergent computation. *Physica D: Nonlinear Phenomena* 42 (1–3): 12–37. https://doi.org/10.1016/0167-2789(90)90064-v.

Latty, T. and Beekman, M. (2013). Keeping track of changes: the performance of ant colonies in dynamic environments. *Animal Behaviour* 85 (3): 637–643. https://doi.org/10.1016/j.anbehav.2012.12.027.

Lehmann, L. and Feldman, M.W. (2009). Coevolution of adaptive technology, maladaptive culture and population size in a producer-scrounger game.

Proceedings of the Royal Society B: Biological Sciences 276 (1674): 3853–3862. https://doi.org/10.1098/rspb.2009.0724.

Lewin, R. (1999). *Complexity: Life at the Edge of Chaos*. University of Chicago Press.

Maschwitz, U., Lenz, S., and Buschinger, A. (1986). Individual specific trails in the *Leptothorax affi- nis* (Formicidae: Myrmicinae). *Experientia* 42 (10): 1173–1174. https://doi.org/10.1007/bf01941298.

Mesoudi, A. (2008). An experimental simulation of the 'copy-successful-individuals' cultural learning strategy: adaptive landscapes, producer-scrounger dynamics, and in formational access costs. *Evolution and Human Behavior* 29 (5): 350–363. https://doi.org/10.1016/j.evolhumbehav.2008.04.005.

Möglich, M. (1979). Tandem calling pheromone in the genus *Leptothorax* (Hymenoptera: Formicidae): behavioral analysis of specificity. *Journal of Chemical Ecology* 5 (1): 35–52. https://doi.org/10.1007/bf00987686.

Möglich, M., Maschwitz, U., and Hölldobler, B. (1974). Tandem calling: a new kind of signal in ant communication. *Science* 186 (4168): 1046–1047. https://doi.org/10.1126/science.186.4168.1046.

Mora, T. and Bialek, W. (2011). Are biological systems poised at criticality? *Journal of Statistical Physics* 144 (2): 268–302. https://doi.org/10.1007/s10955-011-0229-4.

Morand-Ferron, J., Giraldeau, L.-A., and Lefebvre, L. (2007). Wild Carib grackles play a producer scrounger game. *Behavioral Ecology* 18 (5): 916–921. https://doi.org/10.1093/beheco/arm058.

Morand-Ferron, J., Wu, G.-M., and Giraldeau, L.-A. (2011). Persistent individual differences in tactic use in a producer-scrounger game are group dependent. *Animal Behaviour* 82 (4): 811–816. https://doi.org/10.1016/j.anbehav.2011.07.014.

Mottley, K. and Giraldeau, L.-A. (2000). Experimental evidence that group foragers can converge on predicted producer-scrounger equilibria. *Animal Behaviour* 60 (3): 341–350. https://doi.org/10.1006/anbe.2000.1474.

Moynihan, M. and Hall, M.F. (1955). Hostile, sexual, and other social behaviour patterns of the spice finch *(Lonchura punctulata)* in captivity. *Behaviour* 7 (1): 33–75. https://doi.org/10.1163/156853955x00021.

Nicolis, S.C. and Deneubourg, J.-L. (1999). Emerging patterns and food recruitment in ants: an analytical study. *Journal of Theoretical Biology* 198: 575–592. https://doi.org/10.1006/jtbi.1999.0934.

Nicolis, S.C., Fernandez, J., Perez-Penichet, C. et al. (2013). Foraging at the edge of chaos: internal clock versus external forcing. *Physical Review Letters* 110 (26). https://doi.org/:10.1103/PhysRevLett.110.268104.

Pavlic, T.P. and Pratt, S.C. (2013). Superorganismic behavior via human computation. In: *Handbook of Human Computation* (ed. P. Michelucci), 911–960. Springer. https://doi.org/10.1007/978-1-4614-8806-4_74.

Plotnik, J.M., Lair, R., Suphachoksahakun, W., and de Waal, F.B.M. (2011). Elephants know when they need a helping trunk in a cooperative task. *Proceedings of the National Academy of Sciences of the United States of America* 108 (12): 5116–5121. https://doi.org/10.1073/pnas.1101765108.

Popescu, M., Keller, J.M., and Zare, A. (2013). A framework for computing crowd emotions using agent based modeling. In: *2013 IEEE Symposium on Computational Intelligence for Creativity and Affective Computing (CICAC)*. Singapore: IEEE. https://doi.org/10.1109/cicac.2013.6595217.

Pratt, S.C., Brooks, S.E., and Franks, N.R. (2001). The use of edges in visual navigation by the ant *Leptothorax albipennis*. *Ethology* 107 (12): 1125–1136. https://doi.org/10.1046/j.14390310.2001.00749.x.

Pruitt, J.N. (2012). Behavioural traits of colony founders affect the life history of their colonies. *Ecology Letters* 15 (9): 1026–1032. https://doi.org/10.1111/j.1461 -0248.2012.01825.x.

Pruitt, J.N. and Goodnight, C.J. (2014). Site-specific group selection drives locally adapted group compositions. *Nature* 514 (7522): 359–362. https://doi.org /10.1038/nature13811.

Pruitt, J.N. and Riechert, S.E. (2009a). Frequency-dependent success of cheaters during foraging bouts might limit their spread within colonies of a socially polymorphic spiders. *Evolution* 63 (11): 2966–2973. https://doi.org/10.1111 /j.1558-5646.2009.00771.x.

Pruitt, J.N. and Riechert, S.E. (2009b). Sex matters: sexually dimorphic fitness consequences of a behavioural syndrome. *Animal Behaviour* 78 (1): 175–181. https://doi.org/10.1016/j.anbehav.2009.04.016.

Pruitt, J.N., Riechert, S.E., and Jones, T.C. (2008). Behavioural syndromes and their fitness consequences in a socially polymorphic spider *Anelosimus studiosus*. *Animal Behaviour* 76 (3): 871–879. https://doi.org/10.1016 /j.anbehav.2008.05.009.

Reeve, H.K. and Hölldobler, B. (2007). The emergence of a superorganism through intergroup competition. *Proceedings of the National Academy of Sciences of the United States of America* 104 (23): 9736–9740. https://doi.org/10.1073 /pnas.0703466104.

Restall, R. (1997). *Munias and Mannikins*. Yale University Press.

Sasaki, T. and Biro, D. (2017). Cumulative culture can emerge from collective intelligence in animal groups. *Nature Communications* 8: 15049. https://doi.org /10.1038/ncomms15049.

Sasaki, T. and Pratt, S.C. (2012). Groups have a larger cognitive capacity than individuals. *Current Biology* 22 (19): R827–R829. https://doi.org/10.1016/j.cub .2012.07.058.

Sasaki, T., and Pratt, S.C. (2013). Ants learn to rely on more informative attributes during decisionmaking. *Biology Letters* 9 (6). https://doi.org/10.1098/rsbl .2013.0667.

Shaffer, Z., Sasaki, T., and Pratt, S.C. (2013). Linear recruitment leads to allocation and flexibility in collective foraging by ants. *Animal Behaviour* 86 (5): 967–975. https://doi.org/10.1016/j.anbehav.2013.08.014.

Smith, A. (1776). *An Inquiry into the Nature and Causes of the Wealth of Nations*. London: W. Strahan & T. Cadell.

Soon, C.S., Brass, M., Heinze, H.-J., and Haynes, J.-D. (2008). Unconscious determinants of free decisions in the human brain. *Nature Neuroscience* 11 (5): 543–545. https://doi.org/10.1038/nn.2112.

Sumpter, D.J.T. (2010). *Collective Animal Behavior*. Princeton, NJ: Princeton University Press.

Sumpter, D.J.T. and Beekman, M. (2003). From nonlinearity to optimality: pheromone trail foraging by ants. *Animal Behaviour* 66 (2): 273–280. https://doi.org/10.1006/anbe.2003.2224.

Tinbergen, N. (1951). *The Study of Instinct*. Clarendon Press/Oxford University Press.

Vickery, W.L., Giraldeau, L.-A., Templeton, J.J. et al. (1991). Producers, scroungers, and group foraging. *The American Naturalist* 137 (6): 847–863. https://doi.org/10.1086/285197.

Visscher, P.K. (2007). Group decision making in nest-site selection among social insects. *Annual Review of Entomology* 52: 255–275. https://doi.org/10.1146/annurev.ento.51.110104.151025.

Welch, I. (1992). Sequential sales, learning, and cascades. *Journal of Finance* 47 (2): 695–732. https://doi.org/10.1111/j.1540-6261.1992.tb04406.x.

Wilson, R.A. and Keil, F. (1998). The shadows and shallows of explanation. *Minds and Machines* 8 (1): 137–159. https://doi.org/10.1023/a:1008259020140.

Wu, G.-M. and Giraldeau, L.-A. (2005). Risky decisions: a test of risk sensitivity in socially foraging flocks of *Lonchura punctulata*. *Behavioral Ecology* 16 (1): 8–14. https://doi.org/10.1093/be- heco/arh127.

12

Moving Social-Behavioral Modeling Forward: Insights from Social Scientists

Matthew Brashears[1], Melvin Konner[2], Christian Madsbjerg[3], Laura McNamara[4], and Katharine Sieck[5,]*

[1] *Department of Sociology, University of South Carolina, Columbia, SC 29208, USA*
[2] *Department of Anthropology and Neuroscience and Behavioral Biology, Emory University, Atlanta, GA 30322, USA*
[3] *ReD Associates, New York, NY 10004, USA*
[4] *Sandia National Laboratories, Albuquerque, NM 87123, USA*
[5] *RAND Corporation, Pardee RAND Graduate School Santa Monica, CA 90407, USA*

As part of preparing a large volume of contributed papers (Davis et al. 2019), the editors sought some step-back-and-think contributions from noted social scientists who might have suggestions as we move forward in the process of improving social-behavioral models. After all, in our rush to understand people by mining their vast troves of receipts, running paths, texts, search queries, tweets, and *likes*, there is great value to a thoughtful consideration of what we are doing, why we are doing it, and how we intend to do it. The editors and I (Sieck) brought together four researchers who might have such suggestions for computational modeling based on their decades of work serving as translators and brokers between the social sciences and other disciplines: engineering, modeling, medicine, data science, and so on. These authors have spent their careers crossing the lines between qualitative and quantitative or explanatory and predictive efforts to understand human motivation and behavior. The goal in this paper is to share some of the critical lessons learned in these endeavors. Our tone is deliberately conversational, in part to avoid exclusionary disciplinary jargon. Our goal is to widen the conversation, build stronger bridges to computational modeling, and find ways to partner more effectively with modelers.

We begin by asking why people do what they do, noting that it is a very old question. We then move into a discussion of whether the *big data revolution* is, in fact, truly revolutionary (or *something old that is new again*). As we move

---

* Panel moderator.

Social-Behavioral Modeling for Complex Systems, First Edition.
Edited by Paul K. Davis, Angela O'Mahony, and Jonathan Pfautz.
© 2019 John Wiley & Sons, Inc. Published 2019 by John Wiley & Sons, Inc.
Companion website: www.wiley.com/go/Davis_Social-Behavioralmodeling

forward, the lessons of past efforts to quantify and model human behavior should not be far from mind. We then delve into why these efforts have so often failed by exploring how people think and behave, which is inherently *social, not just complex*. We look at how the use of models for research is different for their use in decision-making (i.e. using models depends on the *stakes*). Finally, we conclude by considering the implications of how people make sense of the world around them (*sensemaking*) for verification and validation. This provides a first glimpse into the complexities of *context* within behavioral modeling.

Why Do People Do What They Do?

This question has inspired philosophers, artists, theologians, and writers for centuries. Their musings on the nature of humanity have provided much fodder for reflection, especially as they are so often contradictory of each other. Between Hume and Descartes or Locke and Rousseau, we are often left feeling like we are all blindly touching different parts of the same elephant, but never seeing the whole.

That frustration, and these same musings, inspired subsequent generations of scientists and researchers. Is the *fault* of our circumstances due to ourselves or our stars? Do we *know* because we think or because we feel? What are the limits of individuality and social conditioning? Historically, a constraining factor on our answers has been data. These big questions require a broad reach across the globe in order to tease out the range of contextual variables that might lead to different behavioral pathways in different communities.

The digital age, however, provides new opportunities to understand what people do, when they do it, how they do it, and, often, who they are with when they are doing things. From geolocation data to social media postings, transaction receipts to health trackers, and smart homes to search histories, we now have an overwhelming amount of data about people's lives. When laddered against traditional data sources, we can know a great deal about the rhythms of individual lives. There are important gaps, and we will talk of those shortly. But the hope is that this new trove of *big data* will provide unprecedented opportunities to reopen the debates about human behavior, allowing us to explore answers in an efficient, effective, and holistic manner.

Everything Old Is New Again

But is this *really* a new frontier? Is this *big data* world really something we have never seen before? The short answer is: not really. As Mel Konner reminds us, it is worth recalling that this is not the first time we have held out hope that a new scientific approach can answer perpetual questions about humanity. Nor is it the first time we have ignored some of the gaps.

Medicine and related fields such as human biology, and genetics have put forward multiple attempts to understand human behavior as emergent from distinctive features and aspects of the human body. For example, the Greek philosopher and founder of modern medicine, Hippocrates (460 BCE–370 BCE) theorized that four substances in the human body accounted for the differential temperament and behaviors of people. Galen (c. 131–c. 201), the revered Roman philosopher and clinician, extended the original theory by including the influence of place on the relative balance of humors and thus on temperament and behavior. While we may dismiss such efforts through the lens of modern medicine, these practitioners were methodical and rigorous in data collection and keen observers of the cases before them. Their theories continue on even today – we now talk about *stable temperaments*, not humors, but the concepts are closely linked. They were, in many ways, data scientists for their times, looking for patterns and generating potential causal explanations.

So too were the many practitioners of craniometry, the popular eighteenth- and nineteenth-century science of measuring heads and cranial features in order to assess intellect, *civilizability*, and a range of other valued cognitive traits. As Stepan (1982:33–34) argued "that the skull housed the brain, the organ of mind, was no doubt the chief reason for focusing attention on the skull." This endeavor emerged during an important historical moment. On the one hand, Linnaeus had included humans alongside other primates on his classificatory charts, inspiring questions and concerns regarding whether we might be unique from these species. On the other hand, global conquest and slavery were premised on the differential status of human groups on the *great chain of being* (see Lovejoy 1964), and people went to great lengths to justify these differences through references to physical characteristics. Within the heated debates shaping our understanding of humanity, craniometry sought to bring data to our discussions of uniqueness, differences, and potential (see Gould 1996 for a full discussion of this work).

The advent of Mendelian genetics pushed efforts to understand behavior into discussions of inheritance, with behavior often tied to one's parentage and wider culture. At the time, with what we knew, this was state-of-the-art science – linking Mendel's work with the theories of Charles Darwin, Herbert Spencer, and Thomas Malthus in order to understand undesirable behavioral patterns. There was widespread concern that traits such as criminality, indebtedness, and promiscuity were inherited in the same ways as eye color and physical stature (Konner 2002). With the discovery of DNA, the work became more nuanced but continued nonetheless. The Human Genome Project offered hope that, among other things, we might identify the biological antecedents of behavioral differences through the shared language of the genetic code. With enough data and enough profiles, a genome database could be sorted to identify the patterns and outliers, moving us toward an ever-refined understanding of human actions and choices.

Like the glow surrounding *big data* today, this same aura of possibility defined public perceptions about these various endeavors to explain human behavior: can we identify gene sequences that are linked to depression, to risk taking, to greater sociality, to focus and concentration, and so on? Yet it is worth noting that this remains more a dream than a reality. As Gould (1996:32) remarked *more than 20 years ago and more than 40 years after Watson and Crick's discovery*:

> We are living in a revolutionary age of scientific advance for molecular biology. From the Watson-Crick model of 1953 to the invention of PCR and the routine sequencing of DNA – for purposes as varied as O.J. Simpson's blood signature to deciphering the phylogeny of birds – we now have unprecedented access to information about the genetic constitution of individuals. *We naturally favor, and tend to overextend, exciting novelties in vain hope that they may supply general solutions or panaceas – when such contributions really constitute more modest (albeit vital) pieces of a much more complex puzzle.* We have so treated all great insights about human nature in the past, including nongenetic theories rooted in family and social dynamics… If insightful nongenetic theories could be so egregiously exaggerated in the past, should we be surprised that we are now repeating this error by overextending the genuine excitement we feel about genetic explanation? [emphasis added]

As Konner notes, human behavior remains a complex phenomenon influenced by multiple streams: "anthropology, biology, medicine, evolution, the rain, childhood, history, and culture," to cite a few (Konner 2018). And just as there are limits on our ability to predict the weather even when we understand it, there may be limits to our ability to offer precise empirical predictions regarding human behavior, even when we understand it. This shifts how we think of the value and purpose of models themselves, pushing us more toward models-as-thinking-tools, vs. models-as-predictive (see Konner 2015 for a more detailed discussion).

Nonetheless, we push forward in our data collection efforts regarding climate and weather patterns. Similarly, advancements in our understanding of human behavior will only happen when we have relevant advancements in these and other fields, and this will only happen when we have systematic, rigorous data collection, with careful attention to the gaps: to the people, places, and things we cannot or did not track. In the glee over big data, it is important to remember that not everyone is a participant, nor are we always clear about how the data was collected. These sources of uncertainty should temper our modeling, just as the uncertainty of *inheritance* and genetic expression should have tempered the worst abuses of the eugenics movement. Attention to the quality of the data, how it can be used, and what it reveals is as important as knowing that data is available, and for these judgments, modelers should partner closely with social scientists.

Perhaps one of the lasting lessons of Konner's notably interdisciplinary career is that crosscutting work is essential but challenging – necessary, and not impossible, but not nearly as easy as stating that "X behavior means Y." Those who venture into these woods would do well to study the reasons why we have not yet devised the magic box. It is not because of lack of information, or effort, or smarts. It requires a different way of asking questions and framing problems: collaborative, iterative, broad-reaching, attentive to contextual factors, and attentive to limitations. Studying the history of these past efforts to quantify and model human behavior will provide insights regarding the pitfalls of exuberance with each new technological advancement.

This undoubtedly raises the question as to *why* modeling behavior is so challenging. Our conversation turned at this point to Matthew Brashears.

Behavior Is Social, Not Just Complex

Those who study human behavior are keenly aware of how infuriating it can be to try to understand why people do what they do. Anyone who has survived to age eighteen has likely had this experience. But for those who make it their calling and careers, armed with protocols and theories and measurement tools, answers often remain elusive. Just when you think you have understood the dynamic, something changes.

Matthew Brashears has made this challenge the cornerstone of his career. As he explains, *the core issue is not that social systems are complex, but that their operations vary depending on the decisions and perceptions of the persons within them.* For example, individuals do not simply cut off contact to random others as systems grow larger, but make choices based on how category membership appears relevant at the time. Moreover, these perceptions are not simply random, but are highly patterned according to social processes and, themselves, change over time. For example, a spreading religious doctrine not only recruits new members into an organization but also can cut them off to a degree from future behavioral diffusion, changing the nature of the landscape over which organizations compete. Successful modeling of social systems therefore requires the ability to incorporate the changing and contingent nature of entities and processes directly into the fabric of the simulation.

To cope with these cognitive demands, humans use a variety of psychological *tricks* to process social information more efficiently. In particular, humans use *compression heuristics* (Brashears 2013) or rules that allow social networks to be reconstructed from partial information, allowing most of the information on specific connections between individuals to be discarded. Beginning with a nonsocial example, if an individual is asked to remember a sequence of numbers, such as 2, 4, 6, 8, 10, and so on, they would simply remember the first number, 2, and a rule that the sequence increases by an interval of two. These two pieces of information are sufficient to reproduce the sequence in its entirety and are quite easy to learn and remember. In contrast, a series of numbers with no such clear pattern, for example, 7, 2, 76, 21, and 99, can still be recalled but only through the use of repetition (i.e. brute force). Applying compression heuristics to social information follows the same logic. If we can discard recall of specific connections among group members in favor of particular rules, then we can recall a larger amount of social information in the same brain mass. A series of innovative studies have tested this proposition, finding that triads and kin relations function as compression heuristics (Brashears 2013; Brashears et al. 2016); that while network recall is flexible, the default unit of relationship encoding appears to be the triad (Brashears and Quintane 2015); that females exhibit a significant advantage in network recall relative to males (Brashears et al. 2016), and that affective balance operates as a compression heuristic (Brashears and Brashears 2016).

In general there are two classes of compression heuristics: structural and cultural. *Structural heuristics* are those that derive from the properties of a social network graph and thus can be inferred without additional social information being given. Microstructural features of networks such as triadic closure (i.e. a group of three individuals who all share relations) and four cycles (i.e. a group of four individuals with four relationships, but no closed triads) can be deduced simply by examining a graph, even if one is ignorant of the labels attached to particular actors (e.g. mother, second cousin, best friend). While structural heuristics are likely learned as children mature and encounter group life, structurally reducible network features are not culturally specific, and thus these heuristics should appear in many, if not all, human societies.

Cultural heuristics are normative expectations for how particular types of actors *should* be connected. For example, if John and Sue are siblings, and Jake is Sue's nephew, then we can infer that John is likely Jake's dad. These types of relations are not easily derivable from the structure of the graph itself because the same graphs are consistent with many different relationships. For example, John, Jake, and Sue as mutual friends and John, Jake, and Sue as parent/child, siblings, and aunt/nephew both create a closed triadic structure. As a result, cultural heuristics must be learned and then activated when the situation seems appropriate. However, these cultural heuristics permit significantly larger compression ratios by encoding larger amounts of information into a handful of terms. For example, the term *aunt* implies something about relative age, sex, and common family relationships. Moreover, these heuristics allow individuals to infer the presence of additional relationships not otherwise given, as in the above example of Jake, Sue, and John. As such, while cultural heuristics are more challenging to study due to their cultural variability, they offer significant information processing advantages. However, inappropriately activated schemas (e.g. using a triadic heuristic for a network containing only open triads) also have the potential to worsen recall, since the heuristic leads individuals to expect connections that do not exist (e.g. Brashears 2013; Freeman 1992).

When taken seriously, compression heuristics have important implications not just for how people think about networks, but for how realized social networks grow. For quite some time, the most dangerous threat to a group of humans has been another group of humans, with larger groups generally triumphing over smaller groups (DeScioli and Kurzban 2009). This implies that humans were under significant selection pressure to adopt network structures that permitted larger groups to be achieved. Compression heuristics provided one response to such pressure. Groups that conform to these heuristics are easier to cognitively manage, which suggests that members will make fewer social errors when interacting. In that they reduce the cognitive difficulty of managing social networks, compression heuristics should also permit the development of larger groups. As a result, we should expect that some of the regular features of observed human social networks are a result of not simply our cognitive limits, but of the strategies we adopted to circumvent those limitations.

Compression heuristics in general, and cultural heuristics in particular, therefore present an interesting challenge to the modeler: individuals make choices based on their perceptions, but perceptions are shaped by the particular heuristics that they activate, and these heuristics are, themselves, frequently social products. Moreover, decisions made on the basis of a selected heuristic will influence behaviors shaping the observed structure of a social group, thereby reinforcing some heuristics, as well as making others more/less relevant, thereby triggering a new round of heuristic (in)activation. Finally, developing a model for such complex processes must proceed in conjunction with identifying the existence of the heuristics in question. In other words, we lack a complete list of all of the things we need to be modeling, and the list itself is changing continually.

The problem above likely now seems insurmountable. Indeed, our difficulties in understanding social behavior likely resemble nothing so much as those a Victorian scientist would encounter in trying to understand a personal computer connected to the internet. They might, with strenuous effort, be able to correlate inputs to outputs and develop a working understanding of what the machine does, only to inadvertently open a new application that changes all the rules, connect to a multitude of resources not located anywhere they can easily grasp, or even accidentally reboot the machine into a different operating system. Likewise, human behavior is dependent on complex routines specific to particular tasks (e.g. seduction vs. polite collaboration), on social rules and understandings not fully resident in a specific individual (e.g. norms resident in the social *cloud*), and on systems for evaluating events and circumstances (e.g. definitions of the situation) that shape how we perceive inputs and how we select behaviors to enact. Yet, the difficulty is not insurmountable once modelers can wrap their heads around it. Efforts to specify social entities and processes precisely fail not because social science is imprecise, but because social processes are more dynamic and contingent than any other domain yet studied. Thus, in order to model them successfully, it is necessary to develop entirely new approaches to modeling itself.

But the modeling we develop will depend in large part on what we hope it will yield and how we intend to use it. For some reflections on this, we turned to Laura McNamara.

What is at Stake?

Extending Brashears' compression metaphor, modeling human behavior is akin to transcribing an orchestral performance into a musical score. It is the act of taking a live, contextually rich, multi-actor experience and translating it into a standardized, shared code.[1] As any musician knows, the score is a framework, but every subsequent playing of that piece will have subtle (and not-so-subtle) differences depending on the skills of the players, the preferences of the conductor, the acoustics of the room, the size of the audience, and a multitude of other factors. Hence what gets coded into the score matters tremendously if the point is a general guide vs. an instruction manual for how to repeat a specific rendition of a piece. The latter case requires a far greater level of detail to account for the many contextual factors that influence sound in a particular event, whereas the former would be a traditional score.

In building computational models of human behavior, this matters because *why* we model is equally as important as *what* and *how* we model. As the resident anthropologist within a nuclear laboratory, the question of why people do what they do is pivotal to McNamara's work. When people make bad choices in nuclear facilities, the consequences can be devastating. But potentially more catastrophic is when people make a decision using a tool that is not suited for those purposes. Thus for McNamara, understanding why we model behavior can mean the difference between a conceptual exercise gone sideways and a devastating event.

Modeling as a research practice is exploratory. It helps teams to understand what happened in the past and what might happen under hypothetical future conditions. Its purpose is to extend the range of ways that we think about

---

1 Thanks to Todd Richmond, faculty at Pardee-RAND Graduate School, for the comparison between computational modeling and musical scores. I have extended his metaphor here, but the insight was his.

problems and opportunities. Modeling in these cases can help generate future research questions. For example, research in behavioral genetics indicates some heritability of temperament dimensions such as sociality, activity, shyness, adaptability, attention/persistence, and overall emotionality (Saudino 2005). Yet as with most genetic studies, these findings are based on a limited, primarily Euro-American data set. If we were to extend the data more broadly, including vastly different cultural contexts, we could begin to model more sophisticated questions as to when, why, and how genes influence temperament across the life course. Researchers could include longitudinal work to understand how cultural and environmental factors might influence gene expression or look at the prevalence of genetic markers across different populations, thereby generating new questions regarding differential prevalence rates of, say, attention disorders across communities.

In this work, the researchers have more flexibility to be wrong. The purpose is to see how models behave in order to generate questions for continued refinement in the research. It is a tool for improving and assisting in learning, with an outcome that is intended to be an indicator, not an answer. Therefore, the standards we apply for validation and verification would be somewhat more relaxed given that the goal is to inspire more thinking.

The danger comes in when modeling as a research practice becomes *modeling as an applied decision-making tool.* Unlike the exploratory roaming of the former, the latter is focused. In these cases, models are designed to help teams think through multiple scenarios and select one that holds the most promise for *achieving a desired outcome.* The goal is action oriented. Models as decision-making tools are intended to give reliable, accurate, and valid results because they will be used to inform choices that have real-world consequences.

Slippage between the two rationales happens all too often and often with problematic consequences. Konner (2002) reminds us that this very shift from exploratory to decision-making led to the worst abuses of the eugenics movement, to decades of forced sterilizations, to lobotomies, to discriminatory practices in classrooms and workplaces, and to institutionalizing people – all because exploratory models indicated potential social problems. All are examples where society linked "biological determinism to some of the oldest issues and errors of our philosophical traditions," including reductionism, reification, and hierarchy (see Gould 1996:27).

Big data has not necessarily improved the situation and, in fact, can be seen to exacerbate it through an aura of *rigor* and *science* that it may not warrant. Consider the case in which judges used the proprietary COMPAS model to make sentencing decisions based on predicted risk of recidivism – including denying parole to offenders whose behavior would otherwise merit release. Prior to its implementation, this technology had not been through any robust, public, and transparent process of verification or validation, but for all intents and purposes, some judges seem to have *accredited* COMPAS for sentencing decisions.

The outcomes are very disturbing – namely, it is "no more accurate or fair than predictions made by people with little or no criminal justice expertise" (Dressel and Faird 2018), and it is not clear the judges really understand how to evaluate the correctness of the model they are using.

Think about the difference between the same algorithm used in a court-room and being used as a research methodology. In the latter environment, the consequences are quite different. COMPAS might be referenced in papers and discussed in graduate student research seminars; it might raise community-level questions about the quality of the data being used; it may be even lead policy-makers to pass legislation supporting the collection and curation of better data sets to refine what we know about recidivism at the population level. It is probably easy for most of us in the research community to imagine how COMPAS would serve as a research tool to inform academic and policy debates about incarceration, rehabilitation, and the opportunities facing parolees after release.

As a sentencing tool, the same technology is being used to predict outcomes at the level of the individual. Yet it is not clear that the creators of this tech-nology are assuming any responsibility for the application of what is probably a population-level model to prediction of individual future behaviors. Are judges applying this technology asking questions about the quality of the data, the correctness of the software implementation, or the validity of the model? Do law school graduates have the mathematical literacy or ethical sensitivity to recognize the danger of applying what is likely a population-level model to pre-dict individual outcomes? What is clear is tat COMPAS is having a strong and perhaps frightening impact on individual lives (see Wexler 2017).

For McNamara, it is critical to remember that all models and simulations ultimately run within a social domain. Understanding the actors, motivations, and contexts that shape the tool and why it was requested are pivotal if we are to design modeling scenarios that achieve the desired outcomes.

And how do we know if our models have *achieved* a desired outcome? Con-sidering the advice from the panelists, how do we think about verification and validation in the context of behavioral modeling? Our final discussant, Chris-tian Madsbjerg, turned his considerable energy toward this issue.

Sensemaking

Continuing with the COMPAS model used for predicting recidivism, one of the academic reviewers of the algorithm noted the following:

> The problem isn't necessarily that COMPAS is unsophisticated, says Farid, but that it has hit a ceiling in sophistication. When he and Dressel designed more complicated algorithms, they never improved on the

bare-bones version that used just age and prior convictions. *"It suggests not that the algorithms aren't sophisticated enough, but that there's no signal,"* he says. Maybe this is just as good as it gets. Maybe the whole concept of predicting recidivism is going to stall at odds that are not that much better than a coin toss.

(Yong 2018)

Madsbjerg lives in a world of finding signals or what he terms *sensemaking*. As the managing director of a consulting firm that marries deep qualitative and ethnographic work with massive data sets in order to help clients make critical business and policy decisions, his teams are tasked with finding the signal amid all the noise of daily life. Their sensemaking process (see Madsbjerg 2017 for details) begins with understanding the cultural contexts in which work will occur (as McNamara recommended), prioritizing human intelligence gleaned from real-world observations to define the relevant factors, and then back translating those insights into identifiable patterns in massive data sets.

One way to begin improving the algorithms is to rethink how we collect and choose the information that will inform the model, as well as rethinking the results of an algorithm once formed. There are four main problems with algorithms that could be addressed by social scientists:

1. *The underlying models and algorithms risk being biased by the opinion of the stakeholders and data scientists developing them, which can lead to models based on incorrect assumptions about people or overlook the particularities of a culture or market.* To avoid or mitigate such outcomes, modelers should partner with social scientists who have a deep familiarity with the on-the-ground complexity of a particular topic. These partnerships will help unpack the compression heuristics described by Brashears so that we can see the full range of nuance and complexity and model appropriately.

2. *The data sets used for the algorithms are based more on availability than on accuracy.* Models can easily be limited to the data sets that are most readily available, which may or may not include the most relevant indicators or proxies of behavior to be working with. This harkens back to Konner's warnings about past efforts to explain behavior through cranial measurements or inheritance. Relying solely on what is available does not necessarily tell you whether this data is suited for the challenge of modeling human behavior. All data sets have gaps and weaknesses, and taking time to identify what these are, to consider how it will impact the modeling, and determining whether there are other ways to bolster or support the work are essential steps to ensuring a valid and verifiable model. But this means a critical dive into the data: Is it current? Is it from the right population? What other data might support or contradict these findings?

3. *A flawed algorithm may only serve to reinforce existing faulty ways of thinking*. Beyond the COMPAS example cited above, we have seen multiple examples in which flawed algorithms perpetuate false assumptions (see Baer and Kamalnath 2017; Knight 2017). As algorithms are increasingly used to assess credit, provide access to job opportunities, and assess work performance, we run the risk that constitutionally problematic human biases such as those based on gender and race are reinforced if the biases are not corrected in the underlying data. To the extent that a bias is pervasive in a culture, it is unlikely to be questioned or challenged in the design of the algorithm.

4. *A flawed algorithm creates a false sense of security, when, in fact, it fails to solve the problem it was intended for*. When the results of the algorithm are not observed to be solving the problem, we need to return to the drawing board. But we need sufficient perspective and insight to recognize when the algorithm itself is failing.

The key challenge for successful computational modeling of human behavior is knowing which kinds of data are meaningful – and how the combinations of different data points, once contextualized, make sense and can be used to explain broader patterns of behavior. In contrast to using *common sense* or *hard* variables as the initial inputs to an algorithm, qualitative data-informed variables of human behaviors, essentially preliminary assumptions about why people do what they do, are more likely to generate data that mirrors reality.

In *Silo Effect*, Tett (2015) provides an account of how New York City had to rethink its approach to predicting which buildings were likely to become firetraps when its algorithm failed to predict accurately. In the decade before 2011, there were an average of 2700 structural fires resulting in 85 deaths per year in NYC. Despite the 20 000 reported complaints of dangerous housing per year, inspections revealing unsafe conditions occurred only 13% of the time, and with only 200-odd inspectors to monitor four million properties, in one million buildings, the magnitude of the challenge was clear. The team knew that the root cause stemmed from properties that had been illegally subdivided to enable landlords to extract more rent.

An initial hypothesis was that reports of illegally converted building from the city's 311 line could help, but it was found not to be a good predictor as the majority of calls came from lower Manhattan, and not from the outer boroughs, where most fires or illegal conversions happened. Instead, the city government spent time with firemen, policemen, and inspectors from various municipal departments to learn clues for the existence of problems. Based on their conversations with these people, they found several warning indicators: buildings constructed before 1938 (when codes were looser) and were located in poor neighborhoods; homeowners delinquent on mortgage payments; and the buildings had typically received complaints about issues like vermin. Once the team was able to collate the disparate data sets that stood as proxies for the factors they found to be the most relevant, the new algorithm yielded unsafe conditions 70% of the time – even if nobody had complained about problems; something many residents, mostly poor immigrants, were reluctant to do because of their wariness toward interacting with state authorities.

What Tett's story shows is that computational models and algorithms are only as good as the hypothesis and data on which they are based. The key question is: how can we stop ourselves from overlooking an important aspect of the equation? One answer is to incorporate social science methods in order to identify the most relevant factors. The ways in which social scientists mine and analyze data can provide new theories and hypotheses for why people, communities, and societies behave the way they do, based on empirical observation, deep appreciation of the social context, and informed analyses.

Final Thoughts

Modeling human behavior is no small challenge, yet it is a critical component of everything from algorithmic platforms that predict the risk of child maltreatment and terrorist activity to those that assess credit worthiness and job suitability to those that recommend movies and social contacts. As researchers and practitioners move forward in these efforts, they would be wise to partner with social scientists who have a deep understanding of the communities impacted by these algorithms.

We know these partnerships are challenging. It can feel at times like learning a foreign language on the fly when all you really want are some basic directions. In order to facilitate these efforts, we recommend starting with the following questions and considerations:

1. *Pay attention to history*: When have we tried something like this in the past? What did we learn from those efforts? Are our tools realistically suited to this challenge? What part of the puzzle will this be? Why will this effort improve our understanding?
2. *Cross-train*: Computational modelers should read the social sciences literature around the subjects they model. There is good writing about people and cultures – taking an active interest in this will make modelers more successful in their efforts. So too should social scientists embrace the possibilities and work of modeling. It is imperative that we have, at minimum, a passing understanding of how modeling works, of the different types of models, and of the insights possible through well-crafted simulations and a keen appreciation for how our work intersects with and informs modeling.
3. *Understand the context*: How do we design models that have as much agility and fluidity as human behavior? What are the suite of contextual factors

that influence the ways in which humans make choices? Of note is that the choices may seem endless, but the contextual factors are not.

4. *Clarify purpose*: Is the modeling for exploration or learning? How do we deliberately and thoughtfully transition between the two (and not just slip into decision-making)? What is *good enough* when it comes to verification and validation of models used for decision-making? Should validation include addressing issues of ethics and privacy?

5. *Guide the translation*: Let social scientists understand the phenomenon first, and then ask them what each finding might look like in data. For example, if the goal is to model radicalization in a fringe community, what would change in social networks, in online searches, in spending behavior, in geolocation patterns, and in the tone of social exchanges? And what would we not see in large data sets? The questions are designed to connect the behavioral insights of social scientists to documented data sets where modelers could tease out relevant factors such as the frequency, intensity, and direction of behavioral change.

The challenge and opportunity inherent in building computational models of human behavior is simultaneously enticing and overwhelming. Our goal herein was to offer a few steps toward building a connective thread between the teams that specialize in the modeling work and those that specialize in understanding human behavior. The most good will come from patience, openness, and continued effort down a collective path. Because what is clear from our panelists is that siloed work on these efforts will be useless at best, destructive at worst.

References

Baer, T. and Kamalnath, V. (2017). *Controlling Machine Learning Algorithms and Their Biases*. McKinsey & Co. Insights, November 2017. https://www.mckinsey .com/business-functions/risk/our-insights/controlling-machine-learning-algorithms-and-their-biases.

Brashears, M.E. (2013). Humans use compression heuristics to improve the recall of social networks. *Nature Scientific Reports* 3: 1513.

Brashears, M.E. and Brashears, L.A. (2016). The enemy of my friend is easy to remember: balance as a compression heuristic. In: *Advances in Group Processes* (ed. S. Thye and E. Lawler), 1–31. Bingley, UK: Emerald Insight.

Brashears, M.E. and Quintane, E. (2015). The microstructures of network recall: how social networks are encoded and represented in human memory. *Social Networks* 41: 113–126.

Brashears, M.E., Hoagland, E., and Quintane, E. (2016). Sex and network recall accuracy. *Social Networks* 44: 74–84.

Davis, P.K., O'Mahony, A., and Pfautz, J. (eds.) (2019, this volume). Improving Social-Behavioral Modeling. *Social-Behavioral Modeling for Complex Systems*. Hoboken, NJ: Wiley.

DeScioli, P. and Kurzban, R. (2009). The alliance hypothesis for human friendship. *PLoS One* 4: https://doi.org/10.1371/journal.pone.0005802.

Dressel, J. and Faird, H. (2018). The accuracy, fairness, and limits of predicting recidivism. *Science Advances* 4 (1); 17 Jan 2018. DOI: https://doi.org/10.1126/sciadv.aao55801.

Freeman, L.C. (1992). Filling in the blanks: a theory of cognitive categories and the structure of social affiliation. *Social Psychology Quarterly* 55: 118–127.

Gould, S.J. (1996 [1981]). *The Mismeasure of Man*, 2e. New York: WW Norton & Co.

Knight, W. (2017). Biased algorithms are everywhere, and no one seems to care, *MIT Technology Review*. https://www.technologyreview.com/s/608248/biased-algorithms-are-everywhere-and-no-one-seems-to-care/ (accessed 12 July 2017).

Konner, M. (2002). *The Tangled Wing: Biological Constraints on the Human Spirit*, 2e. New York: Henry Holt & Co.

Konner, M. (2015). The weather of violence: metaphors and models, predictions and surprises. *Combating Terrorism Exchange* 5 (3): 53–64.

Konner, M. (2018). Personal communication with K. Sieck; April 2016.

Lovejoy, A. (1964 [1936]). *The great chain of being: a study of the history of an idea*. Cambridge, MA: Harvard University Press.

Madsbjerg, C. (2017). *Sensemaking: the power of the humanities in the age of the algorithm*. New York: Hachette Book Group.

Saudino, K.J. (2005). Behavioral genetics and child temperament. *Journal of Developmental and Behavioral Pediatrics* 26 (3): 214–223.

Stepan, N. (1982). *The Idea of Race in Science: Great Britain 1800–1960*. London: Macmillan.

Tett, G. (2015). *The Silo Effect*. New York, NY: Simon & Schuster.

Wexler, R. (2017). When a computer program keeps you in jail, *New York Times*, Opinion Section. https://www.nytimes.com/2017/06/13/opinion/how-computers-are-harming-criminal-justice.html (accessed 13 June 2017).

Yong, E. (2018). A popular algorithm is no better at predicting crime than random people, *The Atlantic*. https://www.theatlantic.com/technology/archive/2018/01/equivant-compas-algorithm/550646/ (accessed 17 Jan 2018).

Part III

Informing Models with Theory and Data

13

Integrating Computational Modeling and Experiments: Toward a More Unified Theory of Social Influence
Michael Gabbay

Applied Physics Laboratory, University of Washington, Seattle, WA 98105, USA

Introduction

Social influence is a central element of many behavioral areas, such as public opinion change, radicalization, and group decision-making – all of concern to public policy. It affects the process by which people form attitudes toward their governments, other population groups, or external actors. Group decision-making applications span political, military, economic, and legal domains. If computational social science is to aid in understanding, anticipating, and shaping such real-world contexts, then the development of accurate and broadly applicable models of social influence is essential. This chapter proposes that a deliberate and concerted integration of experimental investigation and computational modeling is needed to develop these models, an effort that will also advance the fundamental knowledge of social influence dynamics.

Within social psychology, social influence has largely been studied via the experimental testing of discrete theoretical hypotheses that express how a dependent variable responds to a change in an independent variable in qualitative language (e.g. increases, decreases, inverted U-shape). Quantitative reasoning is often employed in the rationale for theoretical propositions, and, to a much lesser extent, formal or computational models are also used to motivate them. While this process has been successful in revealing and extensively probing individual phenomena, it has been less effective at synthesizing and reconciling concurrent and competing processes. Such synthesis would better inform the development of computational models as to the relative strengths of different processes and their interaction. It is especially important if one seeks to apply a model to anticipate the behavior of a particular group of interest. For instance, the group polarization effect might predict that a group will pursue an extreme policy, whereas majority influence points toward a

Social-Behavioral Modeling for Complex Systems, First Edition.
Edited by Paul K. Davis, Angela O'Mahony, and Jonathan Pfautz.
© 2019 John Wiley & Sons, Inc. Published 2019 by John Wiley & Sons, Inc.
Companion website: www.wiley.com/go/Davis_Social-Behavioralmodeling

moderate policy. Consequently, guidance as to how those two processes play out together is necessary in order to model the group's overall behavior. That guidance, of course, would need to be context dependent. As will be seen, the failure of group polarization theory to be integrated with broader social influence phenomena leads it to predict that every group of like-minded members will become more extreme, regardless of their initial opinion distribution. This is not so for the frame-induced polarization theory to be described below that does integrate group polarization with majority influence and consensus pressure: the theory can explain a systematic tendency for like-minded groups to become more extreme while being able to predict that individual groups will not.

The difficulty of synthesis in traditional social influence research has not, however, deterred a surge of modeling research across a range of disciplines. This activity has not been an unalloyed good for computational social science as much of this work has proceeded with little regard for empirical support. Sizable and divergent streams of research have arisen around particular modeling approaches with murky domains of validity. This proliferation casts doubt upon the empirical relevance of the associated behavioral findings and complicates model selection and evaluation for applications.

The integrated approach advocated here calls for experiments designed with the explicit purpose of quantitatively testing computational models against data. It will help restore the balance between modeling and experimental validation. The development of computational models in conjunction with experiment will force researchers to reckon more intently on combining concurrent effects in order to make quantitative predictions. That is, a more unified theory will have effects be caused by multiple factors that earlier work associates with separate hypotheses. A greater orientation of experiments toward testing models rather than seeking new effects will encourage replication efforts and so place empirical findings on more solid ground. The increased focus on synthesis and the inevitable failures of previously successful models as they are tested in new regimes will spur theoretical innovation as well. Eventually, this integrated modeling-experiment approach will lead to convergence upon a set of social influence models that have substantial experimental support and so can be confidently extended to larger-scale systems or included within more complex simulations of particular application contexts.

This chapter first presents a brief survey of social influence and related group decision-making research, along with a discussion of how standard hypothesis testing and also quantitative modeling have been employed. The second section provides an overview of opinion network modeling. Next, the quantitative testing of computational models on experimental data is illustrated using recent work on group polarization (Gabbay et al. 2018; Gabbay forthcoming), which also shows how the modeling goal of synthesizing concurrent effects can lead to new and more unified theory. The fourth section then sketches the envisioned integration of modeling and experiment and discusses its potential benefits.

Social Influence Research

The term social influence is broadly applicable to both attitudinal and behavioral effects of human interactions. We focus upon research involving attitudes, opinions, and judgments here, mental constructs that often guide behavior. Classic social influence phenomena include conformity – the tendency and pressures toward consensus in groups, majority and minority influence – the ability of majorities and minorities to sway group opinions, and group polarization – the tendency of discussion among like-minded people to make positions more extreme. Two primary types of influence routes have often been invoked as explanations of such behaviors, normative and informational. Normative influence refers to the operation of group and broader social norms in setting expectations as to appropriate opinions and the value of consensus. Informational influence is the acceptance of information from others as evidence about the reality of the subject under consideration. Informational influence typically involves the alignment of one's private and publicly expressed judgments, whereas public agreement need not imply private acceptance under normative influence.

The body of social influence research above has been established through a process of hypothesis testing via laboratory experiments. Groundbreaking studies were conducted in the 1950s and 1960s (see Eagly and Chaiken 1993, Chapter 13). In a classic experiment by Asch, a substantial proportion of subjects suspended the clear evidence of their senses when faced with a majority of experimental confederates who stated that a clearly shorter line was longer, thereby demonstrating the power of conformity to induce public compliance. Inverting the direction of influence as the process under investigation, Moscovici and collaborators found in a color discrimination task that minorities who advocated consistent positions were more effective in swaying subjects than inconsistent minorities, leading to a theory that majorities primarily exert normative influence, while minority influence occurs mostly via the informational route. Group polarization, the tendency of discussion among like-minded individuals to lead to more extreme opinions, is another element in the social influence canon. It has both informational and normative influence explanations and will be discussed in detail below.

Research on group decision-making in contexts that allow for interpersonal persuasion also involves social influence. One strain of decision-making research considers the performance of groups in comparison with individuals (Kerr and Tindale 2004). For example, the wisdom-of-the-crowds hypothesis holds that simply aggregating individual judgments over many individuals yields greater accuracy than the judgments of individual experts under the assumption that the members of the pooled population make independent judgments whose uncorrelated errors cancel. Arguments have been made that social influence can either impair this performance by inducing correlated errors or improve it when greater individual confidence tends to be associated with greater accuracy (Becker et al. 2017).

The vast majority of experiments on social influence and decision-making have been aimed at testing discrete qualitative hypotheses. A hypothesis is proposed concerning the direction of an effect on the dependent variable, increase or decrease, due to the variation of an independent variable, which is then tested statistically. If the amount of change in the dependent variable is in the theorized direction and improbably attributed to the null hypothesis of no effect, then the proposed hypothesis is said to be supported by the data.

Within the social influence and group decision-making domain, research affiliated with the literature on social decision schemes (SDS), which are essentially mathematical rules for combining group member initial preferences into a final decision, has most consistently pursued a model-based quantitative approach. Although initially concerned with juries and binary (innocent/guilty) decisions, the SDS program grew to include decisions concerning quantitative judgments such as monetary awards and budgets (Hinsz 1999). While it has been very successful with respect to its original jury concentration (Devine 2012), a shortcoming of the SDS program is that its focus on testing an array of aggregation rules has come at the expense of deeper theoretical and model development with respect to a specific opinion change process. This absence of a theoretical impetus inhibits generalization of the results to broader contexts – for example, when subgroup and social network structure is important or for general opinion dynamics in populations not associated with a focal decision point.

Opinion Network Modeling

Although the experimental study of social influence has been conducted by social psychologists and, to a much lesser degree, in other social sciences such as political science, sociology, and economics, the modeling of social influence dynamics has extended beyond social science to fields including physics, applied mathematics, computer science, and electrical engineering (Crandall et al. 2008; Castellano et al. 2009; Proskurnikov and Tempo 2017). While the primary goal of opinion network models is to predict final opinions from initial ones, the models typically describe a process that occurs over time. This section briefly discusses approaches to opinion network modeling and their empirical application.

Many models of opinion change have been developed in the fields noted above and beyond, involving a great diversity in methodological and substantive choices. One major methodological division involves whether outcomes are produced deterministically or stochastically. A fundamental substantive division involves the way opinions are mathematically represented. A binary representation is clearly applicable to situations that ultimately involve a decision over two alternatives such as a political election. Alternatively, a continuous representation can account for gradations of opinion on an issue or for decisions involving either explicit numerical quantities such as budgets or that can be approximated as a spectrum of options ordered along some dimension (e.g. the extent of an escalatory military response). Binary (or discrete) opinion models tend to have stochastic interactions; continuous opinion models usually (but need not) employ deterministic interactions. The choice of opinion representation also constrains the basic process that governs how opinions change when nodes (a term for individuals within a network) interact. A binary opinion must either remain the same or flip when a node interacts with other nodes. In the voter model, a dyadic interaction is assumed whereby a node adopts the opinion held by a network neighbor selected at random, whereas the majority rule model proceeds by selecting subgroups at random with all member nodes adopting the majority opinion in the subgroup (Castellano et al. 2009). Continuous opinion models, on the other hand, allow for incremental shifts in opinion where the amount of change is a function of the distance between node opinions. The DeGroot and Friedkin–Johnsen models, as well as the consensus protocol (popular in the engineering literature on control), use a linear dependence in which the shift is proportional to the opinion difference (DeGroot 1974; Olfati-Saber et al. 2007; Friedkin and Johnsen 2011). Bounded confidence models assume a hard opinion difference threshold, within which nodes interact linearly but beyond which interaction produces no change (Lorenz 2007). The nonlinear model of Gabbay (2007) uses a soft threshold so that, rather than vanishing completely, the interaction decays smoothly with distance.

The vast majority of papers on opinion network models make no contact with empirical data. They start with a model, reasoned to be plausible (sometimes on the basis of social psychology research but sometimes on an appeal to common sense), and then generate simulation results, often in combination with mathematical analysis, on phenomena such as how the time to reach consensus scales with system size, the conditions conducive to the formation of camps of rival opinions, or the ability of extremists or influential individuals to shift opinions. Usually, the focus is on large systems taken to be representative of population-scale behavior. Consequently, such empirical connections as are reported are usually on the level of noting that model-generated curves exhibit qualitatively similar shapes to relationships observed in naturally occurring data from large population systems (Crandall et al. 2008; Düring et al. 2009; Török et al. 2013). However, some models have been shown to quantitatively reproduce empirical relationships such as the distribution of votes in proportional elections (Fortunato and Castellano 2007; Burghardt et al. 2016).

Application of opinion network models to laboratory experiments remains mostly confined to testing models developed within traditional fields of human behavioral research rather than from the physical sciences and engineering. Friedkin and Johnsen (2011) conducted experiments in which they manipulated network topology for small groups and measured initial opinions, thereby enabling the quantitative comparison of experimental and model results. Although the communication topology was controlled, the network weights assessing interpersonal and self-influence in the model had to be calculated for each group separately on the basis of subjects' post-discussion ratings of interpersonal influence, thereby limiting predictive capability. However, their work remains the most extensive experimental investigation of an opinion network model. More recent work has employed agent-based modeling to qualitatively support and extend experimental results (Mäs and Flache 2013; Moussaïd et al. 2013, 2015).

Integrated Empirical and Computational Investigation of Group Polarization

This section provides an illustration of how experiment and opinion network modeling can be integrated as applied to group polarization and serves as a prelude to the description of the integrated approach in the next section. Recent research is described that demonstrates how a modeling-oriented approach can synthesize previously disjoint phenomena, generating a novel theoretical explanation of a classic social influence phenomenon, which furthermore predicts an effect unanticipated by existing theory (Gabbay et al. 2018). The basic theory is implemented in a simple aggregation model that integrates group polarization with the fundamental social influence processes of majority influence and conformity. Further, a model of opinion network dynamics shows how this basic process can arise from a lower-level attitude change framework that considers how persuasive messages shift both opinion and its associated uncertainty (Gabbay forthcoming). Both models not only qualitatively agree with the results of an online discussion experiment but, in accord with the proposed integration of modeling and experiment, are in quantitative agreement with the data as well.

Group Polarization Theory

The question of how groups shift toward more extreme positions has been a focus of both traditional social influence research and opinion network modeling although the explanatory mechanisms favored by each are disconnected. Group polarization is said to occur when, in a group composed of individuals already on the same side of an issue, the post-discussion mean opinion shifts further in support of that side as compared with the pre-discussion mean (Myers 1982; Brown 1986; Isenberg 1986; Sunstein 2002). Note that, contrary to common parlance, *polarization* here refers to movement toward one pole rather than divergence toward opposite poles. The seminal experiments in the 1960s focused on *choice dilemmas* in which subjects were presented with hypothetical scenarios involving the choice between a risky but higher payoff option over a safer, lower payoff one (Brown 1986). Subjects were asked to choose the minimum odds of success they would accept in order to pursue the riskier option. For most choice dilemma items, discussion led groups to choose lower odds of success as measured by the difference in the group pre- and post-discussion means. The effect, therefore, was originally coined the *risky shift*. However, some choice dilemma items tended to produce shifts toward greater caution, while others produced no shift in either direction. Cautious items were marked by a very large stake such as someone's health or marriage, whereas risky items tended to offer a large potential gain for a small stake. Experiments on group betting involving real rather than hypothetical stakes also have shown a mix of risky and cautious shifts (Isenberg 1986).

Beyond the risk context, discussion among similarly inclined individuals was found to cause more extreme social and political attitudes (Myers and Bishop 1970; Schkade et al. 2010; Keating et al. 2016). Manipulation of the evidence presented to mock juries exhibited discussion-induced shifts to lower presumed guilt and softer sentences in cases where the evidence was weaker and higher presumed guilt and harsher sentences for stronger evidence (Myers 1982). Similarly, jury damage awards exhibit polarization (Schkade et al. 2000). In general, the contexts in which group polarization occurs are on the judgmental side of the intellective–judgmental spectrum in which purely intellective tasks, such as math problems, have demonstrably correct solutions, whereas purely judgmental tasks are matters of personal taste or aesthetics (Laughlin and Ellis 1986). Most real-world decision contexts such as forecasting and policy-making are characterized by both intellective and judgmental aspects; they may draw on a body of knowledge (e.g. expertise on a country's political system), yet judgments must be made as to significant uncertainties (e.g. the intentions of political leaders).

Corresponding to the two main pathways of social influence, informational and normative processes underlie the two main explanations of group polarization within social psychology. In the informational route, known as persuasive arguments theory, group members expose each other to new information in favor of that side. In the normative route, known as social comparison theory, a group norm associated with the broader culture or that particular group's identity defines a preferred direction on an issue so that opinions shift in the direction of the norm; a norm toward risk-taking, for instance, would lead groups to make riskier choices as a result of discussion. While the informational and normative mechanisms for group polarization have received robust experimental support, they have never been integrated with strong social influence phenomena such as consensus pressure and majority influence. Relatedly, neither explanation has been developed with respect to a clear formal model at the individual group level. Although the informational and normative processes occur at the group level, these theories were operationalized mainly with respect to a population of groups with random initial opinion distributions, over which majority influence could be assumed to cancel out. As a result, group polarization theory is effectively silent as to whether a particular group with a specific initial distribution of opinions will become more extreme. Alternatively, one could make a strong reading of either persuasive arguments theory or social comparison theory that neglects other processes in which they always predict polarization for homogeneous groups (for sufficiently judgmental issues). Either alternative – silence or a uniform prediction of polarization – limits the ability of existing group polarization theory to address real-world contexts such as whether, in the face of a foreign policy crisis, discussion among a country's leadership will induce a shift toward a more extreme course of action.

Opinion network models do not suffer from an inability to go from initial to final opinions since that is their fundamental purpose. The dominant approach to modeling extremism within this literature has been to attribute higher network weights to nodes with more extreme initial opinions (Deffuant et al. 2002; Friedkin 2015). This *extremist-tilting* approach is necessitated by the fact that in most continuous opinion models the mean opinion in networks with symmetric weights (i.e. the strength of influence is the same from node i to j as from node j to i) remains constant at its initial value, therefore preventing the shift in mean exhibited in group polarization. Consequently, extremists must be assigned greater influence over moderates than vice versa in order to shift the mean. Psychologically, this move is attributed to extremists' greater confidence, commitment, or stubbornness. Extremist tilting is not widely accepted within the literature dedicated to group polarization, however, and has received only mixed experimental support (Zaleska 1982).

Frame-Induced Polarization Theory

This section discusses the frame-induced theory of group polarization introduced by the author and Zane Kelly, Justin Reedy, and John Gastil (Gabbay et al. 2018). This theoretical mechanism is complementary to and can operate simultaneously with the mechanisms of standard polarization theory (shorthand for both persuasive arguments theory and social comparison theory). However, the frame-induced mechanism provides an explanation of group polarization that, unlike standard polarization theory, is integrated with consensus pressure and majority influence, thereby enabling prediction given the group initial opinion distribution. The theory is developed specifically with respect to a quantitative policy under debate, although it should prove applicable to opinions more generally. Examples of quantitative policies include budgets, investment amounts, interest rates, jury damage awards, or military operation sizes. In its emphasis on how the policy is discussed, the theory takes into account the basis of policy positions and not just the policy value alone. For the particular context discussed here, this basis is grounded in the theory of decision-making under risk and uncertainty (Pleskac et al. 2015) and so constitutes a further theoretical element that is integrated within the frame-induced theory.

Crucial to the frame-induced theory is the recognition of a distinction between the policy under debate and the *rhetorical frame* by which it is discussed. In general, one would expect the rhetorical frame to be a substantive aspect of the policy for which there is substantial disagreement among group members. The frame could represent a key uncertainty or differences in how group members value the outcomes associated with the policy. A given issue may admit multiple frames if there are different dimensions of comparable disagreement. Yet, a single dominant rhetorical frame may emerge due to group-specific dynamics such as deliberate efforts to focus a debate as occurs in political framing (Chong and Druckman 2007). Persuasion, and hence agreement, is driven by proximity along the rhetorical frame not the policy itself. The shape of the *rhetorical function* that maps policy positions into positions along the rhetorical frame (*rhetorical positions*) plays an essential role in frame-induced polarization theory.

Focusing on when uncertainty is the source of the rhetorical frame, uncertainty can generate disagreement if group members have different estimates of the probability of either an unknown variable or an impending outcome. A simple but important example involves a policy that depends on the outcome of a binary gamble so that the likelier one estimates the outcome to be: the more stake one is willing to risk on its occurrence. For instance, one would prefer to invest more in a defense technology company (the policy), the greater one's subjective probability that the pro-defense spending candidate in an election is likely to win (the rhetorical frame). The use of the subjective probability of a binary outcome as the rhetorical frame is also relevant to the experiment described below.

Two important behaviors that impact group polarization arise from the distinction drawn between policy and the rhetorical frame: (i) *distribution reshaping*, which preferentially facilitates the formation of extreme majorities and so generates group polarization, and (ii) *heuristic frame substitution*, which can enhance polarization on one side of the issue and suppress it on the other. Distribution reshaping arises when a nonlinear rhetorical function causes the relative spacing between group member rhetorical positions to be different than between their policy positions. Consequently, the distribution of group member rhetorical positions will be reshaped with respect to the distribution of policy positions. Such reshaping may reduce the rhetorical distance within some subgroups relative to others as compared with their distances directly along the policy itself, thereby affecting the composition of the majority that emerges during deliberations. Specifically, for a concave (downward curvature) rhetorical function, rhetorical position increases more slowly as the policy becomes more extreme. This causes the rhetorical distance between more extreme members to be compressed relative to the distance between more moderate ones. This compression favors the emergence of a majority at the extreme, which then drives consensus to a policy that is more extreme than the mean of the initial policy distribution. For the case where the policy, such as a wager or investment, arises from the subjective probability of an outcome in a binary gamble, Gabbay et al. (2018) show that a concave rhetorical function is expected using the theory of decision-making under risk and uncertainty (Pleskac et al. 2015).

As an illustration of distribution reshaping, consider a group of three military planners in wartime tasked with deciding whether to increase or decrease the size of the force allocated to defend a certain territory. The policy is then the change in the number of troops, positive or negative, from the current level. Take a planner's preferred force level to be a function of their estimate of the probability of an enemy offensive against this territory and their assessment of its worth relative to other territories. If there is little disagreement as to worth yet there is fundamental uncertainty as to enemy intentions, then the subjective probability of an enemy offensive is expected to be the dominant source of disagreement and, hence, the rhetorical frame. The rhetorical function is then the transformation that maps a given change in force level to the corresponding subjective probability of an enemy offensive. For instance, assume that the three planners are all inclined to boost the force level and that their respective preferences for the increase in troops are (500, 1500, 2500), which are mapped by the rhetorical function to subjective probability estimates of (0.55, 0.65, 0.70). These values indicate that the rhetorical function is nonlinear: a policy difference of 1000 between the first two members corresponds to a change in probability of 0.10, while a 1000 difference between the second and third members yields a probability change of only 0.05. More specifically, it is concave in that the subjective probability goes up at a slower rate as the increase in force level becomes more extreme. While the policy distribution consists of one member at the mean of 1500 with the other two an equal distance below and above it, the rhetorical position distribution has one member below its mean of 0.633 and two above; a symmetric policy distribution has been reshaped into an asymmetric rhetorical one. As will be described presently, the theorized opinion change process assumes that the two more extreme members are likely to form a

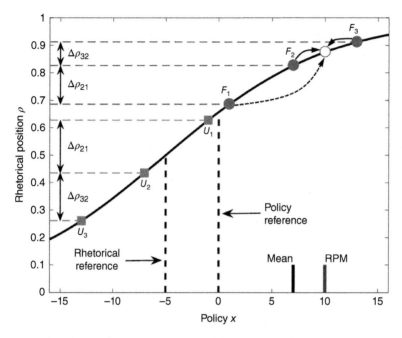

Figure 13.1 Illustration of distribution reshaping, RPM process, and reference point shifting due to heuristic frame substitution. Rhetorical function relating rhetorical frame position to policy is solid curve. Solid curved arrows indicate formation of the RPM pair by F_2 and F_3; dashed curved arrow indicates conformity of F_1 to form final consensus. Short dark gray line at the bottom is the mean of the initial F group policies; short light gray line is RPM process prediction for the F group consensus policy.

majority, converging at their midpoint of 2000, which yields the ultimate group consensus.

To yield a systematic tendency for homogeneous groups to shift toward the extreme, distribution reshaping must be linked to an opinion change process. The rhetorically proximate majority (RPM) process specifically treats consensus formation. Most group polarization experiments have required consensus, and it was the most common outcome in our experiment. Figure 13.1 illustrates how distribution reshaping combines with the RPM process to generate group polarization. Two separate three-person groups, the F and U groups (the reason for these designations will be revealed below), are depicted. The policy reference at $x = 0$ is the boundary between the opposing policy sides. If a group is entirely on one side of the policy reference, then it is said to be homogeneous. The F and U groups are seen to be homogeneous on the positive (pro) and negative (con) policy sides, respectively. Both groups have symmetric policy distributions given by $(x_1, x_2, x_3) = (\pm 1, \pm 7, \pm 13)$ where the $+(-)$ sign corresponds to the $F(U)$ group.

The F group is used to describe distribution reshaping and the RPM process. The rhetorical function $\rho(x)$ maps the policy x to the rhetorical position ρ. The rhetorical position scale goes from 0 to 1 and so can correspond to the choice of subjective probability as rhetorical frame. Although the policy of the centrist F_2 is exactly midway between those of the moderate F_1 and the extremist F_3, the F group members are arrayed along the shoulder of the rhetorical function so that, with respect to the rhetorical frame, F_2 is closer to F_3 than to F_1; the rhetorical distance $\Delta\rho_{32}$ is much less than $\Delta\rho_{21}$. As agreement is driven by proximity of rhetorical positions, it is therefore more likely that F_2 will join with F_3 to form the RPM pair than with F_1. There remains the question of where the RPM pair will converge. Assuming F_2 and F_3 have equal influence on each other, then they should converge at either the midpoint between their rhetorical positions, which is then transformed back to the corresponding policy position, or midway between their policies. Either choice will lead to a policy more extreme than the mean x_2, but the latter process is more direct than the former. In addition, the policy positions are explicitly numerical, whereas the rhetorical ones need only be expressed qualitatively. It is therefore more plausible that the RPM pair forms midway between their policies (at $x_{23} = (x_2 + x_3)/2 = 10$). The convergence of the RPM pair on this point is indicated by the solid arrows leading to the open circle. As indicated by the dashed arrow, majority influence then causes F_1, now in the minority, to conform to the F_2, F_3 position. The result is a consensus policy that is more extreme (i.e. further from the policy reference) than the initial policy mean – in accord with definition of group polarization.

Although distribution reshaping due to a concave rhetorical function in combination with the RPM process can explain group polarization, it predicts only a systematic tendency for groups to shift toward the extreme. It is clear from Figure 13.1 that if F_2's initial policy were moved sufficiently close to F_1, then those two would form the RPM pair, which would yield a consensus policy below the mean. This ability to predict that individual groups can depolarize against an overall polarization tendency stands in contrast to standard theory's strong prediction of polarization for every homogeneous group.

Note that the references corresponding to the policy and rhetorical frame are offset along the horizontal axis in Figure 13.1. This misalignment between the policy and rhetorical references can arise from heuristic frame substitution in which a simpler heuristic rhetorical frame is discussed in place of a more complex frame that directly corresponds to the policy. In the binary gamble context, such substitution can occur when there are two distinct gambles that depend on the same random variable but with different thresholds: the policy gamble that directly determines whether one's policy choice is successful and a heuristic gamble that is more intuitively accessible. In a stock investing scenario, for example, the policy gamble could be whether or not the return of the stock over a given period of time would exceed that of a fixed return asset such as a bond.

If the stock's return is greater than the bond's return, then the investment is successful. The proper rhetorical frame would then be the subjective probability of that outcome. If it is greater than 0.5, then one should invest in the stock and not the bond. However, discussions about the investment might focus on the more intuitive heuristic gamble of whether the stock's price will rise or fall. Both of these gambles depend on the same random variable, the stock's return, but with different thresholds – the fixed return for the policy gamble, zero for the heuristic gamble – and so they have distinct subjective probabilities. If a probability of 0.5 is taken as the neutral reference for both subjective probabilities, then they will be related to different policy reference points by the rhetorical function.

The U group in Figure 13.1 illustrates the effect of reference point shifting due to heuristic frame substitution. The reference point of the rhetorical function is shifted left from that of the policy itself so that the U group members, who are all on the same (negative) side of the policy axis, straddle the rhetorical reference point (U_1 is to the right, while U_2 and U_3 are to the left). Because they are arrayed along the roughly linear part of the curve, the U group members are subject to weak distribution reshaping. One would expect therefore that the U group would be less prone to polarize than the F group. Strict application of the RPM process, however, would, for the case illustrated, lead to the formation of a U_2 and U_3 majority as the rhetorical distance $\Delta\rho_{32}$ is slightly less than $\Delta\rho_{21}$, which would then yield a substantial shift toward the negative extreme. But a small shift in rhetorical positions due to noise or uncertainty could readily flip which distance is smaller, causing the (U_1, U_2) majority pair to form, which would lead to depolarization instead. Whether the U group will polarize is consequently much harder to predict than for the F group. Considering a population of similar U groups, about equal numbers will become more moderate as become more extreme and so there will be no systematic polarization. In contrast, the offset of the rhetorical reference places the F group further along the shoulder and so enhances systematic polarization on the positive policy side.

Accept-Shift-Constrict Model of Opinion Dynamics

Two mathematical models have been presented in connection with frame-induced polarization theory. The RPM process described above is formulated as the RPM model, which determines a consensus policy by a weighted average of the policies of the majority of group members whose rhetorical positions span the least range (Gabbay et al. 2018). Network structure is accommodated by weighting policies by relative node degrees. Rather than static aggregation, the second model describes the opinion change process over time as a result of dyadic-level interactions. This accept-shift-constrict (ASC) model makes two innovations beyond existing continuous opinion network models (Gabbay forthcoming). First, it makes a distinction between policy (or opinion, more generally) and rhetoric in accord with the theory above. Second, it incorporates a novel uncertainty reduction mechanism that does not require that node uncertainties be visible to others.

The ASC model assumes an underlying dyadic process in which one node sends a message to a receiver node in an effort to persuade the latter. The message can impact both the receiver node's policy and its uncertainty interval around that policy. Conceptually, the model proceeds in distinct *accept, shift*, and *constrict* phases (although all occur simultaneously in the mathematical formulation). The accept and shift phases occur in the equation that governs the rate of change of the node's policy. In the accept phase, the ASC model assumes that the probability that the receiver node will accept the message as persuasive decreases as a Gaussian function of the *rhetorical* distance between the sender and receiver nodes. The uncertainty of the receiver's position is taken to be the standard deviation parameter in the Gaussian. If a message is accepted, then, in the shift phase, the receiver shifts its policy in the direction of the sender's by an amount proportional to their *policy* difference. The constrict phase is governed by a second equation for the rate of change of a node's uncertainty, modeling a process in which interaction with others with close positions reduces uncertainty. If the sender's rhetorical position is within the uncertainty interval of the receiver, then the receiver decreases its uncertainty but not below a certain minimum value. Accordingly, unlike other models that involve uncertainty dynamics (Deffuant et al. 2002), it is the difference in (rhetorical) positions among dyad members rather than their difference in uncertainties that drives uncertainty change. The network weights in the ASC model represent the influence of one node upon another due to factors such as communication rate and expertise; they need not be symmetric. The ASC model is implemented in terms of coupled nonlinear ordinary differential equations, with two equations for each group member, one for the policy and one for the uncertainty.

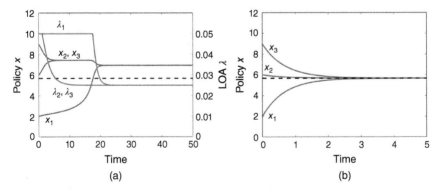

Figure 13.2 Evolution of policy positions and uncertainties for completely connected triad. (a) ASC model. (b) Consensus protocol. Dashed lines indicate initial mean policy.

A sample simulation of the ASC model as applied to a triad is shown in Figure 13.2a. All the nodes are connected and the network weights between nodes are symmetric and all equal. The initial policies are set such that x_2 is closer to x_3 than x_1. We observe that x_2 and x_3 form a majority and their uncertainties (λ_2, λ_3) quickly reach their minimum values, while that of the minority member (λ_1) stays at its initial value. Consequently, x_1 is more open to accepting messages from the majority pair than vice versa, so that x_1 essentially comes up to the majority position, resulting in a consensus policy that is shifted upward from the mean. This ability for a majority to emerge and persist in the face of minority influence is a crucial part of the frame-induced mechanism of group polarization. This dynamic is not present in linear opinion network models such as the DeGroot, Friedkin–Johnsen, and consensus protocol models. As shown in Figure 13.2b for the consensus protocol (also known as the Abelson model, a continuous time equivalent of the DeGroot model), no interim majority emerges as all nodes converge simultaneously on the initial mean.

Experiment and Results

In the experiment reported in Gabbay et al. (2018), triads of knowledgeable fans of the National Football League (NFL) wagered on the outcomes of upcoming NFL games. In accordance with standard NFL betting practice, the wager did not concern simply which team would win the game but rather the margin of victory. Professional oddsmakers set an expected margin of victory, known as the point spread, by which the favorite team (the team deemed likely to win the game) is expected to win over the underdog. A bet on the favorite is successful if the favorite wins by more than the spread; otherwise, a bet on the underdog is successful (neglecting actual practice of returning bets when the favorite's victory margin equals the spread). The spread is set so as to estimate the median margin of victory if the game were repeated many times and, empirically, the chances of favorites and underdogs with respect to the spread are effectively equal. As a consequence, the payoff is the same regardless of which team one bets on.

Several days before the selected game, subjects drawn from an online labor pool were asked in a survey to choose which team they expected to win against the spread and how much they would wager from $0 to $7 on their choice. Triads were then assembled into online discussion groups according to three manipulated variables: (i) policy side conditions of *favorite* or *underdog* corresponding to the team that all group members had chosen (there were no groups of mixed team choice), (ii) disagreement level conditions of *high* ($7) and *low* ($3, 4, 5) corresponding to the difference between the highest and lowest wagers in the group, and (iii) network structure conditions of *complete* in which all group members could send messages to each other and *chain* in which the intermediate wager person was the middle node and the low and high wager individuals were the ends. Groups discussed the game for 20–30 minutes after which members selected their final individual wagers. The winnings from successful wagers were donated to a specified charity.

Analyzing only groups that made consensus wagers (the vast majority of outcomes), the polarization metric is the difference between the mean post- and pre-discussion wagers of each group. If the post-discussion mean is higher, the group is said to have displayed a risky shift. Statistically significant results were found for all three manipulated variables: (i) favorite groups displayed a large risky shift, whereas underdog groups showed a small shift not statistically distinguishable from zero; within the favorite groups, a greater risky shift was observed for (ii) high disagreement groups than low and (iii) complete networks than chains.

These results are not readily explained by either standard polarization theory or existing opinion network models. The differential polarization behavior in particular stands at odds with the informational and normative explanations, which predict that a risky shift should occur for both the favorite and underdog conditions. Since both favorite and underdog groups were homogeneous with respect to policy side, group members have more novel arguments in support of their team choice, and so persuasive arguments theory predicts polarization for both favorite and underdog groups. Similarly, given the low stakes of the task, social comparison theory predicts that a norm favoring risk taking should be present in both groups and so both should exhibit a risky shift. This differential polarization result is also counter to the extremist-tilting explanation of opinion network modeling because, presumably, a high bet on the underdog is equally as extreme as the same amount bet on the favorite and so the level of extremist tilting is the same for both sides. With respect to the results for disagreement and network structure, their effects have been under-theorized and under-explored in the literature. The only previous experimental investigation of network structure in group polarization found no effect of topology (Friedkin 1999).

The experimental results, however, are in qualitative agreement with frame-induced theory. The differential polarization by policy side arises from heuristic frame substitution in which the question directly related to the wager policy – who will beat the spread? – is replaced by the heuristic one – who will win the game? The subjective probability of the favorite winning the spread is the proper rhetorical frame but is replaced by the subjective probability of the favorite winning the game. While professional gamblers may be able to think directly in terms of the spread victor probability, the game victor probability is a much more natural one for most knowledgeable fans to consider and so constitutes the rhetorical frame operative in the discussion. This substitution also entails a shift of reference point since both gambles depend on the same random variable, the margin of victory, but with different thresholds for their resolution. The reference margin of victory for the spread victor gamble is the point spread, whereas the game winner gamble has a reference of zero points. These different references for the margin of victory yield different policy and rhetorical reference wagers. The rhetorical reference is obtained by considering the wager for a subjective probability of the favorite winning the game equal to 0.5. Believing that the game is a toss-up implies that one estimates the margin of victory to be zero and so one should bet on the underdog if the oddsmakers have set a nonzero spread. Therefore, the rhetorical reference equates to some wager on the underdog. If positive and negative wagers are used to represent favorite and underdog bets, respectively, then the rhetorical reference corresponds to a negative wager. Accordingly, the F and U groups in Figure 13.1 can now be seen as analogous to the favorite and underdog groups in the experiment.

That favorite groups with high disagreement are expected to show a greater risky shift is a consequence of the RPM process described in connection with Figure 13.1. Expanding the difference between x_1 and x_3 while keeping x_2 fixed implies that the F_2 and F_3 will form the RPM pair at a more extreme policy since x_3 is more extreme. The greater polarization for complete networks vs. chains is due to the greater relative communication rate of the center node in the chain along with its intermediate wager. Rather than forming at their policy midpoint as in the complete network, the greater influence of the chain center node causes the RPM pair to form at a policy that is closer to x_2 and so implies a lesser shift to the extreme.

As a visual comparison between the data and models, Figure 13.3 displays the observed and simulated pre- to post-discussion shifts in the group mean wager as a function of the wager difference (averaged over all groups at each difference) where favorite and underdog groups are shown, respectively, on the positive and negative sides of the horizontal axis. Groups are simulated using their actual wagers and spreads. The weights in the complete and chain networks are set by a priori considerations of the topological effects upon communication rates. In the complete network, all weights are equal, whereas the middle node in the chain has twice the weight of the end nodes (these expectations are in approximate agreement with the measured communication rates). The free parameters in the models, the level of risk aversion plus the initial and minimum uncertainties (for ASC), were chosen so as to minimize the total χ^2 error over both networks between the observed and simulated data. The data displays the observed greater polarization for favorite groups, high disagreement level, and complete topology. The RPM and ASC simulations also display these behaviors demonstrating qualitative agreement between the experiment and simulations. That the simulation results mostly pass through the error bars further suggests quantitative agreement whose testing we now discuss.

In general, when statistically testing the fit of a model, one assesses whether its predictions are consistent with the data in the sense that it is reasonably probable that the model could have produced the data given the presence of

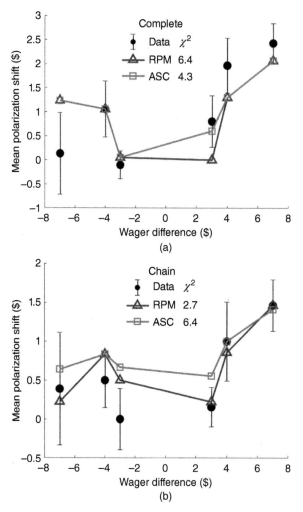

Figure 13.3 Comparison of experimental data and simulations of ASC and RPM models. (a) Complete network. (b) Chain. Positive and negative sides of the x-axis correspond to favorite and underdog groups, respectively. Error bars are standard errors. Simulation results are rounded upward to nearest dollar.

unmodeled noise. The null hypothesis is that the model is correct, and so support for the model is found if the null hypothesis cannot be rejected. Note that this criterion is opposite to that used in the testing of qualitative effects where one seeks to reject the null hypothesis in order to claim support for the theorized relationship. If the model passes the test, one can only claim that it is consistent with the data, not that it is the true model, whereas failure to pass the test indicates that the model can be rejected as false. Free parameters – those that cannot be determined without using the dependent variable – are fit so that the error statistic is minimized. Using more free parameters, however, has the effect of decreasing the maximum allowable error beneath which one can claim support for the model. This makes it harder for models with more parameters to pass the test. However, if two models – a parameter-lean one and a parameter-rich one – both pass the test, one cannot claim more support for one than the other on the basis of the test itself. While the parameter-lean model might be preferred by virtue of its parsimony, the parameter-rich model may be preferred if it has more general scope beyond the experiment under study or if there are more grounds for its causal process in the relevant literature.

The RPM and ASC models were evaluated using a χ^2 goodness-of-fit test. A χ^2 goodness-of-fit test uses the sum of the squared errors between the observed and predicted data points (normalized by the variance at each point) as its test statistic. A threshold of $Q \leq 0.2$ was chosen for rejecting the null hypothesis that the model is correct. This threshold is conservative with respect to the standard significance threshold (p-value) of 0.05 used in testing qualitative hypotheses in that a higher Q-value makes it more difficult for the model to pass the test. The RPM model (one free parameter) has $Q = 0.61$, and the ASC model (three free parameters) yields $Q = 0.3$. Accordingly, both models were found to be consistent with the data. On the other hand, several alternative models, such as the median and a proximate majority model based directly on the policy (not the rhetorical frame), did not pass. Consequently, this test provides a statistical basis for rejecting these models as explanations of the experimental data.

While the agreement between the experiment and the models is encouraging, more experimentation is needed to judge the validity of the theory. Of greater relevance for present purposes is that this work illustrates some of the themes of the integrated approach. First, it demonstrates that quantitative agreement between computational models and experiment is possible. Plots such as Figure 13.3 comparing data points with error bars against model predictions as a function of an independent variable are common in the physical sciences, but not so in social influence research or social science more generally. The work also shows that the modeling goal of predicting outcomes for groups with specific initial conditions drives a heightened concern for the synthesis of concurrent effects, as done here for group polarization and majority influence. This concern for synthesis can lead to a new theory, which, in turn, can predict fundamentally novel phenomena such as differential polarization due to frame substitution. In addition, the synthesis driven by modeling can yield clear predictions for the effects of variables left ambiguous in qualitative theory, such as disagreement level and network structure.

Integrated Approach

The proliferation of opinion network modeling research that is largely uncoupled from empirical support may be impeding rather than advancing computational social science. Particularly in fields outside of social science such as physics, computer science, and engineering, an initial model can be subjected to ever more sophisticated analysis or variation, thereby achieving prominence incommensurate with its level of empirical support. This proliferation makes it hard for application-oriented researchers to choose among various models as model popularity need not indicate empirical validity. Likewise, it is difficult for prospective operational consumers of simulations that incorporate opinion network models to evaluate their validity.

For its part, social psychology is not wanting of experimental investigation of social influence. Commonly, a new behavior is discovered experimentally and further experiments are aimed at generalizing and elaborating upon the origins and circumstances of its occurrence. Experiments can extend the behavior to new contexts and winnow down competing theoretical explanations, but much less effort is dedicated to synthesizing it with other social influence phenomena. This lack of synthesis hampers application to real-world contexts. Many experiments have been conducted on group polarization since the 1960s, but the results have not been firmly integrated with majority influence, consensus pressure, or attitude change research bearing on disagreement. As applied to a natural group decision-making context for an issue with a substantial judgmental element, the normative influence theory, for example, would advise identifying whether a culturally salient norm is present that would push groups further to the extreme. But then, given such a norm, it would always predict polarization. Another example is how the hidden profiles paradigm (Schulz-Hardt and Mojzisch 2012) has not been reconciled with the persuasive arguments theory of group polarization: the former emphasizes how group discussion centers on pieces of information held in common to the detriment of the sharing of unique information held by individuals, while the latter hinges upon group members exchanging their unique information.

An approach that centers on integrating experiment with the quantitative testing of computational models of social influence will place models on more solid empirical footing. Of particular importance to this approach are complex systems models (e.g. the ASC model), in which the variables of interest, such as group member opinions, evolve from initial conditions as a result of endogenous feedback with each other and perhaps exogenous signals (Gabbay 2014). Unlike standard statistical models, complex systems models do not directly posit some generic functional form, often linear or quadratic, that expresses an overall relationship between independent and dependent variables. Rather, the overall functions relating input and output are determined by the unfolding of the processes in the model and so need not result in a function that is expressed in a simple analytic form. However, the predicted relationship is more specific than simply saying, for instance, that there is an interaction effect between variables as in regressions. That complex systems models must deal with specific group initial conditions, rather than an overall population of groups, provides a greater motivation to synthesize different effects than do statistical approaches.

Group polarization research provides an example of how seeking to model specific initial conditions can drive synthesis. One can statistically analyze the overall group polarization in a population of homogeneously inclined groups with random initial preference distributions without worrying about majority influence. Since groups with a majority preference above the mean would be roughly balanced by those whose majority was below the mean, (preference-based) majority influence could not be the cause of any observed net shift toward the extreme. However, majority influence strongly affects whether a particular group will polarize or not as it can operate counter to polarization when a group has a majority below the mean. This implies that majority influence cannot be ignored when one seeks to predict whether individual groups will polarize. Moreover, the focus on making predictions for specific within-group initial preference distributions, rather than over a population of groups with random distributions, helped spur the frame-induced theory's reconceptualization of majority influence as operating over the rhetorical frame, thereby becoming integral to the group polarization process.

Implementation of an integrated approach will entail changes for both the experimental and modeling sides. For the former, the major change is that experiments will be designed from the outset with the objective of quantitatively testing computational models, not just qualitative hypotheses. For hypothesis testing, one often coarse grains the values of an independent variable in order to construct experimental cells corresponding to, say, a binarization of that variable, as we did with disagreement level above. Quantitative testing of computational models, however, typically requires greater resolution, and experiments should be designed so that a sufficient number of variable values are present to conduct a goodness-of-fit test. As complex systems models can utilize specific experimental initial conditions for individuals or groups instead of treating them as random error, greater attention should be given to the distribution of initial conditions in the design than is needed when the goal is simply to populate binarized condition cells. Beyond a focus on initial and final states, the testing of complex systems models would also benefit from experimental measurements over time, when such measurement is feasible and does not unduly interfere with the process under study.

For the modeling side, a greater focus on developing models capable of being estimated from experimental data is needed. Making quantitative contact with experimental results requires more disciplined consideration of parameters than when more simply endeavoring to show a qualitative correspondence with the data. The temptation to develop rich models must be tempered against the need to estimate parameters from the data itself if they cannot be independently determined. Discretion should be exercised when considering the addition of parameters beyond those directly related to the variable of primary concern. The essential parameters in continuous opinion network models will involve distance in the opinion space.

Another important element is that models be capable of prediction. This requires that, once parameters are estimated on an in-sample population, models are then capable of predicting cases held out of the sample or from a new experiment. As noted above, the network weights in the Friedkin–Johnsen model have primarily been calculated using post-discussion ratings of influence by the group members themselves. Such a procedure precludes prediction. However, it would be possible to test the extremist-tilting explanation of group polarization in conjunction with the Friedkin–Johnsen model if one were to fit the function relating persuasion resistance to opinion extremity.

The focus on a relatively tight number of parameters that can be determined a priori or estimated from the data will make models more robust and applicable across experiments falling into the same broad context. A more ambitious goal of an experimentally oriented modeling program would be to bridge different experimental contexts, such as problem solving, forecasting, policy-making, and ideological attitudes. As an example, the distinction between intellective and judgmental tasks is an important one in group decision-making. At opposite ends of the intellective–judgmental spectrum, purely intellective tasks like math puzzles have solutions that are demonstrably correct, whereas purely judgmental tasks are matters of personal taste. Forecasting problems, for instance, lie in between, having not only intellective elements that are demonstrably right or wrong, such as the record of a football team or what party has the most registered voters in a given district, but also judgmental ones involving the factors likely to be most important in a particular circumstance, such as motivational differences between teams or the impact of national-level political considerations on a local election (Kerr and Tindale 2011). Where a task lies on the intellective–judgmental spectrum affects the relative importance of social influence effects such as minority or majority influence. Rather than simply categorizing a context as intellective or judgmental and choosing a model accordingly, it would be preferable to define a parameter that gauges the balance between intellective and judgmental factors and therefore the weight of the dynamical mechanisms at play in a given context. Experiments could then test whether models integrated using the parameter could successfully make predictions for various tasks along the intellective–judgmental spectrum. As an illustration, one might conjecture that with respect to group polarization, the frame-induced mechanism might best suit forecasting problems, whereas political ideologies, falling further on the judgmental side, might best be modeled by extremist tilting. Policy-making might fall in between the two, and a parameter reflecting that balance could then be used to weight relative strengths of the frame-induced and extremist-tilting mechanisms.

A potential hazard of orienting experiments toward model testing is that experiments may be treated as primarily data-fitting exercises in which researchers test a raft of different models driven more by the various mathematical or simulation possibilities rather than by theory. This tendency will lead to models that are narrow in scope and not readily generalized to new circumstances. The SDS literature suffered from this tendency, as mentioned above, and never produced a compelling account of group polarization. However, the increasing prevalence of requirements among journals to make data sets available may help counter the lack of convergence caused by the tendency to seek and emphasize the best-fitting models for experiments in isolation. A new model, which best accounts for the results of a particular experiment, can now also be tested against data from previous experiments. The growth of a norm toward testing models against new and old data will encourage the development of more general models.

Initial successes in model development and testing will eventually lead to the emergence of a self-sustaining research community dedicated to the integration of modeling and experiment (an example of a new and virtuous epistemic culture as described in Chapter 2). In the short term, a comprehensive program aimed at providing experimental data to test and develop a range of models would help generate the nucleus of such a community as well as advancing social influence research itself. The goal of the program would be to develop general models that integrate different social influence phenomena over a range of contexts rather than the current practice that investigates behaviors in divergent research streams. To effect such synthesis, the program could unfold in phases in which experiments and models initially focus on relatively narrow phenomena with later phases becoming successively more integrative. This would encourage the development of more general and robust models and counter the tendency toward one-off model fitting. Follow-on research could test the models on real-world contexts of interest. Ideally, experiments would be conducted by separate teams of researchers who would then share the data with modeling teams. However, as there is little tradition in social psychology (and social science more broadly) of publishing experimental results without at least some theoretical embellishment, experimental teams could be allowed to develop their own models or included as coauthors on initial publications using their data.

While the goal of quantitatively testing computational models of social influence is challenging, the rewards for doing so would be high. On a scientific level, it would make the study of social influence more synthetic and cumulative. The bar would be raised for evaluating competing theories: a theory implemented in terms of a model that provided a quantitative account of experimental results would be preferable to one that only provided a qualitative account. In addition, the present practice of testing hypotheses experimentally based on the coarse graining of variables results in nominal categorizations that can be difficult to extend to more general conditions and synthesize when competing effects and nonlinear interactions are present. Standard group polarization theory, for example, only applies when all group members have common inclinations. Although a more relaxed condition of *most* group members is often stated, that is too vague to enable prediction in situations where groups have members with opposing preferences. Effectively, the reference point is a wall beyond which standard polarization theory is silent. By allowing for gradations of effects, computational models are less constrained by such ambiguous categorizations; for instance, the rhetorical reference point in the ASC model is simply a parameter that affects the rhetorical distance between group members, a distance that can be calculated regardless of whether it spans the reference point or not. As a result, the ASC model can treat the combined F and U groups in Figure 13.1, while standard theory cannot. This freedom from dependence upon categorizations implies that models can be more readily extended to variable and parameter regimes not yet explored experimentally. In combination with the ability to probe the effects of nonlinear interactions via mathematical analysis and simulation, models can therefore be used to reveal novel, potentially counterintuitive behaviors not anticipated by qualitative theorizing.

Greater incentive to perform replication experiments would be another scientific benefit if experiments were to become more oriented toward computational models. Historically, there has been little incentive for social scientists to perform replications of previously reported effects and for journals to publish them. In the physical sciences, however, better measurements of model parameters, such as physical constants, are valued even if no new effect is reported, as such measurements improve the accuracy and precision of model predictions. Similarly, social influence experiments aimed at testing models would yield improved parameter estimation and so hold more value than merely replicating an effect. New experiments could repeat earlier ones but with higher resolution or an extended variable range, thereby enhancing the precision and robustness of parameter estimates. It is also possible that systematic deviations from model predictions could be observed pointing the way toward new theory and model development. A greater ability to publish such discrepant experimental data as valuable in its own right (without theoretical explanation) would therefore allow social scientists to learn more from data than is presently the case. Fundamental advances in physics have occurred because of the publication of experimental findings that ran counter to accepted theoretical models. A pivotal event in the genesis of quantum mechanics, for example, was the discovery of the photoelectric effect, a phenomenon at odds with classical physics, which Einstein eventually explained. The concentration on developing general models that minimize the number of free parameters will also discourage data dredging practices in which researchers sift through a large number of covariates in order to find statistically significant, albeit likely spurious, relationships.

The approach outlined above is intended to develop a community within social influence research in which theory, modeling, and experiment proceed in a fashion similar to the physical sciences, albeit with reduced expectations of predictive power. It is not a call to end the traditional paradigm for the investigation of social influence. Although we have argued that the integrated approach will lead to new theories and discoveries, taking it further by demanding that novel theories be implemented formally before experimental testing would, on net, likely hamper the discovery of new behaviors, given the richness of social systems. A more desirable outcome would be for the model-oriented and traditional approaches to work in tandem. The standard testing of qualitative hypotheses could explore variables and effects not yet incorporated within quantitative models. This exploratory role would identify promising areas that could benefit from modeling and facilitate model development by narrowing the range of viable theoretical explanations. It is also possible that some behaviors will not be amenable to quantitative modeling and so remain in the province of qualitative theory in which modeling continues to play its more usual historical role in support of hypothesis generation.

Conclusion

Much of the recent surge of activity in computational social science has revolved around the analysis of massive amounts of data available from naturally occurring activity on the Internet and social media involving large networks consisting of thousands or millions of individuals. However, such studies do not shed light on the small group context, which is central to decision-making in leadership groups as well as political attitude change among ordinary citizens. Accordingly, the agenda put forth here emphasizes experiments with human subjects. Given their ability to control conditions, experimental studies can more directly test opinion models than can data from online networks or other observational sources. Network topology and initial opinion distributions can be controlled, the latter enabling testing of the core objective of modeling how opinions change from their initial values, rather than predicting final distributions on the basis of assumed initial conditions. Moreover, experimental results can provide a sounder basis for application of opinion network modeling to large systems. Such applications typically use models based on dyadic or other local interactions, and if those models cannot predict the results of small group experiments or be derived from

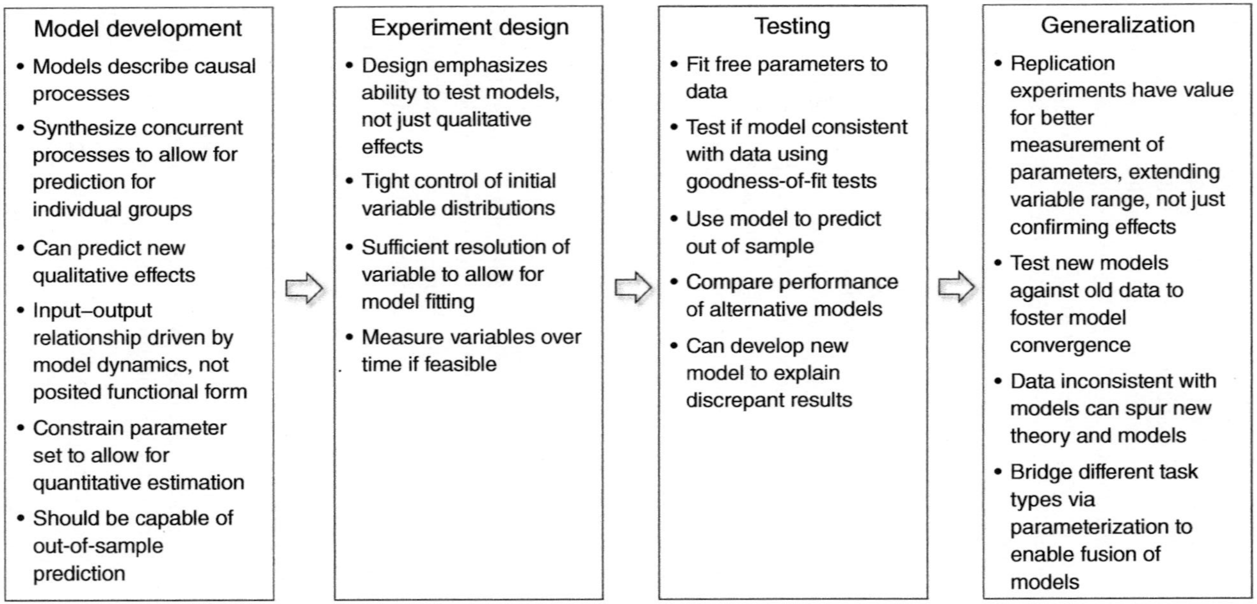

Model development

- Models describe causal processes
- Synthesize concurrent processes to allow for prediction for individual groups
- Can predict new qualitative effects
- Input–output relationship driven by model dynamics, not posited functional form
- Constrain parameter set to allow for quantitative estimation
- Should be capable of out-of-sample prediction

Experiment design

- Design emphasizes ability to test models, not just qualitative effects
- Tight control of initial variable distributions
- Sufficient resolution of variable to allow for model fitting
- Measure variables over time if feasible

Testing

- Fit free parameters to data
- Test if model consistent with data using goodness-of-fit tests
- Use model to predict out of sample
- Compare performance of alternative models
- Can develop new model to explain discrepant results

Generalization

- Replication experiments have value for better measurement of parameters, extending variable range, not just confirming effects
- Test new models against old data to foster model convergence
- Data inconsistent with models can spur new theory and models
- Bridge different task types via parameterization to enable fusion of models

Figure 13.4 Overview of integrated modeling-experiment approach.

approximations of models that can, then little rationale exists for their use on large systems.

Figure 13.4 summarizes the integrated modeling and experimental approach proposed in this chapter. The goal of quantitatively testing computational models of social influence is no doubt an ambitious one. The approach advocated in this chapter centers upon the conduct of experiments explicitly designed to test the quantitative predictions of models rather than the standard experimental paradigm of testing qualitative hypotheses. Its aim is the development of models that can account for a range of phenomena and experimental results. Elements of this approach include: exercising discipline and discrimination with respect to model parameters, conducting goodness-of-fit tests, more highly resolved initial variable conditions, more deliberate control of initial opinion distributions, measuring opinions or other variables over time, greater use of out-of-sample prediction, testing models on new and old data to foster model convergence not proliferation, and parameterizing the nature of group tasks along a spectrum rather than ambiguously assigning them to nominal categories such as intellective or judgmental.

While experimentation with human subjects is much more expensive and laborious than modeling and simulation, a greater emphasis on their integration will enhance both the influence of computational social science and the science of social influence. A major advantage of this integrated approach is an improved ability to synthesize different effects. Since opinion network models make predictions for specific groups, they must take more serious account of effects concurrent to the one under study, which otherwise might be assumed to wash out in a population of groups. Models can synthesize multiple effects more readily than combining different, often ambiguous, categorizations of conditions. The bar will be raised for the evaluation of rival theories with higher precedence given to theories whose associated models are in quantitative accord with experiment. Stronger incentive to conduct experiments for the purpose of providing better measurement or expanding the range of model variables – not just to test hypothesized relationships or competing theories – will be fostered under this approach. Greater replicability will ensue as will the ability to publish anomalous findings, thereby spurring new theory and model development. Ultimately, on an applications level, the integration of quantitative model testing and experiment will raise the confidence and scope with which models can be applied to natural situations for purposes of both prediction and designing interventions to shape outcomes.

Acknowledgments

This research was supported by the Office of Naval Research under grants N00014-15-1-2549 and N00014-16-1-2919.

References

Becker, J., Brackbill, D., and Centola, D. (2017). Network dynamics of social influence in the wisdom of crowds. *Proceedings of the National Academy of Sciences* 114 (26): E5070–E5076. https://doi.org/10.1073/pnas.1615978114.

Brown, R. (1986). *Social Psychology*, 2e. New York: Free Press.

Burghardt, K., Rand, W., and Girvan, M. (2016). Competing opinions and stubborness: connecting models to data. *Physical Review E* 93 (3): 032305.

Castellano, C., Fortunato, S., and Loreto, V. (2009). Statistical physics of social dynamics. *Reviews of Modern Physics* 81 (2): 591–646.

Chong, D. and Druckman, J.N. (2007). Framing theory. *Annual Review of Political Science* 10 (1): 103–126. https://doi.org/10.1146/annurev.polisci.10.072805 .103054.

Crandall, D., Cosley, D., Huttenlocher, D. et al. (2008). Feedback effects between similarity and social influence in online communities. Paper presented at the Proceeding of the 14th ACM SIGKDD international conference on Knowledge discovery and data mining, Las Vegas, Nevada, USA.

Deffuant, G., Amblard, F., Weisbuch, G., and Faure, T. (2002). How can extremism prevail? A study based on the relative agreement interaction model. *Journal of Artificial Societies and Social Simulation* 5 (4).

DeGroot, M.H. (1974). Reaching a consensus. *Journal of the American Statistical Association* 69 (345): 118–121. https://doi.org/10.2307/2285509.

Devine, D.J. (2012). *Jury Decision Making: The State of the Science*. New York: NYU Press.

Düring, B., Markowich, P., Pietschmann, J.-F., and Wolfram, M.-T. (2009). Boltzmann and Fokker–Planck equations modelling opinion formation in the presence of strong leaders. *Proceedings of the Royal Society A: Mathematical, Physical and Engineering Science* 465 (2112): 3687–3708. https://doi.org/10 .1098/rspa.2009.0239.

Eagly, A.H. and Chaiken, S. (1993). *The Psychology of Attitudes*. Fort Worth, TX: Harcourt College Publishers.

Fortunato, S. and Castellano, C. (2007). Scaling and universality in proportional elections. *Physical Review Letters* 99 (13): 138701.

Friedkin, N.E. (1999). Choice shift and group polarization. *American Sociological Review* 64 (6): 856–875.

Friedkin, N.E. (2015). The problem of social control and coordination of complex systems in sociology: a look at the community cleavage problem. *IEEE Control Systems* 35 (3): 40–51. https://doi.org/10.1109/MCS.2015.2406655.

Friedkin, N.E. and Johnsen, E.C. (2011). *Social Influence Network Theory: A Sociological Examination of Small Group Dynamics*. Cambridge, UK: Cambridge University Press.

Gabbay, M. (2007). The effects of nonlinear interactions and network structure in small group opinion dynamics. *Physica A* 378: 118–126.

Gabbay, M. (2014). Data processing for applications of dynamics-based models to forecasting. In: *Sociocultural Behavior Sensemaking: State of the Art in Understanding the Operational Environment* (ed. J. Egeth, G. Klein and D. Schmorrow), 245–268. McLean, VA: The MITRE Corporation.

Gabbay, M. (forthcoming). Opinion Network Modeling and Experiment. In: *Proceedings of the 5th International Conference on Theory and Applications in Nonlinear Dynamics* (ed. V. In, P. Longhini and A. Palacios). Springer.

Gabbay, M., Kelly, Z., Reedy, J., and Gastil, J. (2018). Frame-induced group polarization in small discussion networks. *Social Psychology Quarterly* 81 (3): 248–271. https://doi.org/10.1177/0190272518778784.

Hinsz, V.B. (1999). Group decision making with responses of a quantitative nature: the theory of social decision schemes for quantities. *Organizational Behavior and Human Decision Processes* 80 (1): 28–49. http://dx.doi.org/10.1006/obhd.1999.2853.

Isenberg, D.J. (1986). Group polarization: a critical review and meta-analysis. *Journal of Personality and Social Psychology* 50 (6): 1141–1151. http://dx.doi.org/10.1037/0022-3514.50.6.1141.

Keating, J., Van Boven, L., and Judd, C.M. (2016). Partisan underestimation of the polarizing influence of group discussion. *Journal of Experimental Social Psychology* 65: 52–58. http://dx.doi.org/10.1016/j.jesp.2016.03.002.

Kerr, N.L. and Tindale, R.S. (2004). Group performance and decision making. *Annual Review of Psychology* 55 (1): 623–655.

Kerr, N.L. and Tindale, R.S. (2011). Group-based forecasting?: a social psychological analysis. *International Journal of Forecasting* 27 (1): 14–40. http://dx.doi.org/10.1016/j.ijforecast.2010.02.001.

Laughlin, P.R. and Ellis, A.L. (1986). Demonstrability and social combination processes on mathematical intellective tasks. *Journal of Experimental Social Psychology* 22 (3): 177–189. https://doi.org/10.1016/0022-1031(86)90022-3.

Lorenz, J. (2007). Continuous opinion dynamics under bounded confidence: a survey. *International Journal of Modern Physics C: Computational Physics and Physical Computation* 18 (12): 1819–1838.

Mäs, M. and Flache, A. (2013). Differentiation without distancing. Explaining bi-polarization of opinions without negative influence. *PLoS One* 8 (11): e74516. https://doi.org/10.1371/journal.pone.0074516.

Moussaïd, M., Kämmer, J.E., Analytis, P.P., and Neth, H. (2013). Social influence and the collective dynamics of opinion formation. *PLoS One* 8 (11): e78433. https://doi.org/10.1371/journal.pone.0078433.

Moussaïd, M., Brighton, H., and Gaissmaier, W. (2015). The amplification of risk in experimental diffusion chains. *Proceedings of the National Academy of Sciences* 112 (18): 5631–5636. https://doi.org/10.1073/pnas.1421883112.

Myers, D.G. (1982). Polarizing effects of social interaction. In: *Group Decision Making* (ed. H. Brandstatter, J.H. Davis and G. Stocker-Kreichauer). London: Academic Press.

Myers, D.G. and Bishop, G.D. (1970). Discussion effects on racial attitudes. *Science* 169 (3947): 778–779. https://doi.org/10.2307/1729790.

Olfati-Saber, R., Fax, J.A., and Murray, R.M. (2007). Consensus and cooperation in networked multi-agent systems. *Proceedings of the IEEE* 95 (1): 215–233. https://doi.org/10.1109/JPROC.2006.887293.

Pleskac, T.J., Diederich, A., and Wallsten, T.S. (2015). Models of decision making under risk and uncertainty. In: *The Oxford Handbook of Computational and Mathematical Psychology* (ed. J.R. Busemeyer, Z. Wang, J.T. Townsend and A. Eidels), 209–231. Oxford: Oxford University Press.

Proskurnikov, A.V. and Tempo, R. (2017). A tutorial on modeling and analysis of dynamic social networks. Part I. *Annual Reviews in Control* 43 (Supplement C): 65–79. https://doi.org/10.1016/j.arcontrol.2017.03.002.

Schkade, D., Sunstein, C.R., and Kahneman, D. (2000). Deliberating about dollars: the severity shift. *Columbia Law Review* 100 (4): 1139–1175. https://doi.org/10.2307/1123539.

Schkade, D., Sunstein, C.R., and Hastie, R. (2010). When deliberation produces extremism. *Critical Review* 22 (2–3): 227–252. https://doi.org/10.1080/08913811.2010.508634.

Schulz-Hardt, S. and Mojzisch, A. (2012). How to achieve synergy in group decision making: lessons to be learned from the hidden profile paradigm. *European Review of Social Psychology* 23 (1): 305–343. https://doi.org/10.1080/10463283.2012.744440.

Sunstein, C.R. (2002). The law of group polarization. *Journal of Political Philosophy* 10 (2): 175–195.

Török, J., Iñiguez, G., Yasseri, T. et al. (2013). Opinions, conflicts, and consensus: modeling social dynamics in a collaborative environment. *Physical Review Letters* 110 (8): 088701.

Zaleska, M. (1982). The stability of extreme and moderate responses in different situations. In: *Group Decision Making* (ed. H. Brandstatter, J.H. Davis and G. Stocker-Kreichauer). London: Academic Press.

14

Combining Data-Driven and Theory-Driven Models for Causality Analysis in Sociocultural Systems

Amy Sliva[1], Scott Neal Reilly[1], David Blumstein[1], and Glenn Pierce[2]

[1] *Charles River Analytics, Cambridge, MA 02138, USA*
[2] *School of Criminology and Criminal Justice, Northeastern University, Boston, MA 02115, USA*

Introduction

One of the key aspects of nearly all social science research questions is to validate or discover an explanation for some type of causal relationship. These causal explanations provide a greater understanding of complex social and political dynamics, with the long-term goal of influencing policy interventions or producing theories that can facilitate better decision-making by policy-makers. The growing accessibility of digital information promises many exciting research opportunities for expanding the breadth and depth of causal analysis. For example, studies of the relationship between conflict and natural resource constraints can pull economic data from any number of online sources (e.g. World Bank, World Trade Organization), records of violent incidents from event databases, political sentiments posted in online social media, in-depth interviews from refugees provided through social science data exchanges or aid organizations, and even climate and water quality trends from environmental biologists, all with a few clicks of a mouse.

However, exactly how to leverage this *big data* environment remains a challenge for social scientists, who typically have two distinct methodologies at their disposal: quantitative and qualitative. While quantitative data-driven approaches to causal analysis can provide empirical insights and discover previously unknown relationships, these models can be difficult to fully explain or may overfit existing data sets, producing models that contain spurious statistical relationships that are not truly causal or are not robust to real-world social variations. Adding normative theoretical causal approaches to data-driven analyses can help generate hypotheses and provide the necessary framing for interpreting and understanding data-driven results. Analyzing

Social-Behavioral Modeling for Complex Systems, First Edition.
Edited by Paul K. Davis, Angela O'Mahony, and Jonathan Pfautz.
© 2019 John Wiley & Sons, Inc. Published 2019 by John Wiley & Sons, Inc.
Companion website: www.wiley.com/go/Davis_Social-Behavioralmodeling

causality, then, often requires a diversity of multidisciplinary models – both quantitative and qualitative – that enable researchers to achieve the greater scale promised by readily available data as well as the deeper understanding necessary to make these results useful and applicable. Data can provide new avenues for discovering causal relationships or validating (and invalidating) causal assumptions in many contexts, but theoretical knowledge is needed to make sense of possible interactions and their applicability to particular situations.

In this chapter, we explore the nature of this combination of data-driven and theory-driven approaches to causal analysis. The role each method plays in an analysis depends on the research question, the goals of the analysis, and the type of data available. Using these factors, we have identified two general ways that data-driven and theory-driven approaches can interact with one another to better analyze causality in social systems: (i) existing theory describes relationships that constrain or direct data-driven models, which can validate these theoretical understandings, and (ii) data-driven approaches discover alternative relationships that can be explained by or enrich existing normative theories. Within these two basic methodological approaches, there is still quite a bit of room for variation. To generalize this space, we have developed a framework for constructing mixed-methods causal ensembles, leveraging not only the diversity of data but also the rich diversity of methods and theoretical insights that can be combined to achieve a more comprehensive causal analysis. Finally, we will present four case studies based on our prior work (Sliva et al. 2013; Sliva et al. 2016; Sliva et al. 2017) as examples of different balances between data-driven and theory-driven combinations for causal analysis.

Understanding Causality

Before we explore different approaches for combining data- and theory-driven analyses of causality in social systems, we first need to better understand causality and how it is explored in social science. Causality is notoriously difficult to analyze, a challenge amplified in complex social systems. When looking at physics at a Newtonian or even quantum scale, many questions of causality are addressed by mathematical models that are close representatives of the underlying objective truth. Reality at a larger scale, especially social systems, is much more complex, making the objective truth of causal relationships and processes difficult to define, observe, and measure. We have identified two major factors contribute to this difficulty (Sliva and Neal Reilly 2014; Sliva et al. 2015; Sliva et al. 2017): (i) the lack of a universally accepted definition of causality that applies to all research questions and (ii) the complexity and uncertainty inherent in the social and political phenomena in question.

Goldthorpe (2001) identifies three possible definitions or approaches to working with causality that have been prominent in social science research. The simplest (and most methodologically contentious) view is causality as robust dependence, where causation merely refers to some amount of observed association or correlation with strong predictive power. However, while prediction may be important in some contexts, this definition of causality is not applicable to situations where explanation, rather than forecasting, is the goal – researchers can discover statistically significant *causal* relationships, but explaining how and why these relationships develop requires additional theoretical insight. Another prominent approach to causality views causation as consequential manipulation. That is, causation means that there are different consequences when the causal variable is manipulated or varied (even if only theoretically). In this approach, researchers verify a causal relationship through experimentation – if X is manipulated, it produces systematic effects in the dependent variable Y. Finally, a growing field of study views causality as a generative process whereby the relationship between X and Y involves more than just time and precedence, but is determined by some underlying social process or mechanism, which itself may be unobservable. This generative process concept of causality prevails in many fields in both the natural and social sciences, focusing on the question of *how* the relationship between variables is actually produced, where the underlying process may consist of relationships and latent variables that cannot be directly measured. The intent is to explain and characterize interactions instead of simply quantifying them. In general, these three definitions represent increasing amounts of knowledge. Understanding a generative process implies the ability to manipulate causes and effects, which supports characterization of dependence.

Many data-driven models adhere to the first definition of causality and describe relationships that are highly correlated with a strong predictive component. For example, because causality has an inherently temporal component – the effect does not precede the cause – techniques that exploit this can help analyze and extract causal models from time-series data. The most basic model of predictive power is simple correlation, such as the classic Pearson correlation measure; by adding a temporal offset to the correlation to mimic the sequencing of cause and effect, Pearson can successfully capture many causal/predictive relationships in time-series data. Granger causality (Granger 1969, 1980), a technique familiar to many social science researchers, takes strong predictive correlation further. Granger causality was originally introduced for time-series analysis in economics and assumes a temporal order of cause and effect and also assumes that the causal variables provide information about the effect that can otherwise be unavailable.

Dynamic time warping is a similar approach that was developed by the authors for reasoning over arbitrary time-series data (Myers and Rabiner 1981; Salvador and Chan 2007). It was originally created for gait recognition in computer vision, where temporal offsets may be inconsistent over time (similar to how the distance between steps may change depending on whether a person is walking or running, an increase in crime might happen anywhere from 6 to 12 months after an uptick in unemployment). Convergent cross mapping (CCM) (Sugihara et al. 2012; Clark et al. 2015; Ye et al. 2015), a recent advance in biological studies, can identify even deeper feedback relationships among time series, enabling scientists to model cyclic causal relationships, such as the feedback relationship between poverty and conflict known as the conflict trap (Collier 2003; Collier and Hoeffler 2004; Sliva et al. 2015, 2016) (see section "Choosing Data-Driven Approaches Using Theory"). These approaches can be very useful for identifying causal/predictive relationships in time-series data sets, for example, validating models relating drops in unemployment with successive increases in criminal activity. However, this temporal data only represents a small proportion of the available social science data sets – which might also include a variety of cross-sectional, relational, or event-based data – and analysis methods. In addition, these approaches focus more on correlation and predictive power, making it difficult to verify truly robust causal relationships that not only enable prediction of a social phenomenon but suggest policy directions or possible interventions.

Experimental intervention, Goldthorpe's second definition of causality, is especially difficult in social systems, where it tends to be limited by cost, time, and ethical constraints. Increasingly in the social sciences, researchers are using *natural experiments* that can provide insights into causal mechanisms – such as work examining voter turnout under varying conditions of electoral redistricting (Brady and McNulty 2004) or analysis of the economic impact on civil conflict given the occurrence of different economic shocks caused by weather or natural disasters (Miguel et al. 2004). In these situations, researchers regard the subjects according to a standard of *as if* random assignment where causes are assigned randomly (or as if randomly) among the population; that is, people are not self-selecting into electoral districts or areas of economic shock in ways that would confound the causal analysis (Dunning 2008, 2012). However, natural experiments, while useful in causal analysis, can often fall victim to the counterfactual problem – because we cannot go back in time and rerun history with a different set of parameters, it is impossible to prove in a perfectly controlled experiment that the outcome in one situation would have been different if certain causal variables were changed. Many of the analysis techniques used to actually assess the causal influences in these natural experiments fall into either the first definition of causality previously described or Goldthorpe's third definition, where they are used to capture the complex underlying generative process by which causes produce effects, attempting to mitigate the counterfactual problem and moving beyond mere correlational or predictive relationships.

In this third category are several qualitative theory-driven approaches common in social science, as well as several computational approaches that are gaining prevalence in the social science community. One of the most prominent qualitative approaches to causal analysis is process tracing, whereby researchers enumerate the possible causal mechanisms (i.e. processes), as well as the other events, actions, or intervening variables that may enable them to diagnostically link the processes to the observed effects or outcomes (Bennett and George 1997; Bennett 2010). In one approach to process tracing, the researcher tests whether the processes and lines of influence in a case match existing normative theories – does the causal mechanism appear to operate as expected or diverge in some way? Another approach takes a more inductive view, identifying potentially new causal influences indicated by the data in the case study by examining the impact of different values of some potentially causal factor or working backward from unexpected effects to determine possible causes. In both cases, process tracing looks at the data empirically using a series of qualitative tests (e.g. identify variables that are necessary vs. sufficient for causal interactions) to confirm or reject theoretically driven causal hypotheses (Bennett 2010). Qualitative comparative analysis (QCA) is another qualitative technique designed to identify underlying causal processes (Ragin 1989). QCA is based on Boolean logic, looking at the possible combinations of all variables in a case and using formal logical inference (or, in more recent variations, fuzzy set theory [Ragin 2009]) to determine which causal relationships logically explain the observations. QCA is focused not only on identifying the causal variables related to outcomes of interest but specifically the combinations or sequences of variables that jointly produce some effect, enabling the identification of rich causal explanations.

In addition to these qualitative approaches, relatively recent advances in the uncertainty in artificial intelligence (UAI) community provide computational models that also seek to represent the structure of some underlying process that is responsible for generating observed outcomes, capturing the direct and indirect causal relationships that exist in social systems. For example, rather than simply indicating that there is a statistically strong relationship between data on unemployment and data on crime, these approaches will explicitly model the assumed causal chain from lack of jobs in the legitimate economy to unemployment to economic hardship to criminal behavior, capturing the entire process that links the observed rise in unemployment with the subsequent rise in crime. Unlike the causal/predictive methods described above, the observational data used to construct and validate these models can be nontemporal, such as survey data (e.g. household development indicators surveys) or cross-sectional records (e.g. election outcomes for different regions) often encountered in social science research.

Graphical probabilistic models, such as Bayesian networks (Pearl 2014), have provided the foundation for work in causal analysis since Pearl's *Causality:*

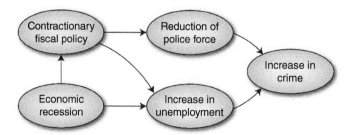

Figure 14.1 Graphical causal model illustrating the causal chain from economic recession to increases in crime.

Models, Reasoning, and Inference (Pearl 2002). These models have proven useful as formal tools for describing and reasoning about causality for many fields of inquiry (e.g. epidemiology, policy, and social sciences). Graphical models mimic naturalistic ways of thinking about causality by explicitly representing the *structure* of causal relationships. Figure 14.1 shows an example of a graphical model illustrating the causal links between economic factors and increases in criminal activity. With their basis in formal mathematics (e.g. probability theory, system dynamics), these models also facilitate rigorous reasoning and inference; however, the intuitive graphical structure means that it is often straightforward to reason visually over these graphs and make complex assessments about causal interactions in real-world situations quite easily and naturally, often without needing to resort to algebraic analysis. The graphical model structure can be provided either by human experts or through automated structure learning techniques, such as heuristic search or constraint satisfaction (e.g. inductive causation, IC; [Verma and Pearl 1991] or the PC algorithm [so called after the first names of the main authors, Peter Spirtes and Clark Glymour; Spirtes et al. 2000]). However, these automated structure learning methods depend on large quantities of data from all variables in the system, which in many cases may be unavailable or researchers have not identified all of the necessary variables present in the system.

Mooij et al. (2016) have developed a technique for determining the directionality of a causal relationship between only two variables that does not rely on any additional structural knowledge, enabling structured causal analysis from more limited observational data. For example, if we know that there is some causal relationship between crime and unemployment, we can determine the directionality of this link simply from observational data about these two variables, rather than requiring knowledge of other factors in the graph. This approach assumes that there is noise in any causal relationship and that by evaluating the effects of some standard noise models on causal relationships, it is possible to tease apart cause and effect in some situations. This has proven effective in many practical cases (see section "Case Studies: Integrating Data-Driven and Theory-Driven Ensembles").

Cutting across all of these definitions of causality are particular challenges unique to the social sciences. The complex indeterminate causation in a social system can be characterized by structural and relational features that make it especially challenging to establish the underlying causal truth (Hoffman et al. 2011). Structural complexity – such as a convergence of causes into a single effect, indirect causation or influence on more direct causes, long chains of cause and effect, temporal discrepancies (e.g. the cause may persist beyond the duration of some of its effects), and distinctions between the individual and aggregate system (e.g. is social behavior an aggregation of many individual behaviors, or are there higher-level social causes as well?) – make it extremely difficult, if not impossible, to establish measurable variables and controlled, repeatable experiments (even natural experiments) to establish an understanding of these systems. Further relationship complexities, such as counter causes, enabling conditions that are not in themselves causal (i.e. that are necessary, but not sufficient factors for inducing some effect), and emergent or evolving effects, introduce an uncertainty to this complex environment that further defies precise measurement and experimentation. In this complex environment, many important events are so singular they will occur only once. For an extreme example, Taleb argues that some of the most impactful events in history are *black swans* that were highly unlikely and not broadly anticipated (Taleb 2007), making them more amenable to theory-driven qualitative analysis than any larger-scale statistical technique.

The wide range of methods described above – by no means an exhaustive list – indicate that there is no single causality analysis technique that fully addresses all of the unique challenges of understanding social systems. Instead, social scientists have at their disposal a suite of both theory-driven and data-driven causal/predictive analysis approaches, each with strengths and weaknesses. We believe the real strength of this analysis suite lies not in finding the perfect technique for a particular system, but in combining these approaches in unique ways to produce a more comprehensive causal model (Sliva and Neal Reilly 2014; Sliva et al. 2015, 2016, 2017) that takes advantage of both theoretical and data-driven insights. The next section will explore different methods for managing these combinations and balancing the insights of normative theory and available data.

Ensembles of Causal Models

Regardless of which view of causality they adopt, researchers share the common goal of attempting to model or make sense of complex social systems.

Because of the large quantity of potential information, we see the availability of diverse large-scale data sets as a way to enrich the breadth and depth of theories of causation – there are more possible causal variables or outcomes that can be measured, more samples and cases to analyze, and more variations in context. However, as already discussed, fully taking advantage of these data sources is likely to require a hybrid approach to causality that mixes data-driven and theory-driven techniques in a mixed-methods approach. At the basic level, mixed-methods research is exactly what it sounds like: "the combination of qualitative and quantitative approaches in the methodology of a study" (Tashakkori and Teddlie 1998). This type of reasoning is actually very intuitive, and in fact, these techniques occur in everyday life in news reports, sports coverage, documentaries, etc., where statistical trends are supported by individual stories and cases (Creswell and Clark 2007). From this perspective, mixed-methods research is not just a way of combining analytical traditions but also encompasses the idea of seeing the world in multiple ways (Creswell and Clark 2007). This multidimensional view of the world maps neatly onto the challenge of causal analysis in social systems, where understanding causality can not only use different methods to address a research question from many different angles but actually requires a multiperspective approach to be addressed effectively.

In statistics and computer science, this type of model combination is known as ensemble reasoning. Ensemble reasoning has proven extremely useful in the computational modeling and the machine learning communities to provide capabilities beyond those provide by any individual technique. Also, when using machine learning to, for instance, learn a classifier, it is possible to create ensembles that combine various learning techniques (e.g. support vector machines, neural networks, naïve Bayes classifiers, combinations of decision trees) – each with its own strengths and weaknesses – to provide better results than any individual approach (Opitz and Maclin 1999). We extend this concept to the domain of causal reasoning to combine multiple causality analysis techniques, each with strengths and weaknesses, that more thoroughly and accurately model causality in real-world systems. In each of these ensembles, theory-driven and data-driven models can interact with each other in two ways.

First, theories can propose relationships that constrain or direct data-driven models. In this situation, theories provide the initial insights for quantitative studies in terms of which variables should be included in the analysis and the important relationships to be modeled, including temporal sequencing of interactions, identification of multiple causal pathways, and determination of the best statistical approach given the characteristics of the relevant variables (Collier et al. 2010). Second, we can use data-driven approaches to suggest alternative relationships that may be explained by normative theories not originally under consideration or, perhaps more importantly, can provide new information that can enrich or enhance existing theoretical assumptions. In this combination, the deeper understanding provided by theories can help explain the patterns discovered across a data set. Such explanations may point in the direction of future work (e.g. new data sets or variables that can be incorporated or related research questions that can be explored).

Within these two paradigms, there are many possible structures for combining multiple causal reasoning methods into ensembles and for joining multiple ensembles into larger ensembles. We call the building blocks of our approach data ensembles, chain ensembles, technique ensembles, and nested ensembles (Sliva et al. 2017). Figure 14.2 shows these graphically.

Data ensembles: There are qualitatively different types of causal relationships to reason about (e.g. causality when there are hidden variables, effects with multiple causes) and qualitatively different types of available data to learn from (e.g. continuous, categorical, noisy). Just as different machine learning techniques can learn different kinds of things from different kinds of data, different causal analysis techniques are needed to extract causal models based on the nature of the causal relationship present and the types of available data. For instance, Granger causality works on temporal data; process tracing works on nontemporal data.

Chain ensembles: Most causal analysis techniques are very specific about what kinds of knowledge they extract, so they need to be composed with other specific approaches that perform other types of reasoning to provide the necessary analysis. For instance, correlation analysis (with appropriate human supervision) can be used to find pairs of likely causal variables, but not the direction of the causation. Then additive noise methods (e.g. those described in Mooij et al. (2016) and Peters et al. (2014)) can be used to identify the direction of the causation given two causal parameters, but does not identify those parameters. By combining the two, we provide the ability to find pairs of cause–effect parameters in data.

Technique ensembles: There are different ways to extract similar kinds of results from similar kinds of data, combining the results in ways that produce better results than any individual method can provide. For instance, Granger causality, dynamic time warping, Pearson correlation, and offset correlation are all mechanisms to find possible causal links in temporal data (Sliva and Neal Reilly 2014; Sliva et al. 2015, 2016), though all are known to have limitations. By combining the results, we can provide an ensemble approach that is more accurate than any of the individual methods.

Nested ensembles: Ensemble causal analysis can enable us to combine causal models that derive from different data sources. This approach abstracts away from the raw data, which may come in a variety of forms (e.g. qualitative observations, surveys, time series) that are very challenging to analyze jointly, to identify linkages that can combine causal models from diverse data sources. For example, a time-series causal analysis method, such as

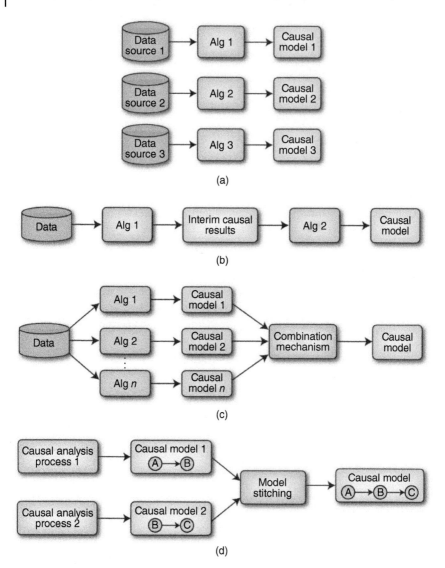

Figure 14.2 The building blocks of ensemble causal reasoning. (a) Data ensembles. (b) Chain ensembles. (c) Technique ensembles. (d) Nested ensembles.

Granger causality, may indicate that increasing unemployment leads to increased crime. A human survey study might indicate that a higher level of crime results in increased preference for increased police spending. Using an ensemble approach, we can combine these two causal models from data sets that cannot be linked using standard quantitative or qualitative methods to achieve a more complete picture of the causal process that leads from unemployment to police spending.

Case Studies: Integrating Data-Driven and Theory-Driven Ensembles

In this section, we present several examples of prior work on ensemble causal reasoning that combine theory-driven and data-driven insights in different ways. We first look at technique ensembles using primarily data-driven methods, but with room for the application of theory to suggest initial variables of interest and interpret the results. We next look at cases of chain ensembles, where theoretical insights direct the exploration of data or where data is used as input to parameterize a theoretically based model structure. Finally, we look at an example of nested ensembles where data-driven and theory-driven causal models interact in more complex iterative ways to form a complete analysis of a social system.

Letting the Data Speak: Additive Noise Ensembles

Our first case study illustrates the use of a primarily data-driven technique ensemble to assess causal relationships in nontemporal observational data. In this case, we ran an experiment using a combination of several parallel additive noise models to determine the directionality of causal relationships that have been hypothesized or assumed by theory. Additive noise models (Peters et al. 2014; Mooij et al. 2016) are a new technique in the UAI community that can identify causal relationships in purely observational data, i.e. data without a clear temporal component. Assessing causality in observational data was previously thought to be an impossible task, but these models have shown that under certain conditions, it is possible to tease out the direction of a causal interaction from this data. In an additive noise model of causality where X causes Y, we define Y as

$$Y = f(X) + E_y \quad \text{where } E_y = Y - \exp(Y \mid X) \text{ is independent of } X$$

Intuitively, this states that Y, the effect, is some function of the cause X with the addition of a noise term E_y that is independent of X, indicating that X has a direct influence on the value of Y, but some amount of additive random noise is allowable without disrupting the causal relationship. To use these models to identify causal relationships, we apply regression on Y and see if the residual noise terms are independent of X. The regression method and the specific noise models can vary, as long as the noise is additive relative to $f(X)$.

For this experiment, we used the CauseEffectPairs (CEP) benchmark data set used in previous studies (Peters et al. 2014; Mooij et al. 2016) of additive noise models of causality. This data consists of 99 different *cause–effect pairs* taken from data sets from a variety of different domains (e.g. meteorology, biology, medicine, engineering, economy, etc.). For example, the data contains pairs of altitude and temperature measurements from weather stations across Germany (here, altitude is clearly the cause, and temperature the effect) and pairs of CO_2 emissions and energy use compiled by the United Nations for 152 countries between 1960 and 2005 (energy use is the cause of CO_2 emissions). Each cause–effect pair consists of samples of a pair of statistically dependent random variables, where one variable is known to cause the other one based on experiential, theoretical, or empirical knowledge, providing us with ground truth for evaluating the data-driven technique ensemble.

The ensemble approach proved very effective when dealing with this nontemporal data. We implemented a variety of state-of-the-art additive noise models from Mooij et al. (2016). Using machine learning techniques, including support vector machines, naïve Bayes, and logistic regression, to learn a weighted combination of these additive noise models, we created an ensemble that was much more effective at identifying causal relationships than any single model. Individually, the best additive noise models were only capable of 60% accuracy when determining the direction of the cause–effect pairs in the data set, while most models had around a 50% accuracy rate, making them no better than flipping a coin. Using our learned ensemble, this improved dramatically to around 90% accuracy. This result indicates that an ensemble can identify important features of a data set, such as the amount and type of noise, and determining which combination of additive noise models will be most successful.

Choosing Data-Driven Approaches Using Theory

In our next case study, we developed a chain ensemble where theoretical insights into a particular research question are used to constrain the data-driven approaches that are applied (Sliva et al. 2015, 2016). Here, we demonstrate a representative exploration of the causal/predictive relationship between poverty and conflict. A large body of literature exists that explores the *conflict trap* – the process whereby countries get stuck in a repeated pattern of violent conflict and economic underdevelopment (Collier 2003). There have been several studies evaluating the causal/predictive link between these two features using standard statistical approaches, with some finding evidence for poverty driving societies into conflict (Collier and Hoeffler 2004; Braithwaite et al. 2016), while others (Djankov and Reynal-Querol 2010) indicate that civil conflict may be the cause of depressed economic growth.

For this problem, the choice of data is itself a challenge requiring theoretical insights, as the complex and abstract concepts of *poverty* and *conflict* are difficult to represent as measurable variables. To measure conflict we use the Uppsala Conflict Data Program/Peace Research Institute Oslo (UCDP/PRIO) data set (Themnér and Wallensteen 2014), which tracks the incidence and intensity of global armed conflict between 1946 and 2013. To capture the notion of poverty, which is not merely a measure of income, but also of relative well-being, we use two variables from the World Bank World Development Indicators data set (World Bank 2013) – infant mortality rate, measured as the number of infants per thousand live births that die each year, and gross domestic product (GDP), to measure the overall level of development. We consider two operationalized conflict variables from PRIO: (i) a categorical variable ranging from 0 to 3 indicating the intensity of a conflict in a given year and (ii) a numerical value with counts of the battle deaths due to conflict within a country. We focused on the time frame from 1960 to 2013 as both data sets were more complete for this time period.

While the literature has some debate about the direction of a causal relationship between conflict and poverty, most work on the conflict trap implies a feedback or cyclic relationship, as illustrated in Figure 14.3. Rather than attempting to analyze our time-series data using standard time-series analysis approaches, such as Granger causality (Granger 1969, 1980), this theoretical insight can direct us to another type of causal analysis, CCM (Sugihara et al. 2012; Clark et al. 2015; Ye et al. 2015), which was originally developed in ecological biology to help analyze cyclic causal relationships. This method can be adapted to model social and political systems, providing a new opportunity to generalize cyclic causal theories. To use CCM, we assume some underlying dynamic generative process that produces the observable variables of interest, X and Y. This process can be *projected*, or visualized, as a set of vectors for variables X and Y called the *shadow manifolds*, essentially estimating *how* the unobservable process generates the observed values. For a time-series variable X, the shadow attractor manifold M_X consists of points $x(t) = (X(t), (X(t - \tau), X(t - 2\tau), \ldots, X(t - E\tau))$ where τ is a sampling time lag and E is the maximum manifold dimension we want to explore. For subsets of time series X and Y of length L, we can construct manifolds M_X and M_Y, generating the structures shown in Figure 14.4. CCM will then determine how well local *neighborhoods* – small regions of M_X – correspond to neighborhoods in M_Y. If X and Y are causally linked, there will be a one-to-one mapping between points in M_X and M_Y. To create this cross mapping, we use a neighborhood in M_X to predict the values of contemporaneous points in M_Y and compute the correlation ρ between the predicted values. If a causal relationship exists, predictions

Figure 14.3 The hypothesized causal relationship between conflict and poverty.

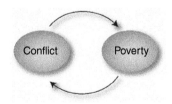

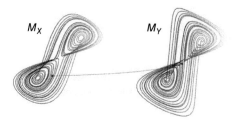

Figure 14.4 Shadow attractor manifolds for CCM neighborhoods of *X* to neighborhoods of *Y* to identify causal relationships.

of Y from X (and vice versa) will improve as the amount of data (L) increases, i.e. the mapping of X and Y will converge to perfect predictability $\rho = 1$.

Because CCM examines the relationships between projections of the time series, we normalized the data for our analysis of the conflict trap to measure the percent change at each time point to account for the vastly different scales of conflict casualties, infant mortality, and GDP. We observed convergence in 80% of countries in the PRIO data set, supporting the hypothesis that conflict causes poverty, and 12% for poverty leading to conflict. However, these results may be skewed by the countries that experienced no conflict during the time period under study. In these cases, where conflict is always zero, the underlying generative process is perfectly stable, so mappings from poverty to conflict start off exhibiting perfect predictability (i.e. regardless of the neighborhood in the poverty manifold and the amount of data, it can predict that there will always be zero conflict), and there is no room to illustrate the asymptotic behavior that defines causal convergence in CCM. As such, this lack of convergence is not really conclusive, and a cyclic relationship may still exist.

Figure 14.5 shows an example of convergence to support the hypothesis that conflict causes poverty in Comoros. In this graph, we show how the ability to predict poverty (i.e. GDP) increases with the amount of conflict data we use for this prediction; that is, each additional time point of conflict data is adding to our ability to predict GDP. Eventually, the predictive capability starts to converge toward 1 as more data is used, which is the indicator in CCM that this is a causal relationship. Whereas previous investigations of the relationship between conflict and poverty relied on linear models, this study leveraged theory – the idea that these phenomena are mutually reinforcing rather than unidirectional – to help us better align the data-driven analysis techniques with the data in order to answer the desired research question.

Parameterizing Theory-Driven Models Using Data

Our next case study provides an example of a type of chain ensemble where data is used to generate parameters for a theory-driven structural model of intergroup conflict. The prevalence of intergroup conflict throughout the world over the past century has been widely examined from a broad range

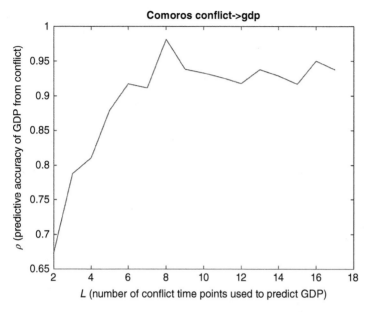

Figure 14.5 Results from CCM analysis illustrating convergence indicative of a causal link from conflict to poverty.

of substantive perspectives. An extensive empirical literature exists, which examines the political, structural, and economic factors associated with the incidence, character, and duration of conflict between groups within a political entity and across political entities. Some investigators have incorporated a variety of these substantive perspectives into game-theoretic frameworks. This line of theoretical and empirical research has yielded insights into some of the factors associated with the incidence and persistence of intergroup conflict (see, for example, [Fearon and Laitin 1996, 2003; Gagnon 1994]). Despite this extensive body of work, however, uncertainty exists concerning the factors that cause the persistence of intergroup conflicts in situations that are clearly at odds with the well-being of the respective groups of the general populations that are in conflict.

Unlike most political, structural, and economic factors associated with intergroup conflict, psychological factors are potentially far more volatile and, as a result, potentially more likely to produce abrupt impacts on public perceptions within fairly brief time spans. Research at the social psychological level has found that public attitudes may be especially susceptible to change under conditions of threat – or perceptions of threat – from outside groups (Rosenblatt et al. 1989; Greenberg et al. 1990, 1992). For example, analysis of public opinion and casualties arising from the Israel–Palestine conflict found that support for military operations on both sides was highly correlated with the level of conflict-related casualties, as shown in Figures 14.6 and 14.7 (Kohentab et al. 2010). Figure 14.6 shows the declining Israeli support for the peace process as Israeli casualties increase, and a similar pattern is seen on the Palestinian side in Figure 14.7; these correlations potentially indicate a relationship between an increase in perceived threat (i.e. more casualties) and the propensity to support military conflict vs. a peace process. This research also found that support for military operations against Israel among Palestinians doubled from 35% in a May 1999 survey to 72% in a December 2000 survey, following the start of the Second Intifada in late September 2000. Importantly, Palestinian casualties also showed a dramatic rise following the Second Intifada.

However, this data, while indicative of a potentially interesting relationship, cannot tell the whole story on its own. Many critical variables (e.g. competing political priorities, cultural narratives surrounding the conflict) are not captured by comparing only public opinion and conflict-related casualties. Instead, this data can be used to contribute to a theory-driven model of intergroup conflict, providing concrete parameters where available to help assess and validate the theoretical framework.

We developed an intergroup conflict model designed to incorporate the psychological effects of perceived threat on decision-making, without necessarily contradicting rational choice theory (Sliva et al. 2013). The model is designed to enable analysis of the degree of *commitment to conflict* – that is, support for continued conflict vs. negotiation – of a population under different conditions of perceived or real threat. This model has variables that take into consideration the potential exposure to threat-related information (e.g. news reports about casualties), the preexisting vested interest in the conflict (e.g. cultural narratives or identities deeply tied to the conflict state, financial stakes in continued conflict, political motivations, etc.), and the different groups that might be competing for decision-making control in this situation (e.g. the Israeli government vs. hardline fringe elements in Parliament).

The model combines traditional game-theoretic concepts with mathematical representations of the above psychological factors in a discrete dynamical system whose state space consists of sets of utility functions. Moving through the system is comparable with moving through a set of separate, but closely related games, and in this manner the system is formulated in terms of game theory. Actors behave according to the utility functions defined in the current state. For each group, we represent a main group and a fringe group, capturing that there are different levels of vested interest in an ongoing conflict even within a given *side*.

Using data-driven correlations and empirical evidence from studies, such as the Israeli–Palestinian examples in Figures 14.6 and 14.7, we can parameterize our decision model, enabling us to provide realistic values for the starting degree of commitment to conflict of different main and fringe groups. We can also simulate the occurrence of actual shocks to the system, such as the

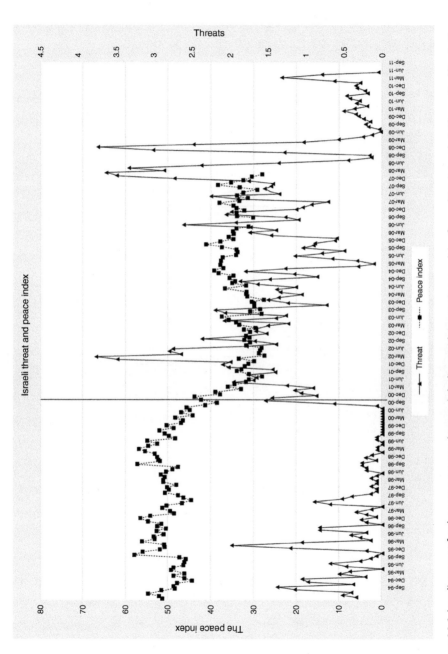

Figure 14.6 Israeli support for the peace process (squares) declines with increasing number of casualties (triangles). The vertical line in the middle of the graph indicates the start of the second Intifada. The Oslo Accords were officially signed in Washington, DC on 13 September 1993. The violence in the Second Intifada started in September of 2000.

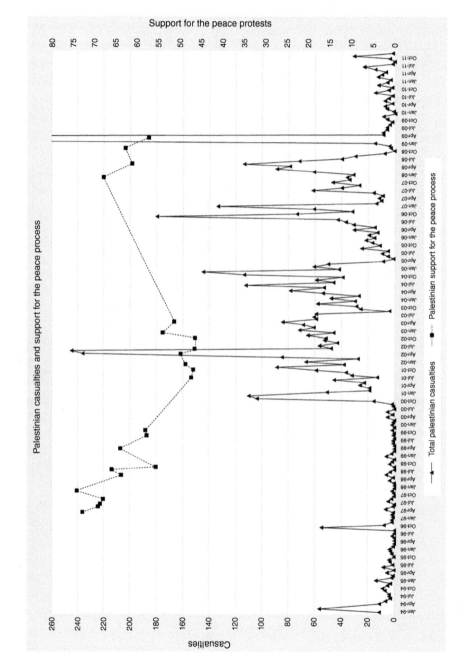

Figure 14.7 Palestinian support for the peace process (squares) declines with increasing casualties (triangles).

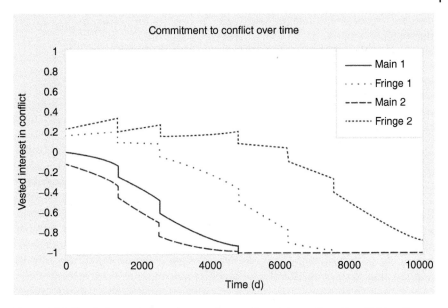

Figure 14.8 Simulation run where the overall commitment to continued conflict trends downward for all groups after a decrease in threat message amplification.

huge spike in casualties that occurred after the Second Intifada. With these parameters set to values representative of a real scenario, it is possible to validate the theoretical model (i.e. does the outcome conform to real-world occurrences?) and enable further exploration of variable manipulation.

Further, this model also enables counterfactual exploration based on real-world values, helping to identify potential policy interventions that may ultimately reduce conflict over time. For example, starting our model using commitment to conflict values based on Israeli and Palestinian public opinion polls, we can then experimentally manipulate other aspects of the model – such as the degree to which threat perception may be amplified by news reporting or social media – to identify changes that lead to overall decreasing conflict over time. This exploration provides interesting insights into not only the causal mechanisms in the model but also potential policy interventions that might lead to more desirable outcomes (e.g. counteracting messages of threat and causalities in the news with positive messages of cooperation to mitigate threat amplification). An example of the outcome of this type of counterfactual analysis is shown in Figure 14.8.

Theory and Data Dialogue

Our final case study looks at a broader example for constructing nested ensembles. In this case, the interaction between theory-driven and data-driven analysis is more of a dialogue – an iterative process for constructing a causal model representative of the scenario that leverages the strengths of both theory and data. The workflow is based on a general procedure for causal analysis in social science described by Goldthorpe (2001): (i) establish a phenomenon of interest, (ii) hypothesize causal relationships, and (iii) test the hypotheses of the causal process. At each step, hypotheses can be driven by either normative theory or discoveries in the data but are then tested against the alternate knowledge source (i.e. data-driven findings are compared with theoretical expectations, and theory-driven hypotheses are validated or enriched through data-driven discoveries).

To demonstrate how ensemble models can help identify causal relationships to characterize social systems, we applied this approach to modeling violence in Iraq (Sliva et al. 2017). Currently, Iraq is plagued by continuing and increasing violence from the terror organization that calls itself the Islamic State of Iraq and the Levant (ISIL), and militant activity is growing throughout Iraq, Syria, and parts of Lebanon. World leaders have been debating how best to address this increasing threat and stabilize the region. Many experts have turned to the recent past to better understand and model the present Political, Military, Economic, Social, Infrastructure, Information (PMESII) effects, looking at parallels to the situation in Iraq's Anbar Province in 2006 when the region was under increasing control by Al-Qaeda militants. Using a variety of causal analysis approaches and diverse data, we developed a rich nested ensemble that employs data, chain, and technique ensembles to capture the dynamics of the situation, illustrating how decision-makers might use these approaches to understand a complex situation and the possible impacts of different policy options.

For our analysis, we used data from the Empirical Studies of Conflict (ESOC) data from Princeton University (2016). ESOC contains a variety of different types and sources of data regarding Iraq that we used for this evaluation: (i) time series of violent events in Iraq yearly, monthly, and weekly at the province and district levels from February 2004 to February 2008 as compiled from the significant activity (SIGACT) reports by coalition forces; (ii) time-series reconstruction data on all projects executed in the country from 2003 to 2008, including the type of project and the amount and source of funding; and (iii) nontemporal demographic data based on a national census in 2008. Each of these data sources can provide insight into a different aspect of the situation in Anbar Province in 2006, capturing the militant activity – particularly activity targeting the US troops – as well as sociocultural and sectarian issues. In addition, the choice of data for this experiment was itself motivated by a causal theory presented by human experts, indicating that there is in fact some relationship between socioeconomic conditions and the propensity for violence in the region, provided that basic security issues could also be provided by US military presence. In order to make use of all this data, including expert social science theories, it was necessary to employ a variety of the causal analysis methods discussed in section "Understanding Causality." Using a chain ensemble, we combined human expertise in identifying potential causes and effects with both additive noise and time-series analyses to determine the actual direction and nature of the causation. Because we have these different types of data sources, including both temporal and nontemporal data, we used a technique ensemble approach to combine additive noise results and time-series analysis into a single coherent causal model. For intuitive analysis by policy-makers, the resulting model was represented as a graphical model with a structure similar to Pearl's (Verma and Pearl 1991; Pearl 2002, 2014). Given all of these factors, we constructed a complex nested ensemble, enabling us to combine human expertise with the additive noise and time-series ensembles developed for the prior two experiments in a graphical model representation.

Figure 14.9 shows the final model that was constructed using this ensemble of ensembles over the ESOC data. Looking at the time series in the reconstruction data and the SIGACTS data, our temporal ensemble of Granger causality, offset Pearson correlation, and dynamic time warping identified several causal relationships where specific types of reconstruction projects (e.g. military facilities, democracy building, infrastructure improvements, etc.) influence (either increase or decrease) the number of violent acts, or vice versa. From the nontemporal demographic data, the additive noise ensemble discovered causal links from the size of the Shia, Sunni, and mixed populations in a district to the number of SIGACTS (i.e. violence was higher in places

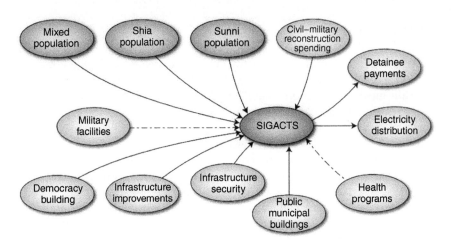

Figure 14.9 Causal model of the impact of socioeconomic and demographic factors on militant activities in Anbar Province, extracted from real-world data and expert theoretical knowledge using a nested iterative ensemble.

with a more mixed population across sectarian lines). In addition, the causal ensemble was also able to identify specific theory-driven hypotheses that were not discovered by the data-driven approach. In the structure proposed by human experts, reconstruction efforts targeting courts and schools were considered important factors in reducing the violence; however, subsequent analysis in the ensemble by data-driven time-series methods did not support this assumption. When considering the strong impact of the demographic conditions, the human experts concluded that the underlying security situation in many regions was too volatile and the government too weak to make these aspects of civil society impactful in this situation. Based on the recommendation of the data-driven analyses, these factors were removed from the final model.

This case study demonstrates the power of ensemble combinations for analyzing causality in social systems, enabling us to combine various different types of data-driven discoveries and theoretical knowledge into a single analysis that provides richer and more accurate results than any single approach. By combining causal analysis techniques, policy-makers weighing different approaches to a situation, such as stabilizing Anbar in 2006 or dealing with ISIL in 2016, can employ the variety and richness of information available in different types of data sets, ultimately producing a model of the social system. For example, using the model in Figure 14.9, a decision-maker can determine the types of reconstruction projects that might be most influential for stabilizing the region, which demographic features contribute most to unrest and violence, and identify new or unexpected causal linkages.

Conclusions

Social science provides an opportunity for researchers and policy-makers to gain a deeper understanding of the world around them and facilitate decision-making based on causal models of social dynamics. However, due to the inherent complexities of human behavior and the inability to conduct carefully controlled experiments in most social systems, causal analysis remains extremely challenging. In this chapter, we have presented a mixed-methods approach to combining data-driven and theory-driven analysis techniques in ensembles, enabling researchers to leverage the increasingly available large-scale data resources while also maintaining an ability for deep understanding and causal explanation. We provided several case studies that illustrate the efficacy of this mixed-methods approach, each demonstratively achieved the essential ambition promised by the field of ensemble machine learning: combinations of multiple diverse techniques outperforming individual approaches.

References

Bennett, A. (2010). Process tracing and causal inference. In: *Rethinking Social Inquiry*, 2e (ed. H. Brady and D. Collier), 207–220. Rowman and Littlefield.

Bennett, A. and George, A.L. (1997). Process tracing in case study research. In: *MacArthur Program on Case Studies*. Washington, DC.

Brady, H.E. and McNulty, J.E. (2004). The costs of voting: evidence from a natural experiment. Annual Meeting of the Society for Political Methodology, Palo Alto, CA.

Braithwaite, A., Dasandi, N., and Hudson, D. (2016). Does poverty cause conflict? Isolating the causal origins of the conflict trap. *Conflict Management and Peace Science* 33 (1): 45–66.

Clark, A.T., Ye, H., Isbell, F. et al. (2015). Spatial convergent cross mapping to detect causal relationships from short time series. *Ecology* 96 (5): 1174–1181.

Collier, D., Brady, H.E., and Seawright, J. (2010). Outdated views of qualitative methods: time to move on. *Political Analysis* 18 (4): 506–513.

Collier, P. (2003). *Breaking the Conflict Trap: Civil War and Development Policy*. World Bank Publications.

Collier, P. and Hoeffler, A. (2004). Greed and grievance in civil war. *Oxford Economic Papers* 56 (4): 563–595.

Creswell, J.W. and Clark, V.L.P. (2007). *Designing and Conducting Mixed Methods Research*, 2e. Los Angeles: SAGE Publications, Inc.

Djankov, S. and Reynal-Querol, M. (2010). Poverty and civil war: revisiting the evidence. *The Review of Economics and Statistics* 92 (4): 1035–1041.

Dunning, T. (2008). Improving causal inference: strengths and limitations of natural experiments. *Political Research Quarterly* 61 (2): 282–293.

Dunning, T. (2012). *Natural Experiments in the Social Sciences: A Design-Based Approach*. Cambridge University Press.

Fearon, J.D. and Laitin, D.D. (1996). Explaining interethnic cooperation. *American Political Science Review* 90 (4): 715–735.

Fearon, J.D. and Laitin, D.D. (2003). Ethnicity, insurgency, and civil war. *American Political Science Review* 97 (1): 75–90.

Gagnon, V.P. (1994). Ethnic nationalism and international conflict: the case of Serbia. *International Security* 19 (3): 130–166.

Goldthorpe, J.H. (2001). Causation, statistics, and sociology. *European Sociological Review* 17 (1): 1–20.

Granger, C.W. (1969). Investigating causal relations by econometric models and cross-spectral methods. *Econometrica* 37: 424–438.

Granger, C.W. (1980). Testing for causality: a personal viewpoint. *Journal of Economic Dynamics and Control* 2: 329–352.

Greenberg, J., Pyszczynski, T., Solomon, S. et al. (1990). Evidence for terror management theory II: the effects of mortality salience on reactions to those who threaten or bolster the cultural worldview. *Journal of Personality and Social Psychology* 58 (2): 308–318.

Greenberg, J., Solomon, S., Pyszczynski, T. et al. (1992). Why do people need self-esteem? Converging evidence that self-esteem serves an anxiety-buffering function. *Journal of Personality and Social Psychology* 63 (6): 913.

Hoffman, R., Klein, G., and Miller, J. (2011). Naturalistic investigations and models of reasoning about complex indeterminate causation. *Information Knowledge Systems Management* 10 (1–4): 397–425.

Kohentab, S., Pierce, G., and Ben-Perot, G. (2010). *A Preliminary Analysis of the Impact of Threat on Public Opinion and Implications for the Peace Process in Israel and Palestine (Working Paper)*. Institute for Security and Public Policy, College of Criminal Justice, Northeastern University.

Miguel, E., Satyanath, S., and Sergenti, E. (2004). Economic shocks and civil conflict: an instrumental variables approach. *Journal of Political Economy* 112 (4): 725–753.

Mooij, J.M., Peters, J., Janzing, D. et al. (2016). Distinguishing cause from effect using observational data: methods and benchmarks. *Journal of Machine Learning Research* 17 (1): 1103–1204.

Myers, C. and Rabiner, L. (1981). A level building dynamic time warping algorithm for connected word recognition. *IEEE Transactions on Acoustics, Speech, and Signal Processing* 29 (2): 284–297.

Opitz, D.W. and Maclin, R. (1999). Popular ensemble methods: an empirical study. *Journal of Artificial Intelligence Research* 11: 169–198.

Pearl, J. (2002). Causality: models, reasoning, and inference. *IIE Transactions* 34 (6): 583–589.

Pearl, J. (2014). *Probabilistic Reasoning in Intelligent Systems: Networks of Plausible Inference*. Morgan Kaufmann.

Peters, J., Mooij, J.M., Janzing, D., and Schölkopf, B. (2014). Causal discovery with continuous additive noise models. *Journal of Machine Learning Research* (15): 2009–2053.

Princeton University (2016). Empirical studies of conflict database. Retrieved from https://esoc.princeton.edu/country/iraq (accessed 9 September 2018).

Ragin, C.C. (1989). *The Comparative Method: Moving Beyond Qualitative and Quantitative Strategies*. University of California Press.

Ragin, C.C. (2009). Qualitative comparative analysis using fuzzy sets (fsQCA). In: *Configurational Comparative Methods: Qualitative Comparative Analysis (QCA) and Related Techniques*, vol. 51 (ed. B. Rihoux and C. Ragin), 87–122. Los Angeles: SAGE Publications, Inc.

Rosenblatt, A., Greenberg, J., Solomon, S. et al. (1989). Evidence for terror management theory: I. The effects of mortality salience on reactions to those who violate or uphold cultural values. *Journal of Personality and Social Psychology* 57 (4): 681.

Salvador, S. and Chan, P. (2007). Toward accurate dynamic time warping in linear time and space. *Intelligent Data Analysis* 11 (5): 561–580.

Sliva, A., Malyutov, M., Pierce, G., and Li, X. (2013). Threats to peace: threat perception and the persistence or desistance of violent conflict. In: *Intelligence and Security Informatics Conference (EISIC), 2013 European*, 186–189. IEEE.

Sliva, A. and Neal Reilly, S. (2014). A big data methodology for bridging quantitative and qualitative political science research. Annual Meeting of the American Political Science Association (28–31 August 2014). Retrieved from https://papers.ssrn.com/sol3/papers.cfm?abstract_id=2454237.

Sliva, A., Neal Reilly, S., Blumstein, D. et al. (2017). Modeling causal relationships in sociocultural systems using ensemble methods. In: *Advances in Cross-Cultural Decision Making*, 43–56. Springer. Retrieved from http://link.springer.com/chapter/10.1007/978-3-319-41636-6_4.

Sliva, A., Neal Reilly, S., Casstevens, R., and Chamberlain, J. (2015). Tools for validating causal and predictive claims in social science models. *Procedia Manufacturing* 3: 3925–3932.

Sliva, A., Neal Reilly, S., Chamberlain, J., and Casstevens, R. (2016). Validating causal and predictive claims in sociocultural models. In: *Modeling Sociocultural Influences on Decision Making: Understanding Conflict, Enabling Stability* (ed. J.V. Cohn, S. Schatz, H. Freeman and D.J.Y. Combs), 315. CRC Press.

Spirtes, P., Glymour, C.N., and Scheines, R. (2000). *Causation, Prediction, and Search*. MIT Press.

Sugihara, G., May, R., Ye, H. et al. (2012). Detecting causality in complex ecosystems. *Science* 338 (6106): 496–500.

Taleb, N. (2007). *The Black Swan: The Impact of the Highly Improbable*, 2e. Random House.

Tashakkori, A. and Teddlie, C. (1998). *Mixed Methodology: Combining Qualitative and Quantitative Approaches*, vol. 46. Thousand Oaks, CA: Sage Publications.

Themnér, L. and Wallensteen, P. (2014). Armed conflicts, 1946–2013. *Journal of Peace Research* 51 (4): 541–554.

Verma, T. and Pearl, J. (1991). Equivalence and synthesis of causal models. In: *Proceedings of the Sixth Annual Conference on Uncertainty in Artificial Intelligence*, 255–270. New York, NY: Elsevier Science Inc. Retrieved from http://dl.acm.org/citation.cfm?id=647233.719736.

The World Bank (2013). *World Development Indicators*. Washington, DC: The World Bank http://data.worldbank.org/data-catalog/world-development-indicators.

Ye, H., Deyle, E.R., Gilarranz, L.J., and Sugihara, G. (2015). Distinguishing time-delayed causal interactions using convergent cross mapping. *Scientific Reports* 5.

15

Theory-Interpretable, Data-Driven Agent-Based Modeling
William Rand

Department of Marketing, Poole College of Management, North Carolina State University, Raleigh NC 27695, USA

The Beauty and Challenge of Big Data

The beauty of big data is all around us and has the potential to be a huge boon to the social sciences. Right now billions of people are carrying cell phones, posting content on social media, driving smart cars, recording their steps, and even using smart refrigerators (Riggins and Wamba 2015). Data is being generated at an extraordinarily fast rate, and many of our traditional methods of analyzing this data face challenges.

The hope and promise for the social sciences is that these vast data sets will give us new insight into the basic way that people interact and behave. There are many different types of data that are available as part of the big data revolution. First, there is the digitization of traditional administrative data (Kitchin 2014). This includes everything from income tax records to parking violation data. This data has always been collected (though not always stored) but now is much more amenable to analysis due to the digitization of the data and the development of ways to get access to this data. For instance, the open government data movements (Ubaldi 2013) have the stated goal of making and increasing the amount of data available to be analyzed by private citizens. Hackathons have been one way to take advantage of this data and to create apps that the average individual can use to make better decisions using this data (Matheus et al. 2014). Though the use and the analysis of this data may be new and different, the basic data has existed for a long time, and so this chapter will not dwell too much on this type of data.

Another form of big data is trace data. Trace data is the data left by an individual as they move through the world sometimes purposefully recording data about themselves, e.g. social media data, and sometimes being unaware that they are creating the data, e.g. the GPS traces stored on a user's cell phone.

Social-Behavioral Modeling for Complex Systems, First Edition.
Edited by Paul K. Davis, Angela O'Mahony, and Jonathan Pfautz.
© 2019 John Wiley & Sons, Inc. Published 2019 by John Wiley & Sons, Inc.
Companion website: www.wiley.com/go/Davis_Social-Behavioralmodeling

Some researchers have derided this type of data as digital exhaust (Watts 2013), but this data is more than that since it can sometimes give us insights into what people are doing (Eagle and Pentland 2006) or even their political beliefs (Golbeck and Hansen 2011). However, even if this data is nothing more than the social equivalent of the exhaust from a car's tailpipe, that still would potentially give us new insights into human behavior. After all a car's exhaust if we could track it would tell us where the car is and how well the car was running. In the same way, tweets from a user on Twitter may tell us where that user is (Backstrom 2010) and whether or not they are mentally healthy (De Choudhury et al. 2013). Regardless, this data is clearly valuable, since it was essentially trace data in the form of liking of pages that enabled Cambridge Analytica to target users with political campaign ads and potentially affect the 2016 presidential election (Grassegger and Krogerus 2017). Since this form of data is much newer in the social science toolkit, I will concentrate on this form of data throughout this chapter. However, many of the principles discussed apply equally well to administrative data and trace data.

The fact that we have millions of traces of individual-level data at a time resolution as precise as a tenth of a second or even higher is beautiful, but it is also a challenge. Never before in the history of social science have researchers had access to such a wealth of data, and as a result the academic community does not really know how to process such data. There have already been a few missteps. The most classic case might be Google Flu Trends, which attempted to predict the prevalence of flu based on examining hundreds of millions of Google searches (Cook et al. 2011). Flu Trends worked well for a while but then started wildly over- and underpredicting the prevalence of flu. The Google Flu Trends approach, and many other approaches to big data, uses an averaging approach of some sort, but the true beauty of big data is the richness of individual-level data. Traditionally social scientists have had to depend on samples or surveys of data (Bertrand and Mullainathan 2001), but the presence of big data means that we have a more complete picture of individual behavior than ever before. Though there is probably an optimal level of data resolution for any given problem, for many questions that are determined by individual-level heterogeneity, it would be useful to have a methodology that could automatically account for the differences between individuals as recorded in big data.

We want computer models that can be built or constructed from data. Many theory-driven models are never really compared to data. In other words, many models have been inspired and generated from theory or speculation, and therefore are somehow taken to be correct representations of the phenomenon at hand, but have never been shown to apply to data, or potentially have just been tested on some data set that also inspired the original theories, i.e. they have not been tested in an out-of-sample situation. Cathy O'Neil in her book, *Weapons of Math Destruction*, warns about the reliance on models (O'Neil 2017), for instance, that make predictions about teacher performance based solely on theories about how teachers should add value to student learning. However, if these models have never been validated (Wilensky and Rand 2015), i.e. showing that they do in fact predict some inherent aspect of teacher performance, then these models are not valid tools of reasoning. As a community, we need to be able to create valid models from big data.

A modeling approach has been used in the social sciences that does give the researcher the ability to account for all of the individual-level heterogeneity that might be present. This approach is called agent-based modeling (ABM), and it is defined by creating a computational representation of every agent or individual (Wilensky and Rand 2015). This gives an agent-based model (ABM) the power to capture a rich variety of heterogeneity in the underlying system. However, historically many social science ABMs have been developed purely from theory, and not from data. Some of these theories have been grounded in pure speculation, based on casual observations and intuitions, and have often focused on one magical cause. Other theories have been driven by more in-depth thinking with at least some consideration of interactions. Still other theories have initially been grounded by examining data, but no serious effort has been made to test them outside of the original data. Instead, I would urge the development of data-driven ABMs. There have been cases of data-driven ABMs in the past in the social sciences (Axtell et al. 2002), but they are the exception rather than the rule.

What if it was possible to derive an ABM directly from data? Theoretically, such a system would not suffer from the problem of being dependent on averages, since the agents would manifest the rich heterogeneity of the underlying individuals. Moreover, if the system was trained using individual-level data, then it would not suffer the problems that arise when a system is based solely on averages. In order to achieve this goal, I will begin by discussing a framework that will guide the construction of big data, social science models that relies on the idea that a good model needs theory, and then building on this framework examine two methods for the creation of large-scale, data-driven ABMs. The first method is parameter optimization (PO), which involves taking an ABM and altering its parameters until the output of the model fits the data, and the second method is rule induction (RI), which involves inducing rules from big data that are directly used to create an ABM.

In order to temper expectations, it is necessary to realize that the vision presented in this chapter is not complete, and we do not yet have a large-scale, dynamically rich ABM that has been derived directly from data. However, the components are being put together, and progress is being made. The goal of this chapter is not to provide the perfect solution for big data, social science models, but rather to start to discuss how such models should be created and what some first steps toward their creation currently look like. Thus, this chapter is not the end of a small research project by a small team of investigators, but rather the beginning of a community-wide project over years.

A Proposed Unifying Principle for Big Data and Social Science

Many of the problems of big data are often related to the fact that the computer and statistical big data models are not connected to theory, meaning that there is no explanation as to why the inputs are connected to the outputs beyond correlation. It is worth noting that in many disciplines the words theory and model are often considered equivalent words, but in some of the social sciences, especially the managerial sciences, there is a distinction that is often drawn between a theory, i.e. a causal explanation, and a model, i.e. a statistical or computational model (Shmueli 2010). I will use this distinction throughout this chapter. Thus, model in this chapter can be read as statistical or computational model relating inputs to outputs, and theory can be read as explanatory framework or causal theory, where such a framework could be anything from a simple one-variable causal hypothesis to a multivariate causal theory to a procedural description that explains behavior. Regardless of its format, the theory should provide some explanation of the phenomenon not just a black-box relationship between inputs and outputs.

Given this distinction the problem with many big data approaches is that without a theory we have not created a grounded model, but rather just a particular prediction of the future. For instance, with regards to Google Flu Trends, many arguments have been advanced as to why it stopped working, but it is hard to assess the cause of the inaccuracies because there was never a theory as to why it was working in the first place beyond the idea that somehow a user's searches on different subjects may be correlated with whether or not they have the flu. Without a theory it is hard to justify an explanation as to why something went wrong, since the theory provides the reason why it might work in the first place. With a theory, if something stops working, then it can be explored whether (i) the theory is wrong, i.e. it has been falsified, or (ii) the assumptions of the theory have been violated, i.e. the theory no longer applies.

This problem is particularly profound in the social sciences, where the basic elements of the systems, i.e. humans and human-created organizations, can change and adapt over time, meaning that we truly need to understand how the system works and not just what the end result of the current set of inputs will be. This does not mean that we need to necessarily know exactly what theory best fits the data before we start building models. Sometimes the goal of modeling is to generate a bunch of different models based on different theories and compare and contrast based on how well they fit the data or predict future outcomes. In fact, we may generate millions of different models, many of which we are just going to discard based on fits to data, but if we have a few that actually match well with data, then we need the ability to explore those models from a theoretical perspective. A model that is constructed in such a way that it is essentially a black-box model, such as many of the deep learning models that are popular right now (LeCun et al. 2015), cannot ever really be compared to theory since there is no easy way to tell why the model is generating its results[1] (Yosinski et al. 2015).

Sometimes the distinction here is referred to as black box vs. white box. Black-box models just take inputs and give outputs, while white-box models allow inspection to determine why they are generating the outputs. Fundamentally, this means that black-box models are usually not very useful from a social science perspective, because it is hard to gather additional knowledge about human processes when it is not possible to explain how the inputs are related to the outputs. Moreover, a black-box model that may not be useful outside the data is trained on, because it is not possible to understand under what assumptions it works. This makes it difficult to determine that a black-box model that works now will also work next year, which seems to have been the case in the Google Flu Trends. This is somewhat related to the Lucas critique of macroeconomics (Lucas 1976), which states that no macro-model is useful for predicting what would happen as a result of a change of macroeconomic policy since the model was trained on data where that policy was not the case. In much the same way, the application of a black-box model is not useful if the goal is to understand how society will respond to a change in a structural element of society, since the model was not developed in a world where the structure had changed.

The result of all this is that we need social science models that are at least potentially amenable to a theoretical analysis. However, we also need theories that are useful as well. Though the focus of this chapter is on models and not theories, it is important to remember that a theory that does not lend itself to be interpreted from the perspective of a modeling framework is also problematic, since the theory is not testable. If the theory is not testable, then it has the same problems that a big data-driven correlational analysis does since there is no way to determine why the theory does not explain a particular set of data. Given these concerns, I propose the following principle for all theories and models that use big data in the social sciences.

Principle. *Theory-Interpretable Models and Model-Interpretable Theory (TIMMIT) Principle: All social science models using big data should be interpretable from a theoretical perspective, i.e. they should not be black box, and all social science theories applied to big data should be interpretable from a modeling perspective, i.e. they should be falsifiable.*

---

1 There are recent efforts to make deep learning neural nets more understandable. This agenda is sometimes referred to as explainable AI.

As mentioned, this does not require that the model be built from first principles, i.e. by hand, but it does require that it is potentially possible to inspect the model and then compare the model behavior to theory. Let us examine a quick example. Imagine that a researcher is trying to understand how information spreads on social media and they are examining Twitter. In particular, they want an individual-level model that predicts whether or not a focal user will retweet a particular piece of content. They could accomplish this by gathering a bunch of features about the user, including all of the time series of all the users that the focal users follow, and feeding this into a large wide and deep neural network. The resulting model might be very predictive, but if there is no way to understand why it is making the decisions, i.e. it is not theory interpretable, that is potentially problematic. On the other hand, the researcher could take the same inputs and create a Markovian-type model from the data. Though this model is not necessarily developed from theory, it is possible to inspect the model and potentially explore why the model is producing the outputs that it is by examining the states and how the user transitions through those states as time goes on. Even though there was no causal theory that was used to generate this model, the model was created using a framework that enables the researcher to explore each and every action that the agent in the model takes from a mechanistic point of view. These mechanisms can then be compared to theory, and the researcher can determine if the model makes sense from a causal theory point of view. In this particular example, it turns out that many traditional theories of information diffusion, such as the threshold (Granovetter and Soong 1983) and cascade theories (Goldenberg et al. 2001), can be written in this same modeling form. This means it may be possible to compare the model directly to theory and see how much the fully data-driven model gains you versus a potentially more restricted theory-based model. This is what I mean by a potentially theory-interpretable model.

Data-Driven Agent-Based Modeling

So now that we have a principle to guide our creation of social science models, generated from big data, the next question is how do we implement this principle. In the rest of this chapter, we will explore one particular solution that involves using machine learning to take an ABM and fit it to data. There may very well be other approaches that work and still meet the goals of the *TIMMIT* principle, but we will focus on this particular solution here. The ABM solution has a number of important benefits. First, ABM is flexible, i.e. it can incorporate many different modeling frameworks for the individual agents. Second, ABM is interdisciplinary, i.e. it applies equally well across the spectrum of the social sciences and has been used in just about all of them. Third, ABM potentially provides an answer to the Lucas critique. If we can understand low-level decision rules that drive human behavior, then theoretically we should be able to explore how changes in their environment or policies affect their behavior. In macroeconomics, the Lucas critique resulted in a shift of research effort within that field to a better understanding of micro-foundations for the same reason. If you can understand the micro-principles, then even if you change the world, the model should adapt appropriately. This does not guarantee that the model will be correct, since it may still be the case that some hidden variables were not included in the model, but it is closer to building a correct model.

Within the space of machine learning and ABM, we will examine two different approaches to developing an ABM that fits to data and is theoretically interpretable: (i) PO and (ii) RI. PO starts with a theoretically grounded model and then modifies the parameters of the model until the output of the model matches some empirical data set (Bonabeau 2002). RI, on the other hand, attempts to induce rules of behavior directly and then asks whether the combined output of all of the agents acting together matches the empirical data. We will explore each of these approaches in turn.

Parameter Optimization

PO has been around since at least the first computer simulations, and even the use of machine learning to optimize parameters (Weinberg and Berkus 1971) has been studied for some time. The basic method for PO is that the researcher first constructs an ABM based on first principles and theory and then does their best to manually match the parameters to measured data about the world based on previous research, a process, which is sometimes called input valida-tion (Rand and Rust 2011). However, it is usually the case that there is some uncertainty or unknown aspects to these parameters (Smith and Rand 2018). Given that, and if the goal of the simulation enterprise is to create a model that is descriptive and potentially predictive of the real world, one way to choose a final set of parameters is to compare the output of the model to an empirical data set and then tune the parameters until the model output and the real-world data are in close alignment.

In previous work, with Forrest Stonedahl, we described this more formally (Stonedahl and Rand 2014). We start by assuming a real-world data set, R, and a model, $M(P, E)$, with parameters, P, and environmental variables, E. The parameters, P, are the input values to the model that we do not know and are trying to optimize. Though we discuss this in the context of finding particular point values for P, it could also be that P is a set of distributions or even a choice of behaviors. The environmental variables, E, are the input variables that vary from particular context to particular context, and are not being optimized. For instance, they could be the input variables that control the geography the model is operating in. Given this notation, we can formalize PO in the following way. We split R into data sets: R_{train} that is the data set we

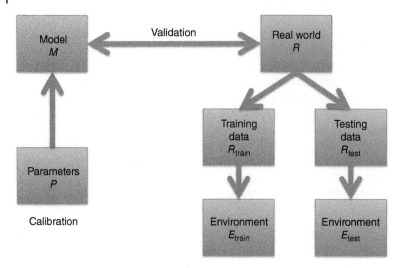

Figure 15.1 An illustration of the parameter optimization process.

will optimize the model with and R_{test} that is the data set that we are using to test the optimization. The environmental variables may differ between these two data sets, so we can denote them E_{train} and E_{test}. Given this notation, we can start the PO process by calibrating on the training data, which is simply a matter of identifying a set of parameter, P^*, such that some error measure $\epsilon(R_{\text{train}}, M(P^*, E_{\text{train}}))$.[2] This becomes a search problem to some extent, and any number of machine learning methods can be employed to optimize the system. Once the model has been calibrated, we can then determine if the model is valid by computing $\epsilon(R_{\text{test}}, M(P^*, E_{\text{test}}))$. If this value is less than some threshold T, then we can say that model has been validated. This approach is illustrated in Figure 15.1.

This provides a general framework by which an ABM that is driven by theory can be compared to and calibrated by big data. One way to use this framework would be to exactly find the best P values to match the underlying data as well as possible. However, it would also be possible to adjust what was within P and how ϵ was defined to instead explore both an estimate of the mean values and the uncertainty ranges around those values. Moreover, this framework enables the comparison of multiple models that have been generated from theory. For instance, if we have two models constructed from competing theories, say, M_1 and M_2, we can use the above procedure to calibrate each model to data and then examine the resulting parameter sets, P_1^* and P_2^*, as well as the minimal error that is achievable given the model and the error measure. If one of

---

2 In our original paper, we showed that the choice of ϵ is critical since the error measures directly affect the way in which the model is calibrated to data.

the models fits better to the data than another, then that can help adjudicate between the theories (Claeskens and Hjort 2008).

Alternatively, if both models fit the data fairly well, then the parameter sets can be investigated, and if it is determined that one of the parameter sets is more realistic than another, then that can also help to adjudicate between the models. It could also be the case that one of the parameter sets is very different from the other one, but not enough is known about the real world to adjudicate the distinction. This could be used as fodder for additional research into this area and helps isolate the distinction between the models.

Finally, it is possible that both models fit the data very well, and there is not much difference between the parameter sets. In this case, there are a number of possibilities: (i) the training and testing data is not sufficiently large enough to explore the differences in the model, and more data is needed; (ii) the environmental parameters might not be sufficiently varied to illustrate the differences between the models, and additional circumstances should be examined; or (iii) the models are both potentially good explanations of the underlying phenomenon, and it is necessary to reassess the differences between the models and see if the differences really are that significant.

This is an abstract explanation of how to use big data to drive the calibration and construction of an ABM. In the next two subsections, we will explore two specific examples. The first is in the context of examining a news consumption model, and the second is in the context of calibrating an ABM of urgent diffusion on social media.

News Consumption

This example is primarily drawn from previous work that developed the PO framework (Stonedahl and Rand 2014). The Internet is quickly and dramatically changing a number of different industries. One industry that has been particularly hard-hit by the rampant growth of the Internet is the news industry. However, news is considered more vital than ever to a well-functioning democracy. Well before the current #fakenews cycle, we investigated this phenomenon using an ABM. There is a desire by a number of individuals to figure out a way to remonetize news (Schmidt 2009), but to do so we need an understanding of what kinds of revenue platforms would work online. Fortunately, the takeover of the news industry by the Internet has resulted in huge, large, and at times nearly unmanageable trace of how individuals are consuming news. This data is eminently trackable, but researchers do not really have a good understanding of how people consume news online, which prevents the analysis of revenue models. In order to start to solve this problem, it would be useful to first construct an ABM that represents user behavior.

The goal of this project was therefore to create a model that matches the data patterns as close as possible. In this case the data was represented by clickstream data, i.e. for a panel set of the same users we could observe over time,

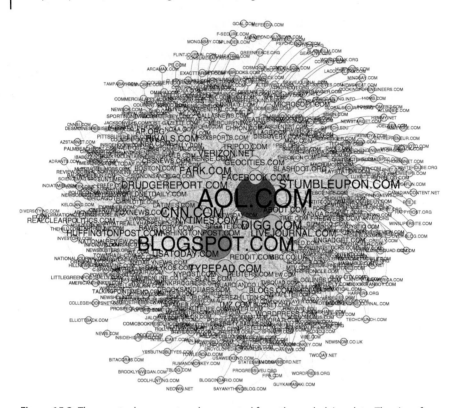

Figure 15.2 The eventual news network generated from the underlying data. The size of the node indicates the amount of traffic that node received. Image Credit: Forrest Stonedahl.

which links the user clicked on. This data contained 90 000 users during the entire year of 2007. In order to organize the data in a proper framework, we took one month of data (January) to use for training and one month of data (December) to use for testing. We also filtered the data to only contain clicks on news-oriented websites, e.g. cnn.com and nytimes.com. We also recorded all of the incoming traffic to these websites. We created a network by linking nodes that individuals had clicked from and to. This resulted in a network of 422 nodes in January and 417 nodes in December. The network is visualized in Figure 15.2.

We then built an ABM in NetLogo (Wilensky 1999) where we constructed a representative average Internet surfer who started at a random node in the network and then would decide which node to click on or to end its Internet session. The hypothesis is that users decide to click on different nodes based on their reputation and that the reputation of the node can be proxied by various network properties of the node. Therefore, the way the user decides which node to click on was decided based on a utility function where the utility of a node was determined by (i) randomness – a pure random element, (ii) in-degree – number of nodes that directly link to this node, (iii) out-degree – number of nodes that this node directly links to, (iv) in-component – number of nodes that can reach this node, (v) out-component – number of nodes reachable from this node, (vi) PageRank score (Page et al. 1999) – the original Google ranking algorithm, (vii) Hits-Hubs score (Kleinberg 1999) – a measure of how much this node serves as a hub, (viii) Hits-Authorities score (Kleinberg 1999) – a measure of how much this node serves as an authority, (xi) clustering – how clustered the connected nodes are, (x) betweenness – the betweenness centrality of the node, and (xi) eigenvector centrality – the eigenvector centrality measure of the node. In this case the answer to this question would provide us with a potential explanation of the underlying user choices observed even if it did not provide us with a causal theory.

The weights that were associated with each of these 11 measures were the variables of concern. In addition to the weights of these 11 utility components, there were two additional parameters. The first parameter, *random-restart*, controlled how often the model restarted, i.e. it was a probability that the user restarted their search at any node. The second parameter, *no-backtrack*, was a Boolean flag that controlled whether the user was allowed to go back to the node they had just visited. Essentially, the model was run with a set of 13 different parameters for each of these variables for a number of user visits equivalent to the number of user visits observed in the actual data, and then the number of clicks on each website node was recorded. This distribution of website traffic was then compared with the actual website traffic using a number of different error measures, including correlation and a number of different L-norms, which included a right or wrong distance measure, Euclidean distance, Manhattan distance, and the L^{∞}-norm (Stonedahl and Rand 2014). The model was trained on the January data, and then the results were evaluated on the December data.

This can be placed into the PO framework. The real-world data set R is the whole set of clickstream data. The model, $M(P, E)$, is the ABM, where P is the set of 11 parameters describing the agent's utility function and the two additional parameters and E is the number of clicks observed in the corresponding set of data; this varies slightly between the January and December data. R_{train} is the January data set, and R_{test} is the December data set. The error measure, ϵ, is the corresponding set of measures, corr, L^0, L^1, L^2, and L^{∞}. The model was then calibrated on the January data (R_{train}) using the BehaviorSearch package in NetLogo (Stonedahl and Wilensky 2010) using a genetic algorithm (Holland 1975).

In this discussion rather than exploring the actual model implications for news consumption, since those results were largely inconclusive, let us concentrate on the methodological results, which directly bear on this chapter's goal of exploring the utility of PO. One central question was whether one of the error functions that was used to train the model would be superior to the other error functions. For instance, would using the correlation-based error measure also optimize the L^1 (Manhattan) error measure or some other error measure? Three clear findings stood out. First, all of the error measures achieved similar performance on the testing data for the L^0 (the number of matching elements) error measure, which means if this measure is a goal, the choice of how you train the model is largely irrelevant. Second, the correlation error measure was pretty good at optimizing its own measures on the testing data set, and did okay on the other error measures, but none of the other error measures did a very good job of achieving high performance on the correlation measure in the testing data. Finally, the L^2 (Euclidean) measure actually achieved high performance on the L^1, L^2, and L^∞ measures, meaning that it is a potentially robust error measure that should be considered more in the future, but none of the measures was strictly dominant.

Urgent Diffusion

The study of news consumption essentially resulted in a finding that was methodologically relevant, but did not have much practitioner value. In a different study, we focused more on the practical applications of calibrating an ABM to data (Yoo et al. 2016). In this study, we examined the role of social media in spreading information about disasters. An area of information diffusion is called urgent diffusion (Rand et al. 2015), since unlike traditional information diffusion, there is a time-critical nature to the diffusion process. The fundamental question of interest was: how can humanitarian organizations use social media to spread information better in urgent diffusion scenarios, such as those that occur during disasters?

We did this by first modeling the process of diffusion using a standard implementation of an agent-based information diffusion model, namely, the independent cascade model (Goldenberg et al. 2001). This model has two parameters, an internal influence parameter q (governing how quickly information spreads within the social media platform) and an external influence parameter p (governing how external sources of information influence the spread of information). We also constructed a social media network that was based on the longest observed cascade (i.e. series of retweets) observed in the actual data. Once we had constructed this model, we then calibrated the model to a number of different diffusions that had been observed in social media data. However, unlike the news consumption model, we were not interested in creating a predictive model, but rather in creating the best descriptive model that we could from the data. As a result, we did not separate into training and testing data, but instead merely fit the model as best as possible to all of the data. This is because we were actually going to use inferred values from the model as the dependent variable for another model, and so validation of the predictive capability of this model was not relevant (Shmueli 2010).

As a result, though the purpose is slightly different, we can still place this model in the PO framework. The real-world data set R is the set of different diffusion patterns on Twitter, essentially being a set of time series where each point of time was the number of new people who had tweeted about the disaster. The model, $M(P, E)$, is the ABM, where P is the two parameters p and q, describing the agent's influence function, and E was the observed social media network. The error measure, ϵ, was mean average precision error, or MAPE, which was a measure of how different the observed time series was from the model time series. The model was then calibrated using all of the data in R using the BehaviorSearch package in NetLogo (Stonedahl and Wilensky 2010) using a simulated annealing algorithm (Kirkpatrick et al. 1983).

In the end, the goal with this chapter was to identify the p and the q values that were most likely have given rise to the observed diffusion patterns. Once these were identified the ratio of the two values $\frac{q}{p}$ was used to express how much social media affected the diffusion. We could then regress properties of the diffusion process to identify factors that led to high rates of social media diffusion. Our results indicated that diffusion events were more likely to spread quickly on social media if (i) they were started by influential individuals, (ii) they were posted earlier in the overall timeline of the disaster, and (iii) the original information was posted repeatedly over time. These are all actionable items from the perspective of a humanitarian organization. However, interestingly we also found that fake news was more likely to spread on social media and that diffusion events containing content that promoted situational awareness (i.e. information about what is going on in the environment) did not significantly affect the spread of news.

Rule Induction

RI is not as old an idea as PO, but it has been explored in the past. The basic idea is that rather than trying to construct the rules that an agent will follow from scratch, the modeler tries to infer rules of behavior directly from data. To some extent this could be used as a large-scale form of PO. Where PO attempts to alter and tweak the parameters of a system to get it to fit better, the goal of RI is to identify the best rule that fits the data. However, there is a significant qualitative difference. The focus in most PO is altering micro-level, meso-level, or even macro-level parameters of the system to get macro-level model outputs that correspond well with actual data. However, in the RI context the goal is to start with empirical data at the micro level and then develop the best agent rules that match that real-world data. In other words, PO starts with altering micro-parameters to generate macro-patterns that match macro-level data. RI works by starting with micro-level data and then generating rules that create macro-level patterns. As a result, both the types of data (micro vs. macro) are different, and the rules/parameters being generated are also often at different levels of modeling complexity.

Commuting Patterns

One example of RI that came about almost 10 years ago was work by Lu et al. (2008) to investigate the use of public transportation by commuters in the Chicago area. In this context, they were interested in investigating how different public policies would affect the decision by commuters to use one particular mode of transit over another. In particular, is it possible to reverse the trend toward car dependence in a large city?

In order to carry out this investigation, they created a synthetic population based on statistics of Chicago's actual population. All agents then had to make two decisions: (i) transit mode choice, i.e. do they use public transit or drive a private car?, and (ii) are they happy with their current residential location, or should they move? For the purposes of this chapter, we will focus on the first decision, because they used a machine learning approach known as (class association rules) CAR to derive the rules of behavior for agents with regard to transit mode choice. They attempted to identify one of five modes of transportation that an agent might use to get to work: (i) walking, (ii) driving, (iii) passenger in a private car, (iv) transit that had to be driven to, and (v) transit that did not have to be driven to. The CAR then used a number of different features to make this decision: household income, household size, number of vehicles in the household, age, gender, employment status, total travel time, access time to public transit, number of public transit transfers required, distance from work, and whether work was located in the central business district. All of this data was derived from actual data about real commuters that were part of the Chicago Area Transportation Survey. This resulted in 404 rules that were then embedded into each agent. The overall structure of this approach is illustrated in Figure 15.3.

It should be noted that this process is almost a mini-calibration exercise in that the rules of behavior were calibrated to real-life behavior, and in fact a training and testing approach was used in order to determine the best rules to use.

The interesting result of this approach is that agents can now make decisions contingent upon their situation. For instance, in this work, Lu et al. took the fully calibrated model and then asked what would happen if you changed the distribution of residential locations and the location of transit options. Since the agent rules were calibrated at the agent level and not at the aggregate level, it was possible to observe how changes in the underlying population and urban form affected agent decisions without having to actually build the city where those patterns existed, creating the possibility for a policy flight simulator (Holland 1992; Sterman 2001). The results showed that though mode choice is

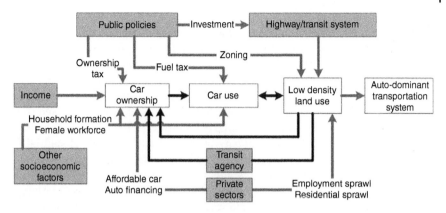

Figure 15.3 A description of the framework in the Lu model. Source: Courtesy of Yandan Lu.

sticky, i.e. people do not switch often, it is possible to design a city that has a high transit share.

Social Media Activity

In the commuting mode choice example, the decision rules were determined for a large population and then embedded contingently in each agent, and the rules for every agent were essentially the same, but the actions differed based on the context the agents found themselves in, i.e. location, number of children, income, etc. However, sometimes this level of heterogeneity is not enough to adequately describe an individual's behavior. In some cases, it may be necessary to create not just different actions for each agent to take, but also different rules of behavior that govern those actions. In these cases it may make sense to learn the behavior of every individual directly from the observable data. This approach would require a large amount of data for every individual and so may not be practical in many cases. For instance, in the commuting mode choice example, it would take a long time to accumulate enough data to determine how each individual chooses different modes of commuting and, in some cases, may not even be practical since you cannot easily alter the income an individual has and then observe how that affects their behavior.

However, in other circumstances it may well be that it is possible to get large amounts of data about the way an individual makes decisions in a limited context. In fact, in the case of the explosion of big data, that is often what we have. We have large-scale trace of individuals using apps, devices connected to the Internet of things, social media, and websites. In these situations we can build models of the individual's interaction with these platforms at the individual level, and then we can embed these individual-level models in an ABM and observe the interactions between them.

We did this recently (Darmon et al. 2013; Harada et al. 2015; Ariyaratne 2016) using a technique known as causal state modeling (CSM). For the purposes of this discussion, the CSM approach, also called computational mechanics or ϵ-machine approach, creates a hidden Markov model-like representation of a time series that is minimally complex and maximally predictive, i.e. has the fewest number of states necessary to predict the time series as well as possible (Shalizi and Crutchfield 2001). In each of the chapters where we used this approach, we took Twitter data and inferred a model of behavior for the individual users of Twitter. The complexity of these models varied substantially. In one case, 12.8% of the inferred models had one state, 58.8% had two states, 4.4% had three states, 3.3% had four states, and 20.7% had more than four states (Darmon et al. 2013). This indicates that the heterogeneity of the underlying population was substantial. The four most common structures that we observed are illustrated in Figure 15.4.

Once we built these models, we were able to predict whether or not a user would tweet in the near future fairly accurately based on these models. We compared the use of CSMs with echo state networks, a recurrent neural network architecture, and found that the model performed fairly similarly at predicting the behavior of an individual (Darmon et al. 2013). Moreover, we

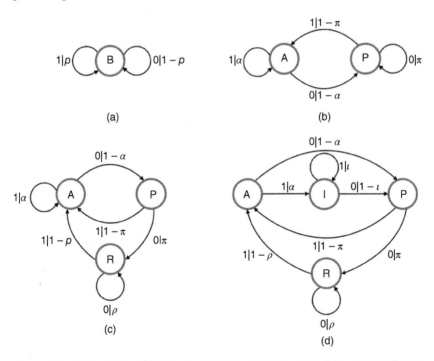

Figure 15.4 An illustration of the four most common CSM structures in the Twitter data set, ranging from the simplest (a) to the most complex (d). Source: Reproduced with permission of David Darmon.

showed that when the behavior of these models was aggregated across large groups of individuals, we were also able to accurately predict the aggregate patterns of behavior of a group of Twitter users. In fact, we performed as well as an aggregate-level autoregressive moving average model and outperformed a pure seasonality model. This last result is interesting, because our model was never trained to perform well at the aggregate level, but rather it was trained at the individual level, and the aggregate-level results are a by-product of the agent-based approach (Harada et al. 2015).

One of the reasons why we have chosen to use CSMs as opposed to neural network architectures or some other data mining approaches is that causal state models are potentially interpretable. By that we mean that once the CSM is learned, a practitioner could interpret the states of the model and label them. This enables comparison to social science theory about how users will behave. In one case it was pretty clear that over half of the users of Twitter had an active and passive state, which corresponds to the idea that users are either actively engaged with Twitter or have stepped away from their phone or the app. Moreover, we can also retroactively generate explanations for why the model makes the prediction that it does, which is contrary to a black-box model where it is difficult to interrogate the model. This transparency of the model gives it a quality that is necessary when pursuing social science explanations and not just curve fitting.

However, there is still work to do in this space. So far the models we have constructed that have been successful have been stand-alone agent models where there was minimal to no interaction between the agents. When we attempted to construct a model where there was social interaction between the agents, we found that the system performed adequately for a few minutes but then stopped predicting the overall state of the system accurately (Ariyaratne 2016). We feel that this is because over long periods of time, seasonality effects have a large role in interactions, and the framework that we constructed did not account for this. Nonetheless, we feel that the CSM approach can be adapted to take a clock as an input that would account for seasonality.

Conclusion and the Vision

ABM and large-scale individual-level data are a great match. The data can provide the insight we need to generate individual-level models of human behavior. This should never be an exercise in curve fitting, since if it is, the individual-level models will probably do no better than the aggregate models. Instead, big data and ABMs should be combined under the TIMMIT principle, and the models should be theory interpretable, and the theories should be model interpretable. This enables the creation of models that are constantly being validated by comparing them with real-world data. If done well then these models that are specified at the level of the individual and not at the level of the aggregate pattern of data may well be able to overcome the Lucas critique (Lucas 1976), since the models of the individual specify an individual's beliefs, goals, and actions, and therefore these models should be able to adapt to changes in the macro-level policies governing individual actions. Thus, the promise of data-driven ABM provides one of the few approaches that may actually work for examining individual responses to new incentive structures imposed by macroscopic policy changes.

In this chapter, I have presented two basic concepts of how this could be done. One approach involves taking a model completely developed from theory and then calibrating it to fit real-world data, i.e. PO. The other approach involves taking a set of data and inferring rules of behavior for that data, often informed by theory development and compared with theoretical models, i.e. RI. I have highlighted two examples of each approach that seem to work very well, but there are many other ways that these processes can be carried out using additional machine learning or statistical inference approaches. Many of them have been explored in previous work (Zhang et al. 2016).

One vision for this work would be the eventual creation of an ABM directly from data. Especially, in the context of the RI approach, it might be possible to specify some basic features with relation to the basic form of an ABM and then give a set of data to a computational data processing pipeline and have that pipeline automatically spit out a fully realized ABM in a common modeling language, such as NetLogo (Wilensky 1999) or Repast (North et al. 2006). Moreover, this model could be updated on a continuous basis without much additional effort by continuing to feed new data into it, creating a dynamically created realistic model of a complex system.

For instance, imagine an ABM of a city or urban landscape that was constantly fed data from social media, e.g. Twitter, administrative data feeds, e.g. police feeds, and satellite or aerial imagery. This model might adequately represent the focal city at enough level of detail that it could then be used as a policy flight simulator (Holland 1992; Sterman 2001) and used to assess the effect of street closures, new crime policies, and changes in the zoning laws. In fact, if carried to the extreme, and developed at the right resolution, such a model might enable detailed predictions about the future state of our fictional urban center. This could lead to the creation of a new science, such as Asimov's psychohistory, where history, psychology, and mathematical statistics (or in this case ABM) are combined to make predictions about how large populations will act in the future (Asimov 1951).

Acknowledgments

I would like to thank all of my coauthors and the researchers who worked on the projects showcased above, especially, Forrest Stonedahl, Uri Wilensky, Eunae

Yoo, Jeffrey Herrmann, Brandon Schein, Neza Vodopivec, Yandan Lu, Moira Zellner, Kazuya Kawamura, David Darmon, Jared Sylvester, and Michelle Girvan. Most of these colleagues also gave invaluable feedback on early versions of this chapter. I also want to thank Paul Davis and Angela O'Mahoney for making this opportunity possible and for providing advice on how to clarify this chapter.

References

Ariyaratne, A. (2016). Modeling agent behavior through past actions: simulating twitter users. Master's thesis. College Park, MD: University of Maryland.

Axtell, R.L., Epstein, J.M., Dean, J.S. et al. (2002). Population growth and collapse in a multiagent model of the Kayenta Anasazi in long house valley. *Proceedings of the National Academy of Sciences of the United States of America* 99 (Suppl. 3): 7275–7279.

Backstrom, L., Sun, E., and Marlow, C. (2010). Find me if you can: improving geographical prediction with social and spatial proximity. In: *Proceedings of the 19th International Conference on World Wide Web*, 61–70. ACM.

Bertrand, M. and Mullainathan, S. (2001). Do people mean what they say? Implications for subjective survey data. *American Economic Review* 91 (2): 67–72.

Bonabeau, E. (2002). Agent-based modeling: methods and techniques for simulating human systems. *Proceedings of the National Academy of Sciences of the United States of America* 99 (Suppl. 3): 7280–7287.

Claeskens, G. and Hjort, N.L. (2008). *Model Selection and Model Averaging*. Cambridge Books.

Cook, S., Conrad, C., Fowlkes, A.L., and Mohebbi, M.H. (2011). Assessing Google flu trends performance in the United States during the 2009 influenza virus A (H1N1) pandemic. *PLoS ONE* 6 (8): e23610.

Darmon, D., Sylvester, J., Girvan, M., and Rand, W. (2013). Predictability of user behavior in social media: bottom-up v. top-down modeling. In: *2013 International Conference on Social Computing (SocialCom)*, 102–107. IEEE.

De Choudhury, M., Gamon, M., Counts, S., and Horvitz, E. (2013). Predicting depression via social media. *ICWSM* 13: 1–10.

Eagle, N. and Pentland, A.S. (2006). Reality mining: sensing complex social systems. *Personal and Ubiquitous Computing* 10 (4): 255–268.

Golbeck, J. and Hansen, D. (2011). Computing political preference among twitter followers. In: *Proceedings of the SIGCHI Conference on Human Factors in Computing Systems*, 1105–1108. ACM.

Goldenberg, J., Libai, B., and Muller, E. (2001). Talk of the network: a complex systems look at the underlying process of word-of-mouth. *Marketing Letters* 12 (3): 211–223.

Granovetter, M. and Soong, R. (1983). Threshold models of diffusion and collective behavior. *Journal of Mathematical Sociology* 9 (3): 165–179.

Grassegger, H. and Krogerus, M. (2017). The data that turned the world upside down. Vice Magazine (30 January). https://motherboard.vice.com/en_us/article/mg9vvn/how-our-likes-helped-trump-win.

Harada, J., Darmon, D., Girvan, M., and Rand, W. (2015). Forecasting high tide: predicting times of elevated activity in online social media. In: *Proceedings of the 2015 IEEE/ACM International Conference on Advances in Social Networks Analysis and Mining 2015*, 504–507. ACM.

Holland, J.H. (1975). *Adaptation in Natural and Artificial Systems: An Introductory Analysis with Application to Biology, Control, and Artificial Intelligence*, 439–444. Ann Arbor, MI: University of Michigan Press.

Holland, J.H. (1992). Complex adaptive systems. *Daedalus* 121 (1): 17–30.

Asimov, I. (1951). *Foundation*. New York: Gnome Press.

Kirkpatrick, S., Gelatt, C.D., and Vecchi, M.P. (1983). Optimization by simulated annealing. *Science* 220 (4598): 671–680.

Kitchin, R. (2014). The real-time city? Big data and smart urbanism. *GeoJournal* 79 (1): 1–14.

Kleinberg, J.M. (1999). Hubs, authorities, and communities. *ACM Computing Surveys (CSUR)* 31 (4es): 5.

LeCun, Y., Bengio, Y., and Hinton, G. (2015). Deep learning. *Nature* 521 (7553): 436.

Lu, Y., Kawamura, K., and Zellner, M.L. (2008). Exploring the influence of urban form on work travel behavior with agent-based modeling. *Transportation Research Record: Journal of the Transportation Research Board* 2082 (1): 132–140.

Lucas, R.E. Jr. (1976). Econometric policy evaluation: a critique. In: *Carnegie-Rochester Conference Series on Public Policy*, vol. 1, 19–46. Elsevier.

Matheus, R., Vaz, J.C., and Ribeiro, M.M. (2014). Open government data and the data usage for improvement of public services in the Rio de Janeiro city. In: *Proceedings of the 8th International Conference on Theory and Practice of Electronic Governance*, 338–341. ACM.

North, M.J., Collier, N.T., and Vos, J.R. (2006). Experiences creating three implementations of the repast agent modeling toolkit. *ACM Transactions on Modeling and Computer Simulation (TOMACS)* 16 (1): 1–25.

O'Neil, C. (2017). *Weapons of Math Destruction: How Big Data Increases Inequality and Threatens Democracy*. Broadway Books.

Page, L., Brin, S., Motwani, R., and Winograd, T. (1999). The Pagerank Citation Ranking: Bringing Order to the Web. Tech. Rep. 422. Stanford InfoLab.

Rand, W. and Rust, R.T. (2011). Agent-based modeling in marketing: guidelines for rigor. *International Journal of Research in Marketing* 28 (3): 181–193.

Rand, W., Herrmann, J., Schein, B., and Vodopivec, N. (2015). An agent-based model of urgent diffusion in social media. *Journal of Artificial Societies and Social Simulation* 18 (2): 1.

Riggins, F.J. and Wamba, S.F. (2015). Research directions on the adoption, usage, and impact of the internet of things through the use of big data analytics. In: *2015 48th Hawaii International Conference on System Sciences (HICSS)*, 1531–1540. IEEE.

Schmidt, E. (2009). How Google can help newspapers. The Wall Street Journal (1 December). https://www.wsj.com/articles/ SB10001424052748704107104574569570797550520.

Shalizi, C.R. and Crutchfield, J.P. (2001). Computational mechanics: pattern and prediction, structure and simplicity. *Journal of Statistical Physics* 104 (3–4): 817–879.

Shmueli, G. (2010). To explain or to predict? *Statistical Science* 25 (3): 289–310.

Smith, E.B. and Rand, W. (2018). Simulating macro-level effects from micro-level observations. *Management Science*. In press.

Sterman, J.D. (2001). System dynamics modeling: tools for learning in a complex world. *California Management Review* 43 (4): 8–25.

Stonedahl, F. and Rand, W. (2014). When does simulated data match real data? In: *Advances in Computational Social Science*, 297–313. Springer.

Stonedahl, F. and Wilensky, U. (2010). Behaviorsearch [computer software]. In: *Center for Connected Learning and Computer Based Modeling*. Evanston, IL: Northwestern University. http://www.behaviorsearch.org.

Ubaldi, B. (2013). Open government data: Towards empirical analysis of open government data initiatives. OECD Working Papers on Public Governance, (22):0_1.

Watts, D.J. (2013). Computational social science: exciting progress and future directions. *The Bridge on Frontiers of Engineering* 43 (4): 5–10.

Weinberg, R. and Berkus, M. (1971). Computer simulation of a living cell: Part I. *International Journal of Bio-Medical Computing* 2 (2): 95–120.

Wilensky, U. (1999). *NetLogo: Center for Connected Learning and Computer-Based Modeling*. Evanston, IL: Northwestern University.

Wilensky, U. and Rand, W. (2015). *An Introduction to Agent-Based Modeling: Modeling Natural, Social, and Engineered Complex Systems with NetLogo*. MIT Press.

Yoo, E., Rand, W., Eftekhar, M., and Rabinovich, E. (2016). Evaluating information diffusion speed and its determinants in social media networks during humanitarian crises. *Journal of Operations Management* 45, 123–133.

Yosinski, J., Clune, J., Fuchs, T., and Lipson, H. (2015). Understanding neural networks through deep visualization. In: *ICML Workshop on Deep Learning*.

Zhang, H., Vorobeychik, Y., Letchford, J. and Lakkaraju, K. (2016). Data-driven agent-based modeling, with application to rooftop solar adoption. *Autonomous Agents and Multi-Agent Systems* 30 (6): 1023–1049.

16

Bringing the *Real World* into the Experimental Lab: Technology-Enabling Transformative Designs

Lynn C. Miller[1], Liyuan Wang[2], David C. Jeong[2,3], and Traci K. Gillig[2]

[1] Department of Communication and Psychology, University of Southern California, Los Angeles, CA 90007, USA
[2] Annenberg School of Communication and Journalism, University of Southern California, Los Angeles, CA 90007, USA
[3] CESAR Lab (Cognitive Embodied Social Agents Research), College of Computer and Information Science, Northeastern University, Boston, MA 02115, USA

Understanding, Predicting, and Changing Behavior

A myriad of studies in the social sciences aim to understand, predict, and change individual behavior to improve public health and national security. These studies are often laboratory-based experiments (i.e. systematic designs), which allow for causal inferences (Cook and Campbell 1979; Trochim and Donnelly 2006). However, a critical weakness of laboratory-based studies is that they are rarely representative of the real-world environments and scenarios faced by those whose behavior is to be influenced (Cook and Campbell 1979; Dhami et al. 2004; Fiedler 2017). This raises serious questions about both causality and generalizability. This chapter describes a new paradigm and how to exploit it.

The basic idea is to design virtual experiments that are more representative of *real-world* challenges faced by those whose behavior is to be influenced. The results from such experiments about the ability to influence behaviors can yield causal inferences generalizable for use in computational models. We illustrate the process and challenges of doing so for three domains: preventing disease, mitigating harm during crises and disasters, and reducing terrorism.

Today's technologies (e.g. virtual environments and games, ecological momentary assessments (EMAs), and intelligent agents) allow us to create and validate the representative environments in question. These can provide artificial but sound *default control groups*. Representative environments can also provide a foundation for experimental studies: that is, using the default

Social-Behavioral Modeling for Complex Systems, First Edition.
Edited by Paul K. Davis, Angela O'Mahony, and Jonathan Pfautz.
© 2019 John Wiley & Sons, Inc. Published 2019 by John Wiley & Sons, Inc.
Companion website: www.wiley.com/go/Davis_Social-Behavioralmodeling

control as a base, researchers can make systematic manipulations and compare results. That is, the approach can provide generalizable testbeds for examining and validating cause–effect relationships in representatively designed scenarios pertaining to social domains of interest. In this chapter we draw on our own research experience with virtual representative designs to illustrate the ideas; we then discuss how such research can be extended to other domains with important public health and national security implications.

Social Domains of Interest

In the United States, many thousands of premature deaths annually result from unhealthy and/or risky behaviors, failure to adopt safety precautions in environmental crises/disasters, and acts of terrorism (U.S. Centers for Disease Control and Prevention 2017). Below, we provide an overview of these three pressing issues, considering the challenges these contexts pose for research that attempts to untangle cause–effect relationships and suggest effective interventions.

Preventing Disease

In the United States alone, at least 35% of 900 000 annual premature deaths are preventable with changes in individual-level decision-making (e.g. diet, physical activity, substance use, safety precautions, sexual behavior) (US National Research Council and Institute of Medicine 2015). Meta-analyses suggest that behavioral interventions can reduce premature death by causing related changes in individual behavior (Lussier et al. 2006; Dutra et al. 2008; Head et al. 2013). Further, behavioral interventions prompting problem-solving behaviors in response to challenges in everyday life can increase healthy behavior (Lara et al. 2014). Such interventions can be accomplished in the laboratory with so-called *serious games* (i.e. videogames dealing with real-world scenarios and health outcomes) (Baranowski et al. 2008, 2011). Although exceptions exist, relatively few existing serious games create environments representative of real-world challenges and scenarios and, at the same time, allow substantial variation. Recreational games can in some respects be quite realistic, but are not designed for the purpose of analysis and only allow for some kinds of manipulation (see other papers in this volume, Lakkaraju et al. (2019) and Guarino et al. (2019)).

Individuals at risk for one health behavior may be at risk for other behaviors due to common underlying variables. We need to better understand common causal factors that, if prevented, can mitigate risk behaviors that have broad social implications. One such underlying variable is emotional regulation – skills and competencies that allow individuals to adapt to stressors and challenges (Thompson 1994; Gross 1998; Eisenberg 2000). Enhancing the emotional skills of at-risk individuals could have positive downstream consequences. For example, one affective factor affecting emotional regulation is shame, a painful feeling that results from believing that one's situation is stable (unchangeable) and global (due to one's self as a whole, that is, that one is a bad person) (Tracy and Robins 2004, 2006) and has failed to live up to one's ideal self (Lazarus 1991). Shame is caused by experiencing and internalizing stigma (Hatzenbuehler et al. 2012). Shame and stigma are related to risk behavior in many health and mental health domains (Hatzenbuehler et al. 2012). Can shame be reduced and does that reduce risky behavior? Initial work is promising in this regard: sexual shame can be reduced with correspondingly reduced risky sexual behavior of men who have sex with men (MSM) (Christensen et al. 2013).

Harm Mitigation in Crises

Next consider how risky decision-making manifests in crisis situations, potentially putting individuals, groups, and communities more broadly in harm's way. Mitigating loss of life due to such man-made or environmental crises as pandemics, earthquakes, hurricanes, or flood preparation often depends on individual behavior. For example, the US Centers for Disease Control and Prevention (CDC) released mitigation guidelines to mitigate influenza effects in the United States (Qualls et al. 2017). The guidelines rely heavily on personal protective measures (e.g. hand hygiene, voluntary home isolation) and additional measures during pandemics (e.g. face masks, quarantines). As another example, 45 states and territories in the United States are at some risk for earthquakes (American Red Cross 2017). In cases of severe earthquakes, individual preparations such as taking steps to avoid gas explosions, bolting down furniture, and storing adequate post-quake supplies could dramatically mitigate loss of life and injuries.

Individual behaviors differ significantly on such matters. In documented disaster cases, some individuals actually traveled *into* not *away from* risk zones, perhaps to learn what was happening (Drabek 1969). After warning of a real tsunami, 15% of individuals moved toward rather than away from risk zones (Helton et al. 2013). Similarly, people vary on whether they do or do not leave an earthquake-affected city (Helton et al. 2013). Computational models and agent-based simulations of disaster responses should reflect this heterogeneity (Chien and Korikanthimath 2007).

Among the reasons for heterogeneity is that some individuals panic or otherwise respond irrationally in crisis situations (Walters and Hornig 1993), are driven by personal beliefs or motives (Drabek 2013), or have differential thresholds for perceiving threats (Helton et al. 2013). Other reasons include diversity of personal histories and related differences in how crisis-time cues activate goals and emotions.

To elaborate, consider that despite years of educational efforts to enhance earthquake preparation, preparation levels generally remain low (Becker et al. 2012). Individuals may not have thought through or experienced what might happen in an emergency or experienced lifelike simulations. Thus, the crisis-time cues may not activate those emotional, cognitive, and social processes that lead to preventative actions. If so, perhaps fear coupled with plans and capabilities for action in a broader social context would motivate better individual preparation (Becker et al. 2012). Virtual games might be ideal for enhancing such sensitivities and propensity to prepare for emergencies. Indeed, some such training interventions exist (e.g. serious games for crisis management). These, however, tend to focus on emergency response personnel. Further, their effectiveness has not typically been evaluated carefully with controlled trials (Di Loreto et al. 2012).

Terrorism Reduction and Lone Actors

Although Al-Qaeda's attack on 9/11 was obviously planned, since 2006 lone actors have caused 98% (156) of the terrorist deaths in the United States (Institute for Economics and Peace 2016). This domestic terrorism (i.e. involving citizens or legal permanent residents, many entering the United States as children) is a *home-grown*, preventable phenomenon (Nowrasteh 2016). Another form of terrorism involves school shooters: this century, there have been 211 school shootings, including one in January 2018 resulting in the deaths of 2 students and an additional 18 individuals injured (CNN and Yan et al. 2018). As we were revising the current chapter, an additional school-shooter incident took place in Parkland, Florida, further increasing the number of school-related incidents involving death and injury.

Existing research offers insight into how to approach the problem of domestic terrorism. First, radicalization of opinions appears quite different from radicalization of behavior. Only 1% of people with radicalized ideas act on those ideas, and those whose behaviors become radicalized do not typically radicalize their opinions. This puts a premium on the pyramids and processes and cues associated with escalation of radicalized behaviors (McCauley and Moskalenko 2017). Regarding radicalization of behavior, one key set of contextual factors involves humiliation, a combination of shame and anger. In a related domain, among criminals, guilt and shame are negatively associated with delinquency (Spruit et al. 2016). However, existing studies exploring these relationships are correlational in design and do not provide causal evidence that increasing shame can decrease criminal behavior. Indeed, interventions trying to induce guilt (Spruit et al. 2016) or shame (Jones 2014) that seek to reduce recidivism have been unsuccessful or have backfired. Furthermore, shame promotion has been clinically discouraged because of its potential to trigger aggression (Schalkwijk 2015). We know little about circumstances that produce humiliation (and its concomitants of shame and aggression) and whether humiliation is indeed critical to understanding radicalized behaviors (McCauley and Moskalenko 2017).

Some research suggests that individuals who exhibit violent behavior directed toward their peers may first experience bullying (Unnever 2005; Stein et al. 2006; Wike and Fraser 2009) or teasing/marginalization (Verlinden et al. 2000). Other problems that are associated with bully victims include poor social adjustment (Nansel et al. 2001), social isolation (Juvonen et al. 2003; Veenstra et al. 2005), alcohol use (Nansel et al. 2004; Stein et al. 2006), mental health symptoms (Kaltiala-heino et al. 2000; Juvonen et al. 2003), behavior disorders (Kokkinos and Panayiotou 2004), personality difficulties (Kaltiala-heino et al. 2000), and relationship difficulties (Kumpulainen and Räsänen 2000). Consistent with the above, other research suggests that a subgroup of shooters experience psychological disorders such as depression, suicidal ideation, and/or various psychiatric disorders and/or were loners who felt isolated and separated from peers (Gerard et al. 2016).

Ioannou et al. (2015) conducted a systematic content analysis of 18 variables involving the offender characteristics of 40 *rampage* school-shooting cases, undertaken by students or former students from 1966 to 2012 (95% male, mean crime age of about 16). Three major themes emerged using the smallest space analysis, suggesting that the school shooter (i) was psychologically disturbed (e.g. on psychiatric medication, had been diagnosed with a psychiatric disorder, had been bullied, had suicidal thoughts, was a loner, watched violent media, or wrote about violence), (ii) was rejected (e.g. as evidenced by a recent breakup of a relationship, past suicide attempts, home abuse, or recent suspension from school), and/or (iii) had engaged in a pattern of past criminal behavior.

Experimentally assessing the circumstances under which emotions (e.g. shame, rejection) or social behavior (e.g. bullying, isolation) activate aggression for particular people, as well as finding potential ways to mitigate these processes, would fill important gaps in our knowledge. Interestingly, in a 19-year longitudinal study, the existence of emotional regulatory problems in children 6–7 in age – as assessed by teachers and parents – was the single best predictor of later adult convictions for violent offenses at age 25–27 (Young et al. 2016). In that study, high *emotional regulatory problems* were signaled by parental/teacher endorsement of items that included worrying about many things: being unhappy, tearful, or distressed, being fearful of new situations, being fussy, and being unresponsive or apathetic. These findings at least suggest the possibility that interventions targeting improving emotional self-regulation and skills development, including communication skills, in handling fears and new situations in a more age-appropriate way might mitigate the potential for subsequent teen and adult violence, including that of school shooters.

Another key, as suggested above pertaining to school shooters, may be understanding the link between mental disorders and individual involvement in terrorism (Gill and Corner 2017). A review of the literature outside of school shooters indicates no common mental health profile for terrorists. However, as with school shooters, higher levels of some psychological characteristics may be contributing factors. Such characteristics account for less than half of all individuals in any terrorist subsample (Gill and Corner 2017). Subsets of individual factors (e.g. suicidal ideation, alcohol and substance abuse, underlying mental health factors) may be prevalent prior to terrorist involvement or exacerbated during involvement (Bubolz and Simi 2015), as is also the case for school shooters.

Recently, researchers have acknowledged the complexity of terrorist involvement and have advocated moving from *profiles* to understanding *pathways* that elucidate how individuals are engaging and disengaging from involvement with terrorist activity (Horgan 2008). Additional individual factors that may predict terrorist action may include early experience (e.g. risk taking, reduced social contact) (Taylor and Horgan 2006), personal victimization (McCauley and Moskalenko 2011), and displaced aggression (Moghaddam 2005). Just as complex dynamics may lead to behavioral radicalization, deradicalization may involve grappling with a range of affective responses (e.g. burnout, regret, distress, fear, experienced victimization) through means such as psychological, social, and/or familial counseling (Gill and Corner 2017).

Taking a broader look at the literature exploring mental health and antisocial/violent teen and adult outcomes, chief predictors appear to be childhood difficulties with emotional regulation and/or abuse (Juvonen and Graham 2001; de Roy van Zuijdewijn and Bakker 2016). Indeed, childhood and adolescent emotional regulation capabilities are negatively predictive of subsequent internalizing and externalizing issues and may be critical underlying causal factors across a range of mental health and other outcomes (e.g. alcohol and abuse disorders, self-harm, etc.) (Berking and Wupperman 2012; Marusak et al. 2014). Interesting, adult individuals with early life history indicators (e.g. early abuse, violent delinquency) in a prospective longitudinal study were at risk for a variety of criminal offenses including violence (Lansford et al. 2007).

Researchers must better understand this *pyramid of behavior* and its underpinnings. What cues/indicators might predict radicalization and deradicalization of behavioral risk? Understanding the role of individual and contextual factors that may dynamically operate to produce emergent radicalization of lone actors is an important target of future work (Gill et al. 2014; Gill 2015; Gill and Corner 2017). Many of the same underlying factors (e.g. problems with emotional self-regulation) that may provide a pathway to terrorism may also provide a pathway to other negative outcomes for individuals and society (e.g. violent and other criminal externalizing behavior, mental health disorders, self-harm, alcohol and drug abuse).

An additional problem is the asymmetrical nature of these risks. Thus, for example, many teens and young adults (a) are rejected, (b) have mental health problems, or (c) have a history of criminal problems yet don't become school shooters: if a, b, and/or c, the behavior (d: school shooting) is a low probability event. But given d behavior, the probability that the school shooter exhibited a, b, and/or c is relatively high. On the other hand, similar interventions (e.g. involving emotional regulation and communication skills training) might be effective for a broad segment of adolescents and young adults with high levels of rejection, mental health problems, and criminal historical patterns in reducing their propensity for negative outcomes, including relatively rare terrorist events (e.g. shooting rampages). Later, we return to a consideration of an example from the terrorist domain, identification of children who may have the emotional regulation difficulties that might be mitigated to reduce school-scooter propensities. Next, we discuss how to use an approach called Socially Optimized Learning in Virtual Environments (SOLVE) to develop a representative game with virtual choices as in everyday life and also incorporate an intervention to change behavior of interest for a given target audience.

The SOLVE Approach

Our work with the SOLVE approach involved some of the first successful attempts to reduce shame through an intervention and to show that reducing shame subsequently reduces risky behavior. In the following, we describe how we used virtual representative design to model real-world situations experienced by our target population (i.e. young at-risk MSM). We also discuss the results of our research using the representative virtual world and implications of those findings for future research designs.

Overview of SOLVE

The original method of SOLVE was a way to study the potential value of virtual interventions in changing risky sexual behaviors for MSM to reduce the spread of HIV (Miller et al. 2011, 2016; Christensen et al. 2013). SOLVE was designed to capture the real-world challenges MSM face and the behavioral options they had in sequences and narratives that lead to risky sex. Indeed, whether the SOLVE intervention used interactive video (Read et al. 2006) or, more recently, intelligent agents and a game platform (Miller et al. 2011), the intervention was found to reduce risky sexual behavior for men in the game condition (vs. control) and, based on self-reporting, in the real world as well. When we designed SOLVE for the application of changing MSM's risky sexual behaviors, we intervened in the decision-making of SOLVE users in numerous ways. For example, we included a *virtual future self* (VFS). That is, users

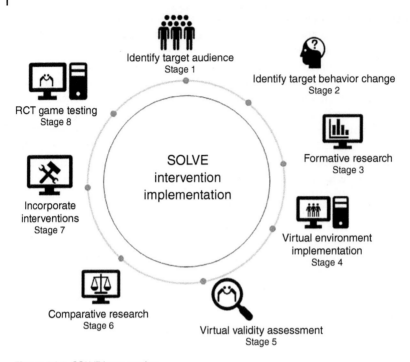

Figure 16.1 SOLVE intervention.

developed an avatar that was then *aged* to represent their future self, which delivered intervention messages. Within the virtual world of SOLVE, the VFS intervened after the user made a risky choice and delivered a message (prior pilot work suggested) might be effective in changing their decision-making (i.e. they would choose to engage in a safer behavior). Users then could revise their behavior choice before action in the game proceeded. In this way, SOLVE modeled social and communication skills for negotiating safer sex. See Figure 16.1.

Shame Reduction as a Key Intervention

As we worked with this population of risky men starting in the late 1980s, it became apparent that sexual shame was a contributing factor to risk behavior. Indeed, we have found among our risky MSM that sexual shame is correlated with past sexual risk (Christensen et al. 2013). To reduce shame in our interventions, as well as normalize men's sexual desire for other men, our guide messages used what we called an ICAP (Interrupting risky choices; Challenging those choices; Acknowledging men's desires, emotions, and motives; and Providing a means to achieve those in a less risky fashion). The SOLVE intervention effectively reduced shame, and subsequently sexual risk behavior, over three months, compared with a control group (Christensen et al. 2013), becoming one of the first interventions to reduce shame and risky choices. Of note, across four game studies, the participants' real-world 90-day sexual behavior was significantly related to virtual sexual choice behavior, highlighting the relevance of virtual behavior to real-world behavior.

Intelligent Agents in Games

In designing environments to be representative of the contexts within which individuals make decisions, the other social actors are an important part of context. Although one can create other social actors as a director would in choosing actors for roles, it is also possible to differentially set parameters of intelligent agents (e.g. their beliefs, their goals) to simulate the types of potential partners that may differentially result in risky or safer choices for our target population. That is what the SOLVE-IT (SOLVE using intelligent technologies) intervention did, using a software tool called PsychSim (Marsella et al. 2004), which is a multi-agent-based tool that simulates and predicts human social behavior. PsychSim is theoretically grounded in theory of mind and allows construction of agent simulations that include mental models about others. Considerable formative research found MSM who took sexual risks disproportionately included men whose attachment style was apt to be avoidant/dismissive, fearful, and anxious. Thus, we created a similar set of potential virtual partners whose propensities (e.g. goals, beliefs, verbal responses) would be representative for our target population. The result was conversational sequences and interactions that were realistic for these men (Miller et al. 2011). Obviously, one could manipulate the parameters of intelligent agents for many experimental conditions examining controlled social interactions between humans and agents (and agents with one another) (Feng et al. 2017; Jeong et al. 2017).

Generalizing Approach: Understanding and Changing Behavior Across Domains

Up to this point we have argued that it is possible to create representatively designed virtual environments for a target population for behavior of interest. These could serve as a *control* or default condition against which to compare experimental conditions, which are the default plus additional systematic manipulations of variables within that environment. Intelligent agents in games offer another rich source of experimental manipulations. Together, a default control and experimental conditions, built onto that default environment, provide a new way to combine systematic and representative designs into a new integrative design: systematic representative designs. If we standardized our approach regarding default control game conditions and agents in them, this could also open possibilities for complex *model systems* that might be interoperable for social science research and applications and provide a better evidence basis for multi-agent simulations of the dynamics of social behavior in a variety of domains of interest. In the following sections, we discuss how our experience can inform future research designs in a variety of pressing health and security contexts.

Across domains of behaviors that have potential risk for health and safety in the United States, research has examined contextual, affective, and social factors influencing behavior. For example, affective factors such as shame and humiliation may be especially important mechanisms to understand in health contexts. Other affective responses, such as fear and threat avoidance are key in contexts such as emergencies. Emotional regulation skills have positive effects for individuals and society across diverse outcomes, and deficits in emotional regulation skills are problematic. Social factors, including the perceived or actual responses of others, and our beliefs about others, may play key roles in individual decision-making (e.g. deciding whether to trust authorities in creating social challenges that may lead to riskier choices). But, exactly how and when these factors (one's own or the inferred goals, affect, and beliefs of others) exert causal influence, and for whom, is unclear. Details regarding the causal sequences and choice points that lead up to the behavior of interest – key points for intervention – are largely unknown.

Experimental Designs for *Real-World* Simulations

Our understanding of decision-making is informed by both theory and methodology. One of the major challenges for modeling social behavior is that our classic systematic experimental designs are primarily focused on establishing internal validity, often at the expense of generalizability (Mook 1992). However, an alternative – representative design – can contribute to both types of validity. Below we discuss the two types of designs – systematic and representative – that historically have been used in psychological experiments, and we suggest how they could be integrated to afford experimental designs for virtual environments and real-world simulations.

Standard Systematic Designs and Representative Designs: A Primer

The dominant modus operandi in social scientific theory testing is standard *systematic design*: a design that affords the potential for causal inference (subject to well-known problems, such as hidden variables). Systematic designs involve measuring one or more participant outcomes (i.e. the dependent variable(s)) after first varying (e.g. manipulating) one or more preconditions (i.e. the independent variable(s)) to which participants are exposed. These designs afford the opportunity to test the causal effect of a manipulation of an independent variable (e.g. treatment vs. control) or combination of independent variables (e.g. treatment A and B vs. alternative control conditions) on one or more dependent variables (Shadish et al. 2001). A goal of systematic design is to ensure the internal validity of the experiment and therefore permit at least cautious causal inference. An experiment is said to have internal validity if variability in the dependent variable across conditions can only be attributed to researcher manipulation(s) of prior condition(s).

Key to classic systematic designs is random assignment of subjects to conditions. Researchers randomly assign participants to experimental conditions to reduce a threat to internal validity (i.e. self-selection to condition might introduce an alternative source of variability that might be responsible for differences across conditions other than the manipulated variable). Yet, the impact of random assignment and self-selection (as in everyday life) on the generalizability of research results to the real world warrants further attention in social science research.

First introduced by (Brunswik 1955), representative design is founded on the logic of inductive inference and sampling theory. These designs sample from the situational features to which one wishes to generalize (Hammond and Stewart 2001). Situational features include both individual, social, and other factors, as well as environmental conditions. If sampling mechanisms pertaining to the environment are incorporated, then it can be argued that representative designs may have a higher capacity to produce generalizable results and increase external validity for conclusions about causality (Cook and Campbell 1979).

Systematic Representative Virtual Game Designs

Systematic designs have historically had an advantage in enabling researchers to make cause–effect claims, assuming an absence of threats to internal validity. Typically, however, systematic designs do not generalize to real-world situations as representative designs can. We believe that virtual games can afford the benefits of both approaches if the following conditions are met: (i) a *control or default* condition is designed, (ii) *hooks* for experimental manipulations within the default condition that afford alternative *experimental groups* against which to compare participants in the control/default condition are created, and (iii) participants within a given target audience are randomly assigned to representative control and experimental virtual game groups. Once randomly assigned to condition (representative control or experimental group(s) that are control conditions with experimental manipulations), participants within conditions are self-selecting themselves into various paths, as they do in real life.

What Is a Default Control Condition?

We can design a representative virtual game for the behaviors of interest for a target audience of interest. This is a naturally occurring representative design or a kind of background or *control* condition, which for the specific target population and behaviors of interest brings the real world into the lab. The control group provides a correlational laboratory, enabling us to examine differences in how individuals respond, which might produce safer or riskier choices. That is, the default control provides a testbed for understanding correlational patterns potentially within persons as well as between persons. Details regarding how to create such an environment are provided below.

What Are *Hooks* and Experimental Alternatives?

When designing a virtual environment that is representative of everyday life, the choices provided, including the sequences within which they are embedded, are designed to be representative of those for a given target audience (e.g. MSM) with a given behavior of interest (e.g. risky sex). In the virtual world, the behavioral options at choice points can be expanded, narrative sequences can be changed, and additional interventions can be built in (e.g. after a user makes a choice, he/she is given an intervention message and has a second chance to decide how to continue). These *hooks* for alternatives within the game allow for alternative *experimental conditions* (e.g. systematically manipulating alternatives of interest to be compared against the control/default condition).

This enables us to have a systematic representative design that combines a *background* (i.e. a control condition representative of real-life decisions in a domain for a given target audience) that is held constant while manipulating variables of interest across conditions. This manipulation allows for the comparison of experimental condition results against the control condition (i.e. the *background*). Indeed, we could think of these control groups or *default* conditions as *model systems* that have been demonstrated to have validity (e.g. virtual validity involving significant correlations between target individuals' virtual and real-life past behavior). Because these *model system* control groups are given in games and experimental conditions are added (or changed) in them, replications are much easier whether for basic or applied research.

Although to date, some game interventions have been developed using representative designs, they have not to our knowledge been built combining representative design and interventions. This is because when many of these game interventions were first developed, cost significantly precluded designing and developing game *default* controls. Nonetheless, we believe it is possible to pull these functions apart to create a systematic representative design approach. But questions remain: how does one create representative environments, what are their challenges, and are they feasible? What evidence suggests their promise? We explore such questions below.

Creating Representative Designs for Virtual Games

There are challenges in designing representative virtual games for behaviors of interest in the target audience of interest, including: (i) measuring naturalistic behavioral occurrences over time, (ii) identifying decision points for risky vs. safer behaviors, (iii) creating a sampling frame of challenges to enable representative sampling of the situations to which researchers wish to generalize, (iv) coding/structuring sequences as in everyday life (in part since the meaning of a challenge [stimulus] situation and responses to it may depend on where it is embedded in the sequence), (v) considering naturally covarying factors in situations, (vi) understanding options available for individuals in the population at each choice point, (vii) determining how and when things go differently to produce risky or safer choices, (viii) adding more detail regarding precipitating cues/triggers and reactions (e.g. affect, cognitions, motives) and their relationship to challenging moments, and (ix) evaluating the effectiveness of creating desired representative design. In our work, we used a range of tools (e.g. interviews, surveys, focus groups) to identify factors (e.g. obstacles to safer sex) that could produce the behavior of interest (e.g. risky sex) for our target population (e.g. MSM). Today's technologies make this process more feasible. In the forthcoming sections, we provide a guide to designing virtual representative environments, drawing from lessons learned in our own work.

Measuring Occurrence of the Behavior of Interest (BoI) in Time

Before one can create a sampling frame, one has to identify when and under what circumstances the behavior is occurring. EMA (Stone and Shiffman 1994; Shiffman et al. 2008) is one useful tool. EMA allows participants to report their real life *in the moment* of naturalistic experiences. Typically, researchers know their ultimate behavioral target of interest. They can track these by asking participants to self-report this information close to the behavioral occurrence (e.g. daily; every minute) in diaries or using EMA technologies. Smartphone technologies and similar devices may eventually enable the automatic trace recording of (or probabilistic inference about) behaviors without (or with less) conscious input from target audience members. For example, wearable wrist biosensors (for electrodermal activity, skin temperature, acceleration) on emergency department patients have been used to assess the ability of researchers and clinicians to detect drug (e.g. opioid, cocaine) relapse in naturalistic settings (Carreiro et al. 2015; Carreiro et al. 2017). A recent review (Adams et al. 2017) indicates how biomarkers pertaining to psychiatric functioning (e.g. stress, anxiety, substance abuse) are being used in conjunction with geo-positioning system data and self-reports to assess behavioral occurrence (when, where). Another review further highlights how sensors (both wearable and digestible) and social media applications can be used in specific domains for evaluating target behavior (Carreiro et al. 2017).

Beyond the When of BoI: Identifying Challenges and Preconditions

Representative designs require sampling from the stimulus condition to which one wishes to generalize. To do so requires creating a sampling frame of stimuli/challenges and preconditions that can result in BoI. Before creating a sampling frame, researchers must identify and be able to measure the potential challenges and preconditions of interest. Several approaches may facilitate this. One approach is identifying possible challenges based on self-reports. For example, in formative unpublished work, we identified over 100 obstacles to safer sex, as well as the frequency of their occurrence, as reported by a large target population sample. Another approach is to measure thoughts, feelings, motives, events, or behaviors in real time at regular intervals (e.g. daily if the behavior might occur daily) over a long enough period of time (e.g. months) to establish possible contextual cues or contingencies prior to a BoI (e.g. feelings of rejection or shame immediately precede risk 90% of the time, and the absence of such feelings results in safe choices 100% of the time). Learning more about the contextual factors that precipitate the risk (e.g. what type of rejection or shame from whom) and cues (e.g. what was said, done, sensed) that precipitated this reaction (e.g. precipitated rejection or shame) could be part of identifying contextualized challenge points for inclusion in the game.

Creating a Sampling Frame of Challenges

Once a thorough listing of potential challenge points that could lead to risky behavior is generated, researchers may consider other factors to determine how to sample these potential challenge points. For example, how frequently is a specific challenge encountered by the target audience? How impactful is this challenge (e.g. when confronted with this challenge, a high percentage of target individuals exhibit risk, but do not do so in the absence of the challenge)? How extensive are the research team's resources (e.g. for animation, game design) in developing challenge points within the intervention (i.e. fewer resources may suggest more of a focus on the most frequent and high impactful challenges)? Of course, the more one samples from the population of challenges to which one wishes to generalize, the more likely game experiences and choices will mirror real-life choices.

Coding/Structuring Sequences as in Everyday Life

Challenges and responses are not random in individuals' sequences of actions. Rather, their meaning likely is affected by the events (including one's own and others' behaviors) that precede the challenge. The meaning of events is often structured in causal sequences called *scripts*. Scripts represent a structured event sequence, or knowledge structure, that organizes expectations regarding one's own and others' behaviors in fulfilling roles. For example, in the context of a visit to a restaurant, various roles (e.g. customer, waiter) are fulfilled in different scenes (e.g. enter, order, eat, pay). Entry and exit conditions to the next scene (i.e. branches and alternative tracks in the *story*) may include leaving the restaurant to walk to a movie theater in this context (Schank and Abelson 2008). Scripts are retrieved or activated when a precondition is met (e.g. one is hungry and has money, and there is a restaurant) and there are available props (cues) to activate the start of a restaurant script entry scene (e.g. maître d' asks if you wish a table for two). Scripts that have been activated afford inferences that are not stated (e.g. how one might acquire food), and these scripts help activate roles that are relevant (e.g. waiter).

As we developed our game for reducing risky sexual behaviors, we identified different types of places where our target population most frequently met a prospective partner. For example, one of these was a bar/club. Within a bar or club, we had to identify frequent *pick-up script* bar scenarios. A common structure in a bar *pick-up* scenario (as identified in our formative research) involved the following: looking around to determine potential availability of partners of interest, creating means to *meet* this other, chatting up the other and getting to know the other, escalating intimacy, testing the waters, and sealing the deal. Within a given scene, some situations are possible, while others are not. Some situations are possible across scenes but their instantiation will be different. For example, one social challenge is grappling with potential conflict (e.g. giving someone a brush-off). This could occur during any scene, but the specific behavioral options involved should differ by phase of the interaction. A *brush-off* can involve simply ignoring one's potential partner's gaze early on in a *pick-up* scenario or involve more (e.g. explicitly refusing a partner's advances).

Identifying script structures is needed to understand the entry and exit requirements into different phases of the interaction and how that may constrain the situation, roles, or behavioral options available to participants. Formative research must include understanding typical structures so that once challenges are instantiated within a game, participants will find the game engaging and realistic. Those of us who build games must build these scenes in a game just as a director builds the scenes of a movie or theatrical production. A difference here is that the interactive experience is not linear. At every point where the user selects options, scenes must be prepared for multiple ways in which the action might proceed. Whichever sequence the user takes, the scenes must dovetail naturally to be believable from the user's perspective.

How are the structures underlying social scenarios and sequences determined? In creating SOLVE, we first conducted focus groups and asked participants to rank possible actions (i.e. suggest their frequency). From these we determined (based on levels of agreement) where there were natural segments or scenes in the sequence (e.g. of a bar pick-up script). We also examined which behaviors (e.g. brush-offs) could be enacted in all scenes. Then, these results were subjected to other tests with participants (e.g. interviews) to determine their validity.

Naturally Covarying Factors/Cues in Situations

Interviews were used by our teams to walk users through their own and potential more general scenarios/scripts, asking target audience members as we slowly proceeded to describe what they saw, smelt, felt, or heard as well as what their thoughts, feelings, and motives might be as we *walked through* a telling or rough set of visual cue cards for possible scripts. These enabled us to create a more *fleshed-out* scenario that we could present to others (focus groups, interviewees) to get additional feedback (for additional inclusions we missed or suggested frequent deletions or issues). For example, in some SOLVE scenes, we chose physical features (e.g. for and within bars, club scenes, bed-room scenes, sounds, actors, etc.) that, based on target audiences, would likely co-occur in these settings (e.g. certain alcohol types and brands, colorings, other potential cues/triggers). The goal was to optimize the representation of stimulus–response sequences as these are naturally embedded in larger sequences for both the target (risky) and non-target (non-risky) audiences.

Options Available in the Game

For every possible challenge point, there had to be behavioral options that included options that 90% or more of the target audience would choose at such a choice point. This required formative research at a conceptual level (e.g. give a compliment now, or ask a question about the potential partner's background) as well as at a detailed instantiation level (the specific compliments that would be offered given the situation). With each option we also needed to know (and conduct formative research to assess) how the action would follow from that point. Also, we needed to be mindful of options that would lead to both safer and riskier sequences of choices (again based on additional formative research).

Determining When and How Things Go Differently to Produce Riskier or Safer Choices

We conducted additional formative research to examine what choices were predictive of riskier or safer outcomes in terms of the behavior of interest (i.e. condomless anal sex). For example, we had MSM who varied in how sexually risky they were indicate the extent to which they experienced various obstacles to safer sex and correlated these with choices of interest that might precipitate risk based on the literature (e.g. alcohol use, methamphetamine use) as well as the major behavior of interest (e.g. condomless anal sex).

More Detail Regarding Precipitating Cues

As we developed our storyboards for scenes and started to make them more concrete, we also had additional process evaluation focus groups and interviews to make sure, based on feedback to our developing materials, we were including the sorts of cues, triggers, reactions, and needed challenges that precipitated risk.

Evaluations of the Effectiveness in Creating Representative Designs

By collecting questionnaire responses (now EMA could be used for this) regarding participants' behaviors over the past 90 days, we could correlate that real-life behavior (e.g. condomless anal sex, sexual position preference, alcohol use, methamphetamine use) with the same participants' behavior and similar choices (e.g. condomless anal sex, sexual position preference, alcohol use, methamphetamine use) within the game. Generally, especially when the game was not directly trying to change the behavior of interest, we expected positive correlations between individuals' real-life and virtual choice behaviors. We refer to this correlation as virtual validity. Indeed, across four virtual environments, we have consistently found that participants' past behavior was significantly related to their virtual choices (e.g. pertaining to sexual and drug choices).

Default Control and Experimental Condition Alternatives

Above, we mostly focused on what we would have to do to create the *default* control condition. But, as mentioned earlier, we could simultaneously build in mechanisms to modify the *model default system* in creating an experimental or intervention condition.

Applications in Three Domains of Interest

Earlier we described three domains of interest (preventing disease, mitigation of man-made and natural disaster effects, and terrorism). The SOLVE intervention is an implementation of one of these domains (e.g. reducing premature death by reducing risky sexual behaviors).

Representative design controls or defaults could be developed for a range of target behaviors for many other specific target audiences. In each of these domains, affective/motivational, cognitive, social, and contextual factors are apt to play a significant role in affecting behavior. For example, shame and feeling *less than*, feeling diminished in status by others, and feeling humiliated by others may be key in internalizing risk behaviors (continuing unhealthy practices) or externalizing behavior, as with acting out on others when anger is a key part of the affective reaction. Our team has successfully reduced shame with virtual games and found that doing so significantly reduces real-world risk behavior for MSM (Christensen et al. 2013; Park et al. 2014). Perhaps similar shame or humiliation targeted interventions could enhance self-regulation by violence-prone and/or fearful individuals. Generally, however, emotional self-regulation and communication skills may be important in reducing risky behaviors that adversely affect individuals and society. The potential even exists for effective interventions to be accessed over the web free to the public (as seen in the SOLVE intervention). Given their potential for scaling up, advancing the opportunities for such evidence-based interventions and linking them to experiments might rapidly advance cumulative social science and the well-being of the public.

With this background, Table 16.1 illustrates succinctly how systematic representative designs could be developed, using examples from the new domains of interest we indicated at the outset: preventing disease, mitigating harm in crisis, and reducing school-shooter terrorism.

Conclusions

A systematic effort to develop the theory- and evidence-informed virtual systems we have discussed could transform our basic social science paradigm and, in the process, advance replicability and generalizability. We have used our SOLVE game in neuroscience studies to provide insight into the cognitive and affective dynamics underlying risk taking and in randomized-controlled trials to reduce such sexual risk taking. In this work we have suggested ways to extend the method into other domains in which behavioral changes are sought.

We believe such *model systems* could materially change social science. First, *default control* conditions uniquely afford a way to examine the virtual validities or relationship between real-life behavior and virtual lab behavior, providing a statistical way to ensure the generalizability of the laboratory game environment. Sufficiently complex *default conditions* in games, as in real life, could provide *model systems*, each for a target population for a specific behavior of interest.

Second, because one can use these *default controls* as a starting place for the experimental conditions of choice, one can hold variables constant except for those being manipulated experimentally: this enables the experimentalist to systematically experimentally manipulate one or more factors against a known generalizable environment. That means that the experimental manipulations

Table 16.1 Guidance for developing a systematic representative design.

		Examples		
		Preventing disease	Harm mitigation	Terrorism reduction
Step	Steps content	Weight reduction	Earthquake preparation	School-shooter mitigation
1	Identify specific target audience	Teens (14–18 years old) at or greater than 95th percentile for age and sex in body mass index (BMI)	Homeowners/neighborhoods at risk for earthquake-related fires/being ill prepared following earthquakes	Male high school students 14–19 years in age who exhibit or have exhibited one or more risk patterns involving mental health disorders, rejection/isolation, criminal or violent behavior
		Comparison: teens with normal BMI	*Comparison: Those unprepared vs. those prepared*	*Comparison: students without such patterns*
2	Identify desired behavior change in *real-world individuals or groups*. Conduct baseline/formative research to get ideas on how changes can occur (e.g. by comparing behaviors of at-risk vs. not-at-risk individuals using different risk indicators)	Over a month, reduce caloric consumption by 20% and increase caloric expenditure by 20% Observe via self-reports with smartphones or sensors, and s school clinic measurements Compare baselines for samples (treatment vs. normal)	Increase risk mitigation in homes and neighborhoods: three-day supplies of water and food; skill in turning off gas lines; other measures found valuable by successful neighborhoods	Increase skills for emotional regulation and mitigating humiliation
3	Use formative research to identify common sequences or paths leading to the point where relevant choices occur (e.g. healthier or safer behavior, vs. unhealthier or more risky behavior)	Interview normal vs. obese teens (90% coverage across youth) • What leads up to food/caloric choice and what choices are posed? • What leads up to decision to exercise/move (or not)?	Interview neighbors in target and control neighborhoods about earthquakes (knowledge of how to prepare; obstacles to preparation; motivations to prepare; neighborhood ties; build ties with neighbors)	Conduct studies with interviews and other means to identify (i) when school bullying occurs, (ii) what leads up to a perceived sense of shame/humiliation, (iii) what obstacles to self-control or triggers for problematic reactions (aggressive behaviors) exist, and (iv) ways that normal children interact with peers that produce better results

(Continued)

Table 16.1 (Continued)

		Examples		
		Preventing disease *Weight reduction*	**Harm mitigation** *Earthquake preparation*	**Terrorism reduction** *School-shooter mitigation*
Step	**Steps content**			
4	Represent such paths in virtual environments as *default controls* (i.e. representative paths and choices for the people to be influenced). Use intelligent agents if social interaction is key. Give agents ability to spark representative dialogue or nonverbal interactions with user	Example: Set up default supermarket point-of-sale virtual environment where user has shopping cart can fill with products from aisles and ensure choices similar to those of target audience	Example: Game on decisions regarding how to prepare for an earthquake in one's home (e.g. gas shut-off process; water heater strapping; water storage, etc.) and how to build neighborhood cooperation for an emergency	Example: Set up default school settings, within which relational/physical bullying may occur, e.g. school bullies, name callings from girls, physical bullies in the classroom
5	Assess degree to which virtual-world behavior (and adaptation) are mirrored in real-world behavior	Measure real-life food/caloric choices and correlate these with virtual food/caloric choices	Measure real-life water/food/other supply levels and preparation for gas-risk mitigation Correlate users' real-world with virtual-world choices in game	Measure real-life school experiences, and correlate those with virtual choices

6	Use formative research to identify potential levers for altering behavior (e.g. levers related to shame, loyalty, or awareness of consequences)	Conduct formative studies suggesting that after risky/safer virtual choice what best intervention (e.g. messages; emotional regulation/shame reduction) choice is that leads to subsequent safer choices over a year	Conduct formative studies to identify effective neighborhood strategies and messages as judged by one-year results	Conduct formative studies regarding what are the best strategies, messages, or emotion regulation choices that can lead to less problematic coping in response to bullying triggers
7	Incorporate interventions with hooks into default control to create virtual experimental comparison Incorporate optional interventions into virtual world, thereby allowing systematic experimentation	Incorporate shame reduction and other messages into experimental game	Incorporate steps for effective individual and neighborhood preparation efforts in game for similar neighborhoods	Incorporate interventions to enhance emotion regulation and social interaction with peers into the experimental game
8	Within virtual-world representative of target's challenges, conduct randomized control trials (RCTs)	Conduct RCT with high school students who meet target audience criteria in a six-month trial with DV: weight loss and caloric intake in virtual and real life; revise. Follow with a new larger RCT	Conduct RCT with similar homeowners using game; revise based on results; and conduct larger RCT with new similar homeowner audience	Conduct RCT with at-risk high school males using game; revise based on results; and conduct larger RCT with similar at-risk males

are probably more likely therefore to be more context specific within sequence: if the experimental condition yields a desired change in behavior, the researcher knows when, how, and what to change to affect the behavior. If the researcher can also use sensor technologies to detect when users are in similar contexts in real life, there is the additional possibility of just-in-time interventions in real time at just the right moment to change just the right behavior for just the right target audience. Also, from a computational perspective, data scientists could use that data to know how changes in the system might likely causally change behavior when, for whom, and why.

Third, this approach bridges basic and applied work because one can rapidly go from a default control to an experimental group and conduct initial basic tests of theory and then possible intervention effectiveness. Then, one can rapidly scale up an intervention, if promising, to be conducted over the web. Fourth, this approach allows us to operate across many levels of scale. For example, the basic and experimental conditions and findings from applied work could also generate hypotheses that could be tested at lower levels of scale (e.g. via neuroscience studies) in the lab, while users are behaviorally responding in the virtual game in the scanner. Fifth, experimental (e.g. social psychological) and correlational approaches (e.g. more common in personality psychology) are brought together in that researchers from both traditions can examine the relationships within and between virtual and real-life environments, assessing in finer detail the nuances of situations and contexts that in combination with individual differences affect and change behavior.

All of these possibilities afford a potential revolution in computational social science. First, the design affords the potential for complex environments where few and many environment manipulations can afford causal inference. Second, at the same time researchers and data scientists can know the extent to which the behavior *in context* for a given target population is apt to be generalizable to real life. That should provide data scientists with a much better idea of how to use such evidence-based research to computationally model behavior while also providing a quantitative index for the level of confidence they should have regarding its generalizability.

References

Adams, Z.W., McClure, E.A., Gray, K.M. et al. (2017). Mobile devices for the remote acquisition of physiological and behavioral biomarkers in psychiatric clinical research. *Journal of Psychiatric Research* 85: 1–14. https://doi.org/10.1016/j.jpsychires.2016.10.019.

American Red Cross (2017). Earthquake safety. Retrieved 17 July 2017, from http://www.redcross.org/get-help/how-to-prepare-for-emergencies/types-of-emergencies/earthquake.

Baranowski, T., Buday, R., Thompson, D.I., and Baranowski, J. (2008). Playing for real: video games and stories for health-related behavior change. *American Journal of Preventive Medicine* 34 (1): 74–82.e10. https://doi.org/10.1016/j .amepre.2007.09.027.

Baranowski, T., Baranowski, J., Thompson, D., and Buday, R. (2011). Behavioral science in video games for children's diet and physical activity change: key research needs. *Journal of Diabetes Science and Technology* 5 (2): 229–233. https://doi.org/10.1177/193229681100500204.

Becker, J.S., Paton, D., Johnston, D.M., and Ronan, K.R. (2012). A model of household preparedness for earthquakes: how individuals make meaning of earthquake information and how this influences preparedness. *Natural Hazards* 64 (1): 107–137. https://doi.org/10.1007/s11069-012-0238-x.

Berking, M. and Wupperman, P. (2012). Emotion regulation and mental health: recent findings, current challenges, and future directions. *Current Opinion in Psychiatry* 25 (2): 128–134. https://doi.org/10.1097/YCO.0b013e3283503669.

Brunswik, E. (1955). Representative design and probabilistic theory in a functional psychology. *Psychological Review* 62: 193–217.

Bubolz, B.F. and Simi, P. (2015). Leaving the world of hate: life-course transitions and self-change. *American Behavioral Scientist* 59 (12): 1588–1608. https://doi .org/10.1177/0002764215588814.

Carreiro, S., Smelson, D., Ranney, M. et al. (2015). Real-time mobile detection of drug use with wearable biosensors: a pilot study. *Journal of Medical Toxicology* 11 (1): 73–79. https://doi.org/10.1007/s13181-014-0439-7.

Carreiro, S., Chai, P.R., Carey, J. et al. (2017). Integrating personalized technology in toxicology: sensors, smart glass, and social media applications in toxicology research. *Journal of Medical Toxicology* 13 (2): 166–172. https://doi.org/10 .1007/s13181-017-0611-y.

Centers for Disease Control and Prevention (2017). National Health Report. Retrieved 16 February 2018, from https://www.cdc.gov/healthreport/.

Chien, S.I. and Korikanthimath, V.V. (2007). Analysis and modeling of simultaneous and staged emergency evacuations. *Journal of Transportation Engineering* 133 (3): 190–197. https://doi.org/10.1061/(asce)0733-947X(2007)133:3(190).

Christensen, J.L., Miller, L.C., Appleby, P.R. et al. (2013). Reducing shame in a game that predicts HIV risk reduction for young adult men who have sex with men: a randomized trial delivered nationally over the web. *Journal of the International AIDS Society* 16: 18716. https://doi.org/10.7448/ias.16.3.18716.

Committee on Population, Division of Behavioral and Social Sciences and Education, Board on Health Care Services, National Research Council, & Institute of Medicine (2015). *Measuring the Risks and Causes of Premature Death: Summary of Workshops*. Washington, DC: National Academies Press (US). Retrieved from http://www.ncbi.nlm.nih.gov/books/NBK279971/.

Cook, T.D. and Campbell, D.T. (1979). *Quasi-Experimentation: Design & Analysis Issues for Field Settings*. Boston, MA: Houghton Mifflin.

CNN, Yan, H., Stapleton, A., and Murphy, P.P. (2018). Kentucky school shooting: 2 students killed, 18 injured. Retrieved 29 September 2018, from https://www.cnn.com/2018/01/23/us/kentucky-high-school-shooting/index.html.

Dhami, M.K., Hertwig, R., and Hoffrage, U. (2004). The role of representative design in an ecological approach to cognition. *Psychological Bulletin* 130 (6): 959.

Di Loreto, I., Mora, S., and Divitini, M. (2012). Collaborative serious games for crisis management: an overview. In: *2012 IEEE 21st International Workshop on Enabling Technologies: Infrastructure for Collaborative Enterprises*, 352–357. Toulouse, Cedex 04, France: IEEE https://doi.org/10.1109/WETICE.2012.25.

Drabek, T.E. (1969). Social processes in disaster: family evacuation. *Social Problems* 16 (3): 336–349. https://doi.org/10.2307/799667.

Drabek, T.E. (2013). *The Human Side of Disaster*, 2e. Hoboken, NJ: CRC Press. Retrieved from http://www.crcnetbase.com/isbn/9781466506862.

Dutra, L., Stathopoulou, G., Basden, S.L. et al. (2008). A meta-analytic review of psychosocial interventions for substance use disorders. *American Journal of Psychiatry* 165 (2): 179–187. https://doi.org/10.1176/appi.ajp.2007.06111851.

Eisenberg, N. (2000). Emotion, regulation, and moral development. *Annual Review of Psychology* 51 (1): 665–697.

Feng, D., Jeong, D.C., Krämer, N.C. et al. (2017). "Is it just me?": evaluating attribution of negative feedback as a function of virtual instructor's gender and proxemics. In: *Proceedings of the 16th Conference on Autonomous Agents and MultiAgent Systems*, 810–818. Richland, SC: International Foundation for Autonomous Agents and Multiagent Systems. Retrieved from http://dl.acm.org/citation.cfm?id=3091125.3091240.

Fiedler, K. (2017). What constitutes strong psychological science? The (neglected) role of diagnosticity and a priori theorizing. *Perspectives on Psychological Science* 12 (1): 46–61. https://doi.org/10.1177/1745691616654458.

Gerard, F.J., Whitfield, K.C., Porter, L.E., and Browne, K.D. (2016). Offender and offence characteristics of school shooting incidents. *Journal of Investigative Psychology and Offender Profiling* 13 (1): 22–38.

Gill, P. (2015). *Lone-Actor Terrorists: A Behavioural Analysis*. London, New York: Routledge.

Gill, P. and Corner, E. (2017). There and back again: the study of mental disorder and terrorist involvement. *American Psychologist* 72 (3): 231–241. https://doi.org/10.1037/amp0000090.

Gill, P., Horgan, J., and Deckert, P. (2014). Bombing alone: tracing the motivations and antecedent behaviors of lone-actor terrorists. *Journal of Forensic Sciences* 59 (2): 425–435. https://doi.org/10.1111/1556-4029.12312.

Gross, J.J. (1998). The emerging field of emotion regulation: an integrative review. *Review of General Psychology* 2 (3): 271.

Guarino, S., Eusebi, L., Bracken, B., and Jenkins, M. (2019, this volume). Using sociocultural data from online gaming and game communities. In: *Social-Behavioral Modeling for Complex Systems* (ed. P.K. Davis, A. O'Mahony and J. Pfautz). Hoboken, NJ: Wiley.

Hammond, K.R. and Stewart, T.R. (2001). *The Essential Brunswik: Beginnings, Explications, Applications*. Oxford, New York: Oxford University Press.

Hatzenbuehler, M.L., Phelan, J.C., and Link, B.G. (2012). Stigma as a fundamental cause of population health inequalities. *American Journal of Public Health* 103 (5): 813–821. https://doi.org/10.2105/ajph.2012.301069.

Head, K.J., Noar, S.M., Iannarino, N.T., and Grant Harrington, N. (2013). Efficacy of text messaging-based interventions for health promotion: a meta-analysis. *Social Science & Medicine* 97: 41–48. https://doi.org/10.1016/j.socscimed.2013.08.003.

Helton, W.S., Kemp, S., and Walton, D. (2013). Individual differences in movements in response to natural disasters: Tsunami and earthquake case studies. *Proceedings of the Human Factors and Ergonomics Society Annual Meeting* 57 (1): 858–862. https://doi.org/10.1177/1541931213571186.

Horgan, J. (2008). From profiles to pathways and roots to routes: perspectives from psychology on radicalization into terrorism. *The ANNALS of the American Academy of Political and Social Science* 618 (1): 80–94. https://doi.org/10.1177/0002716208317539.

Institute for Economics & Peace (2016). *Global Terrorism Index: Measuring and Understanding the Impact of Terrorism*. Institute for Economics & Peace.

Ioannou, M., Hammond, L., and Simpson, O. (2015). A model for differentiating school shooters characteristics. *Journal of Criminal Psychology* 5 (3): 188–200. https://doi.org/10.1108/JCP-06-2015-0018.

Jeong, D.C., Feng, D., Krämer, N.C. et al. (2017). Negative feedback in your face: examining the effects of proxemics and gender on learning. In: *Intelligent Virtual Agents* (ed. J. Beskow, C. Peters, G. Castellano, et al.), 170–183. Springer International Publishing.

Jones, C.M. (2014). Why persistent offenders cannot be shamed into behaving. *Journal of Offender Rehabilitation* 53 (3): 153–170. https://doi.org/10.1080/10509674.2014.887604.

Juvonen, J. and Graham, S. (2001). *Peer Harassment in School: The Plight of the Vulnerable and Victimized*. Guilford Press.

Juvonen, J., Graham, S., and Schuster, M.A. (2003). Bullying among young adolescents: the strong, the weak, and the troubled. *Pediatrics* 112 (6): 1231–1237. https://doi.org/10.1542/peds.112.6.1231.

Kaltiala-Heino, R., Rimpelä, M., Rantanen, P., and Rimpelä, A. (2000). Bullying at school—an indicator of adolescents at risk for mental disorders. *Journal of Adolescence* 23 (6): 661–674. https://doi.org/10.1006/jado.2000.0351.

Kokkinos, C.M. and Panayiotou, G. (2004). Predicting bullying and victimization among early adolescents: associations with disruptive behavior disorders. *Aggressive Behavior* 30 (6): 520–533. https://doi.org/10.1002/ab.20055.

Kumpulainen, K. and Räsänen, E. (2000). Children involved in bullying at elementary school age: their psychiatric symptoms and deviance in adolescence. *Child Abuse & Neglect* 24 (12): 1567–1577. https://doi.org/10.1016/S0145-2134(00)00210-6.

Lakkaraju, K., Epifanovskaya, L., States, M. et al. (2019, this volume). Online games for studying behavior. In: *Social-Behavioral Modeling for Complex Systems* (ed. P.K. Davis, A. O'Mahony and J. Pfautz). Hoboken, NJ: Wiley.

Lansford, J.E., Miller-Johnson, S., Berlin, L.J. et al. (2007). Early physical abuse and later violent delinquency: a prospective longitudinal study. *Child Maltreatment* 12 (3): 233–245. https://doi.org/10.1177/1077559507301841.

Lara, J., Evans, E.H., O'Brien, N. et al. (2014). Association of behaviour change techniques with effectiveness of dietary interventions among adults of retirement age: a systematic review and meta-analysis of randomised controlled trials. *BMC Medicine* 12 (1): 177. https://doi.org/10.1186/s12916-014-0177-3.

Lazarus, R.S. (1991). *Emotion and Adaptation*. Oxford University Press.

Lussier, J.P., Heil, S.H., Mongeon, J.A. et al. (2006). A meta-analysis of voucher-based reinforcement therapy for substance use disorders. *Addiction* 101 (2): 192–203. https://doi.org/10.1111/j.1360-0443.2006.01311.x.

Marsella, S.C., Pynadath, D.V., and Read, S.J. (2004). PsychSim: Agent-based modeling of social interactions and influence. *Proceedings of the International Conference on Cognitive Modeling* 36: 243–248.

Marusak, H.A., Martin, K.R., Etkin, A., and Thomason, M.E. (2014). Childhood trauma exposure disrupts the automatic regulation of emotional processing. *Neuropsychopharmacology* 40 (5): 1250–1258. https://doi.org/10.1038/npp.2014.311.

McCauley, C.R. and Moskalenko, S. (2011). *Friction: How Radicalization Happens to Them and Us*. Oxford, New York: Oxford University Press.

McCauley, C. and Moskalenko, S. (2017). Understanding political radicalization: the two-pyramids model. *American Psychologist* 72 (3): 205–216. https://doi.org/10.1037/amp0000062.

Miller, L.C., Marsella, S., Dey, T. et al. (2011). Socially optimized learning in virtual Environments (SOLVE). In: *Interactive Storytelling*, Springer Lecture Notes in Computer Science (LNCS) (ed. M. Si and D. Thue), 182–192. Heidelberg, Berlin: Springer-Verlag.

Miller, L.C., Godoy, C.G., Christensen, J.L. et al. (2016). A Virtual Integrative Science of our Interactive Ongoing Nature (VISION): Advancing Psychological Science in a Technological Age. Los Angeles, CA.

Moghaddam, F.M. (2005). The staircase to terrorism: a psychological exploration. *American Psychologist* 60 (2): 161–169. https://doi.org/10.1037/0003-066X.60.2.161.

Mook, D.G. (1992). The myth of external validity. In: *Everyday Cognition in Adulthood and Late Life*, (1 paperback ed) (ed. L.W. Poon, D.C. Rubin and B.A. Wilson), 25–43. Cambridge: Cambridge University Press.

Nansel, T.R., Overpeck, M., Pilla, R.S. et al. (2001). Bullying behaviors among US youth: prevalence and association with psychosocial adjustment. *JAMA* 285 (16): 2094–2100.

Nansel, T.R., Craig, W., Overpeck, M.D. et al. (2004). Cross-national consistency in the relationship between bullying behaviors and psychosocial adjustment. *Archives of Pediatrics & Adolescent Medicine* 158 (8): 730–736.

Nowrasteh, A. (2016). *Terrorism and Immigration: A Risk Analysis*. Cato Institute. Retrieved from https://search.proquest.com/docview/1846441243.

Park, M., Anderson, J.N., Christensen, J.L. et al. (2014). Young men's shame about their desire for other men predicts risky sex and moderates the knowledge – self-efficacy link. *Frontiers in Public Health* 2: https://doi.org/10.3389/fpubh.2014.00183.

Qualls, N., Levitt, A., Kanade, N. et al. (2017). Community Mitigation Guidelines to Prevent Pandemic Influenza-United States, 2017. MMWR Recommendations Report No. 66(No. RR-1), 1–34. Centers for Disease Control and Prevention. Retrieved from http://dx.doi.org/10.15585/mmwr.rr6601a1.

Read, S.J., Miller, L.C., Appleby, P.R. et al. (2006). Socially optimized learning in a virtual environment: reducing risky sexual behavior among men who have sex with men. *Human Communication Research* 32: 1–34. https://doi.org/10.1111/j.1468-2958.2006.00001.x.

de Roy van Zuijdewijn, J. and Bakker, E. (2016). Analysing personal characteristics of lone-actor terrorists: research findings and recommendations. *Perspectives on Terrorism* 10 (2). Retrieved from http://www.terrorismanalysts.com/pt/index.php/pot/article/view/500.

Schalkwijk, F.W. (2015). *The Conscience and Self-Conscious Emotions in Adolescence: An Integrative Approach*. Hove, East Sussex, New York: Routledge.

Schank, R.C. and Abelson, R.P. (2008). *Scripts, Plans, Goals and Understanding: An Inquiry into Human Knowledge Structures (Repr)*. New York, NY: Psychology Press.

Shadish, W.R., Cook, T.D., and Campbell, D.T. (2001). *Experimental and Quasi-Experimental Designs for Generalized Causal Inference*. Boston, MA: Houghton Mifflin.

Shiffman, S., Stone, A.A., and Hufford, M.R. (2008). Ecological momentary assessment. *Annual Review of Clinical Psychology* 4 (1): 1–32. https://doi.org/10.1146/annurev.clinpsy.3.022806.091415.

Spruit, A., Schalkwijk, F., van Vugt, E., and Stams, G.J. (2016). The relation between self-conscious emotions and delinquency: a meta-analysis. *Aggression and Violent Behavior* 28: 12–20. https://doi.org/10.1016/j.avb.2016.03.009.

Stein, J.A., Dukes, R.L., and Warren, J.I. (2006). Adolescent male bullies, victims, and bully-victims: a comparison of psychosocial and behavioral characteristics.

Journal of Pediatric Psychology 32 (3): 273–282. https://doi.org/10.1093/jpepsy/jsl023.

Stone, A.A. and Shiffman, S. (1994). Ecological momentary assessment (EMA) in behavioral medicine. *Annals of Behavioral Medicine* 16 (3): 199–202.

Taylor, M. and Horgan, J. (2006). A conceptual framework for addressing psychological process in the development of the terrorist. *Terrorism and Political Violence* 18 (4): 585–601. https://doi.org/10.1080/09546550600897413.

Thompson, R.A. (1994). Emotion regulation: a theme in search of definition. *Monographs of the Society for Research in Child Development* 59 (2–3): 25–52.

Tracy, J.L. and Robins, R.W. (2004). Putting the self into self-conscious emotions: a theoretical model. *Psychological Inquiry* 15 (2): 103–125.

Tracy, J.L. and Robins, R.W. (2006). Appraisal antecedents of shame and guilt: support for a theoretical model. *Personality and Social Psychology Bulletin* 32 (10): 1339.

Trochim, W. and Donnelly, J.P. (2006). *The Research Methods Knowledge Base.* Cengage Learning.

Unnever, J.D. (2005). Bullies, aggressive victims, and victims: Are they distinct groups? *Aggressive Behavior* 31 (2): 153–171.

Veenstra, G., Luginaah, I., Wakefield, S. et al. (2005). Who you know, where you live: social capital, neighbourhood and health. *Social Science & Medicine* 60 (12): 2799–2818. https://doi.org/10.1016/j.socscimed.2004.11.013.

Verlinden, S., Hersen, M., and Thomas, J. (2000). Risk factors in school shootings. *Clinical Psychology Review* 20 (1): 3–56. https://doi.org/10.1016/s0272-7358(99)00055-0.

Walters, L.M. and Hornig, S. (1993). Profile: faces in the news: network television news coverage of Hurricane Hugo and the Loma Prieta earthquake. *Journal of Broadcasting & Electronic Media* 37 (2): 219–232. https://doi.org/10.1080/08838159309364217.

Wike, T.L. and Fraser, M.W. (2009). School shootings: making sense of the senseless. *Aggression and Violent Behavior* 14 (3): 162–169. https://doi.org/10.1016/j.avb.2009.01.005.

Young, S., Taylor, E., and Gudjonsson, G. (2016). Childhood predictors of criminal offending: results from a 19-year longitudinal epidemiological study of boys. *Journal of Attention Disorders* 20 (3): 206–213. https://doi.org/10.1177/1087054712461934.

17

Online Games for Studying Human Behavior

*Kiran Lakkaraju[1], Laura Epifanovskaya[2], Mallory Stites[1], Josh Letchford[2],
Jason Reinhardt[2], and Jon Whetzel[1]*

[1] *Sandia National Laboratories, Albuquerque, NM 87185, USA*
[2] *Sandia National Laboratories, California, Livermore, CA 94551, USA*

Introduction

Games and simulations have been used as experimental platforms for centuries, particularly in the form of war games used for planning purposes and for reen-actment of historical battles (Sabin 2014). In these instances, they are used as a type of virtual laboratory to test hypotheses about what will happen (or in the case of historical battles, what might have happened) when a given deci-sion is made under a certain set of conditions. Such games can be highly sim-plistic or exquisitely detailed; they can be board, card, or Internet games; or they can be highly choreographed simulations, sometimes involving multiple (in many cases high-level) personnel from military and government. While the use of games as a didactic and exploratory tool is widespread, they are used less frequently to gather data on player actions for subsequent statisti-cal meta-analysis. Some reasons for this are the small size (in statistical terms) of the participating groups, the differences in the games and simulations them-selves from game to game, and the difficulty in acquiring and storing analyzable data in real time.

Online games offer the potential to address some of these difficulties. Online games are designed and developed to be hosted and accessed via the Internet, allowing a large and diverse player pool to participate under sets of conditions that are both tunable and recordable by researchers. One particular category of online games, massively multiplayer online games (MMOGs), is especially intriguing as they can provide data on a large number of players interacting in a shared, persistent world over an extended period of time.

Social-Behavioral Modeling for Complex Systems, First Edition.
Edited by Paul K. Davis, Angela O'Mahony, and Jonathan Pfautz.
© 2019 John Wiley & Sons, Inc. Published 2019 by John Wiley & Sons, Inc.
Companion website: www.wiley.com/go/Davis_Social-Behavioralmodeling

In this chapter we discuss the value of online games,[1] including MMOGs, as experimental platforms that might augment or even displace some other data-gathering methodologies in specific areas of research and inquiry. Experimental techniques are a powerful means of identifying causal relationships, which are critical for designing interventions in a system.

We argue that for studying national security issues, where data is sparse, it is difficult to experiment: behaviors can be complex and varied, and online games can serve as a unique and powerful tool to experimentally understand causal relationships.

We begin with a discussion of the potential of online games and the specific applicability to help address national security issues. We outline a proof of concept analysis that we performed using data collected from an existing MMOG, which herein will be referred to as Game X to preserve the anonymity of the game, in order to compare conflict phenomena within the game to data from the real world and a scientific analysis of the real-world data selected from the academic international relations literature.

We end with a discussion of considerations to have when using games as experiments. Particular issues arise when developing a game as a vehicle for studying human behavior. The fundamental problem is the need to engage subjects in the game for extended time periods (on the order of hours to years). In order to naturally encourage this (as opposed to paying subjects), games must be designed in particular ways. These design choices have ramifications on the subject behavior and analysis of data from a game. We outline some research considerations on this topic.

Online Games and Massively Multiplayer Online Games for Research

Online games can span a wide range of characteristics (Laamarti et al. 2014; De Lope and Medina-Medina 2017). There are many types of online games, ranging from simple single player games such as Fruit Ninja, to more social games such as Words with Friends or Farmville, to tactical heavy action games such as Call of Duty, to highly complex and social role-playing games such as World of Warcraft and Eve Online. The latter games are often referred to as massively multiplayer role playing games (MMORPGs); however, we will use the term MMOG to refer to all games (whether role playing or not) in which players interact for extended periods of time in a shared, persistent world.

MMOGs are online games that attract players from around the world of all ages, genders, and educational backgrounds to a shared virtual world (Yee 2006). The diversity and size of the player base, which for some games

---

1 See also Guarino et al. (2019) in this volume on online games.

is numbered in millions, is an especially attractive advantage of this type of game as a data-gathering platform. Data on millions of actions performed by a large and diverse sample of people lends itself well to statistical analysis, better than surveys and laboratory experiments with much smaller sample sizes taken from a more homogeneous group (e.g. college students, who frequently participate in academic human research studies (Gosling et al. 2010; Henrich et al. 2010). They also offer the opportunity to see how different types of players respond under different circumstances, to interrogate differences among players, and also, in war gaming, to uncover novel strategies that would not have occurred to personnel typically involved in these games.

There are many types of MMOGs, ranging from simple browser-based games such as Farmville to highly complex and realistic role-playing games such as World of Warcraft and Eve Online. The latter may be of particular interest as an experimental platform, especially in the social sciences, because of their realism, complexity, and degree of player involvement. The use of games as experimental platforms for scientific and other researches will, of course, be criticized as providing data that is only meaningful within the game context and that cannot be used to draw conclusions about the real world (Williams 2010). The legitimacy of game data for research purposes is a real concern, and one must considered both in the experimental design phase and in analyzing and appropriately caveating the results of MMOG experiments. However, research has shown that player behavior in complex and realistic role-playing games may be representative of behavior in the real world because of the investment in time, effort, and reputation made by participants in their player avatar (Castronova 2008; Lu et al. 2014). In addition, this type of MMOG is often based in a virtual world composed of highly complex economies and social structures, some of which evolve organically and not as a result of a rule set governing the game platform. Because of the complexity and realism of this type of game, in-game behaviors may mimic real-world economic and social behaviors.

Additionally, it is worth noting that many existing methods of data gathering have their own associated problems. In social science research, for example, surveys and questionnaires are used that may collect incomplete and/or biased information; questions may be misunderstood by respondents or may be formulated in such a way as to bias the respondents' answers. As with all scientific data, the analysis of and conclusions drawn from MMOG data must be understood appropriately, and the inherent limits of such games acknowledged. Findings from a statistical analysis of game data, for example, are easily validated using results from future game play; however, this only demonstrates that the findings are valid within the game context. Conclusions drawn from player actions in an MMOG will be much more difficult to validate in the real world (Williams 2010). Game data can still be used for statistical analysis and compared with results in the real world; the work that we will describe below is an example of this. As in the real world, statistical results from game data must also be understood not to provide "the answer" to a given research question, but rather a range of likely answers, a distribution, a mean, and outlying data points which in and of themselves may be interesting to researchers.

Where Is the Benefit?

Online games can provide exceptional value to research efforts that meet any of the following three criteria: data is sparse and difficult to obtain, it is difficult or impossible to build a real-world experimental laboratory with built-in controls to perform the research, and the research is interested in not only finding averages and distributions of player behavior but also exploring a wide range of possible behaviors and examining results that don't fall neatly within a given distribution (i.e. "distillation games"; Perla et al. 2005). Of course, these types of unexpected results are always the most interesting to science; online games provide a method of capturing and quantifying conditions that creates these results in a way typically unavailable in some realms of scientific inquiry.

Areas where these conditions apply and online games might be leveraged include the social and behavioral sciences, war studies and planning, and international relations research. The last, international relations, meets the criteria because of the sparsity of data usable for quantitative analysis. As will be discussed in detail below, academics often use the military and interstate disputes (MIDs) data collected by the Correlates of War project to mathematically evaluate the effects of various military, economic, and other parameters on the likelihood of conflict between sovereign states. The MIDs data span instances of conflict occurring over less than a century; they do not capture all variables of potential importance to the analysis of conflict; and, crucially, it is not possible to gather significantly more data over the few years funded by a typical academic research grant. As such, this last area is one in which MMOGs can potentially contribute tremendous value as experimental platforms that can generate large amounts of data in condensed timeframes from games engineered to answer specific research questions. Our work described below shows a first attempt to take data from an MMOG, *operationalize* it so that it is captured in a way approximately equivalent to the MIDs data and the economic variables from an international relations journal article that we chose as a comparison study, and perform the same statistical analysis on the game data as was used in the journal article in order to compare the results.

War Games and Data Gathering for Nuclear Deterrence Policy

War studies and planning meet all of the criteria used to evaluate online games as research tools that were enumerated in the Introduction: data from wars that can be used in statistical meta-analysis is sparse and difficult to obtain; it is difficult or impossible to create a real-world experimental laboratory with built-in controls to perform the research; events that create new data naturally are undesirable; and the research is interested in exploring a wide range of possible behaviors and examining results that fall outside of the normal distribution. Games and simulations have been used for more than a century to study important historical battles and to envisage future ones. Because they are referred to as *games*, they are often disparaged as serious tools of scholarship, although there have been cases where the game contains more historically accurate detail about a given battle than narrative works on the subject (Sabin 2014). However, war gaming and simulations are taken quite seriously by military planners and policy-makers. During the Cold War period, Nobel Prize-winning economist Thomas Schelling, working at the RAND Corporation, was instrumental in designing and executing war games for military planning purposes. Schelling believed that war gaming was essential in filling a gap in war planning: analysts don't know what they don't know, and they won't think of every possible future battle or crisis contingency. War games allow scenarios to unfold without a human mastermind planning and predicting every step, which means that they can sometimes enter new and unusual territory. Reid Pauly summarized this point nicely in what is perhaps the only meta-analysis of United States politico-military war games published to date, in recounting events surrounding the Cuban Missile Crisis of 1962. As Pauly tells it: "During the Cuban Missile Crisis, a participant in the office of John McNaughton remarked, 'This crisis sure demonstrates how realistic Schelling's [war]games are.' Another responded, 'No, Schelling's games demonstrate how unrealistic this Cuban crisis is.'" In the same meta-analysis, Pauly examines historical war games in order to probe the attitudes of *strategic elites* – that is, military and policy professionals with experience and education relevant to combat and nuclear weapons – toward willingness to use a nuclear weapon in combat. This work followed a publication by Scott Sagan and Benjamin Valentino citing the surprising willingness of the American public at large to use a nuclear weapon against an adversary state, in violation of a hypothesized *nuclear taboo* (Press et al. 2013).

One enormous potential benefit of using an MMOG to answer a similar question about the willingness or unwillingness of particular groups of people to deploy a nuclear weapon (or other weapon of war) in combat is the large, diverse group of people that participate in this type of online game. If some basic data are gathered on the player participants (such as occupation and level of educational attainment), then data from game play can be post-processed to evaluate the decisions of various groups and compare them against each other. In fact, it would be fascinating to compare such an analysis of MMOG data to Sagan and Valentino's finding that the public is not overwhelmingly averse to using nuclear weapons in combat and Pauly's finding that strategic elites are averse to their use. Would we find the same difference in attitudes in player behavior in the game setting?

Attitudes toward nuclear use may be a factor in effective – or ineffective – nuclear deterrence. The United States and its allies rely in part on deterrence, in a communicated willingness to retaliate in kind, to protect them from nuclear aggression by nuclear armed adversaries. As Schelling remarked in his landmark 1966 book Arms and Influence: "The power to hurt is bargaining power. To exploit it is diplomacy – vicious diplomacy, but diplomacy." In order for deterrence to work, that is, in order for it to influence an adversary to avoid the proscribed action, the threat of nuclear retaliation must be deemed to be credible. The Cold War era logic of the threat of mutual annihilation as the foundation of deterrence is summarized neatly by the acronym MAD, or mutual assured destruction, a phrase coined in Cold War strategist Herman Kahn's Hudson Institute (Deudney 1983). The deterrent power of MAD, in part, drove the United States and the Soviet Union to invest hugely in their respective nuclear arsenals in order to maintain nuclear parity with each other and preserve global strategic stability.

Today's political and technological environments are markedly different than those of the Cold War. The Soviet Union collapsed and splintered in the 1990s, leaving only a nuclear-armed Russia. China, India, and Pakistan have acquired nuclear arsenals. In addition to the change in the nuclear landscape, however, the world has also seen changes in political and economic relationships and the rise of new technologies that may affect strategic stability in new ways. Whereas fear of nuclear destruction may have driven strategic military planning during the Cold War, for example, fear of economic devastation may serve as a strong deterrent in today's economically interconnected environment.

Given the (fortunate) paucity of data on nuclear use in conflict, war games are a valuable means of generating and analyzing scenarios involving nuclear exchange. Nuclear standoffs have been evaluated using war games for decades; the additional value that the online game brings to the study of these scenarios is the ability to record massive amounts of data from hundreds of players, for later analysis over multiple instances of game play. A chat feature will allow researchers to gather data on player's thoughts and motives surrounding in-game decisions. In addition, the game allows capture data not only on nuclear use but also on the conditions of the game environment at the time of use. These metadata can later be evaluated to determine if there are other heretofore unexamined factors that significantly influence nuclear use (or the likelihood of simple conventional conflict, for that matter).

In the section that follows, we will discuss how MMOG data can inform efforts in the academic areas of international relations and policy studies to understand the influence of multiple different factors on the likelihood of conflict between states. These factors include military, economic, and political elements, and their effect on conflict is analyzed using the MIDs variables first discussed in the Introduction. The data collected using an MMOG may be combined with the real-world MIDs data to better study the effects of these variables. The MMOG allows the collection of significant amounts of additional data in years to come, much more (hopefully) than the data that will be generated by real-world wars. In addition, the game can be designed to gather data on variables that may be relevant to conflict but which are not currently in the MIDs canon. The utility of the MMOG to further academic debate on the subject of economic interdependence and conflict will be examined in detail in the following section.

MMOG Data to Test International Relations Theory

The relationship between economic interdependencies among states and likelihood of conflict between them has been explored using data in the academic fields of political science and international relations, with divergent results. For decades, academics have performed regression analysis on conflict data from the Correlates of War project to establish linkages between trade volumes and conflict likelihood. The results of these analyses range from evidence that trade increases instances of conflict between states (the view of the *realist* school of international relations theory) to findings that trade decreases conflict likelihood (the view of the *liberal* school) to work testing the assumption that the impact of trade relations on conflict is more complex than the realist/liberal dichotomy would suggest. To date, this dispute in the academic literature has not been resolved.

The research question of whether economic interdependence positively or negatively affects conflict is one that lends itself to study using an MMOG since additional data is desirable for the type of statistical regression analysis typically performed in the literature but is hard to come by in the real world. Data from game play of an online game such as the one described above may be useful in further elucidating the nature of the relationship between economic interdependencies and their effect on the willingness of nations to go to war with each other. As a proof of concept, however, we took data from an extant online serious game with a steady user base, data which members of our project team had collected and used earlier in unrelated research. Game X contains the elements necessary for comparison to real-world research on economic interdependence and conflict: there are guilds within the game (comparable with nation-states in the real world) that trade and wage war with each other, and there are other quantifiable factors within the game that correlate with those captured by the MIDs variables. These parameters include contiguity (whether states are located next to each other geographically; a strong predictor of conflict in the real world), alliance, and capability ratio. The capability ratio variable in the MIDs data set is a relative measure of the Composite

Table 17.1 Mapping between MIDs variables and Game X variables.

MIDs variable	Game X variable
Dyadic trade	Trade between two guilds
Contiguity	Share a border
Alliance	Is Foe $\neq$ True (Is Foe is a designation voluntarily selected by a guild to describe another guild)
Capability ratio	Combat strength ratio, economic strength ratio, size ratio

Index of National Capabilities (CINC) scores of the states within a dyad. The CINC score is based on the population, urban population, iron and steel production, energy consumption, military personnel, and military expenditure of a state. In Game X, we used combat strength ratio, economic strength ratio, and size ratio, all quantities tracked within the game as a measure of player score, as proxies for capability ratio. The MIDs variables and comparable Game X variables are shown below in Table 17.1.

Frequently in the international relations literature, linear regression analysis is performed on one or more economic variables to discern a correlation between them and instances of conflict within a dyadic pair. The variables listed in Table 17.1 are used as controls. We performed a similar analysis on 739 days of Game X data *operationalized* to populate the economic, conflict, and control variables for direct comparison of our analysis to the academic literature. We began by taking a sample journal article (Barbieri 1996) that regresses the MIDs data against three different economic variables defined by the author: salience, symmetry, and interdependence. The variables are defined mathematically as follows:

$$\text{Trade Share}_i = \frac{\text{Dyadic Trade}_{ij}}{\text{Total Trade}_i} \tag{17.1}$$

$$\text{Salience} = \text{Sqrt}(\text{TradeShare}_i * \text{TradeShare}_j) \tag{17.2}$$

$$\text{Symmetry} = 1 - |\text{TradeShare}_i - \text{TradeShare}_j| \tag{17.3}$$

$$\text{Interdependence} = \text{Salience} * \text{Symmetry} \tag{17.4}$$

The purpose of our work was not to validate or invalidate the variable definitions or the conclusions of the article, but simply to directly compare statistical analysis of serious game data to similar analysis of real-world data reported in the literature and to compare results of the analysis and examine how and why they might be different.

To operationalize the economic and conflict data in Game X, we started by measuring combat and trade between guilds over the entire 739-day period.

Guilds change over time; few were present and unchanged over the entire measurement period. As such, we divided the time period in 25 consecutive 30-day month periods and calculated the trade and combat measures on a month-to-month basis. Dyads were the main unit of measurement, as defined by a pair of guilds that engaged in trade, combat, or both at any point within a month period. The resulting unit of dyad month is similar to the dyad year unit used in the Correlates of War database to store the MIDs data. There were a total of 297 guilds and 13 079 unique dyad pairs over the entire time period. We excluded dyads that included the game itself (i.e. game-controlled entities or *non-player characters*) and guilds engaging in trade or conflict within itself. This resulted in 47 748 observations. On average, only 18% of dyads measured engaged in conflict over the 25 periods of data gathered.

Measurements of trade between members of a dyad were modeled after (Barbieri 1996). First, the trade share for each member of the dyad was calculated by taking the amount of trade with the dyad partner (imports and exports) divided by the total amount of trade conducted by that dyad member with all trading partners (including trade with other guilds, trade with players not belonging to guilds, and trade with the game). See Eq. (17.1). Trade Share is bounded by 0 and 1; its value approaches 1 as trade from $Guild_i$ becomes a larger fraction of overall trade conducted by $Guild_j$. From here, three economic variables were calculated: salience, symmetry, and interdependence, defined by Eqs. (17.2)–(17.4). Salience was formulated by the journal article author (Barbieri) in order to capture the importance of the trade relationship to the dyadic partners; a high salience score should indicate that the relationship is important to at least one of the partners in the dyad. In fact, a highly asymmetric trade relationship might obfuscate the importance to one trading partner (a very small fraction of one multiplied by a larger fraction of one), but our intention here was not to contest the measurement methods employed by the article but simply to replicate them. The variable salience turned out to be very small on average in the Game X data set (mean = 0.006, range = [0, 0.09], median = 0.0001). The dyadic trade share was often a very small fraction of overall trade for a given guild.

The variable symmetry is defined in Eq. (17.3). Using this definition, values close to 1 should indicate a relatively symmetrical trade relationship, while very small values should indicate an asymmetric relationship. In the Game X data set, most trade share values were very small, as noted above. Because of this, most symmetry values were close to 1 (mean = 0.998, range = [0.3, 1]). Interdependence, defined as in Eq. (17.4), reflects the problems inherent in both the salience and symmetry values, and because it multiplies the symmetry values, which approach but are still fractions of 1, by the small salience (approaching 0) numbers, it results in each case in very small interdependence scores (mean = 0.0006, range = [0, 0.074]).

Conflict was defined in broad terms for purposes of this analysis; a dyad was coded as having engaged in conflict if any of its members engaged in any sort of combat with the other dyadic guild within the month period. This method of encoding does not capture whether the conflict was part of a larger effort coordinated by the guild or simply a *one-off* battle between two individuals. As such, the variable of conflict in Game X may not correlate well with the conflict variables from the MIDs data set, which capture state-level disputes and wars (Table 17.2).

Table 17.2 Game X conflict data summary.

	Count of dyads engaging in conflict and/or trade in each month period		
	Conflict present		Conflict absent
Month ID	Trade present	Trade absent	Trade present
1	3	2	11
2	73	66	398
3	120	97	786
4	145	142	1016
5	123	124	1271
6	104	160	1421
7	149	118	1566
8	130	162	1589
9	124	133	1731
10	153	230	1817
11	133	172	1772
12	153	186	1951
13	159	195	1981
14	147	193	1977
15	209	705	2000
16	155	398	1944
17	127	203	1971
18	177	155	2013
19	147	138	2061
20	218	433	2137
21	183	494	1971
22	182	255	2213
23	199	189	2249
24	195	129	2083
25	108	77	1558

Analysis and Results

Analysis was performed on the data using linear mixed-effects regression models run with the lme4 package for R, which allows the inclusion of multiple observations of the same dyad over the full-time period. Random intercepts were included for each dyad, such that the overall estimated intercept was adjusted slightly for the error variance due to each dyad's likelihood to engage in combat. This allowed us to look at the effect of the economic predictor variables, also known as the fixed effects, while accounting for variability in the baseline likelihood of conflict between individual dyads. The dependent variable predicted by the logistic regression model was likelihood of conflict between two dyads (bounded by 0 and 1) during the time period analyzed. Logistic (also known as logit) regression was used because of the categorical nature of conflict (either it occurred or it didn't, there is no combat continuum), resulting in a binary dependent variable that can take either a value of 0 or 1. The fit to the data from the logit model generates an equation that estimates a probability, or likelihood, of the dependent variable occurring (having a value of 1) with a given value of the independent variable. The equation includes an intercept and a coefficient applied to the independent variable, which are shown in the results tables that follow. The equation takes the form:

$$p(x) = \frac{1}{1 + e^{-(\beta_0 + \beta_1 x)}} \tag{17.5}$$

where x is the dependent variable and p is the function of x that estimates the probability. The regression coefficients β, shown in the odds ratio column in each table, indicate the strength of each economic indicator in predicting the dependent variable. These coefficients are a calculation of the odds ratio:

$$\beta = \left(\frac{p(\text{conflict})}{1 - p(\text{conflict})} \right) \tag{17.6}$$

Three different analyses were performed: the first on all guilds over the full-time period, the second only on dyads that include one big guild (defined for purposes of this analysis as guilds with more than 30 members), also over the full-time period, and the third on data gathered only during the periods of stability between a period of two large wars, which took place in months 17–19. These three different analyses were performed in part to examine whether big guilds behave differently than small guilds and whether economic predictors behave differently during peacetime and wartime.

Analysis 1: All Guilds, Full-Time Period

The data set for this analysis included all dyads across the entire time period. The three economic indicator variables are not actually independent of each other, since all are a function of dyadic trade share. Fortunately, Barbieri does separate regressions against each variable in addition to her full model, which includes all three variables together. We replicated her methodology, using four different models, which we will designate M1, M2, M3, and M4. M1 regresses against only against salience as the economic predictor variable, M2 only against symmetry, M3 against interdependence, and M4 against all the economic predictor variables together (equivalent to the full model). The model results can be seen in Table 17.3 for all guilds.

These results appear to indicate that the symmetry variable is correlated strongly and positively with likelihood of conflict. It is possible that only symmetry registers as statistically significant because it is the only indicator that is not a vanishingly small number, as are salience and interdependence. Simply for purposes of comparison directly to the academic literature, the results of the M4 analysis are shown side by side with the full model results from [1] in Table 17.4. As mentioned above, the variables salience, symmetry, and interdependence are not independent of each other, and the M4 results are given here simply for purposes of comparison. Barbieri's work shows a negative correlation between both those variables and likelihood of combat, while our analysis shows a positive correlation. On the other hand, Barbieri shows a positive correlation between interdependence and conflict, (ours is negative), where interdependence is simply a product of the two other economic variables. It is likely that the nonindependence of the economic variables is confounding the statistical analysis and is complicating comparison of our results.

Analysis 2: Large Guilds, Full-Time Period

The data set for the second analysis included only large guilds (guilds with 30 or more members) across the entire time period. The results of all four models run on this data set are shown in Table 17.5.

Models M1–M3 show no significant effect of the economic indicators on conflict likelihood. There is, however, a positive correlation between strength and size ratios and combat likelihood (combat is more likely if there is a power asymmetry).

Large Guilds, Interwar Period

The data set for analysis 3 included only dyads with at least one large guild and was restricted to the three months between large wars. The results of this analysis can be found in Table 17.6. Here we see the same correlation between power asymmetries and conflict likelihoods that we saw in analysis 2 and the same lack of significance in the economic variables.

Table 17.3 Model results for all guilds, full-time period.

Predictor and control variables	Models											
	M1: Salience			M2: Symmetry			M3: Interdependence			M4: All		
	Odds ratio	SE	x	Odds ratio	SE	p	Odds ratio	SE	p	Odds ratio	SE	p
Fixed parts												
(Intercept)	0.15	0.06	<0.001	0.00	0.00	<0.001	0.15	0.06	<0.001	0.00	0.00	<0.001
Salience	0.00	0.00	0.005							Inf	Inf	<0.001
Symmetry				56 887.24	161 762.56	<0.001				8 933 765.75	38 675 407.04	<0.001
Interdependence							0.00	0.00	<0.001	0.00	0.00	<0.001
Economic strength ratio	1.51	0.64	0.339	1.44	0.62	0.391	1.51	0.65	0.337	1.42	0.62	0.420
Combat strength ratio	0.76	0.13	0.110	0.76	0.13	0.110	0.76	0.13	0.110	0.76	0.13	0.110
Size ratio	0.54	0.05	<0.001	0.52	0.05	<0.001	0.54	0.05	<0.001	0.52	0.05	<0.001
Contiguity (IS contiguous)	0.24	0.01	<0.001	0.24	0.01	<0.001	0.24	0.01	<0.001	0.24	0.01	<0.001
Alliance (IS Foe)	12.83	0.68	<0.001	12.86	0.68	<0.001	12.82	0.68	<0.001	12.76	0.65	<0.001

Regression coefficients expressed as the odds ratio, SE is the standard error, and p is the significance value estimated with Wald's Z using the sjt.lmer package in R.

Table 17.4 Game X M4 analysis compared with Barbieri's full model results.

Variable	Game X			Barbieri		
	β	SE	p	β	SE	p
Salience	463.27	131.95	≤ 0.001	−22.64	6.69	≤ 0.01
Symmetry	16.01	4.33	≤ 0.001	−4.46	0.80	≤ 0.01
Interdependence	−490.88	140.78	≤ 0.001	26.60	7.28	≤ 0.01

β is regression coefficient from the logistic regression, expressed as an odds ratio, SE is the standard error, and p is the significance value estimated with Wald's Z using the sjt.lmer package in R.

Caveats

These experimental findings must be interpreted in light of several caveats. One important note is that the definition of the economic factors was defined in such a way that many dyads had extreme values (either at the very low or very high end of the scale) and were all derived in some way from trade share (so they were nonindependent predictors), both of which could have skewed the results. No interactions were included between any of the control variables and the economic predictors, so we cannot say whether, for example, symmetry is a more important predictor for contiguous versus noncontiguous guilds. Future work could address these questions in more detail.

Operationalizing MMOG Data

There were several issues as we operationalized the real-world variables into Game X.

In Game X the closest analogue to countries was guilds, which were player created and managed. However guilds have no physical boundaries and can vary greatly in size. Countries and states in the real world have strict boundaries. This made the operationalization of the *contiguity* control variable difficult as it relied on geographic distance between countries.

The lack of publicly documented agreements had a significant effect on operationalizing the *alliance* control variable. Since we primarily used the *Is Foe* variable, we were aggregating positive relationships that span multiple types, from merely neutral to strongly positive. In the real world treaties and other agreements provide a formal method to assessing the strength of alliances.

Many guilds in Game X have exhibit organizational properties, including roles and a hierarchy (Lakkaraju and Whetzel 2013). However, there is no public process to codify this organization. In contrast, in the real world, constitutions and laws are drawn up and publicized. This allows observers and the general public to gain information about the organization and can be important for

Table 17.5 Model results for large guilds, full-time period.

Predictor and control variables	Models										
	M1: Salience			M2: Symmetry			M3: Interdependence			M4: All		
	Odds ratio	SE	p	Odds ratio	SE	p	Odds ratio	SE	p	Odds ratio	SE	p
Fixed parts												
(Intercept)	1.02	0.84	0.983	Inf	Inf	0.254	1.02	0.84	0.983	Inf	Inf	0.071
Salience	0.00	0.00	0.040							Inf	Inf	0.004
Symmetry				0.00	0.00	0.255				0.00	0.00	0.071
Interdependence							0.00	0.00	0.214	0.00	0.00	0.004
Economic strength ratio	0.10	0.09	0.012	0.10	0.09	0.012	0.10	0.09	0.012	0.11	0.10	0.017
Combat strength ratio	0.98	0.33	0.961	1.00	0.34	0.992	0.98	0.33	0.960	1.03	0.35	0.941
Size ratio	0.95	0.19	0.785	0.95	0.20	0.807	0.95	0.19	0.785	0.97	0.20	0.896
Contiguity (IS contiguous)	0.43	0.03	<0.001	0.42	0.03	<0.001	0.43	0.03	<0.001	0.44	0.03	<0.001
Alliance (IS Foe)	4.68	0.44	<0.001	4.73	0.44	<0.001	4.68	0.44	<0.001	4.61	0.42	<0.001

Regression coefficients expressed as the odds ratio, *SE* is the standard error, and *p* is the significance value estimated with Wald's Z using the sjt.lmer package in R.

Table 17.6 Model results for large guilds, interwar period.

Predictor and control variables	Models											
	M1: Salience			M2: Symmetry			M3: Interdependence			M4: All		
	Odds ratio	SE	p	Odds ratio	SE	p	Odds ratio	SE	p	Odds ratio	SE	p
Fixed parts												
(Intercept)	0.00	0.00	<0.001	0.00	0.00	0.643	0.00	0.00	0.471	0.00	0.00	<0.001
Salience	0.00	0.00	0.370							0.00	0.00	<0.001
Symmetry				Inf	Inf	0.649				Inf	Inf	<0.001
Interdependence							0.00	0.00	0.795	Inf	Inf	<0.001
Economic strength ratio	0.08	0.00	<0.001	0.12	1.37	0.851	0.04	0.45	0.784	0.14	0.00	<0.001
Combat strength ratio	6.57	0.01	<0.001	0.39	1.67	0.826	12.78	57.87	0.574	1.16	0.00	<0.001
Size ratio	3.08	0.00	<0.001	1.69	3.49	0.799	3.30	7.09	0.578	1.45	0.00	<0.001
Contiguity (IS contiguous)	0.19	0.00	<0.001	0.23	0.18	0.053	0.20	0.16	0.041	0.17	0.00	<0.001
Alliance (IS Foe)	7.39	0.00	<0.001	6.65	6.52	0.053	7.68	8.28	0.058	7.50	3.65	<0.001

Regression coefficients expressed as the odds ratio, SE is the standard error, and p is the significance value estimated with Wald's Z using the sjt.lmer package in R.

understanding conflict (Barbieri 1996) used the control variable of *joint democracy* as a way of capturing the type of government of a country. We could not operationalize that in our current analyses.

Even if a guild establishes laws and codified them into a public document, it is not clear if we can define what a *democracy* is, and whether real world definitions can apply.

MMOGs, at their core, are games that are meant for entertainment. An important part of entertainment is the ability to explore and make mistakes in an environment with little consequence. Clearly such behavior can be exhibited by players, especially early on in the game. We must account for this in our analysis. We suspect that as players stay in the game longer, they are more attached to their character and will act in a way to protect their character. We must be aware of this and sample the data to try to avoid the exploration phase of the player's behavior.

It is impossible to draw any real conclusions about the effects of economic ties and combat likelihood in the game, given the way that the economic variables are defined. However, the purpose and focus of this work was not to do so but to demonstrate a proof of concept: that data from an online serious game (Table 17.5) can be operationalized in the form of economic, political, and military variables and that statistical analysis can be performed on that data for direct comparison with academic studies of real-world data (Table 17.6). Through this exercise, we have identified some of the key issues with using game data for national security research, which we describe in more detail below. With a carefully designed game such as the proposed MMOG described above, we believe that we can populate a database with conflict data to use in future academic research. The work on Game X described here is an example of how MMOGs can be used as experimental platforms and contribute to research efforts in policy areas where data are otherwise sparse.

Games as Experiments: The Future of Research

As discussed and demonstrated in the sections above, games as experiments have both a long history and a bright future. While games have historically been used to probe possibilities in a simulated space and recreate in detail history as it happened, the future of research with games may include gathering large amounts of data for analysis of multiple kinds, not only the traditional analysis of interesting trajectories that the game play took due to unexpected player actions but also statistical and other quantitative analysis on the potentially massive amounts of data that can be gathered using a game with a large player base such as a MMOG. This type of game and the accompanying data set will allow for large-scale analysis of actions taken over time and across multiple games; a type of analysis that has heretofore been extremely difficult to perform. The ability to operationalize game data that we demonstrated above will allow researchers to compare game play results directly to those from the real world and may enable them to augment real-world data sets with MMOG data. Analysis such as the logistic analysis performed above can help inform researchers and policy-makers of the potential impact of particular trade, military, or diplomatic relationships on future outcomes, such as the likelihood for conflict that we tested in our analysis of Game X, which correlated positively with power asymmetries in our analysis.

As we consider the utility of online games for research, we must still be aware of the difference in intention: games are meant, fundamentally, to engage a player through entertainment. To maintain engagement, the design of games may cause issues when used for experimentation. We posit the following considerations when analyzing game data.

Simplification

Entities and processes in the game world are simpler than in the real world. This is for multiple reasons. First, it is to reduce the cognitive burden of learning the game. Second, it is to provide an environment that is focused on the core purpose of the game. The simplification makes the mapping between real-world entities (our actual target of interest) to game world entities difficult, as we saw in the mapping of guild to nation states above.

Option Abundance

Game choices (while certainly simplified from the real world) are often provided in plenty. To maximize engagement, players need to be able to explore and discover new things. However, in experimental contexts one often wants to limit the number of options for a subject in order to study the underlying relationship better.

Event Shaping

The game may push the player to make certain choices or experience certain events. For instance, games often encourage conflict through resource manipulation. When considering the correlation between in-game behavior and real-world behavior, one must be careful to account for the potential forces that are driving in-game behavior.

Another factor is that players may be focused on exploring the world initially, and the game may encourage that by providing simple initial environments.

Final Discussion

MMOGs will never be perfect representations of the real world; no experimental laboratory is, nor are they intended to be. What both MMOGs and real-world laboratories offer is the possibility to test hypotheses in a controlled setting and to manipulate the controls in order to observe how differences in controls affect experimental results. The potential for MMOGs to serve as experimental platforms for various types of research is enormous given the large sample size that they provide for analysis and the ability to engineer the online environments to address specific research questions.

The exercise we detailed in this chapter, to operationalize variables from an existing MMOG, served a useful purpose in highlighting potential issues that can arise when studying game data, especially for national security issues. Understanding, and addressing, how simplification, option abundance, and event shaping can influence data analysis and interpretation of results is an important future work.

Acknowledgments

Sandia National Laboratories is a multimission laboratory managed and operated by National Technology & Engineering Solutions of Sandia, LLC, a wholly owned subsidiary of Honeywell International Inc., for the US Department of Energy's National Nuclear Security Administration under contract DE-NA0003525. This document is numbered: SAND2018-6296 B.

References

Barbieri, K. (1996). Economic interdependence: a path to peace or a source of interstate conflict? *Journal of Peace Research* 33 (1): 29–49.

Castronova, E. (2008). A Test of the Law of Demand in a Virtual World: Exploring the Petri Dish Approach to Social Science. SSRN Scholarly Paper ID 1173642. Rochester, NY: Social Science Research Network.

De Lope, R.P. and Medina-Medina, N. (2017). A comprehensive taxonomy for serious games. *Journal of Educational Computing Research* 55 (5): 629–672.

Deudney, D. (1983). *Whole Earth Security: A Geopolitics of Peace, Worldwatch Paper Series*. Worldwatch Institute.

Gosling, S.D., Sandy, C.J., John, O.P., and Potter, J. (2010). Wired but not WEIRD: the promise of the internet in reaching more diverse samples. *The Behavioral and Brain Sciences* 33 (2–3): 94–95.

Guarino, S., Eusebi, L., Bracken, B., and Jenkins, M. (2019, this volume). Using sociocultural data from online gaming and game communities. In:

Social-Behavioral Modeling for Complex Systems (ed. P.K. Davis, A. O'Mahony, and J. Pfautz). Wiley.

Henrich, J., Heine, S.J., and Norenzayan, A. (2010). The weirdest people in the world? *The Behavioral and Brain Sciences* 33 (2–3): 61–83; discussion 83–135.

Laamarti, F., Eid, M., and Saddik, A.E. (2014). An overview of serious games. *International Journal of Computer Games Technology* 2014: 11.

Lakkaraju, K. and Whetzel, J. (2013). Group roles in massively multiplayer online games. *Proceedings of the Workshop on Collaborative Online Organizations at the 14th International Conference on Autonomous Agents and Multiagents Systems.*

Lu, L., Shen, C., and Williams, D. (2014). Friending your way up the ladder: connecting massive multiplayer online game behaviors with offline leadership. *Computers in Human Behavior* 35: 54–60.

Perla, P.P., Markowitz, M., and Weuve, C. (2005). *Game-Based Experimentation for Research in Command and Control and Shared Situational Awareness.* CNA.

Press, D.G., Sagan, S.D., and Valentino, B.A. (2013). Atomic aversion: experimental evidence on taboos, traditions, and the non-use of nuclear weapons. *American Political Science Review* 107 (01): 188–206.

Sabin, P. (2014). *Simulating War: Studying Conflict through Simulation Games.* London, Oxford, New York, New Delhi, Sydney: Bloomsbury Academic. reprint edition.

Williams, D. (2010). The mapping principle, and a research framework for virtual worlds. *Communication Theory* 20 (4): 451–470.

Yee, N. (2006). The demographics, motivations, and derived experiences of users of massively multi-user online graphical environments. *Presence: Teleoperators and Virtual Environments* 15 (3): 309–329.

18

Using Sociocultural Data from Online Gaming and Game Communities

Sean Guarino, Leonard Eusebi, Bethany Bracken, and Michael Jenkins

Charles River Analytics, Cambridge, MA 02138, USA

Introduction

In social science studies, laboratory experiments enable a high degree of control over conditions and independent variables, but the conditions may be artificial. Also, this comes at the cost of small or unrepresentative samples, priming effects, and situation-specific constraints that can limit the generalizability of research findings. Field studies are an alternative for observing populations in their natural social environments, but managing interactions can be difficult (although *natural experiments* are occasionally possible). The advent of online communities and mobile technologies has created new opportunities for social, behavioral, and economic (SBE) research. Commercial games, in particular, provide a rich opportunity for SBE researchers to perform a different type of field study, observing populations in the *natural* social environment that is provided by the game. This is often a more controlled environment than real-world habitats – defined by the rules and characteristics of the game – but still provides an opportunity for observing human behavior outside of a laboratory environment (albeit with many of the behavioral implications of online anonymity and game behavior). Communities evolve within games (e.g. players interacting in massively multiplayer online games (MMOGs)) and in the emergent communities surrounding them (e.g. forums focused on game discussion or popular streaming communities). Using games and surrounding game communities, researchers can observe behaviors reflective of other online SBE interactions, as well as limited aspects of face-to-face interactions. For example, games and surrounding communities often include their own forms of law enforcement (i.e. trusted community members responsible for enforcing rules and addressing misbehaviors) and provide a platform for a host of intricate social functions, from powerful

Social-Behavioral Modeling for Complex Systems, First Edition.
Edited by Paul K. Davis, Angela O'Mahony, and Jonathan Pfautz.
© 2019 John Wiley & Sons, Inc. Published 2019 by John Wiley & Sons, Inc.
Companion website: www.wiley.com/go/Davis_Social-Behavioralmodeling

friendships to online marriages (Shi and Huang 2004; Wu et al. 2007; Bates IV 2009). Many games have independent player-driven economies (Taylor et al. 2015; Drachen et al. 2016), some of which maintain larger economies than some real-world countries (Nazir and Lui 2016). Game communities can have a significant impact on player behavior within the games because players evolve their approach based on community discussions and, in many cases, real-world economic implications (e.g. streamers earning a comfortable living through donations from their followers [Gandolfi 2016; Johnson and Woodcock 2017]).

A strength of research on human behavior in games and game communities is the sheer size of the available populations. Large-scale game populations provide more opportunities to observe behavioral phenomena and ultimately to increase the statistical power that can be realized in field or experimental studies. Game populations that number in the millions provide an opportunity to run studies with populations on par or larger than real-world field studies (e.g. observing people in natural interactions), but within a more controlled context where game rules guide interactions, and, in some cases, can be manipulated to study specific phenomena. For example, Riot Games' League of Legends (LoL) has over 67 million players that use the game each month and over 7.5 million simultaneous players using the game regularly at peak hours. Riot Games collects a vast amount of data, including player behaviors and decisions, player communications in public and private communities, and player characteristics. Much of this is made available through a public application programming interface (API). In collaboration with academic institutions, Riot has run a variety of studies to understand what motivates players, what behaviors are observed in effective teams, and how problematic behaviors (e.g. bullying) can most effectively be discouraged (Blackburn and Kwak 2014; Conway and deWinter 2015; Kwak et al. 2015; Kim et al. 2016).

Surrounding communities also provide vast populations for study. Twitch.tv – a key social media framework that enables live video streaming of gameplay to large audiences – is the fourth most trafficked site in the United States, with over 100 million monthly users (15 million daily active users) and more than 2.2 million unique content creators each month. Twitch reaches a live, engaged, and concentrated Millennial population that has been constantly growing since its inception (Gargioni 2018). Nearly half of users spend more than 20 hours each week viewing content, and streaming has become a lucrative business for popular content creators. At a minimum, further exploration of large-scale games and surrounding game communities can provide an unparalleled opportunity for field studies, observing human behavior in partially constrained environments with large populations. To the extent that researchers can collaborate with game companies and associated game communities, these online settings can provide an even more powerful capability to manage experimental research with these large-scale populations (e.g.

manipulating game rules in collaboration with game companies, manipulating incentive mechanisms when working with community providers, such as Twitch or associated Twitch data collection services).

The remainder of this review describes the characteristics of in-game behavior and surrounding game community behavior, with the goal of inspiring future SBE research that focuses on these communities. We first describe the types of social interactions, behaviors, and economic interactions that can be observed in these game environments. We then describe the various data sources that are available in the gaming community and the challenges for accessing and interpreting this data for SBE research objectives. We review three case studies in which we collected and analyzed user behavior in games and community environments, performing limited-scope SBE research to demonstrate the value of these data sources. Finally, we provide conclusions summarizing the potential of gaming as a platform for SBE research and recommendations for improving these communities to better support future studies.

Characterizing Social Behavior in Gaming

While researchers are still learning how best to interpret behavior in games and game communities, these settings clearly enable extensive observation of many SBE phenomena. Table 18.1 provides examples comparing the context of experiments and behaviors in live environments with those of games and game communities. Many modern online games require significant human interaction and often involve player teams ranging from two-player partnerships to large-scale coordinated raids with dozens of players. Players cooperate to overcome game-provided content and to implement potential game strategies. In some games, players compete with others, either as a primary objective of the game (e.g. in player-vs.-player (PVP) strategy games) or as a potential additional challenge while overcoming other game objectives. Collecting data around these in-game behaviors can provide research insights into teaming and leadership behaviors, incentive mechanisms, friendships, and differences between positive and negative player behaviors. Furthermore, SBE researchers can study how the rules and characteristics of the game affect behaviors across games and teams.

Many games have rich and complex social and economic features as well. In MMOGs, for example, players socialize as a regular part of their gameplay. This socialization can include political interactions where different players compete for control of social groups (e.g. to control teams and guilds). It leads to social structures that drive how players interact (e.g. through leadership and support roles). Within games, players discuss their personal lives, bully other players based on disagreements, and form lasting relationships (e.g. in-game marriage). While many of these constructs may not be identical to parallel

Table 18.1 Contextual comparisons between live SBE research environments and game and game community environments.

Concept/behavior	Live context	Game and game community context
Participants	Laboratory experiments use size-limited and often demographic-limited populations	Games and game communities provide large-scale populations in more controlled environments than many live field studies
	Field studies are done with live populations usually acting in an unconstrained social setting; natural experiments are possible and powerful when available, but are difficult to organize and manage	Some opportunities exist to manipulate existing environments, although they require close collaboration with developers who are focused on business objectives
Study timelines	Laboratory experiments are generally short-term events, limiting the interactions that can be studied	Game populations can be anonymous and difficult to track in a longitudinal way
	Field studies collect data over short or long periods with no promises of instances of targeted phenomena	Significant potential exists for multi-instance data collection across many games and population groups, but it is difficult to reproduce exact scenarios and conditions across game instances
	Both aspects provide a good opportunity for managing known populations in longitudinal studies	Games and game communities provide extensive access to historic data
Data	Data collection is limited to available or designed instruments or coding mechanisms; historical data does not exist in laboratory environments and is limited in field studies	Games and game communities often incorporate built-in data collection
	Instruments can be designed for easy interpretation	Game data are not designed for behavioral interpretation; interpretation can be challenging and can rely on detailed game knowledge
	Coding of observations can be subjective in nature; it is challenging to develop objective metrics	Community data requires heavy text interpretation
		Human coding can be just as subjective as live content coding
Behavioral context	Real-world behaviors include wide variety of psychological interactions that can be manipulated	Game behaviors are expressed by actions in the game that must be interpreted based on game knowledge with respect to behaviors of interest
	There are a range of environment and personal moderators that have a strong impact on behavior, including (but not limited to) culture, education, experience, priming effects, stress, and fatigue	Game community behaviors are social in nature
		Online context has a significant impact on how people behave (e.g. anonymity effects) and can be impacted by the full range of moderators

Table 18.1 (Continued)

Concept/behavior	Live context	Game and game community context
Social context	Real-world social interactions are strongly impacted by affect Social norms have a significant impact on interactions Social behavior is impacted by the full range of moderators Face-to-face interactions can be rapid, but may not be as private as in-game or community-based interactions	Online behavior – particularly game behavior – can be strongly impacted by the feeling of anonymity Limited ability exists to communicate affect (although a rich language of emojis is developing to address this issue) Text creates a different speed of interaction but enables a variety of simultaneous independent interactions and private communications
Economic context	Real-world economies strongly impact one another, making it difficult to tease international from national effects Real-world economies have direct and clear ties to populations' livelihood and survival, so in most cases can be considered more meaningful	Game economies are often tightly constrained within the context of the game Interactions with the outside world are often limited in nature, and many players act in the economy with no outside world interaction Some game community economies (e.g. Twitch) directly interact with live economies (e.g. donating real money, providing salaries to streamers)

constructs in live interactions, these games provide a unique opportunity for observing similar social interactions in partially constrained settings. Many games also have economic context with players acquiring and selling in-game content to access or succeed at new parts of the game (or even just to impress other players). Game economics sometimes carry over to the outside world, with players purchasing in-game resources using real-world money and some players even using this mechanism to earn a real-world salary through gameplay (Knowles et al. 2015; Patel 2016). Using these games, economic researchers can potentially investigate and manipulate limited game economies with constrained interactions in real-world economies.

Game communities also provide opportunities for studying human behavior, social interaction, and economic decision-making. Surrounding communities can be split into two categories: asynchronous and synchronous. Asynchronous communities, such as forums or reviews of recorded content, are characterized by interactions over long periods of time with delays between initial posts and responses. Synchronous communities, including live streaming and chatting, are characterized by real-time interactions that can lead to real-time changes in content (for example, many Twitch streamers will select their game or approach based on requests from live viewers). Both contexts exhibit political, social, economic, and behavioral interactions. Highly skilled players or popular personalities gain significant respect and rise to positions of power and profit that make them trusted contributors. Community members form friendships, which may extend outside of the original social media setting. In some ways, the surrounding game communities can provide more context for studying SBE phenomena than in-game behaviors, as players are not as distracted by the actual gameplay. A deeper understanding of game community data can enable SBE research to investigate a wide range of interactive behaviors across a variety of online behavioral phenomena.

Game-Based Data Sources

In this section, we review gaming data sources that can be used to study SBE phenomena. We categorize these into four classes. The first two focus on data sources for in-game behavior, treating the game as an online environment for SBE research. Direct in-game data is often collected by the game developer and accessed by APIs when they are made available (e.g. LoL at developer.riotgames .com; Guild Wars 2 (GW2) at wiki.guildwars2.com/wiki/API:Main; Eve Online at community.eveonline.com/support/api-key). In some cases, another option for accessing this kind of data is third-party software that extracts similar data as the game executes (e.g. TorchCraft for StarCraft 2 data). Meta-game data, which includes a host of meta-information that external users collect about the game, describes successful and unsuccessful game strategies, skills, and behaviors. This data can be found in a host of community and third-party sites describing the game, where avid players work to gather data about how the game is played (e.g. what strategies and approaches are used).

The next two classes focus on data from the community of players surrounding games. Here, the focus is not on studying game behavior, but is rather on studying how the player community behaves and interacts outside of the game. Asynchronous community data includes forums, podcasts, and videos, along with the commenting and discussions that surround this content. Players are not interacting in real time, but rather are interacting asynchronously as they observe and react to new content. Synchronous community data includes a variety of streaming content sources (e.g. Twitch.tv, YouTube Live) and the real-time interactions involving those sources.

In-Game Data Sources

Many online games record a significant portion of the events and actions within each game session, primarily for developers to use in improving the game experience. Some games, such as LoL, GW2, and Eve Online (Eve), make this data available to the public or to registered third-party developers for use in developing websites or other products designed to improve the broader game experience. Some developers (e.g. Riot Games, developers of LoL) may agree to expose this data to researchers if the research is relevant to their interests. Typically, the data available consist of the most important actions within a game, including combat or other active decisions, strategic choices, and player-to-player transactions. For example, in LoL, the data API provides a *match detail* object that can be collected for known matches. Each match detail object contains data on the summoners (player profiles) involved in the match (which team they were on, which champion [character type] they selected, aggregated statistics for the match, and whether their team won). The objects also contain key match events, including when and where champions, buildings, or major neutral minions were killed; the items purchased by each champion; and each champion's position every 60 seconds. This level of detailed data allows researchers to examine how game players behave and interact within the constraints of the game rules and strategic objectives. In our first case study (see section "Case Study 1: Extracting Player Behavior from League of Legends Data"), we used LoL data to analyze links between specific in-game behaviors and performance.

Even when game companies make data available, that data is often constrained. Few game companies make their data easily available to researchers, limiting the pool of possible sources. APIs that do exist can often be burdened with licensing fees or usage restrictions. These licenses are crafted to protect not only the personal data of the players but – most importantly perhaps – the company's intellectual property and commercial aims. Companies are sensitive to perceptions of improper data use, and gamer communities are uncomfortable with clandestine activities that might affect the integrity of the game (such as asking study participants to act according to out-of-game objectives). This makes it difficult or impossible to design studies that manipulate play in some way and then measure the effects of these manipulations. While there have been some examples of game companies working with researchers to investigate game manipulations (most notably, work by Riot Games to investigate bullying in LoL (Blackburn and Kwak 2014; Conway and deWinter 2015; Kwak et al. 2015; Kim et al. 2016)), in most cases it is difficult or impossible to manipulate game characteristics.

For this reason, most studies using game data can best be compared to field studies, where game players are observed in the game environment and comparisons can be made across players performing in different game instances and modes. Researchers must work within the constraints of planned game updates and game features provided by the game company. For example, the most common forms of competitive play in LoL requires random matchmaking, which would restrict researchers from ensuring study participants are teaming together or playing under a targeted condition or constraint. While small-scale studies can bring participants into a lab and use assigned matches, these studies do not take advantage of the vast populations that could otherwise be exploited in these games. In GW2 or Eve, typical behaviors involve many players in an open, persistent world, creating situations where manipulations will affect and be impacted by nonparticipants who happen to be present at or tied to the events that are manipulated. Avoiding this *collateral participant* issue requires clever design of manipulations that affect only a single team, isolate participants from nonparticipants, or obtain consent from unplanned participants.

While game data is made available in some instances, chat data that is common to many online games (e.g. for in-game communications and coordination) is rarely, if ever, made available to researchers through game APIs. Many game companies collect this data when it is done in-game, but the perception of potential privacy infringement prevents them from sharing that data with researchers or third-party developers. While the data can be manually collected by researchers or their proxies, such collection is limited. For example, researchers performing field studies in which they observe chat data within the game will be unable to view any private channels or messages. Researchers can require study participants to share their chat data, but this can limit studies to small populations of self-selected participants, potentially creating biased results and removing the ability to exploit large-scale game populations. Another approach may be to collect chat data using companion apps (e.g. apps that players can run to automatically record chat while playing the game). However, while these apps may be easier to use, they ultimately have the same constraints on study populations as requests for participants to share their chat data.

There are also challenges interpreting in-game data, stemming from several confounding factors. In-game strategies are complex and give significant contextual meaning to different behaviors. Aggression and risk-taking behaviors, for example, manifest in very different ways in LoL, GW2, and Eve. A metric tuned to identify aggressive play in LoL might look at how often a player attacks other players over non-player character (NPC) adversaries. In Eve, this metric would be less meaningful, since most situations are either purely PVP or player-vs.-environment (PVE). In GW2, there is some blend of the two, but the metric would be harder to interpret because of the need to determine the context. Tuning metrics of this sort requires deep knowledge not only of the meaning of the variables recorded by the game but also of the strategic and meta-game context of those variables, as well as the way people actually play or expect each other to play. In games with deeper social aspects, social norms (such as expectations of leaders or group members) can also affect the meaning of these variables.

Another challenge is that the data provided by game companies is shared at a level of resolution decided by the company. Having no obligation to share any data with research communities, companies will often constrain how they share the data, both because they may not wish to share detailed behaviors with potential competitors and because they are trying to minimize data sharing costs. For example, in LoL, position data within the game is provided at very low precision (every minute). These limitations can create significant uncertainty in interpreting the actual motion of players' characters in the game and lead to missing entire events that occur between data points.

Despite these constraints, when in-game data is provided – particularly for games with very large populations, such as LoL – it provides an extensive source for studying online behavior and social interaction. Researchers have been able to directly observe in-game interactions across tens or hundreds of thousands, or even millions of players, and can extract how these interactions impact a variety of cooperative and competitive social behaviors. The data can provide insights into how strategic decisions and activities are linked to performance, how these factors impact current and future teaming decisions, and how teaming impacts retention and participation. Where relevant, data can provide economic insights as well, enabling researchers to study how much players are willing to pay for certain goods within the game, and, in some cases (e.g. Eve Online), tracking goods exchange in a detailed manner within a thriving economic system (Papagiannidis et al. 2008; Shukla and Drennan 2018; Patel 2016). While in-game data sources limit researchers' ability to make experimental manipulations, they allow for extensive field study that can help to make significant advances in the SBE research and modeling community.

Meta-Game Data Sources

Gameplay and in-game actions are not the only useful source of data describing in-game behaviors. There are many data sources that revolve around the meta-game of popular games – that is, the strategies, information, goals, competitions, and other factors that the game's fans track, discuss, and consider between game sessions (King et al. 2010; Harviainen and Hamari 2015; Sköld et al. 2015). Thought and discussion around these topics make up a significant portion of the time that many players spend on the game. This is especially true for highly competitive players but can even be observed with casual fans who want to know what the professional players are doing or want to maximize the effectiveness of their limited time to play the game. Meta-game information is collected, disseminated, and interpreted across numerous fan sites, blogs, YouTube videos, and other websites, creating a social ecosystem that surrounds many games. For example, for Blizzard's *Hearthstone* – a card-based online strategy game in which players construct decks to compete against one another – this ecosystem includes (i) the official website (blizzard.net), which provides news on game updates and details; (ii) major podcasts (e.g. the Angry Chicken) providing commentary on rising game strategies; (iii) major fan sites (e.g. HearthPwn, LiquidHearth, and tempostorm) that provide strategy reviews and articles; (iv) aggregated data sites (e.g. HearthstoneTopDecks, https://www.hearthstonetopdecks.com/decks/) that provide historic gameplay data; and (v) in-game aggregated data (e.g. collected by vS Data Reaper). These data sources are typically freely available to all visitors or sometimes restricted to visitors who create an account or pay a small subscription fee. These sources provide essential information to understand the game, which is critical for interpreting in-game data. But they also provide a useful data source on their own for assessing game behavior. For example, sites like Hearthstone Meta Stats (see Figure 18.1) collect data on the popularity of specific strategies, how they are built and applied, and their success rates. This data can be used to track trends in the game community, as well as reactions to game updates, new strategies, and public tournaments. It can identify when game approaches become too popular (indicating that they may be overpowered) or lose popularity (indicating they may be too weak). It can provide insights into the popularity of new game features and capabilities, helping to identify which features should be stressed in research studies.

While some of these sites may provide an API for accessing data, most provide data that is unstructured or unprepared for automated consumption; these must be scraped and searched to extract useful information. Analysts

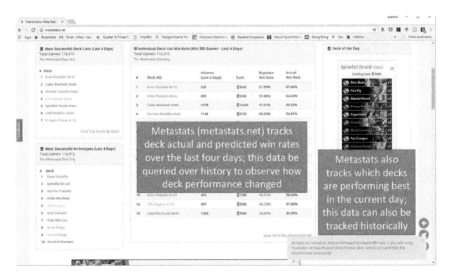

Figure 18.1 Information on popularity and performance of common decks (strategies). Hearthstone Meta Stats (http://metastats.net) provides detailed information on popularity and performance of common decks (strategies) in the game, providing a powerful source for investigating strategic trends in the player community.

can pull out text or tables captured on a webpage, snapshots of graphics, and other processed information, though copyright concerns apply and researchers should ensure that they are following the terms of use provided on the website. For each data source, the challenge lies in first finding ways to extract and aggregate the data (e.g. through crawling the Web, subscribing to RSS feeds, or using existing APIs) and then developing metrics for interpreting behaviors and activities within the data. The data is not inherently designed to study SBE phenomena; rather, researchers need to identify the patterns or terms that represent events in the context of a particular game and then must extract those from the data source. For example, in studying information spread in Hearthstone (see Case Study #2, section "Case Study 2: Extracting Popularity Patterns from Hearthstone Community Data"), we made significant progress by identifying key terms reflective of Hearthstone strategies using the MetaStats data (e.g. cards and deck names). We explored each of the possible Hearthstone meta-game data sources and prioritized those that provided the most useful and simple to extract data. We were then able to use statistical methods to measure the popularity of particular game strategies based on the frequency of their appearance on these sites.

Data from these sources can require automated interpretation or hand coding of video content, audio content, images of graphs, or strategy pages and articles. There are three main forms of meta-game data that must be interpreted:

(i) Aggregated, static breakdowns that detail strategies or popular patterns of game behaviors that change slowly – This data is typically updated regularly but may grow stale if it is not updated frequently enough.

(ii) Strategy articles, reviews of game changes, or descriptions of major events – This data mostly applies to a specific moment in time but represents a well-expressed and organized expert view of that moment.

(iii) Forum threads, discussions, and posts – This data is also situated in time but is typically less organized, is more colloquial, and frequently contains oblique references to current events.

Useful content from the latter groups will often have lasting appeal and be referred to repeatedly in later discussions. For example, in *Magic: The Gathering (M:tG)*, Flores (1999) produced a strategy article *Who's the Beatdown?*; this article captures a central aspect of deck-based strategy games and is often referred to by not only M:tG players but also Hearthstone and other card game players. Another challenge in using these sources is that some of them may be flawed, incomplete, or even incorrect, requiring some attention to data validation or verification. Even when these articles are incorrect, it can be interesting to study how the community goes about discovering (or not discovering) those errors. Finally, many meta-game sources rely on accepted context and jargon from the game, often without direct reference to its origin or definition. This context is dynamic, growing as new game features are released, and often changing meaning entirely based on changes to the game (e.g. when a particular feature is minimized or overpowered, its meaning can change dramatically). This can make interpretation of these data sources difficult to automate; it is not only an unstructured data problem but also a challenging process of data discovery and verification.

Nevertheless, meta-game data can be particularly useful for studying SBE trends in a game community, ranging from tracking key decisions and evolutions of game strategies to observing economic histories of the game (e.g. monthly economic reports for *Eve Online*). The data can often be accessed for free, with website developers providing an aggregated outcome that researchers can download and interpret. Many of these sites are driven by game players and are well supported by the game developers who see the sites as a good means to maintain player interest and participation. Meta-game sites still suffer from the same limitations regarding experimental control but provide a powerful field research tool for understanding how game communities are behaving within the game. In the context of many games, they can be considered parallel to census reports or popular opinion polls for game participants.

Asynchronous Community Data Sources

Moving beyond in-game data, many games are surrounded by a thriving community that itself can be an interesting subject of SBE research. The category of asynchronous community data includes any of a wide variety of sites that players and fans of a game use to socially interact outside of the game, with one another, with popular personalities or skilled players within the game, and with game developers. Community behaviors – or behaviors of individuals within the community – can provide opportunities to trace how information spreads, understand how information is repeated, altered, and embellished in that spread, and understand how different community members contribute to information attaining high popularity. Asynchronous community data sources are generally found in three forms: (i) forums that contain posts about in-game activities, meta-game shifts, game news, and other broad discussions between community members (e.g. Reddit threads, official forums, major fan site forums); (ii) video-based feeds that contain both in-game actions and often a broader discussion around those in-game actions with attached comments, chat rooms, and viewers (e.g. YouTube videos, Twitch videos on demand [VODs], and other recordings of play by skilled players and personalities); and (iii) podcasts focused on the game, which contain information about the game's community and events. Across all of these forms, the culture of a game community can be studied, providing an SBE research opportunity than is often more social than the games themselves. Researchers can learn which topics, goals, and actions are considered important or taboo. They can study the language used within the community, including unique jargon and novel uses of emoji. They can track how events are managed by the game community and by the game developers – including both player-driven events such as raids or in-game weddings and developer-driven events such as major game releases. They can understand how groups are formed and leaders chosen. They can understand how information spreads or cascades within a community and how it changes based on user interactions. Furthermore, in studying all of these topics, researchers have the potential to participate – that is, to post information in a community forum or create their own forum with different characteristics or rules to produce manipulations for potential experiments.

Data collected from forums, videos, and podcasts is typically subject to license agreements with the website that owns the content, including copyright concerns for the content producers. In most cases (for major sites, like Reddit, YouTube, or Twitch), these agreements protect the personally identifiable information of the users. For public personalities, it can be difficult to strip the identity of the content creator, as the creator's name itself may be a form of shorthand for a particular style of play or commentary (for example, key Hearthstone personalities can be recognized by their playstyle). Despite these constraints, text data from these sites will often be freely available and quite comprehensive, though researchers may need to develop software to scrape it from the website. The data can be voluminous, with many language artifacts that can be difficult to effectively parse and interpret (e.g. abbreviated terms or references, newly invented terminology, careless typos and spelling errors). Content from videos and podcasts poses its own challenges, such as speech-to-text translation in a domain with jargon and many nonstandard proper nouns. Many videos have consistently positioned elements, such as the camera feed of the content creator or parts of the game's display (e.g. a minimap or heads-up display), which can be used to extract information from the video given a sufficient computer vision algorithm (e.g. posture, facial expressions, in-game locations, basic game stats).

Game community interactions pose unique challenges for interpretation. From a linguistic perspective, the jargon used in these communities evolves as the game is updated and changed. Further, it contains community-specific valence words, such as references to past events or game elements that have been viewed positively or negatively, and thus convey that sentiment; and this valence can evolve as well, as new events occur and change the associated sentiment. Games often contain a host of proper nouns that are specific to the content of the game, including names of players, characters, or personalities; game elements or equipment; competitions or raids; game levels or maps; and many others. Even with a strong understanding of the game's jargon and content, understanding events or information spread requires a multi-source, multi-format data-extraction procedure. Correlating sources and timing of events is also a key challenge. Media produced on one day might receive comments across weeks or months, and those comments could refer to one of dozens of things occurring in the media. Articles might be updated and redistributed, without clear indications of what has been changed. Forum threads might be revived after long periods of inactivity, and new posts may refer to early statements from much earlier in the thread. Nevertheless, many of these technical challenges can be addressed to analyze behavioral interactions in these communities, without the need to overcome the harsher data restrictions common to in-game data sources. For example, in Case Study #2 (section "Case Study 2: Extracting Popularity Patterns from Hearthstone Community Data"), after identifying terms of interest reflecting common cards and decks used in Hearthstone, we were able to use Reddit discussions to explore information and topic spread surrounding these topics.

Synchronous Community Data Sources (Streaming Sources)

Synchronous community data sources – most notably, streaming sources – are similar to asynchronous data sources in that they represent an opportunity to study social behavior in the community surrounding the game, rather than behavior within the game itself. The major difference found in streaming data sources is that they represent a real-time interaction between community members. Whether it be video streaming or real-time chat across games, these sources provide a view into how game players and observers interact and manipulate each other in real time. Perhaps most notable among these is Twitch.tv, a highly popular streaming video platform in which game players, or streamers, play a game for an audience of subscribers and viewers. The audience observes the stream live, discussing and donating in chat rooms based on what the streamer is doing. This live interaction – which is found on a variety of similar streaming services in other countries – is uniquely different from other game-based data sources, as it provides an opportunity for content producers (streamers) to customize their behavior based on the demands of the crowd (viewers).

Working directly with streaming websites or data aggregators for those websites, significant data is readily available. As with many other data sources, there are licensing terms that must be followed when accessing this data; however, the data that can be made available includes not only aggregate statistics but detailed chat logs as well. While there remains a strong level of sensitivity around how the community will perceive research, companies such as Twitch.tv and Muxy have a strong interest in participating in SBE research to demonstrate the impact of their data.

Existing aggregation tools make two kinds of data available. First, a variety of tools provide statistical data about the streams, assessing the behaviors of the viewership during streaming events (e.g. How does the population change over time? What is the donation or subscription behavior of the population? How often does the population shift to new channels?). Second, several tools provide access to chat data from the viewers and subscribers (e.g. the actual text that they provide).

While the statistical data is relatively simple to extract and interpret, other data poses significant challenges for interpretation and use in SBE research. Chat data poses many of the same interpretation complexities found in forums – the text is often characterized by excessive use of ever-evolving jargon, valence terms, misspellings, and other difficult text (Olejniczak 2015). Because users are often typing quickly, this problem is even more excessive in chat rooms, where shorthand is used as much as possible. Furthermore, communities use an extensive library of emoji that can have unique interpretations in game contexts. Twitch, in particular, introduces a myriad of new game-specific emoji (Barbieri et al. 2017), including *kappa* to refer to sarcastic reactions to game behaviors (e.g. a sarcastic reference to a good move), *PogChamp* to refer to nice plays within a given game instance, and SMOrc to refer to particularly aggressive game styles. The use of these terms carries significant in-game context as well, for example, bringing a knowledge of what it means to make a nice play or be aggressive in specific games such as Hearthstone or LoL.

A key challenge in using this data is that the data contained in the actual stream is not effectively interpreted for use with other data sources. Currently, interpreting this data would require either a time-consuming and intricate process of manual coding or some combination of advanced video or imagery analysis (Pan et al. 2016), as well as speech-to-text interpretation and text analysis to pull out the events and activities within the stream. Furthermore, it would require a significant effort to align events extracted from the stream with the timeline captured in the chat data and meta-statistics surrounding the stream. Currently, this kind of analysis has not been done for Twitch; building an automated system to perform this analysis would greatly enhance the usefulness of Twitch as a data source.

More so than the asynchronous game community data, the streaming community provides an opportunity to study a unique online cultural environment. The real-time interactive nature of streaming communities provides a different dynamic than any other social media platform, with unique opportunities to study emergent entertainment behaviors. Researchers can learn how streamers build and maintain the interest and donations of their fans and how these behaviors differ across subcommunities within Twitch and other streaming environments. They can understand how the economic framework of Twitch works and what kinds of behaviors lead streamers to be able to be successful within this framework. They can understand how information spreads or cascades within a community, particularly if they can investigate how it might be spread by a streamer or how its spread might interact with other community environments. Case Study #3 (section "Case Study 3: Investigating Linguistic Indicators of Subcultures in Twitch") illustrates an analysis of Twitch chat data to understand community subcultures and metadata attributes of the behavior of those subcultures.

Case Studies of SBE Research in Game Environments

In this section, we present several case studies of our research investigating the use of game-based resources to study online behavioral phenomena. We first look at a behavioral analysis of in-game activities extracted through the public data API provided for Riot Games' LoL. We next look at the use of online forums (specifically, Reddit) and online meta-stat sites to explore social and behavioral phenomena in the context of Blizzard's Hearthstone. Finally, we look at a linguistically based subculture analysis of viewer chat behavior in Twitch.tv.

Case Study 1: Extracting Player Behavior from League of Legends Data

Our first case study illustrates the use of in-game data provided through a game API. Game companies that manage online games regularly record a host of behavioral data capturing the activities of the player community within the game. While most companies primarily use this data internally to assess and revise game balance and discover exploits, some companies make parts of this data available to the research community for study. Riot Games, in particular, provides a public API (as well as example data sets) for LoL, the most popular online battle arena video game. This API provides access to a host of aggregate and behavioral data from the game, including static information about game elements (e.g. items, maps, game types, etc.), match and tournament information (e.g. a record and timeline of the games played and maps used, performance statistics by individuals and teams), and detailed match data (e.g. selected classes, development and progression decisions, and even movement records showing where players are on the game map at each time step). While there is no data for direct social interactions (e.g. chat or audio data) provided by the API, it nevertheless provides access to a host of data that enables assessment of players' SBE interactions. In our analysis, we focused on extracting player behaviors from raw match data, assessing how behavioral patterns such as aggression and consistency were linked to success or failure in the game.

While this match data is extensive, it is provided in a raw format that does not inherently provide insight into the specific activities of players. High-level statistics can be readily extracted and assessed (e.g. performance, teaming patterns, etc.), but specific interpreted game behaviors are not provided. Rather, pre-analysis must be done based on an understanding of game tactics and activities to extract behavioral patterns of interest. For example, in LOL, there is a distinct role to be played in managing combat *lanes* in the game (these are paths by which an adversary can choose to attack a team's base). A key aspect of player behavior is the degree to which they effectively perform a lane management role within a given team and game. However, there is no game data to inform analysts of the role each player is performing, nor is there existing game data to inform the analyst of what behaviors are standard, defensive, or aggressive within those roles (e.g. how often they should be in the given lane, how combative they should be, how they should shift to support teammates). Rather, to effectively analyze player behavior, one must bring a deep understanding of the game to develop extraction techniques and assessments that analyze the player's location throughout the game. This can provide early insight into the role the player appears to be pursuing and how consistent or inconsistent the player is in pursuing that role. Once roles are characterized in behavioral patterns, other characteristics can be assessed in the context of these roles (e.g. is the player achieving appropriate *kills* in the game, is the player in the lane a sufficient period of time, does the player maintain the defense of the lane effectively). Analysts can then begin to understand what constitutes desirable and undesirable behaviors in the game. Table 18.2 provides several examples of specific observable strategies that we extracted from LoL match data. Based on these strategies, we attempted to build more detailed metrics for assessing laning behaviors (e.g. assessing whether players who maintained their role were more likely to win a game) and aggression (e.g. assessing whether aggression toward other players and toward controlling territory led to a higher probability of win). Ultimately, while we observed different behaviors across different character types and roles, we saw no evidence for an impact of laning behaviors on performance, while our metrics of aggression seemed to indicate that higher aggression led to better performance. It was unclear, however, whether these were real findings or simply an artifact of limited definitions of these behaviors. Player and territory aggression, in particular, were likely biased in late game stages as winning teams took control of the game; it may be that these metrics would have worked better had they been restricted to earlier stages of the game.

Importantly, this behavioral analysis of game-specific activities is a challenge in any use of in-game data. The game data is unlikely to be organized around the behavioral patterns that emerge from gameplay. Rather, SBE researchers wishing to study these behaviors will need to bring a deep understanding of the game to bear and will need to develop computational methods for extracting these behavioral patterns from the raw data that is collected from the game.

Table 18.2 Observable strategies in League of Legends.

Strategy	Description	Detection
Laning roles	The champions should sort themselves into lanes in an appropriate way to maximize gold and experience gains. This typically means one player in the top lane, one player in the middle lane, two players in the bottom lane, and one free-roaming player between the lanes. Players should be getting appropriate minion kills in their lane to develop four highly experienced champions across the team	Set benchmarks for minion kills and ensure that appropriate players are exceeding benchmarks Observe player activity and ensure that players are appropriately distributed across objectives
Vision	Early-game vision means placing wards near key areas, such as the Dragon (a high-XP monster), blue and red buff locations, and likely locations for adversary attacks. Mid-game targets include the Dragon and Baron, as well as a greater sense of where the enemy is (map coverage). Late-game vision is about covering the area where most of the combat action is taking place	Check ward placement in specific regions in early and mid-game Check level of champion activity near ward placement in late game
Ganks	A coordinated attack where one champion moves out of position to surprise an enemy champion while the normal opponent is harassing them. The intent is to coordinate so the enemy cannot escape	Check for instances of multiple champions targeting one opposing champion, outside of the bottom lane Check for kill rate on those instances for assessment of coordinated attacks
Defense/ counter-ganks	When a champion is being ganked, the others should try to help out. This is especially true in the mid-game when champions tend to be more mobile. A good team will often save its attacked champion or secure one or two kills in retaliation	Check for successions of multiple champions attacking one opposing champion (e.g. *gank detection*) Determine rate of assist and/or movement toward assists
Dragon/Baron kills	In the mid-game and especially early game, it is not possible to take out these objectives alone, and it is dangerous to go after them even in a group, because it invites an attack while players are recovering from the battle. Good groups will coordinate an attack when they have an advantage in numbers and work together to speed up the combat	Detect when someone is attacking these objectives and analyze with respect to (i) game status (early, mid, late), (ii) coordination with allies (e.g. close allies, arrival of assistance), (iii) advantage over adversaries (e.g. exploiting downtimes on adversaries), and (iv) counterattack impacts by adversaries

Table 18.2 (Continued)

Strategy	Description	Detection
Team fights	By the mid-game and especially late game, it is dangerous to be caught too far from allies. Champions who do so might be attacked and killed and might leave their allies vulnerable to a 5v4 attack. Good players will still roam the map, but they will be aware and able to quickly get back to the rest of the group. The group is said to *collapse on* some objective, like the Baron, a turret, or an attack	Define *team fights* (e.g. 7+ champions in close proximity), and assess when those fights are uneven Assess when team members are alive but not near the team fight

While this analysis can be challenging, if done correctly, it can provide deep insights into how different playstyles affect performance and impact teaming patterns. Within LOL, these behaviors can provide critical insights into several key topic areas, including:

- *Investigate how play patterns contribute to group/team stability*: To what degree do players stay together in groups as they start new games or shift around to new groups? How is this shift impacted by specific behaviors in the games? Do certain behaviors lead to more group shifting or more team consistency? Can we observe patterns of play that shift players to taking on different roles within teams?
- *Investigate how in-game behaviors contribute to game experience*: Are there behaviors in game that contribute to or detract from game experience (e.g. win rate, teaming rate, etc.)? Can we correlate specific in-game behaviors to game success? Are there patterns that are so egregious as to lead to people quitting mid-game?
- *Characterize success or failure of communications based on behavioral patterns*: What patterns indicate coordinated activity or failure to coordinate? Based on these patterns, can we predict success or failure? Do patterns indicative of coordination predict success? Do patterns indicative of failed coordination predict team breakups between sessions?
- *Characterize observable patterns of protest in game*: Are there specific patterns of *quiet protest* that show that a player is not happy with the current team? For example, killing oneself in game, staying at home base, leaving one's role, or making poor strategic decisions?

Future research in this area should focus on investigating ways to code and compute behaviors from in-game data and assess how those behaviors reflect

SBE phenomena. In LoL, there are opportunities to explore a wide variety of phenomena. Social phenomena can be found in detailed analyses of teaming and coordination. Behavioral phenomena can be found in detailed analyses of other game behaviors, ranging from individual protests to bullying to activities centered on following custom and crowd directives. Economic phenomena can be found in detailed analysis of the purchasing patterns for upgrades to the champions players are controlling (e.g. how they choose particular upgrades, what costs make those upgrades worth pursuing). Some of these factors can be directly investigated within the data provided by the LoL API, though it often takes significant interpretation and translation to properly define these factors in the context of the game. As more games expose this data, SBE research communities can benefit by researching phenomena across a wider range of game contexts.

Case Study 2: Extracting Popularity Patterns from Hearthstone Community Data

Our second case study focused on the use of data from the community and meta-game websites surrounding a game. Many modern games are surrounded by an active meta-game and online social community discussing the game to build or better understand game strategies and features. These websites provide a powerful opportunity to study constrained interactions across large populations of individuals with similar interests. Sources that can be used for this type of analysis are broad and often publicly accessible, meaning that the data can be readily downloaded or scraped for use and that users have little expectation for the privacy of that data (since it is already available publicly online). Source types include:

- Community forums discussing particular games or related topics.
- Major fan sites that focus on game or content reviews and ratings; many of these sites aggregate statistics either directly from the game or from third-party observations of the games.
- Strategy sites that focus on providing guides for players to improve their game.
- Official websites that provide game news, release information, and other game data.
- Podcasts that provide commentary on rising trends in the game.
- Streaming sites that provide video content of successful gameplay, often with commentary and related discussion.

In our second case study, we acquired and analyzed online data available from communities surrounding Blizzard's Hearthstone. As previously mentioned, Hearthstone is a card-based strategy game where online players build and compete with strategic decks combining these cards. We explored a variety of meta-game and asynchronous community data sources associated with Hearthstone, seeking those that gave indicators of player behaviors in card and deck selection.

A key challenge in using these data sources lay in addressing the problem of extracting and aggregating the data, which is rarely conveniently stored for research use. SBE researchers will need to use a variety of software tools to acquire such data. They can access it using some combination of subscriptions, Web crawlers, and manual downloads, transferring it into a database for manipulation and analysis. To translate it into a format that can be analyzed, they may need to extract further information from it. For example, extracting commentary from videos in Twitch or YouTube may require a time-consuming coding process that can be challenging to keep consistent (particularly across coders). In most cases, these channels incorporate some form of unstructured text or language and therefore require some level of text parsing capability for interpretation. Depending on the sophistication of the desired analysis, this could range from a simple extraction of key terms and phrases (in our analysis, we focused on deck and card names from Hearthstone) to an interpretation of the valence or content of posts or articles. If more detailed interpretation is desired, this can pose a difficult problem as the language is often characterized by difficult text interpretation challenges (e.g. game community slang, shortened references to key concepts, pronoun use, typing errors).

In our Hearthstone case study, we focused on three data sources:

- *MetaStats class and archetype usage data*: We collected user data for class use and archetype representation in competitive Hearthstone matches from 1 April to 26 June 2017. The user data shows what classes and deck styles the community had used immediately following an expansion, providing a starting point for determining trends in community behavior. This allowed us to create forecasts about future expansions or generalizations to other communities.
- *Reddit Hearthstone community data*: We extracted all discussions for the Hearthstone and competitive Hearthstone communities and computed metrics assessing the characteristics of those discussions. Each file includes post time, score (i.e. popularity), user ID, replies, and discussion bodies. We maintain a hierarchy of the Reddit posts and discussions that followed. We use this data to try to identify gatekeepers (Meraz and Papacharissi 2013) and discussion trends before they occur within the MetaStats data. High scoring posts with many replies are likely to have been seen by many viewers, potentially influencing MetaStats. We also explored valence data for all posts on both Reddit communities, enabling an analysis of community reception surrounding particular messaging.
- *HearthPwn community data*: Similar to Reddit, we collected all forum discussions from the General Discussion, Card Discussion, and Deck Building

forums, as well as meta-information surrounding those discussions. This community provides a more focused look at shifts in the game strategy, with discussions on why cards or deck styles work or fail in the current state of the game. HearthPwn also provides deck building tools within which users share, discuss, and rate decks based on their experiences. We scraped this data to assess how decks evolved.

Using this data, we set forth to assess information spread within the Hearthstone community. We began by using the MetaStat data to identify a set of key terms for referring to deck archetypes. We then used these as discussion topics in the Hearthstone community, assessing the frequency of their mention over time and across community websites and forums. Using this data, we analyzed the relationships between potential gatekeepers in the Hearthstone community, focusing primarily on high-profile game players who are known for their skill at the game and their ability to construct and define new decks and strategies. We found that several deck types – specifically, the *Midrange Shaman* and *Pirate Warrior* – saw an increase in popularity as they were pursued by multiple popular figures, most notably including the Hearthstone personality, *Lifecoach*. While we were able to identify some possible gatekeepers after the fact, we were unable to identify data-driven indicators of gatekeepers before a deck became popular (beyond the knowledge that known game personalities were more likely to make a deck popular than unknown players).

Next we explored potential instances of information spread, in which a topic's use expands rapidly through a community, and information cascades, in which the sentiment around a topic significantly shifts without a clear cause (Cha et al. 2008; Cheng et al. 2014). We began this analysis by plotting the popularity of key deck types across each of the Hearthstone archetypes using MetaStats data. We identified a number of sudden rises in popularity of particular decks, providing a starting point for exploring potential events. Figure 18.2 shows several examples observed in the period surrounding the quarterly Hearthstone update on 4 April 2017. Specifically, we observe (a) the rise of the Aggro Token Druid and the fall and subsequent rise of the Ramp Druid; (b) the belated rise of the Burn Mage and the initial rise, drop, and then significant rise of the Secret Mage; (c) the mid-tier rise of the Control Paladin, belated rise of the Aggro Paladin, and significant rise of the Midrange Paladin; and (d) the significant rise of the Quest Rogue and the drop and spike of the Miracle Rogue.

These examples illustrate common behavioral patterns observed when regular updates occur in games. Players realize that their current strategies are no longer effective (or have become less effective) and rapidly stop using those strategies. Players quickly begin to try new strategies (new decks in the case of Hearthstone), in some cases simply to explore the new content and in other cases because that content is more effective than previous content. As the new content loses its luster, some of it drops off in use (e.g. the content that was

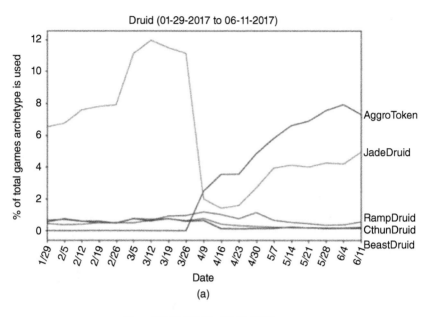

(a)

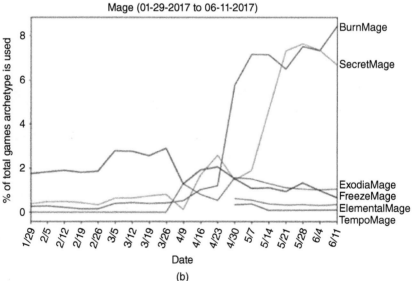

(b)

Figure 18.2 Popularity for a variety of specific types within each class, showing the percentage of total games in which the deck is used (*y*-axis) by date (*x*-axis). (a) Druid deck popularity. (b) Mage deck popularity. (c) Paladin deck popularity. (d) Rogue deck popularity.

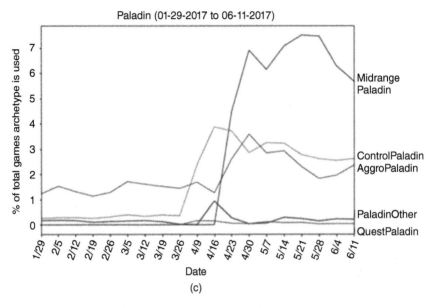

(c)

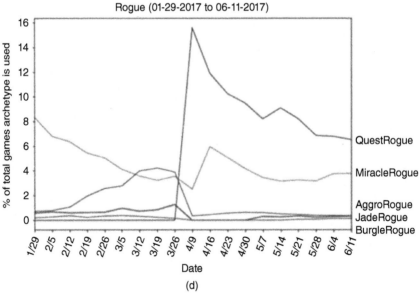

(d)

Figure 18.2 *(Continued)*

simply exciting because it was new), while some of it sees a more consistent rise (e.g. the content that may, perhaps, be overpowered in terms of game balance). These kinds of patterns can be observed across numerous games beyond Hearthstone and represent essential insights for game designers to drive future decisions in game updates to maintain the interest of the player communities.

Based on the instances of popularity increases that we saw – and theorizing that these increases arise from initial information cascades on each particular deck – we next assessed discussions surrounding several key decks on Reddit. We chose not to focus on the decks that were predicted to be successful before the game patch. For example, based on prerelease rumors and advertising material, the *Control Paladin, Aggro Paladin*, and *Midrange Paladin* were each predicted to be powerful, providing a clear explanation for their rise in popularity. Instead, we focused on decks that surprisingly rose in popularity, with the idea that we might be able to observe an information cascade that led to that popularity rise. As shown in Figure 18.3 we explored potential increases in discussion around several decks, including a variety of *Quest* decks (Rogue, Hunter, Druid, and Warrior), as well as the *Secret Mage* and *Control Priest*. While there was a significant increase in discussion around the *Quest Rogue* as it became popular, most of the decks we explored did not see any significant increase in discussion. Furthermore, the discussions surrounding the *Quest Rogue* did not follow the pattern of an information cascade – rather, they showed a more standard information spread pattern as players realized the deck's effectiveness. This pattern is best explained by the deck performing well on aggregated data collection sites, such as MetaStat, rather than a deep and rapid shift in community opinion.

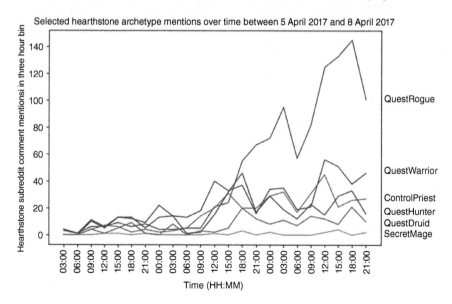

Figure 18.3 Mentions over time for selected Hearthstone decks on Reddit.

Future research in this area should focus on investigating ways to more effectively extract events and link them across the variety of community sources that can be used to investigate these games. The surrounding community for many games represents an extensive online social environment with a variety of sources that are used in tandem by large groups of players. Meta-game sources provide essential context for interpreting interactions in this social environment. Social phenomena can be studied in the interactions occurring regularly in forums, chat communities, and streaming video websites (e.g. Twitch). Behavioral phenomena, ranging from gatekeepers and information spread to full-blown information cascades, are likely to exist throughout these communities. While it was not a focus of our research, many communities incorporate subscription services that provide opportunities for exploring economic phenomena as well. Developing methods to interpret community interactions and behavior can provide a large source of data for analyzing online behavior, particularly from a social and behavioral perspective.

Case Study 3: Investigating Linguistic Indicators of Subcultures in Twitch

Our third and final case study focused on cross-game synchronous community behavior, rather than a specific game, using Twitch.tv. Twitch is a highly popular streaming video platform in which game players, or streamers – many of which are highly skilled and popular online personalities – stream their game activity for an audience. The audience observe the stream live and often discuss and donate in chat rooms based on those observations as the streamer continues to play. Twitch and other similar streaming services – such as Japan's *Live Tube* and China's *Huya* – provide a live interaction dynamic that is uniquely different from other social media services, in that viewers are interacting, in real time, with a performer (streamer) who is trying to maintain their attention and contribution. Streamers often customize their game selection and in-game behaviors to address interests raised by the audience, in an attempt to better engage that audience and draw more donations. This behavior represents a complex social interaction across a variety of users with different roles, dynamics, and behaviors. Further, there is an interactive economic relationship, as many of the streamers rely on contributions from their viewers as a primary source of income and are highly successful in doing so. As previously mentioned, Twitch also provides an advantage for SBE research in that it provides a truly vast population to observe, with over 100 million unique monthly users. Opportunities for researching SBE phenomena within this community are plentiful.

Twitch makes an API available to partner vendors for extracting chat data, as well as aggregate data capturing the popularity of streams and the shifting of viewers across those streams. Several online vendors (e.g. http://muxy

Table 18.3 Muxy data characteristics.

Table	Characteristics
General information	Key information about the stream, including frames per second (FPS), language, maturity, delay, game, viewer count, follower count, chatter count, mentions, and other information
YouTube data	Likes, dislikes, favorites of streams that are shared in YouTube (Twitch streams are often cross-referenced in YouTube; this data is not available from Twinge)
Chat data	Channel list, observers, and user messages (this data is not available from Twinge)
Panels	Title, image, description
Links	Linked data from the Twitch panels
Users leaving	When users leave and/or enter different channels

.io, http://twinge.tv) aggregate and maintain a history of this data, providing online services and commercial applications for using the data. For example, Muxy helps streamers understand what activities lead to more or less support from viewers, leading to better and more profitable decisions in streaming content. Muxy has been collecting Twitch chat data and viewership statistics for several years now (see Table 18.3), including roughly 45 million lines of chat per day. Partnering with these vendors provides a unique opportunity for SBE researchers to access this vast source of online behavioral data.

Working with Muxy, we acquired limited windows of Twitch data, which we used to perform a study using linguistic analysis to identify cultural clusters of related channels within the Twitch community. We focused our analysis on a subset of the messages surrounding donations, which are, in many cases, the more important chat messages within a channel (i.e. the messages that are producing a profit for the streamer and therefore likely impacting streaming content the most). Using Python's Natural Language Toolkit, we cleaned the text, removing punctuation, numeric characters, and stop words and reducing words to their roots. Next we used term frequency–inverse document frequency (TFIDF) encoding (Ramos 2003; Zhang et al. 2011) to rate the importance of each term within each channel, characterizing a linguistic feature set for all of the channels in our data window. We then used k-means clustering (Kanungo et al. 2002) to effectively split these channels into ten coherent clusters with similar linguistic features, summarized in Figure 18.4 and Table 18.4.

There were a number of interesting features to observe across these clusters. While some clusters appear to be focused on particular games (e.g. cluster 3 on *World of Warcraft*), many games clearly were not a differentiating factor for

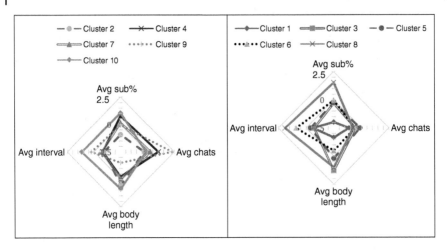

Figure 18.4 Activity features of clusters of channels. Figures are formed based on words or emotes used in bits messages. They show how these two different methods of characterizing channels point to the existence of different Twitch subcultures.

a linguistic analysis (e.g. games such as *Call of Duty: Black Ops III*, *Minecraft*, and *Friday the 13th: The Game* were repeated across many clusters). Most of the clusters were relatively small (several hundred channels), though there were several outliers (clusters 4 and 5, in particular). Average viewers per channel and donations per channel were significantly different across a number of clusters; this, in turn, has significant implications for the nature of the communications in some of those channels (e.g. clusters that average 200+ viewers per channel will likely portray very different behavior than those with less than 100 viewers per channel; clusters with donations exceeding an average of 200 bits are likely very different from those with significantly lower donations).

Next, we characterized these clusters based on meta-information describing behavioral elements of the chat in these clusters, shown in Figure 18.4. These plots show the z-score of each cluster's average value for subscriber percentage (fraction of all chat messages observed that were sent by subscribers, as a measure of how insular the cluster is), number of chats observed in our sample set, length of chat message bodies, and interval between chats. These plots highlight different patterns across the clusters we identified, providing further evidence that at least some of these clusters may clear behavioral differences in the Twitch community – behavioral differences that may be indicative of different Twitch subcultures. We believe that this work provides a foundation for analyzing and identifying potential avenues for information sharing and

Table 18.4 Breakdown of clusters identified through linguistic analysis, identifying the key games played in the cluster, the number of channels in the cluster, the average viewers per channel in the cluster, and the average bit donations per channel in the cluster (a form of payment that viewers make to streamers).

No.	Games streamed (arbitrary order)	Channel count	Average viewers/ channel	Average bit donation/ channel
1	Smite, Elite Dangerous, Call of Duty: Black Ops III, World of Warcraft, NBA 2K17, Arma 3, Minecraft	278	71.2	223.9
2	Friday the 13th: The Game, Darkest Dungeon, Dead by Daylight, Minecraft	871	61.0	134.0
3	World of Warcraft, ARK	429	81.3	228.9
4	Rocket League, Black Desert Online, RuneScape, Diablo III, World of Warcraft, Minecraft, Hearthstone	2687	203.1	214.7
5	Friday the 13th: The Game, Call of Duty: Black Ops III, Minecraft	3556	81.0	188.2
6	M.U.G.E.N, Diablo III, Call of Duty: Black Ops III, The Sims 4, Terraria, Fallout 4, Hearthstone, Arma 3, Heroes of the Storm, World of Tanks	205	116.8	182.5
7	Halo 5: Guardians, RuneScape, Final Fantasy XIV Online, Diablo III, Call of Duty: Black Ops III, 7 Days to Die, Paladins, Arma 3, Minecraft, World of Tanks	323	107.5	112.0
8	Persona 5, Legend of Zelda: Breath of the Wild, Pokémon Trading Card Game, MapleStory, Dark by Daylight, 7 Days to Die, Minecraft, Call of Duty: Infinite Warfare, Black Desert Online, Blade & Soul, Life is Strange, Mario Kart 8	217	94.0	124.6
9	Dead by Daylight, Darkest Dungeon, Dota 2, Blade & Soul, Black Desert Online, Battlefield 1, Dark Souls III, Minecraft, Super Mario Maker	358	465.6	220.6
10	Halo 5: Guardians, Retro, Elite Dangerous, Pokémon Sun and Moon, Call of Duty: Black Ops III, Friday the 13th: The Game, Paladins, Pinball, Train Simulator 2017, MechWarrior Online, Legend of Zelda: Ocarina of Time	228	139.0	248.5

gatekeeping. If these clusters are, indeed, indicative of related subcultures within Twitch, then it stands to reason that information is more likely to spread within these clusters than across them. Further research investigating the details of these clusters – in particular, who the streamers are and how those streamers interact with viewers – would be needed to better characterize the underlying cultural differences across the Twitch community.

This linguistic analysis represents only a limited subset of the type of analysis that can be done with Twitch data. We also performed several studies investigating the potential of information spread within the Twitch community, investigating how rumors and topics spread over time across streams. While there are clearly extensive communications occurring across the streaming community, we were unable to demonstrate evidence of specific information spread within Twitch. There were a number of confounding factors in this analysis. For example, there are clear confounds in the linguistic patterns of the community that make it difficult to track key terms. Community members often use slang that can be difficult to predefine or use pronouns in response to others discussing the topic. Behavioral patterns also confound the interpretation of this data. For example, we observed that many viewers remained in a single channel for the majority of each session (on average, across the community, viewers join 1.3 channels per session, despite sessions lasting, on average, 45+ minutes); once viewers were watching a particular stream, they were unlikely to switch channels and therefore unlikely to be the source of information spread across channels. However, despite this constraint, many viewers – particularly power users – remain active in many chat channels at once using third-party chat tools. Because of this external ability to observe and participate in chat, it seems likely that there are a variety of connections across channels that we are unable to track using the Twitch data alone.

Finally, a key limitation of Twitch data (and other streaming video communities) is limited insight into the content of the actual streaming video. While popular streams remain available online for long periods after they occurred, they are not coded into text or events, and they are not aligned with any other data surrounding the discussions regarding the stream. There has been some limited work extracting image-based events from streams where possible (Pan et al. 2016), but there has been no comprehensive attempt to encode stream content. This is important because the stream content could very well be a mechanism of information spread (e.g. the streamer could mention a topic or rumor to the viewers), and the stream itself can provide essential context to interpret the discussion occurring in chat. Future work developing automated tools to encode stream content and align detected events with the timeline of the chat data could create significant opportunities for performing more advanced behavioral research studying the interactions between streamers and subscribers.

Conclusions and Future Recommendations

Games and game communities provide a powerful platform for pursuing SBE research. While there is a clear need for further research to better understand behavioral differences between games and live settings, multiplayer games express extensive SBE interactions across often large player communities that surpass populations that can be accessed in research settings. Behaviors include players teaming to achieve game objectives, competing against other teams of players, and interacting socially and economically between play sessions and events. Researchers can use a variety of data sources to study in-game behaviors, including direct access to limited game data from APIs provided by a handful of game companies, indirect access to in-game data provided by observations made in public game spaces (e.g. markets or team formation areas in MMOGs), and a host of websites and forums that collect meta-game statistics capturing the popular strategies and approaches used within the games. While there are clear challenges in extracting and interpreting the data from these sources (interpreting game behaviors; parsing text, audio, and video; addressing gaps in data collection), they nevertheless provide a capability to analyze how people behave in game environments.

A key area for future research to enhance in-game behavior study would be finding ways to enhance data collection and availability. Partnering with successful game companies to share more extensive data in exchange for feedback useful to their game improvement can greatly enhance the potential of research in this area. Developing learning algorithms to aggregate meta-game information and use that information to interpret in-game data and events could streamline behavioral analysis. Another approach might be to work with emerging game companies directly, funding those companies from an early stage to ensure that they make data available to SBE researchers when the game is complete.

Surrounding game communities also express extensive SBE interactions across large player communities. Behaviors include players discussing and spreading information about the game, sharing and developing ideas and strategies, learning from more skilled players and streamers, and earning profit by satisfying and engaging player communities. Researchers can readily gain access to a variety of data sources in this area, including public forums and content sharing sites, as well as detailed chat and streaming statistics from real-time sources such as Twitch. This data can be easier to access than in-game data and is more suited to analyzing online behavioral phenomena (e.g. gatekeeping, information spread and evolution, information cascade). However, it does not involve the detailed real-time interactions and structure that can be found with in-game data. As with in-game data, there are challenges in extracting and interpreting this data, mostly challenges of interpreting text. Nevertheless, clear and extensive social communities exist for study, some of which continue to grow and evolve. Twitch, in particular, has continued to grow since its inception and has been evolving its content beyond game communities with creative, real-life, and other types of streaming content.

Future research areas for enhancing the study of game communities should focus on two constraints in current data sources. First, research should focus on enriching existing data to address under-characterized or under-interpreted elements of the source. For example, Twitch analyses would greatly benefit from automated or crowdsourced tools to code events and behaviors within the streams, ideally aligning those events with activities in the chat data. This would enable more robust analysis of the interactions occurring in Twitch, as researchers could better understand what the viewers are reacting to and how the streamers are reacting to viewers. Similarly, studies using podcasts and videos as data sources could benefit from better automated tools for coding events and behavior within that media. Second, research should focus on aligning social behavior in these communities with activity in other social media sources. We found that, in many cases, it was unclear how information spread was occurring within the social communities that we were investigating. Aligning events across a variety of information sources – Twitch, Reddit, Twitter, and news services – can create a fundamentally better ability to determine how information spread is occurring across what is, ultimately, a highly interconnected social media environment. Finally, while SBE phenomena can clearly be observed within games and game communities, more research is needed to understand how these observations can be used as a basis to understand analogous behaviors in live environments.

Acknowledgments

The work described in this paper was funded by DARPA/I2O and performed under AFRL contract number FA8750-17-C-0081. The authors thank Dr. Jonathan Pfautz, Dr. Jennifer Roberts, Dr. Brian Dennis, and Mr. Steven Drager for their engagement and technical support throughout the project. The authors also thank Mr. Peter Bonanni of Muxy for his support in investigating Twitch as a data source for SBE research investigating game communities. Any opinions, findings, and conclusions or recommendations expressed in this material are those of the authors and do not necessarily reflect the views of the US Air Force or DARPA.

References

Barbieri, F., Espinosa-Anke, L., Ballesteros, M. et al. (2017). Towards the understanding of gaming audiences by modeling Twitch emotes. http://repositori.upf.edu/handle/10230/33289 (accessed 10 September 2018).

Bates, M.C.B. IV (2009). Persistent rhetoric for persistent worlds: the mutability of the self in massively multiplayer online role-playing games. *Quarterly Review of Film and Video* 26 (2): 102–117. https://doi.org/10.1080/10509200600737770.

Blackburn, J. and Kwak, H. (2014). STFU NOOB!: predicting crowdsourced decisions on toxic behavior in online games. In: *Proceedings of the 23rd International Conference on World Wide Web*, 877–888. New York, NY: ACM https://doi.org/10.1145/2566486.2567987.

Cha, M., Mislove, A., Adams, B., and Gummadi, K.P. (2008). Characterizing social cascades in flickr. In: *Proceedings of the First Workshop on Online Social Networks*, 13–18. New York, NY: ACM https://doi.org/10.1145/1397735.1397739.

Cheng, J., Adamic, L., Dow, P.A. et al. (2014). Can cascades be predicted? In: *Proceedings of the 23rd International Conference on World Wide Web*, 925–936. New York, NY: ACM https://doi.org/10.1145/2566486.2567997.

Conway, S. and deWinter, J. (2015). *Video Game Policy: Production, Distribution, and Consumption*. Routledge.

Drachen, A., Riley, J., Baskin, S., and Klabjan, D. (2016). Going out of business: auction house behavior in the massively multi-player online game. ArXiv:1603.07610 [Cs, Stat]. Retrieved from http://arxiv.org/abs/1603.07610.

Flores, M. (1999). StarCityGames.com - who's the beatdown? Retrieved from http://www.starcitygames.com/magic/fundamentals/3692_Whos_The_Beatdown.html (accessed 21 April 2018).

Gandolfi, E. (2016). To watch or to play, it is in the game: the game culture on Twitch.tv among performers, plays and audiences. *Journal of Gaming & Virtual Worlds* 8 (1): 63–82. https://doi.org/10.1386/jgvw.8.1.63_1.

Gargioni, A. (2018). 10 reasons why you should not ignore marketing on Twitch. [Blog post]. Retrieved from https://www.greengeeks.com/blog/2018/02/13/10-reasons-why-you-should-not-ignore-marketing-on-twitch/.

Harviainen, J.T. and Hamari, J. (2015). Seek, share, or withhold: information trading in MMORPGs. *Journal of Documentation* 71 (6): 1119–1134. http://doi.org/10.1108/JD-09-2014-0135.

Johnson, M.R. and Woodcock, J. (2017). 'It's like the gold rush': the lives and careers of professional video game streamers on Twitch.tv. *Information, Communication and Society* 1–16. https://doi.org/10.1080/1369118X.2017.1386229.

Kanungo, T., Mount, D.M., Netanyahu, N.S. et al. (2002). An efficient k-means clustering algorithm: analysis and implementation. *IEEE Transactions on Pattern Analysis and Machine Intelligence* 24 (7): 881–892. https://doi.org/10.1109/TPAMI.2002.1017616.

Kim, J., Keegan, B.C., Park, S., and Oh, A. (2016). The proficiency-congruency dilemma: virtual team design and performance in multiplayer online games. In: *Proceedings of the 2016 CHI Conference on Human Factors in Computing*

Systems, 4351–4365. New York, NY: ACM https://doi.org/10.1145/2858036 .2858464.

King, D., Delfabbro, P., and Griffiths, M. (2010). Video game structural characteristics: a new psychological taxonomy. *International Journal of Mental Health and Addiction* 8 (1): 90–106. https://doi.org/10.1007/s11469-009-9206-4.

Knowles, I., Castronova, E., and Ross, T. (2015). Virtual economies: origins and issues. In: *The International Encyclopedia of Digital Communication and Society* (ed. P.H. Ang and R. Mansell), 1–6. American Cancer Society https://doi .org/10.1002/9781118767771.wbiedcs046.

Kwak, H., Blackburn, J., and Han, S. (2015). Exploring cyberbullying and other toxic behavior in team competition online games. In: *Proceedings of the 33rd Annual ACM Conference on Human Factors in Computing Systems*, 3739–3748. New York, NY: ACM https://doi.org/10.1145/2702123.2702529.

Meraz, S. and Papacharissi, Z. (2013). Networked gatekeeping and networked framing on #Egypt. *The International Journal of Press/Politics* 18 (2): 138–166. https://doi.org/10.1177/1940161212474472.

Nazir, M. and Lui, C.S.M. (2016). A brief history of virtual economy. *Journal of Virtual Worlds Research* 9 (1): https://doi.org/10.4101/jvwr.v9i1.7179.

Olejniczak, J. (2015). A linguistic study of language variety used on Twitch.tv: descriptive and corpus-based approaches. *Redefining Community in Intercultural Context* 4 (1): 329–334.

Pan, R., Bartram, L., and Neustaedter, C. (2016). *TwitchViz: A Visualization Tool for Twitch Chatrooms*, 1959–1965. ACM Press https://doi.org/10.1145/2851581 .2892427.

Papagiannidis, S., Bourlakis, M., and Li, F. (2008). Making real money in virtual worlds: MMORPGs and emerging business opportunities, challenges and ethical implications in metaverses. *Technological Forecasting and Social Change* 75 (5): 610–622. https://doi.org/10.1016/j.techfore.2007.04.007.

Patel, R. (2016). Computational market dynamics of virtual economies. Master thesis in Computational Engineering. Lappeenranta University of Technology.

Ramos, J. (2003). Using TF-IDF to determine word relevance in document queries. In: *Proceedings of the First Instructional Conference on Machine Learning*, vol. 242, 133–142.

Shi, L. and Huang, W. (2004). Apply social network analysis and data mining to dynamic task synthesis for persistent MMORPG virtual world. In: *Entertainment Computing – ICEC 2004* (ed. M. Rauterberg), 204–215. Berlin, Heidelberg: Springer https://doi.org/10.1007/978-3-540-28643-1_27.

Shukla, P. and Drennan, J. (2018). Interactive effects of individual- and group-level variables on virtual purchase behavior in online communities. *Information and Management* https://doi.org/10.1016/j.im.2018.01.001.

Sköld, O., Adams, S., Harviainen, J.T., and Huvila, I. (2015). Studying games from the viewpoint of information. In: *Game Research Methods* (ed. P. Lankoski and

S. Björk), 57–73. Pittsburgh, PA: ETC Press. Retrieved from http://dl.acm.org/citation.cfm?id=2812774.2812781.

Taylor, N., Bergstrom, K., Jenson, J., and de Castell, S. (2015). Alienated playbour: relations of production in EVE online. *Games and Culture* 10 (4): 365–388. https://doi.org/10.1177/1555412014565507.

Wu, W., Fore, S., Wang, X., and Ho, P.S.Y. (2007). Beyond virtual carnival and masquerade: in-game marriage on the Chinese internet. *Games and Culture* 2 (1): 59–89. https://doi.org/10.1177/1555412006296248.

Zhang, W., Yoshida, T., and Tang, X. (2011). A comparative study of TF*IDF, LSI and multi-words for text classification. *Expert Systems with Applications* 38 (3): 2758–2765. https://doi.org/10.1016/j.eswa.2010.08.066.

19

An Artificial Intelligence/Machine Learning Perspective on Social Simulation: New Data and New Challenges

Osonde Osoba and Paul K. Davis

RAND Corporation and Pardee RAND Graduate School Santa Monica, CA 90401, USA

Objectives and Background

There is a growing demand to develop social and behavioral models competent to inform decision-making in such diverse domains as counterinsurgency, political polarization, adversary propaganda campaigns, and public health behaviors. Success will depend on effective use of empirical information drawn from observation of both online and physical network human behavior. The objectives of this discussion are:

- To characterize the current infrastructure of data relevant to behavioral modeling.
- To describe progress on methods relevant to behavioral modeling that come from research on artificial intelligence (AI) and machine learning (ML).
- To identify shortcomings and challenges with current modeling approaches.
- To suggest where advances are needed to address them.
- To identify some mechanisms for doing so.

An earlier study conducted over three years by the National Research Council (Zacharias et al. 2008) presented a comprehensive review of social and behavioral modeling. Most of that analysis remains solid and apt today. Thus, we focus primarily on selected developments over the last 10 years.

Relevant Advances

Overview

Two trends relevant to this chapter have been notable over the last decade, (i) the burgeoning of data, data sources, and data infrastructure and (ii) AI/ML methods:

Social-Behavioral Modeling for Complex Systems, First Edition.
Edited by Paul K. Davis, Angela O'Mahony, and Jonathan Pfautz.
© 2019 John Wiley & Sons, Inc. Published 2019 by John Wiley & Sons, Inc.
Companion website: www.wiley.com/go/Davis_Social-Behavioralmodeling

Data: social media more than MMOGs: The 2008 NRC report highlighted massive multiplayer online game (MMOG) platforms, surveys, and ethnography as predominant sources of behavioral data for training behavioral models. Some recent work has explored MMOGs for understanding observed social behaviors (e.g. identifying patterns observable in the play of Pokémon Go; Althoff et al. 2016, 2017). More generally, however, it seems that the usefulness of MMOG platforms for behavioral insight is less than was imagined. MMOGs are interesting troves of data, but the insights appear not to add much to the foundations of behavioral modeling research. This observation makes sense in hindsight. MMOGs are specific ecosystems of behaviors and contrived ones at that. Researchers can distill out interesting patterns of behavior (Tomai et al. 2013), but these are not necessarily generalizable or informative outside the contrived environment. Arguably, surveys and ethnography (including micro-narratives) are better for eliciting insights on more general behavioral patterns, although they can be unwieldy and intrusive. Ubiquitous *social media* (SM) platforms seem to be popular as sources of relevant behavioral data for now. We shall discuss some of this in a later section.

AI/ML methods: As for advances in AI/ML methods, we construe the topic broadly to include:

- Adaptive statistics- and optimization-based methods for teaching computers to identify or exploit regularities in signals (ML or pattern recognition).
- Expert-, rule-, or logic-based methods for planning, problem solving, or knowledge representation.
- Models for representing or imitating cognition and decision-making (human or otherwise).

These topics fall under the useful general definition of AI as the discipline "concerned with intelligent behavior in artifacts.[1]"

The term *ML* often refers to the more statistically flavored subfields like supervised, unsupervised, and reinforcement learning (RL). Other AI subfields rely more on symbolic-based and rule-based methods for tasks like knowledge representation and automated planning. Some earlier versions of these were known as expert systems and knowledge-based systems in the 1980s. These include AI approaches like automated planning solvers, fuzzy cognitive maps (FCMs) (Amirkhani et al. 2017), and tree-based methods for parsing semantics

---

1 There have been numerous other attempts to define AI canonically. This description is due to Nilsson (1998). McCarthy (2007) defines intelligence as "the computational part of the ability to achieve goals in the world." Minsky (1961) gave an enumeration of functions required for such intelligence: search, pattern recognition, learning, planning, and induction (or generalization from observed examples). Any artificially constructed or software-based system performing combinations of these functions to achieve goals in the world will qualify as AI for the purpose of our discussion.

and ontologies. Other non-ML strands of AI research include cognitive modeling architectures like BDI (Tambe et al. 1988), Soar (Georgeff et al. 1998), ACT-R (Anderson 1996), and EPIC (Rubinstein et al. 2001). These have been useful for enabling tasks like team-based collaboration in robots. The varied nature of social and behavioral modeling requires the full diversity of AI methods. We see it as important to consider all of these methods (i.e. to include what some refer to as both strong and weak strands of AI).[2]

Numerous relationships exist among what are sometimes treated as different methods. The authors of the 2008 NRC report generated one depiction of the various methods and how they relate to each other, as indicated in Figure 19.1. Although useful for drafting and structuring the large and complex NRC report, Figure 19.1 uses a fine-grained disaggregation of modeling approaches, which the report discussed as individual, organizational, and societal (IOS) modeling tools. For our purposes, such a disaggregation downplays how deeply interconnected these approaches sometimes are and sometimes should be. For example, the optimization node (near bottom left) stands alone in the figure. However, a central theme in current ML methods is learning as optimization (see dashed line 1); much of modern ML relies on optimization procedures like stochastic gradient descent during training. The distinction between ML and statistics is also not nearly wide as the figure suggests (see dotted line 2). Another problem with such a disaggregation is that it may obscure opportunities for innovation. An important recent innovation in AI/ML, generative adversarial networks (GANs), combines game theory and ML to improve unsupervised learning tasks. Other current innovations include the use of statistical, ML, and social network analysis methods to infer behavioral patterns (Sapiezynski et al. 2016).

The subsequent sections discuss advances in data sources and AI/ML in more detail.

Advances in Data Infrastructure

New Sources

Many new data sources have emerged in the last decade as important ecosystems for exhibiting and recording behavioral patterns. Each data source has blind spots. On the one hand, a larger ecosystem might be expected to increase the chances of capturing key behavioral information. On the other hand, a smaller data ecosystem would require less modeling effort and might be more cost efficient because infrastructure costs burgeon with ecosystem size, perhaps faster than any benefits. Nonetheless, from a purely modeling

---

2 Strong AI aspires to computer programs that represent human cognition and achieve significant aspects of humanlike intelligence. AI researchers are strongly divided about the degree to which strong AI is feasible.

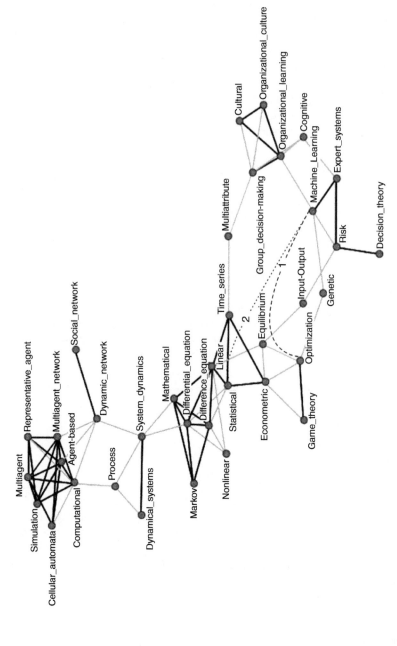

Figure 19.1 A similarity network of modeling methods. Source: Adapted from NRC report Zacharias et al. (2008), p. 93.

perspective, one might think that more data from more sources should – other things being equal – increase capacity for behavioral modeling.

Some novel data streams have already shown value for behavioral modeling. For example, SM data has become ubiquitous and is now an important analytic tool in national politics. SM platforms also serve as tools for influencing discourse and for measuring behaviors/influence. They measure individual data and also relational or network data. The ascendance of SM platforms has led to a rise in methods and tools for dealing with relational/network/graph information.

Cellular data records (CDRs), including metadata, are also valuable, especially for inferring spatial behavioral preferences. Recent research indicates that CDRs are very useful for identifying spatial behaviors relevant to, e.g. migration/disaster response (Bengtsson et al. 2011), shopping patterns (De Montjoye et al. 2015), and personal network affinities. And even limited metadata on CDRs can reveal useful patterns (e.g. how much talking between two contacts).

Other novel data streams include:

- Financial records from banks, retail records, and financial technology firms (FinTech).
- Public and private video surveillance from street cameras and cell phone recordings.
- Voice records from AI personal assistants (Siri, Alexa, Google Now).

The growth in behaviorally relevant data streams mirrors a growing reliance on tools and devices for mediating behaviors (e.g. GPS for navigation, music streams for mood management, SM platforms for expression). It is now common to have records or signals of an individual's plans, intentions, and mental states in digital form, especially in affluent cultures with high smartphone adoption. Clark and Chalmers (1998) used the term *the extended mind* to refer to the extension of mental deliberation and cognition outside the (as yet) unobservable confines of the human mind. Extended minds with *accessible* digital data exhausts can be potentially revolutionary for behavioral modeling. Data ecosystems that include extended minds enable finer-grained behavioral models that may be able to capture the effects of intention, self-censorship, or even ambivalence in a way that models based solely on actual observed behavior cannot. The data ecosystem seems to be growing in that direction.

Evaluating the Data Ecosystem

Whatever its primary function may be, for our purposes a data ecosystem serves at least a measurement function. Its value in this role depends on at least the following:

- *Representativeness of the measured population*: The measured population (the proxy population) should be representative of the background population about which we want to infer behavioral patterns. Unfortunately, SM platforms generally exhibit significant *proxy population mismatch* (Ruths and Pfeffer 2014). Proxy population mismatch has been implicated as the primary reason for the errors observed in SM polls (O'Connor et al. 2010; Chung and Mustafaraj 2011; Gayo-Avello 2013).

- *Signal fidelity and resolution*: The value of a measurement tool depends on its ability to indicate signals of interest faithfully. Can the data unambiguously indicate behavioral signals of interest? At reasonable effort? Linguistic interactions (e.g. on SM platforms), for example, may not recognize behavioral signals such as sarcasm without considerable effort to parse the data. Also, the ability to usefully indicate location or geographical behavior depends on the resolution of the location sensors. Signal fidelity also includes questions of *misrepresentation* in observed signals. The ability of agents (human or organizational) to harbor unrevealed or unstable preferences and engage in game-theoretic behaviors means that dishonest signaling can be prevalent in behavioral contexts. Behavioral models will need to account for misrepresentation in signals.

- *Systemic selective nonresponse*: Silence – i.e. the absence of data – is sometimes extremely informative, especially for active voluntary interactions. Unfortunately, the tendency in analysis is to emphasize signal presence over absence. This tendency has been implicated as a potential cause of excessive polarization in SM interactions. Studies of political discourse on blogs and SM platforms highlight the tendency for hyperpolarized minorities to commandeer discourse and thereby skew platform data streams away from underlying population preferences (Tumasjan et al. 2014).

Trends
The data ecosystem is likely to keep growing, perhaps even exponentially as the Internet of things (IoT) takes off. The goal is not just to obtain larger quantities of data, but also to obtain more raw/unfiltered qualitative and quantitative data, free of the self-report and interpretive biases that often afflict surveys and ethnographies. This ecosystem skews heavily toward observational data as opposed to controlled trial data that would identify causal connections more readily. This limits the fitness of the data ecosystem for some purposes.

Questions of data quality, fitness for use, and access will be key. Much of the data ecosystem exists under different jurisdictions, access/legal restrictions, and quality levels. There will likely be a growing demand for methods and practices for *fusing* data sources and patching blind spots in the ecosystem. Advances in AI/ML, discussed below, may help. Besides those discussed below, the research community is also developing important tools like *quasi-experimental approaches* for inferring causality from observational data and *fusion methods* for fusing heterogeneous data/knowledge/information sources. These could be invaluable in time.

Although we do not dwell on the matter here, *substantial* questions and concerns exist about the rules and regulations governing the use of such data, e.g. questions of data privacy, data representativeness, and legal access to data (Levendowski 2018; Balebako et al. 2019).

Advances in AI/ML

Many of the advances in AI/ML have occurred in AI subfields that use statistical learning concepts (pattern recognition including advanced regression, clustering, classification). Expert systems, rule-based, and solver-based AIs continue to be important for problems like planning and knowledge representation, but progress there has not gotten as much popular attention. We mention them nonetheless. For example, we present an expert system, a Fuzzy Cognitive Map (FCM) (Osoba and Kosko (2017)), for modeling behaviors later in this chapter. Fuzzy-based expert systems also have robust mechanisms for knowledge fusion. Such fusion approaches hold promise for addressing the issue of limited model federation (which we highlight later). Other AI subfields like knowledge representation may be useful for specifying underlying behavioral ontologies.

Deep Learning

Deep learning (DL) is the most touted recent trend in ML. Traditional connectionist ML models, by definition, connect nonlinear processing (neural) units configured in shallow hierarchies or layers to solve classification, regression, or dimension reduction tasks. Shallow networks are limited in the complexity of features (or combinations of input variables) they can find and use. But deeper hierarchies are harder to train. DL leverages advances in computational power and statistical learning theory to update the standard connectionist learning models with deeply stacked layers of processing units. The use of deep stacks allows the model to identify complex features in the data that can be useful for improving the model's performance.

Depth in learning architectures is an idea that has been considered for much of the history of ML research. The main hurdle preventing the exploitation of deep architectures has been limits in usable computing power and previously limited access to relevant training data. ML models solve an optimization problem in the process of learning. The optimization problem is a function of the number of tunable parameters in the model. Neural network models (both shallow and deep) typically rely on the backpropagation algorithm for parameter tuning. Shallow models have fewer parameters than deep models. The difficulty of the optimization task grows exponentially with the number of parameters. So deep models can be prohibitively computationally intensive. Deep models often also require larger data sets for training. The key factors that make DL models feasible are *the existence of large application-relevant data sets and massive computing power.* Our previous discussion already highlighted the growth of the ecosystem of behavior-related data. Available computational power has also grown explosively.

More recent DL advances include the use of time-varying layer weights that allows models to represent temporal or sequential dependence. These are called *recurrent neural nets* (RNNs). These models are useful for modeling time series with temporal correlations. Depth in these models refers to the length of time dependence in the signal, not number of layers. *Long short-term memory* (LSTM) models are a popular type of RNN model. LSTMs maintain an internal input-dependent hidden state that modulates what transformation produces the output signal. LSTMs are used for handwriting recognition in some Windows systems. They have also shown good results in the generation of text for dialog (e.g. chatbots). LSTMs are especially useful for sequence-to-sequence learning tasks (Sutskever et al. 2014) (e.g. machine language translation like the *Google Neural Machine Translation*, GNMT; Wu et al. 2016).

The DL community has also made widespread use of the *convolutional neural network* (CNN) architecture for video, image, speech, and text tasks. CNNs are neural networks with layers that apply biologically inspired weighted local averages (or convolutions) to input signal fields. These weighted averaging schemes are one way of incorporating sequential dependence. CNNs have proven especially effective at image and audio tasks. A recent empirical analysis (Bai et al. 2018) argues that CNN-based models may be better than RNN-based models as default models for handling sequence learning tasks.

The value of DL for social and behavioral modeling lies mainly in the ability of DL models to convert a larger portion of the data ecosystem into meaningful behavioral signals. The semantics of images and videos used in social interactions become more accessible using appropriate DL models.[3]

Language data is behaviorally meaningful. But language has historically been hard to model. The next section talks about *natural language processing* (NLP). Much of the recent advances in NLP are due to the application of DL models to language tasks. DL has also fostered the development of computing architecture for scalable computation on large data sets. But DL models themselves have not had a large footprint so far as tools for modeling social behavior directly.

Natural Language Processing (NLP)

The most important AI improvement relevant to behavioral modeling over the last half decade has probably been the maturation of NLP. This includes text mining, topic modeling, sentiment inference, speech recognition, machine

---

3 DL is also appealing because it is often possible to implement portfolios (or federations) of modeling subtasks entirely within the DL paradigm.

translation, etc. These are important because they enable the quantitative analysis of textual data. Behavioral cues in language use are now observable. Caliskan-Islam et al. (2015) demonstrate an example of behavioral modeling (identifying telltale patterns of language use) based on NLP.

Topic modeling (Blei and Lafferty 2006; Blei 2012) has been particularly useful for measuring trends in what is otherwise unstructured data. Blei gives an example of the use of NLP methods to track publishing behaviors in academic fields (Blei 2012). Topic modeling tools have also been indispensable in the analysis of micro text corpora from SM platforms (e.g. short tweets). Some SM studies use latent Dirichlet allocation (LDA) for NLP (Gross and Murthy 2014).

Other modes of text analysis are needed. Topic modeling is a strictly statistical analysis of text with limited use of semantics or lexical structure. It treats text as pure symbols with no meaning separate from the statistical co-occurrence patterns learned from corpora. Significant insight can be gained from this purely symbolic analysis (as its current use in text mining indicates), but topic modeling results still require significant human interpretation to identify and understand context-dependent meanings. Automated tools for semantic, lexical, and lexicographic analysis of larger bodies of discourse can improve social and behavioral (SB) models. The rise of NLP suites of tools like *word2vec* and *GloVe* represents increasing capacity for the algorithmic comprehension of text meaning. They are currently good at solving analogical questions – arguably the minimal task required to indicate semantic comprehension.

Automated machine translation (AMT) is another NLP domain with significant innovation. Google recently introduced the GNMT (Wu 2016) for language translation using an end-to-end neural network framework trained in a purely data-driven fashion (no pre-coded language rules) (Sutskever et al. 2014).

Adversarial Training for Unsupervised Learning

The discovery of adversarial training approaches may hold the most promise for future social-behavioral modeling and related AI. Goodfellow et al. first formalized stable adversarial training in their development of GANs (Goodfellow et al. 2014; Radford et al. 2015). The training approach takes inspiration from game theory and minimax decision-making.

The standard statistical learning approach begins with training data and a parameterized model (e.g. regressions, perceptions, support vector machines, or neural networks). The goal is to learn a desired behavior encoded in the training data. Changing the model parameters changes the behavior of the model. Learning algorithms encode the behavior to be learned as critical points of an objective function over the model's parameter space. Statistical learning (clustering, fitting, classification, regression being the major forms) thus often reduces to stochastic optimization – See Vapnik's (2013) formalization of the empirical risk minimization (ERM) framing for supervised and unsupervised learning tasks. This mode of automated learning works well for problems in which the desired behavior is reasonably articulable as objective functions, e.g. image/facial recognition and some aspects of NLP. This is essentially a teacher–student learning model.

Learning to generate subtle but sometimes crucial behaviors from data sets can be much harder. For example, generating believable super-resolved images, social networks, conversations, or even facial expressions is difficult. The key theme in these tasks is the need to learn *implicit models* of data generation (Diggle and Gratton 1984; Mohamed and Lakshminarayanan 2016). These are models for which the generative structure of the data is apparent post hoc but not easy to articulate a priori. And thus, the process of learning the structure consists of *imitative* model adaptation in response to a repeated game of discriminating between examples.

Adversarial training proceeds as a game between two agents: the generator and the discriminator. The generator's goal is to learn to produce synthetic samples that are representative of the training data. The discriminator's goal is to learn to discriminate between the generator's output and samples from the training data. Both agents escalate in the course of training to the point where the generator has learned to behave indistinguishably from the training data. The learning model is related to actor-critic RL model (Pfau and Vinyals 2016). Adversarial training provides a way for AI systems to do better in games-against-nature scenarios or in adversarial scenarios (e.g. individuals are seeking to hide information or even mislead). Such training could enable agents in social science models to learn behaviors that are not easy to encode or explain (e.g. identifying fake news or recognizing patterns in the presence of efforts to hide them or deceive).

Reinforcement Learning

RL is a branch of statistical ML focused on teaching agents how to act to achieve goals in an uncontrolled environment (Figure 19.2) (Sutton and Barto 1998 [second ed., forthcoming]). It has its origins in research on control theory, robotics, automated planning, and behavioral psychology. RL's key defining features are the explicit modeling of the environment, the built-in emphasis on exploration, the sparsity of evaluative feedback to guide the agent's learning, and the learning of action policies from experience or data. The rise of RL is a response to the inadequacies of supervised learning for planning-style tasks in which the value of real-time actions derives from their downstream effects rather than from immediate rewards. Planning is an integral part of human social behavior. And we need models that can capture such behaviors.

The maturation of RL enables the development of large-scale agent-based models (ABMs) with adaptive intelligent agents possibly in adversarial

Figure 19.2 Standard reinforcement learning framework.

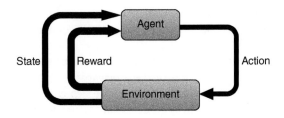

(red-vs.-blue) scenarios. Developments in RL may allow for more adaptive ABM models of intelligent human behavior in complex adaptive systems.

Emulating Human Biases and Bounded Rationality

Another important area of work on AI/ML deals with methods for learning true or revealed preferences. This is important for behavioral modeling because self-report data and behaviorally revealed preferences often diverge (Rudder 2014). Biases in preferences drive many social behaviors. Research on cognitive biases, as reviewed in Kahneman (2011), shows that biases perform useful functions even if they sometimes lead users astray. More specifically, recent work by Pita et al. (2010) on security games shows the importance of accurately modeling human bounded rationality (Simon 1996 [first edition, 1969]; Simon and Newell 1962). Eliminating heuristic biases (or even irrational aspects of decision-making) from social-behavioral simulation models may reduce their descriptive accuracy. The social simulations need to reflect true biases and preferences.

Researchers are beginning to develop simple games and other methods to discover hidden preferences. These may help address the NRC concern about reproducing nonrational agent behaviors in models (Zacharias et al. 2008, p. 359). It is also important, as discussed in the literatures on deterrence (National Research Council 2014, p. 37) and *wicked problems* (Rittel and Noble 1989) to recognize that humans (all of us) often do not actually have the stable utility functions postulated by both rational actor theory and usual versions of bounded rationality theory. We often *discover* or *develop* values in the processes of human interaction and experience. *Limited rationality* is sometimes used to refer to problems that go beyond those of bounded rationality (Davis 2014b, p. 6). Such problems include mental health problems (as when leaders are perhaps depressed and using alcohol or drugs and are sometimes paranoid) and emotion-driven irrational acts (*op cit*).

Trends

Most of the AI/ML methods highlighted seem to have more promise as tools for comprehending novel data streams than as new approaches to social and behavioral modeling. The application of RL to ABMs may prove to be the exception in time. But the addition of heterogeneous data comprehension and fusion capabilities to the behavioral modeling repertoire is substantial boon, if nothing else.

Perhaps a key limitation of many currently popular AI/ML methods is the overrepresentation of feed-forward model structures. This limits the use of such methods to subtasks of the behavioral modeling enterprise (e.g. NLP). Training statistical models that feature feedbacks can be daunting or unstable. But many interesting behaviors involve significant amounts of feedback. The feed-forward emphasis also inhibits the representation of chaotic dynamics that can often occur in behavioral models (e.g. in simple predator–prey models; Schaffer 1985), especially complex ones. The introduction of ML models like adversarial training and sequence learning marks a growing emphasis on simple feedback in ML models. Some less popular AI models like FCMs for modeling causal networks also model feedback since causation often features feedback loops.

Data and Theory for Behavioral Modeling and Simulation

Prefacing Comments on Fundamentals

Certain foundational questions must be addressed if we are to model and simulate social behavior. These include how to represent aspects of behavior, how to understand and establish *validity*, and how to compose models from smaller models (*model federation*). We do not address all such questions here. Here, we focus on the question of how to relate theory and data.

The relationship between theory and data demands further scrutiny because of the advent of *big data*. Advances in AI/ML, driven in part by the glut of big data sources, have predisposed modelers to focus on data as the primary foundation for building and validating models of behavior. This data focus is evident in the emphasis on *predictive power* in much of modern ML approaches to modeling. The growth of ML modeling approaches is arguably due to the ability to convert measures of predictive validity into concrete mathematical objective functions (e.g. classification accuracy for classification, mean squared error for regressions) that are directly amenable to optimization methods. *Explanatory power*, if considered at all, has been a secondary goal – sometimes with the rationale that increases in predictive power *surely* occur due to increases in the intrinsic or effective explanatory power of the model. Modelers sometimes (implicitly) resort to this rationalization even if that increased explanatory power is difficult to verify because the model is complicated or opaque.[4] That rationalization is neither valid nor acceptable. Explanatory power is extremely important for social and behavioral modeling, especially when that is to be used to inform decision-making (Davis et al. 2017). A purely data-driven modeling approach would be inadequate and potentially misguided.

For Want of Good Theory...

Researchers have multiple examples highlighting the inadequacies of purely data-driven models in complex domains. Recent discussion (Jonas and Kording 2017; The Economist 2017) argues that many data-driven neuroscience techniques aimed at uncovering brain architecture and function are misguided. One example involves the common approach of localizing function in the brain based on observations of function loss or impairment due to localized brain lesions. Jonas and Kording apply modern data-driven neuroscience techniques to an older simpler microprocessor. The goal was to reconstruct the known architecture of the microprocessor by analysis of measured signals and localized impairments or interventions – the way a neuroscientist might seek to reconstruct the architecture of brain. The methods failed to identify high-level structures (e.g. the arithmetic and logic unit, ALU) fundamental to explaining the microprocessor's function.[5] The failure of the methods at identifying crucial structures in a simpler system raises questions about the ability of these methods to *make sense* or *explain* brain functions physiologically.

An older critique of heavily data-driven models of behavior is Chomsky's critique (Chomsky 1959; Fodor 1965) of B.F. Skinner's model of learned verbal behavior. Skinner had put forward a theoretical model of language acquisition in which infants learn language entirely from experience filtered through a sparse operant conditioning learning framework. Chomsky argued that Skinner's theory of verbal behavior was wrong because the stimulus–response mechanism it hinged on was too sparse to explain the speed of language acquisition or the observed complexities of language use (e.g. *latent* learning in no-reward scenarios). Chomsky pointed out that studying verbal behavior solely based on input (stimuli and reinforcements) and output (verbal utterances) observables is incomplete since it ignores complex *selective mechanisms* that mediate verbal behavior. The infinite variation and novelty in language use makes it unlikely that verbal behavior is the product of *generalization*

---

4 A number of model selection instruments (e.g. the Akaike information and the Bayesian information criteria) try to incentivize simpler (and therefore less opaque?) models by penalizing models with larger parameter spaces.

5 Mathematically, a core problem is that statistical methods do not *discover* model fragments that are not part of the specification used.

from limited observation examples. He further argued that the human facility with language likely exists because of an innate grammar acquisition model (a theory of behavior) more complex than operant conditioning.

Chomsky's criticisms arguably apply to other models of complex human behavior besides verbal behavior. Efforts at building social and behavioral models are subject to similar misspecification risks. Data-driven behavioral models, without the benefit of strong hints from plausible social science theories of behavior, may have poor explanatory power – especially in nonlinear systems. The statistical models may not even be useful for postdiction or post hoc sensemaking. To be sure, fully theory-driven modeling can also be seriously misguided.

> Usual statistical analysis use *specifications*.

In an ideal situation with complete relevant theory and perfect and complete data, the data validates the theories, and the theories provide an explanatory/interpretive frame for the data. The current situation, however, is very different: we have a lot of data, but it is often imperfect, incomplete, and/or biased, and we have a great many social science *theories*, which are often narrow, fragmentary, unvalidated, and certainly not settled.[6]

> **Key Questions:** How do we design models that are valid for purpose and representative of the relevant reality when the data and theories are unsettled? What is the right balance of theory and data focus in models? And what are the best practices on weaving the two together effectively?

The Scope of Theory and Laws for Behavioral Models

Some earlier literature on the validity of modeling and simulation (M&S) often supposed that complete systems understanding is the key to good models. In other words, good models should have structural validity. This is not necessarily true; *good* theories are necessary, but they often do not need to be perfect, complete, or structurally isomorphic to the real system. This point will

---

6 It is notable that in the social sciences, *theory* may correspond to nothing more than a single-variable hypothesis, such as "More of X should tend to increase Y," whereas in the physical sciences, *theory* often (but not always) is regarded as that which pulls together a great many considerations coherently. Thus, in conversation, a physical scientist may use the word *theory* in a very positive way, whereas a social scientist may regard someone else's *theory* skeptically as nothing more than yet another hypothesis to be tested. Miscommunication occurs routinely on this matter.

be important as we attempt to tease out the laws and theories needed for good social-behavioral M&S.

The most persuasive way to argue this is perhaps to appeal to daily experience. We do not use detailed mental models of the world to function successfully. The taxi driver does not need a comprehensive mental model of car mechanics or of petrochemistry to operate his taxi. Similarly, animals do not need a full understanding of physical theories (e.g. Newton's laws or quantum mechanics) to thrive.

So also, an agent's internal models of and behaviors in complex systems need not be perfect. They should, however, reflect *effective theories* (Randall 2017) to guide their behavior in the system over time. These are simplified mental models that have proven sufficiently useful for guiding the agent's actions even though they are not fully accurate or complete. These idealized mental models are *useful untruths* (Appiah 2017).

Key Questions: Is the social science research community able to identify which theories of social behavior are effective theories vs. accurate theories vs. just-interesting-but-invalid theories? Is there a stable mechanism or framework for making this distinction? How useful is data for this purpose?

The Model-Inquiry Gap

This value of effective theories for useful modeling suggests a diagnostic concept for modeling practice: *the model-inquiry gap*. This is the conceptual gap between the level of a model's representation of the world and the level of analysis the model is meant to inform. It is one aspect of estimating how well a model matches its purpose. One example of a large model-inquiry gap would be using a quantum mechanical model in an attempt to answer questions about planetary motion. It would simply be inappropriate. An example of a small model-inquiry gap (i.e. a good match) might be using a social network model to study individual influence in an organization. Where a broad gap exists, using the model (if it can be used at all) will require vastly more effort, to include extensive calibration if good and comprehensive data exists, than if a suitably simpler model were used. Effective theories act as tools for reducing the gap with lower effort. This is also the reason that recent discussion of Department of Defense's (DoD's) modeling for defense planning decried overdependence on detailed models and urged greater emphasis on simpler models (with more detailed models used to study selected issues in more detail, which is often crucial). The lesson is not to choose simple rather than complex, but to have the right family of models for both broad analysis and for in-depth analysis when necessary (Davis 2014a).

The use of effective theories is a way to give agents some aspects of *bounded rationality*[7] (Cioffi-Revilla 2014, p. 132) as they try to act in complex environments. With effective theories it is possible to create models and simulations of poorly understood systems that are valuable and valid for a purpose (Zacharias et al. 2008). Comparison of game-theoretic models in Stackelberg games (used as a proxy model for critical infrastructure defense) and empirical data showed that humans deviate significantly but predictably from rational expectations. Failing to incorporate bounded rationality in behavioral models leads to suboptimal or invalid models (Pita et al. 2010).

> **Key Questions:** What are the standard or best-practice approaches for equipping behavioral models with the kinds of heuristic decision-making mechanisms that humans demonstrate? When is it important to do so?

Consider the preceding discussion as a discussion on the *depth* of a model's theories. What about the *breadth* of a model? What is the relevant scope of a model? Most systems of interest are subsystems of larger systems. There will be interactions within the hierarchy of systems. Identifying which interactions and systems are relevant to the modeler's interest is not always obvious. Casting too wide a net leads to large unwieldy models. An overly parsimonious model may be too incomplete to be valid for purpose.

Herbert Simon argues that most systems are nearly decomposable federations of subsystems (Simon 1996 [first edition, 1969]; Simon 2002). This is because there are fitness benefits to abstracting away or encapsulating functions into subsystems rather than having a single, entangled monolithic system of functions (Simon 2002). Near-decomposability tends to improve adaptability and reduce fragility. This suggests that composite social-behavioral models need to be well designed so that they have the benefits of near-decomposability.

We are much less sanguine about the role of federated models than was the NRC report (Zacharias et al. 2008). Model composition will be important for particular purposes, but the tendency to adopt *approved* model federations with *approved* input data should be fiercely resisted because the problems of interest tend to be dominated by uncertainty. Analysis in such cases needs to be uncertainty sensitive.

> **Key Questions:** When should federated models be encouraged, and, in those cases, how should they be designed and used?

---

7 Herbert Simon's *bounded rationality* refers to making decisions under constraints like imperfect or incomplete information, limited or imperfect computational capacity, and limited time – i.e. the kinds of constraints that almost always apply in real life.

The Scope of Data for Behavioral Models

What about the system data? System data here refers to measurement of the system's state. The typical roles of system data in modeling are tuning models, validating models, generating hypotheses of underlying system behavior, and finally establishing the appropriate inputs for a given application of simulation. System data in our context includes behavioral or social data.

The emerging *big data* ecosystem is creating a steady stream of behavioral data (Ohm 2010). A significant body of data-driven work exists on mobility behavior (De Montjoye et al. 2013), financial behavior (De Montjoye et al. 2015), and personal networks (Sapiezynski et al. 2016). The goal of these is to help make sense of how people behave or make decisions. Such insights would be commercially useful for improving targeted advertising and other applications that deal with transitions from user intent to user action. Much of this work leverages newer data-driven methods to highlight patterns. But they tend to fall short on making sense of the patterns they find.

This sensemaking is important if behavioral models are to reach their potential. Social and behavioral simulations will need to bridge the theory–data gap. Characterizing behavior without a theoretical frame is often not enough to determine proper interventions. And positing theories of user behavior without empirical support is not enough.

> **Key Questions:**   Which data-driven methods are most useful for sensemaking? Which methods have limited sensemaking value? What would a research program focused on developing methods for sensemaking look like?

Some of the issues have been discussed recently in a volume summarizing results of DoD's human, social, cultural, and behavioral (HSCB) modeling program (Egeth et al. 2014) (see, e.g. Chapter 10).

Causal inference is another consideration that motivates the need for a theory frame around behavioral data. Good causal theories, however, must include variables and relationships that are often omitted. For example, some theories and formalisms do not include feedback cycles. Also, the variables used for hypothesis testing statistical work are often *not* very appropriate for causal explanation because they are poor proxies for the *real* variables of interest and because they sometimes lock in a simplistic framework (e.g. linear dynamics and trivialized rational actor decision-making).[8]

---

8 Attempting to use quantitative social science theories to inform policy-level issues is criticized in Davis (2011, p. 326ff), drawing on criticisms from within the social science community itself (Sambanis 2004; Kalyvas 2008). A major problem is the tendency for the quantitative work to be aggregated and too little informed by factors known to be important from more micro-level case studies.

Key Questions: The current data ecosystem is strongly skewed toward the collection of observational data. What is the state of scalable methods (quasi-experimental methods) for extracting causal relationships from observational data? Are the limits of causal inference on observational data going to diminish with a larger ecosystem? Or are these limits fundamental? Are there robust scalable ways of eliciting causal insight from experts?

Bridging the Theory–Data Gap

Initial Observations

Models are sometimes starkly dichotomized models as theory driven vs. data driven, but it is better to think in terms of a spectrum:

- On one end, the theory-driven approach is said to rely on *deductive reasoning* – using general system laws to infer behavior in specific instances.
- At the other end, the data-driven approach is said to rely more on *inductive reasoning* – using a collection of specific observations to abstract out governing laws and relationships.
- *Abductive reasoning* is a somewhat intermediate mode of reasoning. Abductive reasoning starts from a portfolio of general hypotheses or laws and proceeds to rank them based on how well they match observed data. Abduction is concerned with making *inferences to the best explanation*. This bridges the gap between theory-focused and data-focused approaches.[9]

The approach we urge is shifting the balance in social-behavioral search from a statistics- and-data-driven dominance toward a healthier abductive approach in which theory informs data analysis and data analysis informs theory in a continuous dynamic tension. This corresponds as well to urging a relatively greater emphasis on *causal* modeling rather than correlational work.[10]

Thinking now of M&S, an agent reasoning in an abductive mode would observe state and have a number of hypotheses to *explain* it. It would then act using a hypothesis (or theory) that is as simple as possible for explanation, but no simpler (an expression of Occam's razor attributed to Einstein). The first part (as simple as possible) encourages models that are more likely to be generalizable to cover new observations (Pearl 1978). The second part (but no simpler) is important because, especially when data is incomplete and imperfect, it may be very important to persist in using some model

---

9 Charles S. Peirce discussed abductive inference (Peirce and Buchler 1940).

10 We recognize, of course, that many papers that have been written about how, allegedly, cycles within system models make the concept of causality untenable, how *everything* is correlational because we do not know the ultimate *true* theory if it exists, and so on. This is not the place to discuss such matters.

complexities that are rooted in knowledge or persuasive theory.[11] Data may need to *catch up*. This full process, from observations to best explanation, is a better description of what scientists do when they seek out governing laws of nature. Abductive reasoning is a less structured process that draws on both data and theory; this lack of structure has inhibited its automation.

> **Key Questions:** Are there modeling approaches that enable flexible exploration of alternative explanations of observations? Which analytic techniques better represent the abductive process? What could it mean to foster a capacity of programmable abductive model building?

The preceding sections discuss criteria for evaluating social simulations. It may be useful to examine currently implemented social simulations through the lens of these criteria. These simulations are rudimentary, but they can serve as useful caricatures of how to use data to train simulation models in a theory-informed manner.

Example 1: Modeling Belief Transmission: Memes and Related Issues at the Micro Level

The Model The transmission of beliefs is a social behavior of considerable interest. Often transmission occurs in "memes," units such as ideas, behaviors, or styles (Dawkins 2016 [first ed., 1976]). Memes include information, conspiracy theories, aspects of individual identity, and even language. The flexible and structurally valid model of belief transmission can have significant policy relevance for questions about the transmission of health, voting, radicalization, and other types of behaviors.

Behaviors and incentives for behaviors can evolve spontaneously over time and in response to an individual's information diet. An important explanatory model of belief transmission sees an individual or agent's beliefs and actions as a function of the beliefs and actions of members of the agent's social networks. Social behavior or culture propagates through social networks. Every person belongs to multiple networks, but infectiousness of such memes depends on

---

11 Some twentieth-century physicists noted candidly that their theoretical work, the stuff of their Nobel Prizes, sometimes was driven by the *beauty* of the mathematics they articulated rather than at-the-time empirical information (Dirac 1939; Weinberg 1994). At a more mundane level, some aspects of social-behavioral theory are intuitively very persuasive. If so, they should not be omitted without strong empirical disconfirmation. Interestingly, in DoD's modeling of combat, many operations researchers have often omitted the factor of morale because it is difficult to measure and *subjective*. Senior officers, historians, and some modelers have recognized such practice as absurd.

the network in which an agent encounters the meme. Teenagers are more likely to adopt behaviors observed in their peer network. We can illustrate the phenomenon on a simple model based on social networks equipped with a provisional theory of transmission.

Our simple model draws on theories of *belief fixation* and *the intersectional nature of identity*. Consider a model that is concerned with only a single binary belief. Belief fixation measures how strongly an agent's belief state resists changing when confronted with new information.[12] We represent doubt about the belief using Bayesian priors. Fixation refers to how strongly the agent resists changing its beliefs in light of new information. *Inquiry* is the process by which the agent moves from doubt to certainty. We operationalize this using the concept of intersectionality: an agent resolves doubt by examining the beliefs of neighboring agents in a set of personal networks (e.g. family, friends, workplace, and community). The strength of an agent's impulse to switch belief is a weighted function of the fraction of each such network with belief contrary to the agent's. This models Peirce's *appeal to authority* mode of changing belief; the *authority* here is the weighted plurality of beliefs in an agent's local networks. It also reflects the function of social influence in opinion formation. In a simple agent-based simulation, then, an agent's belief regarding the issue in question (the meme) depends on its prior belief one time step earlier and the beliefs of the other agents in his network.[13] If the initial beliefs of all agents are specified and the simulation is then executed, the fraction of the agents believing the meme will change over time as illustrated in (Figure 19.3).

For the illustration, we assumed that each agent was a member of four networks, as shown, the networks being of different characters. This particular simulation shows a particularly rare belief/behavior (initial prevalence at 1%) spreading through the population rapidly and then randomly oscillating in frequency around 45% (Figure 19.4). The illustrative simulation shows the meme *caught on*. This is the result of the parameter values assumed in the model. With other choices, the meme would never catch on, or might propagate for

---

12 Peirce's conception (Peirce 1877; Peirce and Buchler 1940) was that an agent's main goal was certainty, with the state of doubt being inherently repugnant.

13 In equation terms,

$$p_k(t) \sim Beta\left(\vec{\alpha}_k(t) \cdot \vec{w} + \lambda \cdot \delta(X_k(t-1)), \vec{\beta}_k(t) \cdot \vec{w} + \lambda \cdot \delta(1 - X_k(t-1))\right)$$

where $X_k(t) \sim Bernoulli(p_k(t))$,

$\alpha_{k,j}(t)$ = number of agents connected to k in the jth network who believe
$\beta_{k,j}(t)$ = number of agents connected to k in the jth network

who do not believe

w_j = relative weight or leverage of belief in the jth network
λ = inertia of prior belief

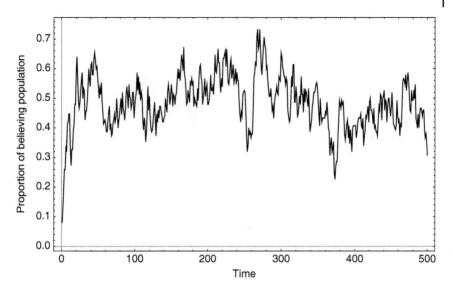

Figure 19.3 Population proportion of belief as a meme propagates (assuming 1% initial belief).

a while and then dissipate again. Thus, even this very simple model can – by varying parameters – generate quite a range of behaviors. Conversely, if behavior is observed, then the model parameters can be inferred.[14]

Evaluating the Model The belief transmission model and similar such models can be useful for social and behavioral M&S. This is an example of a *micro-level model*, specifically a micro-level ABM. Since it revolves around a single belief, it might prove simplistic when comparing it with full models of human decision-making. In some cases, however, it might be useful for exploring and building intuition about belief transmission about a particular meme.

The question of validity is a primary concern for such models. And validity has different dimensions (Davis et al. 2012). The first dimension is the question of *structural validity*: does the model actually mimic the key processes in the transmission of memes in the real world? The simplicity of the model suggests that it most likely does not capture all relevant features of the real process. It is an open question whether it captures enough of the relevant features. Another dimension is the question of *replicative validity*: does the model replicate overall behavior under identical initial conditions? The data ecosystem may be becoming robust enough to tackle questions of replicative and structural

---

14 The illustrative model is an example of Markov random field (MRF) equipped with Bayesian temporal update dynamics. More specifically within MRFs, it is an *Ising model* (Brémaud 2010).

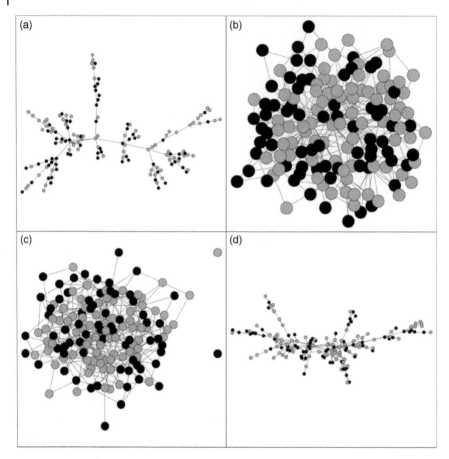

Figure 19.4 The multiple networks to which agents belong.

validity for this model. SM platforms and CDRs can provide useful estimates of both multimodal social network relationships and the memes propagating through these networks. Using these data sources could also enable the model to account for dynamic networks and shock events. It is thus possible, in theory and with careful calibration, to compare meme transmission in this model with transmission in the real world.

The questions of structural and replicative validity are *not* synonymous for ABMs. An ABM may mimic system processes closely and still fail to replicate observed macro-level system behavior. And conversely, an ABM may replicate observed macro-level behavior based on internal dynamics that are not identical to the true system's internal dynamics. Such mismatches may be attributed to incorrect/incomplete specification of model dynamics, hidden/latent/unaccounted system variables, or issues of *emergence* (chaotic or otherwise) more broadly. The question of validity for ABMs is an area of continuous research effort.

Example 2: Static Factor-Tree Modeling of Public Support for Terrorism

The Model Behavioral models may also work at a more macroscopic level. Consider, for example, a model intended to examine the factors that promote or inhibit the *public support for insurgency and terrorism* (PSOT). The macroscopic nature of the question suggests that the effective theories governing the model should be macro level to achieve a low model-inquiry gap. A large 2009 DoD study drew on a comprehensive review of social science relating to counterterrorism to construct a composite qualitative model in the form of a *factor tree* (a kind of primitive but broad static causal model) of, e.g. public support for terrorism (Paul 2009). A later study tested it empirically (qualitatively) and refined it slightly (Davis et al. 2012) (Figure 19.5). The work was synthetic across different fragmentary social science theories.

A next step (Davis and O'Mahony 2013) went beyond the purely pictorial description to build a computational version of the factor tree but with numerous degrees of freedom to accommodate different possibilities about how the factors in fact combine. That is, the model incorporated not just different levels of resolution as indicated in the tree, but also different structures for the combining relationships. The intent was to enrich the ability to discuss causal phenomena *at a point in time*. The price paid was deliberate suppression of dynamics. Although simple enough to present in a single page, the factor tree integrates a great deal of knowledge – moving discussion away from what the alleged primary reason for public support is to the many factors that affect it, with the relative significance of factors varying from one context to another as expected. The same factors appeared in subsequent case studies (i.e. the qualitative theory had significant generality), but – as predicted – the relative significance of the factors varied with case (Davis et al. 2012).

The computational version of the model allowed generating broad outcome maps showing the circumstances (contextual variables) under which public support would be expected to be very low, low, medium, high, or very high (Figure 19.6) (Davis and O'Mahony 2013).

Adding Dynamics with Fuzzy Cognitive Maps Osoba and Kosko (2017) extend the simulation capacity of the Davis–O'Mahony model by adapting the factor tree into an FCM. FCMs are a type of expert system for capturing and simulating causal knowledge from experts and data. Both factor trees and FCM models

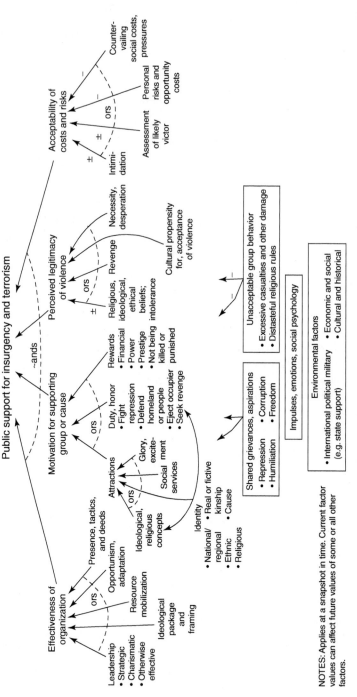

Figure 19.5 A factor tree for public support for insurgency and terrorism. Applies at a snapshot in time. Current factor values can affect future values of some or all other factors.

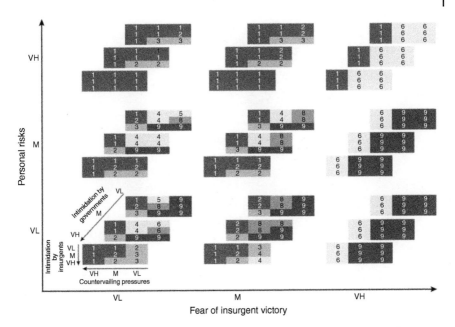

Figure 19.6 An illustrative outcome map showing public support vs. five contextual variables. Public support (model output) is indicated by a cell's color and number with scores of 1, 3, 5, 7, and 9 corresponding to green, light green, yellow, orange, and red, respectively. High scores indicate strong public support for insurgency. Colors are not legible in gray-scale versions of figure.

score high on important modeling concerns like producing interpretable and *multiresolution* representations of systems and concepts. But FCMs have the added benefits of being able to:

- Flexibly represent causal dependences (even feedback or cyclic dependence).
- *Fuse* knowledge sources (e.g. experts and data).
- Perform data-driven *automatic hypothesis generation.*
- *Simulating* static and dynamic what-if scenarios over short or long time horizons.

These added benefits greatly increase the capacity for *abductive model building.* Analytic methods like FCMs can be useful for bridging the gap between theory and data for M&S.

Figure 19.7 shows the graphical depiction of FCM-PSOT, the FCM adaptation of the Davis–O'Mahony PSOT model. The FCM-PSOT is an example of an AI method (an expert system) that is directly applicable for social and behavioral modeling. It is a directed graph (digraph) as opposed to the tree structure

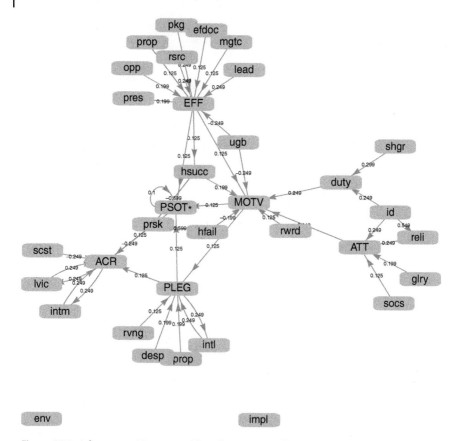

Figure 19.7 A fuzzy cognitive map adding dynamics to a factor-tree model.

of the PSOT. Cycles are easy to incorporate into digraphs as the figure shows. The digraph's adjacency matrix combined with an appropriate signal squashing function at the nodes (e.g. the logistic function) allows the analyst to simulate causal progression by single time steps or all the way to convergence to a fixed point or limit cycle. Progression in time on FCMs is a simple matrix-vector multiplication composed with a nonlinear squashing operation on the output vector.

The FCM's digraph structure also enables the fusion of multiple maps, e.g. from separate experts on the same topic. FCM fusion amounts to weighted combinations of adjacency matrices (augmented or zero padded if necessary). Figure 19.8 shows the graphs and adjacency matrices in an example of such knowledge combination. The figure shows elicited maps from two experts on the blood-clotting process known as *Virchow's triad* combining into one FCM.

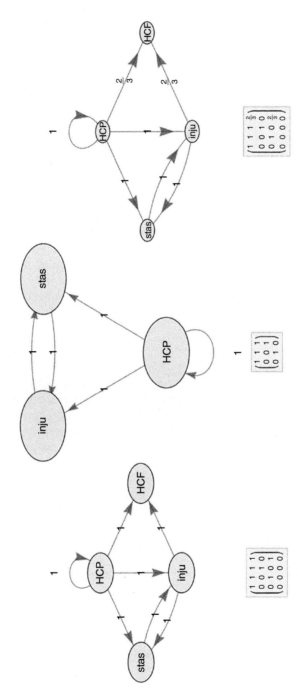

Figure 19.8 Fusing or combining fuzzy cognitive maps.

Evaluating the PSOT Models

PSOT and FCM-PSOT are examples of macro-level models of behavior (compared with the belief transmission model). The relevant entities are populations of people, not individual agents or people. More specifically, the models operate on the population-level prevalence of beliefs and the interactions among these beliefs. The model's focus on common beliefs makes the validation process more complex. The interaction between beliefs and the more observable actions or events requires careful calibration in this context (and in more general contexts). The existing data ecosystem currently does not (and most likely will not) inform most of the model variables. We would need careful surveys and ethnographies to measure the pervasiveness of the relevant beliefs or factors. Thus, tests of validity (replicative or predictive) would be hard to implement especially for the dynamic FCM-PSOT model.

The value of these models is mainly as tools to characterize and simulate scenarios based on carefully vetted expert knowledge about causal links. The correctness of the elicited expertise wholly underpins the structural validity of both models. The fusion capability for FCMs means probabilistic limit laws guarantee the structural validity of the final FCM if a large number of independent experts present individual FCMs for fusion.

Conclusion and Highlights

This discussion has outlined key changes and trends in the data and modeling environment since the NRC report of 2011. The methods highlighted focused on the AI/ML domain. The data conversation examined the emerging ecosystem of data streams with content relevant to behavioral modeling. SM data stands out as a high-leverage stream. It is just one part of the ecosystem though. The extended mind trend (in which individuals' cognition and mental states leave digital footprints) suggests that the behavioral data ecosystem may be able to record formerly unobservable useful signals.

The AI/ML discussion highlighted a series of innovations that could be of value to behavioral modeling. The innovations seem to hold the most promise for unlocking or combining information from complex or heterogeneous data streams (e.g. language, images, video). We demonstrate the value of the FCM AI approach for developing an expert system model of macro-level behavior. This suggests that older expert-system-style AI methods may hold promise for behavioral modeling. Alternatively, RL may be of value to adaptive or dynamic ABMs of behavior. And adversarial training may be useful for capturing or simulating patterns that are difficult to articulate.

The rest of the chapter focused on more fundamental questions about the interplay between theory and data in efforts to model and simulate human behavior. We identified a series of questions that will need answers if capacity for behavioral M&S is to grow. We identified questions on the following:

The interplay between data and theory:

1. How do we design models that are valid for purpose and representative of the relevant reality when the data and theories are unsettled? What is the right balance of theory and data focus in models? And what are the best practices on weaving the two together effectively?

2. Is the social science research community able to identify which theories of social behavior are effective theories vs. accurate theories vs. just-interesting-but-invalid theories? Is there a stable mechanism or framework for making this distinction? How useful is data for this purpose?

3. Are there modeling approaches that enable flexible exploration of alternative explanations of observations? Which analytic techniques better represent the abductive process? What could it mean to foster a capacity for programmable abductive model building?

Modeling practice:

4. What are the standard or best-practice approaches for equipping behavioral models with the kinds of heuristic decision-making mechanisms that humans demonstrate? When is it important to do so?

5. When should federated models be encouraged, and, in those cases, how should they be designed and used? We are skeptical about how broad the value of model federation.

6. Which data-driven methods are most useful for sensemaking? Which methods have limited sensemaking value (DL)? What would a research program focused on developing methods for sensemaking look like?

Causal inference:

7. The data ecosystem skews strongly toward observational data collection. What is the state of scalable quasi-experimental methods for extracting causal relationships from observational data? Are the limits of causal inference on observational data going to diminish with a larger ecosystem? Are there robust scalable ways of eliciting causal insight from experts? We believe the need for interpretable models of behavior to guide intervention makes the development of more capable causal modeling approaches key for useful behavioral modeling.

Our discussion ended with an exploration of rudimentary models and simulations of behavior at different levels of abstraction. These served as simple illustrations of the interplay between theory- and data-driven perspectives on social and behavioral modeling. They also serve as a concrete canvas on which to test our growing sophistication in evaluating behavioral models and simulations.

Acknowledgments

Part of this work was funded under a project for the Defense Advanced Research Projects Agency.

References

Althoff, T., White, R.W., and Horvitz, E. (2016). Influence of pokémon go on physical activity: study and implications. *Journal of Medical Internet Research* 18 (12): e315.

Althoff, T., Horvitz, E., White, R.W., and Zeitzer, J. (2017). Harnessing the Web for Population-Scale Physiological Sensing: A Case Study of Sleep and Performance. arXiv, preprint arXiv:1710.07083.

Amirkhani, A., Papageorgiou, E., Mohseni, A., and Mosavi, M. (2017). A review of fuzzy cognitive maps in medicine: taxonomy, methods, and applications. *Computer Methods and Programs in Biomedicine* 42: 129–145.

Anderson, J.R. (1996). Act: a simple theory of complex cognition. *American Psychologist* 51 (4): 355.

Appiah, K.A. (2017). *As If: Idealization and Ideals*. Cambridge, MA: Harvard University Press.

Bai, S., Zico Kolter, J., and Koltun, V. (2018). An empirical evaluation of generic convolutional and recurrent networks for sequence modeling. arXiv:1803.01271v2 [cs.LG]. 19 April 2018.

Bengtsson, L., Lu, X., Thorson, A. et al. (2011). Improved response to disasters and outbreaks by tracking population movements with mobile phone network data: a post-earthquake geospatial study in haiti. *PLoS Medicine* 8 (8): e1001083.

Blei, D.M. (2012). Probabilistic topic models. *Communications of the ACM* 55 (4): 77–84.

Blei, D. and Lafferty, J. (2006). Correlated topic models. *Advances in Neural Information Processing Systems* 18: 147.

Brémaud, P. (2010). *Markov Chains: Gibbs fields, Monte Carlo Simulation, and Queues*. New York, NY: Springer Science+Business Media.

Caliskan-Islam, A., Harang, R., Liu, A. et al. (2015). De-anonymizing programmers via code stylometry. Proceedings from the 24th USENIX Security Symposium (USENIX Security), Washington, DC.

Chomsky, N. (1959). A review of B. F. skinner's verbal behavior. *Language* 35 (1): 26–58.

Chung, J.E. and Mustafaraj, E. (2011). Can collective sentiment expressed on twitter predict political elections? *AAAI* 11: 1770–1771.

Cioffi-Revilla, C. (2014). *Introduction to Computational Social Science: Principles and Applications*. London: Springer-Verlag.

Clark, A. and Chalmers, D. (1998). The extended mind. *Analysis* 58 (1): 7–19.

Davis, P.K. (ed.) (2011). *Dilemmas of Intervention: Social Science for Stabilization and Reconstruction.* Santa Monica, CA: RAND Corporation.

Davis, P.K. (2014a). *Analysis to Inform Defense Planning Despite Austerity.* Santa Monica, CA, RAND Corporation.

Davis, P.K. (2014b). Toward theory for dissuasion (or deterrence) by denial: using simple cognitive models of the adversary to inform strategy. *Working Paper WR 1027.*

Davis, P.K. and O'Mahony, A. (2013). A Computational Model of Public Support for Insurgency and Terrorism: A Prototype for More General Social-Science Modeling. Santa Monica, CA: RAND Corporation.

Davis, P.K., Larson, E., Haldeman, Z. et al. (2012). Understanding and Influencing Public Support for Insurgency and Terrorism. Santa Monica, CA: RAND Corporation.

Davis, P.K., O'Mahony, A., Gulden, T. et al. (2017). Priority challenges for social-behavioral research and it's modeling. Santa Monica, CA. RAND Corporation.

Dawkins, R. (2016). *The Selfish Gene: 40th Anniversary Edition,* Oxford Landmark Science, 4e. Oxford University Press.

De Montjoye, Y.-A., Hidalgo, C.A., Verlesen, M., and Blondes, V.D. (2013). Unique in the crowd: the privacy bounds of human mobility. *Scientific Reports* 3 (1): 1376.

De Montjoye, Y.A., Radelli, L., and Singh, V.K. (2015). Unique in the shopping mall: on the reidentifiability of credit card metadata. *Science* 347 (6221): 536–539.

Diggle, P.J. and Gratton, R.J. (1984). Monte Carlo methods of inference for implicit statistical models. *Journal of the Royal Statistical Society, Series B* 46: 193–227.

Dirac, P.A.M. (1939). The relation between mathematics and physics. *Proceedings of the Royal Society of Edinburgh* 59 (Part II): 122–129.

Egeth, J.D., Klein, G.L., and Schmorrow, D. (eds.) (2014). *Sociocultural Behavior Sensemaking: State of the Art in Understanding the Operational Environment.* McLean, VA: MITRE.

Fodor, J.A. (1965). Could meaning be an RM? *Journal of Verbal Learning and Verbal Behavior* 4 (2): 73–81.

Gayo-Avello, D. (2013). A meta-analysis of state-of-the-art electoral prediction from twitter data. *Social Science Computer Review* 31 (6): 649–679.

Georgeff, M., Pell, B., Pollack, M., and Tambe, M. (1998). The belief-desire-intention model of agency. In: *International Workshop on Agent Theories, Architectures, and Languages,* 1–10. Berlin, Heidelberg: Springer.

Goodfellow, I., Pouget-Abadic, J., Mirza, M. et al. (2014). Generative adversarial nets. *Advances in Neural Information Processing Systems* 2672–2680.

Gross, A. and Murthy, D. (2014). Modeling virtual organizations with latent dirichlet allocation: a case for natural language processing. *Neural networks* 58: 38–49.

Jonas, E. and Kording, K.P. (2017). Could a neuroscientist understand a microprocessor? *PLOS Computational Biology* 13 (1): e1005268.

Kahneman, D. (2011). *Thinking, Fast and Slow*, 1e. New York: Farrar, Straus and Giroux.

Kalyvas, S.N. (2008). Promises and pitfalls of an emerging research program: the microdynamics of civil war. In: *Order, Conflict, and Violence*, 397–421. Cambridge University Press.

Levendowski, A. (2018). How copyright law creates biased artificial intelligence's implicit bias problem. *Washington Law Review* (93): 579.

McCarthy, J. (2007). What is artificial intelligence. http://www-formal.stanford .edu/jmc/whatisai.pdf (accessed 30 September 2018).

Minsky, M. (1961). Steps toward artificial intelligence. *Proceedings of the IRE* 49 (1): 8–30.

Mohamed, S. and Lakshminarayanan, B. (2016). Learning in Implicit Generative Models. arXiv, preprint arXiv:1610.03483.

National Research Council (2014). *U.S. Air Force Strategic Deterrence Analytic Capabilities: An Assessment of Methods, Tools, and Approaches for the 21st Century Security Environment*. Washington, DC: National Academies Press.

Nilsson, N.J. (1998). *Artificial Intelligence: A New Synthesis*. San Francisco: Morgan Kaufmann.

O'Connor, B., Balasubramanyan, R., Routledge, B.R., and Smith, N.A. (2010). From tweets to polls: linking text sentiment to public opinion time series. *ICWSM* 11: 122–129.

Ohm, P. (2010). Broken promises of privacy: responding to the surprising failure of anonymization. *UCLA Law Review* 57: 1701.

Osoba, O. and Kosko, B. (2017). Fuzzy knowledge fusion for causal modeling. *Journal of Defense Modeling and Simulation* 14 (1): 17–32.

Paul, C. (2009). How do terrorists generate and maintain support (April 8, Trans.). In: *Social Science for Counterterrorism: Putting the Pieces Together* (ed. P.K. Davis and K. Cragin), 113–209. Santa Monica, CA: RAND Corporation.

Pearl, J. (1978). On the connection between the complexity and credibility of inferred models. *International Journal of General Systems* 4: 255–264.

Peirce, C.S. (1877). The fixation of belief. *Popular Science Monthly* 12: 1–15.

Peirce, C.S. and Buchler, J. (eds.) (1940). *Philosophical Writings of Peirce*. Dover Publications.

Pfau, D. and Vinyals, O. (2016). Connecting Generative Adversarial Networks And actor-Critic Methods. arXiv, preprint arXiv:1610.019452016.

Pita, J., Jain, M., Tambe, M. et al. (2010). Robust solutions to Stackelberg games: addressing bounded rationality and limited observations in human cognition. *Artificial Intelligence* 174: 1142–1171.

Radford, A., Metz, L., and Chintala, S. (2015). Unsupervised Representation Learning with Deep Convolutional Generative Adversarial Networks. arXiv:1511.06434.

Randall, L. (2017). *Effective Theory*. Edge. Retrieved from https://www.edge.org/response-detail/27044.

Rittel, H. and Noble, D. (1989). Issue-based information systems for design. *Working Paper 492*. Berkeley, CA: Institute of Urban and Regional Development, University of California.

Rubinstein, J.S., Meyer, D.E., and Evans, J.E. (2001). Executive control of cognitive processes in task switching. *Journal of Experimental Psychology: Human Perception and Performance* 27 (4): 763.

Rudder, C. (2014). *Dataclysm: Who We Are When We Think No One is Looking*. Crown.

Ruths, D. and Pfeffer, J. (2014). Social media for large studies of behavior. *Science* 346 (6213): 1063–1064.

Sambanis, N. (2004). Using case studies to expand economic models of civil war. *Perspectives on Politics* 2 (02): 259–279.

Sapiezynski, P., Stopczynski, A., Wind, D.K. et al. (2016). Inferring Person-to-Person Proximity Using Wifi Signals. arXiv, preprint arXiv:1610.04730.

Schaffer, W.M. (1985). Order and chaos in ecological systems. *Ecology Letters* 66 (1): 93–106.

Simon, H.A. (1996). *The Sciences of the Artificial*, 3e. Cambridge, MA: The MIT Press.

Simon, H.A. (2002). Near decomposability and the speed of evolution. *Industrial and Corporate Change* 11 (3): 587–599.

Simon, H.A. and Newell, A. (1962). Computer simulation of human thinking and problem solving. *Monographs of the Society for Research in Child Development* 27 (2): 137–150.

Sutskever, I., VInyals, O., and Le, Q.V. (2014). Sequence to sequence learning with neural networks. *Advances in Neural Information Processing Systems* 3104–3112.

Sutton, R.S. and Barto, A.G. (1998). *Reinforcement Learning: An Introduction*. MIT Press.

Tambe, M., Kale, D., Gupta, A. et al. (1988). Soar/PSM-E: investigating match parallelism in a learning production system. *ACM SIGPLAN Notices* 23 (9): 146–160.

The Economist (2017). Through a glass darkly: testing the methods of neuroscience on computer chips suggest they are wanting. *The Economist* (21 January). Retrieved from https://www.economist.com/science-and-technology/2017/01/21/tests-suggest-the-methods-of-neuroscience-are-left-wanting.

Tomai, E., Salazar, R., and Flores, R. (2013). Simulating aggregate player behavior with learning behavior trees. Proceedings from 22nd Annually Conference on Behavior Representation in Modeling and Simulation, Ottawa, 2013.

Tumasjan, A., Sprenger, T.I.O., Sandner, P.G., and Elpe, I.M. (2014). Predicting elections with twitter: what 140 characters reveal about political sentiment. *ICWSM* 10 (1): 178–185.

Vapnik, V. (2013). *The Nature of Statistical Learning Theory*. Springer Science & Business Media.

Weinberg, S. (1994). *Dreams of a Final Theory: The Scientist's Search for the Ultimate Laws of Nature*. New York: Vintage.

Wu, Y., Schuster, M.I., Chen, Z. et al. (2016). Google's Neural Machine Translation System: Bridging the Gap Between Human and Machine Translation. arXiv, preprint arXiv:609.08144.

Zacharias, G.L., MacMillan, J., and Van Hemel, S.B. (eds.) (2008). *Behavioral Modeling and Simulation: From Individuals to Societies*. Washington, DC: National Academies Press.

20

Social Media Signal Processing

Prasanna Giridhar and Tarek Abdelzaher

Computer Science Department, University of Illinois at Urbana–Champaign, Champaign, IL 61801, USA

Social Media as a Signal Modality

Posts on social networks collectively comprise a new type of indexing into physical reality, social beliefs, concepts, biases, and ideas (Levy 2013). A logical question becomes: can one develop an instrument to browse this reality, a new *macroscope* into world state? One purpose of such a device would be to reliably observe physical and social phenomena at scale, as interpreted by the collective intelligence of social media users.

This chapter describes computational insights that give rise to information processing algorithms for such an instrument. Engineers have long investigated how signals propagate through noisy channels. For example, engineers study how AM/FM radio transmissions propagate through air, walls, and metal, how vibrations travel through terrain, and how acoustic waves travel through physical matter. Signal loss and distortions are introduced. Such loss and distortions change or bias the received signal. An understanding of signal emission and propagation properties can simultaneously help reconstruct both a good approximation of the original transmitted signal as well as a model of the introduced perturbation or bias. Can one apply the same wisdom to social sensing to reconstruct both *physical reality* and *human biases* from posts propagating on social media?

The analogy between humans and physical media might at first seem too superficial if not outright offensive. After all, humans are much more intelligent, sophisticated, and unpredictable compared to physical things. They do not obey strict laws of nature, they have hidden agendas, and they engage in behaviors that are substantially more complex than air molecules or earth particles. In view of such drastic differences, how can one possibly borrow from engineering, a science geared toward the study of physical artifacts?

Social-Behavioral Modeling for Complex Systems, First Edition.
Edited by Paul K. Davis, Angela O'Mahony, and Jonathan Pfautz.
© 2019 John Wiley & Sons, Inc. Published 2019 by John Wiley & Sons, Inc.
Companion website: www.wiley.com/go/Davis_Social-Behavioralmodeling

The relevance of engineering foundations comes from the intuition that, at an appropriate level of abstraction, the outcomes of human behavior can be modeled by a finite set of choices. For example, in elections, the available choices are *vote yes, vote no,* or *abstain.* Similarly, when propagating information on Twitter, the choices are *retweet an existing tweet, write a new tweet,* or *stay silent.* While it is hard to predict the behavior of an individual, in aggregate, the overall proportions of different outcomes are easier to predict. For example, one may predict with some accuracy whether a given gun control law will meet community approval, given a high-level classification of the community as a whole, such as liberals vs. conservatives. The same applies to other behaviors, such as deciding to propagate a post that espouses a given political point of view (i.e. *retweet* it), or not. Predictions can be described by probability distributions over the set of possible outcomes. Predictions at the community level (e.g. results of a vote) are easier to compute than predictions of individual human choices.

We should also distinguish between the *what* and the *why* questions. The question of "*what* the result of a (conservative or liberal) vote is going to be" is a much easier question to answer compared with the question of *why* individuals will vote that way. The rationalization is usually a lot more complex, rooted in origins that remain subject to active research and debate, such as moral foundations theory (Graham et al. 2013) and the Schwartz value system (Schwartz and Bilsky 1987). Indeed some social scientists suggest that humans are prone to making decisions intuitively based on simple rules wired by evolution (Kahneman 2011; Graham et al. 2013), then consciously rationalizing the intuitive decisions, as opposed to using rational thinking to determine decision outcomes in the first place (Haidt 2001). If such is indeed the case, then collective human decisions may be easier to predict than their accompanying rationalizations. This chapter is exclusively focused in the *what* question. We focus on algorithms that model and exploit (the statistical distribution of) behavior *outcomes* as opposed to behavior *reasons* and *rationales.* We call the observable outcomes on the social medium the social (media) *signal.*

The above discussion leads to the core of our analogy: much in the way physical objects induce distinguishable signals in their physical environment that can be detected by observing the physical medium, socially relevant events induce distinguishable signals in their social environment that can be detected by observing the social medium. A study of the way these signals are emitted, the way they travel, and the way they are perturbed by the medium can lead to algorithms that help reconstruct both reality in the physical world and the perturbation introduced by the medium. The analogy opens up a novel research field, where human-centered sciences meet research on physical, computing, and engineered systems to better characterize the collective properties of social media signals, with the goal of exploiting social media to understand both physical reality and human/cultural biases.

We view the aforementioned reconstruction problem as a collaboration between humans and machines. The machine automates data preprocessing by observing data (propagation) patterns on social media and using those patterns to convert the firehose of emitted data into a more structured representation. The added structure separates descriptions of different events, separates reality from falsehoods, and separates text based on different source biases. This structure helps humans select the specific events they are interested in studying, set appropriate alert triggers (when events of interest occur), understand a notion of truth (vs. falsehoods) about these events, and learn biases of the different communities involved. The human should not have to aid the machine in doing its preprocessing. The premise of the social sensing analogy is that this preprocessing can be accomplished automatically by an appropriate signal processing algorithm much the way the signal processing algorithm of an acoustic sensor array might automatically eliminate echoes, account for sound reflections and other imperfections, and ultimately detect the true location of a tank. It remains to be seen how far this approach can go, but initial results, presented later in this chapter, suggest that there is promise.

Interdisciplinary Foundations: Sensors, Information, and Optimal Estimation

We view individuals on social media as agents stimulated by physical world events to share or relay information. These agents are imperfect. They may produce reports (e.g. tweets) that are biased and may disseminate willful misinformation. In doing so, they act as noisy sensors and/or noisy communication channels. Two problems can be defined on this context, inspired by physical sensors. The first, referred to in engineering literature as *signal detection*, is to reconstruct reality in the physical world from the shared reports by automatically detecting and correcting for misinformation and bias. The second, referred to as *channel estimation*, is to characterize information distortion properties of the reporting medium (i.e. the community of individuals involved). For example, we show that one may characterize the prevalent biases in the community from the manner in which the community propagates or filters information. In turn, such characterization may shed light on underlying cultural norms, political affiliations, or moral foundations. Three key concepts have been developed in embedded systems to address the above two problems analytically. They are described below:

Maximum likelihood estimation: There is a circular dependency between signal detection and channel estimation. Accurate signal detection needs a statistical model of introduced distortion in order for effects of distortion to be undone. Hence, it needs results of channel estimation. Conversely, channel estimation needs accurate measurements of the signal in order for the nature of distortion to be determined. Hence, it needs results of signal detection. Fortunately, prior literature developed iterative algorithms for joint channel estimation and signal detection, in which both problems are solved simultaneously from scratch. A popular solution approach in this context is *maximum likelihood estimation*, which can be accomplished using the expectation–maximization (EM) algorithm (Moon 1996). It proceeds by solving the signal detection problem given a channel estimate, solving the channel estimation problem given presently detected signals, and iterating until convergence is reached. Later in this chapter, we use EM to simultaneously detect misinformation on the social medium (signal detection) as well as estimate biases of different sources (channel estimation).

Information gain: Another key concept is to separate important vs. unimportant signals. Intuitively, important signals carry more information on the probability distribution of variables of interest. Information theory (Cover and Thomas 2012) offers a formal definition wherein the *amount* of information received is measured by the reduction in the number of bits needed to encode the possible values of a variable as a result of receiving the information on that variable. More precisely, it is measured by the reduction in the shortest possible average length of lossless compression encoding of the data. This reduction is formally called *information gain*. We shall use information gain in lieu of semantic notions of information to decide on importance of signals received. To illustrate, consider an investigation into the color of an escape vehicle used in a recent bank robbery. Assume the color was initially determined to be either black, blue, red, or green, with equal probability. Since there are four equally probable colors, conveying the actual vehicle color will take two bits. Now, assume that additional information was received that eliminates the color red and makes blue twice as likely as either of the remaining two colors. Hence, the probabilities now are 50% for blue, 25% for black, and 25% for green. An optimal encoder saves bandwidth by giving more probable values fewer bits; for example, it might denote blue by the code word *0*, black by the code word *10*, and green by the code word *11*. The expected number of bits transmitted to convey the color is the weighted sum of color probabilities, each multiplied by the corresponding code word length, or $0.5 \times (1 \text{ bit}) + 0.25 \times (2 \text{ bits}) + 0.25 \times (2 \text{ bits}) = 1 \text{ bit}$. The received information therefore reduces the average number of bits needed to convey color from 2 bits to 1 bit. The information gain is $2 - 1 = 1$ bit. Information theory offers convenient general mathematical expressions to compute information gain that allow comparing different signals. For example, if information was received that make one outcome of a vote significantly more likely, the resulting information gain can be formally quantified (as in the above example). In general, one can associate information gain with any *spike* or change in the probability distribution of a signal. We shall use information gain expressions to determine which social media signals are more important.

Event detection: In engineering sciences, a common model of physical media is the input/output model. It describes the statistical properties of the medium's reaction (or output) when exposed to a given stimulus, event, or action (the input). For example, when a tank moves, it produces an input stimulus on its environment, causing the emission of an output signal (e.g. sound) whose statistical properties are characteristic of the tank. From the perspective of an engineer, one way to model the *signal* generated on text-based social media, such as Twitter, is by changes in the *joint* probability distribution of emitted words and phrases. The joint probability distribution of words and phrases simply refers to the probability of occurrence of their different *combinations*. This distribution can be estimated empirically from the observed *frequency* of occurrence of these combinations on the medium. Individual words and phrases are indeed the fundamental building blocks of more complex text patterns. When events occur in the physical world, their occurrence changes the *joint* probability distribution of emitted words and phrases on the medium. We can use information gain to determine how much *information* (in the information-theoretic sense described above) a given change carries, which can be used as a foundation for detecting *important* events. The larger the change in the joint probability distribution, the higher the information gain.

Below, we describe how the above analytical insights can be used to solve the intertwined problems of signal detection and channel estimation on social media.

Event Detection and Demultiplexing on the Social Channel

Event detection refers to recognizing that a particular event has occurred. Event demultiplexing is the act of separating signals attributed to different events. Before we discuss event detection and demultiplexing on social media, it is important to define the input stimulus more precisely. The input stimulus is the event that causes reaction on the medium. It is the event we want to detect. For the purposes of this chapter, an *event* refers to an *incident that (i) occupies continuous limited time and space and (ii) is observable by humans*. Examples include individual demonstrations, flash-crowd events, car accidents, tornadoes, explosions, or parades. Let us consider a Twitter-based example. Consider the task of detecting the occurrence of tornadoes in the United States from their signature on Twitter. A simple approach might be to cut time into windows and inspect the received tweets for the number of occurrences of the word *tornado* in each window. The resulting probability distributions are compared across successive windows. A distribution change that has a high information gain can then be indicative of tornado-related event, such a tornado warning or a tornado sighting.

This basic approach needs to be extended to demultiplexing. On Twitter, signal demultiplexing is simply the challenge of separating tweets pertaining to one event from those pertaining to another. For example, we want to separate the tweets about one tornado from the tweets about another concurrent tornado. A key question is: can we do such separation automatically by a machine that does not understand the language? Interestingly, not only is the answer in the affirmative, but also the event separation algorithm turns out to be very simple, if proper intuitions are exploited.

When signals are represented in a sufficiently sparse feature space, different sources give rise to unique signal features (e.g. sets of signal *spikes* unique to each source) that allow one to distinguish among the sources. As mentioned earlier, on Twitter, we consider a lexical feature space, where words and phrases are the observed features. Events change the joint probability distribution of combinations of words and phrases (that are related to those events) on the medium. By the principle of Occam's razor, let us consider the simplest such combinations first, namely, combinations of two words. The lexicon of commonly used words in a language, such as English, may contain around 10 000 words. There are 100 million possible combinations of two words (i.e. work pairs). This is several orders of magnitude larger than the number of event instances we might need to distinguish (demultiplex) at any given time. Hence, events in the feature space of word combinations are indeed sparse. The probability of overlap between the frequent word pairs used in describing *different* events is very small. Note that common phrases such as *Bay Bridge* or *New York* are treated as a single word or token for the purposes of this discussion. To detect new events today, one merely needs to look for new frequent combinations that were not prevalent on the medium yesterday.

To illustrate, consider tracking car accidents. A particular car accident involving a drunk truck driver killing a dog on a bridge might be described by tweets containing such words as *drunk*, *dog*, and *bridge*, leading to a spike in the co-occurrence of new pairs of words such as *drunk* and *dog* in tweets describing the event. These words do not normally co-occur. A spike in their joint frequency of use in today's tweets indicates the occurrence of a new event. A question becomes: how do we know what pattern of new event-specific co-occurring words to look for in order to detect a new event instance?

The answer is: we do not have to know it ahead of time. We simply use information gain. More precisely, we cut time into windows and count all combinations of two words in the tweets emitted in each window. Hence, a tweet of 10 words, for example, would contain $(10 \times 9)/2 = 45$ different two-word combinations. We then identify those pairs that occur disproportionately more frequently in the current time window, compared with their normal frequency. The information gain metric, discussed earlier, offers a statistically well-founded measure of significance of frequency deviations of a word pair from the norm. High information gain word pairs are identified in the current window of observation. Each is then associated with a separate *bin* of tweets. Tweets containing both of the words of an identified high information gain word pair (in any order) are placed in the corresponding bin. Bins with largely similar tweets are then consolidated into one. The result is that tweets on the most informative events (i.e. events that resulted in high information gain changes in the joint word probability distribution on the medium) are collected into corresponding bins. They are automatically separated into a single bin per event, which constitutes demultiplexing.

Table 20.1 shows examples of tweets about different detected protests after demultiplexing. In this example, tweets were collected that contain the word *protest*. The stream of collected tweets was then automatically separated into bins using the technique described above. In the table, tweets on three different detected protests are shown that occurred around 26 December 2017. The high information gain automatically detected word pair corresponding to each bin is also shown. It can be seen that the automatic separation of tweets by the protest they describe is accurate despite the fact that the text uses slang, broken English, and ad hoc abbreviations. Indeed, the power of the signal processing approach, described in this chapter, lies in that the machine never interprets the text. It does not need to know proper vocabulary, slang, or specific abbreviations. It simply counts co-occurrences of word combinations in tweets with no interpretation of what the words represent. No prior training or labeling is needed. The significance of this approach for distinguishing different event instances lies in eliminating the human effort for machine training and labeling. The approach applies regardless of the language used (because it does not need to understand it) and is thus insensitive to slang, abbreviations, grammatical errors, and other imperfections that may pose problems for approaches that know something about vocabulary or grammar. (The latter approaches are usually sensitive to the language, dialect, and spelling that they were trained with.) The technique presented above is valuable as a preprocessing step to impose structure on data by separating it into different event bins.

The wealth of information posted on social media offers promise as a mechanism for understanding human behavior. The event detection and demultiplexing approach described in this section allows the human information processing effort to be reduced. For example, it becomes easier to automatically filter out less important events, generate alerts when important events occur, or focus exclusively on particular event instances, without manually going through all tweets to distinguish ones that belong to those specific instances. We are therefore hopeful that solutions, such as the above, can empower and accelerate social science research.

Table 20.1 Examples of demultiplexed protests.

(High information gain) word pair	Tweet bins
Karachi, Teachers	We all newly h.ms condemn govt of sindh brutial action against nts teachers peacefull protest in karachi, govt must regularized immediately
	#BREAKING: Teachers hold protest outside press club in #Karachi, claims to march toward #Sindh's Chief Minister house… (https://t.co/1nXnT5gpUp)
	_ Karachi Press Club Pe Hazroon Teachers Ka ApNe Ke Liye Protest jari, MPA Nusrat Seher Abbasi & PTI Ke Arif Alvi Be Maujood A_Baqi Jakhrani
	Miss please do a show for nts teachers who are road's of Karachi for protest
Mahadayi, Farmers	Mahadayi River row: Farmers gather to protest outside BJP office in Bengaluru.(UTTAR BANDH) Part-01: https://t.co/Gxvn2Lpjxg via @YouTube
	Bengaluru: Mahadayi farmers protest – details from Reporter (https://t.co/zOP2Lhcgqd) via @YouTube
	Vatal Nagaraj Meets & Joins Protest With Farmers Over Mahadayi Issue (https://t.co/gO0c5uxkMI) via @YouTube
	B S Yeddyurappa Decided Not To Meet Mahadayi Protest Farmers | ಸುದ್ದಿ ಟಿವಿ Suddi TV #BSyeddyurappa #Mahadayi… (https://t.co/piHREfiCEm)
Rockets, Clippers	The Rockets filed a protest over the Clippers game lmaoooo, they might be worse than the Warriors when it comes to being sore losers
	#ClippersNation #Clippers #LAC Rockets File Protest Over Officiating Error in Loss to Clippers (https://t.co/5nr3IZ0o2H)

Filtering Misinformation

Another key signal processing task for a sensor is to detect and reduce noise (or, in the case of social sensing, reduce misinformation). Our social medium comprises imperfect human observers who act as information sources and relays. How well does the collective output of those observers represent reality in the physical world? The reception of more reports about an alleged occurrence of some event does not necessarily lend more credibility to that event. Some sources may simply repeat information they heard from others without independent verification. Similarly, internal bias of sources makes them predisposed to believing information that confirms their bias and relaying it without questioning. Hence, misinformation may be propagated, leading to highly correlated widely spread beliefs that misrepresent reality.

It is useful at this point to distinguish factual statements from opinions. By factual statements we refer to concrete claims made about conditions of the physical world that a reliable observer can easily determine to be either true or false. For example, the claim that a bomb was detonated on Main Street in a given city at a given time can be factually verified to be either true or false. Opinions, in contrast, are not easily verifiable via direct observation. For example, whether a particular leader is *good* or *bad* is usually a matter of opinion. The opinion may depend on an individual's value system, loyalties, and rationale. Let us first consider factual statements only. Later we shall extend the discussion to consider opinions.

Social media posts cast as factual statements constitute *misinformation* if they contradict ground truth. One question to consider is: can we automatically detect and flag misinformation on the social medium? Clearly, if the reliability of each source on the social medium is known, the problem of detecting misinformation will be much simpler. Unfortunately, source reliability is generally unknown. Hence, we jointly compute the truth value of each reported statement (which is a signal detection problem), together with computing the reliability of each source (which is a channel estimation problem) using algorithms for joint signal detection and channel estimation.

Conceptually, we represent the reported observations (tweets) about an event by a graph of sources and claims that we call the source–claim network. Sources simply refer to the IDs of devices that report the corresponding data. Claims can be thought of as abstract statements reported by the sources. In this case, we take the tweets as claims. In the resulting source–claim network, a link between a source and a claim indicates that the source asserted that claim (i.e. made that tweet). Hence, when multiple individuals make the same claim (e.g. retweet the same statement), they are connected to the same claim node in the graph. Similarly, all claims made by the same individual are connected to the same source node. The source–claim network is a general representation of reported data that enables cleaning. Another data structure is the empirically observed correlations among different pairs of sources (measured by the percentage of time the pair makes the same claim). These are used to estimate probability of correlated errors. Once the source–claim graph and the source correlation graph are formed, the joint channel estimation and signal detection problem can be solved. Recent literature reported different ways to jointly estimate source reliability parameters, together with the true/false value of each claim, using the source–claim graph, SC, and source correlation graph, SD, as input (Wang et al. 2014; Yao et al. 2016). These techniques require no prior labeling of data; no one needs to tell the machine what reality is or tag any tweets as true or false. Results show that the algorithms automatically determine the truth values of tweets with up to 80–90% accuracy with no prior training or labeling.

The idea behind these techniques is depicted in Figure 20.1, where circular nodes represent sources and square nodes represent claims. Arcs between

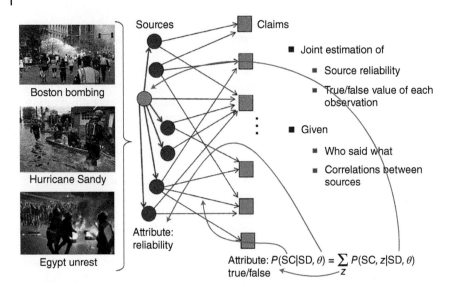

Figure 20.1 The error correction problem formulation.

sources represent correlations, whereas arcs between sources and claims represent who said what. The likelihood function is also depicted as a conditional probability of observing the graph, SC. Hence, for example, a square on the right with many incoming edges represents a tweet repeated by many sources. A pair of circles on the left that are connected represents a pair of correlated sources, perhaps one that follows the other on Twitter. Intuitively, a claim made by a larger number of more independent (i.e. not connected) sources is more likely to be true. Similarly, a source who makes more claims that are deemed true is likely to be more reliable. The observation suggests an approach where source reliability and claim correctness can be jointly determined in an iterative fashion with no prior knowledge of ground truth and no prior reputation scores.

Informally, the estimation algorithm randomly initializes reliability parameters for each source and truth/falsehood values for each claim. An iterative approach is used to adjust source reliability parameters and claim correctness values in a direction that maximizes the likelihood. The iterations are derived by applying a maximum likelihood estimation algorithm, called EM. It converges to a maximum likelihood estimate, which in this case yields the most likely source reliability values and the most likely true/false classification for each tweet. Moreover, variance of the resulting estimator could be obtained using the Cramer–Rao bound (Cramer 1946). In turn, this variance allows computing a confidence interval in results, which separates the aforementioned approach from prior work on media exploitation and fact-finding. As a result, it is possible for the machine to assess automatically which statements are likely to be true of the physical world and which statements are likely to be false without prior training or supervision. The algorithm augments work described earlier on event detection and demultiplexing by a capability to assess veracity of individual claims (tweets) made about the events in question.

Human Bias, Opinions, and Polarization

Detecting misinformation, as described in the previous section, is useful when the social medium is leveraged as a *sensing* instrument of physical reality. Equally important, however, is the use of the social medium to understand human opinions and beliefs. Following our signal processing approach, we would like to infer beliefs from the way signals propagate on the medium, much in the way propagation of acoustic waves can be used to infer properties of terrain in which the waves propagate (e.g. map the ocean floor).

Modeling Signal Propagation

Let us first understand the impact of trust relations among sources and the impact of source bias on information propagation. Toward that end, we conducted an experiment on Amazon Mechanical Turk, where individuals were asked whether they would relay and whether they would contradict each of multiple statements describing different experiences. Three treatments were considered, depending on whether the experiences in question were (i) presented as witnessed firsthand by the individual, (ii) reported to the individual by a source they consider reliable, or (iii) reported by an unreliable source. Table 20.2 presents the statements that the participants were presented with. One should add the caveat that what participants claimed they would do is not necessarily the same as what they would do in reality. Nevertheless, as we show later, results inspired effective bias separation algorithms.

Observe in particular that statements 6–10 described a conflict between two imagined families in the imagined city of the participant (which was named *Zalawera*). Each of the three treatments mentioned above was thus further divided into two groups. In one, participants were told that they were a member of one of the families (the Clays). In the other, participants were not given a family affiliation. The purpose was to measure the impact of bias (to protect family name) on choices made by the participants in relaying or denying information.

The experiment included 759 participants (limited for logistical reasons to the US citizens). All participants included in the study were 18 years of age or older, with over 50% of participants' ages ranging between 18 and 34 years of age. Sex was represented equally within the sample (50% female and 50% male). Participants were randomly assigned to one of the six (i.e. three by two) experimental conditions manipulating group bias and trust in information source. Participant responses on (i) how likely they were to relay information

Table 20.2 Statements presented to participants for propagation analysis.

1.	A very strong earthquake shook the city of Zalawera
2.	After days of intense rain, the river of Zalawera is flooding nearby neighborhoods
3.	Severe thunderstorms causing major power blackouts in Zalawera
4.	The local Zalawera hospital catches fire after a lightning strike and burns to the ground
5.	Category 4 tornadoes level entire neighborhoods in the outskirts of Zalawera
6.	Violent riots and looting erupt, carried out by the Clays on the Main Street in Zalawera. (The Clays are one of two warring clans in the city)
7.	Unmanned drones hover in the air above Zalawera, carrying the Clay Clan logo
8.	Fire shots are exchanged between the Clay Clan and the Lions Clan, the two warring Zalawera factions
9.	The local Zalawera government militia shoot at a peaceful Clay gathering causing several fatalities
10.	A huge explosion rattles the Clays' place of worship in Zalawera
11.	A major car accident is blocking all lanes of Highway 66, the main Highway of Zalawera
12.	Severe traffic congestion around Zalawera Stadium because of the football game
13.	Hazardous debris partially blocking the left lane on Shore Blvd of Zalawera
14.	Stuck in severe traffic on Mission Avenue because of a construction zone
15.	Zalawera police are asking shoppers to evacuate Mall's parking lot because of a bomb threat

and (ii) how likely they were to contradict information were provided on a 7-point Likert scale, ranging in increasing numeric order from *very unlikely* to *very likely*.

While the detailed findings of this study are reported elsewhere (Roy et al. 2016), two key observations are noted. First, the study found a statistically significant dependence between predisposition to relay information and the level of trust in source. Specifically, groups who were told that they witnessed the stated events firsthand were more likely to relay the statements than those who were told that they received the information from another source they trust. Similarly, the latter groups were more likely to relay information than those who were told they received the information from a source not known for reliability.

Second, among groups who were given the Clays affiliation, there was a statistically significant correlation between predisposition to contradict the information and the degree of damage the information posed to the Clay name. More specifically, looking at questions 6–10, it is easy to see that they are sorted from most damaging to Clays to most empathetic with Clays. The average predisposition to contradict a statement increased monotonically between questions 10 and 6. No such monotonic relation was observed in groups that were given no affiliation, as well as in answering other questions (not featuring a conflict with Clays). These results offer evidence that in-group bias produces an information filtering effect, whereby groups are less likely to relay (and more likely to contradict) information that conflicts with their bias. In contrast, in contexts involving no conflicts, no such filtering effect exists.

While the above study was limited to the US participants, it would be interesting to compare the results for different cultures and backgrounds. Empirical evidence suggests that the collective filtering introduced by conflicted groups creates a valuable means to infer their respective biases, as explained below.

Opinion Separation and Polarization Detection

The informal intuition is that individuals will propagate information they agree with. Let us draw a graph where individuals are nodes and where the thickness of an edge between two nodes represents how many times they agreed on the same post (e.g. tweeted/retweeted the same thing). The more frequently they agree, the thicker the edge. Now, let us grow clusters of nodes by merging nodes connected by thick edges into the same cluster. The output of such clustering is a partitioning of the community by opinions shared by different subgroups. Recent empirical work has shown that such partitioning is especially effective at separating supporters of different sides of a debate as well as their opinions (Al Amin et al. 2017). For example, consider the US political system. Republicans will more often propagate content that is favorable to Republicans, whereas Democrats will more often propagate content that is favorable to Democrats. Hence, thicker edges will exist among Democrats as well as among Republican, while thinner edges will exist across the two groups. The clustering algorithm will therefore result in two big clusters of nodes: one of a Democrat disposition and the other of a Republican one. The clusters reveal who belongs to each group as well as what opinions (posts) they espouse.

Table 20.3 illustrates the result of applying this process to the separation of supporting and opposing claims tweeted after the controversial removal of the former Egyptian president, Mohamed Morsi, from power. Pro-Morsi tweets are shown in the left column, whereas anti-Morsi tweets in the right. The table shows that the separation is very accurate despite the fact that the machine does not understand the text. Indeed, the left column depicts Morsi and his supporters as victims, whereas the right column depicts Morsi and his supporters/party (the Muslim Brotherhood) as victimizers. The two sides are separated based on text propagation patterns. In a polarized scenario, claims would generally propagate along one of two major pathways, depending on which group they are favorable to, making separation possible.

Observe that separating supporting and opposing claims is different from sentiment analysis. For example, the fourth claim under the *pro* column in Table 20.3 has a strong negative sentiment (mentioning arrests, beatings, and denial of rights). Yet, the claim is clearly pro-Morsi. Indeed, most tweets on

Table 20.3 Separating support from opposition to a person or cause.

Pro-Morsi	Anti-Morsi
Sudden Improvements in Egypt Suggest a Campaign to Undermine Morsi (http://t.co/0yCjbKGESr)	Prayers for the Christian community in Egypt, facing violent backlash for opposing the Muslim Brotherhood (https://t.co/O5X7BwUjCI)
Saudi Arabia accused of giving Egypt $1B to oust Morsi (http://t.co/d4ZQNntCH)	Egypt's Coptic Christians, under attack for supporting overthrow of Muslim Brotherhood, need continued prayers (http://t.co/dW0gdcielb)
Before Morsi's Ouster, Egypt's top generals met regularly with opposition leaders (http://t.co/LbdHKJF508) via @WSJ	Islamic extremists reportedly attacking Egypt's Christian community over Morsi ouster – *Fox News* (http://t.co/VMMN2m49Sw)
#Egypt: #Morsi supporters denied rights amid reports of arrests and beatings – Amnesty International (http://t.co/koVRHlmdWk)	In Egypt, the death toll in the clashes between police and pro-Morsi supporters in Cairo has risen to 34
Crowds March in Egypt to Protest Morsi Detention (http://t.co/Hp9566xyfB)	Amnesty International – Egypt: Evidence points to torture carried out by Morsi supporters (http://t.co/8hgAHrNoWd)

both sides have a negative sentiment as they heap blame on the other party for alleged injustice and atrocities. The technique reported above can be used to automatically separate human attitudes toward controversial issues. In turn, such an automatic separation can be used to map political affiliations, underlying moral foundations, or prevalent religious beliefs in communities of interest.

The techniques described above can serve as preprocessing tools (e.g. to help an analyst prepare a situation report for a commander or help an anthropologist study the biases of a culture). Machine errors will occur. For example, supportive statements of a cause might, on occasion, be put in the opposing bin. The human user should understand that the machine is not perfect.

Online Tools

The above ideas were implemented in a social sensing toolkit, called Apollo. Apollo uses Twitter and Instagram APIs (or rather a reseller of Instagram content) to collect tweets and pictures. A parser parses tweet content into a source ID and claim body, inserting the tuple in a file. For simplicity, lexically similar or identical claims are considered to be the same one. A network of sources and claims is thus constructed. Claims are demultiplexed into event streams and passed to the module responsible for misinformation detection. An interface offers user access to distilled information after processing.

In the Apollo Social Sensing Toolkit, a data collection and analytic pipeline is called a task. A task can be started by the user to collect data on a particular topic. Once the relevant topic information is entered, users are presented with options that allow them to choose different processing modules to run on the collected data, such as veracity analysis and event detection. Experiences with Apollo suggest that it often detects new events before they are reported by the mainstream news media. While some events are reported by news media first, before they are relayed on social networks (e.g. statements by political figures and exclusive interviews with celebrities), others appear on social networks before they are picked up by news media. Examples include reports of traffic accidents, demonstrations, terrorist acts, or damage from natural disasters. More on the Apollo tool can be found in the authors' prior work (Roy et al. 2017). An in-depth dive into the analytical foundations of the approach can be found in the author's recent book (Wang et al. 2015). At the time of writing, techniques described in this chapter have been used by researchers on a daily basis to monitor protests, natural disasters, civil unrest events, acts of terror, social movements, and other natural and man-made phenomena around the globe.

The approaches discussed in this chapter differ from what is traditionally referred to as *machine learning* or *artificial intelligence* (AI) solutions. Typical machine learning/AI solutions need labeled data in order to learn how to classify inputs by categories of interest. This need is a source of great power as well as great pain. Clearly, exposure to large amounts of labeled data leads to improved solution accuracy. On the disadvantage side, the labeling effort is often prohibitive. For example, Project Maven (Cukor 2018) operationalizes machine learning in military contexts by using deep learning for image recognition. The most expensive component of the project lies in the labeling effort to train the machine. This training is domain sensitive. Indeed, despite the availability of large amounts of labeled training imagery on public media, Maven has to do its own labeling because public media images differ from military images. The algorithms need labeled data from the right domain. In contrast, the solutions described above do not require machine training. No labeled data are used. These solutions, therefore, generalize easily across domains, languages, topics, and dialects. Current efforts are focused on improving resilience with respect to subversive behavior, collusion, and other efforts targeted at misdirecting these algorithms. Preliminary work suggests that low-cost collusion detection techniques are possible, based on correlation analysis across sources, combined with anomaly detection. In general, detecting bad online actors is an arms race. Techniques will continue to evolve as attacks as well as defenses increase in sophistication over time. Discussion of such techniques, however, is beyond scope for the current chapter.

Conclusions

This chapter illustrated an analogy between digital data processing techniques, applied to the extraction of signals from noisy inputs, and new algorithms for social network data analysis. The analogy demonstrates the feasibility of building signal processing libraries for social network data analysis that are founded in techniques, borrowed from the engineering discipline, such as information-theoretic signal detection and estimation-theoretic EM. The work calls for a novel research agenda on developing robust data fusion algorithms and theory for social sensing, not unlike those for acoustic, magnetic, or optical sensing, based on estimation-theoretic and information-theoretic concepts. These algorithms and theory may accelerate social science research by automating the preprocessing of large social media data sets for scientists by adding structure to data toward a better understanding of physical and social phenomena via the lens of social media data.

Acknowledgment

Research reported in this chapter was sponsored in part by the Army Research Laboratory under Cooperative Agreements W911NF-09-2-0053 and W911NF-17-2-0196 and in part by DARPA under award W911NF-17-C-0099. The views and conclusions contained in this document are those of the authors and should not be interpreted as representing the official policies, either expressed or implied, of the Army Research Laboratory, DARPA, or the US government. The US government is authorized to reproduce and distribute reprints for government purposes notwithstanding any copyright notation hereon.

References

Al Amin, M.T., Aggarwal, C., Yao, S. et al. (2017). Unveiling polarization in social networks: a matrix factorization approach. In: *INFOCOM 2017-IEEE Conference on Computer Communications, IEEE*, 1–9. IEEE.

Cover, T.M. and Thomas, J.A. (2012). *Elements of Information Theory*. Wiley.

Cramer, H. (1946). *Mathematical Methods of Statistics*. Princeton University Press.

Cukor, D. (2018). Project Maven: operationalizing machine learning. In: *Proceedings of the SPIE Defense and Security*. Orlando, FL: SPIE.

Graham, J., Haidt, J., Koleva, S. et al. (2013). Moral foundations theory: the pragmatic validity of moral pluralism. In: *Advances in Experimental Social Psychology*, vol. 47, 55–130. Academic Press.

Haidt, J. (2001). The emotional dog and its rational tail: a social intuitionist approach to moral judgment. *Psychological Review* 108 (4): 814.

Kahneman, D. (2011). *Thinking, Fast and Slow*. Macmillan.

Levy, P. (2013). *The Semantic Sphere 1: Computation, Cognition and Information Economy*. Wiley.

Moon, T.K. (1996). The expectation-maximization algorithm. *IEEE Signal Processing Magazine* 13 (6): 47–60.

Roy, H., Bowman, E.K., Kase, S.E., and Abdelzaher, T. (2016). Investigating social bias in information transmission: experimental design and preliminary analyses. *Proceedings of the 21st International Command and Control Research and Technology Symposium (ICCRTS)*, London, UK.

Roy, H., Abdelzaher, T., Bowman, E.K., and Al Amin, M.T. (2017). Information flow on social networks: from empirical data to situation understanding. In: *Next-Generation Analyst V*, vol. 10207, 1020702. SPIE.

Schwartz, S.H. and Bilsky, W. (1987). Toward a universal psychological structure of human values. *Journal of Personality and Social Psychology* 53 (3): 550.

Wang, D., Abdelzaher, T., and Kaplan, L. (2015). *Social Sensing: Building Reliable Systems on Unreliable Data*. Morgan Kaufmann.

Wang, D., Amin, M.T., Li, S. et al. (2014). Using humans as sensors: an estimation-theoretic perspective. In: *Information Processing in Sensor Networks, IPSN-14 Proceedings of the 13th International Symposium on IEEE*, 35–46. ACM.

Yao, S., Hu, S., Li, S. et al. (2016). On source dependency models for reliable social sensing: Algorithms and fundamental error bounds. In: *Distributed Computing Systems (ICDCS), 2016 IEEE 36th International Conference on IEEE*, 467–476. IEEE.

21

Evaluation and Validation Approaches for Simulation of Social Behavior: Challenges and Opportunities

Emily Saldanha[1], Leslie M. Blaha[2], Arun V. Sathanur[3], Nathan Hodas[1], Svitlana Volkova[1], and Mark Greaves[1]

[1] Data Sciences and Analytics Group, National Security Directorate, Pacific Northwest National Laboratory, Richland, WA 99354, USA
[2] Visual Analytics, Pacific Northwest National Laboratory, Richland, WA 99354, USA
[3] Physical and Computational Sciences Directorate, Pacific Northwest National Laboratory, Seattle, WA 98109, USA

Overview

Broad Observations

Humans often behave in ways that defy simple explanations. For this reason, complex behavior models may be used to draw together what we know and believe about humans and to help us model and predict events in the real world. For example, we may like to know the effect of a public health intervention on diabetes rates (Jones et al. 2006), forecast human migration during a refugee crisis (Edwards 2008), or understand how online content becomes viral (Hodas and Lerman 2014). As we advance the science of modeling and simulation of complex social systems, we are challenged to grapple with complexity at multiple levels of the system. Individual decision-making behaviors can be modeled, with complexity of behavior emerging from basic cognitive mechanisms. The interpersonal interactions between individuals then give rise to the complex, dynamic behaviors of social systems. It follows that the development of new techniques for modeling social systems requires careful consideration and potential advances in the methods for verification, validation, and accreditation of those models to both inform the model development process and to determine usability for desired applications.

This paper is written to explore validation issues in computational social science research that is primarily data-driven with massive amounts of automatically collected data, such as data from social media. Social media is

Social-Behavioral Modeling for Complex Systems, First Edition.
Edited by Paul K. Davis, Angela O'Mahony, and Jonathan Pfautz.
© 2019 John Wiley & Sons, Inc. Published 2019 by John Wiley & Sons, Inc.
Companion website: www.wiley.com/go/Davis_Social-Behavioralmodeling

redefining approaches to social science because it creates novel social environments that interact with the real world while producing massive amounts of data that we seek to understand and simulate. Additionally, we seek to understand the interplay of the online and offline social landscape reflected in social-behavioral data. To do this, we must have validated models producing trustworthy simulated data to explore the phenomena of interest. Thus, this chapter focuses on how we can best evaluate and validate models in this challenging problem space. Our perspective emphasizes data-driven evaluation, meaning we will explore how modeling of patterns of observed data can inform the larger process, and we emphasize the validation challenges in this space.

In what follows, we explore current practices in validation of models of social behavior and challenges in performing this validation. We suggest best practices in validation given the growing opportunities that large-scale, online social data sets provide.

All system models involve inevitable assumptions and simplifications. To quote George Box, "all models are wrong but some are useful" (Box 1979). How can we be confident that a model is valid enough to capture phenomena of interest? One facet of confidence is finding consistent patterns when comparing the outputs of the models with the corresponding empirical observations. We focus on this element throughout the current chapter. Elsewhere in this volume, Davis and O'Mahoney (2019) discussed six dimensions for evaluating model validity, and Carley (2019) discusses some of the major challenges to validation.

As simple as this sounds, execution of this validation involves many challenges. First, to determine how well a model captures phenomena of interest, validation can incorporate both qualitative and quantitative components. *Qualitative* evaluation by subject matter experts (SMEs) is a common approach for model evaluation and may provide insight into a model's basic trends. In this case, it is imperative that the SMEs developing the model not be those evaluating it, and common standards should be established a priori for consistent SME evaluation. This can be fraught with subjective judgment pitfalls (Young 2003; Campbell and Bolton 2005). If two models both show the same qualitative trends, the use of quantitative comparison may provide a way to distinguish the two. *Quantitative* metrics also provide a mechanism for incorporating machine learning or other automated model tuning and calibration techniques and provide a means to determine if a model's performance evolves over time. Quantitative metrics indicating the ways in which a model differs from empirical observations are critical for determining how a model should be improved.

A second challenge is that we may not always have a way to articulate the precise metric we seek to optimize. As Alfred Korzybski stated, "the map is not the territory" (Korzybski 1958). For example, the discrepancy between predicted unemployment vs. actual unemployment may be a clear, quantitative metric, but does it reflect whether or not an entire economy is being accurately modeled? Do we say our model of social media is correct if we precisely capture the temporal correlation of user posts? What metric or metrics do we choose if we want to validate a model of public response to building a hospital? When working with complex systems like social systems, we will often need multiple metrics to capture the multifaceted behaviors. How should we weigh the importance of each metric to the overall effort?

Finally, we may not possess the data necessary to compare important model predictions with our observations in a robust manner. A presidential election in the United States occurs every four years, and systematic polling has only been present for a fraction of past elections. Some models require data that may not be directly observable (latent factors), such as an individual's political preference, so we might turn to rough proxies such as the individual's registered party. Lastly, we may only have data aggregated at a resolution too coarse or too fine for our models, such as election outcome per state instead of per zip code or per neighborhood block.

Before delving into validation practices, we specify some terminology. In the present work, *theory* refers to a conceptual explanation of some phenomena of interest; it supports conceptual explanation and exploration of those phenomena. A *model* is a formal instantiation of a theory, usually mathematical or computational in nature. A theory, therefore, could have multiple models if different approaches are used to develop the formal instantiations. A *simulation* is at least one experiment conducted with a model, in which code executes or calculations are completed. It produces behaviors consistent with the theoretical perspective and model instantiation. This data can then be compared with real-world observations as a test or realization of the model or the theory.

Online Communication in Particular: A Valuable Venue for Validation

All of the factors described above make model validation challenging. Enter social media. Online social platforms, such as Twitter, GitHub, or Reddit, provide a public forum for millions of people to exchange information. They provide an unprecedented volume of data. As of 2017, Facebook had approximately 2 billion active monthly users.[1] While social media activity takes place on a computer or phone, it is also very much the *real world*. People meet, form groups, exchange knowledge, create friendships, etc. These large-scale data sets provide a unique opportunity in social-behavioral modeling to explore a full range of validation metrics on a wide variety of human actions, addressing many of the above concerns.

Using social media as a window into social behavior, we can see how different modeling approaches might influence different evaluation metrics. Studying

---

1 https://techcrunch.com/2017/06/27/facebook-2-billion-users

the interactions of billions of people might be unfeasible for agent-based modeling but very easy in a mean-field approach (Weidlich 2006; Helbing 2010). Conversely, understanding microscopic interactions in small groups would be accessible to an agent-based model but not to an approach operating at a population level. Therefore, modeling approaches necessarily drive evaluation metrics. Social media, with millions of users and perhaps billions of events, provides a venue for exploring and validating both microscopic and macroscopic models of social behaviors.

Simulation Validation

Validation for modeling and simulation is the process of determining the extent to which the behavior produced by the model is an accurate representation of the real-world phenomena it attempts to reproduce (Roache 2009; Sargent 2011; U.S. Department of Defense 2011). Validation allows for conclusions to be drawn about the utility of the simulation for its intended use. We note that validation methods are often discussed together with verification and accreditation processes, and it is important to distinguish these processes. Verification is the process of evaluating the modeling and simulation *implementation*; that is, it addresses whether the model and its associated data are implemented accurately, reflecting the developer's concept and specifications (Roache 2009; U.S. Department of Defense 2011). This is the process whereby software bugs are found, errors in equations or logic are corrected, etc. Accreditation is a formal certification process that must be accomplished before a model can be deployed into operations (U.S. Department of Defense 2011). This typically occurs for mature models that have been vetted through verification and validation already. In the present chapter, we emphasize only the validation process, because this is the critical process by which the social science research community will compare and contrast models, seeking to advance the research field.

Model validity can be broken down into concept validity, or the adequacy of the underlying conceptual or theoretical model in characterizing the real world, and operational validity, or the adequacy and accuracy of the computational model in matching real-world data (Sargent 2011). Additionally, application validity captures the degree to which a model is valid for its intended use (Campbell and Bolton 2005). There are many possible techniques for probing a model's validity. Face validity is evaluated by SMEs judging the *reasonableness* of the simulation outputs. Turing tests also entail SME evaluation, as experts attempt to distinguish simulation from reality. Quantitative approaches include graphical comparisons between simulation and observation and numerical quantification (e.g. goodness-of-fit statistics) of the difference between the model predictions and ground truth.

Compared with validation of simulations in the physical sciences and engineering, validation of simulations of social behavior both online and offline has relied extensively on face validity and qualitative comparisons. That is, the emphasis has been on expert judgment that a model is parsimonious with the expectation for how the system should work. This can be explained by the unique data collection challenges in the social sciences, including systems that cannot be observed directly and systems with many interdependencies that cannot be measured simultaneously. And although the validation needs in social science simulation are in many ways similar to other fields, a key difference is particularly evident when considering online social systems: model development and validation are concurrent with the efforts to define and understand the very phenomena to be modeled. Older fields, like physics, engineering, and behavior representation modeling, benefited from years of empirical work in which models were developed and evaluated for construct validity prior to computational simulation research (Campbell and Bolton 2005). The consequence is that social simulation work must validate the theoretical constructs together with modeling and simulation approaches and it must incorporate the ability to adapt all these tools with our changing understanding of social behavior and emerging phenomena in evolving social environments. However, as novel data sources have increased the available volume and variety of ground truth observations of social behaviors, we are increasingly able to leverage the quantitative techniques developed in other fields to both aid the characterization of the data and evaluation of models. In this chapter, we focus on the application of quantitative operational validity for comparative validation of social and behavioral models.

Simulation Evaluation: Current Practices

To explore common practices in current social science modeling and simulation validation efforts, we examine the methods in recent use for simulation of information diffusion in online systems. Modeling of information propagation in social networks and other online ecosystems has typically focused on a descriptive approach to diffusion phenomena. For example, some (e.g. Kwak et al. 2010) study the network properties of Twitter and the temporal diffusion properties of trending topics. Others (e.g. Ferrara and Yang 2015) study how sentiment affects information diffusion on Twitter, and yet others (e.g. Adamic et al. 2016) study how memes evolve during a diffusion cascade on Facebook. Such *descriptive studies* provide key insights for properties of information diffusion that should be replicated by simulations and models of these phenomena.

Validation of the simulation of information diffusion through online social systems is typically performed either through *qualitative evaluation* of observed diffusion phenomena or through *quantitative comparison* of the observed propagation of specific messages. This quantitative comparison can be performed on properties of the full diffusion path of a message or on the propagation of messages through specific nodes in the social network. Examples of qualitative evaluation of simulation results include the visual comparison of the cumulative distribution of the total size, depth, and structural virality of cascades through online social networks (Del Vicario et al. 2016; Goel et al. 2016).

Some examples of metrics that quantitatively evaluate the full diffusion process of a message include the root-mean-square error (RMSE), relative error, or mean error of the total number of message shares (de C Gatti et al. 2013; Jin et al. 2013), the absolute difference in number of users of who agree or disagree with a particular message (Serrano and Iglesias 2016), and the relative absolute error of the proportion of users within a certain graph distance that receive the message (Wang et al. 2012).

Approaches that focus more on quantitative node-level evaluation include predictive metrics such as the use of accuracy, precision, recall, and F1 score of whether individual users will propagate a message and the mean square error (MSE) and mean absolute error (MAE) of the predicted rate at which they will do so (Galuba et al. 2010; Gomez Rodriguez et al. 2013; Wang et al. 2014; Hu et al. 2017).

Most existing approaches focus on model fit statistics rather than performing cross-validation, which limits validation to studying only how well a model fits a particular sample of data. Cross-validation using a held-out data set is desirable because it evaluates the ability of the model to predict or generalize across data sets (Pew et al. 2005). Validation is sometimes performed on a metric other than the optimization criteria used for the model fitting, which can add an additional dimension for model performance evaluation and comparison. The use of separate train and test sets is more common for approaches that leverage predictive metrics, with some approaches using a time-based train–test split (Galuba et al. 2010) and others splitting on users (Hu et al. 2017).

Measurements, Metrics, and Their Limitations

When evaluating simulation results, the goal should be to replicate the behaviors and phenomena found in the observed ground truth data. Ground truth data is a sample of real-world data containing the phenomena of interest; data selection or design is driven by the questions of interest for the modeling and simulation activities. In any complex social system, there will be a multitude of *simultaneously* occurring phenomena observed in the same set of social interactions. *Measurements* are designed to directly probe these properties in the data. For example, in the simulation of online information diffusion, measurements used in previous work have included the total volume of the diffusion cascades (Jin et al. 2013), the transmission rate of a particular network edge (Wang et al. 2014), and the number of shares as a function of time (Matsubara et al. 2012).

Given a set of such measurements, one must then evaluate how well the measured behavior corresponds with observations. For this purpose, *metrics* allow you to compare the results of modeling or simulation with ground truth through quantitative comparison of their observed measurements. A large range of possible metrics are available; choices depend on the scaling properties of the measurements being compared and the goals of the comparisons. Metrics could include distribution comparisons such as Kullback–Leibler divergence, ranked list comparisons such as Spearman correlation, and sequence comparisons such as Levenshtein distance.

Most existing approaches for modeling online social behavior in social networks rely on a limited set of measurements, usually focused on a specific *phenomenon* (e.g. information cascades or gatekeepers), a specific *problem* (e.g. predicting meme propagation), or a specific social media platform (e.g. Twitter or Instagram). In current practice, model *evaluation depends on the specific modeling approach*, with different approaches being used for machine learning vs. epidemiological models.

Moreover, current evaluation approaches suffer from difficulty with properly measuring *causal relationships in the data*, sensitivity of the existing measurements to *initial conditions and model assumptions*, and failure to account for *uncertainty*. Additionally, comparison among simulation and model approaches is inhibited by the lack of common computational environments, common input/output formats for data, and common measurements and metrics. The impact of this is a challenge for generalizability to new contexts, inability to reproduce the validation approach for new models, and difficulties in interpretation of validation results.

Lack of Common Standards

Previous social simulation efforts have applied a variety of validation strategies and metrics. However, because each effort applied different validation techniques, comparison of the results among the approaches is not feasible. At a high level, simulation approaches can be thought of as differing along the granularity of simulation, while at a deeper level the approaches could differ in the nature of the techniques employed. For example, the approaches could be simulating individual, group-level, or population-level dynamics, while two individual-level approaches could differ in the nature of the agent models used. Granularity may also vary along dimensions like time or geospatial resolution (e.g. municipality vs. nation), which can affect the patterns observed in individual-, group-, or population-level approaches.

Common data sets and challenge problems can play a critical role in model comparison and validation. Understanding which approaches capture the observed behavior and phenomena at a specific resolution with greater fidelity would allow better understanding of the relative success of each approach. However, because the choices of measurements and metrics are often limited by the chosen modeling approach, it can be challenging to select a common set of validation techniques to apply broadly.

Let us assume that our test panel consists of three simulation approaches. S_1 is a population-level system dynamics simulator, and S_2 and S_3 are two agent-based approaches. S_2 consists of agents modeled via statistical distributions and S_3 via cognitive models. Testing against only the population-level time-series measurements diminishes the ability to see the value of the finer-resolution models S_2 and S_3. Testing against only the individual traces means that S_1 is bound to fail because the levels of relevant dynamics are mismatched. Nonetheless, testing against the individual traces is needed to reveal the parallels and discrepancies between the S_2 and S_3.

Selection of Appropriate Measurements and Metrics

Even when the stated goal of a simulation or modeling effort is relatively straightforward, such as predicting the proportion of a population that will adopt a certain behavior after a public health campaign, selecting appropriate measurements and metrics to compare the agreement between the model and observations is not always clear-cut. For example, should the proportion of adoption be measured as a function of time or only after a fixed period? Should the metrics measure relative or absolute error? Does the model performance vary across different types of campaigns, different communities within the population, or different volumes of campaign outreach? In the case of more complex simulation scenarios, such as *sandbox* simulations used for exploration and hypothesis testing, the selection of appropriate validation criteria becomes even more of a challenge. In such complex scenarios, validation may be less about success or failure of a model, but instead about defining a set of metrics that elaborate on the multivariate nature of the modeling results to understand the ways the model did or did not capture aspects of the scenarios under study.

Additionally, when employing measurements and metrics for cross-model comparison and validation, the selection of metrics that can appropriately distinguish the relative performance of the models is crucial. For example, choosing metrics on which all models fail completely or perform perfectly does not provide the ability to make a useful comparison. Nor would such approaches be informative about how to improve the models or the validation process. Indeed, as Pew et al. (2005) suggested, when complexity of the modeling increases, we are in danger of learning less about human behavior representation requirements. We can carefully avoid this by emphasizing model comparison through a broad set of relevant validation metrics.

Correlations, Causation, and Transfer Entropy

While a lot of big data analyses revolve around finding meaningful correlations, when we are dealing with social data sets, it is much more useful to find evidence of causality. This is because causal influence can produce very interesting phenomena such as cascades and viral spreading of behaviors. Therefore, it becomes important to develop evaluation criteria that can probe the causal relationships present in the observed social phenomena. Understanding causality between observed entities, and differentiating direct causality from latent factors, allows us to determine whether the simulation models the causal relationships. If it does, we can be more confident that the model will replicate the empirically observed effects of causal relationships.

Distinguishing causality (influence) from other confounding factors such as homophily in social settings is an extremely hard problem as discussed in Shalizi and Thomas (2011) and Anagnostopoulos et al. (2008). Anagnostopoulos et al. (2008) also pointed out that statistical tests can be designed to identify the presence of influence when the time series of users' actions are available such as in the case of online social media. Influence patterns between time series of entities can help characterize significant social phenomena. Therefore such analysis needs to be part of the evaluation strategy, particularly to help identify if the measured influence is causal or the by-product of another latent factor or phenomenon. Potential analyses seeking insights into this question include measures of synchrony between two entities, such as cross-correlation, cross-spectral coherence, cross-recurrence quantification analysis, or average mutual information. We note that most of these analyses rely on finding patterns within a single level of granularity in the data by comparing time series. This can be useful for comparing data from two simulations, two sets of real-world data, and between real-world data and simulated data.

One specific, useful model-free statistic for measuring causal influence (or information transfer) is transfer entropy (TE) introduced by Schreiber (2000). TE is rooted in information-theoretic analyses of source and target time-series data and does not assume any interaction models (Vicente et al. 2011). TE has acquired increasing popularity in the measurement of influence between time series corresponding to social media entities (Ver Steeg and Galstyan 2012; Borge-Holthoefer et al. 2016). TE, however, comes with its own set of issues. While a model-free measure, TE still needs some parameters to be fixed for computation. This process is nontrivial. The statistical significance of TE is also difficult to quantify. It is challenging to form an intuitive understanding of the strength of the causal influence given a certain numeric value for TE. Finally, spuriously high TE values can occur when two processes X and Y

have a common origin Z, but there is a predictable one-sided delay between X and Y (Ver Steeg and Galstyan 2012). Without the knowledge of the underlying cause Z, it appears that the leading process (X) is influencing the lagging (Y) even though the same is not true. Thus, careful analysis of possible relationships between factors and probing for missing knowledge is needed.

Initial Conditions and Model Assumptions

Evaluation of modeling and simulation must be done in context. Context, in this case, encompasses articulation of (i) the problem under study, (ii) the theoretical framework in which both the problem is being studied and the formal models are grounded, (iii) the mathematical or computational language(s) with which the model is formalized, and (iv) the initial conditions for the simulations. Implicit in this list is that context includes both the assumptions that inform both the model structure and the conditions in which the model-based analyses can be applied. Our use of *context* is very similar to Ören and Zeigler (1979) definition of experimental frame, which defines the circumstances under which a model is tested and evaluated. Experimental frames consist of observational variables (independent and dependent measures), model and simulation initialization settings and input schedules, termination conditions, and data collection and compression specifications. In many traditional modeling approaches, the mathematical theory dictates the key assumptions, if sufficient theory has been developed prior to modeling and simulation efforts. We note that the evaluation context has elements that are critical to both the verification and validation processes. For example, choice of a programming language may dictate the types of common bugs that must be identified and corrected through verification and may dictate performance constraints that will manifest in validation metrics.

For many large-scale social network simulations, the theoretical foundations are currently in development. Consequently, articulating models within the theoretical frameworks is not fully possible. We must rely on observation and the current state of the data to inform the assumptions of the modeling and simulation approaches. Clear articulation of the model assumptions is even more important in this case, so that the assumptions can be tracked between simulation studies, between use cases, and between data sets. Model assumptions will usually reflect to some degree the theoretical perspective of the researchers. Knowing the assumptions is critical to determine the appropriateness of a researcher's choices about analyses and model interpretations (Axelrod 1997).

Output behaviors can additionally depend on the initial conditions of the simulation (Axelrod 1997). In dynamic systems, future behaviors that change markedly as a function of the starting values of the simulation are termed *sensitive to initial conditions*. Chaos theory is one well-known case of systems highly sensitive to initial conditions, because seemingly random or divergent outputs arise from small changes to initial conditions. We note that a common criticism of modeling is that assessment of sensitivity to initial conditions or input parameters often considers either only extreme parameter values, seeking extreme behaviors, or midrange values, seeking representative or average behaviors (Mehta 1996). Validation should seek to capture the full range of behaviors on both retrospective analysis and predictive evaluations.

There are some unique ways in which social networks may be sensitive to assumptions and initial conditions relative to other dynamical systems. Network structure and dynamics coevolve (Farajtabar et al. 2015). As networks grow over time, diameters shrink, meaning that the shortest distance between the furthest nodes is getting shorter (Leskovec et al. 2005). Network distance is computed by counting the number of links to move through the network between the two entities. So the network diameter is the largest number of links to move between any two entities in the network. Diameter is used to gauge the overall size of a social network, and shrinking diameters can mean that people are closer to each other and information made propagate faster. This can potentially affect cascade depth distributions and suggests that behaviors on a network may be highly sensitive to initial structure and size at the time of observation. Simulations exploring the evolution of network structures and dynamics will need to provide periodic assessments to capture those sensitivities. This implies that we cannot evaluate the social behaviors of networks without also evaluating the structure of the network itself.

Uncertainty

When comparing a single realization of a real-world social system with a single realization of a simulation aiming to reproduce the dynamics of that system, we must consider how sources of uncertainty from both the data and the modeling approach can affect the agreement between the two. This is especially crucial to properly quantify risk as a component of application validation when employing the simulations in decision-making contexts.

Data uncertainty affects the training data used to calibrate the models, the initial conditions used to seed them, and the testing data used to evaluate them. Understanding the level of uncertainty in training data and initial condition data is key to understanding the forward uncertainty propagation of these errors to the final output of the model. Uncertainty in the testing data must be understood for proper validation of model results. For online social systems, data uncertainties can arise due to many sources including population subsampling, inferred network structure, inferred node or edge properties, and missing data.

In addition to uncertainty in the data, uncertainty arises in model outputs due to uncertainty in the model parameters as well as stochastic mechanisms in the model. Nondeterministic models of complex systems will produce a range of simulated output even when given the same input data and initial conditions. A deterministic model dependent on some uncertain inputs will similarly give a range of outputs. Thus, a single model run with a single set of assumed inputs or a single run of a stochastic model does not provide a full picture of the expected behavior of the system. Instead, enough runs are needed to quantify both the expected behavior and its uncertainty given uncertain assumptions and stochastic considerations. When multimodal output behavior is generated from the model, this increases the difficulty of assessing whether the observed reality is consistent with the simulation.

We note that there is an important distinction between error and uncertainty that is often lost in the process of data analysis, due to some computations of variability being termed *error*. Error may be considered actual or recognizable deficiencies in the modeling and simulation process that is not attributed to lack of knowledge, such as a wrong assumption, an approximation, or errors in computation (Axelrod 1997). Errors of this nature should be caught and corrected in the verification process. Uncertainty, on the other hand, captures the potential deficiencies due to lack of knowledge (Axelrod 1997). Uncertainty has many sources in modeling and simulation, as noted above. Computations of *error bars* and confidence intervals are statistical techniques for attempting to quantify such uncertainty. Goodness-of-fit statistics and hypothesis testing are validation techniques that can take uncertainty into account. The challenge is that even though we have some ways of accounting for uncertainty, those do not usually support developing an understanding of the sources of uncertainty (data or model), as they are often conflated in observed data and measurement.

Generalizability

Because models are often developed to study a particular phenomenon or scenario, it is difficult to understand how the model will perform given new contexts. In many cases, simulation is desired to explore certain counterfactual scenarios for purposes such as instruction, future decision support, and system design under hypothetical conditions. However, in the absence of real-world data for training and evaluating these models, the validity of the resulting outputs for previously unobserved scenarios cannot be readily ascertained. As researchers then, we want to ensure that the model has strong validity with respect to underlying principles or basic phenomena that might arise across scenarios. This will help support determination of the appropriateness of a model for generalization to novel scenarios.

Interpretation

While the performance of a model can be measured according to certain validation metrics, the question of how to interpret the performance results is a key component of the evaluation process. The purpose of validation analysis includes answering a number of key questions. Is the simulation fidelity sufficient for the desired application? Under what conditions is the approach valid? How does the simulation performance of each modeling approach compare with performance of other modeling approaches or formalisms? How should the model be modified to target improved performance? How can multiple modeling approaches be combined to leverage their relative strengths? Each dimension of interpretation may use different metrics or aspects of the system performance. This suggests that a single metric will not be adequate to support thorough interpretation of modeling and simulation results.

Proposed Evaluation Approach

While many of the challenges and issues noted in the above section are open questions without any single or easy solution, we propose a set of best practices for comparative, reproducible validation of varied social science simulation approaches against real-world data. By adhering to common evaluation practices, the social modeling and simulation community can more easily share knowledge and adapt to new forms of data and novel modeling techniques. In what follows, we outline these proposed best practices and highlight how these evaluation procedures can address many of the current challenges outlined above.

Considering the Goal of the Simulation

Because, by definition, a simulation cannot be successful unless it achieves its goal for its intended use, understanding the stated aim and purpose of a particular simulation effort is key to designing an appropriate evaluation strategy. As in Ören and Zeigler (1979) experimental frame, a model is expected to perform well within the space of intended use and may perform poorly outside that space. Defining the goals and related constraints on the expected performance limitations is therefore key for evaluating application validity. In particular, to understand the goal of simulation, we can ask the following:

1. *What is the general use case of the simulation?* The intended use of the modeling and simulation output will dictate the types of metrics and other evaluation strategies that are needed. For example, if the simulation is intended to support high-consequence decision-making, the accuracy or precision requirements of the validation may be different than for a simulation designed for scientific exploration of competing theories. In high-consequence uses, metrics providing risk assessment through uncertainty analysis techniques become a more crucial component of the validation.

2. *What phenomena is the simulation designed to capture?* While there may be a multitude of simultaneously occurring behavioral phenomena in a social system, simulations may aim to study the dynamics of particular phenomena that can guide the selection of measurements. The phenomena of interest should be selected to be relevant and informative to the intended use of the simulation.

3. *Under what different experimental conditions should this behavior be replicated?* There are limits to model capabilities that may render the model invalid or inappropriate if applied under incompatible conditions. For example, if the simulation approach will be used for prediction of message spreading in select English-speaking communities, the validation domain can be significantly reduced compared with a model designed to be applied across global, multilingual populations.

4. *What are the particular output variables of interest and how will they be used?* In the case of targeted simulations, such as those for decision support, the nature of the information the decision-maker needs to receive as output from the model should be the focus of the evaluation measurements and metrics. This includes not only the set of necessary output behaviors but also the measurement scale on which they are reported (e.g. categorical output like positive or negative value or a specific continuous real number).

We should remember that at the core, the goal of every simulation is to *reproduce* certain dynamics, behaviors, and phenomena of the system it seeks to model. Therefore, many common principles can be applied across modeling efforts even while they pursue different goals and aims.

Data

The current online social environment provides not just unprecedented volumes of data, but also unprecedented opportunities to study complete systems rather than samples. It also offers unprecedented opportunities to study the connections among different systems, both online and offline, as information can now be tracked and linked as never before.

While large data volumes and cross-domain links can create new technical challenges for modelers, these unique capacities can be leveraged by validation strategies that can comprehensively specify the environment in which the social behavior occurs. This serves to mitigate issues related to the incomplete specification of initial conditions that arises from treating the simulation environment as a *closed system* where behavior may differ between observation and simulation due to unmeasured environmental conditions.

When data is collected for a simulation effort, it should be divided into three key components: a set of training data to be used for model calibration and refinement, a set of initial conditions data used to seed the simulations, and a set of testing data used to evaluate the model performance.

For robust evaluation of a model's capacity for generalization, it is crucial that validation be performed on a held-out test set that was not observable by the model during training and calibration. Such cross-validation allows testing of predictions and adaptability to new sources of variance rather than relying solely on goodness of fit. Like the training data, the test set should be representative of the whole system. We differentiate this from a test set that might be derived from a different system, which would be a novel test set rather than a held-out test set. Yet, in both cases, the test sets should adhere to consistent experimental frames and contain the phenomena of interest. In the context of predictive simulation, where past or current observations are used to predict future behaviors, a held-out data should be selected to temporally follow all observed data in the training set to provide two types of evaluation. One type of test examines the model's capability to make appropriate or accurate predictions about future unseen scenarios. This is appropriate if a phenomenon of interest has not changed from the training time period. The other type of test helps to capture changes in phenomena, like concept drift, by mapping the ways in which the model must change to continue to predict the phenomena of interest. For example, changing event dynamics within a system may be captured by changes of a parameter, and the rate and range of those parameter changes can be informative about the rate of change in the system itself.

Because of these temporal considerations, a traditional cross-validation strategy in which multiple training and test sets are selected randomly from the data is often not applicable in social simulation scenarios. However, a rolling time window evaluation approach can be used to produce multiple training and test scenarios for more robust evaluation. We recommend that people take time to characterize their training and test sets to understand each set's patterns and characteristics. This is crucial for correctly interpreting simulation outputs.

We also note that time is not the only dimension along which training and test sets can be created. Communities within a system can be defined by different demographics, like geolocation, language, or self-defined user communities. Similar considerations must be made for training and test sets defined by dimensions other than time, in that researchers should characterize the sets to see if the phenomena of interest are present and that the sets continue to adhere to consistent experimental frames.

Depending on the simulation scenario, the initial condition data may take a variety of forms. For example, they could include a sample of recently observed activity in the system, a set of environmental conditions, the properties of a particular set of entities of interest, or some combination of these. The choice of initial condition data should replicate the desired application of the simulation.

Modeling Assumptions and Specifications

The preceding section on measurements and metrics outlined the difficulties with applying evaluation criteria selected based on the simulation approach. The remedy is to accommodate the multifaceted nature of the approaches by means of multidimensional evaluation criteria. The results of the evaluation need understanding along different dimensions. Accommodation of different techniques to allow fair comparison should be a priority when designing evaluation strategies. With the above considerations, applying a common evaluation framework to different approaches will enable improved comparison between methods as well as potential for learning how to best combine different modeling strategies when considering specific scenarios. A particular research direction on the multi-criteria evaluation could be the design of data-driven measurements in the form of low-dimensional representations of entire ground truth or simulated trace sets that allow for a quick model-free comparison.

In our example of three different simulation approaches from the section on measurements and metrics (S_1 = population level, S_2 = agent-based statistical, S_3 = agent-based cognitive), the evaluation strategy in its simplest form should contain both the population-level time-series measurements (P) and the individual-level activity measurements (I). Individual-level activity measurements for the top k individual users (I_k), which is a less noisy version of I, can also be included as part of the evaluation. With such an evaluation panel, on the same hardware, it is possible that we end up with an evaluation matrix as follows:

| Approach | Measurement | | | |
	P	I	I_k	Time (h)
S_1	Good	—	—	1
S_2	Moderate	Poor	Good	12
S_3	Good	Moderate	Good	48

Thus, the multidimensional nature of the evaluation strategy can point out the complete set of trade-offs between the capabilities of the simulation approaches, the computation time, and the performance of the corresponding simulation compared across different measurements and metrics. It is apparent that the multi-criteria evaluation should incorporate, at a minimum, the union set of the indicators at which each modeling approach excels with additional criteria included for more insights. For example, we can add community-scale metrics in the above example to gain insight about group behavior. The following sections elaborate on the various aspects of multi-criteria evaluation.

Measurements

There are many factors that influence the choice of measurements to compare between simulation outputs and ground truth observations. While a prime consideration is the targeted purpose of the simulations, it is often beneficial to probe the performance of a model beyond its direct purpose to enable both improved interpretation of the results and improved understanding of how the model might generalize to new contexts. While typically validation is performed on only a single observed phenomenon, we propose probing multiple facets of the simulated behavior using a set of varied quantitative measurements to probe different scales, data attributes, and behavior types.

An evaluation strategy that leverages these multiple facets of evaluation can provide increased interpretability of performance results. Greater insight into the success of a modeling approach can be achieved by measuring performance across multiple *scales* including across network scales (e.g. node level, group level, and population level), activity scales (e.g. more vs. less active entities), and temporal scales (e.g. batch properties vs. temporal trends; shorter vs. longer elapsed intervals after the initial conditions). Second, by faceting the evaluation across data *properties*, such as population demographics, we can study how performance varies across conditions, and the relative success in different conditions tells us about the appropriate *operating conditions* of the model. Finally, we can study the generalizability of the approach by performing evaluation of multiple *phenomena* such as cascades, recurrence, gatekeepers, and persistent groups. This ensures the approach can capture multiple facets of behavior in any complex social system.

It is also important to select measurements for evaluating system behaviors that are agnostic to the initial conditions. Measurement and analysis of the outputs should emphasize the behaviors of interest, defined without special consideration of the specific model definitions. Then, we might define a consistent set of measurements to apply across models and initial conditions for direct comparisons. In this way we can define measurements about the distributions of output behaviors, statistics about the top k performing agents in the network, or metrics about the individual agent behaviors. This enables insights at and potentially across multiple levels of system resolution.

While many typical measurements probe the emergent patterns of behavior in a simulation, particular measurements such as TE can be designed that probe causal mechanisms. Some of the issues with characterization of causal influence via TE that we enumerated earlier can be mitigated by careful experimentation and interpretation as elaborated next. The issues related to parameter selection in the computation of TE have been investigated in detail in Sathanur and Jandhyala (2014) and Wibral et al. (2013). Wibral et al. (2013) proposed sweeping the TE value with delay and using the maximum value of the resulting parametric sweep. The significance of TE value can be understood by comparing the actual TE with TE between the source and various randomly shuffled versions of the target time series and testing for statistical significance (Wollstadt et al. 2014). The strength of causal influence can be understood, for example, by characterizing the TE between two Gaussian processes or Bernoulli processes with a known delay and correlation. Misinterpretation of spurious high TE values can be somewhat mitigated by computing TE values for entity pairs that are related to each other. For example, it makes sense to compute TE values between members of the same friends group, team, and community as opposed to random pairs of individuals.

Metrics

In combination with the multifaceted measurement approach, a set of metrics for comparison of the measurements between simulation and ground truth must be employed. A strong set of metrics allows both the determination of the validity of an individual simulation approach and the robust comparison of the relative performance among different approaches. Given a single measurement, such as the activity level of users within a social network, three separate levels of metrics can be employed to probe the agreement from the macro to the micro level.

Distributional
Comparisons of distributions between the simulation and the ground truth provide a mechanism for probing the broader trends observed in the scenario. Distributional comparison can be used to evaluate the range of behavior observed in an individual, group, or population. Examples of metrics that can be employed for this comparison include Kullback–Leibler divergence and the Kolmogorov–Smirnov test.

Rankings
To probe the simulation results in more detail, we can move beyond distributional comparisons and evaluate the relative properties of specific entities in the scenario. For example, we can move beyond comparing the distribution of the number of actions across a population and determine if the simulation captured those individuals responsible for high-volume activity and those who were relatively inactive. Possible metrics for this purpose include the Spearman rank-order correlation and the rank-biased overlap (Webber et al. 2010).

One to One
At this level, metrics delve even deeper into individual-level behavior, by determining if the simulation captured the specific activity of individuals. For example, we can perform one-to-one comparison of activity volume on a per user basis in a social network. Metrics that perform one-to-one comparison

between entities in the simulation and the ground truth include the RMSE, R^2 coefficient, and Pearson correlation coefficient.

By employing metrics that probe all three levels of evaluation, we can ensure the ability to understand the relative level of performance of different approaches. If the general distributions of behavior prove to be trivially predicted by all approaches, more fine-grained comparison can distinguish model performance at the micro level. Alternatively, if simulation of exact individual behavior proves too challenging in a given context, then comparison at the level of distributions or rankings may provide the appropriate scale for relative evaluations.

Because simulation of complex social systems must capture the dynamics of the emergent behavior in the system, the evaluation of temporal dynamics of a simulation should receive special attention. Temporal patterns can be compared in several formats, including event sequence data and time-series measurements. Sequence comparison metrics, such as the Levenshtein distance, can answer questions about whether events occur in the correct order without concern over the inter-event timing, which can be a proxy for causal patterns in the data. When comparing numeric time-series measurements rather than event sequences, we can use error metrics such as RMSE or more flexible metrics such as dynamic time warping (Müller 2007). By probing both exact match and trend validity, we can perform evaluation of qualitative trends using rigorous quantitative methods.

Evaluation Procedures

A robust evaluation procedure will include the use of the training data for model calibration and training, while the initial condition and test data are reserved until the models are finalized. Once the single or multiple sets of initial conditions are released, the resulting simulated behavior can be compared with that observed in the testing data. Any performance gains achieved through model modifications after observations of the validation results on the test data may not be indicative of the true potential of the model to generalize to new situations but instead may indicate overfitting to the particular test cases used.

To properly assess model uncertainty and sensitivity to initial conditions, we must build on the basic evaluation procedure of simply comparing a single run of the simulation with the observed ground truth. Evaluation of stochastic models can instead be performed on a population of simulation runs rather than a single iteration to study both the expected simulation variability and the modeling uncertainty.

Understanding how the output behaviors of a system vary with initial conditions requires model exploration. This is particularly true in agent-based systems where complexity of the output behaviors emerges from the interactions of agents over time. Complexity is not necessarily coded into the agent-based model, as it might be in a coupled nonlinear dynamic system model. Thus, exploration is needed to find both the expected and unexpected system behaviors. Systematic exploration of the initial conditions, such as grid search through input parameter space, is the foundation of the study of sensitivity to initial conditions. In an environment where data is now collected at large enough scale that observations can be made of the same phenomenon under numerous environmental perturbations, often differing only slightly, we are able to perform an empirical study of this sensitivity. Employing a range of initial condition scenarios each observed in the ground truth data allows a systematic comparison between initial condition sensitivity as observed in the real world and the sensitivity exhibited by the models, a technique that is supported by the increasingly complete and complex available data sets.

Because initial condition sensitivity may especially arise in the presence of nonlinear effects with high path dependence, we can gain additional insight into this effect by measuring simulation performance as a function of time elapsed since the initial conditions. Understanding the dynamics of model performance provides key input into properly limiting the domain of applicability for the simulation.

Interpretation

There are three primary objectives to the interpretation of a set of evaluation results: (i) determining whether a given approach is valid for its stated objectives and under what specific conditions, (ii) comparing between multiple simulation approaches to identify their relative success, and (iii) identifying targeted improvements to modeling techniques. While a great deal of focus is typically placed on the first two objectives, less attention has been paid to the third that would allow us to answer *why* a certain approach fails, *when* it works well, and *how* to make improvements (Campbell and Bolton 2005).

By probing different resolutions, scales, population subsets, and phenomena, we can ascertain the conditions where a model performs well and where it falls short. When performing a comparative evaluation of multiple approaches, we can gain insight by observing the correlation in performance among the multiple metrics. That is, which groups of metrics tend to rise or fall together? For example, does increased performance on predicting the timing of recurring activity bursts correlate with increased performance predicting the volume of those bursts? This can provide crucial feedback on potential improvements to the approach.

On the other hand, are there trade-offs to consider among the various dimensions of evaluation? For example, do modeling techniques that achieve high-fidelity prediction of the distribution of sentiment toward a messaging campaign tend to have reduced performance of predicting the particular users who will propagate the campaign messages? This may point to the need to employ different approaches to target the types of behavior and phenomena these metrics are capturing. Further, to meet standards of validation toward model accreditation, we must ensure that the approaches selected predict behavior in a way that is relevant and appropriate for the intended use of the model (U.S. Department of Defense 2011).

Finally, by identifying approaches that perform well on different groups of metrics, one can leverage the particular strengths of these different techniques by selecting the appropriate model for a given application or by applying a combination of models using an ensemble (Tebaldi and Knutti 2007) or modular (Portegies Zwart et al. 2012) approach.

Conclusions

The ever-increasing volumes and varieties of data available to support the modeling and simulation of social-behavioral phenomena provide both challenges and opportunities for model validation. We now have increased capability to perform rigorous, repeatable, quantitative evaluation of social-behavioral modeling at scale through comparison with observed ground truth and qualitative interactions. These new opportunities for rigorous evaluation can best be leveraged by the community through the adoption of common standards and approaches for data selection, selection of validation criteria, and implementation of evaluation procedures. However, there are many ongoing unsolved validation challenges that should receive continued attention in future research. These include the development of novel measurements and metrics to support model-agnostic comparison and improved methods for interpreting validation results to support insights into model improvements and methods for leveraging the strengths of varied available approaches. Indeed, a key contribution of validation efforts is a reduced opacity of models through improved capabilities to inspect and interpret them (Pew et al. 2005). Ultimately, there is much to be gained through the development of novel techniques for comparative and reproducible validation of computational social models, because such models, if they can be meaningfully and empirically tested, will provide unprecedented insights into human behavior.

References

Adamic, L.A., Lento, T.M., Adar, E., and Ng, P.C. (2016). Information evolution in social networks. In: *Proceedings of the 9th ACM International Conference on Web Search and Data Mining*, 473–482. New York, NY: ACM.

Anagnostopoulos, A., Kumar, R., and Mahdian, M. (2008). Influence and correlation in social networks. In: *Proceedings of the 14th ACM SIGKDD International Conference on Knowledge Discovery and Data Mining*, 7–15. ACM.

Axelrod, R. (1997). Advancing the art of simulation in the social sciences. In: *Simulating Social Phenomena*, Lecture Notes in Economics and Mathematical Systems, vol. 456 (ed. R. Conte, R. Hegselmann and P. Terna). Berlin, Heidelberg: Springer.

Borge-Holthoefer, J., Perra, N., Gonçalves, B. et al. (2016). The dynamics of information-driven coordination phenomena: a transfer entropy analysis. *Science Advances* 2 (4): e1501158.

Box, G.E. (1979). Robustness in the strategy of scientific model building. In: *Robustness in Statistics* (ed. R.L. Launer and G.N. Wilkinson), 201–236. Elsevier.

Campbell, G.E. and Bolton, A.E. (2005). HBR validation: integrating lessons learned from multiple academic disciplines, applied communities, and the AMBR project. In: *Modeling Human Behavior with Integrated Cognitive Architectures: Comparison, Evaluation, and Validation* (ed. K.A. Gluck and R.W. Pew), 365–395. Mahwah, NJ: Lawrence Erlbaum Associates.

Carley, K.M. (2019, this volume). Social-behavioral simulation: key challenges. In: *Social-Behavioral Modeling for Complex Systems* (ed. P.K. Davis, A. O'Mahony and J. Pfautz). Hoboken, NJ: Wiley.

Davis, P.K. and O'Mahoney, A. (2019, this volume). Improving social-behavioral modeling. In: *Social-Behavioral Modeling for Complex Systems* (ed. P.K. Davis and K. Pfautz). Hoboken, NJ: Wiley.

de C Gatti, M.A., Appel, A.P., dos Santos, C.N. et al. (2013, December). A simulation-based approach to analyze the information diffusion in microblogging online social network. In: *2013 Winter Simulations Conference (WSC)* (ed. R.R. Hill and M.E. Kuhl), 1685–1696. Washington, DC: ACM.

Del Vicario, M., Bessi, A., Zollo, F. et al. (2016). The spreading of misinformation online. *Proceedings of the National Academy of Sciences* 113 (3): 554–559.

Edwards, S. (2008). Computational tools in predicting and assessing forced migration. *Journal of Refugee Studies* 21 (3): 347–359.

Farajtabar, M., Wang, Y., Rodriguez, M.G. et al. (2015). Coevolve: a joint point process model for information diffusion and network co-evolution. In: *Advances in Neural Information Processing Systems*, 1954–1962. Cambridge, MA: MIT Press.

Ferrara, E. and Yang, Z. (2015). Quantifying the effect of sentiment on information diffusion in social media. *PeerJ Computer Science* 1: e26.

Galuba, W., Aberer, K., Chakraborty, D. et al. (2010). Outtweeting the Twitterers – predicting information cascades in microblogs. In: *Proceedings of the 3rd Conference on Online Social Networks*, 3–11. Berkeley, CA: USENIX Association.

Goel, S., Anderson, A., Hofman, J., and Watts, D.J. (2016). The structural virality of online diffusion. *Management Science* 62 (1): 180–196.

Gomez Rodriguez, M., Leskovec, J., and Schölkopf, B. (2013). Structure and dynamics of information pathways in online media. In: *Proceedings of the 6th ACM International Conference on Web Search and Data Mining*, 23–32. New York, NY: ACM.

Helbing, D. (2010). *Quantitative Sociodynamics: Stochastic Methods and Models of Social Interaction Processes*. Springer Science & Business Media.

Hodas, N.O. and Lerman, K. (2014). The simple rules of social contagion. *Scientific Reports* 4: 4343.

Hu, W., Singh, K.K., Xiao, F. et al. (2017). Who will share my image? Predicting the content diffusion path in online social networks. CoRR, abs/1705.09275.

Jin, F., Dougherty, E., Saraf, P. et al. (2013). Epidemiological modeling of news and rumors on twitter. In: *Proceedings of the 7th Workshop on Social Network Mining and Analysis*, 8:1–8:9. New York, NY: ACM.

Jones, A.P., Homer, J.B., Murphy, D.L. et al. (2006). Understanding diabetes population dynamics through simulation modeling and experimentation. *American Journal of Public Health* 96 (3): 488–494.

Korzybski, A. (1958). *Science and sanity: an introduction to non-Aristotelian systems and general semantics*. Institute of GS.

Kwak, H., Lee, C., Park, H., and Moon, S. (2010). What is Twitter, a social network or a news media? In: *Proceedings of the 19th International Conference on World Wide Web*, 591–600. New York, NY: ACM.

Leskovec, J., Kleinberg, J., and Faloutsos, C. (2005). Graphs over time: densification laws, shrinking diameters and possible explanations. In: *Proceedings of the 11th ACM SIGKDD International Conference on Knowledge Discovery in Data Mining*, 177–187. New York, NY: ACM.

Matsubara, Y., Sakurai, Y., Prakash, B.A. et al. (2012). Rise and fall patterns of information diffusion: model and implications. In: *Proceedings of the 18th ACM SIGKDD International Conference on Knowledge Discovery and Data Mining*, 6–14. New York, NY: ACM.

Mehta, U.B. (1996). Guide to credible computer simulations of fluid flows. *Journal of Propulsion and Power* 12 (5): 940–948.

Müller, M. (2007). Dynamic time warping. In: *Information Retrieval for Music and Motion*, 69–84. Berlin, Heidelberg: Springer.

Ören, T.I. and Zeigler, B.P. (1979). Concepts for advanced simulation methodologies. *Simulation* 32 (3): 69–82.

Pew, R.W., Gluck, K.A., and Deutsch, S. (2005). Accomplishments, challenges, and future directions for human behavior representation. In: *Modeling Human Behavior with Integrated Cognitive Architectures: Comparison, Evaluation, and Validation* (ed. K.A. Gluck and R.W. Pew), 397–414. Mahwah, NJ: Lawrence Erlbaum Associates.

Portegies Zwart, S.F., McMillan, S.L.W., van Elteren, A. et al. (2012). Multi-physics simulations using a hierarchical interchangeable software interface. *Computer Physics Communications* 184 (3): 456–468.

Roache, P.J. (2009). *Fundamentals of Verification and Validation*. Socorro, NM: Hermosa Publishers.

Sargent, R.G. (2011). Verification and validation of simulation models. In: *Proceedings of the Winter Simulation Conference*, 183–198. Winter Simulation Conference.

Sathanur, A.V. and Jandhyala, V. (2014). An activity-based information-theoretic annotation of social graphs. In: *Proceedings of the 2014 ACM Conference on Web Science*, 187–191. ACM.

Schreiber, T. (2000). Measuring information transfer. *Physical Review Letters* 85 (2): 461.

Serrano, E. and Iglesias, C.A. (2016). Validating viral marketing strategies in Twitter via agent-based social simulation. *Expert Systems with Applications* 50 (1): 140–150.

Shalizi, C.R. and Thomas, A.C. (2011). Homophily and contagion are generically confounded in observational social network studies. *Sociological Methods & Research* 40 (2): 211–239.

Tebaldi, C. and Knutti, R. (2007). The use of the multi-model ensemble in probabilistic climate projections. *Philosophical Transactions of the Royal Society of London, Series A: Mathematical, Physical and Engineering Sciences* 365 (1857): 2053–2075.

U.S. Department of Defense (2011). *VV&A Recommended Practices Guide*. Washington, DC: Defense Modeling and Simulation Coordination Office. Retrieved 4 April 2018 from https://vva.msco.mil.

Vicente, R., Wibral, M., Lindner, M., and Pipa, G. (2011). Transfer entropy – a model-free measure of effective connectivity for the neurosciences. *Journal of Computational Neuroscience* 30 (1): 45–67.

Ver Steeg, G. and Galstyan, A. (2012). Information transfer in social media. In: *Proceedings of the 21st International Conference on World Wide Web*, 509–518. ACM.

Wang, F., Wang, H., and Xu, K. (2012). Diffusive logistic model towards predicting information diffusion in online social networks. In: *Proceedings of the 2012 32nd International Conference on Distributed Computing Systems Workshops*, 133–139. Washington, DC: IEEE Computer Society.

Wang, S., Hu, X., Yu, P.S., and Li, Z. (2014). MMRate: Inferring multi-aspect diffusion networks with multi-pattern cascades. In: *Proceedings of the 20th ACM SIGKDD International Conference on Knowledge Discovery and Data Mining*, 1246–1255. New York, NY: ACM.

Webber, W., Moffat, A., and Zobel, J. (2010, November). A similarity measure for indefinite rankings. *ACM Transactions on Information Systems* 28 (4): 20:1–20:38.

Weidlich, W. (2006). *Sociodynamics: A systematic approach to mathematical modelling in the social sciences.* Courier Corporation.

Wibral, M., Pampu, N., Priesemann, V. et al. (2013). Measuring information-transfer delays. *PloS ONE* 8 (2): e55809.

Wollstadt, P., Martínez-Zarzuela, M., Vicente, R. et al. (2014). Efficient transfer entropy analysis of non-stationary neural time series. *PloS ONE* 9 (7): e102833.

Young, M.J. (2003). Human performance model validation: One size does not fit all. In: *Summer Computer Simulation Conference*, 732–736. Society for Computer Simulation International.

Part IV

Innovations in Modeling

22

The Agent-Based Model Canvas: A Modeling *Lingua Franca* for Computational Social Science

Ivan Garibay[1], Chathika Gunaratne[2], Niloofar Yousefi[1], and Steve Scheinert[3]

[1]*Department of Industrial Engineering and Management Systems, College of Engineering and Computer Science, University of Central Florida, Orlando, FL 32816, USA*
[2]*Institute for Simulation and Training, University of Central Florida, Orlando, FL 32816, USA*
[3]*Department of Industrial Engineering and Management Systems, University of Central Florida, Orlando, FL 32816, USA*

Introduction

When (Grimm et al. 2006) first published their Overview, Design Concepts, and Details (ODD) framework, they recognized both the growing use of computational models in the social sciences, particularly agent-based models (ABMs), and the difficulty in describing ABMs with sufficient clarity for other researchers to understand and replicate the model design components and construction. When Ostrom (2005) published the Institutional Analysis and Development (IAD) framework, she recognized the difficulty in describing complex institutions and the spaces in which they operate. The framework provides a tool for organizing inquiry and description of institutions by providing researchers with a set of questions to answer, which, when all are answered, provide a thorough description and analysis of that institution. Both identified a common challenge for complex computational models: describing models is itself a difficult and complicated task. Both responded to this challenge by providing a framework to guide researchers' efforts in meeting this challenge. Ostrom's framework guides the design of conceptual models, while Grimm et al.'s framework guides the description of computational models. Neither framework guides the translation of theories and conceptual models into computational models. Like the specification of statistical or econometric models (Kennedy 2003), this process is left largely for the researcher to determine and apply. An additional framework, one that can help bridge the gap between conceptual, modeling, and technical languages used in different fields, is now needed to support the growing use of ABMs in the social sciences.

Social-Behavioral Modeling for Complex Systems, First Edition.
Edited by Paul K. Davis, Angela O'Mahony, and Jonathan Pfautz.
© 2019 John Wiley & Sons, Inc. Published 2019 by John Wiley & Sons, Inc.
Companion website: www.wiley.com/go/Davis_Social-Behavioralmodeling

Social science researchers, in a wide array of social scientific fields, have long used a variety of computational methods. Econometrics was sufficiently developed to be recognized, and its meaning debated more than half a century ago (Tintner 1953). Game theory undergirded the US strategies during the Cold War (Schelling 1960, 1966). Network analytic techniques are now regularly used in a variety of fields, particularly including sociology, political science, and policy science (Borgatti et al. 2009; Comfort et al. 2011; Koliba et al. 2011). The growing use of network analysis and simulation, along with a clear articulation of an epistemology for applying ABMs' generative capabilities to the social sciences (Epstein 2006), is helping to drive the emergence of a new field, computational social science (CSS), that brings together researchers with computational backgrounds and researchers with social scientific backgrounds to apply ever more sophisticated computational analytic techniques to social scientific research questions.

It is the interdisciplinary nature of CSS that enables and necessitates collaboration between social and computational scientists, many of whom now identify as *data scientists*. In utilizing modeling and simulation approaches, this collaboration has the potential to make monumental changes in addressing society's most challenging problems in ways not previously possible. Opportunities resulting from these collaborative efforts are many (Epstein 2008). Social scientists provide substantive theories, guiding the development of hypotheses on whether, when, and under what conditions certain behaviors occur. Computational scientists and statisticians test these hypotheses with advanced techniques analyzing empirical and model-generated social scientific data to identify what and when certain social scientific phenomena will take place. Although this collaboration can lead to breakthrough innovations that no discipline could produce single-handedly, just as with any interdisciplinary team project, a primary obstacle is establishing a common language for effective communication among researchers with diverse backgrounds. Though we use the term *language*, it is not necessarily that this be a full-fledged natural or computer language, but a common ground for scientific reasoning. What matters is that the concepts and terms that each member of a CSS research team uses are understood by the other members of the team. Without such a common language, the collaboration is unlikely to succeed.

The Stakeholders

Three stakeholder groups are important for the success of a CSS research effort:

Modeling and simulation specialists: Modeling and simulation specialists have the necessary knowledge for constructing computer models and simulation to replicate complex system phenomena. They understand the capabilities and limitations of modeling and simulation tools and which tools are appropriate under the given problem context. Importantly, they recognize the requirement for a model to have enough detail to be computationally implemented and understand how to analyze simulation output. For example, *starting from a generic ABM simulation, experts are able to put forth questions that help define how a model must be constructed so that it can provide an answer to the research question.*

Social scientists: Social scientists in this context are the domain experts. Without the input from social scientists, the models built by simulation experts and data scientists might fail to leverage the existing body of knowledge from the social sciences. As a result, these models, in turn, might fail to answer questions grounded in social science theory. Despite the comprehensiveness of the ABM implementation tools and even the availability of design documentation (ODD/ODD+D), domain experts will often have a difficult time producing ABMs to suit their work. Social scientists often struggle in implementing ABMs as a computer program without the computational background to mathematically define and describe the concepts and relationships contained in theories. For example, rules for movement of agents are commonly overlooked: *what does a movement in the horizontal direction mean in relation to a movement in the vertical direction in a social context?*

Data scientists/machine learning experts: Data scientists and machine learning experts specialize in the knowledge and skills in using advanced analytic techniques that can be used to test the theories proposed by the social scientists. Though many social scientists regularly perform their own analyses, they often cannot stay up to date with the plethora of methodological developments or the scripting languages upon which the newest tools and techniques rely. For example, *in a model of social media, the domain expert may realize the importance of influence, yet not have the expertise to extract influence from the big data resulting from social media.* In turn, the data scientist may not know which variables to look at in their search for user influence. This is where the communication between the data expert and the domain expert is crucial.

Need for a *Lingua Franca*

The answers to today's most exciting research questions lie buried in large data sets. These data sets are often unstructured and have complex relationships, requiring specialized, cutting-edge computational analyses. As will be fully developed in the next section, the different stakeholder groups rely on diverging sets of concepts, methods, and lexicons. In many cases, the same specific term may have vastly different meanings in each stakeholder group. Building effective interdisciplinary research teams means bridging this language gap between social scientists, data scientists, and modeling and simulation experts. The absence of a common language or a *lingua franca*, spoken by all three types experts, is then a key concern and requirement for effective computational social scientific research.

The Agent-Based Model Canvas

We propose such a *lingua franca*: the Agent-Based Model Canvas (ABMC). The ABMC aims to bring the collective effort of modeling and simulation experts, social scientists, and data scientists/machine learning experts toward the successful completion of a CSS research collaboration. The ABMC is inspired by the Business Model Canvas (Osterwalder et al. 2014). Using a similar approach, the ABMC visualizes all key CSS project building blocks into a single paged canvas. This representation allows the multidisciplinary discussion, design, and iterative refinement of the CSS project. A completed ABMC is able to completely define the requirements of a CSS research effort, including the guiding theories; the concepts, variables, and relationships; the data required for hypothesis building; the construction of the ABM itself; the expected target output; and the data and methods for the validation and output analysis. The Canvas consists of nine building blocks that a CSS research team must define in order to fully specify an ABM-guided CSS project. It acts as a framework, similar to ODD and IAD, for guiding the construction of ABMs using steps drawn from the scientific method that is common to all scientists.

ABM-based CSS research is the result of two, usually iterative, processes. The ABMC is highly visual so that it can easily encapsulate and communicate these processes to researchers from all fields. First, during the hypothesis building process, the input of domain and data experts is drawn upon to construct meaningful theory- and data-verified testable hypotheses for the model to test. The hypotheses building process ensures that the ABM is grounded in relevant social scientific theories. Second, the model construction and experimentation process uses the hypotheses for model construction. The model construction and experimentation process ensures that the models perform as desired, accurately and effectively implementing the relevant theories and producing output that can be used to test the chosen hypotheses. Both processes require data though for separate purposes. The first process relies on empirical data gathered about the context that is being modeled and uses those data for filling gaps in substantive the knowledge of theories and the modeling context. The second process relies more heavily on model-generated data for model validation and output exploration, often comparing model-generated data against empirical data.

The Language Gap

Languages, by which we mean the varying tools that represent concepts and variables for analysis, are the key tools for scientific research. They are the mediums through which theories and concepts are expressed, data are structured and analyzed, and results are interpreted, all in the service or testing hypotheses and providing evidence for and against theories. Social scientists, modeling and simulation experts, and data scientists/machine learning experts have different languages for communication within their respective fields. These languages help with knowledge representation and dissemination by building a common lexicon with highly refined definitions that allow quick, easy, and succinct expression of complicated ideas. However, in the case where the research outcomes are the result of interdisciplinary teams working toward transdisciplinary outcomes, these multiple communication streams render the lexicons uninterpretable, and communication breaks down. Jargon from one camp is misunderstood by the other, and the shared mental model is never truly communicated across the various experts. In this section we explore some existing efforts of research communication in the three disciplines considered in the CSS efforts and emphasize the disparity between them.

The Modelers

Agent-based modeling has become a common approach to answer social science research questions through simulation. ABMs deal with large orders of agents, each activated at every discrete time step. Within a given time step, an agent must plan and perform actions that can range in computational complexity. Due to the massive amount of computations, agent-based simulation experiments must make use of the computing power of modern computers. Agent-based modeling toolkits encapsulate and implement functions common across ABMs such as the scheduler, which determines the order in which agents act per time step. In doing so, they provide a further level of abstraction upon which model developers can skip the detail necessary to implement such a project directly on a lower-level language such as Java or C++, by providing a scripting interface. For example, Repast Simphony, written in Java, and RepastHPC, written in C++, are abstracted through ReLogo; NetLogo, which provides NetLogo scripting language, is an abstraction of Java and Scala implementations; and AnyLogic, again an implementation in Java, is abstracted as a visual programming language, which allows the developer to organize blocks of encapsulated agent behaviors.

Thus, the entirety of the ABM is captured through the programming language used to implement it. Yet, it is usually unreadable to domain experts.

Concepts and key relationships between variables are now expressed in deep logic that is not easily interpretable through the eyes of social scientists or data scientists/machine learning experts. To address this issue, ABM documentation standards have emerged such as ODD and the human decision-making specific version of ODD, ODD+D (ODD + Decision) (Grimm et al. 2006, 2010). The ODD standards help frame the ABM itself highlighting:

Purpose, Entities, state variables, and scales, Process overview and scheduling, Design Concepts (Basic principles, Emergence, Adaptation, Objectives, Learning, Prediction, Sensing, Interaction, Stochasticity, Collectives, Observation), Initialization, Input data, and sub models In setting this standard, ODD has given the modeling community a standard for communicating complex models that is easily understood throughout the community. By doing this, ODD provides not a tool for modeling but a tool for communication.

Social Scientific Languages in CSS

Certain communication standards are prevalent in the social science literature, though many lack the broad standards of practice that make their manner of use consistent across fields. Concept maps are common artifacts embodying social science concepts and research outcomes (Novak 2010). However, no standard exists (Roth and Roychoudhury 1992; Kinchin et al. 2000; Kamsu-Foguem et al. 2014; Raei and Kardan 2015; Dipeolu et al. 2016; Schwendimann and Linn 2016; Haymovitz et al. 2017; Fu et al. 2017).

Frames (Minsky 1975) and factor trees (Davis and O'Mahony 2013) offer social scientific tools that are closer in their execution to the tools used by modelers and data scientists. Frames have helped the artificial intelligence community define knowledge representation and flow. Knowledge representations are common to both the social sciences and artificial intelligence. Good knowledge representations can be directly converted into models that can be programmed into a computer to simulate reality and may have symbolic representations or nonsymbolic representations (Ramirez and Valdes 2012). Factor trees were introduced specifically to link individual human behaviors with the factors that influence those behaviors and the choice between them (Davis and O'Mahony 2013). Since trees are developed using a high-level visual programming language, such as Analytica (Davis and O'Mahony 2013), the tree is not only a specification of behaviors and their motivating factors but also an initial implementation of a model. The factor tree should then be easily reviewable and readily reprogrammed into a computational model, regardless of which modeling or programming language may be used.

Natural language narrative is the norm in describing social theories (Czarniawska 2004). That often leads to ambiguous and abstract theory definitions due to a lack of formalism (Casas 2013). Even methodologically rigorous case study techniques, which are ubiquitous throughout the social sciences (King et al. 1994; George and Bennett 2005), including case studies conducted with complexity-friendly frameworks like IAD, are not designed and rarely executed with computational or simulation modeling in mind. This manner of usage presents difficulties for researchers trying to translate case studies into computational models. An additional analytic step becomes necessary to further case study results and conclusions before machine-readable modeling code can be constructed. Several researchers have promoted the use of mathematics-friendly representations such as Petri nets (Köhler et al. 2007; Casas 2013), Discrete Event System Specifications (DEVS) (Zeigler 1989), Unified Modeling Language (UML) (Fowler and Scott 2000), or SysML (Weilkiens 2011). While the social sciences are witnessing a growing use of these methods, the standards for their use as well as the communication of their implementation and results are still evolving and inconsistent, with many researchers preferring to adapt the traditional structures for explaining research methods to include explanations of algorithms and, sometimes, the inclusion of code as an appendix.

Data Analysis Languages

High-level scripting languages, like Matlab/Octave, R, or Python, are common for data analysis or machine learning efforts that happen in computer science, statistics, or mathematics, where they offer a much more direct link to modeling and programming languages. With a script written in a scripting language, the exact detailed steps of an analysis or algorithm can be coded and executed in a highly repeatable format. Algorithms are commonly communicated in the form of pseudo-code (Horowitz et al. 1997) or flowcharts or implemented through scripting language *packages*, which provide complete and vetted code for running complex or highly tailored commands. These are ready to implement, sometimes automatically, particularly in the case of packages. Large-scale softwares designed for such research are described in UML (Fowler and Scott 2000). All of these methods have been established with the direct conversion to computer programming languages in mind. Discrepancies may exist, for example, due to differences among object-oriented programming, functional programming, etc. Less often formal grammars and their many variants (Chomsky 1956) can be used to specify a complete functional set used for computer programming. The benefit of using a scripting language is that it can easily be rendered into advanced computer and modeling code since it is already written in a type of programming language. The challenge is that writing the script often requires the same cross-disciplinary boundaries as designing and implementing an ABM. Indeed, in some ABM platforms, such as RSiena, the tools and procedure for coding a script and a model is identical, and the result largely indistinguishable.

Table 22.1 A comparison of existing languages for knowledge and flow representation by the three stakeholder types of CSS.

Language	Social science experts	Data analysis	Computational modeling
NetLogo (and related software)	✓̸	X	✓
ODD/ODD+D	✓	X	✓
Concept maps	✓	X	✓̸
Factor trees	✓	X	✓
Frames	✓	X	X
Natural language narratives	✓	X	X
UML	✓̸	✓	✓̸
High-level scripting languages (Matlab, R, Python)	X	✓	✓
Pseudo-code, flowcharts, formal grammars	X	✓	✓
Visual programming languages (Vensim, Analytica)	✓	✓	✓

✓̸: adaptable to the domain, ✓: actively used by the domain, X: not applicable.

A Comparison of Existing Languages

In Table 22.1, we compare and contrast some of the qualities of the more commonly used of these research languages to emphasize that there is a requirement for a *lingua franca* across the three disciplines. Here, languages that are adaptable to a discipline but require extra effort on the expert's part are awarded a ✓̸, and languages that are actively used to cater to a certain discipline are marked with a ✓.

Comparing across the three columns, what we notice is that computational modeling shares languages with social sciences due to its penetration as a methodology for the social sciences and with data analysis due to its roots as a computer science domain. However, there is a clear gap between the social sciences and machine learning/data scientists in the languages they use to frame and execute their research.

The Agent-Based Model Canvas

In this section we introduce our *lingua franca* for the constructive collaboration between the three stakeholder types involved in a large-scale CSS research project. Our approach, the ABMC, shown in Figure 22.1, is highly visual, making it easily accessible to researchers in all fields.

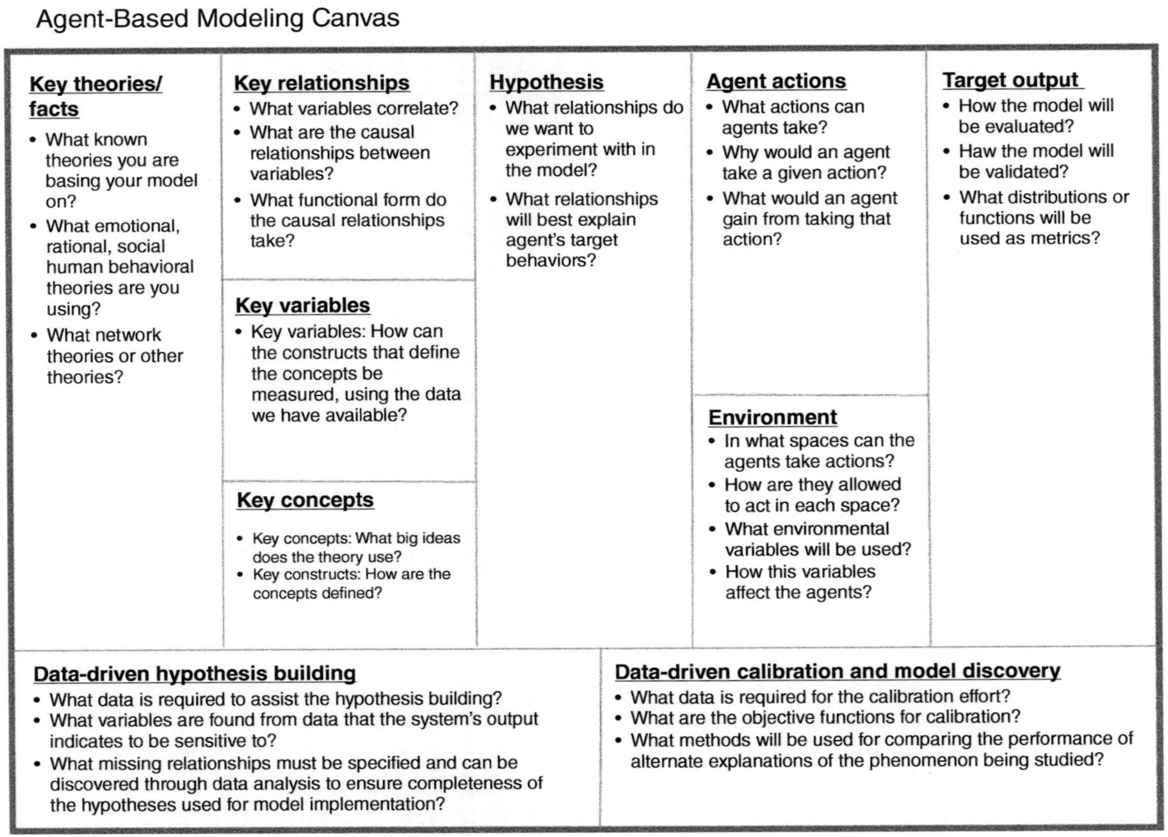

Agent-Based Modeling Canvas

Key theories/facts
- What known theories you are basing your model on?
- What emotional, rational, social human behavioral theories are you using?
- What network theories or other theories?

Key relationships
- What variables correlate?
- What are the causal relationships between variables?
- What functional form do the causal relationships take?

Key variables
- Key variables: How can the constructs that define the concepts be measured, using the data we have available?

Key concepts
- Key concepts: What big ideas does the theory use?
- Key constructs: How are the concepts defined?

Hypothesis
- What relationships do we want to experiment with in the model?
- What relationships will best explain agent's target behaviors?

Agent actions
- What actions can agents take?
- Why would an agent take a given action?
- What would an agent gain from taking that action?

Environment
- In what spaces can the agents take actions?
- How are they allowed to act in each space?
- What environmental variables will be used?
- How this variables affect the agents?

Target output
- How the model will be evaluated?
- Haw the model will be validated?
- What distributions or functions will be used as metrics?

Data-driven hypothesis building
- What data is required to assist the hypothesis building?
- What variables are found from data that that system's output indicates to be sensitive to?
- What missing relationships must be specified and can be discovered through data analysis to ensure completeness of the hypotheses used for model implementation?

Data-driven calibration and model discovery
- What data is required for the calibration effort?
- What are the objective functions for calibration?
- What methods will be used for comparing the performance of alternate explanations of the phenomenon being studied?

Figure 22.1 The Agent-Based Model Canvas.

From Theory to Hypothesis: Human-Aided Data-Driven Hypothesis Building

Several steps are needed to get from a guiding theory to a testable hypothesis. A social theory is a coherent explanation of one or more phenomena observed reliably in a systematic empirical research (Shaughnessy and Zechmeister 1985). Social theories are typically very abstract and aim to explain or interpret social phenomena. Theories often are not stated in a falsifiable manner, meaning that they are rarely amenable to direct or easy testing. A theory, however, can be tested by rendering it into a set of more specific and concrete instances, called hypotheses. In identifying the relevant theories for a research project, and consequently filling this building block of the ABMC, we ask questions such as: *What known theories are we basing our model on? What emotional, rational, social human behavioral theories are we using? What network theories or other theories?*

Once the answers to these questions are identified, for a theory to generate more precise research questions, it must have a clear focus on a set of *key concepts* explained through a set of *key variables*. The concepts help define the theory, making it more specific and more testable by defining the broad ideas upon which the theory is built (Kerlinger and Lee 2000). Variables further refine the definitions of concepts and provide specific and measurable operationalizations of those concepts. The variables are amenable to testing since they can be observed and measured. Questions such as *What big ideas does the theory use? or How are the concepts defined?* can help fill *key concepts* block of the ABMC. Also, in order to identify the *key variables*, a researcher should ask: *What constructs describe and define the concepts? How can the constructs be defined and measured using the data we have available?*

In the next step, the ABMC asks for the *key relationships* interrelating the involved variables. For example, we should ask: *How do the variables correlate? What are the casual relationships between variables? What functional form does the casual relationship take?*

The relationships between variables provide the content needed to define and test *hypotheses*. Hypotheses, according to the scientific method, have been defined as "statement(s) or prediction(s) about a phenomenon that should be observed if a particular theory is accurate (Shaughnessy and Zechmeister 1985). In the context of ABM, these hypotheses will drive agent behavior, which, in turn, will determine the simulation output needed to test those hypotheses. To fill this building block of the ABMC, we try to answer questions such as *What relationship do we want to experiment within the model? or What relationships will best explain agent's target behaviors?*

As an example of how hypotheses can be derived from existing theories, consider Milkman's study of *How Race and Gender Biases Impact Students in*

Higher Education (Milkman et al. 2015), which can be considered as a part of Feagin's theory of systematic racism (Feagin 2013). In Milkman's study, a group of researchers hypothesize that race would play a role in shaping how university professors respond to prospective graduate students who express interest in their research. In order to test this hypothesis, they measured email responses of 6500 professors across 250 of the US top universities to the messages sent by students, who showed admiration to the research of those professors and asked for a meeting. Their study strengthened the hypothesis of "faculties are biased in favor of white men."

Feagin's theory on systemic racism is the relevant theory for this study, which along with its content help fill the theory block of the ABMC. The key concepts are race, racial bias, and interpersonal interactions between professors and prospective graduate students. Operationalizations of these concepts include the identified professor population and the relevant content of their email correspondence with perspective students. Accordingly, the hypothesis being tested by Milkman, based on Feagin's theory, was that *there are differences in how the professors respond to prospective students of different races, measured through differences in the email correspondence.*

From Hypothesis to Model: Data-Driven Calibration and Model Discovery

Hypotheses are clear, specific, and falsifiable statements or predictions. These characteristics make them testable. Similar to the process of translating theories into testable hypotheses, several steps are needed to generate a set of interpretable models from a set of hypotheses. Furthermore, there is often more than one plausible theory explaining a phenomenon. No matter if the existing theories are complementary or competing, it may be necessary to identify and consider all plausible theories that offer explanations of the phenomena of interest. In this case, it might be crucial for a modeling framework, such as ABM, to consider not one but multiple social theories (along with their associated hypotheses) that are competing or cooperating to drive *agent actions.* ABMs enable replicating and predicting socioeconomic systems through characterizing the behaviors and interactions of the agents. The behaviors of the agents, in turn, are indicated by a set of rules derived from the relevant theories. In order to fill this block, we need to identify the following: *What actions can agents take? Why would an agent take a given action? What would an agent gain from taking that action?*

Another entity called the *environment* is also needed to drive the behavior and dynamics of the agents. The environment exists to determine how the model links to the project's context and also to implement the rules for the physical, conceptual, or reified space in which agents act. To be able to identify these elements, we need to answer questions such as the following: *In what spaces can the agents take actions? How are they allowed to act in each space? What environmental variables will be used? How do these variables affect the agents?*

Before using an ABM to perform a forecast on a real system, one needs to ensure that the ABM is relatively capable of recreating observed phenomena. This process is called validation. A crucial step in validating any ABM is calibration, which is typically performed by comparing the resulting output of the constructed model with that of the real system. Note that an ABM-based simulation is often characterized by a set of variables:

1. Parameters of the model driving the simulation dynamics and define agent populations.
2. Initial conditions such as the characteristics of the agent populations and environmental state.
3. The relative weights given to certain parameters.

The calibration process can be achieved by tuning the associated variables for *desired output*. This block of the Canvas concerns *how the model will be evaluated, how the model will be validated, and what distributions or functions will be used as metrics.*

Finding the best set of parameters typically involves an optimization problem with respect to a loss or objective function, capturing the distance between the model's simulated output and the real experimental data. Parameter tuning, most often, requires exploring an extensive parameter space. This motivates automated machine learning-based or computational strategies, which facilitate faster and more efficient explorations of the parameter space. The result of this validation process will provide insight into the:

1. Basins of attraction of the phenomenon being simulated.
2. Plausible archetypes/characteristics of entities in the real world.
3. Importance of factors of behavior.

At the end, if the calibration fails to produce a sufficiently accurate fit to the empirical data, one can conclude that the hypothesis being tested is invalid, that is, the underlying hypotheses cannot explain the phenomena of interest. In other words, ABM rules and populations do not sufficiently match those of the real system. In this case, the modeling team should consider alternative theories and different testable hypotheses. A possible approach to perform this task is to apply an evolutionary model discovery technique. This approach allows the researcher to explore multiple candidate models embodying existing theories of micro-behavior rules for the agents (Gunaratne and Garibay 2017, 2018). Although the need for this type of rule-space exploration has been emphasized for a while (Epstein 1999, 2006), it has not generally been sufficiently investigated.

Two Application Examples

In this section, we demonstrate the usefulness of ABMC by examples. We take two of the most well-studied ABMs and use the ABMC to fully describe them: Schelling's seminal work on segregation modeling (Schelling 1969), a good example of an abstract model of a social science concept that provides intriguing insights, and the Long House Valley project's Artificial Anasazi model (Dean et al. 2000), one of the first CSS models to be compared to data and have computational calibration efforts (Stonedahl and Wilensky 2010b) and model discovery efforts (Gunaratne and Garibay 2018) applied to. The purpose of choosing these examples is to allow future transdisciplinary teams guidance toward the application of ABMC in their own CSS efforts.

Schelling's Segregation Model

The ABMC of Schelling's segregation model (Schelling 1969) is shown in Figure 22.2. The nine building blocks are as follows:

Theory: The theory used is an household-level description of discrimination, one motivated by statistical observations about the impact of racial segregation in the United States.

Key concepts: Segregation is driven by individual-level neighbor preferences. Those of a similar race can require that a minimum fraction of immediate neighbors are of the same race. Until this threshold is met, the individual is unsatisfied and seeks an alternate living situation.

Key variables:
1. *Parameter*: White's threshold of tolerance of immediate neighbors of similar race.
2. *Parameter*: Black's threshold of tolerance of immediate neighbors of similar race.
3. *Initial condition*: Fraction of region's population that is white.
4. *Initial condition*: Fraction of region's population that is black.

Key relationships: $H_i = s_i > k_i$ where H_i is the binary residential satisfaction of the individual, s_i is the ratio of similar neighbors around the individual, and k_i is the desired minimum of neighbors of similar race by the individual.

If happiness is not met, the agent should relocate to a new, free residence.

Hypothesis building data: Data on the correlation of race an economic status provided by Pascal (1967).

Hypothesis:
1. Segregation is the result of individual racial preferences, and segregation patterns can result from even modest preference ratios.
2. There exist a tipping point in the dynamics of segregation, dependent on the similar race wanted parameter, for which the entire neighborhood can be completely occupied by one race or the other.

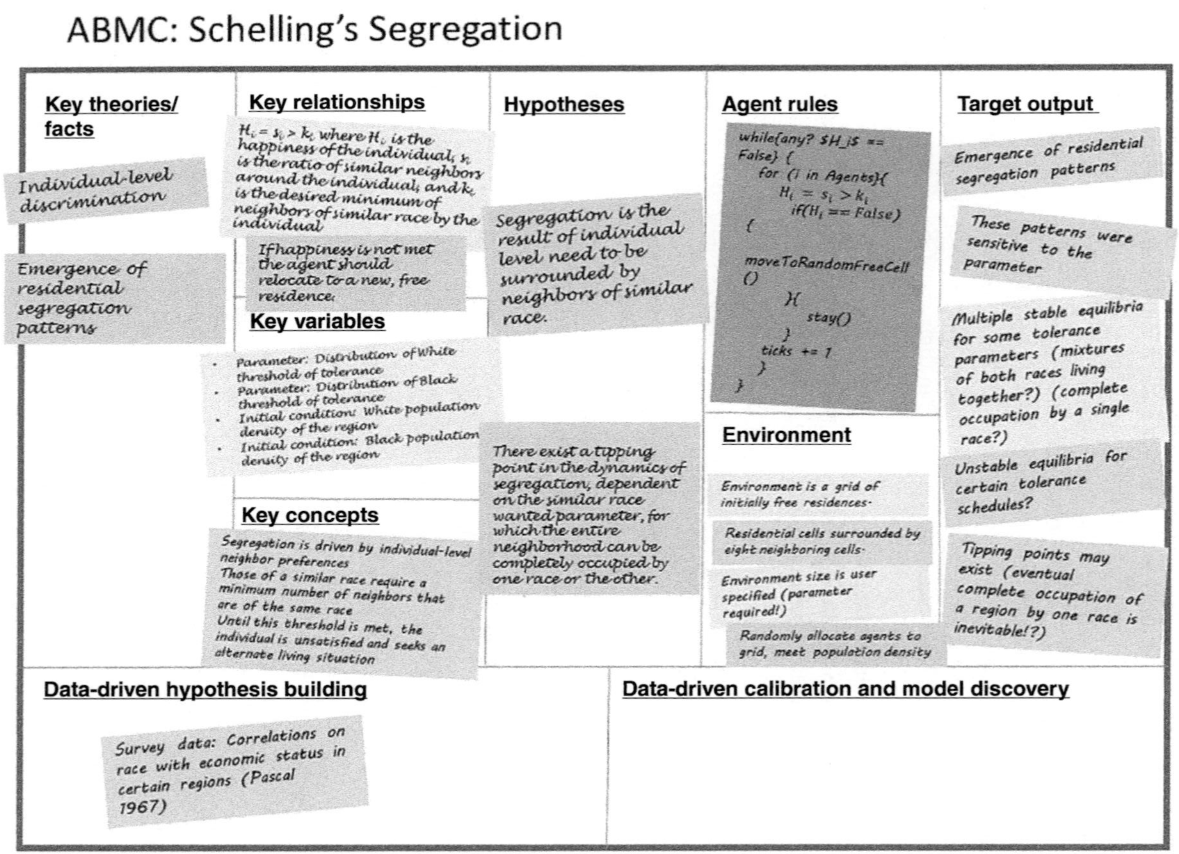

ABMC: Schelling's Segregation

Key theories/ facts

Individual-level discrimination

Emergence of residential segregation patterns

Key relationships

$H_i = s_i > k_i$ where H_i is the happiness of the individual, s_i is the ratio of similar neighbors around the individual, and k_i is the desired minimum of similar neighbors of the individual

If happiness is not met the agent should relocate to a new, free residence.

Key variables

- Parameter: Distribution of White threshold of tolerance
- Parameter: Distribution of Black threshold of tolerance
- Initial condition: White population density of the region
- Initial condition: Black population density of the region

Key concepts

Segregation is driven by individual-level neighbor preferences

Those of a similar race require a minimum number of neighbors that are of the same race

Until this threshold is met, the individual is unsatisfied and seeks an alternate living situation

Hypotheses

Segregation is the result of individual level need to be surrounded by neighbors of similar race.

There exist a tipping point in the dynamics of segregation, dependent on the similar race wanted parameter, for which the entire neighborhood can be completely occupied by one race or the other.

Agent rules

```
while(any? $H_i$ ==
False) {
    for (i in Agents){
        H_i = s_i > k_i
        if(H_i == False)
        {
            moveToRandomFreeCell
        }{
            stay()
        }
    }
    ticks += 1
}
```

Environment

Environment is a grid of initially free residences.

Residential cells surrounded by eight neighboring cells.

Environment size is user specified (parameter required).

Randomly allocate agents to grid, meet population density

Target output

Emergence of residential segregation patterns

These patterns were sensitive to the parameter

Multiple stable equilibria for some tolerance parameters (mixtures of both races living together?) (complete occupation by a single race?)

Unstable equilibria for certain tolerance schedules?

Tipping points may exist (eventual complete occupation of a region by one race is inevitable?)

Data-driven hypothesis building

Survey data: Correlations on race with economic status in certain regions (Pascal 1967)

Data-driven calibration and model discovery

Figure 22.2 The Agent-Based Model Canvas of the Schelling's segregation model.

Agent rules: Agent's behavior according to the following algorithm:

```
while any? Hᵢ == False do
    forall i in Agents do
        Hᵢ = sᵢ > kᵢ if Hᵢ == False then
        |   moveToRandomFreeCell()
        else
        |   Stay()
        end
        ticks += 1
    end
end
```

Environment:
1. Environment is a grid of initially free residences.
2. Each residential cell is surrounded by eight neighboring cells.
3. Environment size is user specified (parameter required!).
4. Initialize grid at each simulation run start randomly allocating agents to the grid until the population density conditions are met.

Validation data: None

Target output:
1. The emergence of residential segregation patterns.
2. These patterns were sensitive to the parameter.
3. There can exist multiple stable equilibria for some tolerance parameters, and they may or may not allow mixtures of both races living together or complete occupation by a single race.
4. There may even exist unstable equilibria for certain tolerance schedules.
5. Depending on tolerance schedules, tipping points may exist beyond which the eventual complete occupation of a region by one race is inevitable.

Artificial Anasazi

In Figure 22.3 we consider ABMC applied to the Artificial Anasazi model (Dean et al. 2000) and some of the calibration (Stonedahl and Wilensky 2010b) and model discovery efforts applied on it (Gunaratne and Garibay 2017, 2018).

Theory:
1. Agriculture was a centerpiece of Pueblo cultures (Dean et al. 2000).
2. The typical household size at the period of study (800 CE–1350 CE) consisted of five individuals on average (from archaeological ethnographic and demographic studies) (Dean et al. 2000).
3. Soil quality is determinant of farming potential and yield, with some zones of the Long House Valley being arable while others were not.

Key concepts: The Kayenta Anasazi were an agriculture-dependent civilization, and fertility of land and availability of water would have affected the carrying capacity of the Valley.

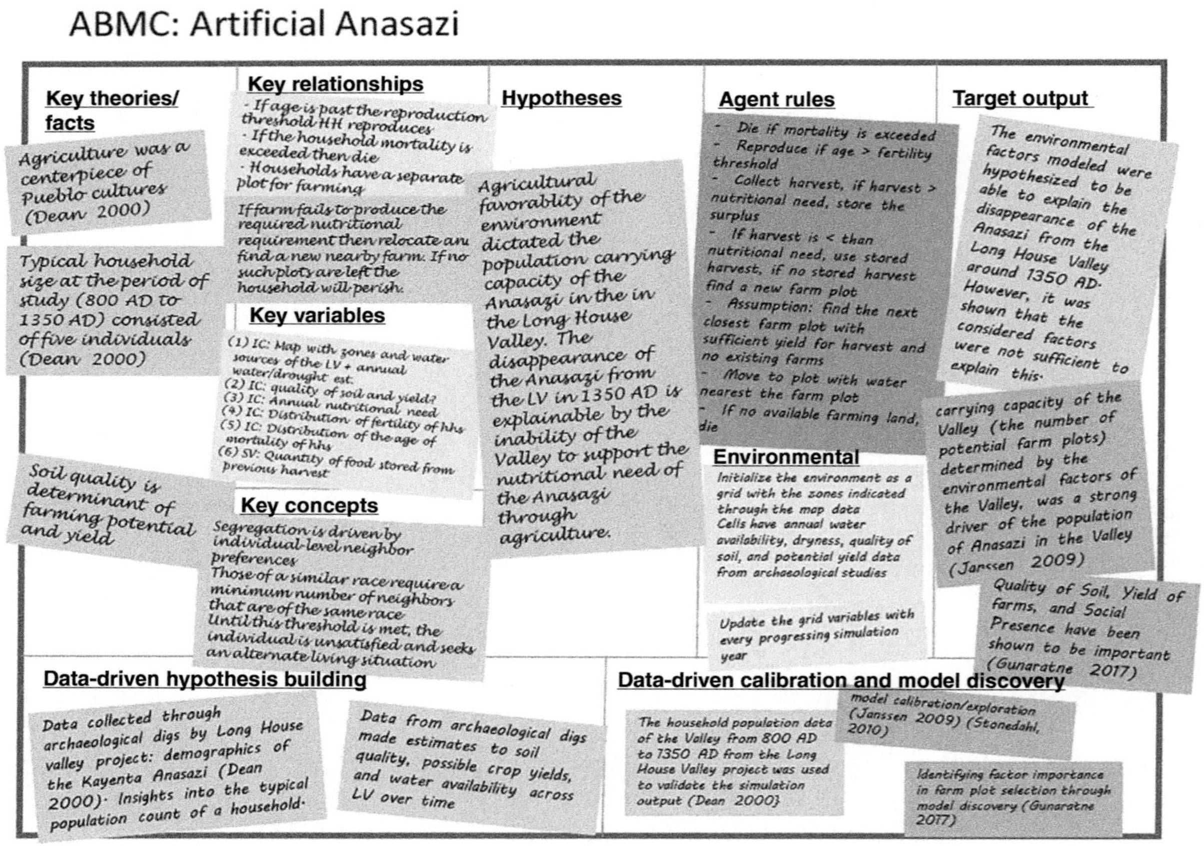

ABMC: Artificial Anasazi

Key theories/facts

Agriculture was a centerpiece of Pueblo cultures (Dean 2000)

Typical household size at the period of study (800 AD to 1350 AD) consisted of five individuals (Dean 2000)

Soil quality is determinant of farming potential and yield

Key relationships

- If age is past the reproduction threshold HH reproduces
- If the household mortality is exceeded then die
- Households have a separate plot for farming

If farm fails to produce the required nutritional requirement then relocate an find a new nearby farm. If no such plot are left the household will perish.

Key variables

(1) IC: Map with zones and water sources of the LV + annual water/drought est.
(2) IC: quality of soil and yields
(3) IC: Annual nutritional need
(4) IC: Distribution of fertility of hhs
(5) IC: Distribution of mortality of hhs
(6) SV: Quantity of food stored from previous harvest

Key concepts

Segregation is driven by individual-level neighbor preferences

Those of a similar race require a minimum number of neighbors that are of the same race. Until this threshold is met, the individuals unsatisfied and seek an alternate living situation.

Hypotheses

Agricultural favorability of the environment dictated the population carrying capacity of the Anasazi in the the Long House Valley. The disappearance of the Anasazi from the LV in 1350 AD is explainable by the inability of the environment of the Valley to support the nutritional need of the Anasazi through agriculture.

Agent rules

- Die if mortality is exceeded
- Reproduce if age > fertility threshold
- Collect harvest, if harvest > nutritional need, store the surplus
- If harvest is < than nutritional need, use stored harvest, if no stored harvest find a new farm plot
- Assumption: find the next closest farm plot with sufficient yield for harvest and no existing farms
- Move to plot with water nearest the farm plot
- IF no available farming land, die

Environmental

Initialize the environment as a grid with the zones indicated through the map data
Cells have annual water availability, dryness, quality of soil, and potential yield data from archaeological studies

Update the grid variables with every progressing simulation year

Target output

The environmental factors modeled were able to explain the disappearance of the Anasazi from the Long House Valley around 1350 AD. However, it was shown that the considered factors were not sufficient to explain this carrying capacity of the Valley (the number of potential farm plots) determined by the environmental factors of the Valley, was a strong driver of the population of Anasazi in the Valley (Janssen 2009)

Quality of Soil, Yield of farms, and Social Presence have been shown to be important (Gunaratne 2017)

Data-driven hypothesis building

Data collected through Long House archaeological digs by Long House archaeological digs Kayenta valley project: demographics of the Kayenta Anasazi (Dean 2000). Insights into the typical population count of a household.

Data from archaeological digs made estimates to soil quality, possible crop yields, and water availability across LV over time

Data-driven calibration and model discovery

The household population data of the Valley from 800 AD to 1350 AD from the Long House Valley project was used to validate the simulation output (Dean 2000)

model calibration/exploration (Janssen 2009) (Stonedahl,

Identifying factor importance in farm plot selection through model discovery (Gunaratne 2017)

Figure 22.3 The Agent-Based Model Canvas applied to the Artificial Anasazi model and later calibration and model discovery efforts performed on it.

Key variables:

1. *Initial condition*: Map of the geography of the various fertility zones and water sources of the Valley along with annual water/drought estimates.
2. *Initial condition*: Variables of the land important to agriculture such as quality of soil and yield.
3. *Initial condition*: The annual nutritional need of households.
4. *Initial condition*: Distribution of the threshold of the fertility of households.
5. *Initial condition*: Distribution of the age of mortality of households.
6. *State variable*: Quantity of food stored from previous harvest by household.

Key relationships:

1. If the household age is past the reproduction threshold of the agent, it may reproduce.
2. If the household mortality is exceeded, the household must die.
3. Households have a separate plot for farming.
4. If household's receive more harvest than their nutritional requirement, the excess is stored.
5. If the household's farm fails to produce the required nutritional requirement, the household must relocate and find a new nearby farm. If no such plots are left, the household will perish.

Hypothesis building data:

1. Data collected through archaeological digs during the course of the Long House Valley project were used to determine the demographics of the Kayenta Anasazi (Dean et al. 2000). The data gave insights into the typical population count of a household (five individuals).
2. Conditions of the Valley itself changed over the course of time. Data from archaeological digs made estimates to soil quality, possible crop yields, and water availability across the Valley.

Hypothesis: Agricultural favorability of the environment dictated the population carrying capacity of the Anasazi in the Long House Valley. The disappearance of the Anasazi from the Valley in 1350 CE is explainable by the inability of the Valley to support the nutritional need of the Anasazi through agriculture.

Agent rules:

1. Die if mortality is exceeded.
2. Reproduce if age exceeds fertility threshold.
3. Collect harvest; if harvest is greater than nutritional need, store the rest.
4. If harvest is less than nutritional need, use stored harvest, and if no stored harvest, find a new farm plot.
   (a) (Assumption in original model) Find the next closest farm plot with sufficient yield for harvest and no existing farms.

    (b) Move household to plot with water nearest the farm plot.

    (c) If no available farming land, die.

Environment:

1. Initialize the environment as a grid with the zones indicated through the map data.
2. Populate grid with annual water availability, dryness, quality of soil, and potential yield data from archaeological studies.
3. Update the grid variables with every progressing simulation year.

Validation data: The household population data of the Valley from 800 CE to 1350 CE from the Long House Valley project was used to validate the simulation output (Dean et al. 2000). This data was used for model calibration (Janssen 2009; Stonedahl and Wilensky 2010b) and for identifying factor importance in farm plot selection through model discovery (Gunaratne and Garibay 2018).

Target output: The environmental factors modeled were hypothesized to be able to explain the disappearance of the Anasazi from the Long House Valley around 1350 CE. However, it was shown that the considered factors were not sufficient to explain this.

In further studies, Janssen (2009) demonstrates that the carrying capacity of the Valley (the number of potential farm plots), determined by the environmental factors of the Valley, was a strong driver of the population of Anasazi in the Valley, in comparison with the agent parameters (fertility, mortality, etc.). In Gunaratne and Garibay (2018) this result is strengthened by discovering and ranking the importance of factors of farm selection through data-driven evolutionary model discovery. It is shown that quality of soil, yield of farms, and social presence (two factors affecting the carrying capacity and the last an emergent result of agent decisions) are actually stronger determinants of the population in the Valley than the other variables.

Conclusion

We are now witnessing the rise of a new paradigm at the intersection of social, computer, and data science. This interdisciplinary field needs a common language for effective communication between researchers with highly varied backgrounds, skill sets, and research cultures. The ABMC offers a form of *lingua franca*, a common framework for both high-level and detailed model design, specification, and building, using mutually understood elements of research methods. The Canvas facilitates rapid interdisciplinary model creation, description, validation, improvement, and evolution by visually organizing the research project into the 9 model building blocks and, in our view, substantially contributing toward the success of any CSS project. A completed canvas facilitates the formation of testable hypotheses and the models used to test the hypotheses. In this way, the proposed framework embodied by the Canvas aids in constructing ABMs by sketching the expected target output and employing data and methods for validation as well as analysis and hypothesis testing.

References

Bersini, H. (2012). Uml for ABM. *Journal of Artificial Societies and Social Simulation* 15 (1): 9.

Bettencourt, L.M., Lobo, J., Helbing, D. et al. (2007). Growth, innovation, scaling, and the pace of life in cities. *Proceedings of the National Academy of Sciences of the United States of America* 104 (17): 7301–7306.

Borgatti, S.P., Mehra, A., Brass, D.J., and Labianca, G. (2009). Network analysis in the social sciences. *Science* 323 (5916): 892–895. https://doi.org/10.1126 /science.1165821.

Casas, P.F.I. (2013). *Formal Languages for Computer Simulation: Transdisciplinary Models and Applications*. IGI Global.

Chomsky, N. (1956). Three models for the description of language. *IRE Transactions on Information Theory* 2 (3): 113–124.

Cioffi-Revilla, C. (2014). *Introduction to Computational Social Science*. London and Heidelberg: Springer.

Comfort, L.K., Wukich, C., Scheinert, S., and Huggins, L.J. (2011). Network theory and practice in public administration: designing resilience for metropolitan regions. In: *The State of Public Administration: Issues, Challenges, and Opportunities* (ed. D. Menzel and H. White), 257–271. M.E. Sharp Inc.

Czarniawska, B. (2004). *Narratives in Social Science Research*. Sage Publications.

Darity, W. (1975). Economic theory and racial economic inequality. *The Review of Black Political Economy* 5 (3): 225–248.

Davis, P.K. and O'Mahony, A. (2013). A Computational Model of Public Support for Insurgency and Terrorism: A Prototype for More-General Social-Science Modeling. Tech. Rep. TR-1220-OSD. Santa Monica, CA: Rand National Defense Research Institute.

Dean, J.S., Gumerman, G.J., Epstein, J.M. et al. (2000). Understanding Anasazi culture change through agent-based modeling. In: *Dynamics in Human and Primate Societies: Agent-Based Modeling of Social and Spatial Processes*, 179–205. New York, NY: Oxford University Press, Inc. ISBN: 0-19-513168-1.

Dipeolu, A., Cook-Cottone, C., Lee, G.K. et al. (2016). A concept map of campers perceptions of camp experience: implications for the practice of family counseling. *The Family Journal* 24 (2): 182–189.

Epstein, J.M. (1999). Agent-based computational models and generative social science. *Complexity* 4 (5): 41–60.

Epstein, J.M. (2006). *Generative Social Science: Studies in Agent-Based Computational Modeling*. Princeton, NJ: Princeton University Press.

Epstein, J.M. (2008). Why model? *Journal of Artificial Societies and Social Simulation* 11 (4): 12.

Epstein, J.M. and Axtell, R. (1996). *Growing Artificial Societies: Social Science from the Bottom Up*. Brookings Institution Press.

Feagin, J. (2013). *Systemic Racism: A Theory of Oppression*. Routledge.

Fiore, S.M., Smith-Jentsch, K.A., Salas, E. et al. (2010). Towards an understanding of macrocognition in teams: developing and defining complex collaborative processes and products. *Theoretical Issues in Ergonomics Science* 11 (4): 250–271.

Fiore, S.M., Wiltshire, T.J., Oglesby, J.M. et al. (2014). Complex collaborative problem-solving processes in mission control. *Aviation, Space and Environmental Medicine* 85 (4): 456–461.

Fowler, M. and Scott, K. (2000). *UML Distilled: A Brief Guide to the Standard Object Modelling Language*, 2e. Boston, MA: Addison-Wesley Longman Publishing Co., Inc. ISBN: 0-201-65783-X.

Fu, Z., Huang, F., Ren, K. et al. (2017). Privacy-preserving smart semantic search based on conceptual graphs over encrypted outsourced data. *IEEE Transactions on Information Forensics and Security* 12 (8): 1874–1884.

George, A.L. and Bennett, A. (2005). *Case Studies and Theory Development in the Social Sciences*. MIT Press.

González-Bailón, S. (2013). Social science in the era of big data. *Policy & Internet* 5 (2): 147–160.

Grimm, V., Berger, U., Bastiansen, F. et al. (2006). A standard protocol for describing individual-based and agent-based models. *Ecological Modelling* 198 (1–2): 115–126.

Grimm, V., Berger, U., DeAngelis, D.L. et al. (2010). The odd protocol: a review and first update. *Ecological Modelling* 221 (23): 2760–2768.

Gunaratne, C. and Garibay, I. (2017). Alternate social theory discovery using genetic programming: towards better understanding the artificial Anasazi. In: *Proceedings of the Genetic and Evolutionary Computation Conference*, 115–122. New York, NY: ACM.

Gunaratne, C. and Garibay, I. (2018). Evolutionary Model Discovery of Factors for Farm Selection by the Artificial Anasazi. arXiv preprint arXiv:1802.00435.

Haymovitz, E., Houseal-Allport, P., Lee, R.S., and Svistova, J. (2017). Exploring the perceived benefits and limitations of a school-based social-emotional learning program: a concept map evaluation. *Children & Schools* 40 (1): 45–54.

Hjorth, A., Head, B., and Wilensky, U. (2015). *Levelspace Netlogo Extension*. Evanston, IL: The Center for Connected Learning and Computer-Based Learning.

Horowitz, E., Sahni, S., and Rajasekaran, S. (1997). *Computer Algorithms c++: C++ and Pseudocode Versions*. Macmillan.

Janssen, M.A. (2009). Understanding artificial Anasazi. *Journal of Artificial Societies and Social Simulation* 12 (4): 13.

Kamsu-Foguem, B., Tchuenté-Foguem, G., and Foguem, C. (2014). Using conceptual graphs for clinical guidelines representation and knowledge visualization. *Information Systems Frontiers* 16 (4): 571–589.

Kennedy, P. (2003). *A Guide to Econometrics*, 5e. MIT Press.

Kerlinger, F.N. and Lee, H.B. (2000). *Foundations of Behavioral Research*, 4e. Harcourt College Publishers.

Kinchin, I.M., Hay, D.B., and Adams, A. (2000). How a qualitative approach to concept map analysis can be used to aid learning by illustrating patterns of conceptual development. *Educational Research* 42 (1): 43–57.

King, G., Keohane, R.O., and Verba, S. (1994). *Designing Social Inquiry: Scientific Inference in Qualitative Research*. Princeton, NJ: Princeton University Press.

Köhler, M., Langer, R., Von Lude, R. et al. (2007). Socionic multi-agent systems based on reflexive petri nets and theories of social self-organisation. *Journal of Artificial Societies and Social Simulation* 10 (1): 3.

Koliba, C., Meek, J.W., and Zia, A. (2011). *Governance Networks in Public Administration*. CRC Press.

Mason, W., Vaughan, J.W., and Wallach, H. (2014). *Computational Social Science and Social Computing*. Springer.

McFarland, D.A., Lewis, K., and Goldberg, A. (2016). Sociology in the era of big data: the ascent of forensic social science. *The American Sociologist* 47 (1): 12–35.

Milkman, K.L., Akinola, M., and Chugh, D. (2015). What happens before? A field experiment exploring how pay and representation differentially shape bias on the pathway into organizations. *Journal of Applied Psychology* 100 (6): 1678.

Minsky, M. (1975). Minskys frame system theory. In: *TINLAP75: Proceedings of the 1975 Workshop on Theoretical Issues in Natural Language Processing*, 104–116. Stroudsburg, PA: Association for Computational Linguistics.

Novak, J. (2010). *Learning, Creating, and Using Knowledge: Concept Maps as Facilitative Tools in Schools and Corporations*. Routledge.

Osterwalder, A., Pigneur, Y., Bernarda, G., and Smith, A. (2014). *Value Proposition Design: How to Create Products and Services Customers Want*. Wiley.

Ostrom, E. (2005). *Understanding Institutional Diversity*. Princeton, NJ: Princeton University Press.

Pascal, A.H. (1967). The economics of housing segregation. Unpublished doctoral dissertation. Elsevier.

Rafiei, M. and Kardan, A.A. (2015). A novel method for expert finding in online communities based on concept map and PageRank. *Human-Centric Computing and Information Sciences* 5 (1): 10.

Ramirez, C. and Valdes, B. (2012). A general knowledge representation model of concepts. In: *Advances in Knowledge Representation*. InTech.

Roth, W.-M. and Roychoudhury, A. (1992). The social construction of scientific concepts or the concept map as device and tool thinking in high conscription for social school science. *Science Education* 76 (5): 531–557.

Schelling, T.C. (1960). *The Strategy of Conflict*. Harvard University Press.

Schelling, T.C. (1966). *Arms and Influence*. Yale University Press.

Schelling, T.C. (1969). Models of segregation. *The American Economic Review* 59 (2): 488–493.

Schwendimann, B.A. and Linn, M.C. (2016). Comparing two forms of concept map critique activities to facilitate knowledge integration processes in evolution education. *Journal of Research in Science Teaching* 53 (1): 70–94.

Shaughnessy, J.J. and Zechmeister, E.B. (1985). *Research Methods in Psychology*. Alfred A. Knopf.

Stonedahl, F. and Wilensky, U. (2010a). Behaviorsearch [computer software]. In: *Center for Connected Learning and Computer Based Modeling*. Evanston, IL: Northwestern University. Available online: http://www.behaviorsearch.org.

Stonedahl, F. and Wilensky, U. (2010b). Evolutionary robustness checking in the artificial Anasazi model. In: *AAAI Fall Symposium: Complex Adaptive Systems*, 120–129.

Tintner, G. (1953). The definition of econometrics. *Econometrica* 21 (1): 31–40. https://doi.org/10.2307/1906941.

Weilkiens, T. (2011). *Systems Engineering with SYSML/UML: Modeling, Analysis, Design*. Elsevier.

Zeigler, B.P. (1989). DEVS representation of dynamical systems: event-based intelligent control. *Proceedings of the IEEE* 77 (1): 72–80.

23

Representing Socio-Behavioral Understanding with Models

Andreas Tolk[1] and Christopher G. Glazner[2]

[1] Modeling, Simulation, Experimentation, and Analytics, The MITRE Corporation, Hampton, VA 23666, USA
[2] Modeling, Simulation, Experimentation and Analytics, The MITRE Corporation, McLean, VA 22103, USA

Introduction

Scientific work in all domains follows a set of common principles and guidelines that ensure high-quality research, dissemination of the results, and the general contribution to the archived body of knowledge that provides a comprehensive and concise representation of concepts, terms, and activities needed to make up a professional scientific domain. Therefore, scientific work requires one to comprehend, share, and reproduce research results to increase knowledge over time. Given the complexity of the world around us, this is often accomplished using models that represent a scientific theory in a more generally comprehensible and testable form. In his work, Goldman (2006) describes the development of science as a series of models that either evolve by integrating new research results or that must be replaced due to revolutionary new insights by new models.

The use of simulation in support of such efforts is increasingly accepted in engineering disciplines (Mittal et al. 2017). Computational sciences allow the execution of hypotheses or theories that are captured in a mathematical model and implemented in a computer program, demonstrating a causal link between theory and behavior. While real-world social systems are far more complex than most physical systems and therefore more difficult to model, we can gain insight from making purposefully simplified representations of social systems and associated hypotheses executable. Multiple models, each bringing a unique perspective, are possible. Further, exploding data collected on social systems are opening the door to the creation of more accurate models of social systems and behaviors (Pentland 2015). Simulation allows researchers to conduct computer-based experiments to evaluate their hypotheses in the

Social-Behavioral Modeling for Complex Systems, First Edition.
Edited by Paul K. Davis, Angela O'Mahony, and Jonathan Pfautz.
© 2019 John Wiley & Sons, Inc. Published 2019 by John Wiley & Sons, Inc.
Companion website: www.wiley.com/go/Davis_Social-Behavioralmodeling

virtual realm that in reality are too expensive, too dangerous, or simply not possible to conduct. While experimentation is critical, the ability also clearly and effectively to communicate and reproduce research is an important benefit of simulation modeling. In this regard, executability is important, as this may become an important advantage for communicating ideas by bringing them to life in a virtual world and make scholars and students *experience* it.

Many simulation pioneers successfully demonstrated the power of simulation, among those Axelrod's paper "Advancing the art of simulation in the social sciences" (1997). However, to this day the use of simulation-based solutions in social sciences and the humanities is often overshadowed by the lack of comprehensibility, shareability, and reproducibility, which led to recent activities in professional societies to address these concerns by encouraging researchers to share simulations used to produce insights presented in journal papers, as discussed by Uhrmacher et al. (2016). The general question is: how can we share our assumptions and constraints within our modeling efforts with other computational scientists?

The following criteria were identified as tenets for useful modeling in social sciences during the expert discussions at the RAND workshop in support of DARPA in Spring 2017. Models are mainly used to *communicate ideas* and may eventually lead to a *computational representation of laws of social science* that can be extracted from peer-reviewed work. As such, *visualization* and other representations of these laws and their effects that integrate all team members, including the customer and sponsor of the work, into the modeling process are as important as *mathematically consistent representation*. The resulting models must be understandable, reproducible, replicable, reusable, and credible. Models developed by computational social sciences shall ultimately be used to communicate and archive the social science body of knowledge in computably accessible and executable form.

Within this chapter, we will first address the philosophical foundations for model-based science and their applicability for the various schools of social science. We will then present a set of selected modeling methods, well knowing that such a list can neither be complete nor exclusive. It is also likely that computational social scientists will need to use a set of orchestrated tools and data, such as recommended, among others, by Levis (2016). The use of models and simulations within social science leads to a recommended way forward. Within all of this, the focus lies on using models to document knowledge and research and communicate it unambiguously with other members of the broader research community.

Philosophical Foundations

Modeling in Support of Scientific Work

In the Introduction, we already stated that models can be seen to be the essence of science. According to Rosen (1998), modeling is the essence of science and the habitat of all epistemology! In his method on the physical sciences, John Von Neumann (1955) states that "sciences do not try to explain, they hardly even try to interpret, they mainly make models." This statement is surely meant to emphasize the importance of models to communicate insights, as in particular in the inductive–empirical world, we make real-world predictions rooted in experience captured in models. However, is such a euphoric view of models applicable to computational social sciences as well, as they are more often represented by the deductive–rational view of the world? Since Sir Francis Bacon (1561–1626) introduced the inductive–empirical method, accompanied by Rene Descartes (1596–1650) introducing the deductive–rational method, we perceive ourselves applying the scientific method to gain knowledge. The inductive–empirical method evolved into scientific realism, while the deductive–rational method can be interpreted as the foundation for the development of constructionist views.

Philosophers of science contributed many directly related insights in recent publications that all support the role of models and increasingly the role of simulation, in support of scientific work. Gelfert (2016) compiled an overview of philosophical views on the use of models in science. Magnani and Casadio (2016) provided an overview of model-based reasoning research, followed by Magnani and Bertolotti (2017) compilation of model-based science, both written from the perspective of humanities and philosophy. Humphreys (2004) provided a philosopher's perspective on how modeling and simulation can fit into the philosophy of science. If our models represent the objective knowledge sufficiently, they can be used to extend our abilities to take full advantage of computational benefits. Like we use microscopes and telescopes to extend the ability of our visual abilities, simulations help us to extend our mathematical skills to understand the dynamic behavior of complex systems. Providing a simulationist's viewpoint was attempted by Tolk (2015), similarly making the point that if we have a theory, understood as the collection and consistent representation of guiding laws of the discipline, a valid model represents this theory, and the resulting verified simulation system can now be used to gain insight into the represented system, produce quasi-empirical data to spawn further research, and so forth. Validity in this context means alignment with the theory, and verification ensures the lossless transformation of the model into computer code.

Gelfert (2016) closes his evaluation on the scientific use of models from the philosophical perspective with the following conclusion.

> Whereas the heterogeneity of models in science and the diversity of their uses and functions are nowadays widely acknowledged, what has perhaps been overlooked is that not only do models come in various forms and shapes and may be used for all sorts of purposes, but they also give unity to this diversity by mediating not just between theory and data, but also between the different kinds of relations into which we enter with the world. Models, then, are not simply neutral tools that we use at will

> to represent aspects of the world; they both constrain and enable our
> knowledge and experience of the world around us: models are mediators,
> contributors, and enablers of scientific knowledge, all at the same time.
>
> (Gelfert 2016, p. 127)

If we use models to capture empirical insight or if we use them to construct our current understanding, models are accepted as enablers for scientific knowledge. In summary, the case for model-based science to capture knowledge in comprehensible, shareable, and reproducible form has been established. Representing our concepts and understanding about their relations and behavior in a form that allows gaining more knowledge in the form of models has been argued for from different philosophical standpoints. This small literature research, which can neither be complete nor exclusive, motivates that models seem to be the best option to enable translational social sciences.

Epistemological Constraints for Computational Science

Computational sciences use their models to develop simulation systems that help to conduct simulation-based experiments. Whenever a real-world experiment is too expensive, is too dangerous, or simply cannot be conducted due to the absence of necessary components, such as systems that are still under development, such simulation-based experiments can help to gain insight even in the absence of empirical experiments.

Kleijnen (2008), Law (2014), and Zeigler et al. (2000) agreed on the basic principle that in a simulation-based experiment, the system of interest is replaced by a valid simulation thereof, which means that by variation of the free input parameters and observation of the same bound input parameters, the same output parameters with the same temporal behavior should be observed as if the experiment were conducted with the real system. An experimental frame binds the overall system, sufficient parametric variations allow for a good understanding of solutions and their sensitivity, and uncertainty can be captured by choosing and calibrating the appropriate probability functions. Traoré and Muzy (2006) showed the duality of experiment frame and simulation system.

The experimental frame is sometimes compared with the experimental setting used to exclude unwanted influences from the natural setting, but epistemologically, there is a significant difference, as the experimental frame bounds the reality to what is known of the simulation, while the experimental settings simply exclude known but unwanted influences. A real-world experiment can reveal something we do not know, while a simulation system can only reveal what we already know. As captured by Chaitin (1977), computer programs including simulation systems are purely transformational; they simply map input parameters to output parameters using computable functions. Whatever we find in the output parameters has either been in the program or in the input parameters. Nothing new is created in this process; information is simply transformed.

On the other side, recent success stories have demonstrated that computational sciences are a powerful tool that allow evaluation of a scientific theory in a detail that otherwise wouldn't be possible, by scanning through vast regions of the solution space by sheer computing power. Ken Wilson's discoveries about phase changes in materials using simulation earned him a Nobel Prize in physics (Wilson 1989). Another recent example is the discovery of the Higgs boson elementary particle (Atlas Collaboration 2012) in which simulation was successfully used to guide experiments to allow for successful observations. The simulation comprised all aspects of the guiding theory and therefore could help to look exactly where the theory predicted certain events to occur so that the scientists could focus their observations of the real world accordingly.

It is, therefore, interesting to focus on the epistemologically interesting characteristics of a model. As compiled in Tolk (2015), *modeling is a task-driven purposeful simplification and abstraction of a perception of reality*. Let's have a closer look at the components of this definition:

- *Task-driven*: A model is generated for a task. It is built to help look for an answer to a scientific question, or it may be built to construct a simulation system that helps to extract information from users. A serious game may be required to educate people or just to provide an immersive virtual environment in which individuals can react and be observed without any risks. The task drives the modeling process.
- *Purposeful*: Modeling is a creative act. The various activities are focused and determined by the task to be supported.
- *Simplification*: Like the experimental settings eliminate all unwanted influences, simplification eliminates all elements that are not important and only distract from the main event. Only what is necessary for the task becomes part of the model.
- *Abstraction*: Abstraction levels are often used to hide details in support of focusing on particular concerns that do not require knowledge about the underlying details. Complex systems may expose different characteristics on a micro, meso, and macro level, the methods of applied sociology that may be used are usually influenced by comparable levels, and therefore such focus on level-specific concerns and methods is important for the modeling process as well.
- *Reality*: The model shall be rooted in empirical data, observations, or logical extensions of valid theories. As discussed before, this can be simplified and abstracted, but the real world builds the foundation of the model.
- *Perception*: The perception of reality is shaped by physical–cognitive aspects as well as other constraints. It represents our current understanding of

the problem. The physical aspect defines what attributes of an object are observable with the sensory system of the observer or, more generally, the information about the object that can be obtained. Cognitive aspects are shaped by the education and knowledge of the observer, their paradigms, and even knowledge of related tools associated with the tasks. Additional constraints are legal or ethical borders that cannot be crossed when collecting data needed for the modeling process. Social factors play a critical role as well.

The result of the modeling phase is a conceptualization. This conceptualization is then transformed over several steps into an executable simulation. This transformation process is also shaped significantly by the resources and abilities of the programmers. Are memory and processing capacities sufficient, is the programming language expressive enough, are numerical approximations needed, and does computational complexity require the use of heuristics instead of formal solutions? Oberkampf et al. (2002) compiled a list of such transformation and implementation challenges that result in errors and uncertainties in simulation experiments.

These mathematical, computational, and epistemological challenges are systemic. Honest and extensive documentation of assumptions, constraints, and applied heuristics and numerical approximations are pivotal to understanding simulation results as well as for providing for reproducibility, reuse, and common research. Modeling methods and paradigms can help to accomplish this goal. From the broader philosophy of science perspective, these results are consistent with the insights provided by McKelvey (1997, 2002).

Furthermore, models – by their very own definition – are providing a facet of our understanding. Some models are unifying, while others are focusing on a special effect of particular interest. If we want to use them to capture our knowledge in total, we have to make sure that the models – and the resulting simulations – can work together. The insights from the field of semiotics can be of tremendous help, as semiotics deals with signs and symbols, their use, and interpretations, up to the definition of artificial and natural languages. The various ideas have been compiled by Tolk et al. (2012) regarding modeling and simulation, showing a close relationship between these approaches. Semiotics differentiates between the syntax (how to provide information structurally), the semantics (how to interpret the provided information), and pragmatics (how to use this information). Similarly, the underlying technical standards define how information is structured in a simulation system, the data or object model provides the interpretation, and the procedures of the simulation system show how this information is used. System engineers may recognize the similarity to the data, information, knowledge, and wisdom hierarchy made popular by Ackoff (1989): data are facts, close to syntax, that become information when put into a common context, giving facts meaning, so belonging to the domain of semantics. Organizing the dynamics of information and processing, it creates knowledge, a pragmatic process. The level of wisdom addresses why certain knowledge is used, adding a level that represents the scientific theory that is used to describe, analyze, and structure the know-what and know-how of the lower three levels into the know-why. All these levels are needed when we bring models together into a common knowledge repository. If the data are not aligned, the processes not harmonized, and the theories not consistent, the result of using two models may offer little scientific value. If the models and simulations are working together, the scientific support can be tremendous, as already mentioned by pointing to Wilson (1989) and the Atlas Collaboration (2012).

However, models are a perception of reality, representing a theory of causality that explains observable correlation in empirical observations (or successfully predicting such observations). As such, models ideally reflect our current best knowledge to explain phenomena. Like theories, as described by Goldman (2006), different models can make different predictions.

Two challenges tightly related with the use of simulation are epistemological and hermeneutical in nature. Scientists make an epistemological mistake when they do not model some phenomenon important for the topic of interest, as discussed above. Hermeneutical mistakes interpret something into simulation results that is not included in the model: scientists unconsciously project their worldview into the observed results, which are potentially biased. In summary, not modeling what is important is an epistemological challenge, while interpreting something into the model that is not in it is a hermeneutical challenge.

Tolk et al. (2013) provided a framework that treats different models as different viewpoints. In highly dynamic systems, such predictions can differ significantly, as is well known from weather models. Providing a common context that allows the display of various forecasts side by side is a useful tool. If, in addition, the different level of trust in certain models is considered, based on experience or scientific rigor of the model, powerful decision support can be provided. However, models must be placed into a common context, and that can be a challenge by itself, in particular, when the originating disciplines use different terms and concepts to express their theories. The next section will present some methods to support overcoming these challenges.

Multi-, Inter-, and Transdisciplinary Research

The terms multi-, inter-, and transdisciplinary are often used interchangeably by researchers not used to working in such a diverse collection of domains, but they define distinct phases and degrees of alignment of collaboration between two fields. Before addressing this, it is worthwhile to address why so many different disciplines with special concepts and terms exist in the first place.

We already addressed the rise of the scientific method to address epistemological growth during the era of enlightenment, such as described by Goldman (2006) and others, in earlier sections. The main underlying idea is first to formulate a hypothesis, then define an experimental frame to support an experiment that allows validation or falsification of the hypothesis. In the case of success, integrate the hypothesis as a new component into the existing theory, or if the new results falsify the current theory, formulate a new theory that combines old and new research insights consistently. The guiding principle is reductionism, derived from the principle of noncontradiction, which allows a researcher to divide all objects into sets defined by the presence or absence of a characteristic attribute. We can divide the whole knowledge domain into various fields and subfields, leading to more and more specialized experts in disciplines that are well defined by the subset of knowledge they are working on. Reductionism has been very successful in understanding more and more detail, for example, in the domain of molecular biology or particle physics.

However, in recent decades, the need for systems thinking and holistic approaches arose from understanding complexity and complex systems better. In complex systems, the principle that "the system is more than the sum of its parts" requires an additional understanding of the interrelations between the components, which are well understood by the specialist. Complexity does not replace reductionism but adds additional value to the relation and interactivities of components. The highly specific and detailed understanding of the fields is still needed, but the seams between these fields and subfields are shifting into the focus of scientific interest as well, requiring a collaboration of researchers that traditionally did not collaborate, leading to multi-, inter-, and transdisciplinary research. The taxonomy and definitions published by Klein (2010) clarify the different degrees of collaboration between different disciplines to address such new challenges.

- *Multidisciplinarity* describes a loose and temporary coupling of disciplines to solve a common problem. It juxtaposes disciplines and their methods. The multidisciplinary team coordinates and sequences their contributions, complementing each other. The definition of terms is aligned with a focus on the collaborative effort, the information exchange is an ad hoc established by the team.
- *Interdisciplinarity* creates a closer linkage between disciplines. Data are aligned and processes are orchestrated, supporting a much higher degree of an integrating approach. The interdisciplinary team focuses on identifying overlapping domains of knowledge and uses integrated solutions in its participating disciplines. By doing so, they are building permanent bridges that link domains together. In addition to the research results, common concepts and common semantics of terms are established that become part of the originating disciplines.

- *Transdisciplinarity* represents the strongest coupling of disciplines. Transcending, transgressing, and transforming the disciplines and specialties defines this approach. Concepts, terms, and activities are not only described in common terms but also systematically integrated, and new interactions are defined across the sectors of original contributions. While knowledge components are used to complement each other in multidisciplinary teams, transdisciplinarity hybridizes the knowledge.[1]

The following figure was developed for the medical community to show not only the need for striving for higher degrees of collaboration, using the definitions of Klein (2010), but also the necessary alignment of methods from M&S needed to allow such degrees of collaboration (Tolk 2016) (Figure 23.1).

The simulation-based support must allow for such collaboration and therefore needs to advance integratability, interoperability, and composability as well. A recent compilation of related work and definitions has been published in Benali and Ben Saoud (2011). More related work can be found in the recent NSF report (Fujimoto et al. 2017). As no single model can provide all the functionality needed and facets required to understand a certain research challenge, we will always have to use an orchestrated set of tools, using different methods, modeling paradigms, etc.

- To support data exchange between supporting simulation methods in *multidisciplinary* approaches, ad hoc agreements are established to ensure the correct interpretation and use of data. Besides using fundamental agreements about the supporting infrastructures providing a common syntax to describe data and common technical means to exchange them, there is no need for further alignment. In the extreme, *swivel chair* solutions – where the experts of one discipline must re-encode the results in a usable form for their simulation method – can and will be applied.
- The more persistent *interdisciplinary* team will establish interoperability solutions that use the team agreement resulting from their integrating and interacting collaboration: the alignment of data and the orchestration of processes. Interoperability allows the exchange of data and their use in the receiving systems of the interdisciplinary team.
- *Transdisciplinary* teams ultimately create a coherent theory that results from the transcending, transgressing, and transforming collaboration. Not only does this allow for the exchange of data and their use, but it also ensures the consistent representation of truth in accordance with the common theory in all participation systems.

---

1 While most researcher distinguish between multi- and interdisciplinary approaches, the differentiation between inter- and transdisciplinary approaches is sometimes omitted. We follow Klein (2010) observations that clearly describe three different degrees of collaboration.

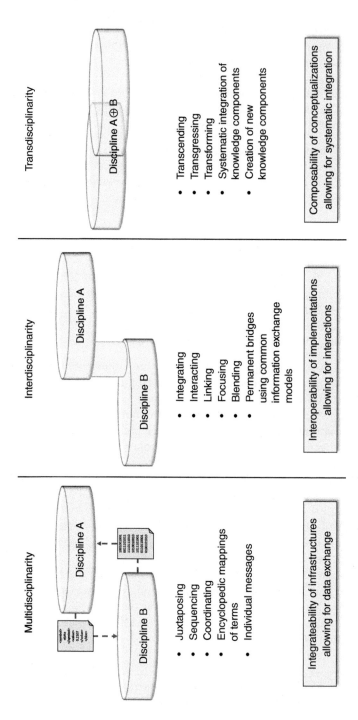

Figure 23.1 Principles of various degrees of collaboration between disciplines. Source: Tolk (2016). Reproduced with permission of John Wiley & Sons.

In summary, multidisciplinarity requires integratable simulation solutions, defined by a common infrastructure and syntax of data. Interdisciplinarity requires interoperable simulation solutions, allowing the exchange of data and meaningful use of data in the receiving systems, based on common semantics and pragmatics. Transdisciplinarity uses composable simulation solutions that are not only interoperable but also supportive of a common theory and therefore ensure the common representation of truth in all participating systems.

When selecting modeling methods, as discussed in the next section, these guiding ontological and epistemological principles need to be taken into consideration to ensure the best possible support of computational social science by simulation-based solutions.

Simulation and Modeling Approaches for Computational Social Scientists

Using simulations becomes a more commonly accepted method to advance social science. In addition to the general ontological and epistemological principles, social scientists must also understand the concrete contributions and limitations of the diverse array of the available modeling and simulation approaches. Like the social systems under study, simulation methodologies vary widely. By employing simulations from different perspectives, we can build a more complete understanding of complex social systems. To advance computational social science, researchers must build an appreciation of these perspectives to ensure that the simulation approach employed will contribute toward building a deeper understanding of a system, be that through inductive, deductive, or even a generative logical approach. As researchers apply these methodologies, they must do so in a way that ensures replicability, ease of sharing, and ease of understanding.

There are multiple ways that simulations can advance our understanding of social systems, employing deduction (derived from principles), abduction (through observation with probabilistic inference), and even induction (observations leading to principles). We can use them in an exploratory fashion to explore the implications of social structures and incentives by generating observed behaviors (simulation *of* social systems); we can also use them as tools to gather data from individuals and organizations that would be too difficult, dangerous, or costly to obtain directly, such as in combat or policy simulations (simulation *for* social systems).

Humans can be *in the loop*, interacting with a simulation, or the simulation can exist in a standalone constructive environment. We can simulate behaviors with a paucity of data such as the collapse of the Anasazi culture 800 years ago by searching for rules that can generate observed behavior (Epstein 2006), or we can use them in data-rich environments such as the collection of biometrics to build a causal simulation of system to investigate patterns discovered through statistical analysis. The space of possible applications of simulation in social science is extensive. Rather than cataloguing every possible application, it is useful to examine the broader classes and approaches of simulation to social and behavioral sciences. We also refer to more detailed publications for further evaluation, such as Diallo et al. (2019) providing examples and perspectives from social science and the humanities and obviously several chapters in this volume.

Simulation OF Social Systems

Perhaps the most common way in which simulation is employed is the simulation *of* social systems, when social systems are modeled, simulated, and studied. This typically follows a deductive approach, where the simulation is used as a tool to explore the logical implications of rules and structures, based on general principles. This allows researchers to understand how different structures, incentives, and neurocognitive models may generate previously observed social behaviors and provide a testbed to study how intervention can alter behavior. The benefits of such an approach are many: reduced cost, faster analysis, cleaner data, greater controllability, and the ability to simulate situations that are too difficult to study in real-world systems. The primary challenge, however, is validity: real-world social systems are extremely complex, and we have no laws governing social interaction in the way that we have laws for thermodynamics. While we may be far away from being able to describe laws of social physics in the way we have for the physics of motion, simulation of social systems can still prove to be very valuable in accelerating our understanding and encourage social science researchers to describe their theories with the rigor necessary for computational examination.

In modeling physical systems, such as an aircraft, we employ a deductive approach: based on our understanding of aerodynamics, we can use simulation to understand how a new aircraft design will behave, and we can have confidence that our simulated aircraft will perform favorably to the actual aircraft. In social systems, our understanding of basic principles is less developed. We have not advanced our theories of social systems such that our understanding of organizational dynamics allows us to simulate a new organizational design and predict its performance with reasonable confidence. We can, however, create higher-level models driven by causal logic as tools to advance our understanding of such systems. Such *top-down* approaches help us to better create and shape social systems, even if they are sufficiently abstract that they miss many of the emergent phenomena, heterogeneity, and other complexities that characterize social systems. Another approach that can be taken to build our understanding is to replicate behavior seen in practice by generating it from low-level interactions of boundedly rational individuals. This *bottom-up* approach to theory development shows tremendous promise in fundamentally advancing our theoretical understanding of social and behavioral systems, yet produces many challenges that researchers continue to struggle with. Both perspectives of simulations of social systems provide value.

Simulation of Social Systems from the *Top Down*

The oldest approaches to social system simulation are equation-based approaches, which model the dynamics of social systems in terms of differential or difference equations. By necessity of the approach, such modeling is done at a very high level, and the structure of the system is determined from the *top down*. Individuals are not modeled; rather, it is aggregates and rates that are modeled in an attempt to understand how quantities change over time as a result of feedback and interaction.

One such early application was the Lanchester equations, first developed in 1916 to model losses in combat between two forces (Lanchester 1956). These equations describing the modern warfare are very simple and can be given as

$$\frac{dA}{dt} = -\beta B$$

$$\frac{dB}{dt} = -\alpha A$$

where A is the initial number of soldiers in one force, B is the initial number of soldiers in a second force, and α and β are the respective rates at which each force can eliminate the other force per unit of time. This model, simple enough to be solved by hand, helped strategists and historians understand the trade-offs between numerical advantage and technical superiority, but did so by treating a social system as a physical system, no different than water moving between two containers. The mathematical formalism of such models meant that they were not commonly used in the social sciences, as they were seen as unapproachable and unintuitive to the social sciences.

Beginning in the 1950s, Jay Forrester at the Massachusetts Institute of Technology (MIT) began to develop a methodology called system dynamics to apply and teach a differential equations approach to modeling using a graphical notation that could be easily taught to students without a rigorous mathematical background. These diagrams of *stocks and flows* could be translated into differential equations and then solved using computers. Using system dynamics, Forrester and his students began applying the technique first to industrial dynamics and then later to social issues such as social policy in urban areas (Forrester 1969) and even limits to global growth using the World 3 model (Meadows et al. 1972).

This simulation approach first defines the structure of systems, with feedback and delays, and produces dynamic behavior. This structure was developed based upon a causal hypothesis: component A's increase *causes* component B's decrease. This structure-driven, continuous time approach meant that it was not individuals that were modeled, but rather aggregates behaving in deterministic response to a designed, but complex, system. This is the essence of a top-down approach, which provides insights into the theoretical performance of idealized social (and other) systems in which actors behave identically and the data that drive the system are population averages, rather than heterogeneous, path-dependent characteristics of individuals. The goal of such models is to identify and quantify policy levers in a system that can influence behavior, such as housing policies or growth targets. Critiques (Nordhaus 1973) of the World 3 model and the accompanying book Limits to Growth (Meadows et al. 1972) slowed the widespread adoption of this approach in the social sciences, although it did develop a following in the study of business strategy and consulting (Senge 1990; Sterman 2000).

Top-down simulation approaches, such as system dynamics or other equation-based approaches, can be very useful in developing understanding in idealized, and designed, systems. They can often be developed rapidly, the causal nature of the underlying logic is unambiguous and easily communicated, and the models themselves are straightforward to rigorously analyze (Rahmandad et al. 2015). These models, however, are highly abstracted and fundamentally do not model the behavior of individuals, the basis of social systems. Any social theory based upon individual behaviors, preferences, heterogeneity, time history, or even position cannot be evaluated using top-down approaches without significant effort. In such examples, another approach is better suited.

Simulation of Social Systems from the *Bottom Up*

In contrast to the *top-down* approach where the top-level structure is modeled and behavior is simulated in the aggregate, *bottom-up* simulation of social systems seeks to understand how social behavior emerges from micro-level interactions. The unit of analysis is the individual and the individuals' incentives, rather than the high-level structure of the systems. Microsimulation, multi-agent simulation, individual-based models, and agent-based models are related simulation approaches that follow the bottom-up modeling paradigm. Due to its ability to generate behaviors that are observable at multiple scales and great flexibility, the bottom-up approach to simulation, and specifically agent-based modeling, is the most widely used simulation approach in social and behavioral research (Boero 2015). Advances in the field are enabling researchers to better communicate and build upon each other's work.

One of the earliest explorations of bottom-up approach to modeling a social system was done manually by Schelling (1969), who employed coins on a paper grid to model to study segregation. Schelling defined a set of rules to govern if a given coin would choose to move to an unoccupied square based on the color of its neighbors and demonstrated that only slight preferences for homophily can quickly lead to segregated communities after a period of time. In this way, he established how "micromotives can drive macrobehaviors" (Schelling 1978). There was no central structure; system-level behavior was shown to emerge from interactions of individuals obeying local incentives in response to their environment. This approach to modeling did not capture a specific neighborhood or attempt to make forecasts, but rather sought to help researchers posit rules and incentives and then see and observe what behavior would emerge if all actors exactly obeyed accordingly.

The structure of Schelling's model made it amenable to simulation and others in many disciplines such as ecology and began to employ computer simulation to study how individual-level modeling could give rise to system behavior. Simulation tools such as StarLogo, Swarm, and NetLogo began to appear in the 1990s, enabling researchers in fields ranging from physics to ecology to build bottom-up models without developing them from scratch. Bonnabeau (2002) provided an excellent overview of the development of the tools and early applications, which often sought to explore the implications on how a globally shared set of incentives and behaviors could give rise to system-level behavior and patterns. Bottom-up simulations in which the agents are homogeneous can be made to very closely approximate top-down simulations, especially as the number of agents increases (Borschsev and Fillipov 2004).

While there is value in simulations that demonstrate how patterns emerge from a given rule set with homogeneous agents, many researchers desired to advance the autonomy and heterogeneity of their agents in ways that reflect the real world, employing bounded rationality, realistic interactions, and networks and placing them in an explicit space. As these models include more of these attributes, they gain the ability to explore a greater number of dynamics. Models possessing these attributes are called agent-based models to distinguish them from the more simplistic, mechanistic approaches that might be termed microsimulations or simply multi-agent systems.

Epstein and Axtell (1996) were early and vocal proponents of agent-based models in the social science, and Epstein (2006) went on to develop the case that ABM represents a new approach, which he termed *generative*, to conducting social science research. The accepted standard of scientific explanation in the social science, he argues, should be the ability to generate observed behaviors. The generative standard, with the motto "if you didn't grow it, you didn't explain it," employs an empirical, deductive approach based upon available data about a social system. Generating behavior in a social system, Epstein argues, is a necessary but not sufficient condition for explanation is social systems. Building upon the Sugarscape model (Epstein and Axtell 1996), the Artificial Anasazi project (Epstein 2006) demonstrated how an agent-based model could explain how the indigenous population of the Long House Valley in northeastern Arizona between 800 and 1350 CE. adapted its behaviors in response to environmental changes. Using the climatic record and developing decision rules for agents in conjunction with anthropologists, the model could generate patterns of behavior consistent with the archaeological record.

A difficult aspect of bottom-up modeling approaches is developing decision and interaction rules for agents. These are often developed through observation and iteration, and it has often been observed that even simple rules can give rise to complex behavior. An early model of flocking in birds, for example, produced strikingly familiar patterns of behavior using only three simple rules for the motion of birds in a flock (Reynolds 1987). These rules were written by a computer scientist looking to build a convincing animation, however, not a biologist looking to develop an understanding of starlings. To make computational approaches more relevant to the advancement of social science, rather than being an enlightening exercise, the tools for computationally modeling social systems from the bottom up need a theoretically grounded framework. The eternal blank canvas for the computational researcher, even if it can generate familiar behaviors, will not become a primary tool of social scientists until it can incorporate and extend existing theory.

The current frontier in bottom-up simulation of social systems is in developing frameworks built on existing social science theory that are understandable to social scientists while still providing the experimental flexibility of simulation. Rather than using perfectly rational and consistent agents whose behavior seeks a mathematically desirable equilibrium, the frontier seeks to develop frameworks to generate behaviors with boundedly rational agents. For example, Thompson et al. (2017) sought to operationalize the theory of planned behavior (Ajzen 1991), which describes how individual beliefs and norms drive behavior as it relates to participating in government. The challenge lies in translating a causal hypothesis from the social science literature into the precise mathematical language of simulations.

Epstein ambitiously sought to move the field forward by developing such a framework called *Agent_Zero*, which he developed to provide neurocognitive basis for agent-based models (2013). The Agent_Zero framework posits that individual behavior in groups is driven by three partially understood and interdependent processes: the emotional, cognitive, and social. The goal of such a framework was to be able to generate original behaviors in populations of cognitive plausible agents that are able to do more than mimic and propagate behavior, generating phenomena such as collective violence and financial panic. This framework draws upon research from cognitive neuroscience but has been developed explicitly for the development of agent-based models. While Epstein has demonstrated a handful of simulations covering topics from the outbreak of the Arab Spring in 2011 to jury processes using Agent_Zero, the development of new social science theory arising from this framework has yet to be seen. There is active effort underway to continue to advance the framework and for others to apply it in useful ways that can contribute toward theory development, and better understanding of how individual behaviors can influence the behavior of social systems at different scales.

Simulation FOR Social Systems

In addition to simulation of social systems for theory development, researchers can also develop simulations FOR social systems, where the simulations exist as stimuli for human subjects. Rather than a computational exercise where all aspects exist in silico and the full experiment can be easily shared, simulations for social systems seek to immerse human subjects in simulated, virtual environments so that their interaction with the simulation and each other can be studied. These *human-in-the-loop* simulations can control for factors that are often too difficult, costly, or otherwise untenable to experiment with directly. For example, a flight simulator is a human-in-the-loop simulation often used to train pilots at significantly reduced cost, placing them in otherwise dangerous scenarios. Such simulations, when used for research rather than training can be highly effective at generating responses from users that are reflective of real-world behaviors.

In research with human-in-the-loop simulation, the human contribution is observed, rather than modeled as it is in simulations of social systems. Applications can vary widely, ranging from studying individual cognition and response in virtual environments to more socially oriented simulation on topics such as war gaming (Harrigan and Kirschenbaum 2016), command and control, or policy development (Mayer 2009). Rothcock and Narayanan (2011) provided a recent overview of many uses of human-in-the-loop simulation for research on human interactions in simulated contexts, examining human involvement in complex systems, and the study of cognitive models. When used for research, these simulations provide a testbed for experimentation that researchers can use as the basis for hypothesis development and validation.

The military community refers to such a mix of human-in-the-loop simulation approaches as live-virtual-constructive simulation: in live simulations, real humans operate real systems in a maneuver; in virtual simulations, real humans operate simulated systems or simulator; and in constructive simulations, simulated humans operate simulated systems (Hodson and Hill 2014). While the technical challenges of such compositions are well addressed, much research still needs to be conducted on effectiveness of the resulting training and the transferability of experiments from the virtual to the real world, as the validity and transferability are often limited by the behavior pattern of humans, which may be different if they know that they are acting in the real or a simulated world.

Simulation *for* social systems, such as human-in-the-loop simulations, is distinct from simulation *of* social systems, such as agent-based modeling, in that it represents an experimental platform to collect data from human subjects, rather than being a fully encapsulated experimental, virtual environment. While it does capture the behavior of actual humans in a controlled environment, it does not provide the same benefits of cleanly articulating, documenting, and sharing repeatable research across the social science research community as is needed to advance the field of social science. There is a need for both approaches, however, as we do not yet have accurate, predictive, and theoretically grounded simulation models of individual behavior (as Agent_Zero aspires to provide a framework for) that are needed for simulations of social systems. We will continue to need to experiment with and understand individuals, so that we can build more accurate, testable, and constructive simulations of social structures and dynamics.

The Way Forward

The examples have shown that there is a growing place for simulation solutions and applications in computational social science and that models and simulations are significant contributions to the body of knowledge. To take full advantage of the potential use of model-based knowledge representations, these models must be conceptually aligned, as described for transdisciplinary research. These activities foremost not only support reaching the objectives of comprehensibility, shareability, and reproducibility of research and results but also support the direct use of computational support by the resulting simulations, as competing theories are recognized on the conceptual level, so researchers know that they should use the resulting simulation to evaluate competing alternatives, and common theories allow for the development of model compositions with complementary views provided by the simulations.

Competing and complementing simulation compositions can be understood as the foundation of computational social science, as these simulation systems are developed to replace observed empirical correlations or assumed theoretic assumptions with causality: implemented as functions on input domain and resulting in output ranges! The computational representation makes them precise, the use of modeling methods as discussed in the previous sections makes them communicable, and their implementation as a simulation makes them easily reproducible. This requires a solid understanding of the various modeling paradigms and their application potential and constraints, mathematically, epistemologically, and hermeneutically. There is a rising interest in topics on simulation solutions in support of social science, as books like Gilbert and Troitzsch (2005), Conte et al. (2013), and Diallo et al. (2019) show.

Despite all these advantages, Denning (2017) provided a critical review of computational thinking, warning against overselling the ideas when claiming that computational thinking will be good for everyone. To take full benefit of the advantages, he observed that computational thinking includes designing the model, not just the steps to control it. This can be generalized into the necessity to understand fully a model to be able to control it. Otherwise, hermeneutical misinterpretations, overselling of results, and other unintended uses of the computational scientific methods may result. In soft sciences like social science and the humanities, the computational support will continue to become increasingly important.

On the other side, simulations enabled significant breakthroughs in computational sciences, with the already earlier mentioned simulation-supported discovery of the Higgs boson elementary particle being one of the most important recent contributions (Atlas Collaboration 2012). In addition, Diallo et al. (2013) showed that simulation can be used not only to simulate known phenomena but also to create insight in the form of building theories in bottom-up approaches, helping to discover new structures of knowledge. While this domain is still in its infancy, the application examples clearly show its potential, including for complex fields like social sciences.

Computational social science will continue to be one of many contributions to the ontology and epistemology of social science. However, to maximize the benefits and support, the principles discussed in this chapter need to be socialized not only in the computational social science community but also in the broader general social science community. This implies the need for easier access to simulation technology for nonengineering disciplines, including better representation of models, simulations, and their results for the nonengineer, as well as openness to such new methods in the social sciences and humanities.

Acknowledgment

We would like to thank our many colleagues and friends for their help and input in writing this chapter, foremost Saikou Diallo, Jose Padilla, Ernie Page, LeRon Shults, Brian Tivnan, and Wesley Wildman.

Disclaimer

References

Ackoff, R. (1989). From data to wisdom. *Journal of Applied Systems Analysis* 16: 3–9.

Ajzen, I. (1991). The theory of planned behavior. *Organizational Behavior and Human Decision Processes* 50 (2): 179–211.

Atlas Collaboration (2012). Observation of a new particle in the search for the Standard Model Higgs boson with the ATLAS detector at the LHC. *Physics Letters B* 716 (1): 1–29.

Axelrod, R. (1997). Advancing the art of simulation in the social sciences. In: *Simulating Social Phenomena*, Lecture Notes in Economics and Mathematical Systems, vol. 456 (ed. R. Conte, R. Hegselmann and P. Terna), 21–40. Berlin, Heidelberg: Springer.

Benali, H. and Ben Saoud, N.B. (2011). Towards a component-based framework for interoperability and composability in modeling and simulation. *Simulation* 87 (1–2): 133–148.

Boero, R. (2015). *Behavioral Computational Social Science*. Hoboken, NJ: Wiley.

Bonnabeau, E. (2002). Agent-based modeling: methods and techniques for simulating human systems. *Proceedings of the National Academy of Sciences of the United States of America* 99 (Suppl. 3): 7280–7287.

Borschsev, A. and Fillipov, A. (2004). From system dynamics and discrete event to practical agent based modeling reasons techniques, tools. Proceedings of the 22nd International Conference of the System Dynamics Society, Oxford, England.

Chaitin, G.J. (1977). Algorithmic information theory. *IBM Journal of Research and Development* 21 (4): 350–359, 496.

Conte, R., Hegselmann, R., and Terna, P. (eds.) (2013). *Simulating Social Phenomena*, Lecture Notes in Economics and Mathematical Systems, 2e, vol. 456. Springer.

Denning, P.J. (2017). Remaining trouble spots with computational thinking. *Communications of the ACM* 60 (6): 33–39.

Diallo, S.Y., Padilla, J.J., Bozkurt, I., and Tolk, A. (2013). Modeling and simulation as a theory building paradigm. In: *Ontology, Epistemology, and Teleology for Modeling and Simulation: Philosophical Foundations for Intelligent M&S Applications* (ed. A. Tolk). Berlin, Heidelberg: Springer-Verlag.

Diallo, S.Y., Wildman, W., Shults, F.L., and Tolk, A. (eds.) (2019). *Human Simulation: Perspectives, Insights, and Applications*. Cham: Springer International Publishing.

Epstein, J.M. (2006). *Generative Social Science: Studies in Agent-Based Computational Modeling*. Princeton, NJ: Princeton University Press.

Epstein, J.M. and Axtell, R. (1996). *Growing Artificial Societies: Social Science from the Bottom Up*. Washington, DC: Brookings Institution Press.

Forrester, J.W. (1969). *Urban Dynamics*. Cambridge, MA: MIT Press.

Fujimoto, R., Bock, C., Chen, W. et al. (eds.) (2017). *Research Challenges in Modeling and Simulation for Engineering Complex Systems*. Cham: Springer International Publishing.

Gelfert, A. (2016). *How to do Science with Models: A Philosophical Primer*. Cham: Springer.

Gilbert, N. and Troitzsch, K. (2005). *Simulation for the Social Scientist*. McGraw-Hill Education.

Goldman, S.L. (2006). *Science Wars: What Scientists Know and How they Know It*. Chantilly, VA: Lehigh University, Teaching Company.

Harrigan, P. and Kirschenbaum, M.G. (2016). *Zones of Control: Perspectives on Wargaming*. Cambridge, MA: MIT Press.

Hodson, D.D. and Hill, R.R. (2014). The art and science of live, virtual, and constructive simulation for test and analysis. *Journal of Defense Modeling and Simulation* 11 (2): 77–89.

Humphreys, P. (2004). *Extending Ourselves: Computational Science, Empiricism, and Scientific Method*. Oxford University Press.

Kleijnen, J.P. (2008). *Design and Analysis of Simulation Experiments*. New York, NY: Springer.

Klein, J.T. (2010). A taxonomy of interdisciplinarity. In: *The Oxford Handbook of Interdisciplinarity* (ed. R. Frodeman, J.T. Klein and C. Mitcham), 15–30. Oxford University Press.

Lanchester, F.W. (1956). Mathematics in warfare. *The World of Mathematics* 4: 2138–2157.

Law, A. (2014). *Simulation Modeling and Analysis*, 5e. McGraw-Hill Education.

Levis, A.H. (2016). Multi-formalism modelling of human organization. In: *Seminal Contributions to Modelling and Simulation* (ed. K. Al-Begain and A. Bargiela), 23–46. Cham: Springer International Publishing.

Magnani, L. and Bertolotti, T. (eds.) (2017). *Handbook of Model-Based Science*. Cham: Springer International Publishing.

Magnani, L. and Casadio, C. (eds.) (2016). *Model-Based Reasoning in Science and Technology*. Cham: Springer International Publishing.

Mayer, I. (2009). The gaming of policy and the politics of gaming: a review. *Simulation Gaming* 40 (6): 825–862.

McKelvey, B. (1997). Perspective – quasi-natural organization science. *Organization Science* 8 (4): 351–380.

McKelvey, B. (2002). Model-centered organization science epistemology. In: *Companion to Organizations* (ed. J.A.C. Baum), 752–780. Thousand Oaks, CA: SAGE Publishing.

Meadows, D.H., Meadows, D.L., Randers, J., and Behrens, W.W. III, (1972). *The Limits to Growth; A Report for the Club of Rome's Project on the Predicament of Mankind*. New York, NY: Universe Books.

Mittal, S., Durak, U., and Ören, T.e. (2017). *Guide to Simulation-Based Disciplines*. Cham: Springer International Publishing.

Nordhaus, W.D. (1973). World dynamics: measurement without data. *The Economic Journal* 83 (332): 1156–1183.

Oberkampf, W.L., DeLand, S.M., Rutherford, B.M. et al. (2002). Error and uncertainty in modeling and simulation. *Reliability Engineering & System Safety* 75 (3): 333–357.

Pentland, A. (2015). *Social Physics: How Social Networks Can Make Us Smarter*. New York, NY: Penguin Books.

Rahmandad, H., Oliva, R., and Osgood, N.D. (2015). *Analytical Methods for Dynamic Modelers*. Cambridge, MA: MIT Press.

Reynolds, C.W. (1987). Flocks, herds and schools: a distributed behavioral model. *ACM SIGGRAPH Computer Graphics* 21: 25–34.

Rosen, R. (1998). *Essays on Life Itself*. New York, NY: Columbia University Press.

Rothcock, L. and Narayanan, S. (2011). *Human-in-the-Loop Modeling*. London: Springer-Verlag.

Schelling, T.C. (1969). Models of segregation. *American Economic Review* 59 (2): 488–493.

Schelling, T.C. (1978). *Micromotives and Macrobehavior*. New York, NY: W.W. Norton & Company.

Senge, P.M. (1990). *The Fifth Discipline*. New York, NY: Doubleday/Currency.

Sterman, J.D. (2000). *Business Dynamics: Systems Thinking and Modeling Tools for a Complex World*. McGraw-Hill Higher Education.

Thompson, J.R., Caskey, T., Dingwall, A.D., and Henscheid, Z.A. (2017). Citizen-centric government services ecosystem: an agent-based approach to the theory of planned behavior. *Journal on Policy and Complex Systems* 3 (2): 196–218.

Tolk, A. (2015). Learning something right from models that are wrong: epistemology of simulation. In: *Concepts and Methodologies for Modeling and Simulation* (ed. L. Yilmaz), 87–106. Cham: Springer International Publishing.

Tolk, A. (2016). Multidisciplinary, interdisciplinary, and transdisciplinary research. In: *The Digital Patient: Advancing Healthcare, Research, and Education* (ed. C.D. Combs, J.A. Sokolowski and C.M. Banks), 225–240. Hoboken, NJ: Wiley.

Tolk, A., Diallo, S.Y., and Padilla, J.J. (2012). Semiotics, entropy, and interoperability of simulation systems: mathematical foundations of M&S standardization. In: *Proceedings of the Winter Simulation Conference*, 2751–2762. Piscataway, NJ: IEEE Press.

Tolk, A., Diallo, S.Y., Padilla, J.J., and Herencia-Zapana, H. (2013). Reference modelling in support of M&S – foundations and applications. *Journal of Simulation* 7 (2): 69–82.

Traoré, M.K. and Muzy, A. (2006). Capturing the dual relationship between simulation models and their context. *Simulation Modelling Practice and Theory* 14 (2): 126–142.

Uhrmacher, A.M., Brailsford, S., Liu, J. et al. (2016). Reproducible research in discrete event simulation – a must or rather a maybe? In: *Proceedings of the Winter Simulation Conference*, 1301–1315. Piscataway, NJ: IEEE Press.

Von Neumann, J. (1955). Method in the physical sciences. *Collected Works* 6: 491–498.

Wilson, K.G. (1989). Grand challenges to computational science. *Future Generation Computer Systems* 5 (2–3): 171–189.

Zeigler, B.P., Praehofer, H., and Kim, T.G. (2000). *Theory of Modeling and Simulation: Integrating Discrete Event and Continuous Complex Dynamic Systems*. Academic Press.

24

Toward Self-Aware Models as Cognitive Adaptive Instruments for Social and Behavioral Modeling

Levent Yilmaz

Department of Computer Science, Auburn University, Auburn, AL 36849, USA

Introduction

The use of computational models in systems engineering is pervasive. However, despite the availability of useful tools that assist modelers in routine aspects of system modeling, other stages in evidential reasoning are not yet so helpful, especially for causal modeling. These phases include (i) the generation and prioritizing of modeling assumptions that are ripe for exploration, (ii) hypothesis-guided automated generation and execution of experiments (Yilmaz et al. 2016, 2017), and (iii) interpretation of results to falsify and revise competing behavioral mechanisms.

A critical obstacle includes the disconnect between model discovery and experimentation. Another is the tendency to delay model justification until after model implementation. We conjecture that addressing these issues requires dynamic coupling of model building and experimentation with continuous feedback between their technical domains. Coupling these domains needs to provide sound explanatory characterization about what alternative and complementary mechanisms are plausible, whether they cohere and under what conditions. This is in contrast with the current practice. Often when experimental results deviate from expected behavior, modelers locate and mitigate the specific issue without looking deeper and more broadly for flaws in the model. This leads to premature convergence to a supposedly tested model that is not robust enough to address variability and uncertainty. However, the ability to retain, experiment with, and relate alternative model mechanisms is critical for developing robust solutions. This is akin to the issue in philosophy of science (Klahr and Simon 1999) of discovering robust theories (Levins 1966) that aim to achieve a balance between generality, precision, coherence, and accuracy.

Social-Behavioral Modeling for Complex Systems, First Edition.
Edited by Paul K. Davis, Angela O'Mahony, and Jonathan Pfautz.
© 2019 John Wiley & Sons, Inc. Published 2019 by John Wiley & Sons, Inc.
Companion website: www.wiley.com/go/Davis_Social-Behavioralmodeling

The process of discovering robust and general theories starts with a question in a domain. By using the background knowledge, scientists hypothesize one or more models to advance alternate explanations. These hypotheses generate testable predictions, which steer experiments to yield new evidence. The observed data then feed into the process to update causal models, resulting in continuous learning that allows adjusting hypothesized mechanisms to fit the data better. The premise of the approach is that scientists accept a theory if it provides the best explanation of the expected behavior, where "best explanation" is evaluated based on an overall coherence judgment. In practice, actual cases of scientific reasoning (Bunge 1998) suggest a variety of factors that determine the explanatory coherence of hypotheses. How much does the hypothesis explain? Are its explanations economical? Is the hypothesis similar to ones that explain similar phenomena? How does the hypothesis compare against alternative hypotheses? Historically, the problem of inference to such explanatory hypotheses was explored in philosophy (Bunge 1998), computational discovery (Darden 2001), psychology (Klahr and Simon 1999), and artificial intelligence (Langley 2000). In philosophy, the acceptance of explanatory hypotheses is called inference to the best explanation (Harman 2008; Thagard 2016).

Motivated by these observations, the objective of this chapter is to put forward a methodological basis, which aims at the following:

- Explore the utility of having a special class of models as adaptive agents that mediate among competing domain theories, data, requirements, principles, and analogies.
- Underline the role of cognitive assistance for model discovery, experimentation, and evidence evaluation to discriminate among competing models.
- Examine alternative strategies for explanatory justification of plausible solutions via cognitive models that explicate coherence judgments.

Our thesis is that model building involves the coherent integration of various ingredients (i.e. domain theory, data, requirements, analogies, principles) to meet specific quality criteria. This thesis is predicated on two sub-theses: the context of discovery and the context of justification (i.e. related to what some refer to as validation). In the context of discovery, mechanism integration plays a vital role in transforming the constituent elements by merging them into a unified mechanism. One aspect of merging is the mechanistic representation, while the other is of calibration: the choice of the parameters in such a way that the model not only fits the data adequately where appropriate but also integrates other ingredients, such as constraints, domain theory, principles, and analogies. The context of justification refers to built-in justification; that is, when the set of ingredients implement the requirements the model is expected to address, then justification is built in.

The rest of the chapter is structured as follows. In the section "Perspective and Challenges," we characterize a view of models that play the special role of mediating. The perspective is used to illustrate challenges for model abstraction and cognitive assistance. The section "A Generic Architecture for Models as Cognitive Autonomous Agents" delineates a reference architecture that defines models as mediators, which are construed as cognitive agents. The mediation process that cognitive models are designed to carry out is outlined in the section "The Mediation Process." The section "Coherence-Driven Cognitive Model of Mediation" presents the role of cognitive computing in the mediation process. The section "Conclusions" concludes by summarizing the main contributions of the chapter. Many authors are researching related matters, and the chapter may provide a framework to see how they fit into the larger conceptual architecture.

Perspective and Challenges

The use of models as exploratory instruments that evolve into a plausible explanatory or predictive model suggests that it is reasonable to view a particular class of models as Dynamic Data-Driven Applications (Darema 2004). Notably, as a mediator between theory and data, the special models in question need to seek coherence in explanatory causal hypotheses that govern a model's behavior.

Models as Dynamic Data and Theory-Driven Mediating Instruments

Figure 24.1 illustrates the mediation role of models between theory and data (see also Davis and O'Mahony 2018). Theoretical principles are leveraged to construct causal relations in the form of behavioral rules. The process of simulated experimentation follows the initial construction of the model. Targeted instrumentation of the model results in observed simulation data that form the basis for learning about the efficacy of the hypothesized mechanistic assumptions that define the behavioral rules. The dynamic and mutually recursive feedback between the model and data refers to adaptive learning. The data gathered through experimentation are analyzed to make decisions about model representation. Model revisions, if there is sufficient consensus, result in mechanism revision. Consequently, the theoretical principles induced from limited data become increasingly accurate in their explanatory and generative power.

Similarly, the Dynamic Data-Driven Application System (DDDAS) paradigm (Darema 2004) promotes incorporation of online real-time data into simulation applications to improve the accuracy of analysis and the precision of predictions and controls. As such, DDDAS aims to enhance applications

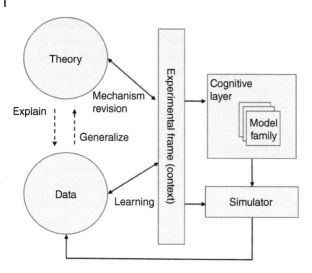

Figure 24.1 Models as mediators.

by selectively imparting newly observed data into the running simulation to make it congruent with its environment. The ability to guide measurement and instrumentation processes is critical when the measurement space is large. By narrowing measurement within a subset of the overall observation space, the methodology reduces both the cost and the time to test while also improving the relevance of data. Viewing coupled model-experiment system dynamics as a DDDAS requires advancements in variability management, model interfaces to instrument the simulation, and facilities to incorporate data-driven inference back into the model's technical space. As such, the provision of run-time models and dynamic model updating is necessary for closing the loop.

Challenges

The adaptive nature of models highlighted above is consistent with using "models as mediating instruments" (Morrison and Morgan 1999). According to this view, a model needs to align theoretical and empirical constructs to converge to an explanation of expected behavior. The integration of the ingredients of the model toward aligning theory and data suggests the autonomous character of a model, measured as a function of the degree of coherence of the competing assumptions, mechanisms, and evidential observations. However, models should not merely mediate existing theoretical and empirical elements. Model developers introduce new features and constructs – not only to represent behavioral mechanisms but also to integrate them in ways that are logical and consistent. Also, to facilitate the acquisition of knowledge, models should be connected with a valid and objective relation to the phenomena while also affording cognitive access to the information it contains to support comprehension and explanation (Ross et al. 2016). Next, we examine challenges in developing model abstractions and cognitive aids to support critical pillars of the perspective outlined.

Model Abstractions

Computer simulations have become instruments of epistemic inquiry in a wide range of domains spanning from natural to social and artificial sciences (Mittal et al. 2017). Further developments in the computational modeling of the information processes that underlie scientific phenomena will require advancements including but not limited to problem formulation, communication, and complexity management. Table 24.1 outlines specific categories of challenges that need to be addressed to better respond to requirements of model-driven scientific activities.

Cognitive Assistance in Modeling

Cognitive computing is concerned with computational models and techniques that study the human mind and simulate thought processes (or proxies), including learning, decision-making, planning, problem-solving, reasoning, and explanation. Such cognitive models amplify modelers' capabilities to formulate questions, generate hypotheses, plan and devise experiments, and analyze results to facilitate learning and revision of both the models and the experiments. Hence, a *cognitive model* is a model that, in the process of experimentation and comparison with data, assists in formulating questions, generating hypotheses, finding casual relationships, planning and devising experiments, and analyzing results to allow improvement of both models and experiments. To this end, we highlight five significant modes of cognitive assistance that can contribute to a model-driven discovery cycle. Support in the form of generative design and execution of experiments as well as model synthesis and revision enables modelers to focus on the objectives of the study rather than platform management.

- *Automation*: Orchestrating simulation experiments and collecting data require transparent access and instrumentation. Experiments should be coupled to the model representation to improve observability while also supporting controllability of simulations. That is, experiments can be defined in a way to facilitate instrumentation and monitoring of simulations seamlessly. Aspect-oriented model programming is a potential strategy to implement such coupling between models and experiments. The data collected through experimentation need to be abstracted into evidence, patterns, and regularities to evaluate hypotheses. Inductive learning and generalization methods, followed by the application of formal deductive methods such as probabilistic model checking (Doud and Yilmaz 2017), can help determine whether or not the hypothesized assumptions lead to the

Table 24.1 Abstraction to assist cognition.

	Issues and challenges	
Based on	**Type of issue**	**Characterization**
Information abstraction	Conceptualization	Algorithmic abstractions of natural entities, relations, and processes need to be based on and generated from the domain terminology
	Communication	Simulation models and the outputs of computational discovery systems need to be transparently communicated to domain scientists
	Formalization	Formal methods such as model checking need to be leveraged to assure that models not only are syntactically well formed but also adhere to semantic constraints of the hypothesized mechanistic behavioral mechanisms
	Understanding	Discovery systems need to produce models that go beyond description to also provide explanations of evidence in terms of generative mechanisms
	Complexity	Scientific phenomena include multiple facets and aspects that require provision of distinct features and formalisms suitable for each aspect
Cognitive assistance	Automation	Execution and orchestration of simulation experiments, possibly on distributed platforms, and collecting data to evaluate hypotheses
	Explanation	Produce explainable models and present results via an explanation interface, which is guided by cognitive and psychological theories of effective explanations
	Decision support	Data collected through simulation experiments need to be abstracted into evidence, patterns, and observed regularities for the purpose of evaluating hypotheses
	Learning	Model learning involves provision of support for model and experiment updating so as to discriminate among competing hypotheses
	Discovery	Mechanistic hypotheses need to be revised, and these revisions, including new behavioral mechanisms, should be transferred into the model's technical space

desired behavior or expected regularity. Exploratory searching for plausible mechanistic hypotheses can leverage heuristic search techniques, as well as abductive, probabilistic (Pearl 2014), and analogical reasoning methods to postulate model revisions.

- *Decision support*: Testing the consequences of hypothesized causal models requires deciding which variables to measure, as well as designing and prioritizing experiments to improve information gain by reducing uncertainty across hypotheses while adhering to principles of reliable and valid design of experiments. Cognitive assistance can also help determine the marginal utility of plausible experiments and provide support for comparing alternatives.

- *Learning*: The discovery system needs to support learning from experimentation to discriminate among competing hypotheses. For instance, statistical machine learning techniques can generate causal probabilistic networks among hypotheses and evidence to provide a quantitative explanatory framework.

- *Explanation*: The output of existing learning techniques is difficult to communicate to disciplinary scientists and other modelers. Advances in cognitive systems are necessary to produce explainable models and present results via an explanation interface, which is ideally guided by cognitive and psychological theories of effective explanations. Alternative machine learning models could be developed to learn structured, interpretable, causal models. Moreover, model induction techniques could be used to infer meta-models that provide approximate explainable models of causal dependencies among the scientific knowledge structures.

- *Discovery*: Following the analysis of the results, if necessary, mechanistic hypotheses could be revised, and these revisions, including new behavioral mechanisms, need to be transferred into the model's technical space to start a new cycle of experimentation. The revision of experiment models brings focus and generates new plans. Alternative experiment models, guided by current goals, could be explored until when new goals emerge due to the revision of hypotheses as well as the availability of new evidence.

A Generic Architecture for Models as Cognitive Autonomous Agents

As an active entity, models with cognitive capabilities provide features that overlay the simulation model and augment the model to support computational discovery. In this view, a model is construed as a family of models, which evolve as learning takes place. As shown in Figure 24.2, models need to be designed with a variability management layer (VML) to support seamless customization and to address a variety of experiment objectives, especially when the target system has multiple facets (Pohl and Metzger 2006).

Besides highlighting the significance of a VML, Figure 24.2 illustrates the building blocks of an active model that is coupled to an experimentation environment to maintain mutually beneficial and adaptive feedback between theory and data. Theoretical constructs are characterized by the features defined in the VML and the models that encapsulate causal hypotheses, principles,

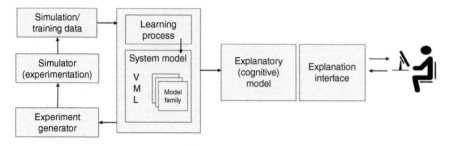

Figure 24.2 Reference architecture.

and constructs underlying the theory. A family of models is defined in terms of features (Kang et al. 2002; Oliveira et al. 2013) that can be configured to synthesize alternative models. Features define mandatory or optional standard building blocks of a family of models along with their interrelationships. Variability management via feature models is a common strategy in the product-line engineering practice.

In support of such a process, a variability management language should enable the specification and weaving of variants in the host simulation modeling language. The selection of such variants in the form of features triggers specific actions that customize a model by adding, removing, or updating model fragments. For instance, FeatureSim (Yilmaz 2017) is a simulation programming language that implements the variation management aspect of the reference architecture. The objective of FeatureSim is to provide an agent simulation environment that allows defining agents in terms of reconfigurable features. By defining features separately and composing them to build a variety of models, experiments can be extended to study the implications of structural and behavioral variations. This opens new avenues, relative to common simulation experiments that focus only on sweeping the input parameter space of models.

As shown in Figure 24.3, a FeatureSim model is encapsulated within an explicit context model, which is constructed by the constraints of the experiment. By making the experiment model a first-class design entity in the overall architecture, simulation model developers are required to explicitly define the constraints under which a model can be explored while making the experiment reusable across a family of models. These similar but distinct models can be generated by updating the resolution model defined as part of the experiment model. The resolution specification refers to the selected features from the feature model that are pertinent to the simulation experiment. The composer evaluates the resolution to synthesize a composite feature that integrates the behaviors associated with the selected features. By varying the resolution model, the simulation can automatically generate a family of models so that the same experiment can be applied to distinct models.

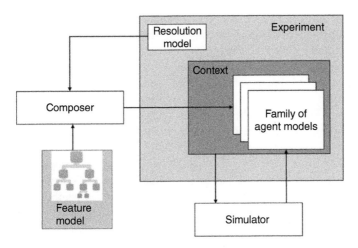

Figure 24.3 FeatureSim component architecture.

In a simulation study, according to the goal of the simulation, there are three major types of variation:

Data-dependent variation: Simulation with dynamic models generates numerous types of data that can be visualized and rendered by using different charts and statistical methods. Moreover, the objective of the experiment dictates which data sources to monitor to align instrumentation of the model and to facilitate derivation of the dependent variables consistent with the goal of the experiment.

Environment-dependent variation: The behavior of a model depends on the independent variables, factors, and their levels, as well as constraints that influence the behavior of agents. Such constraints include the structure of the topology that governs interaction among agents and the signals produced by the environment. Separation of the environment structure and behavior from the model behavior allows testing of the model under alternative interaction topologies and types of selection pressure induced by the environment.

Feature-dependent variation: Besides parametric variations based on environment-dependent factors, a model can be configured with alternative structures and behavioral activity flows (mechanisms) to evaluate the implications of different features or rules associated with agents.

Variable-structure models are necessary to synthesize alternative model representations to address different types of research questions or experiment objectives. Unlike the kinds of variation associated with the goal of a simulation study, which are used to vary the initial context of the model, *context-dependent run-time variation* is based on the recognition of an emergent behavior regarding structural and behavioral regularities exhibited by the model. FeatureSim provides an organizational structure to enable both feature-driven and environment-dependent variation. Feature variation is based on feature modeling strategy, which defines a model in terms features that are visible and prominent behavior, aspect, or quality characteristics of a system.

Extending variant management to online model tuning to support experimentation requires a run-time evaluation mechanism. To this end, an input plug-in transforms raw experiment data into a format that can be analyzed using the Evaluator/Critic component, shown in Figure 24.2. The evaluator can be as simple as a filter that abstracts the data. However, to provide cognitive assistance, a sophisticated model can facilitate coupling a model's technical space to formal methods such as probabilistic model checking (Kwiatkowska et al. 2011) to determine the extent to which the simulation data support the expected behavior. Those mechanisms that lend significant support to the expected behavior are retained, while others are revised or declined from further consideration.

The learning model uses the results of the evaluator to update the confidence levels of the competing hypotheses. For example, a Bayesian net can revise conditional posterior probability estimates using the Bayes' rule in terms of prior probabilities and the observed evidence. Alternatively, cognitive models such as explanatory coherence (Thagard 2016), which is briefly discussed in the section "Coherence-Driven Cognitive Model of Mediation," can be used to acquire, modify, reinforce, or synthesize hypotheses to steer the model-driven discovery process. This incremental and iterative strategy is coordinated by a mediation process that governs the interaction between the technical spaces of models and experiments. Next, we discuss the elements of such a process.

The Mediation Process

Building on (Klahr and Simon 1999), the proposed mediation strategy involves three main components that control the process from the initial formulation of hypotheses, through their experimental evaluation, to the decision that there is sufficient evidence to accept or reject hypotheses. The three major components are search model space, test hypothesis (search experiment space), and evaluate evidence. The output from the Search Model Space component is a set of fully specified mechanistic models, which serves as input to the Test Hypothesis phase that involves simulated experiments, resulting in evidence for or against the hypothesis. The Evaluate Evidence component decides whether the cumulative evidence warrants acceptance, rejection, or further consideration of the current hypothesis.

Search Model Space

Searching the model space requires identifying the general structure and the scope of the hypothesized model mechanism(s), followed by the refinement of the model structure to make it instantiable. In generating model mechanisms, strategies for evoking or inducing a template are needed. Template-based feature-oriented analysis (Meinicke et al. 2016) provides a basis for generative modeling. Specifically, strategies that view models as algebras (Oliveira et al. 2013) allow flexible composition of models in terms of features. Evoking an existing template requires searching a frame to abstract away unnecessary imperative programmatic constructs. In cognitive science, several mechanisms (Harman 2008) have been proposed to account for initial templates and to build the search space incrementally. DSL-guided analogical coherence, heuristic search, and abductive reasoning for evoking mechanisms are plausible strategies. When it is not possible to reuse a template, a new mechanism can be synthesized in the form of a meta-model by generating expected behavioral outcomes and by generalizing over the results to conjecture a mechanism frame.

The second issue that needs to be addressed is the template instantiation process, which takes as input a partially instantiated mechanism and assigns specific parameter values to generate a fully specified hypothesized model. Theory of template-based modeling provides a path for using prior knowledge or particular experimental outcome. If there exist outcomes extracted from previous experiments, they can be reused in the new context. Alternatively, the Generate Outcome goal is used to produce empirical results solely to determine mechanism parameters via sensitivity and boundary analyses, facilitating the refinement of a partially defined hypothesis. Early in the course of experimentation, prior knowledge via analogical coherence can be used to assign values, whereas using experimental outcomes is more likely to be used in the later phases of the iterative discovery process. If all the identified values are tried and rejected, then the mechanism needs to be abandoned, and the process needs to trigger the Generate Mechanism goal.

The Generate Observation/Outcome goal appears multiple times in the goal hierarchy. The first appearance is when simulated outcomes are generated to induce a mechanism template, and the second is when the hypothesized mechanisms are instantiated. Each time the Generate Outcome goal is activated so will the Search Experiment Space goal, which will generate the experiments and should be able to focus on those aspects of the situation that the experiment is intended to elucidate. Discriminating among competing mechanisms is one of the critical functions of the Focus subgoal. One can use coherence maximization to address this issue (see the section "Coherence-Driven Cognitive Model of Mediation"). Once a focal aspect is identified, the Select Strategy subgoal can choose specific independent and control factors according to the search preferences implied by the Focus goal. Means–ends reasoning and heuristic search/optimization strategies are often used to sweep a state space and to steer the search process. The Choose and Set subgoal can assign specific values to independent and control variables to facilitate the application of the Conduct Experiment and Observe subgoals.

Search Experiment Space

This goal aims to generate an experiment that is appropriate for the current set of hypotheses being examined, to make a prediction by running the simulation experiment, and to match the outcome to expected behavior. Once the Search Experiment Space component produces an experiment, conducting the experiment involves execution of the simulation and may require distributing the replications across multiple machines to improve the performance. The Analyze (Compare) goal aims to describe the discrepancy between the expected behavior and the actual outcome. The comparison can involve statistical methods such as ANOVA analysis as well as formal methods that leverage model checking. For instance, the results from simulation replications can generate outcomes that can be generalized as a Markov model that can be formally verified against finite-state verification patterns. When completed, the Test Hypothesis goal generates a representation of evidence for or against the current hypotheses. The Evaluate Evidence component then uses this outcome.

Evaluate Evidence

This component aims to determine whether the cumulative evidence warrants the acceptance or rejection of competing hypotheses. Various criteria can be used to evaluate evidence and hypotheses. These include plausibility, functionality, parsimony, etc. In the absence of hypotheses, experiments can be generated by intelligently navigating the experiment space. During the Evaluate Evidence phase, three general outcomes are possible. The current set of coherent hypotheses can be accepted, rejected, or considered further. In the first case, the discovery process stops. If the hypothesis is rejected, then the system returns to the Search Hypothesis Space phase, which can trigger two possible activities. If the entire mechanism template (frame) is rejected, then the system must attempt to generate a new mechanism. If the Evoke Mechanism goal cannot be satisfied or is unable to find an alternative mechanism, then the system will recourse to the Induce Mechanism subgoal, which requires running simulation experiments to generate outcomes that can be generalized via induction. Having induced a new mechanism frame, or having returned from Evaluate Evidence with a frame needing revised instances with new values, the system resumes with Template Instantiation. If prior knowledge is not applicable or available, here, too, the system may require running experiments to generate outcomes and to make value assignments.

Coherence-Driven Cognitive Model of Mediation

For instilling trust in the cognitive assistance system, the accumulation of evidence over hypothesized models needs to be objectively scrutinized in a transparent, explainable manner. In this context, the explanation model can serve dual purposes. Besides informing the user about the utility of competing model mechanisms, it can serve as a run-time model to guide the revision and selection of hypotheses, identifying which mechanisms to focus, and what experiments to generate to differentiate between competing models. The users of an explainable cognitive model are end users who depend on decisions and recommendations generated by the simulation and therefore need to understand the rationale for the decisions. For example, an engineer who receives recommendations needs to know why the cognitive model recommends specific hypothesized mechanisms over others or particular experiments for further investigation. In seeking a balance between theoretical principles, evidence, system constraints, and mechanisms, a strategy is needed to attain a state of coherent justification that accounts for the recommended mechanisms. Advances are needed in terms of principles and requirements and interactions with such explainable cognitive models, along with strategies, which can contribute to developing explainable cognitive models. The options available include deep learning explanation, interpretable models, and model induction.

Theory of reflective equilibrium: For illustration purposes, consider Figure 24.5, which depicts a connectionist network comprised of observed or expected system behaviors or requirements that aim to explain these observations, and model mechanisms that facilitate addressing the requirements at different levels of resolution. Furthermore, suppose that the system under investigation is a self-organizing system, and the mechanisms are well-known patterns such as gradient-field, digital pheromones, as well as low-level supportive mechanisms such as tag-based, token-based, or market-based coordination (De Wolf and Holvoet 2006) with their specific template structures.

Modeling cognitive coherence: The hypothetical network shown in Figure 24.4 evolves as learning takes place through experimentation. In this example, M_1 and M_2 together facilitate R_1, and therefore, they are compatible, whereas, M_3 provides an alternative to M_1 and hence contradicts it. That is, they do not cohere. Similarly, tag-based coordination mechanism, $M_{1.1}$, and token-based coordination, $M_{1.2}$, are alternative strategies that facilitate or contribute to the higher-level mechanism, M_1. Therefore, they compete. If the network is viewed as a connectionist network with node activations depicting the acceptability of mechanisms and the strength of links representing the degree of coherence among nodes, we can start developing insight into the evaluation of hypothesized mechanisms. The strategy can lend support to providing

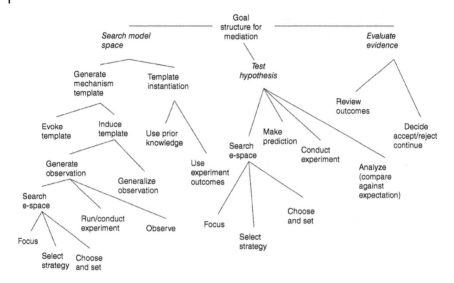

Figure 24.4 The goal structure of mediation.

a coherence-driven justification of decisions made in model discovery and design. If we can view equilibrium in such a network as a stable state that brings design conflicts to a level of resolution, the equilibrium state serves as a coherence account of model justification. As indicated by Daniels (1996) and Thagard (2016), an optimal equilibrium can then be attained when there is no further inclination to revise decisions, because together they have the highest degree of acceptability.

The decisions that one arrives at when the equilibrium is reached can provide an account for the solution under the system requirements examined. In relation to our example shown in Figure 24.5, resolving conflicts among hypotheses from uncertainty to a stable, *coherent state of equilibrium* can be characterized as follows. Consider a finite set of cognitive elements e_i and two disjoint sets, C^+ of positive constraints, and C^- of negative constraints, where a constraint is specified as a pair $(e_i; e_j)$ and weight w_{ij}. The set of cognitive elements is partitioned into two sets, accepted set (AS) and rejected set (RS), and $w(A; R)$ is defined as the sum of the weights of the satisfied constraints among them. A satisfied constraint is defined as follows: (i) if $(e_i; e_j)$ is in C^+, then e_i is in AS if and only if e_j is in AS. (ii) If $(e_i; e_j)$ is in C^- if and only if e_j is in RS. When applied to practical and theoretical reasoning, these elements represent assumptions, goals, and propositions. The coherence problem can then be viewed partly as a parallel constraint satisfaction problem, where the goal is satisfying as many constraints as possible by propagating a wave of activations while taking the significance (i.e. weights) of the constraints into consideration.

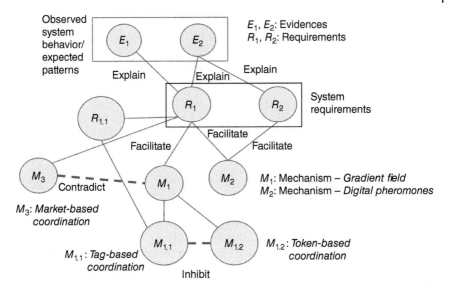

Figure 24.5 Connectionist constraint network.

With these observations, coherence can be adopted in terms of the following premises to support model discovery.

Symmetry: Explanatory coherence is a symmetric relation, unlike conditional probability.

Explanation: (i) A hypothesized model mechanism coheres with what it explains/supports, which can either be evidence/requirement or another hypothesis; (ii) hypotheses that together support some other requirement cohere with each other; and (iii) the more hypotheses it takes to satisfy a requirement, the lower the degree of coherence.

Analogy: Similar hypotheses that support similar requirements cohere.

Evidence/requirement priority: Propositions that describe evidence and requirements have a degree of acceptability on their own.

Contradiction: Contradictory propositions are incoherent with each other.

Competition: If P and Q both support a proposition, and if P and Q are not explanatorily connected, then P and Q are incoherent with each other. P and Q are explanatorily connected if one explains the other or if together they explain something.

Acceptance: The acceptability depends on the coherence of a proposition with the rest of the accepted propositions.

A provisional strategy for determining the state of equilibrium via coherence maximization: Given this characterization of coherence, the activations of nodes in the network are analogous to the acceptability of the respective propositions. This strategy is similar to viewing the state as a unit-length vector in an N-dimensional vector space. Each node receives input from every other node that it is connected with. The inputs can then be moderated by the weights of the link from which the input arrives. The activation value of a unit is updated as a function of the weighted sum of the inputs it receives. The process continues until the activation values of all units settle. Formally, if we define the activation level of each node j as a_j, where a_j ranges from -1 (rejected) and $+1$ (accepted), the update function can be defined as follows:

$$a_j(t+1) = \begin{cases} a_j(t)(1-0) + \text{net}_j(M - a_j(t)), & \text{if net}_j > 0 \\ a_j(t)(1-0) + \text{net}_j(a_j(t) - m), & \text{otherwise} \end{cases}$$

In this rudimentary formulation, the variable θ is a decay parameter that decrements the activation level of each unit at every cycle. In the absence of input from other units, the activation level of the unit gradually decays, with m being the minimum activation, M denoting the maximum activation, and net_j representing the net input to a unit, as defined as $\sum_i w_{ij} a_i(t)$. These computations can be carried out for every node until the activation levels of elements stabilize and the network reaches an equilibrium via self-organization. Nodes with positive activation levels at the equilibrium state can be distinguished as maximally coherent hypotheses. For experimentation purposes, the design of the network can be calibrated or fine-tuned to alter the links' weights, which represent the significance of the constraints.

Conclusions

Following the characterization of the nature of model-driven activities, the issues and challenges in model processing and cognitive activities involved in model-based scientific problem-solving are delineated. These challenges resulted in the formulation of a reference architecture that views mediator models as autonomous adaptive agents with learning capabilities. The premise of the strategy is based on the observed need for mediating between theory and data to facilitate their coherent integration. By aligning the mechanistic hypotheses or assumptions of a model with the empirical evidence or expected regularity, the mediation process aims to facilitate the process of computational discovery by facilitating a search process across the technical spaces of models and experiments. We highlighted that such mediation requires flexibility in generating and evolving multiple models, and the underlying generative process continues until it converges to one or more competing and complementary models that can usefully and convincingly explain the systemic properties of the phenomena of interest.

References

Bunge, M. (1998). *Philosophy of Science: From Problem to Theory*, vol. 1. Transaction Publishers.

Daniels, N. (1996). *Justice and Justification: Reflective Equilibrium in Theory and Practice*, vol. 22. Cambridge University Press.

Darden, L. (2001). Discovering mechanisms: a computational philosophy of science perspective. In: *Discovery Science*, Lecture Notes in Computer Science, vol. 2226 (ed. K.P. Jantke and A. Shinohara), 3–15. Berlin, Heidelberg: Springer http://link.springer.com/chapter/10.1007/3-540-45650-3_2.

Darema, F. (2004). Dynamic data driven applications systems: a new paradigm for application simulations and measurements. In: *International Conference on Computational Science*, Lecture Notes in Computer Science (LNCS), vol. 3038, 662–669. ACM.

Davis, P.K., O'Mahony, A., Gulden, T. et al. (2018). Improving Social-Behavioral Modeling. In: *Priority Challenges for Social-Behavioral Research and its Modeling*. Santa Monica, CA: RAND Corporation.

De Wolf, T. and Holvoet, T. (2006). Information flows for designing self-organizing emergent systems. In: *Proceedings of the Joint Smart Grid Technologies (SGT) and Engineering Emergence for Autonomic Systems (EEAS) Workshop*, 22–29. IEEE.

Doud, K. and Yilmaz L. (2017). A framework for formal automated analysis of simulation experiments using probabilistic model checking. *Proceedings of the 2017 IEEE/ACM Winter Simulation Conference*, Las Vegas, NV (December 3–6), 1312–1323.

Harman, G. (2008). *Change in View: Principles of Reasoning*. Cambridge University Press.

Kang, C.K., Lee, J., and Donohoe, P. (2002). Feature-oriented product line engineering. *IEEE Software* 19 (4): 58.

Klahr, D. and Simon, H.A. (1999). Studies of scientific discovery: complementary approaches and convergent findings. *Psychological Bulletin* 125 (5): 524–543. https://doi.org/10.1037/0033-2909.125.5.524.

Kwiatkowska, M., Norman, G., and Parker, D. (2011). Prism 4.0: verification of probabilistic real-time systems. In: *International Conference on Computer Aided Verification*, 585–591. Springer.

Langley, P. (2000). Computational support of scientific discovery. *International Journal of Human-Computer Studies* 53 (3): 393–410. https://doi.org/10.1006/ijhc.2000.0396.

Levins, R. (1966). The strategy of model building in population biology. *American Scientist* 54 (4): 421–431.

Meinicke, J., Thum, T., Schroter, R. et al. (2016). FeatureIDE: taming the preprocessor wilderness. In: *Proceedings of the 38th International Conference on Software Engineering Companion*, 629–632. ACM.

Mittal, S., Durak, U., and Oren, T. (2017). *Guide to Simulation-Based Disciplines: Advancing our Computational Future*. Springer.

Morrison, M. and Morgan, M.S. (1999). Models as mediating instruments. In: *Models as Mediators*, vol. 52, 10–37.

Oliveira, B.C.T., Loh, A., and Cook, W.R. (2013). Feature-oriented programming with object algebras. In: *European Conference on Object-Oriented Programming*, 27–51. Springer.

Pearl, J. (2014). *Probabilistic Reasoning in Intelligent Systems: Networks of Plausible Inference*. Morgan Kaufmann.

Pohl, K. and Metzger, A. (2006). Variability management in software product line engineering. In: *Proceedings of the 28th International Conference on Software Engineering*, 1049–1050. ACM.

Ross, A.M., Fitzgerald, M.E., and Rhodes, D.H. (2016). Interactive evaluative model trading for resilient systems decisions. *14th Conference on Systems Engineering Research*, Huntsville, AL.

Thagard, P. (2016). Emotional cognition in urban planning. In: *Complexity, Cognition, Urban Planning and Design* (ed. J. Portugali and E. Stolk), 197–213.

Yilmaz, L. (2017). FeatureSim: feature-driven simulation for exploratory analysis with agent-based models. In: *Proceedings of the IEEE/ACM 21st International Symposium on Distributed Simulation and Real Time Applications (DS-RT)* (18–20 October 2017), 163–170. Rome: IEEE.

Yilmaz, L., Chakladar, S., and Doud, K. (2016). The goal-hypothesis-experiment framework: a generative cognitive domain architecture for simulation experiment management. In: *Proceedings of the 2016 Winter Simulation Conference*, 1001–1012. ACM.

Yilmaz, L., Chakladar, S., Doud, K. et al. (2017). Models as self-aware cognitive agents and adaptive mediators for model-driven science. In: *Proceedings of the 2017 IEEE/ACM Winter Simulation Conference* (3–6 December 2017), 1300–1311. Las Vegas, NV: ACM.

25

Causal Modeling with Feedback Fuzzy Cognitive Maps

Osonde Osoba[1] and Bart Kosko[2]

[1] *RAND Corporation and Pardee RAND Graduate School, Santa Monica, CA 90401, USA*
[2] *Department of Electrical Engineering and School of Law, University of Southern California, Los Angeles CA 90007, USA*

Introduction

Social scientific theories can benefit from computational models even when the relevant social scientific concepts and phenomena can be hard to quantify. This chapter describes a powerful tool for modeling causal relationships: fuzzy cognitive maps (FCMs). FCMs offer a practical and flexible way to model and process the interwoven causal structure of policy and decision problems.

Existing tools for causal modeling include Bayesian belief networks (BBNs) and system dynamics (SD) models. FCMs have some notable advantages over the default models of causality: FCMs model causal loops naturally, they combine separate knowledge sources into a single FCM, and they permit fast pattern inference and data-driven adaptation at low computational cost. FCMs also lend themselves particularly well for modeling ambiguous elicited causal beliefs because of their use of fuzz. A flexible and robust modeling approach for capturing causal beliefs is important for social and behavioral models since beliefs are foundational to many observed behaviors.

The next sections describe FCMs briefly. We conclude with two FCM policy applications. The first shows how FCMs can assist in modeling the many causal and policy factors involved in public support for insurgency and terrorism (PSOT). The second shows how an FCM model can give insight into Allison's recent *Thucydides' trap* model of US–China conflict. We draw on a recent journal paper that includes more technical details (Osoba and Kosko 2017).

Social-Behavioral Modeling for Complex Systems, First Edition.
Edited by Paul K. Davis, Angela O'Mahony, and Jonathan Pfautz.
© 2019 John Wiley & Sons, Inc. Published 2019 by John Wiley & Sons, Inc.
Companion website: www.wiley.com/go/Davis_Social-Behavioralmodeling

Overview of Fuzzy Cognitive Maps for Causal Modeling

This section gives a nontechnical overview of FCMs.

FCMs allow users to quickly draw causal diagrams of complex social or other processes (Kosko 1986b). These causal diagrams can have closed loops or paths in them. The closed loops directly model feedback among the causal concepts or nodes (as in Figure 25.1). They are fuzzy because the causal arrows in the diagrams admit degrees or shades of gray. Users can make what-if predictions with a given FCM: What happens given this input policy? The predictions are not numerical predictions. They are pattern predictions or repeating sequences

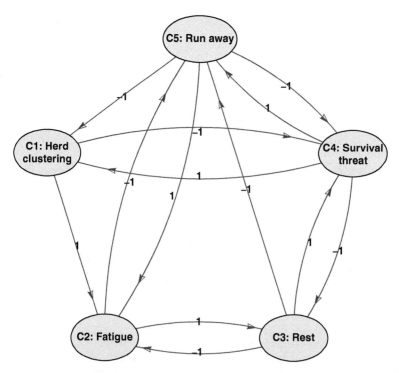

Figure 25.1 Fragment of a predator–prey fuzzy cognitive map that describes dolphin behavior in the presence of sharks or other survival threats (Dickerson and Kosko 1994). The edges in this FCM are trivalent: $e_{ij} \in \{-1, 0, 1\}$. Each nonzero edge defines a causal if-then rule: the dolphin pod decreases its resting behavior if a shark or other survival threat is present. But the survival threat increases if the pod rests more. These two causal links define a minimal cycle or feedback loop within the FCM's causal web. Such feedback cycles endow the FCM with transient and equilibrium dynamics. All inputs produce equilibrium limit cycles or fixed-point attractors in the simplest case when all nodes are bivalent threshold functions and when the system updates all nodes at each iteration. (Dickerson and Kosko 1994). © 1994 by the Massachusetts Institute of Technology, reprinted courtesy of the MIT Press.

of events. Users can also ask *why* questions in some case: Why did these events occur? Different users can also fuse or combine their FCM diagrams into a unified FCM. These FCM techniques are quite general and have led to numerous applications (Glykas 2010; Papageorgiou 2013; Papageorgiou and Salmeron 2013; Amirkhani et al. 2017). A recent notable application uses an FCM to make sense of the impact of "Brexit" on food, energy, and water systems in the UK (Ziv et al. 2018).

An FCM uses an arrow or directed graph *edge* to describe how one concept *node* causally affects another concept node. So an FCM represents causality as a directed edge in graph. An FCM model of military deterrence might include concept nodes for the degree of national military capability and for the degree of a threat's credibility. The directed causal edges among two or more such concept nodes define a directed graph or *digraph*. A key aspect of FCMs is that in general their digraphs contain cycles or feedback loops. This means that most FCMs are not directed *acyclic* graphs (DAGs).

Fuzz

Fuzz is a term of art in logic and engineering. It denotes degree of truth or degree of causality. Fuzzy logic extends binary logic to multivalued logic and allows rule-based approximate reasoning (Zadeh 1965; Kosko and Isaka 1993). FCMs are fuzzy both in their causal edges and often in their concept nodes. Fuzzy causal edges denote partial causality. All-or-none causality can still occur but only as the end points of the spectrum of causal influence. The same holds for the activation of a concept node. Concept nodes are often binary in practice and in the examples below. More sophisticated FCM models use fuzzy gray-scale concept nodes and may also use time lags.

Feedback loops in an FCM imply that an FCM is a dynamical system. Inputs stimulate the FCM system. The causal activation then technically swirls around in the FCM forever or until a new input perturbs the system. Most FCMs quickly reach an equilibrium. The equilibrium serves as the system's what-if prediction or forward inference from the input. FCMs with binary nodes usually converge to a limit cycle or a repeating sequence of states. The cycle of states is a form of pattern prediction.

Comparison with Other Methods

Two other approaches to modeling causal worlds are system-dynamics SD models (Sterman 2000; Abdelbari and Shafi 2018) and BBNs (Pearl 2009).

SD models allow users to represent and simulate causal interactions within relatively complex systems. SD has a long history of modeling practice. The full picture on the relative modeling strengths of FCM and SD models may take some time to emerge. But we can point out a few key advantages of FCMs. For example SD models are statically parameterized. This means they are not easily adapted to available empirical data (although they can use probability distributions for some variables; Granger and Henrion 1992). Domain expertise or random experimentation often chooses the parameters of the subsystems and their interconnections. FCMs admit both data-driven and expert-driven adaptation of the model structure *and* the model parameters. Statistical learning algorithms estimate causal edges from training data. Experts can also state edge values directly. SD models can account for uncertainty by doing sensitivity testing or Monte Carlo analysis. FCMs build uncertainty into their very structure.

BBNs represent uncertain causal worlds with conditional probabilities. A user must state a known joint probability distribution over all the nodes of the directed graph. Forward inference on a BBN tends to be computationally intensive. The directed graph is usually *acyclic* graph and thus has no closed loops. The acyclic structure simplifies the probability structure but ignores feedback among the causal units. This is an important limitation for social and behavioral models that often feature causal loops (such as preferential attachment in social networks). The acyclic constraint also makes it hard to combine BBN causal models from multiple experts because such combination may well produce a cycle. FCMs contain causal loops by design. Combining FCMs only tends to increase the loop or feedback structure and therefore produces richer dynamics. It also tends to improve the modeling accuracy in many cases. FCM forward-inference process is also simple and fast because it uses only vector–matrix multiplication and thresholding. But BBNs do give precise numerical probability descriptions and not the mere pattern predictions of FCMs. So FCMs trade numerical precision for pattern approximation, faster and scalable computation, and richer feedback representation.

FCMs offer substantial value for social scientific modeling and simulation. FCMs can capture the causal beliefs of domain experts. FCM dynamics can reveal global *hidden patterns* in small or large causal webs. These patterns tend to far from obvious when examining a large-scale FCM model. The domain expert expresses his beliefs about how relevant factors causally relate to one another. Some examples of such expert-based FCMs include public support for terrorism (Osoba and Kosko 2017), blood-clotting reactions (Taber et al. 2007), and medical diagnostics (Stylios et al. 2008). An analyst can also convert documents into FCMs. We show below an example translating Graham Allison's textual descriptions of his proposed Thucydides' Trap of international relations into a predictive FCM. FCMs can also apply driving agents' behaviors or decisions in agent-based models (ABMs). Prior work (Dickerson and Kosko 1993) shows how lone or combined FCMs can govern the behavior of agents or virtual actors. These behaviors can be simple or complex depending on the driving FCM. Applying learning laws to the agents' FCMs can simulate the effect of agents learning new behaviors (Figure 25.3).

Inference with FCMs

An FCM tends to have many cycles or closed loops in its fuzzy directed graph. These cycles directly model causal feedback from self-loops to multipath causality. The cycles also produce complex nonlinear dynamics. FCM causal inference maps input states or policies to equilibria of the nonlinear dynamical system. Users can also step through time slices of the FCM dynamical system to at least partially unfold the system in time.

An FCM's feedback structure contrasts with the acyclic structure of BBNs. BBNs form the basis of Pearl's popular model of causal inference (Pearl 2009). Rubin's counterfactual approach to causality is a related statistical model (Rubin 2005; Imbens and Rubin 2015). BBNs for causal inference assign probabilities to a DAG. Their acyclic causal tree structure rules out feedback pathways. This strong acyclic assumption greatly simplifies the probability calculus on such digraphs and may permit finer control when propagating probabilistic beliefs. But the acyclic structure is hard to reconcile with the inherent and extensive feedback causality of large-scale social phenomena from social networks to state-vs.-state wars. These social systems are high-dimensional nonlinear dynamical systems. They have dynamics because they have feedback loops.

Probabilistic inference on a graph computes the conditional probability $P(C_O | C_{Ev})$ of output nodes C_O given a state vector on observable evidence nodes C_{Ev}. This computation involves a complex marginalization operation on general DAGs. It also assumes that the user knows the closed-form joint probability distribution on all the nodes. The computation often requires complex message-passing algorithms such as belief propagation (Yedidia et al. 2001; Murphy 2002) or the more general junction-tree algorithm (Wainwright and Jordan 2008). This probabilistic computation is NP-hard in general (Dagum and Luby 1993; Russell et al. 2016).

Loops or cycles in a causal graph model may render exact probabilistic inference difficult if it is even feasible. Inexact or *loopy* inference schemes can give useful approximations in many cases. These methods include loopy belief propagation, variational Bayesian methods, mean field methods, and some forms of Markov chain Monte Carlo simulation (Beal and Ghahramani 2003, 2006; Wainwright and Jordan 2008; Pearl 2014). But loopy algorithms are still NP-hard and may not even converge.

FCM forward inference uses light computation. It requires only vector–matrix multiplication and nonlinear transformations of vectors. The transformations are often hard or soft thresholds. So FCM forward inference has only polynomial-time complexity. This means that FCMs scale fairly well to problems with high dimension or multiple concept nodes. Simple binary-state FCMs converge quickly to limit-cycle equilibria given an input stimulus (Kosko 1991). FCMs with more complex node nonlinearities can converge to aperiodic or chaotic equilibria if sufficiently complex nonlinearities describe the concept nodes.

FCMs do have at least two structural limitations. The first is that a user may not be able to use some predicted outcomes. A user may find it hard to interpret an FCM's what-if predictions because they are equilibria of a highly nonlinear dynamical system. Simple predictions may be limit cycles that consist of only a few ordered system states. The dolphin FCM below is one such case. One equilibrium output consists of four binary vectors that repeat in sequence. Richer dynamics can produce aperiodic or chaotic equilibria in regions of the FCM state space. Such predicted outcomes may have no clear policy interpretation if we cannot clearly associate the equilibrium attractor's region of the FCM state space with a temporally ordered sequence of FCM states.

A more fundamental limitation is that FCMs do not easily answer why questions given observed outcomes. FCMs do not easily admit *backward inference* from effects to causes because FCM nodes are neural-like nonlinear mappings of causal inputs to outputs. A user may need to test a wide range of random inputs to see which FCM states map to or near a given observed or conjectured output state. We often check all possible input policy states to find all output equilibria. We did this with Thucydides' trap FCM below by *clamping on* input policy variables while the FCM dynamical system converged. We also tested this FCM's pattern predictions against those of a thresholded FCM whose edge values were the trivalent extremes of $-1, 0,$ or 1. The thresholded FCM gave similar equilibrium predictions.

Combining Causal Knowledge: Averaging Edge Matrices

Causal modeling faces a threshold epistemic question when dealing with multiple experts: How do we combine the causal models of multiple knowledge sources? Such knowledge fusion is a key function in many decision-making processes (Kosko 1986c, 1988; Taber 1991; Taber et al. 2007; Davis et al. 2015a,b).

A common answer avoids the question by combining the causal knowledge or expertise before it enters a causal model. Some form of this knowledge preprocessing occurs with AI search trees and other DAG models. Multiple knowledge sources may lead a knowledge engineer to draw or otherwise modify a weighted causal arrow in a model. That differs from first letting each source have its own causal arrow and then combining. This fit-all-in-one-model approach may work well for problems of small dimension or small expert sample size. Even then it may obscure the disparate knowledge that went into the representation. But it can ensure that a causal DAG stays a DAG as it encodes new information. The approach can become more ad hoc and restrictive as the expert sample size m grows. The likelihood of getting a causal cycle only increases with expert sample size and the node count of the model.

FCMs answer the epistemic question directly: average the causal FCMs of each expert (Kosko 1988, 1991; Taber 1991). Preprocessing can still occur. But there is no limit to the number m of FCMs that averaging can combine. FCMs naturally combine into a new FCM. So FCM combination is a closed graph-theoretical operation. And the resulting FCM has all the representative properties of a sample average.

The loop-free structure of BBNs' DAGs makes it hard to do this sort of combination of causal models from multiple experts. Combining DAGs need not produce a new DAG. So combining DAGs is not a closed graph-theoretical operation in general. Some experts will tend to draw opposing causal arrows between nodes. Others will tend to add links that create multi-node closed loops.

The FCM average forms a mixture or convex combination of the causal edges. The user can combine any number of FCMs by adding their underlying augmented adjacency matrices. A group of m experts can each produce an FCM causal edge matrix $\mathbf{E}_k$ that describes some fixed problem domain. Each expert can model different concept and policy nodes. The total number of nodes is n. Augment the edge matrices with zero rows and columns for any missing nodes in an expert's causal edge matrix. Then FCM knowledge fusion or combination takes the weighted average of their augmented causal edge matrices:

$$\overline{\mathbf{E}}_m = \sum_{k=1}^{m} w_k \mathbf{E}_k \tag{25.1}$$

where the weights w_k are convex weights and hence nonnegative and sum to one.

The weights w_k can reflect relative expert credibility in the problem domain. They can reflect test scores or subjective rankings or some other measure of the experts' predictive accuracy in prior experiments. The same weight w_k need not apply to the entire kth FCM edge matrix. Each edge value can have its own weight. So a weight matrix W_k corresponds to each expert's FCM edge matrix. Predd et al. (2008) developed a method for combining expert inputs when the experts abstain or when they are incoherent. Voting schemes (Conitzer et al. 2009; Caragiannis et al. 2013) might also pick the FCM weights and affect the fusion process. We here take the weights as given and use equal weights as a default.

The m edge matrices $\mathbf{E}_k$ in (25.1) must be conformable for addition. So they must have the same number of rows and columns and in the same matrix positions. So we first take the union of all concept nodes from all m knowledge sources. This again gives a total of n distinct concept nodes. Then we zero-pad or add rows and columns of zeros for missing nodes in a given knowledge source's causal edge matrix. This gives a conformable n-by-n signed fuzzy adjacency matrix $\mathbf{E}_k$ after permuting rows and columns to bring them in mutual coincidence with all the other zero-padded augmented matrices.

The strong law of large numbers gives some guarantees about the convergence of this fusion knowledge graph to a representative population FCM if the knowledge sources are approximately statistically independent and identically distributed and if they have finite variance (Kosko 1988; Taber et al. 2007). Then the weighted average in (25.1) can only reduce the inherent variance in the expert sample FCMs. So the knowledge fusion process improves with sample size m. Simulations have shown that the equilibrium limit cycles of the combined FCM tend to resemble the limit cycles of the m individual FCMs (Taber et al. 2007). An expert random sample is sufficient for this convergence result but not necessary. A combined FCM may still give a representative knowledge base when the expert responses are somewhat correlated or when the experts do not all have the same level of expertise or problem-domain focus. Users can also use policy articles or books or legal testimony as proxy experts.

Figure 25.2 shows the minimal combination case where two FCMs fuse into one representative FCM. The mixture or convex combination of FCMs creates a new fused FCM as the weighted averages of the FCMs' augmented signed fuzzy adjacency matrices. Users can add new concept nodes or factors at will. Each new factor converts all m n-by-n edge matrices into $n + 1$-by-$n + 1$ edge matrices. This again amounts to adding a new zero-padded row and column to an edge matrix if its corresponding FCM does not include the factor as a concept node. An expert has a zero row and column for a concept node if the expert impliedly states that concept is not causally relevant.

This fusion averaging technique can reflect bad effects as well as any other effect. The technique can reflect anomalous effects due to active sabotage or extreme variance in expert opinions.

Learning FCM Causal Edges

We turn next to inferring the directed causal edges e_{ij} from time-series data.

Correlation does not imply causation. But some time-lagged correlations *suggest* causation. This is the idea behind the *unsupervised* learning laws below for estimating the directed causal edges e_{ij}. They are unsupervised because there is no teaching signal that the learning process matches against.

We can learn causal edge strengths through the *concomitant activation* among the factor pairs. This approach assumes that events (factor activities) are more likely to involve a causal connection if the events occur together (Hebb 1949; Kosko 1986a, 1991). This suggests the well-known Hebbian correlation learning law (neurons that fire together wire together) for training neural network synaptic weights (Kosko 1991):

$$\dot{e}_{ij} = -e_{ij} + C_i C_j \tag{25.2}$$

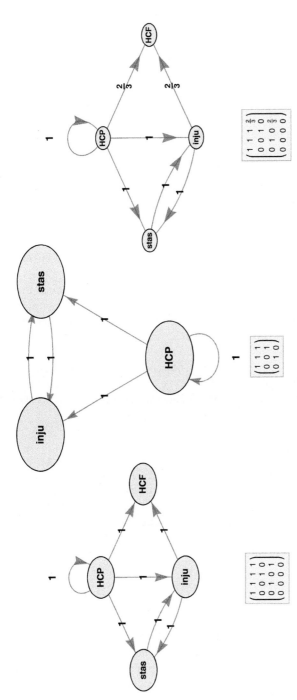

Figure 25.2 FCM knowledge combination or fusion by averaging weighted FCM adjacency matrices. The three digraphs show the minimal case of combining two FCMs that have overlapping concept nodes. The third FCM is the weighted combination of the first FCMs. The expert problem domain is the medical domain of strokes and blood clotting involved in Virchow's triad of blood stasis stas, endothelial injury inju, and hypercoagulation factors HCP and HCF (Taber et al. 2007). Expert 1 has a larger FCM than Expert 2 has because Expert 1 uses an extra concept node. FCM knowledge combination fuses their knowledge webs by averaging their causal edge adjacency matrices with the mixing equation (25.1). This weighted average uses each expert's causal edge matrix $E = \frac{2}{3}E_1 + \frac{1}{3}E_2$ as shown in the combined (third) FCM. The weighting assumes that the first expert is twice as credible as the second expert. Note that Expert 2 ignored the HCF factor. This results in a new row and column of all zeroes in the augmented edge matrix E_2.

where $\dot{x}$ denotes the time derivative of the signal x. The passive decay term $-e_{ij}$ stabilizes the learning in the differential equation model. It also models a *forgetting* constraint that helps the network prune inactive connections. The product term $C_i C_j$ directly models concomitant correlation.

We can instead use *concomitant variation* (Mill 1843) in time between factors as partial evidence of a causal relation between those factors or concepts. Suppose the data show that an increase in C_i occurs at the same time as increase in the C_j. This concomitant increase suggests that the edge value e_{ij} should be positive. Suppose similarly that decreases in C_i occur with decreases in C_j. Then such concomitant decrease suggests a negative causal edge value e_{ij}. Even a slight time lag between the two concept nodes can indicate the direction of causality in practice. Such concomitant variation or covariation leads to the *differential* Hebbian learning (DHL) law (Kosko 1986a, 1988, 1991):

$$\dot{e}_{ij} = -e_{ij} + \dot{C}_i \dot{C}_j \tag{25.3}$$

We use concomitant activation and variation as proxies for causation during unsupervised learning with Hebbian and DHL laws. Hebbian learning tends to learn spurious causal links between any two concept nodes that occur at the same time. This quickly grows an edge matrix of nearly all unity values if most of the nodes are active. DHL correlates node velocities. So it has a type of arrow of time built into it. DHL correlates the signs of the time derivatives. So it grows a positive causal edge value e_{ij} if and only if the concept nodes C_i and C_j both increase or both decrease. It grows a negative edge value if and only if one of the nodes increases and the other decreases.

Both learning laws combine to give a more general version of DHL (Kosko 1988):

$$\dot{e}_{ij} = -e_{ij} + C_i C_j + \dot{C}_i \dot{C}_j \tag{25.4}$$

This hybrid learning law fills in expected values for edge-strength values when there is no signal variation in the factor set (Kosko 1990). The hybrid law takes advantage of the relatively rarer variation events to update the edge weights. It also tends to produce limit cycles or even more complex equilibrium attractors. It can produce fixed-point attractors given some strong mathematical assumptions (Kosko 1988, 1991).

Most applications use discretized versions of the DHL law (Kosko 1996) in (25.3):

$$e_{ij}(t+1) = \begin{cases} e_{ij}(t) + \mu[\Delta C_i(t)\Delta C_j(t) - e_{ij}(t)] & \Delta C_i(t) \neq 0 \\ e_{ij}(t) & \text{else} \end{cases} \tag{25.5}$$

where $\Delta C_k(t) = C_k(t) - C_k(t-1)$.

DHL can infer causal edge values in an FCM if the system has access to enough time-series data. Such data can again come from expert opinion surveys. It can come from direct time-series data on measurable factors. Or it can come from indirect instrumental variables linked to the factors of interest:

social media trends, Google Trends, or topic modeling on news corpuses. Previous explorations (Osoba and Kosko 2017) showed a synthetic example in which DHL learned a close approximation of the *true* causal edge values after only a few iterations. Figures 25.3 shows DHL applied to learning a single causal edge value from empirical data. Figure 25.3 shows the DHL training data for an edge value using Google Trends time-series data on the use of politically charged terms "Black lives matter" (BLM), "All lives matter" (ALM), and Blue lives matter" (BlueLM) in online discourse.

We can also fuse soft and hard knowledge sources through the above averaging technique in (25.1). Let $\mathcal{E}_{data}$ denote the data-driven FCM. Let $\mathbf{E}_{exp}$ denote the expert-elicited FCM. Then the fused causal edge matrix $\mathbf{E}_{fusion}$ is a simple mixture of the two edge matrices:

$$\mathbf{E}_{fusion} = \omega_{data}\mathbf{E}_{data} + \omega_{exp}\mathbf{E}_{exp} \tag{25.6}$$

Then (25.5) or some other statistical learning law can continue the adaptation process by using new numerical data or occasional opinion updates from experts.

We can also learn edge values by taking a cue from the literature on Bayesian networks (Friedman and Koller 2003). This entails putting a prior on a randomized FCM. Assume first that the FCM graph is random. Assume next a prior over the space of amenable graphs. Then use observed node data to update a posterior distribution of compatible FCM graphs. This Bayesian process requires a calibrated understanding of the topology and size of the graph spaces. The process also requires that the user produce an accurate and tractable closed-form prior for the graphs. Fuzzy rules can directly represent these closed-form priors (Osoba et al. 2011).

Learning need not take place only in a stationary causal environment where the underlying causal relations do not change in time. Causal relations are apt to change in large-scale problems of social science. Figure 25.3 gives an example. Adaptive FCMs can still model these nonstationary causal worlds if the causal world does not change too fast and if the FCM learning system has access to enough time-series data that reflects these changes.

FCM Example: Public Support for Insurgency and Terrorism

Our first substantive FCM policy example is to the problem of public support for insurgency and terrorism (PSOT). We based two PSOT FCMs on the factor-tree PSOT analysis of Davis et al. (2012) and Davis and O'Mahony (2013). PSOT has complex sociopolitical causes (Ibrahim 2007; Snow et al. 2008; Davis and Cragin 2009; Davis et al. 2012; Nawaz 2015) that involve numerous factors. Davis's later work (Davis et al. 2015a,b) used the PSOT model to motivate related models of an individual propensity for terrorism.

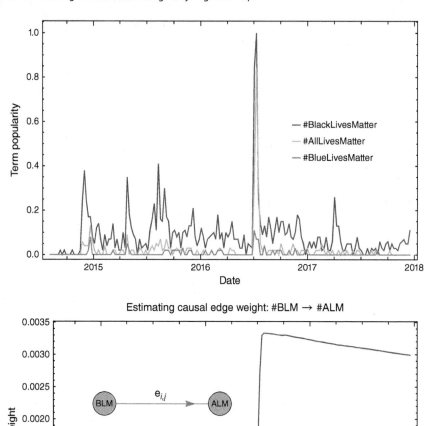

Figure 25.3 Learning FCM causal edge values e_{ij} with Google Trends time-series data for three politically charged phrases: "Black lives matter" (BLM), "All lives matter" (ALM), and "Blue lives matter" (BlueLM). We infer the value of a directed causal edge e_{ij} by applying adaptive inference algorithms such as differential Hebbian learning to relevant Google Trends time-series data. The time series records the weekly popularity of these terms in public Google search activity from January 2014 to February 2017. The time series consisted of 163 ordered samples. The use of BLM-related terms preceded the use of ALM-related terms both in time and in the media narrative. We used this fact to specify the direction for the causal edge.

The PSOT model is a causal factor-tree model because it depicts the degree to which child nodes influence or cause parent nodes. Figure 25.4 and Table 25.1 give more details of the PSOT factor tree. The PSOT nodes represent factors that directly or indirectly relate to the PSOT concept.

Davis's factor-tree models are multi-resolution models (Davis and Bigelow 1998). Major elements have a hierarchical structure that allows users to specify factors at different levels of detail. Each node is an exogenously driven factor, or it fires or activates based on a function of its inputs.

There are also crosscutting factors besides sub-node factors. Crosscutting factors affect multiple factors simultaneously. The ~*and* nodes depend on all fan-in factors being present to a first approximation. The ~*or* nodes depend on any of the fan-in factors being active or on a combination of the fan-in factors being active. There are several top-level factors that directly relate to the general *PSOT* of (Davis et al. 2012): effectiveness of the organization *EFF*, motivation for supporting the group/cause *MOTV*, the perceived legitimacy of violence *PLEG*, and the acceptability of costs and risks *ACR*. Each of these factors has attendant contributory sub-factors.

PSOT edges denote positive influences by default. We denote negative edges with − as with an FCM causal decrease edge. Factor activation along a negative edge reduces the activation of the parent factor. We denote *ambiguous* edges with ±. The ambiguity refers to uncertainty over the edge's direction of influence.

We based our FCM models on the important case of the Al-Qaeda transnational terrorist organization. We augmented the original PSOT with cross-links in the dynamic model to allow richer representation of SD.

Davis et al. (2012) have discussed how the PSOT model explains the public support for Al-Qaeda's mission as follows (paraphrased from Davis et al. (2012)). The organizational effectiveness of Al-Qaeda depends in part on the charisma, strategic thinking, and organizational skills of its leadership (*lead*). Al-Qaeda has framed its ideology to appeal to many Muslims worldwide. Motivation for public support of Al-Qaeda's beliefs comes from shared religious beliefs that stress common identity (*id*) and the sense of duty (*duty*) that such identity fosters. Al-Qaeda also relies on a popular narrative of shared grievances (*shgr*) in the Muslim world. Al-Qaeda stresses the perceived glory (*glry*) of supporting a cause that aims to redress these purported grievances. Religious beliefs and intolerance (*intl*) help increase the perceived legitimacy (*PLEG*) of violence against the West and against the many Muslims who do not share their Salafist views. Countervailing pressure (*scst*) discourages more support for Al-Qaeda. This countervailing pressure occurs in part because much of the public believes that Al-Qaeda will not succeed and thus emerge as ultimate victors (*lvic*). This pressure reduces the acceptability of costs and risks (*ACR*) for Al-Qaeda activities. The parameters of this Al-Qaeda case study determined the relative causal edge weights in our FCM models.

Figure 25.5 shows FCM versions of the old static (acyclic) PSOT model and new dynamic PSOT model.

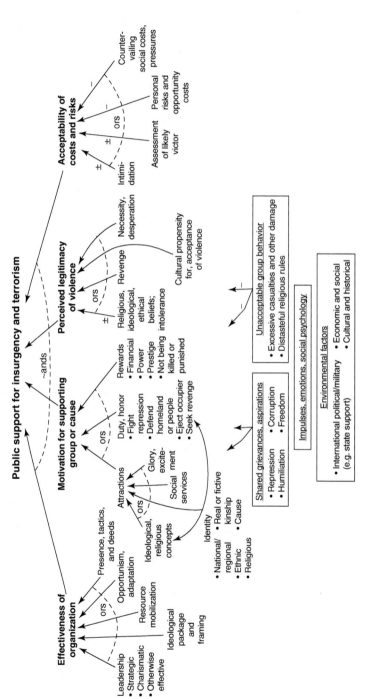

Figure 25.4 PSOT factor-tree model. The figure shows the directed relationships among factors that underlay the public support for terrorism or insurgency model in (Davis and O'Mahony 2013). Note: Applies at a snapshot in time. Current factor values can affect future values of some or all other factor.

Table 25.1 Table of factors in the public support for insurgency and terrorism (PSOT) model.

Label	Full description
lead	Leadership strategic or otherwise
pkg	Ideological package and framing
rsrc	Resource mobilization
opp	Opportunism and adaptation
pres	Presence, tactics, and deeds
EFF	Effectiveness of organization
reli	Ideological religious concepts
socs	Social services
glry	Glory, excitement
ATT	Attractions
duty	Duty and honor
rwrd	Rewards
MOTV	Motivation for supporting group, cause
intl	Religious, ideological, ethical beliefs; intolerance
rvng	Revenge
cprop	Cultural propensity for accepting violence
desp	Desperation, necessity
PLEG	Perceived legitimacy of violence
intm	Intimidation
lvic	Assessment of likely victor
prsk	Personal risk and opportunity cost
scst	Countervailing social costs and pressures
ACR	Acceptability of costs and risks
id	Identity
shgr	Shared grievances and aspirations
ugb	Unacceptable group behavior
env	Environmental factors
impl	Impulses, emotions, social psychology
hsucc	History of successes
mgtc	Management competence
prop	Propaganda, advertising
efdoc	Effectiveness of indoctrination/passing beliefs
hfail	History of failures
PSOT*	Public support for insurgency and terrorism

The starred factor(s) indicate the output node for this FCM.

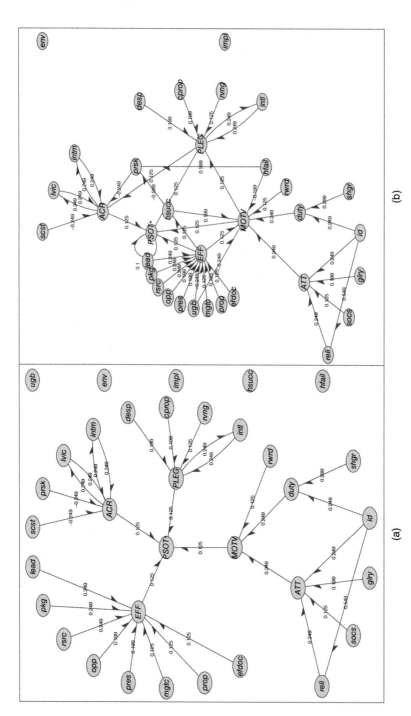

Figure 25.5 Two fuzzy cognitive maps of the PSOT factor-tree model. Panel (a) shows the FCM digraph for the original static (acyclic) PSOT model. Panel (b) shows the FCM digraph of the dynamic PSOT model with cross-links. Table 25.1 gives the key for the concept-node labels in both FCMs. We based the new FCM edges in the right digraph on the findings in Davis and O'Mahony (2013) and on expert input. Figure 6 in Osoba and Kosko (2017) illustrates the causal edge connection matrices **E** (the adjacency matrices) for each FCM using heat maps.

We now outline modifications to the original PSOT model. We first added a weak self-excitation feedback loop to the PSOT concept node because it is the highest-level concept node. This self-excitation loop modeled inertia in aggregate public opinion about insurgency and terrorism. This new feedback source induced a weak serial correlation in time in the PSOT concept node.

The next directed edges connected the top-level factors in (Davis and O'Mahony(2013) from left to right: $EFF \rightarrow MOTV$, $MOTV \rightarrow PLEG$, and $PLEG \rightarrow ACR$. These directed causal edges made explicit an implicit point about O'Mahony and Davis's use of factor trees. Their factor-tree representation assumed a left-to-right dependence of the top-level factors that we have linked (Davis 2011; Davis and O'Mahony 2013). This implicit dependence made their factor tree more readable. The FCM model made this dependence explicit.

O'Mahony and Davis (2013) discuss other dynamic augmentations to the PSOT model. They point to the following new factors. A history of successes or failures can affect motivation and perceived risks. We model this dependence with factors *history of successes* and *history of failures*. These two nodes exert opposing influence on $MOTV$ and *prsk*. We split this history factor because traditional FCM models admit only positive values that represent the degree or intensity to which a concept occurs. The effectiveness of the organization factor EFF partly determines the history of successes: $EFF \rightarrow hsucc$. Unacceptable group behavior *ugb* also influences motivation and effectiveness: $ugb \rightarrow MOTV$ and $ugb \rightarrow EFF$.

US–China Relations: An FCM of Allison's Thucydides Trap

We next use an FCM to model a new conflict dynamic in international relations.

Political scientist Graham Allison calls this dynamic *Thucydides' trap* (Allison 2015, 2017). Allison argues that this dynamic occurs when a new power emerges that challenges the dominance of an older power on the world stage. Superpowers such as the United States and China must avoid Thucydides' trap to avoid war.

Our FCM interpretation of Allison's analysis predicts some type of war pattern in some cases and not in others. A large percentage (most) of the clamped input states led to a war-type outcome. But this was not a probability estimate. It just reflects an exhaustive search of all possible clamped input states. It does not reflect that relative likelihood of the states themselves.

We based the fractional causal edge values e_{ij} for this FCM on Allison's text. We also tested the robustness of this properly fuzzy FCM by thresholding all positive edge values $e_{ij} > 0$ to 1 and all negative edge values $e_{ij} < 0$ to −1. This gave a trivalent FCM that predicted some type of war for the majority of all clamped input states. We stress again that the prevalence of a war outcome in this model does *not* mean that the FCM predicts war with high probability. That would require that all input states are equally likely and they clearly are not. We did not address the issue of which inputs are more or less likely to occur. Our task was to translate Allison's textual claims into a representative FCM causal model and explore its pattern predictions.

The name *Thucydides' trap* stems from a famous political conjecture in Thucydides' *History of the Peloponnesian War* (Thucydides 1998) (Book 1, paragraph 23): "the real though unavowed cause [of the war] I believe to have been the growth of Athenian power, which terrified the [Spartans] and forced them into war." Thucydides expands on his causal theory of war in a speech that an Athenian gives to the Spartan assembly (Thucydides 1998) (Book 1, paragraph 76):

> So that, though overcome by three of the greatest things, honor, fear, and profit, we have both accepted the dominion delivered us and refuse again to surrender it, we have therein done nothing to be wondered at nor beside the manner of men. Nor have we been the first in this kind, but it hath been ever a thing fixed for the weaker to be kept under by the stronger.

Thucydides claimed that three main factors determine how nation-states interact: interest, fear, and honor. The *interest* or profit factor just restates a nation's self-interested actions. Nation-states act against other states to maintain their high-priority national interests within the geographic scope of their power. These interests include national security, economic security, and sovereignty. *Fear* refers to the emotionally charged frames through which a nation views world events. *Honor* refers to the nation's senses of self and entitlement. Examples include the nineteenth-century US *manifest destiny* or China's older concept of *Tianxia* or *all under heaven*.

Allison expands on these factors in his Thucydides' trap model where again the rise of a new power risks war with a dominant power. He argues that fear is the main cause of war between such a dominant power and a new rising power. He looked at 16 such historical power struggles that extend back to the fifteenth century. He found that 12 of these power struggles ended in war. Allison also contends that similar structural dynamics apply elsewhere in international relations.

We parsed Allison's analysis (Allison 2017) to create an FCM of Thucydides' trap for current US–China relations. The FCM follows Thucydides and uses his three main factors of interest, fear, and honor. Auxiliary factors give context to the main factors. The resulting Thucydides' trap FCM has 17 factors. Table 25.2 lists and describes these factors.

Table 25.2 Factors in Thucydides' trap for relations between the United States and China in 2017.

Label	Full description
FEAR	Fear
usd	US military/defense posture
chnd	China military/defense posture
geod	Geographical distance
ENT	Sense of entitlement/honor
uspub	US public resentment
chnpub	Chinese public resentment
dipl	Diplomacy channels and international rules
NUKE	Nuclear power/MAD
ShrdCult	Shared culture
INT	National interests clash
usecon	US economic dominance
chnecon	China economic dominance
econdep	Economic interdependence
ally	Alliance network structural friction
shi	*Shi* or contextual/historical military momentum
WAR*	War, military conflict between the United States and China

The starred factor(s) indicate the output node for this FCM.

Thucydides' trap FCM also uses some of the auxiliary concepts that Allison discussed. One example is how nuclear weapons affect the chance of all-out war. Diplomatic institutions and economic dependencies also affect the chance of war. Treaty and alliance obligations can rapidly induce or expand war as happened in both world wars. The FCM in Figure 25.6 shows the directed causal edges among the concepts and the FCM's causal edge matrix representation **E**.

We surmised the causal edge strengths based on Allison's discussions in (Allison 2017). Below we present the results of thresholding the magnitudes to their binary extremes.

Thucydides' trap FCM predicted war-type patterns between the United States and China more often than it predicted peace-type patterns. An exhaustive search of the space of possible (clamped) scenarios found that only under ∼20% of scenarios led to lasting peace between the dominant power (United States) and the rising power (China). We point out again that these are not representative probabilities because we did not know or estimate the relative probabilities of the input states. We simply assumed that they were all equally likely. The FCM rapidly converged to an equilibrium state where

WAR ∗ was active when the input consisted of US-specific nodes that were stagnant and China-specific nodes that were rising. The key factors present in peaceful accommodations were significant geographic distance, mutual assured destruction (via nuclear weapons posture), a shared culture, economic interdependence, and the presence of diplomatic channels.

We simulated the FCM dynamical evolution from *trap-like* initial conditions (see Figure 25.7 for initial states and evolution trace). Our test scenario consisted of these causal relations:

- The United States maintains a strong military or defensive posture.
- China is economically rising or already dominant.
- US public has high resentment toward China.
- Both sides are economically interdependent.

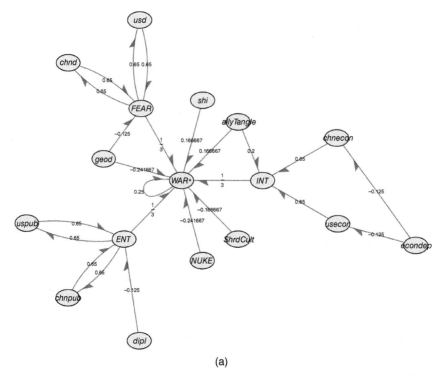

(a)

Figure 25.6 FCM implementation of Allison's *Thucydides' trap* as it is described in (Allison 2017). (a) Causal network representation of Allison's *Thucydides' trap* FCM. (b) Causal edge matrix representation **E** for Thucydides' trap FCM. **E** is the adjacency matrix for the FCM's fuzzy signed directed graph. Each square shows the fuzzy causal edge value e_{ij}. The value e_{ij} is how much the ith concept C_i causes or influences the jth concept C_j. The matrix entries e_{ij} in these FCMs are fuzzy values in the bipolar interval $[-1, 1]$. Uncolored squares indicate the absence of causal influence.

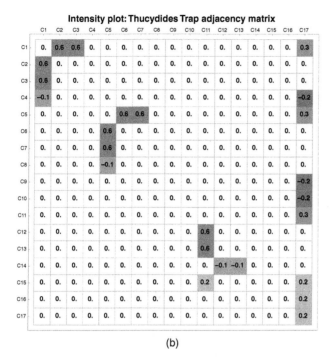

(b)

Figure 25.6 (*Continued*)

- Both sides have enough nuclear capability to pose credible threats to each other (sufficient for deterrence).
- Strong diplomatic channels exist between both sides.

We coded this initial scenario for the FCM concept nodes. Forward inference gave the sequence of states in Figure 25.7.

The FCM converged in four iterations to a fixed-point equilibrium state with the $WAR *$ node active. The FCM's state evolution showed that China's economic dominance led to a clash in national interests while the US defensive posture led to fear. This led China to ramp up its defensive posture. The US public's resentment toward China led to a sense of entitlement. That led in turn to Chinese public resentment. The clash in national interests, fear, and sense of entitlement or national honor combined to activate the $WAR *$ node. This FCM behavior is similar to Thucydides' trap dynamics that Allison described.

The FCM's war prediction was robust against many perturbations of the input state. It persisted despite changes in the activation of nodes like diplomacy and geographic distance. But activating the shared-culture concept node did prevent war. The FCM also fell out of the war equilibrium when we shut off either the concept node for US defense posture or for Chinese

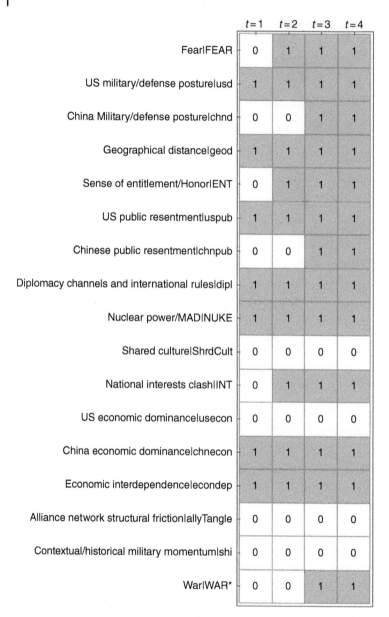

	$t=1$	$t=2$	$t=3$	$t=4$
Fear|FEAR	0	1	1	1
US military/defense posture|usd	1	1	1	1
China Military/defense posture|chnd	0	0	1	1
Geographical distance|geod	1	1	1	1
Sense of entitlement/Honor|ENT	0	1	1	1
US public resentment|uspub	1	1	1	1
Chinese public resentment|chnpub	0	0	1	1
Diplomacy channels and international rules|dipl	1	1	1	1
Nuclear power/MAD|NUKE	1	1	1	1
Shared culture|ShrdCult	0	0	0	0
National interests clash|INT	0	1	1	1
US economic dominance|usecon	0	0	0	0
China economic dominance|chnecon	1	1	1	1
Economic interdependence|econdep	1	1	1	1
Alliance network structural friction|allyTangle	0	0	0	0
Contextual/historical military momentum|shi	0	0	0	0
War|WAR*	0	0	1	1

Figure 25.7 Spreading activation time slices in Thucydides' trap FCM. Each column is a discrete step in time. The FCM converged in four iterations.

economic dominance. These peaceful equilibrium outcomes also appear consistent with Allison's analysis. Figure 25.8a shows the average concept-node activations for initial scenarios that led to peaceful outcomes.

We realize that other analysts would likely surmise somewhat different fuzzy causal edge values $e_{ij} \in [-1, 1]$ given the same source text. We would expect more agreement on the signs of these edges. So we tested whether a thresholded version of our properly fuzzy Thucydides' trap FCM made similar equilibrium predictions. We formed this trivalent Thucydides' trap FCM by replacing all positive edges $e_{ij} > 0$ with 1 and all negative edges $e_{ij} < 0$ with −1. Zero-valued

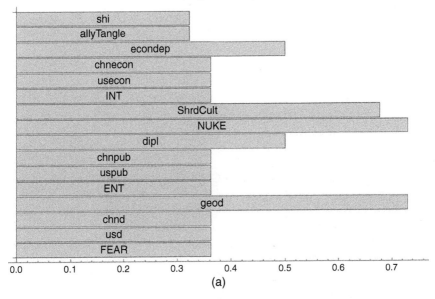

Figure 25.8 Average node activations for input scenarios that converge in peace (for original and thresholded FCM). (a) Average concept-node activations for initial scenarios that did not escalate into war. The bar chart shows that non-escalatory scenarios tended to involve large geographic separation (*geod*), the availability of nuclear weapons (*NUKE*), the presence of diplomatic channels (*dipl*) for resolving issues, and a shared culture between both parties (*shrdCult*). Economic interdependence (*econdep*) was also a common feature of peace. (b) Trivalent Thucydides' trap FCM: Average concept-node activations for initial scenarios that did not escalate into war. The trivalent (thresholded) FCM's average activation patterns for peaceful scenarios were similar to those of the original properly fuzzy FCM. The key concept nodes associated with peace (geographic distance, nuclear posture, shared culture, presence of diplomatic channels, and economic dependence) were the same for both FCMs. 14.7% of all input scenarios resolved peacefully for the quantized FCM compared with 19.3% for the original FCM. This suggests that the Thucydides' trap FCM's behavior was reasonably robust to mis-specified causal edge values e_{ij}.

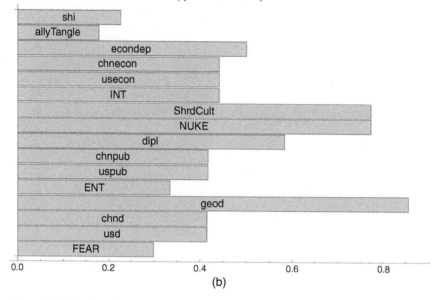

Figure 25.8 *(Continued)*

edges stayed the same. The trivalent FCM still predicted warlike patterns for most clamped input states. It predicted peace for only ~15% of all input scenarios compared with ~20% for the original fuzzy Thucydides' trap FCM. We point out again that we treated all input states as equally likely, and this is clearly not the case in the real world of international relations. Figure 25.8b shows the similar average concept-node activations for the input scenarios that resolved peacefully in the trivalent FCM. This counts as evidence that the properly fuzzy Thucydides' trap FCM was reasonably robust to perturbations in the causal edge value magnitudes.

Analysts may also disagree on structure, not just numerical edge values. Other researchers may surmise different causal links (edges) or even different relevant concept nodes. This may be due to an alternative reading of the source text or due to independently derived domain expertise.

The ability to have these sorts of disagreements actually shows the value of the FCM modeling approach. Our example only demonstrates the basic feasibility of FCM modeling for this application. Domain experts can instantiate concurring or dissenting causal maps. Experts and critics alike can then compare or contrast the different theories in this social scientific domain. Such comparisons extend beyond just comparisons of the basic representation of different social scientific theories (the FCMs themselves). Analysts can also compare the long-term implications of the different theories with more quantitative rigor (the *hidden patterns* and limit cycles in the FCMs' temporal dynamics).

Conclusion

FCMs offer a flexible way to model large-scale feedback causal system and to make forward inferences. Their cyclic structure produces a nonlinear dynamical system that tends to quickly equilibrate to limit-cycle predictions given a causal input or stimulus. Users can also step through FCM transient or equilibrium states and therefore unfold the dynamical system in time. The underlying matrix structure of an FCM's directed causal edges permits natural knowledge fusion through simply adding or mixing the augmented FCM causal edge matrices for any number of experts. Such knowledge combination tends to improve with the number of combined experts.

DAGs lack these features precisely because they are acyclic. Their use of probability to describe causal uncertainty is secondary to their lack of cycles. Cycles describe feedback in directed graphs. And combining directed graphs will in general produce several such cycles. The vector–matrix operations of FCMs also involve much less computation than the probabilistic computations in BBNs. But FCMs cannot produce the precise probability descriptions that DAGs can if the user knows the DAG's corresponding complete joint probability density function and uses the sum–product algorithm. Imposing this or a related probability structure on an FCM is an area for future research.

Current FCM inference and learning have two key limitations that future research also needs to address. The first is that FCMs do not easily permit backward chaining. So they do not in general answer which input caused an observed output effect. Users cannot simply run the FCM in reverse because of the node nonlinearities. We instead must exhaustively test all or nearly all input states to see which inputs map to which output equilibria. This computes the inverse image of each output attractor basin. It carves the FCM state space into attractor regions. We know for a given output only that the input came from an attractor region. Future research should address this limitation with new inferencing or other techniques.

The second limitation is more challenging: How do we infer missing FCM concept nodes? This just asks how we come up with a new causal hypothesis. A new node leads to new causal conjectures for all nonzero edges that connect to the new node. Current adaptive techniques infer and tune the causal edge values only for known concept nodes. An open research problem is to find data-based techniques that infer new or missing concept nodes in large-scale FCM causal models. Solutions may include Bayesian priors or rules over node sets or other statistical techniques for model building.

References

Abdelbari, H. and Shafi, K. (2018). Learning structures of conceptual models from observed dynamics using evolutionary echo state networks. *Journal of Artificial Intelligence and Soft Computing Research* 8 (2): 133–154.

Allison, G. (2015). The Thucydides Trap: are the US and China headed for war? The Atlantic (24 September).

Allison, G. (2017). *Destined for War: Can America and China Escape Thucydides's Trap?* Houghton Mifflin Harcourt. Retrieved from https://books.google.com/books?id=CtmpDAAAQBAJ.

Amirkhani, A., Papageorgiou, E.I., Mohseni, A., and Mosavi, M.R. (2017). A review of fuzzy cognitive maps in medicine: taxonomy, methods, and applications. *Computer Methods and Programs in Biomedicine* 142: 129–145.

Beal, M.J. and Ghahramani, Z. (2003). The variational bayesian em algorithm for incomplete data: with application to scoring graphical model structures. *Bayesian Statistics* 7, 453–464.

Beal, M.J. and Ghahramani, Z. (2006). Variational bayesian learning of directed graphical models with hidden variables. *Bayesian Analysis* 1 (4): 793–831.

Caragiannis, I., Procaccia, A.D., and Shah, N. (2013). When do noisy votes reveal the truth? In: *Proceedings of the 14th ACM Conference on Electronic Commerce,* 143–160. New York, NY: ACM. https://dl.acm.org/citation.cfm?id=2482570.

Conitzer, V., Rognlie, M., and Xia, L. (2009). Preference functions that score rankings and maximum likelihood estimation. *IJCAI* 9: 109115.

Dagum, P. and Luby, M. (1993). Approximating probabilistic inference in Bayesian belief networks is NP-hard. *Artificial Intelligence* 60 (1): 141–153.

Davis, P.K. (2011). Primer for building factor trees to represent social-science knowledge. In: *Proceedings of the Winter Simulation Conference,* 3121–3135 https://dl.acm.org/citation.cfm?id=2431888.

Davis, P.K. and Bigelow, J.H. (1998). *Experiments in Multiresolution Modeling (MRM).* RAND Corporation.

Davis, P.K. and Cragin, K. (2009). *Social Science for Counterterrorism: Putting the Pieces Together.* RAND Corporation.

Davis, P.K. and O'Mahony, A. (2013). *A Computational Model of Public Support for Insurgency and Terrorism: A Prototype for More-General Social-Science Modeling.* RAND Corporation.

Davis, P.K., Larson, E.V., Haldeman, Z. et al. (2012). *Understanding and Influencing Public Support for Insurgency and Terrorism.* RAND Corporation.

Davis, P.K., Manheim, D., Perry, W.L., and Hollywood, J.S. (2015a). *Causal Models and Exploratory Analysis in Heterogeneous Information Fusion for Detecting Potential Terrorists.* RAND Corporation.

Davis, P.K., Manheim, D., Perry, W.L., and Hollywood, J. (2015b). Using causal models in heterogeneous information fusion to detect terrorists. In: *2015 Winter Simulation Conference (WSC)*, 2586–2597. IEEE.

Dickerson, J.A. and Kosko, B. (1993). Virtual worlds as fuzzy cognitive maps. In: *Virtual Reality Annual International Symposium, 1993*, 471–477. IEEE.

Dickerson, J.A. and Kosko, B. (1994). Virtual worlds as fuzzy cognitive maps. *Presence* 3 (2): 173–189.

Friedman, N. and Koller, D. (2003). Being Bayesian about network structure. A Bayesian approach to structure discovery in Bayesian networks. *Machine Learning* 50 (1–2): 95–125.

Glykas, M. (ed.) (2010). *Fuzzy Cognitive Maps*. Springer.

Granger Morgan, M., Henrion, M., and Small, M. (1992). *Uncertainty: A Guide to Dealing with Uncertainty in Quantitative Risk and Policy Analysis*. Cambridge University Press.

Hebb, D.O. (1949). *The Organization of Behavior: A Neuropsychological Approach*. Wiley.

Ibrahim, R. (2007). *The Al Qaeda Reader: The Essential Texts of Osama Bin Laden's Terrorist Organization*. Broadway Books.

Imbens, G.W. and Rubin, D.B. (2015). *Causal Inference in Statistics, Social, and Biomedical Sciences*. Cambridge University Press.

Kosko, B. (1986a). Differential Hebbian learning. In: *AIP Conference Proceedings*, vol. 151, 277–282. New York: American Institute of Physics.

Kosko, B. (1986b). Fuzzy cognitive maps. *International Journal of Man-Machine Studies* 24 (1): 65–75.

Kosko, B. (1986c). Fuzzy knowledge combination. *International Journal of Intelligent Systems* 1 (4): 293–320.

Kosko, B. (1988). Hidden patterns in combined and adaptive knowledge networks. *International Journal of Approximate Reasoning* 2 (4): 377–393.

Kosko, B. (1990). Unsupervised learning in noise. *IEEE Transactions on Neural Networks* 1 (1): 44–57.

Kosko, B. (1991). *Neural Networks and Fuzzy Systems: A Dynamical Systems Approach to Machine Intelligence*. Prentice Hall.

Kosko, B. (1996). *Fuzzy Engineering*. Prentice Hall.

Kosko, B. and Isaka, S. (1993). Fuzzy logic. *Scientific American* 269 (1): 62–67.

Mill, J. (1843). *A System of Logic, Ratiocinative and Inductive: Being a Connected View of the Principles of Evidence and the Methods of Scientific Investigation*, vol. 1. John W. Parker. Retrieved from https://books.google.com/books?id=y4MEAAAAQAAJ.

Murphy, K.P. (2002). Dynamic Bayesian networks: representation, inference and learning. Unpublished doctoral dissertation. Berkeley, CA: University of California.

Nawaz, M. (2015). *Radical: My Journey from Islamist Extremism to a Democratic Awakening*. Random House.

Osoba, O.A. and Kosko, B. (2017). Fuzzy cognitive maps of public support for insurgency and terrorism. *The Journal of Defense Modeling and Simulation* 14 (1): 17–32.

Osoba, O., Mitaim, S., and Kosko, B. (2011). Bayesian inference with adaptive fuzzy priors and likelihoods. *IEEE Transactions on Systems, Man, and Cybernetics Part B: Cybernetics* 41 (5): 1183–1197. https://doi.org/10.1109/TSMCB.2011.2114879.

Papageorgiou, E. (2013). *Fuzzy Cognitive Maps for Applied Sciences and Engineering: From Fundamentals to Extensions and Learning Algorithms*. Berlin, Heidelberg: Springer. Retrieved from https://books.google.com/books?id=S3LGBAAAQBAJ.

Papageorgiou, E.I. and Salmeron, J.L. (2013). A review of fuzzy cognitive maps research during the last decade. *IEEE Transactions on Fuzzy Systems* 21 (1): 66–79.

Pearl, J. (2009). *Causality*. Cambridge University Press.

Pearl, J. (2014). *Probabilistic Reasoning in Intelligent Systems: Networks of Plausible Inference*. Elsevier Science. Retrieved from https://books.google.com/books?id=mn2jBQAAQBAJ.

Predd, J.B., Osherson, D.N., Kulkarni, S.R., and Poor, H.V. (2008). Aggregating probabilistic forecasts from incoherent and abstaining experts. *Decision Analysis* 5 (4): 177–189.

Rubin, D.B. (2005). Causal inference using potential outcomes: design, modeling, decisions. *Journal of the American Statistical Association* 100 (469): 322–331.

Russell, S., Norvig, P., and Davis, E. (2016). *Artificial Intelligence: A Modern Approach*. Pearson Education, Limited. Retrieved from https://books.google.com/books?id=XS9CjwEACAAJ.

Snow, D.A., Soule, S.A., and Kriesi, H. (2008). *The Blackwell Companion to Social Movements*. Wiley.

Sterman, J.D. (2000). Business dynamics: systems thinking and modeling for a complex world (No. HD30. 2 S7835 2000).

Stylios, C.D., Georgopoulos, V.C., Malandraki, G.A., and Chouliara, S. (2008). Fuzzy cognitive map architectures for medical decision support systems. *Applied Soft Computing* 8 (3): 1243–1251.

Taber, R. (1991). Knowledge processing with fuzzy cognitive maps. *Expert Systems with Applications* 2 (1): 83–87.

Taber, R., Yager, R.R., and Helgason, C.M. (2007). Quantization effects on the equilibrium behavior of combined fuzzy cognitive maps. *International Journal of Intelligent Systems* 22 (2): 181–202.

Thucydides (1998). *History of the Peloponnesian War*, Translated by Benjamin Jowett. Oxford.

Wainwright, M.J. and Jordan, M.I. (2008). Graphical models, exponential families, and variational inference. *Foundations and Trends ® in Machine Learning* 1 (1–2): 1–305.

Yedidia, J.S., Freeman, W.T., and Weiss, Y. (2001). Generalized belief propagation. *Advances in Neural Information Processing Systems*. 689–695.

Zadeh, L.A. (1965). Fuzzy sets. *Information and Control* 8: 338–353.

Ziv, G., Watson, E., Young, D., Howard, D.C., Larcom, S.T. and Tanentzap, A.J. (2018). The potential impact of Brexit on the energy, water and food nexus in the UK: A fuzzy cognitive mapping approach. *Applied Energy*, 210: 487–498.

26

Simulation Analytics for Social and Behavioral Modeling

Samarth Swarup, Achla Marathe, Madhav V. Marathe, and Christopher L. Barrett

Biocomplexity Institute & Initiative, University of Virginia, Charlottesville, VA 22904, USA

Introduction

Computational simulations are increasingly being used to study social systems,[1] in order to model a range of phenomena, including infectious disease epidemics, natural- and human-initiated disasters, economic self-organization, online social behavior, and more. These phenomena are behaviorally driven and exhibit coupling between multiple systems and multiple spatiotemporal scales and also offer many opportunities for interventions. These modeling and simulation efforts are giving rise to interesting new questions that need to be addressed through new methods that combine data analytics and simulation science, which we refer to as simulation analytics.

Simulation-based approaches are needed for these complex problems because policy and planning require understanding hypothetical (counterfactual) scenarios, answering what-if and under-what-conditions questions (Davis and O'Mahony 2019), and *discovering* interventions. High-resolution and data-driven simulations of adequate representational complexity provide a natural way of addressing these requirements. For instance, if during an infectious disease epidemic, an epidemiologist wishes to understand the potential consequences of closing a particular school for a certain number of days, it helps to have a model that has a representation of said school and of the behaviors of students on school days as well as nonschool days.

Over the last two decades, social simulations have gotten increasingly larger and more sophisticated. For example, the first TRANSIMS simulation modeled the Dallas–Ft. Worth area in Texas without a detailed representation of daily

1 Eubank et al. (2004), Haas et al. (2012), Zou et al. (2012), Schiller et al. (2015), Axtell (2016), and Lymperopoulos and Ioannou (2016).

Social-Behavioral Modeling for Complex Systems, First Edition.
Edited by Paul K. Davis, Angela O'Mahony, and Jonathan Pfautz.
© 2019 John Wiley & Sons, Inc. Published 2019 by John Wiley & Sons, Inc.
Companion website: www.wiley.com/go/Davis_Social-Behavioralmodeling

activity patterns (Beckman 1997). In 2004, Eubank et al. (2004) described the use of much more refined synthetic populations for high-fidelity urban-scale epidemic simulations. In 2011, Parker and Epstein (2011) developed a platform for global-scale epidemic simulation, albeit without the kind of representational detail described above.

In terms of the level of detail and amount of data used in model design, simulations have also rapidly increased in sophistication. Early stylized models of regions (Batty and Xie 1994) led to models with accurate demographics and geospatial information (Eubank et al. 2004). These, in turn, have led to even more complex simulations, such as the simulation of National Planning Scenario 1, by Barrett et al. (2013), which includes highly detailed, data-driven models of multiple infrastructures and human behavior after a disaster in addition to an accurate population model.

These simulations are intuitive to build and understand in that they require specifying models of *individual* behavior and interactions. The simulation then computes the population-level consequences of the millions of interactions between individuals that occur during the event or phenomenon. Indeed, this is the *raison d'être* of the simulation, since we cannot compute population-level consequences any other way from specifications of the individual behavior, at least once the individual behaviors get reasonably complex. For example, closed-form mathematical modeling of complex behaviors and millions of interacting agents would not be tractable.

High resolution, fidelity, precision, and accuracy of a simulation contribute to its veridicality and therefore its ability to provide actionable information.[2] However, these properties, by making the simulation more complex, also make it more opaque to inspection. With a toy simulation, we can completely explore the parameter space and run large statistical experiment designs, thus obtaining a comprehensive understanding of the phenomena that emerge within the simulation. This experiment design approach is applied to larger simulations also, where we choose a few possible interventions (or perhaps vary some parameters of interest) and run the simulation for each case to examine differences in the outcomes of interest (Halloran et al. 2008).

As we push the boundaries of scale and complexity, however, we encounter two problems: (i) we do not know the right interventions to try ahead of time, i.e. we would like to use the simulation as a means of discovering useful interventions, and (ii) we can run the simulation only a small number of times in a reasonable amount of time. This limits the statistical power we can obtain to distinguish outcomes between the different cells of a statistical experiment design. However, we retain two important advantages over empirical data.

---

2 While validation is not discussed here, it is a crucial step in the design and use of a simulation. It is of course possible to have a lot of high-fidelity detail in a simulation that is ultimately irrelevant, or even erroneous, with respect to the question being studied. Uncertainty quantification and sensitivity analysis are also important in understanding the applicability of a simulation.

First, since a simulation is self-contained, we have complete observability. There are no missing values, and no unobserved variables that could be influencing observed outcomes in the simulation. Second, more fundamentally, we have complete knowledge of the computation being carried out internally by each agent, since we design and implement the agents. This does not mean, however, that we know the effective computation being carried out by each agent, since that also depends on the agent's interactions with other agents and with the environment. In fact, this is a crucial question to be addressed through simulation analytics.

In sum, this is a unique regime of investigation. Sensemaking in this regime is a big data challenge. These simulations produce a large amount of richly structured data and give rise to several novel questions, which require new methods to be developed. We will explore some of these in what follows, especially in reference to social and behavioral modeling.

What Are Behaviors?

In order to design meaningful simulations, we need an operational understanding of the notion of behavior, in the sense of being able to create a computational model that captures the essentials of behavior with respect to a given domain. Such an operational description of human behavior is closely tied to models of bounded rationality, individual agency, social influence, and social structure. It is worthwhile thinking about what each of these terms mean and how they get translated into computational simulations.

The idea of simulating a population of interacting *units* by means of a computer goes back to Orcutt (1957). By the general term units, he indicated the flexibility of the method to model individuals, households, or firms. In his conception, units receive inputs and produce outputs based on probability distributions connecting inputs and outputs. This very general approach is what we term *microsimulation* today (Li and O'Donoghue 2013). Microsimulations are used in many domains and are essential in at least two types of situations.

First, as an illustrative example, consider the problem of estimating population exposure to extreme heat. This can be done by providing a set of individuals with temperature sensors that can log their exposure as they go about their daily routines. This is an expensive approach and is further beset by the problem that the variation in temperature across a large region (such as a US state) makes it difficult to generalize from the sample to the population. However, a microsimulation of the region, which includes typical daily activity patterns of individuals and high-resolution temperature data, can be used to estimate exposure (Swarup et al. 2017). Indeed, it is necessary if we wish to compute population heat exposure without recruiting a very large (and prohibitively expensive) sample.

Second, consider the problem of mitigating an epidemic of an infectious disease such as influenza. Such diseases spread from person to person over a *social contact network*, which is the network induced by physical collocation as people go about their daily activities. The social contact network is impossible to estimate through a survey, even if you could survey every single person, because most people do not know all the other people with whom they are physically collocated on a typical day. We can, however, synthesize the social contact network by integrating the appropriate data sets. This is known as the synthetic population approach where multiple data sets, including the census, travel, and activity surveys, and geospatial data are integrated to construct a representation of the demographics, activity patterns, and activity locations of the entire population of a region, from which a synthetic social contact network can be extracted (Eubank et al. 2004). Microsimulations can then make use of this synthetic social contact network to assess the efficacy of different strategies for epidemic mitigation.

In both the above scenarios, there is a need for a detailed disaggregated simulation as we cannot realistically obtain information about population-level effects otherwise. The simulations include a model of typical daily activities, but these activity patterns are fixed (in the simplest case) and not responsive to changes in the social or physical environment. This lack of adaptiveness at the individual level is what we point to when we say that these simulations lack *behaviors*. In order to incorporate behaviors, we need to go beyond microsimulation to agent-based simulations.

The broad distinction between microsimulation and *agent-based* (or multi-agent) simulation is that the latter contains agents. What distinguishes an agent from *just* a computer program? Franklin and Graesser (1997) addressed this very question and defined an autonomous agent as "a system situated within and a part of an environment that senses that environment and acts on it, over time, in pursuit of its own agenda and so as to effect what it senses in the future." There is thus an implication of teleology, or purpose, in the definition of agency. While this may appear to locate agency within an agent, they also recognize the role of factors external to the agent in determining its agency. They note that, "a robot with only visual sensors in an environment without light is not an agent. Systems are agents or not with respect to some environment." In this sense, agency is also contingent, making it similar to Simon's (1996) definition of artificial systems.

This raises an important question from the simulation perspective: how much of agency is teleological, and how much is contingent? Put another way, how much of agency is individual or agent internal, and how much is determined by interaction? This is not just a matter of philosophical interest; it has implications for the design and scalability of simulations. In a traffic simulation, for instance, drivers may be modeled as sophisticated feedback controllers who adapt to traffic conditions, leading to different observed traffic flow regimes at different traffic densities. Barrett et al. (1996) showed that such an adaptive feedback control mechanism emerges from a much simpler computation in a cellular automaton (CA) model of traffic, through the interactions between simulated vehicles. There are, thus, two equivalent ways of designing traffic simulations. The first is to develop a complex driver model for the agent and put many of these together to simulate traffic. The second is to use the CA approach, where the same driver model emerges from the interactions between the CA rules and the interactions within the system. The two approaches are equivalent in terms of the global phenomena of interest, but the latter is a much more scalable approach in terms of the size of the system that can be simulated.

It is in our interest, therefore, from the perspective of designing scalable simulations, to have the individual agents be as *lightweight* as possible in terms of their individual computational burden. If agent computations can be distributed across interactions within the system, the simulation benefits by having a smaller memory and processing footprint while still achieving the same global objectives, thus allowing scaling to much larger numbers of agents. Barrett et al. (2011) refer to this approach as *unencapsulated agency*. How to do this in general, though, is an open question.

On the other hand, the reader might reasonably object, just because a complex behavior can emerge from interactions in a simpler simulation doesn't mean that that's how it works in the real world. The human brain is a marvelously complex system, and human decision-making is a subtle process, so it could be supposed that locus of agency lies within the individual, and the idea that agency emerges from interactions might seem counterintuitive.

This tension between the individual and the social or between agency and structure (to use the sociological term) has long been recognized. Sewell (1992) referred to agency as *the efficacy of human action*, and wrote that "a social science trapped in an unexamined metaphor of structure tends to reduce actors to cleverly programmed automatons."

Emirbayer and Mische (1998) have a much more complex definition of agency: "the temporally constructed engagement by actors of different structural environments – the temporal-relational contexts of action – which, through the interplay of habit, imagination, and judgment, both reproduces and transforms those structures in interactive response to the problems posed by changing historical situations." This bears unpacking.

They decompose agency into three elements, which they term *iteration, projectivity*, and *practical evaluation*. In their view, an *agent* structures its social environment and creates a stable identity and interactions by selectively reactivating prior patterns of thought and action. This is referred to as iteration. The agent also generates possible future trajectories of action (projection) and chooses among them based on current circumstances as well as practical and normative judgments (practical evaluation) (Emirbayer and Mische 1998).

At first glance, this evokes the standard artificial intelligence concepts of learning and planning where, broadly, the notion of iteration corresponds to learning and projectivity and practical evaluation correspond to planning. It also tracks Franklin and Graesser's definition of an agent, above, in the sense of being situated in a (structural) environment and acting upon it in pursuit of its own agenda to effect what it senses in the future. There are a couple of differences though. Emirbayer and Mische make a commitment to the methods by which an agent acts. For instance, iteration is done by reactivating prior patterns of thought and action; projection is done by generating possible future trajectories. Franklin and Graesser allow for a broader definition since they do not make similar commitments.

A deeper difference, however, is that Emirbayer and Mische emphasize that their's is a fundamentally relational notion of agency. In their conception, "agency [is] always agency toward something, by means of which actors enter into relationship with surrounding persons, places, meanings, and events" (Emirbayer and Mische 1998). While Franklin and Graesser also note the role of context in determining agency, Emirbayer and Mische treat interaction and context as central to their understanding of agency. Indeed, they treat the agent itself as a relational structure and write that, "our perspective, in other words, is relational all the way down" (Emirbayer and Mische 1998).

This view has also been present in the philosophy of distributed artificial intelligence (DAI). Gasser (1991), in laying out a series of foundational principles for DAI, gives the example of an industrialist flying from Tokyo to Los Angeles after initiating business negotiations in Tokyo and having called ahead to inform her associates in Los Angeles. Gasser points out that this industrialist is simultaneously involved in multiple commitments, some of which are not in her control (like the plane landing in Los Angeles), which involve actions by multiple participants in many places. Her agency, in effect, cannot be localized to her physical location, making her a distributed agent.

What are the implications for our understanding of simulations? First, the above discussion shows that unencapsulated agency is not just a means to designing scalable simulations. The non-localization of agency is a fundamental property of agents in the real world. We need to treat this as a foundational design imperative for simulations.

Second, from a simulation analytics perspective, it suggests that what we get out of a simulation might be quite different from what goes in. As we claimed earlier, our reason for building simulations is that we wish to calculate population-level outcomes from descriptions of individual behaviors and interactions. However, the process of interaction causes the influence (and agency) of each agent to spread out through the population, making it tricky, to say the least, to identify causality, or to determine effective interventions. We can then ask the question in the opposite direction: given a simulation, how do we identify the locus of agency? How do we identify the causal structures, and how do we extract meaningful information?

If agency is understood to be relational, how do we understand structure? Sewell (1992) builds upon the prior work of Giddens and Bourdieu to clarify the notion of social structure and develop a theory of the means by which agents are not only constrained by structure but also able to effect structural change. He develops the notions of *schemas*, which refer to the rules governing interactions among agents, and *resources*, which refer to material things that agents can draw upon or use in their interactions. In his view, "To be an agent means to be capable of exerting some degree of control over the social relations in which one is enmeshed, which in turn implies the ability to transform those social relations to some degree."

Sewell cites the analogy made by Giddens (1976), that the relationship of structure to practice is like the relationship of *langue* to *parole* described by Saussure where, broadly, *langue* corresponds to grammar and *parole* to speech. In Sewell's view, structure consists of both virtual things, the schemas, things like norms, patterns of behavior, and roles, and material things, the resources, such as land, tools, and factories. He goes on to discuss how schemas and resources depend upon each other and how each produces the other. In précis, he defines structures to be "sets of mutually sustaining schemas and resources that empower and constrain social action and that tend to be reproduced by that social action" (Sewell 1992).

In simulations, structure is provided by the rules of interaction, the affordances available to agents, and the internal computational models of agents. This structure constrains agent behavior in the simulation. However, while this seems analogous to the *langue/parole* distinction, it doesn't actually reflect the idea of structuration that Sewell elaborates because in his view, crucially, structures tend to be reproduced by social actions. That this is a tendency, and not a rigid determinism, allows for the possibility of structural change. In a simulation, unless we allow for some kind of self-modifying code, the kinds of structure mentioned above are not reproduced by the agents, nor is there any possibility of change to these structures through the actions of the agents. There is therefore an intermediate level of structure we must consider, which is the structure that is emergent in a simulation. For example, the network structure of interactions and the rules of interaction, both of which might be fixed in a simulation, can lead to the emergence of consensus or norms (Fagyal et al. 2010), which is an emergent structure.

From a simulation analytics perspective, the questions this raises have to do with identifying emergent structures. What defines an emergent structure? How do we formalize (intermediate) levels of structure? When do these structures have causal power? Is it possible for emergent structures to have causal power that is more informative than the lower level structures that define the simulation (Hoel 2017)? These are fundamental questions and have been identified (among others) as priority challenges by Davis et al. (2018).

Simulation Analytics for Social and Behavioral Modeling

We present an example of a question for which we have developed a novel method to analyze simulation results. This is barely scratching the surface of the problems in this space, but it shows how taking a simulation analytics perspective allows finding new insights into outcomes of complex simulations.

This example is from a simulation (Barrett et al. 2013) of National Planning Scenario 1, wherein a 10 kT improvised nuclear device is detonated at ground level in Washington, D.C., at approximately 11 a.m. on a weekday. Detailed calculations have previously been done of the physical effects of this event, including the shock wave, the electromagnetic pulse, thermal fluence, radiation levels over time, and the resulting fallout cloud. The explosion will result in damage to the built infrastructure in the area around ground zero, including power loss, loss of communication, disruption of transportation (road damage), and building damage. Our simulation used information about all of these effects in combination with a detailed model of the population and expected behaviors. The simulation modeled the aftermath of the event out to 48 hours post-event, keeping track of population health and mobility.

In this simulation, a synthetic population model of the Washington, D.C., metro area was used to create 730 833 agents, corresponding to an estimate of all the people expected to be in the area we studied at the time of the detonation. The state of each agent was represented by a number of variables, including its demographics, its household relationships, its current geographical location, its health state, and more. Each agent, at each time step, could choose to do one or both of two actions: try to go somewhere and/or try to call someone.

Where the agent decided to go and whom the agent decided to call depended on the agent's current behavior. We modeled six behaviors: household reconstitution, shelter seeking, evacuation, healthcare seeking, worry, and aiding/ assisting others. These behaviors were temporally extended. They were triggered in part by environmental conditions, in part by the health of the agent, in part by the information the agent had about family members, and in part probabilistically. The details of the behavior model are presented elsewhere (Parikh et al. 2013, 2016a). For our current discussion, it is important to note two things: that the agency of these actors, therefore, was distributed across the physical and social environment and that the behaviors were modeled at an intermediate level between the level of actual actions the agents perform and that of the population level outcomes that are observed.

Identifying Causal Connections Between Behaviors and Outcomes

With this complex behavior representation embedded in a complex simulation of multiple physical and social interactions, it is not easy to identify causal influences (Davis et al. 2018). One of the outcomes we ultimately care about is the distribution over health states at the end of the simulation, i.e. how many people die, how many are injured, how many are in good health, etc.; health state in the simulation was divided into eight discrete levels, with level 0 representing death, level 7 representing perfect health, and intermediate levels representing different levels of injury/incapacitation.

We would like to know how behaviors (and other factors) affect health outcomes. Some factors, such as the explosion itself, are obvious. The probability of survival within a half-mile radius of ground zero is very small. However, other factors are not so clear cut. Does searching for family members help or hurt? Is it better to evacuate or shelter in place? There are several such questions. Below we describe a method we have developed (Parikh et al. 2016b) to tease out the causal influences on a given outcome of interest in a simulation. The method is based on the causal state formalism that has been developed by Crutchfield and others, termed computational mechanics (Crutchfield and Young 1989; Shalizi and Crutchfield 2001). We make use of the fact here, as discussed in the Introduction, that we have complete observability in a simulation, i.e. there are no missing data or unmeasured influences on the outcomes.

Suppose that we have a stochastic process, denoted by a sequence of random variables X_t, drawn from a discrete alphabet, $\mathcal{A}$. At time t, we write $\overleftarrow{X}$ to denote the sequence $X_{-\infty} \ldots X_{t-2}X_{t-1}X_t$ the *past* of the sequence, and $\overrightarrow{X}$ to denote the sequence $X_{t+1}X_{t+2} \ldots X_{\infty}$ the *future* of the sequence, following Crutchfield et al. (2009) and Ellison et al. (2009).

Crutchfield and Young (1989) came up with an elegant and simple method for modeling the time series: group all the histories that predict the same future. This gives rise to a state machine that they call an ϵ-machine, defined as (Ellison et al. 2009):

$$\epsilon(\overleftarrow{x}) = \left\{ \overleftarrow{x}' \mid \Pr(\overrightarrow{X}|\overleftarrow{x}) = \Pr(\overrightarrow{X}|\overleftarrow{x}') \right\} \tag{26.1}$$

The states of this ϵ-machine correspond to groups of histories that assign the same probability distribution to the future of the time series. They showed that ϵ-machines have some very interesting and useful properties. For example, $\overleftarrow{X}$ is statistically independent of $\overrightarrow{X}$ given the current causal state, which makes the ϵ- machine process Markovian. ϵ-Machines are also optimally predictive because they capture all of the information $\overleftarrow{X}$ contains about $\overrightarrow{X}$.

An algorithm for learning an ϵ-machine representation from a given time series, known as *Causal State Splitting Reconstruction* (CSSR), was described by Shalizi and Shalizi (2004). Their approach is to learn a function that can predict the next step of the series optimally, a property they call *next-step suffi-ciency* and that can be applied recursively. The idea is that if a function satisfies both these properties, it can be used to compute the entire future of the time series.

CSSR operates incrementally to infer an ϵ-machine as a hidden Markov model (HMM) from a given time series. To start with, the HMM has just one hidden state. More states are added only when a statistical test shows that the current set of states is insufficient for capturing all the information in the past of the time series.

The CSSR algorithm considers increasingly longer past sequences and com-pares the distribution over the next symbol by doing a statistical test. If L is the length of the past sequences evaluated so far and Σ is the set of causal states estimated so far, then in the next step, CSSR looks at sequences of length $L + 1$. If a sequence of the form ax^L, where x^L is a sequence of length L and $a \in A$ is a symbol, belongs to the same causal state as x^L, then we would have (Shalizi and Shalizi 2004)

$$\Pr(X_t | ax^L) = \Pr(X_t | \hat{S} = \hat{\epsilon}(x^L)) \tag{26.2}$$

where $\hat{S}$ is the current estimate of the causal state to which x^L belongs. If these two distributions are statistically significantly different according to a statistical test such as the Kolmogorov–Smirnov test, then CSSR tries to match the sequence ax^L with all the other causal states estimated so far. If $\Pr(X_t | ax^L)$ turns out to be significantly different in all cases, a new causal state is created and ax^L is assigned to it. This process is carried out up to some length $L_{\max}$.

Our approach adapts the causal state formalism to large multi-agent simula-tions. There are two key differences. First, CSSR relies on having a very long, stationary time series in order to be able to estimate the probability distribu-tions. Instead of having a very long time series, we rely on having a very large number of agents. Second, since our simulation is not a stationary process, we construct the optimal set of clusters at each time step, with respect to a final outcome, as explained below.

In our approach, a multi-agent simulation consists of a set of agents, each of which is defined by a k-dimensional state vector $\mathbf{x}(t) = [x_1(t), x_2(t), \ldots, x_k(t)]^\mathsf{T}$, which evolves over time. Let d_i be the number of possible values x_i can take. The simulation proceeds in discrete time steps from $t = 0$ to $t = T$. Let the number of agents be denoted by N.

We use the term state in a broad sense. It can include, e.g. the action taken by the agent at each time step or how many other agents in the agent's neigh-borhood are doing the same action. It can also include historical aggregations of variables, e.g. it might include a variable that tracks if an agent has ever done a particular action or the cumulative value of some variable so far.

Our goal is to compress the trajectory of each agent through state space to a small number of important states that have a significant impact on the outcomes we care about. Let the outcome variable for agent i be denoted by y_i. We assume that y_i is an instance of a random variable Y. Our algorithm for discovering these causally relevant states proceeds as follows.

We divide the agent population into a set of clusters, $C(t) = \{C_1(t) \cup C_2(t) \cup \cdots \cup C_m(t)\}$, at each time step. Initially, all the agents are grouped into just one cluster, i.e. $m = 1$ at $t = 0$. At each subsequent time step, the state of each agent changes because at least one of $x_1, \ldots, x_k$ changes. The number of ways in which $\mathbf{x}$ can change is $d = d_1 \times d_2 \times \cdots \times d_k$.

Consider an arbitrary cluster of agents, $C_i(t)$. At time step $t + 1$, it can split into up to d groups, based on how each agent's state changes. However, not all of these changes may have a significant impact on the outcome variable. We treat each group derived from $C_i(t)$ as a candidate cluster, denoted by $CC_{i,j}(t + 1)$, where $j \in 1 \ldots d$. At each step, we compare $\Pr(Y|C_i(t))$ with $\Pr(Y|CC_{i,j}(t + 1))$ using the Kolmogorov–Smirnov test.

Our null hypothesis (analogous to Eq. (26.2)) is

$$\Pr(Y|CC_{i,j}(t + 1)) = \Pr(Y|C_i(t)) \tag{26.3}$$

We also introduce a parameter δ, which is a threshold on the *effect size*, which we measure as the Kullback–Leibler (KL)-divergence between $\Pr(Y|C_i(t))$ and $\Pr(Y|CC_{i,j}(t + 1))$. If the null hypothesis is rejected at a level α (say 0.001) and $D_{KL}(\Pr(Y|C_i(t))|| \Pr(Y|CC_{i,j}(t + 1)) > \delta$, then candidate cluster $CC_{i,j}(t + 1)$ is accepted as a new cluster at time step $t + 1$. The need for the effect size threshold is explained further below. If none of the candidate clusters at time step $t + 1$ are accepted, then $C_i(t)$ is added to the set of clusters for time step $t + 1$.

Thus, the entire simulation is decomposed into a tree structure of agent clusters. Furthermore, each cluster splits only when the corresponding state change is informative about the final outcome of concern. The trajectory of each agent traces a path through this tree structure. We compress the trajectory by retaining only those time steps at which the cluster to which the agent belongs splits off from its parent cluster. The parameter δ allows us to control how many new clusters are formed at each step and, consequently, how much compression of trajectories we achieve. Setting δ to a high value will retain only the clusters that have a large difference in outcomes from their parent clusters.

The resulting tree structure can then be queried to find information of interest. For instance, the query "For people who started between 0.6 and 1 mile from ground zero, identify top 10 transitions where current health state remains the same but the expected final health state is reduced, order by expected reduction in descending order" gives the results shown in Table 26.1 (Parikh et al. 2016b).

The results pick out a combination of behaviors, time (iteration), health state, radiation exposure, and whether the agent has received an emergency broadcast (EBR) advising to shelter in place, which has a high predictive value in

Table 26.1 Top results for the query.

Rank	Iteration	Health state	EBR	Behavior	Radiation exposure	Treatment	Distance from ground zero
1	9	3	0	HC seeking	Low	0	>1 mile
2	7	7	0	HRO	High	0	<0.6 mile
3	17	7	0	Aid and assist	High	0	<0.6 mile
4	12	3	0	HC seeking	Low	0	>1 mile
5	4	3	0	HC seeking	Low	0	>1 mile
6	9	7	0	HRO	High	0	<0.6 mile
7	8	7	0	HRO	High	0	<0.6 mile
8	4	3	0	Worry	Low	0	>1 mile
9	5	7	0	HRO	High	0	>0.6 mile, <1 mile
10	3	3	0	HC seeking	Low	0	>1 mile

Source: Reproduced from Parikh et al. (2016b)

determining the final health states of the agents. HRO is the "household recon-stitution option," i.e. the behavior where agents are trying to locate their house-hold members. HC seeking is healthcare seeking. The causally relevant states, therefore, are a combination of agency and structure. They are not designed to be this way; they emerge from the interactions in the simulation and are revealed by taking a simulation analytics perspective.

Conclusion

Isaac Asimov, in a collection of short stories that became the novel *Foundation*, invented the idea of a science called psychohistory. This was a science that allowed forecasting the dynamics of large populations, through a combination of psychology, history, mathematics, and statistics. In his stories, the people who used this science used a device called the Prime Radiant to project the immensely complicated equations onto walls, annotate and revise the equations, and extrapolate them forward. They spent their lives in comparing the predictions of the model with reality and in devising careful modifications (interventions) to the model in order to then steer human civilization in a preferred direction.

There are two points implicit here. One is the idea that forecasting accuracy is (supposedly) the ultimate measure of model quality. The other is that the use of the model is for control, i.e. for devising interventions that steer the system in a preferred direction. These two goals are actually at odds with each other. Sup-pose we had the perfect model for a given (social) system, perfect in the sense of being able to forecast any aspect of it very accurately. With such a model to study counterfactual scenarios, in order to determine the consequences of a planned intervention, say, all we would have to do would be to feed the model with different data corresponding to the scenario we wished to study and then run the simulation to observe the outcomes. This approach treats the model and simulation as a black box and concerns itself with just the inputs and outputs to the system.

However, consider the problem of finding good interventions, given such a perfect model. While such a model would offer the affordances to study changes to any aspect of the real world, its complexity would be as much as the real world, by the law of requisite variety (Ashby 1958; Bar-Yam 2004). Finding good interventions requires understanding the causal processes in the system, i.e. opening up the black box, which would be very challenging with such a complex model. In other words, such a model would be very difficult to understand and use. In general, the more realistic a model becomes, the harder it is to work with.

However, even with a complex and realistic model, we retain the advantages of complete observability and complete knowledge of the computation being carried out by each agent, along with the ability to rerun the simulation, as discussed in the Introduction. We do, however, need new simulation analytics methods for sensemaking with such models.

Taking a step back from this dichotomy, there are actually many reasons to do modeling beyond forecasting and control (Epstein 2008), such as explaining, generating new hypotheses, training practitioners, and more. Consider a question like, "how many different types of agents do you have in your simulation?" Generally, this question is asked from a design perspective, e.g. we might have designed civilians and responders into a disaster simulation. However, this question is equally valid, and perhaps more meaningful, from a simulation analytics perspective. Can we look at the outcome of the simulation and derive a taxonomy of agents? We might discover that a finer-grained taxonomy is more meaningful, e.g. the category of civilians might be further refined into those who are seriously injured, those who are trying to evacuate, and those who are moving toward the disaster area (to look for family members, for example). This kind of analysis would clearly help with a high-level understanding of the simulation, and with planning appropriate response procedures. This is just one example of a range of new questions that open up when we take a simulation analytics perspective. The key ideas about this perspective that we have tried to convey in this article are as follows:

- Simulations allow us to compute population-level outcomes from complex individual actions and interactions.
- Large-scaled complex simulations constitute a new regime of investigation.
- Behavior needs to be modeled and understood at a level that is intermediate between individual actions and population-level outcomes.

- A relational view of agency and structure shows how interactions in a simulation result in causality being *diffuse.*
- There are a host of new questions that can be asked in this setting, for which new methods are needed. We term this general area simulation analytics. Discovering interventions is one example.

More generally, our thesis can be summed up as follows: That which we call behavior is a fundamentally distributed, interactionist, and emergent phenomenon. It arises from the interplay between agency and structure. Simulation analytics offers a means of studying it as such.

Acknowledgments

We thank the members of the Network Dynamics and Simulation Science Laboratory for the many useful discussions over several years. This work has been supported in part by Air Force Research Laboratory Contract FA8650-18-C-7826, DARPA Cooperative Agreement D17AC00003, DTRA CNIMS Contract HDTRA1-17-0118, NIH MIDAS Cooperative Agreement U01GM070694, NIH Grant 1R01GM109718, NSF IBSS Grant SMA-1520359, and NSF NRT-DESE Grant DGE-154362. This work was carried out in part when the authors were at Virginia Tech.

References

Ashby, W.R. (1958). Requisite variety and its implications for the control of complex systems. *Cybernetica* 1 (2): 83–99.

Axtell, R.L. (2016). 120 million agents self-organize into 6 million firms: a model of the U.S. private sector. *Proceedings of the 15th International Conference on Autonomous Agents and Multi-Agent Systems (AAMAS)*, Singapore.

Bar-Yam, Y. (2004). Multiscale variety in complex systems. *Complexity* 9 (4): 37–45.

Barrett, C.L., Wolinsky, M., and Olesen, M.W. (1996). Emergent local control properties in particle hopping traffic simulations. In: *Proceedings of the Conference on Traffic and Granular Flow (TGF)* (ed. D.E. Wolf, M. Schreckenberg, and A. Bachem), 169–174. Singapore: World Scientific.

Barrett, C., Eubank, S., Marathe, A. et al. (2011). Information integration to support model-based policy informatics. *The Innovation Journal* 16 (1).

Barrett, C., Bisset, K., Chandan, S. et al. (2013). Planning and response in the aftermath of a large crisis: an agent-based informatics framework. In: *Proceedings of the 2013 Winter Simulation Conference* (ed. R. Pasupathy, S.-H. Kim, A. Tolk et al.), 1515–1526. Piscataway, NJ: IEEE Press.

Batty, M. and Xie, Y. (1994). From cells to cities. *Environment and Planning B: Planning and Design* 21: S31–S48.

Beckman, R.J. (ed.) (1997). Transportation Analysis Simulation System (TRANSIMS): The Dallas-Ft. Worth Case Study. Technical Report. LAUR-97-4502. Los Alamos National Laboratory.

Crutchfield, J.P. and Young, K. (1989). Inferring statistical complexity. *Physical Review Letters* 63 (2): 105–108.

Crutchfield, J.P., Ellison, C.J., and Mahoney, J.R. (2009). Time's barbed arrow: irreversibility, crypticity, and stored information. *Physical Review Letters* 103 (9): 094–101.

Davis, P.K. and O'Mahony, A. (2019, this volume). Improving social-behavioral modeling. In: *Social-Behavioral Modeling for Complex Systems*. Wiley.

Davis, P.K., O'Mahony, A., Gulden, T. et al. (2018). Priority Challenges for Social-Behavioral Research and its Modeling. Technical Report. RR-2208-DARPA. Santa Monica, CA: RAND Corporation.

Ellison, C.J., Mahoney, J.R., and Crutchfield, J.P. (2009). Prediction, retrodiction, and the amount of information stored in the present. *Journal of Statistical Physics* 136 (6): 1005–1034.

Emirbayer, M. and Mische, A. (1998). What is agency? *American Journal of Sociology* 103 (4): 962–1023.

Epstein, J.M. (2008). Why model? *Journal of Artificial Societies and Social Simulation* 11 (4): 1–12.

Eubank, S., Guclu, H., Anil Kumar, V.S. et al. (2004). Modelling disease outbreaks in realistic urban social networks. *Nature* 429: 180–184.

Fagyal, Z., Swarup, S., Escobar, A.M. et al. (2010) Centers and peripheries: network roles in language change. *Lingua* 120 (8): 2061–2079. https://doi.org/10.1016/j.lingua.2010.02.001.

Franklin, S. and Graesser, A. (1997). Is it an agent, or just a program? A taxonomy for autonomous agents. In: *Intelligent Agents III Agent Theories, Architectures, and Languages. ATAL 1996, Lecture Notes in Computer Science (Lecture Notes in Artificial Intelligence)*, vol. 1193 (ed. J.P. Müller, M.J. Wooldridge, and N.R. Jennings), 21–35. Springer-Verlag.

Gasser, L. (1991). Social conceptions of knowledge and action: DAI foundations and open systems semantics. *Artificial Intelligence* 47: 107–138.

Giddens, A. (1976). *New Rules of Sociological Method: A Positive Critique of Interpretive Sociologies*. Hutchinson.

Haas, P.J., Berberis, N.C., Phoungphol, P. et al. (2012). Splash: Simulation optimization in complex systems of systems. *Proceedings of the 50th Annual Allerton Conference on Communication, Control and Computing*.

Halloran, M.E., Ferguson, N.M., Eubank, S. et al. (2008) Modeling targeted layered containment of an influenza pandemic in the United States. *Proc. Natl. Acad. Sci. U.S.A.* 105 (12): 4639–4644.

Hoel, E.P. (2017). When the map is better than the territory. *Entropy* 19 (5): 188.

Li, J. and O'Donoghue, C. (2013). A survey of dynamic microsimulation models: uses, model structure, and methodology. *International Journal of Microsimulation* 6 (2): 3–55.

Lymperopoulos, I.N. and Ioannou, G.D. (2016). Understanding and modeling the complex dynamics of the online social networks: a scalable conceptual approach. *Evolving Systems* 7: 207–232.

Orcutt, G.H. (1957) A new type of socio-economic system. *The Review of Economics and Statistics* 39: 116–123.

Parikh, N., Swarup, S., Stretz, P.E. et al. (2013). Modeling human behavior in the aftermath of a hypothetical improvised nuclear detonation. *Proceedings of the International Conference on Autonomous Agents and Multiagent Systems (AAMAS), Saint Paul, MN, USA.*

Parikh, N., Hayatnagarkar, H.G., Beckman, R.J. et al. (2016a). A comparison of multiple behavior models in a simulation of the aftermath of an improvised nuclear detonation. *Autonomous Agents and Multi-Agent Systems, Special Issue on Autonomous Agents for Agent-Based Modeling* 30 (6): 1148–1174. https://doi.org/10.1007/s10458-016-9331-y.

Parikh, N., Marathe, M.V., and Swarup, S. (2016b). Summarizing simulation results using causally-relevant states. In: *Autonomous Agents and Multiagent Systems: AAMAS 2016 Workshops, Visionary Papers, LNAI*, vol. 10003 (ed. N. Osman and C. Sierra), 88–103. Springer.

Parker, J. and Epstein, J.M. (2011). A distributed platform for global-scale agent-based models of disease transmission. *ACM Transactions on Modeling and Computer Simulation* 22 (1): Article 2.

Schiller, M., Dupuis, M., Krajzewicz, D. et al. (2015). Multi-resolution traffic simulation for large-scale high fidelity evaluation of VANET applications. *Proceedings of the SUMO User Conference – Intermodal Simulation for Intermodal Transport.*

Sewell, W.H. Jr. (1992). A theory of structure: duality, agency, and transformation. *American Journal of Sociology* 98 (1): 1–29.

Shalizi, C.R. and Crutchfield, J.P. (2001). Computational mechanics: pattern and prediction, structure and simplicity. *Journal of Statistical Physics* 104 (3/4): 817–879.

Shalizi, C.R. and Shalizi, K.L. (2004). Blind construction of optimal nonlinear recursive predictors for discrete sequences. In: *Proceedings of the 20th Conference on Uncertainty in Artificial Intelligence* (ed. M. Chickering and J. Halpern), 504–511. Banff, Canada: AUAI Press.

Simon, H.A. (1996). *The Sciences of the Artificial*, 3e. Cambridge, MA: MIT Press.

Swarup, S., Gohlke, J.M., and Bohland, J.R. (2017). A microsimulation model of population heat exposure. In: *Proceedings of the 2nd International Workshop on Agent-based Modeling of Urban Systems (ABMUS).*

Zou, Y., Torrens, P.M., Ghanem, R.G., and Kevrekidis, I.G. (2012). Accelerating agent-based computation of complex urban systems. *International Journal of Geographical Information Science* 26 (10): 1917–1937.

27

Using Agent-Based Models to Understand Health-Related Social Norms

Gita Sukthankar[1] and Rahmatollah Beheshti[2]

[1] *Department of Computer Science, University of Central Florida, Orlando, FL 32816, USA*
[2] *School of Public Health, Johns Hopkins University, Baltimore, MD 21218, USA*

Introduction

Agent-based models (ABMs) have been shown to be valuable for many types of social simulation problems, including predicting the effects of geography, economic fluctuations, and public policy decisions on human populations. Understanding the influence of social norms on human behavior is an important aspect of performing accurate population-level modeling, and forecasting *norm emergence* has attracted research attention in both the agent-based social simulation and multi-agent system communities.

In this chapter, we describe an agent-based simulation that we constructed to model smoking cessation trends at University of Central Florida following the initiation of a smoke-free campus policy. Since social norms have been shown to strongly affect health-related habits such as overeating, binge drinking, and smoking, our simulation focuses on the social, rather than addictive, elements of the smoking cessation problem. Our lightweight normative architecture (LNA) (Beheshti and Sukthankar 2014b) models the impact of personal, social, and environmental factors on recognition, adoption, and compliance with campus smoking norms. When initialized with student survey data, it accurately predicts trends in smoking reduction over a one year timeframe.

One weakness with LNA is that it has a relatively simple internal model of the human decision-making process. To address this issue, we created a general normative architecture, cognitive social learner (CSL) (Beheshti et al. 2015), which is capable of reasoning about any social norm. CSL provides a computational mechanism for transitioning behaviors learned during repeated social interactions into the agent's internal cognitive model of preexisting beliefs, desires, and intentions. By incorporating a more complex normative

Social-Behavioral Modeling for Complex Systems, First Edition.
Edited by Paul K. Davis, Angela O'Mahony, and Jonathan Pfautz.
© 2019 John Wiley & Sons, Inc. Published 2019 by John Wiley & Sons, Inc.
Companion website: www.wiley.com/go/Davis_Social-Behavioralmodeling

reasoning model, CSL not only predicts smoking trends but also accurately forecasts population-level perception on the social acceptability of smoking.

Related Work

Our models in this chapter examine smoking behaviors from a normative point of view. Nonnormative models of smoking behavior already exist; for instance, *SimSmoke* is one of the widely used tobacco control policy simulations. It models the dynamics of smoking use and smoking-attributed deaths in the society of interest, as well as the effects of policies on those outcomes (Levy et al. 2005). Other types of simulations have been used to model the consequences of second-hand smoking (Dacunto et al. 2013). In addition to norms, our proposed approach also simulates network effects as was done in Beckman et al.'s (2011) study on the propagation of adolescent smoking behavior.

Most existing models within the medical and public health community are based on statistical analysis of smoking data (Luo et al. 2015). These methods often focus on a narrow aspect of the problem, such as modeling abstinence due to changes in brain cells. However, some models in the public health domain have been based on system dynamics approaches (Timms et al. 2012). An introduction to this set of techniques can be seen in Homer and Hirsch (2006).

The relationship between social norms and smoking behavior was examined as part of a European Union study on the impact of cultural differences on the emergence of norms in different countries after the commencement of anti-smoking legislation (Dechesne et al. 2013). Our current ABM does not attempt to recreate cultural effects. Rather than studying smoking cessation behavior at the macroscopic level, we adopt a higher fidelity approach in which the daily behavior patterns of individual agents are simulated within an activity-oriented microsimulation.

Lightweight Normative Architecture (LNA)

To construct a normative model for a real-world scenario, we need to define both a norm architecture and the components that are used to recreate the real-world problem.

Each agent has a personal *smoking value* (SV) ranging from 0 to 100 that governs its behavior. As shown in Figure 27.1, our architecture contains three

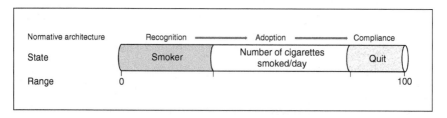

Figure 27.1 A schematic representation of the LNA architecture.

stages: *recognition, adoption,* and *compliance.* In the first stage (recognition), the beliefs of an agent change and develop. During the adoption phase, the agent commences action. Note that the general definition of adoption in normative systems is very consistent with our smoking scenario. During the adoption phase the agent can opt to violate the norm. The equivalent violation in the smoking scenario (recidivism) is quite common in those trying to quit. To quit smoking, a smoker usually decreases the number of smoked cigarettes, which can be considered as another adoption behavior. The compliance phase is used to simulate the time period when the agent seriously attempts to quit smoking.

Cognitive Social Learners (CSL) Architecture

CSL adds an advanced structure of learning and reasoning about norms to the simple decision-making process of the LNA architecture. Figure 27.2 shows a schematic view of CSL. In this architecture, the belief, desire, and intention (BDI) components implement the cognitive aspects of norm formation, while the game-theoretic (GT) interaction and reinforcement learning (RL) recognition parts implement the social aspects. CSL only models the rational and social elements of human health-related decisions; it meant to be a complement to existing biological models of craving and dependency (Gutkin et al. 2006).

The representation used for the BDI components and the norms is based on a simplified version of the framework introduced by Casali et al. (2008) and Criado et al. (2010b) in which a certainty degree is assigned to each representation. For example, $(D^-$payfine, 0.45) designates a negative desire toward paying a fine with a certainty degree of 0.45.

Belief, Desire, and Intention

The CSL architecture follows a classic BDI structure. Like many normative architectures, each agent is initialized with a set of personal values that model innate preferences. In CSL, these personal values are used to create type 1

Figure 27.2 Cognitive social learners (CSL) architecture.

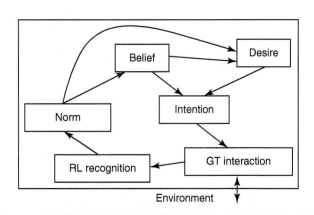

beliefs that have a certainty equal to 1; for instance, $(B[\text{hedonism} = 50], 1)$ indicates that the personal value of the agent regarding hedonism is equal to 50. The second form of belief, type 2, is used to model the agent's actions and is represented as $(B[\alpha]\varphi, \delta)$. For instance, $(B[\text{smoking}]\text{botherRest}, 0.30)$ indicates that the agent believes, with certainty of 0.30, that smoking would bother the other agents.

Desires can be determined independently or based on the agent's beliefs. Desires are represented as $(D^*\varphi, \delta)$, which models the positive or negative $(* = \{-, +\})$ desire of an agent regarding state φ with certainty of δ. An agent may update its desires when its beliefs change. This process is shown in Eq. (27.1); the certainty value of desire D is updated based on function f, which is a user-defined function:

$$((D^*\varphi, \delta_\varphi), (B[\alpha]\varphi, \delta_\phi)) \Rightarrow (D^*\varphi, f(\delta_\varphi, \delta_\phi)) \tag{27.1}$$

Intentions are derived from the set of positive desires, if they have a certainty value higher than sum of the certainty values of all negative desires relevant to the intention. Equation (27.2) shows this:

$$((D^+\varphi_{i_1}, \delta_{\varphi_{i_1}}), \dots, (D^+\varphi_{i_n}, \delta_{\varphi_{i_n}}), (\text{plan}_j, \delta_j))$$
$$\Rightarrow (I_k, f(\delta_{i_1} \dots \delta_{i_n}, \delta_j)) \tag{27.2}$$

while $\Sigma(\delta_{i_1} \dots \delta_{i_n}) \geq \Sigma(\delta_{l_1} \dots \delta_{l_n})$ and l_1 to l_n are indices of negative desires toward effects of I_k. According to this formula, the set of positive desires (from i_1 to i_n) and plan j will determine the intention k based on a user-defined function f. In the smoking case, an agent might have positive desires toward higher happiness, conforming with smoker friends, but negative desires toward becoming ill and being observed by others. In this case, if the sum of certainty values for happiness and consistency is more than the sum of certainty values for becoming sick and being observed (assuming that smoking is part of the agent's current plan), the agent will smoke.

Game-Theoretic Interaction

Instead of deciding its actions based on intentions alone, which is often the case in BDI-based methods, the agent's final action is determined after playing a social dilemma game with (one of) the other agents. The other agent could be a neighbor agent or a friend. The maximum certainty value of available intentions is used to create a two-by-two matrix. The two possible actions are performing or refraining from that action. After calculating the payoff value for an action based on the related intentions, fixed values of α and β are used to increase the value of the elements in the matrices representing coordinated action (the agent and its neighbor selecting the same actions) (Easley and Kleinberg 2010). An example of this matrix for the smoking scenario is shown in Table 27.1. ψ shows the computed payoff value for smoking. ψ' is the payoff for not smoking.

Table 27.1 Example payoff matrix for smoking (S=Smoke, NS=not smoke).

	S	NS
S	$\psi + \alpha$	ψ
NS	ψ'	$\psi' + \beta$

Based on the outcome of games played with this payoff matrix, an agent decides what action to perform. What an agent observes after performing an action may cause an agent to update its personal values (type 1 beliefs) and learned norms, which in turn modifies its behavior in subsequent steps. For instance, in the case of our example scenario, after smoking, if there is an advertisement in its vicinity, its value for the effect of advertisements will increase because of the formed habit.

Norm Recognition Using RL

The goal of this component is to construct a practical way of recognizing/ learning norms while connecting different components of the architecture. Our RL-based recognition component plays the role of a hub among norms and personal values (beliefs) on the one hand and the GT interaction on the other hand.

The combination of GT interaction and RL-based recognition components is used to implement the social learning process, which propagates norms across the agent population. The aim of the social learning framework is different from similar processes in the domain of multi-agent RL. In those the agents play iterative games to learn a policy, resulting in a competitive or coopera-tive equilibrium. Sen and Airiau (2007) note several differences between social learning and multi-agent RL, including the lack of equilibrium guarantees. At every time step, each agent interacts with a single changing agent, selected at random, from the population. The payoff received by the CSL agent depends only on this interaction. We use a basic Q-learning algorithm for recognizing norms in which states are the discretized current values of an agent's payoff matrices. Learning results in modifications to the certainty degree of available norms. Rewards are calculated based on the changes in the personal values.

Norms

The process of recognizing a social norm is modeled by an agent increasing the norm's certainty value to a positive value. The agent updates the certainty values of norms based on its observations after performing an action. Our norms are represented using the format introduced in Criado et al. (2013), $\langle \Delta, C, A, E, S, R \rangle$, in which Δ designates the type of norm, C is the triggering condition, A and E show the activation and expiration periods of the norm, and S and R indicate a reward or sanction. For example, this is an example of a possible norm: $(\langle\text{prohibition}, \text{smoking}, -, -, \text{payfine}, -\rangle, \delta)$, which is always valid since there is no duration on activation, A, and expiration, E.

All of possible norms are initialized at the beginning of the simulation with the certainty value of zero. Agents update their norms by increasing or decreasing the certainty value of each norm after making an observation. For instance, if the agent receives a fine after smoking, it will update its current value of (δ) in the above norm example with $(\delta + \epsilon)$, where ϵ is a user-defined value.

An agent's current norms are used to update its beliefs and desires. The updating procedure is shown in Eqs. (27.3)–(27.5). Here, norms are abbreviated as N instead of $\langle\Delta, C, A, E, S, R\rangle$. If there are any relevant rewards R (or sanctions S), the positive desire D^+ (or a negative desire D^-) will be updated. f functions are user-defined functions:

$$((N_i, \delta_N), (B[\alpha]\varphi, \delta_\phi)) \Rightarrow (B[\alpha]\varphi, f(\delta_N, \delta_\phi)) \tag{27.3}$$

$$((N_i, \delta_N), (D^+\varphi, \delta_\varphi), R \neq \emptyset) \Rightarrow (D^+\varphi, f(\delta_N, \delta_\varphi)) \tag{27.4}$$

$$((N_i, \delta_N), (D^-\varphi, \delta_\varphi), S \neq \emptyset) \Rightarrow (D^-\varphi, f(\delta_N, \delta_\varphi)) \tag{27.5}$$

As an example, if the norm $(\langle\text{prohibition}, \text{smoking}, -, -, \text{payfine}, -\rangle, 0.75)$ exists and a negative desire toward paying fine $(D^-\text{payfine}, 0.55)$, assuming the agent has just paid a fine for smoking $(S \neq \emptyset)$ with $f = \min(\max(0.75, 0.55), 1)$, the resulting updated desire would be $(D^-\text{payfine}, 0.75)$.

Figure 27.3 shows the pseudo-code describing an agent's behavior for one time step in the CSL implementation. The certainty value of beliefs and desires is initialized uniformly at random at the beginning of the scenario.

```
init(blf, des, pln, q-tbl)
repeat
    generateIntention(blf, des, pln)              ▷ Eq. (27.2)
    updatePMatrix(maxIntention)
    if (converged-Qtbl) then
        playGame(pMatrix,neighbors)
        performAction()
        update-qTable(rew, san)
    else
        performAction()
    end if
    update-norms(rew, san)
    update-beliefs(rew, san, norms)               ▷ Eq. (27.3)
    update-desires(rew, san, norms)               ▷ Eqs. (27.1), (27.4), and (27.5)
until agent not selected
```

Figure 27.3 CSL pseudo-code (blf, beliefs; des, desires; pln, plans; rew, rewards; san, sanctions).

Smoking Model

Our smoking model considers three sets of factors that are known to affect human smokers: personal, social, and environmental influences. Considering the complex and challenging nature of modeling smoking behaviors, specifically the addictive property of smoking, we tried to have an inclusive model that contains as many factors as possible.

Personal

Our model includes a set of personal values that are specific to each person and depend on their personality; Dechesne et al. (2013) use a similar set of values within their model of cultural differences that affect smoking behavior. According to the sociological theory of cultural value orientation introduced by Schwartz (2006), three types of values determine cultural differences in societies. These values are defined by three bipolar cultural dimensions that can be used to describe possible resolutions to problems confronting societies. In our model, we adopted two of these values since the third dimension is specifically for cultural differences, which are negligible for our relatively homogeneous undergrad population. The two adopted values are described below:

- *Embeddedness vs. autonomy*: This determines how much an individual's preferences, feelings, and ideas are affected by others through various relationships vs. being cultivated internally.
- *Mastery vs. harmony*: This refers to the dichotomy of being ambitious, daring, and self-assertive vs. being consistent, understanding, and appreciative of the environment.

The first item is referred as *individualism* (ind) and the second one as *achievement* (ach). The third item that is not included in our model is equality. In addition to these two personal values drawn from Schwartz's sociological (or anthropological) model, three other personal values are included:

- *Regret* (rgt): In our scenario, this value shows how much the individual is regretful about smoking and is used to model the phenomenon of addiction. The role of regret in smoking behaviors is described in Conner et al. (2006); it is related to their willingness to quit smoking or decrease their tobacco usage.
- *Health* (hlt): As the name implies, this value shows the extent to which a person is health conscious and pays attention to medical recommendations.
- *Hedonism* (hdn): The pleasure-seeking aspect of one's personality. Health and hedonism were also used in a related study on smoking bans in European Union countries (Dechesne et al. 2013).

Social

The second aspect of our model is used to quantify the effects of the community on the individual. To do this, we create a synthetic friendship network for our simulated community using the method described in Wang et al. (2011) for creating human networks that follow a power law degree distribution and possess homophily, a greater number of link connections between similar nodes. The network generator uses link density (ld) and homophily (dh) to govern network formation. For our smoking model, three elements are defined to determine the homophily of a node: age, gender, and undergraduate major. The nodes of the graph represent the individuals (agents) in the simulation.

Environmental

The third category of factors that affect people's smoking behavior is what they observe or encounter in their surroundings. Four items are considered in this category:

Others (oth): One major factor that affects norm compliance is observing other people's behavior. Seeing other smokers can affect the agents' decisions to obey policies, particularly when complying with smoking cessation rules. Similar behaviors in humans have been shown to exist and are usually referred to as *observational learning*. Various studies have shown the effect of observation on smoking behaviors (e.g. Akers and Lee (1996)).

Signs + butts (sbt): This item is specifically related to the effect of installed *no smoking* signs that advise people to refrain from smoking. A key research challenge here is to simulate the behavior of people in response to this type of notification. A recent study by Schultz et al. (2013) on littering in public locations shows that people tend to obey installed signs when there is no trash around the sign, but when litter exists in the vicinity, the rate of people who do not follow the signs increases significantly. Using a similar approach, we consider signs and cigarette butts together and model the influence of observed cigarette butts on a person's on-campus smoking behavior.

Advertisements (adv): Physical advertisements can also influence smoking behaviors. These advertisements are a major part of the campus smoke-free program. This category refers to tents, fliers, billboards, catalogs, posters, and banners installed permanently in different locations of campus.

Miscellaneous (msc): This category encompasses all of the other factors that might influence a smoker's decisions. One major aspect of this category is nonphysical influences, especially digital, educational, and promotional activities. Also included in this category is the role of different cessation facilities available on campus, such as workshops and nicotine replacement therapy (NRT).

The five elements introduced for the personal values, the social element, and the four environmental factors are all defined as ranging from 0 to 100. The main SV is calculated using this formula:

$$SV = (k_1 * \text{ind}' + k_2 * \text{ach}' + k_3 * \text{rgt} + k_4 * \text{hlt}' + k_5 * \text{hdn}$$

$$+ k_6 * \text{frd} + k_7 * \text{oth} + k_8 * \text{sbt} + k_9 * \text{adv} + k_{10} * \text{msc}) / \sum_{i=1}^{10} k_i$$

$$(27.6)$$

The SV falls between 0 and 100. In this formula, k_1 to k_{10} shows nine coefficients that are assigned to the user. Prime (') means complement, which in this case is equal to 100 -. The friendship value (frd) is determined using the social model.

Agent-Based Model

The original version of the ABM used in this work was built to study the transportation patterns of people and vehicles (Beheshti and Sukthankar 2012, 2014a). Before presenting the new components, we will first describe the function of the base ABM. The model was built to simulate the movement patterns of students at the University of Central Florida. The data for building the model were gathered through an online survey. In the survey, participants were asked to answer questions about the time they arrive and depart campus, locations they visit, and frequency of their visits. A set of statistical distributions was fit to the answers of each question. These distributions were then used to initialize the model parameters, including those that govern the activities of an agent.

Each agent arrives, visits locations on campus, and then leaves campus according to its own personal schedule. Various specialized rules were added to the model to improve the verisimilitude of the whole system. Examples of defined rules include limitations on the number of cars that can enter a parking lot or the hours that shuttle services operate. The accuracy of the ABM was measured in several different ways, including comparing the obtained statistics from the ABM with other independently collected data sources.

To implement the smoking simulation scenario, the proposed smoking model was added to the original ABM. We added two parameters, age and gender, to each agent's parameter set to be used for measuring homophily in the social model. Each agent is initialized as a smoker or nonsmoker at the start of the ABM, based on the number of smokers in the survey data. The smoke-free campus policy is assumed to be in effect immediately after the start of the simulation.

Having a detailed transportation model facilitates implementing the environmental aspects of the proposed smoking model in high fidelity. The assumption is that each smoker agent smokes an average of 15 (for men) and 10 (for women) cigarettes per day. These numbers are based on the reported statistics in Burns et al. (2003). The effect of observing others smoking on campus is incrementally aggregated for each agent through the described RL algorithm. The observation occurs whenever an agent is close to an agent that is smoking at the same time.

The exact location of no smoking signs and physical advertisements is defined in the campus map used in the ABM. Based on our observational study of the campus, cigarette butt locations are marked near the large college buildings, but not general buildings like the student union and library. This trend might occur because of the frequent cleaning of these areas or the tendency of people to avoid smoking in heavily crowded areas. While the agent moves around campus, it passes physical advertisements. Similar to observing others smoking, every encounter with an advertisement increases its chance of affecting the user.

Figure 27.4 shows the user interface of the ABM. In this figure, the location of buildings, routes, and also the advertisements can be seen. The last item of the environmental model (misc factors) is implemented by a random value that represents the aggregation of all other factors.

LNA Setup

Here, we describe the details of implementing smoking factors on the ABM that follows the LNA architecture:

Personal: Personal values were added to the set of parameters possessed by each agent in the ABM. These values are calculated using distributions fitted to the available survey data.

Social: To implement the diffusion of smoking behaviors in the friendship network, a game-theoretic approach (Easley and Kleinberg 2010) is used. Here, a simple two-by-two matrix is defined that contains four different states that can occur in the smoking scenario. Table 27.2 shows this matrix. The descriptions below the table show how the payoffs are calculated. The abbreviations on the right side of the equations relate to being a smoker (s) or nonsmoker (n).

Table 27.2 Payoff matrix governing the diffusion process in the friendship network.

		Node B		
		Smoker	Nonsmoker	
Node A	Smoker	ss+α	sn	ss = ind$'$ + ach$'$ + hlt$'$ + hdn
		ss+α	ns	sn = ind + ach + hlt + hdn$'$
	Non-smoker	ns	nn+β	ns = ind + ach + hlt$'$ + hdn
		sn	nn+β	nn = ind$'$ + ach$'$ + hlt + hdn$'$

Prime ($'$) means complement, which in this case is equal to 100 -. ind, individualism; ach, achievement; hlt, health; hdn, hedonism.

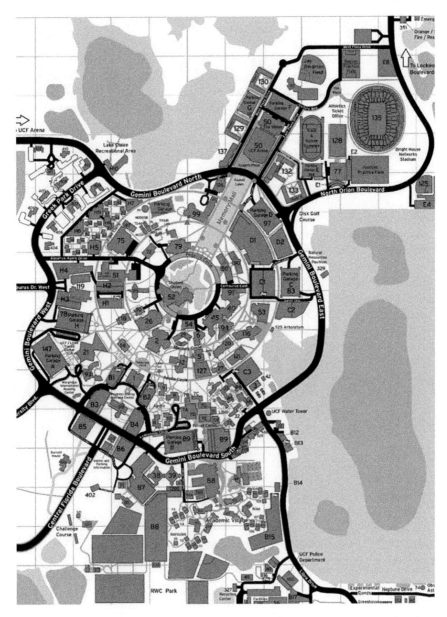

Figure 27.4 Screenshot of the agent-based model. The advertisements (pentagons) and no smoking signs (triangles) are shown on the map.

Each individual is either a smoker or nonsmoker. The payoff for each of four entries of a node is calculated according to three factors: personal values, network neighbors, and whether the subsequent state is similar to the current state. Similar to the mechanism that was described earlier, α and β values are added to the model to show the tendency of people to maintain their current state. These two parameters are constant positive values that make the value of the payoff higher for the cases that the agent remains a smoker or nonsmoker than in the cases that a state transition occurs. The final value for the friendship element of model (frd) is calculated based on the current state of the individual and her friends, using the payoff matrix.

Environmental: Each of the four elements is represented in the model with values ranging from 0 to 100. These factors are modeled as part of agents' beliefs for the CSL architecture. For the implementation of environmental factors on LNA architecture, a simplified version of Q-learning is used to govern the effects of the environmental factors. As Table 27.3 shows, when encountering an environmental factor such as a banner, the state of an agent is defined by the current value of its personal and social elements. The agent can either be affected by the environmental factor or disregard it. In case of the first action, the value of that environmental factor will increase by a fixed amount, but in the second case nothing changes. The reward that agent receives from each action is calculated based on three elements of its personal value vector: regret, health, and hedonism. The reward value falls between -1 and $+1$ and is calculated using the following formula:

$$\text{Reward} = (\text{regret} + \text{health} - 2 * \text{hedonism})/200 \qquad (27.7)$$

A dynamic learning schedule is utilized for the Q-learning, which results in a higher rate of learning at the beginning of the simulation and a lower one afterward.

CSL Setup

Here we describe the elements of the CSL architecture:

Beliefs, desires, and intentions: The two first personal values, individualism and achievement, are implemented as fixed value elements of beliefs (type 1).

Table 27.3 Q-learning definitions for state, actions, and rewards.

States	Current value of personal and social elements
Actions	Pay attention or not
Rewards	Calculated based on the values of regret, health. and hedonism

If the agent does not pay attention, it means that the agent opts to ignore a specific environmental element. Regret and health affects the reward value positively, and hedonism affects it negatively.

The remaining three personal factors, regret, health, and hedonism, plus environmental factors are implemented as variables and part of each agent's beliefs. The certainty values (δ) for beliefs and desires are assigned uniformly at random at the beginning of the scenario. The intentions are determined according to Eq. (27.2). The main desires and intentions defined in this system refer to smoking and not smoking.

Payoff matrices: An agent plays games with both its friends and other agents near to it to determine its actions. For each action, an agent has a two-by-two payoff matrix that determines the agent's decision. The agent picks the intention with the highest certainty value. The values of this payoff matrix are determined by the certainty degree of the selected intention. This means that in our architecture, the intentions do not directly determine agent's actions; instead they define payoff matrix values. The friendship (frd) value in the smoking model is calculated using the payoff matrix values.

Norm recognition: The learning component is implemented using the Q-learning algorithm. Actions are the action performed by the agent: to smoke or refrain. The reward value is assumed to be the same as the reward value defined for the RL and smoking diffusion in LNA. The current values of the payoff matrices determine the states of the Q-table. The selected action modifies the certainty value of norms. After an agent performs an action, it observes the consequences of its action to compute the overall received payoff, which is then used to update the Q-table.

Norms: Norms are created using the same procedure introduced. Only dynamic (variable) parts of beliefs are updated. All possible norms are initialized as having a certainty value of zero. During initialization, we create all of possible norm combinations based on the introduced norm representation: $\langle \Delta, C, A, E, S, R \rangle$. The type of norm and its reward or sanction nature can be determined by the value for C. We assume that all norms are always valid during the experiment, so we do not need to take A and E into account. Thus 12 possible norms are defined for this scenario: |obligation, prohibition, permission|*|smoking, not smoking|*|reward, sanction|.

Data

Our ABM uses data from three surveys of UCF students. In spring 2012, we did an online survey of 1003 students to collect the data used to model campus transportation patterns. The other two surveys were conducted by health services; one of them was done in fall 2011, before the smoke-free policy was instituted, and the second in fall 2012, at the end of the first year of the smoke-free campus. Both of these surveys were performed as part of the annual university ACHA-NCHA reporting process. The student answers to five questions in the first survey were used to determine the numerical values for the five personal values. The personal values and corresponding survey questions are as follows:

- *Individualism*: Do you think breathing smoke-free air on campus is a right?
- *Hedonism*: Do you think smokers have the right to smoke on campus?
- *Achievement*: Would you feel comfortable asking someone to put out their cigarette?
- *Health*: Would a smoke-free campus policy make campus healthier?
- *Regret*: If you smoke, are you interested in attending a smoking cessation program?

The questionnaire was designed using a Likert scale. The personal values in our work were matched to questions after the survey was conducted, and normal distributions fitted to the data were used to initialize the agents' personal values in the ABM. The university administration used the answers to the following three questions to determine the success of the smoke-free campus policy. In our work, the answers to the second and last questions were used to show the accuracy of the proposed model. These three questions are as follows:

- Do you support the campus smoke-free policy?
- Do you smoke?
- Are you likely to take smoking cessation classes?

The other data used to implement the model, including the location of advertisements and installed no smoking signs, were obtained from campus sources.

Experiments

Validation is a major challenge while evaluating ABMs – how to show that the model matches reality. One approach is to evaluate the model by comparing the statistics obtained from the model with other sources of data as indicators of ground truth. Here, the data obtained from the second and third questions of the survey described in the previous section are used to evaluate the model. These two questions show the percentage of smokers among the students and also the percentage of those who are willing to attend smoke cessation workshops.

The ABM is initialized with the same number of smokers and people willing to participate in smoking cessation classes as indicated in the survey data.[1] According to our definition, a smoker is an agent whose SV is below the quitting threshold. Similarly, we use the middle part of the proposed SV range to identify an agent who is willing to attend smoking classes. An agent who is willing

---

1 Since the total number of students is known, the percentage values also determine the numbers; hence we use the terms interchangeably.

Table 27.4 Experimental settings for smoking value (sv).

Agent state	Range
Nonsmoker	90–100
Willing to participate in classes	50–90

to participate in classes has an SV between the two proposed thresholds. The assumption is that the adoption phase in the proposed architecture shows the situation where the agent has not reached the compliance phase. So, assuming that an agent in the compliance mode is willing to attend smoking classes is consistent with the proposed architecture, because attending class is not a clear quitting task, but is a behavior toward quitting (the action phase).

Table 27.4 shows the parameters that are used in the experiments to determine the smoking range. As the table shows, the value 50 is used for the first threshold, and 90 for the second threshold shown in Figure 27.1. In our experiments, the values for the coefficients k_3, k_4, and k_6 in Eq. (27.6) were 3, 3, and 2. The other coefficients were equal to 1. In the next section, a set sensitivity analysis experiments related to these values are presented. For the network generation part, the values for the link density, ld, and homophily, dh, were 0.40 and 0.66.

In addition to the LNA and CSL architectures, we have also implemented the norm–belief–desire–intention (NBDI) architecture (Criado et al. 2010a). The NBDI benchmark does not play the social dilemma game and does not use RL to generate and update norms. In this case, intentions determine actions, and then the norms are updated based on the feedback received from the environment. Note that the way that the norm representation was implemented (by modifying the certainty value of norms) is not part of the original version of NBDI. The norm recognition part in the original NBDI was assumed to work as a black box, and there was insufficient detail about its implementation to recreate it. Hence we simply used the same norm recognition structure for both CSL and NBDI.

Results

Using these assumptions, we ran our agent-based simulation for a period of a year from fall 2011 to fall 2013. In these experiments, we initialized the simulation with the same number of smokers and students willing to go to the classes as the initial survey data and then compared the numbers obtained from the simulation with the final survey data. During this period, the agents commute to campus and follow schedules governed by the transportation model. The proposed smoking model simulates the smoking behavior of students during the year of study. The average simulation error of ten runs of

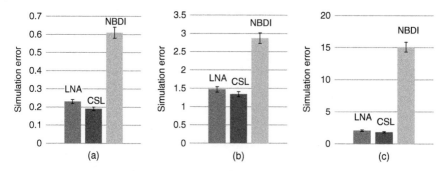

Figure 27.5 Comparison between the performances of different normative architectures. The simulation error refers to the difference between the value obtained by each method and the empirical survey data. (a) Percentage of predicted smokers vs. empirical data for fall 2012. (b) Percentage of predicted smokers vs. empirical data for fall 2013. (c) Percentage of predicted students willing to attend classes vs. empirical data for fall 2012.

the model is reported in Figure 27.5. Simulation error refers to the difference between the values obtained from each method and the real value from the experimental data. The two measures shown here are the percentage of smoker students and the percentage of smoker students who are willing to attend smoking cessation classes. The empirical data for the percentage of smokers was also available for 2013.

Figure 27.5 shows the comparison between the number of students who were smokers and students willing to participate in smoking cessation classes. The performance of CSL at predicting the actual adoption of the smoking cessation norm is comparable with the specialized smoking model (LNA) and superior to NBDI.

A powerful feature of ABMs is their potential ability of predicting future trends, which is not possible if past data is not representative of future data. This can be a great tool for policy-makers who want to analyze the effects of modifying various parameters of a specific model. In Figure 27.6, the predicted percentage of smokers for the period of the years 2011–2016 is shown. The values shown for the years 2011–2013 are the same as Figure 27.5. Both models slightly overpredict the number of smokers identified using the UCF Health Services survey data, with CSL being slightly more accurate. For the year 2013, it was confirmed by the health services department that the reported rate (3.9%) seems a bit lower than what they were expecting based on national and state averages. One possibility is that smoking behavior is underreported by the students or being supplanted by vaping. The current assumption in our model is that different system properties remain the same during the simulated years. A factor that our model does not take into account is the gradual change of the population as students arrive to the school and graduate.

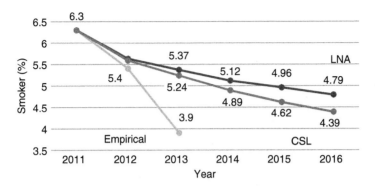

Figure 27.6 Predicted percentage of smokers for future years.

Table 27.5 Standard coefficient (beta) values of the applied linear regression to perceived social acceptability of smoking (independent variable) and quitting intention (dependent variable).

	Beta	*p* level
CSL	0.22	0.001
LNA	0.001	0.007
NBDI	−0.01	0.005

Table 27.5 shows a comparison between the different architectures at predicting the perceived social unacceptability of smoking. This phenomenon is reported in many smoking studies including ((Dotinga et al. 2005; Hammond et al. 2006)) as occurring when smoking bans exist in human cities. Brown et al. (2005) showed that perceived social acceptability of smoking among referent groups is independently associated with both strength of intention to quit and actual quitting behavior.

In our smoking model, it is assumed that an agent has the intention to quit smoking if its SV is within the first and second threshold values. The social unacceptability of smoking across the population of agents is determined using the value for one of the agent's personal characteristics (IND). The value of this factor was initialized based on data from a survey question asking whether the participant believes smoking is acceptable on campus. A linear regression model was used to examine the relationship between these two elements, and the standard coefficient (beta) value of the applied linear regression is shown in Table 27.5. The CSL model produces a positive beta value, which is consistent with the real-world data. This shows that, using CSL, agents are able to reason about the socially perceived unacceptability of smoking behavior and modify their behaviors accordingly. Therefore, CSL is modeling norm emergence in a more realistic manner. On the other hand, the beta values for the

LNA and NBDI architectures are close to zero, which does not accurately reflect the results reported in independent smoking studies.

Additionally, we performed a set of sensitivity analysis experiments on the results that we obtained from the two architectures. Since our models include a number of variables that could directly affect the final behavior of our system, the sensitivity analysis illuminates the effect that each variable can have on the final outcome. Five of the ten coefficients (Eq. (27.6)) plus the two threshold values for determining the three stages of norm formation (Figure 27.1) are used as the independent variables in our sensitivity analysis model. The remaining five coefficients are not shown due to their close relation to the current coefficients. The analysis is performed on one independent value at a time.

Figure 27.7 shows the range of output values for different values that can be assigned to five of the k_i coefficients, and similarly Figure 27.8 shows the output range for the two threshold values. By comparing the results shown in Figure 27.7a with 27.7b, and also 27.8a with 27.8b, we can observe that LNA seems to be more sensitive to noise than CSL. By changing the coefficient values from 0 to 6, the maximum change in the percentage of smokers is close to 4 for LNA and less than 2 for CSL. In the case of the two threshold values (shown in Figure 27.8), LNA's results vary in the range of 3.5, while the length of CSL's range is less than 2.5. Overall, the sensitivity of the model's output to the set of input values is low, and because of type of equation that we proposed, the output range for different values remains linear.

We also study ablated versions of the model that lack one of the three smoking elements (social, environmental, or personal). The results for alternate months during the year of simulation are reported in Figure 27.9. The reported results are, again, averaged over ten runs, and in all cases the initialization

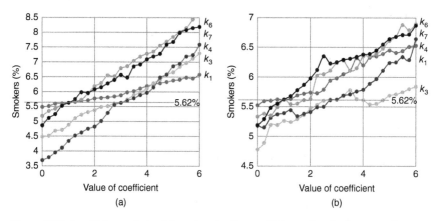

Figure 27.7 Sensitivity analysis of the values for five coefficient values in the regression equation introduced for determining the final smoking value in our models. Horizontal lines show the current values reported by our models. (a) LNA; (b) CSL.

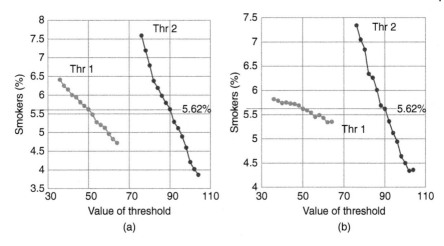

Figure 27.8 Sensitivity analysis of the effects of the two threshold values in our models. Horizontal lines show the current values reported by our models. (a) LNA; (b) CSL.

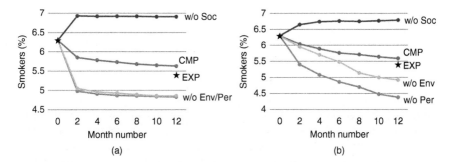

Figure 27.9 The percentage of smoker students in LNA (a) and in CSL (b) during the one year simulation period. The numbers from the survey data are marked by the star icons at the beginning and end of the simulation period (experimental/EXP). The figure shows the predictions of the proposed model (complete/CMP), the model without the personal values, without the social aspect, and without environmental influence. There is a close match between the predicted values of the complete model and the survey data.

configuration is based on the survey data. In Figure 27.9a, b the left star shows the starting value, which is the empirically measured value and is the same for all four experiments. Without the personal and environmental components, the model tends to underestimate results in comparison with the final empirical results. Without the social part, the model overestimates smoking behavior. Based on the size of differences between the empirical results and the other experiments for CSL, it can be concluded that the personal values are the major predictors in determining smoking behaviors. Environmental factors had the lowest impact on predicting smoking behavior.

Discussion

The LNA architecture is based on a simple model of norm adoption that is similar to many of the mechanisms that are currently used for constructing normative multi-agent systems. On the other hand, CSL represents a group of architectures that employ more complex structures for normative reasoning. CSL integrates internal cognitive structures with social interaction mechanisms. Our results demonstrate that CSL provides a richer model of health-related social norms. Specifically, when it comes to modeling the intricacies of humans' behaviors – like the correlation between the unacceptability of smoking in a society and quitting intention – our LNA exhibits a greater sensitivity to design assumptions and produces a lower fidelity model than CSL.

Conclusion

In this chapter, we introduced two architectures (LNA and CSL) for modeling the emergence of norms in ABMs. Our proposed normative architectures focus on the impact of social factors on smoking behavior and are intended to complement biological models of the effects of nicotine addiction. We examine the effect of a richer reasoning model on the overall performance of a model in understanding human behavior. Although both methods demonstrate good prediction abilities at the population level, the agents with more advanced cognitive and learning structures exhibit a higher level of verisimilitude in simulating human behavior. This was a gratifying result because we expect ABM to have this potential. At the same time, results are driven by estimates of a number of parameter values, and more data is needed to permit developing better methods for making such estimates and then updating them as additional information comes in. Doing so would be akin to what statisticians do, but we would anticipate results to be much better with the cognitive and learning structures of the ABM.

Acknowledgments

This research was supported by NSF IIS-08451 and DARPA HR001117S0018-SocialSim-FP-026.

References

Akers, R.L. and Lee, G. (1996). A longitudinal test of social learning theory: adolescent smoking. *Journal of Drug Issues* 26, 317–343.

Beckman, R., Kuhlman, C., Marathe, A. et al. (2011). Modeling the spread of smoking in adolescent social networks. *Proceedings of the Fall Research Conference of the Association for Public Policy Analysis and Management*, Washington, DC.

Beheshti, R. and Sukthankar, G. (2012). Extracting agent-based models of human transportation patterns. In: *Proceedings of the ASE/IEEE International Conference on Social Informatics*, 157–164. Washington, DC.

Beheshti, R. and Sukthankar, G. (2014a). A hybrid modeling approach for parking and traffic prediction in urban simulations. *AI and Society: Journal of Knowledge, Culture and Communication* 30 (3): 333–344.

Beheshti, R. and Sukthankar, G. (2014b). A normative agent-based model for predicting smoking cessation. *Proceedings of the International Conference on Autonomous Agents and Multi-Agent Systems*, 557–564. Paris, France.

Beheshti, R., Ali, A.M., and Sukthankar, G. (2015). Cognitive social learners: an architecture for modeling normative behavior. *Proceedings of the AAAI Conference on Artificial Intelligence*, Austin, TX.

Brown, D.G., Riolo, R., Robinson, D.T. et al. (2005). Spatial process and data models: toward integration of agent-based models and GIS. *Journal of Geographical Systems* 7 (1): 25–47.

Burns, D.M., Major, J.M., and Shanks, T.G. (2003). Changes in number of cigarettes smoked per day: crosssectional and birth cohort analyses using NHIS. *Smoking and Tobacco Control Monograph* 15, 83–99. NIH publication no. 03-5370.

Casali, A., Godo, L., and Sierra, C. (2008). A logical framework to represent and reason about graded preferences and intentions. In: *11th International Conference on Principles of Knowledge Representation and Reasoning*, 27–37.

Conner, M., Sandberg, T., McMillan, B., and Higgins, A. (2006). Role of anticipated regret, intentions and intention stability in adolescent smoking initiation. *British Journal of Health Psychology* 11 (1): 85–101.

Criado, N., Argente, E., and Botti, V. (2010a). A BDI architecture for normative decision making. In: *Proceedings of the 9th International Conference on Autonomous Agents and Multiagent Systems*, 1383–1384.

Criado, N., Argente, E., and Botti, V. (2010b). Normative deliberation in graded BDI agents. In: *Multiagent System Technologies*, 52–63.

Criado, N., Argente, E., Noriega, P., and Botti, V. (2013). Human-inspired model for norm compliance decision making. *Information Sciences* 245: 218–239. Statistics with Imperfect Data.

Dacunto, P.J., Cheng, K.-C., Acevedo-Bolton, V. et al. (2013). Identifying and quantifying secondhand smoke in multiunit homes with tobacco smoke odor complaints. *Atmospheric Environment* 71: 399–407.

Dechesne, F., Di Tosto, G., Dignum, V., and Dignum, F. (2013). No smoking here: values, norms and culture in multi-agent systems. *Artificial Intelligence and Law* 21 (1): 79–107.

Dotinga, A., Schrijvers, C.T.M., Voorham, A.J.J., and Mackenbach, J.P. (2005). Correlates of stages of change of smoking among inhabitants of deprived neighbourhoods. *The European Journal of Public Health* 15 (2): 152–159.

Easley, D. and Kleinberg, J. (2010). *Networks, Crowds, and Markets*, 8. Cambridge University Press.

Gutkin, B.S., Dehaene, S., and Changeux, J.-P. (2006). A neurocomputational hypothesis for nicotine addiction. *Proceedings of the National Academy of Sciences of the United States of America* 103 (4): 1106–1111.

Hammond, D., Fong, G.T., Zanna, M.P. et al. (2006). Tobacco denormalization and industry beliefs among smokers from four countries. *American Journal of Preventive Medicine* 31 (3): 225–232.

Homer, J.B. and Hirsch, G.B. (2006). System dynamics modeling for public health: background and opportunities. *American Journal of Public Health* 96 (3): 452–458.

Levy, D.T., Nikolayev, L., and Mumford, E. (2005). Recent trends in smoking and the role of public policies: results from the SimSmoke tobacco control policy simulation model. *Addiction* 100 (10): 1526–1536.

Luo, S.X., Covey, L.S., Hu, M.-C. et al. (2015). Toward personalized smoking-cessation treatment: using a predictive modeling approach to guide decisions regarding stimulant medication treatment of attention-deficit/hyperactivity disorder (ADHD) in smokers. *The American Journal on Addictions* 24 (4): 348–356.

Schultz, P.W., Bator, R.J., Large, L.B. et al. (2013). Littering in context personal and environmental predictors of littering behavior. *Environment and Behavior* 45 (1): 35–59.

Schwartz, S.H. (2006). A theory of cultural value orientations: explication and applications. *International Studies in Sociology and Social Anthropology* 104: 33–78.

Sen, S. and Airiau, S. (2007). Emergence of norms through social learning. In: *Proceedings of the International Joint Conference on Artifical Intelligence*, 1507–1512.

Timms, K.P., Rivera, D.E., Collins, L.M., and Piper, M.E. (2012). System identification modeling of a smoking cessation intervention. In: *16th IFAC Symposium on System Identification*, 11–13. Brussels, Belgium.

Wang, X., Maghami, M., and Sukthankar, G. (2011). Leveraging network properties for trust evaluation in multi-agent systems. In: *Proceedings of the IEEE/WIC/ACM International Conferences on Web Intelligence and Intelligent Agent Technology, Volume 02*, 288–295.

28

Lessons from a Project on Agent-Based Modeling

Mirsad Hadzikadic[1] and Joseph Whitmeyer[2]

[1] *Department of Software and Information Systems, Data Science Initiative, University of North Carolina, Charlotte, NC 28223, USA*
[2] *Department of Sociology, University of North Carolina, Charlotte, NC 28223, USA*

Introduction

Hadzikadic, O'Brien, and Khouja published an edited volume *Managing Complexity: Practical Considerations in the Development and Application of ABMs to Contemporary Policy Challenges* in 2013 (Hadzikadic et al. 2013), almost five years after the completion of a DARPA-sponsored project called Actionable Capability for Social and Economic Systems (ACSES), which focused on the feasibility of creating a computational social science laboratory. The effort was, at the time, one of the most comprehensive efforts to use agent-based modeling for computation-based quantitative research in social sciences. The contextual framework was the battle for winning the hearts and minds of the population at the height of the ongoing war in Afghanistan. The book outlined the computational platform that was developed during the project and the way that end users were supposed to use it to analyze problems of interest.

However, the book did not address fully some of the interesting issues noticed throughout the project – issues that were simply not within the project's scope – including the complexity of verification and validation of agent-based models, the elusiveness of self-organization and emergence phenomena, and the need to gain the trust and acceptance of agent-based models by the end users, leading to their use of such models for describing, explaining, predicting, and prescribing the states of the domain under their control. This chapter addresses those issues. It also includes a set of recommendations for resolving these issues.

Social-Behavioral Modeling for Complex Systems, First Edition.
Edited by Paul K. Davis, Angela O'Mahony, and Jonathan Pfautz.
© 2019 John Wiley & Sons, Inc. Published 2019 by John Wiley & Sons, Inc.
Companion website: www.wiley.com/go/Davis_Social-Behavioralmodeling

It is first necessary to provide context for the reader. What follows is a summary of the ACSES effort that will be both the basis and the point of reference for the subsequent discussion.

ACSES

The ACSES was a pilot model designed to help a research team evaluate the possibility of creating a computational social science laboratory. The example used to motivate model development was the question of "population allegiance in Afghanistan in the face of the Taliban insurgency, and effort by the Afghan government and coalition forces to combat the insurgency" (Whitmeyer 2013). The purpose in building ACSES was to "demonstrate that a useful ABM model could be built that would: 1) explain or predict a population outcome of interest to policy-makers and strategists, 2) make use of data of various kinds, 3) be testable with real-world data and perform well at prediction, and 4) implement different social theories so that the efficacy of the theories could be tested. The population outcome of principal interest in the ACSES model is citizen allegiance, i.e., the number of citizens supporting the Taliban, supporting the government, or being neutral" (Whitmeyer 2013). The model produces the geographical distribution of citizen allegiance. Below, we discuss the structure of the model and the social-behavioral theories it incorporates.

The model has four kinds of agents: Taliban insurgents, coalition forces, government forces, and citizens. The model implicitly represents ethnic leaders through user-specified allegiance-related orders and leader characteristics, which then affect the citizens' allegiance: "The model incorporates data about the geographic distribution of population, ethnic groups, and economic status. The option exists to calibrate and test the model by comparing model predictions with actual data about violent events in one district of Afghanistan over a three-year span in the 2000s. Model calibration consists principally of choosing the social theory that does best at reproducing the empirical pattern of violence as functions of the above. The model incorporates social theories in such a way as to allow a continuum of different possibilities. To facilitate our analyses, however, we specified 12 distinct theories. In addition, we built into the model four theories of how actors, over time, may change in characteristics important for the social theories" (Whitmeyer 2013).

The ACSES model demonstrates that agent-based models can be useful in explaining or predicting citizens' allegiances and some consequences in situations of insurgency. Here are sample questions (from Whitmeyer 2013) that ACSES may help answer in case, say, that the Taliban would mount an offensive in a specific region of Afghanistan:

(1) How many citizens in the region initially support the Taliban?
(2) How are the allegiance of citizens, their ideology, and their support for the different factions likely to change over time?

(3) What will be the violence level in the region if the coalition forces fight?

(4) How would gaining the support of local leaders for the coalition forces change the answers to questions 1–3?

(5) How would a shift in the support of local leaders to the Taliban change the answers to questions 1–3?

(6) How would sending additional coalition forces into the region change the answers to questions 1–3?

(7) How would increasing or decreasing the resources of local leaders affect the allegiance, ideology, and behavior of citizens?

As mentioned above, the ACSES model can also be used to evaluate a variety of social science theories for their effectiveness in explaining events and predicting the likely range of outcomes. These theories can range across three levels: individual action, social effects of aggregating individuals, and psychological change within individuals. Consequently, ACSES can answer questions such as (Whitmeyer 2013):

(1) Do different social theories produce qualitatively different results (e.g. the answers to questions 1–7 above)?

(2) Do theories at different levels combine to produce new results?

(3) Are the effects of combining theories, whether within or across levels, linear or nonlinear?

The ACSES simulation includes models of the citizen agents at two levels and time scales. One model is at the behavioral level and instantiates social theories that explain allegiance. The theories of allegiance are implemented via a utility function for the citizen agents. The other model changes variables in the utility function and instantiates theories concerning psychological change.

The Social Theories

In social science, the word *theory* typically refers to a hypothesis or set of hypotheses about some relatively narrow category of phenomena.[1] The theory may imply behavioral priorities, i.e. an ordering of behavioral preferences. The key mechanism in the ACSES model that allows implementation of social theories is the utility function. For each considered behavior, the utility function combines the states of the agent on any number of variables into a single quantity, labeled *utility*. Each agent then chooses the behavior with the greatest utility for the next time step. Many different mathematical forms of utility functions exist, with somewhat different mathematical properties. The ACSES model employs the commonly used Cobb–Douglas utility function, which specifies utility as the product of preferences raised to a fractional power.

---

1 Note that this is a different usage than in the natural sciences. It is as if physics referred to *mass theory, friction theory, resistance theory*, and so on. Instantiating many social science *theories* in one model, as with the utility function here, shows they share a common theoretical framework.

Through the utility function, ACSES instantiates three specific theories or hypotheses concerning the obedience of citizen agents to a leader's orders; we call these the *obedience theories*. Two different kinds of social influence may affect citizen allegiance as well. An obedience theory may be tempered by one or both social influence theories.

The utility function (shown below) permits a continuum of preference combinations to affect an agent's behavior. In the ACSES model, specific preference combinations are used that correspond most closely to important theories of allegiance in the social science literature. If other theories involving only the preferences incorporated in the ACSES model are identified, ACSES may be able to accommodate them immediately. For theories involving preferences not in the ACSES model, the modular nature of the Cobb–Douglas utility function would make adding the necessary preferences relatively straightforward.

The ACSES utility function incorporates seven variables likely to affect a citizen agent's allegiance choice. Five of these are measures of an agent's nonsocial environment or its position vis-à-vis its leader: the agent's loyalty to its leader (L), its economic situation (E), its security situation (V), its ideological position (I), and the coercive resources wielded by its leader (C). The other two variables measure aspects of the agent's social environment: the allegiance position of the agent's close associates (F) and absence of repression, a combination of high collective action in the vicinity and little activity to counter it (R).

The following equation gives the specific version of the Cobb–Douglas utility function that is used in ACSES:

$$U = (1 - L)^{W_L}(1 - C)^{W_C}(1 - I)^{W_I}(1 - E)^{W_E}(1 - V)^{W_V}(1 - F)^{W_F}(1 - R)^{W_R}$$

where U is the utility of a behavior and the seven variables, L, C, I, E, V, F, and R, are the deviations from the agent's ideal state (the state of maximum utility) for each of these preferences or motivations, all measured on a 0–1 scale. The weights determine the relative effect of the different preferences on U and, therefore, characterize their relative importance to the agent.[2] Different weight settings correspond to different social theories, as described below. Briefly noting some properties of the Cobb–Douglas function, a weight of zero means the factor takes the value 1 regardless of the state of the agent, i.e. the preference has no effect. Fixing the weight for a given preference, the agent's utility is maximal when the variable is 0, that is, the agent is at its ideal state with respect to that preference. Variables are allowed to approach 1, maximal deviation from ideal, but never to attain 1, because if one variable did, the agent's utility would be 0 regardless of other preferences.

The ACSES model determines allegiance choice for an agent in a given situation for a given setting of the parameters w, then, by calculating U for three

---

2 They are called weights because the logarithm of utility is a sum of terms in which the *w*'s are weights in the normal sense.

different allegiance choices: $A = -1$, supporting one side; $A = 1$, supporting the other side; and $A = 0$, neutral. The choice is whichever yields the highest utility U.

The three obedience theories specifically identified in ACSES are alternative hypotheses about the effects of a leader's order concerning its followers' allegiances. Legitimacy theory states that citizen agents want to follow the orders of the leader if they consider him or her legitimate. This theory is operationalized, therefore, by emphasizing the loyalty component of the utility function, i.e. setting w_L to be large. The coercion theory posits that citizen agents follow the orders of the leader to the extent that they believe that they will be rewarded for doing so and punished for not. This theory is operationalized by setting w_C to be large. The representative theory suggests that citizen agents follow the leader only to the extent that the leader advocates what the citizen agents want him or her to advocate. Essentially, this means that the citizen agents choose allegiance based on other important characteristics of their situation, the ones included in ACSES being their ideology, their economic situation, and their security situation. Thus, for this theory w_I, w_E, and w_V will be large.

The two social influence theories specified in the ACSES model are hypotheses about the effects of social influence on participation in collective action. Social influence theory posits that a citizen agent is more likely to join a collective action if more immediate friends do. The resistance to repression theory encompasses two balancing effects: (i) citizen agents are not likely to join a collective action if there are strong repressive forces around them, and (ii) citizen agents are more likely to join a collective action if a larger segment of the population is engaged in the action. In the ACSES situation, joining a collective action means taking the group's allegiance position. Resistance to repression is taken into account only for citizen agents who are considering helping the weak side in their neighborhood. Given a high setting of w_R, citizen agents are less likely to confront factions that have a relatively strong presence in their neighborhood, especially when that faction has little support among other citizens in the neighborhood. This theory was deemed relevant and included due to the circumstances of Afghanistan in the midst of the Taliban insurgency. Note that these two theories (social influence and resistance to repression) can be used in conjunction with each other and with an obedience theory, but they will necessarily temper the obedience theory, in the sense that the more some kind of social influence matters, the less the obedience considerations will matter.

Table 28.1 summarizes the five theories or hypotheses of citizen allegiance specified in the ACSES model out of the available range of preference combinations. Domain refers to the variables emphasized in the theory, that is, whether they concern a leader or whether they concern an agent's social environment. The emphasis is achieved through noted settings of the parameters w.

Figure 28.1 below presents a schematic diagram of a citizen agent's allegiance choice in the ACSES model, as it has been described above.

Table 28.1 Theories reflected in ACSES.

Theory	Domain	Parameter settings
Legitimacy	Obedience	High loyalty
Coercion	Obedience	High coercion
Representative	Obedience	High economics, violence, ideology
Social influence	Social influence	High influence + obedience theory settings
Resistance to repression	Social influence	High repression + obedience theory settings

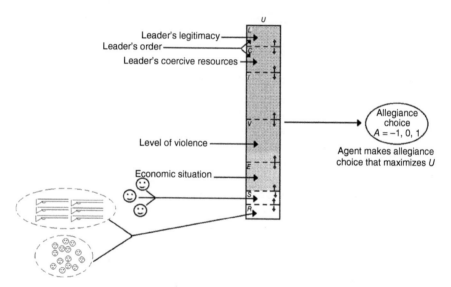

Figure 28.1 Schematic of agent's allegiance choice in ACSES model. Shading indicates the portion of the model relating to obedience theories. The unshaded region refers to the security and absence-of-repression portions.

The specific theories or hypotheses realized in ACSES were drawn from literature in the social sciences concerning political allegiance (Whitmeyer 2013). This literature is, as described in the overview chapter of this volume (Davis and O'Mahony 2019), fragmented. Taken individually, the theories are arguably rather simplistic because it is likely that many factors matter. The ACES framework and its Cobb–Douglas utility function, however, allows for a continuum of parameter settings, as noted above, and for experimentation with composite theories, as well as for the addition of new variables. Thus, ACSES can be used to investigate new theories of citizen allegiance or – given enough data – to discover superior combinations of variables, that is, combinations and weightings of parameters that correspond to no known named theory.

The Adaptation Theories

Over time, the values of the variables that feed into the utility function may change. Some of the variables reflect the agent's external conditions, such as economics and security. Of the seven components of the utility function, two, however, concern variables that are largely internal to the agent. These are loyalty to the leader and ideology (indicating which side of conflict most closely reflects the agent's beliefs).

To accommodate the possibility of these internal variables changing over time, ACSES incorporates four theories of psychological change, labeled (i) cognitive dissonance, (ii) results based, (iii) homophily, and (iv) socialization (Whitmeyer 2013). Cognitive dissonance increases the perceived legitimacy of the leadership if the chosen behavior coincides with the leadership's orders and lowers it if it does not. The result-based mechanism is based on a measure of the agent's welfare comprising the agent's economic situation and security situation. The perceived legitimacy of the leadership is adjusted according to the agent's welfare improvement (or lack of it) over the previous simulation iteration. Homophily adjusts a citizen agent's perception of the legitimacy of the leadership to become more similar to the perception of nearby agents of the same type. Socialization assumes the citizen agent acquires his or her belief in the leadership's legitimacy at some point in the past, and it does not change after that. The name comes from the long-standing sociological idea that as children, people are socialized into the beliefs they will retain throughout their lives (Durkheim 1982 [1895]). Thus, the socialization mechanism is the default psychological change theory if no other theory is invoked. The same four theories of psychological change can be applied to a citizen agent's ideology.

Summary

To summarize, then, ACSES simulates the choice of one behavior (allegiance to the government) and of one class of actor (citizen agents) in the context of Afghanistan during the Taliban insurgency and allows the use and testing of social theories or hypotheses concerning allegiance by implementing them via a Cobb–Douglas utility function. The utility function involves seven preferences that allow agents' allegiance choices to be affected by external conditions, such as security, leaders' orders, and friends' choices, and internal states, such as personal ideology. Because some of the internal states of agents may change over time, ACSES also includes alternative theories of adaptation. Let us now turn to how such a model can be verified and validated.

Verification and Validation

The question in this section is how one should go about verifying and validating agent-based models that simulate real-life complex situations. Many authors have talked about V&V over the years, some in the context of economic systems (Atkinson 1969; Kahneman 2003; Smith 2003; Ormerod 2004; Dosi et al. 2006); others have done so in the context of general social science (Axelrod 1997; Pilch et al. 2000; Ormerod and Rosewell 2004, 2009; Rand and Wilensky 2006; McNamara et al. 2007; Windrum et al. 2007). Leigh Tesfatsion maintains a website that offers a comprehensive treatment of this subject, including references (Tesfatsion 2017). Another chapter in this volume also discusses broad issues of V&V (Volkova et al. 2019).

In this chapter, we discuss some particular V&V issues in agent-based modeling rather than attempting something more comprehensive.

Adding to the complexity of V&V is the fact that models are designed with different functions in mind. The overview chapter of this volume (Davis and O'Mahony 2019) distinguishes among validation for description, explanation, postdiction, exploratory analysis (and coarse prediction), and prediction functions of the model. This chapter does not address those issues, but its observations and suggestions are probably relevant to each of these classes of validation.

Before proceeding with methods for V&V, it is worth noting the magnitude of the V&V challenge if we want to test the model comprehensively. ACSES has 144 different combinations of social theories, the result of having nine behavioral theories, each of which can have any of four theories of change in perceived leadership legitimacy or any of four theories of change in ideology. Also, the relative weights of the seven elements in the utility function can be set to any values between 0 and 1 as long as they sum to 1. Thus, two or more of the obedience theories can be combined, by averaging the weights, or entirely new theories involving these preferences or motivations can be implemented. In a sense, then, a continuum of obedience theories concerning allegiance can be implemented. Lastly it is straightforward, although requiring a bit more work, to add a new motivation or preference to the utility function. This may be necessary to implement some alternative behavioral theory not considered for the ACSES model, especially if the target behavior is changed from citizen allegiance to something else.

In general, an agent-based model can be in any number of states within a high-dimensional space defined by agent attributes, rules of engagement, and spatial movements. In addition, randomness is used often in the model to avoid introducing unnecessary bias into the system/model. Consequently, the search space for agents in an agent-based model is prohibitively large for any realistic problem. This problem of large search spaces is exacerbated by the fact that situations modeled by agents are often rare, unique, complicated, and complex. So, there are not many such situations to use for empirical assessment of the performance of agent-based models. For example, we have not seen hundreds or thousands of wars in Afghanistan of the kind that resulted in the removal of

Taliban from power and its subsequent resurgence. There was only one such a war. Therefore, machine learning methods are not very useful here.

With this sobering background of difficulties, let us now consider some methods that can be brought to bear.

Verification

In general, verification refers to assuring that the program does what the program designer intended it to do. Methods from software engineering can be very useful. Software verification in software engineering usually comprises static and dynamic verification methods. Static methods check the code before it runs, checking for adherence to code conventions, detecting bad practices, and calculating various software metrics associated with code quality and – if a specification exists – more formal verification. Dynamic verification is good for finding faults, also known as software bugs. This chapter focuses mostly on V&V issues specific to agent-based modeling. It takes for granted that the modeling and programming has been done professionally and with good practices (e.g. along-the-way modular testing). If not, verification testing later will probably be frustrating and perhaps hopeless.

Given the huge space of an ABM's possible states, modelers cannot systematically traverse the model instantiations through all the stages to see if the model arrives at a given end state through a *verifiably appropriate* sequence of steps. Based on long experience and the specific experience of the ACSES project, we highlight the following methods:

- *Visualization*: Although informal, this verification method is still the easiest and most frequently used. It is strongly recommended. By using visualization, the modeler can verify whether agents move in the plausible direction, cover the appropriate distance, cluster as expected or possible, change color as expected, or get born or disappear as intended. Also, as the model performs over time, it becomes clear whether specific runs end too early, never end, produce nonsense outcomes, or do not produce expected variability. In practice, a great many *small* bugs introduce visible irregularities.
- *Charts*: Unlike visualizations that show dynamic movement of agents over time, charts show how each key variable changes its value over time. Charts can quickly indicate irregular behavior, especially when paired with other variables that are expected to correlate with it (or not).
- *Gradual implementation*: It is beneficial to start with a minimal model and add additional variables and features only after the existing ones are thoroughly verified.
- *Tracing*: It is also useful to have a trace of the state of a set of randomly selected agents as they act and move in time as part of the model run. Each state of such agents at each tick should be evaluated in the context of the agent's neighborhood.

- *Exploratory analysis*: Ideally, the modelers should run a system with every possible combination of starting conditions, including a sweep of values for agent attributes, their rules of behavior, and ways of spatial movement. This will provide a distribution of outcomes for the dependent variables as a function of initial values for independent variables. Due to combinatorial explosion, it will often be necessary to settle for a careful sampling of the full space.

Validation

Validation is a completely different challenge. Once the model is reasonably verified, the next step is to validate it by making sure that the system reasonably accurately models the phenomenon under consideration for the purposes intended for the model. In the case of ACSES, we had to resort to finding a specific situation that happened on the ground to see if the model was accurately comparing three social science theories that were competing for the best explanation of the real-life events of interest. In general, the range of validation techniques that would be meaningful in the context of a model like ACSES might include:

- *Distribution of possible futures*: This method is useful when we have the distribution of values for each input variable. Starting the system with known initial conditions, proposed rules of behavior, and spatial movement of agents allows the modeler to anticipate the final state of the system after some period of time. Of course, different allocations of values to input variables, based on the known distribution of values for each input variable, may result in a different system outcome. Let us say that we are interested only in two output states of a model warning that a state is failing: Failure Expected = Yes or Failure Expected = No. Then, after a sufficiently large number of simulation runs, the distribution of probabilities for those two possible outcomes will indicate the nature and probability of possible system outcomes.
- *Use cases*: Finding well-documented use cases that can be used to test the validity of the recommendations offered by the model can go a long way toward the model validation in the eyes of the end user. This was the strategy eventually adopted in ACSES. However, this is not sufficient. The model should also be tested with sensitivity analysis (varying the value of one input variable for each system run) or exploratory analysis (varying multiple input variables at the same time) for all variables that matter in the context of the use case at hand. This, of course, is just a use-case-specific variant of the possible future validation strategy.

Self-Organization and Emergence

Definition

Self-organization and emergence are hallmark phenomena of complex systems. They represent a non-directly coded pattern of behavior observed at the agent level (self-organization) or the system level (emergence). They are derived from the fundamental properties of the system.

Many authors have looked into the issue of self-organization from disciplinary perspectives. Krolikowski et al. (2016) provide a good summary of such efforts. Chih-Chun et al. (2007), on the other hand, discuss the issue of emergence in agent-based models. They offer no consensus on either the definition of self-organization and emergence or formal ways for detecting them in computational systems.

The term *emergence* is especially hard to define. Some researchers make a distinction between weak and strong emergence:

> If there are phenomena that are *strongly emergent* [emphasis added] with respect to the domain of physics, then our conception of nature needs to be expanded to accommodate them. That is, if there are phenomena whose existence is not deducible from the facts about the exact distribution of particles and fields throughout space and [time] (along with the laws of physics), then this suggests that new fundamental laws of nature are needed to explain these phenomena.
>
> (Chambers 2002)

> We can say that a high-level phenomenon is strongly emergent with respect to a low-level domain when the high-level phenomenon arises from the low-level domain, but truths concerning that phenomenon are not deducible even in principle from truths in the low-level domain. Strong emergence is the notion of emergence that is most common in philosophical discussions of emergence, and is the notion invoked by the British emergentists of the 1920s. ... We can say that a high-level phenomenon is weakly emergent with respect to a low-level domain when the high-level phenomenon arises from the low-level domain, but truths concerning that phenomenon are unexpected given the principles governing the low-level domain.
>
> (Chalmers 2006)

Most practitioners of agent-based modeling deploy the weak definition of emergence. Holland (1998) offers "that the study of emergence is closely tied to [the] ability to specify a large, complicated domain via a small set of 'laws'."

Similarly, Bar-Yam (1997) states that "a complex system is a system formed out of many components whose behavior is emergent, that is the behavior of the system cannot be inferred simply from the behavior of its components." He also defines the difference between what he refers to as local and global emergence (Bar-Yam 1997):

> The collective behavior is, however, contained in the behavior of the parts if they are studied in the context in which they are found. To explain this, we discuss examples of emergent properties that illustrate the difference between local emergence—where collective behavior appears in a small part of the system—and global emergence—where collective behavior pertains to the system as a whole.

Practice

In this chapter, however, we are mostly interested in the pragmatic aspect of self-organization and emergence – detecting them in agent-based model simulations. In practice, true self-organization and emergence are rarely detected. At least, this is our experience working with students and agent-based modeling practitioners over the years. Their solutions are often prescribed, almost as in algorithmic implementations. For example, the student who designed a system for modeling genocide claimed that the fact that the model always ends in genocide (only the time scale changes, but the outcome is always the same) constitutes an example of an emergent property. However, when the student changed some of the elements of agent behavior in the model, the outcome of the system changed significantly, which indicated that it was a characteristic of agents that caused the observed outcome rather than the nature of the interactions among agents. Thus, the cause of the problem was more of an algorithm nature (a bug if you will) rather than the nature of an agent-based model (interactions among agents).

This may be due to the fact that only few programmers who learn modeling with procedural languages make a seamless shift to the agent-based modeling paradigm. But it is also possible that modelers do not evaluate their models appropriately. For example, sensitivity/exploratory analysis is rarely done in agent-based modeling projects, even though it is critical to do it for a complete understanding of agent-based models. Sensitivity/exploratory analysis implies a thorough sweep through all possible starting states and randomly induced external shocks to the system over sufficiently long runs to detect all possible types of system behavior, including emergent properties that appear with some degree of probabilistic consistency. Of course, this is just a speculation. Other explanations might be even more plausible.

Self-organization is usually interpreted as an un-programmed spatial clustering phenomenon in the runs performed during the sensitivity analysis. If the agent-based model is implemented as a network, one could even use network community algorithms to automatically detect existence of agent clusters that form and dissipate in time. Many experts believe that self-organization is a manifestation of certain kind of emergence and that its detection should be covered within the mechanisms for detecting emergence itself. We believe that this approach is not operationally helpful. Our distinction between self-organization (as a property of a collection/subset of agents) and emergence (as a property of the whole system/model in a holistic sense) is probably more helpful in finding ways to detect each. However, we realize that this is not a generally accepted view of the distinction between self-organization and emergence.

In the ACSES project we could not look for spatial clusters of agents simply because our agents were spatially aggregated in the 2×2 mile blocks. There were simply too many of them to do it otherwise. Meaningful simulations were done on samples of 5000, 10 000, and 30 million agents. Visual representation of them was not practical. However, we could have attempted to look at the self-organization of aggregated agents, the 2×2 blocks. However, due to the tight schedule of the project, we did not have the time to define superagents as the aggregate of agents occupying spatial elements of the environment. Perhaps this is something we could do in the future.

Emergence, on the other hand, is much harder to detect, see, look for, or anticipate. It can happen in any aspect of system performance. It can be in the end state the system finds itself most often in, in the type of agents that survive or succeed, in the distribution of population characteristics, or in the longevity of the system, among many other things.

In the case of ACSES, the system was complicated. It had over 500 variables, and its purpose was to evaluate the utility of three specific social science theories. Even though we performed sensitivity analysis, it was done in the context of which theory was the most plausible one in explaining the use case at hand. Unfortunately, there were not many realistic use cases that the project team had access to. Therefore, our testing was limited in nature. This, in the end, proved to be the most uncertain aspect of the project, as it prevented the DARPA program manager to form a conclusive opinion about the potential utility of the project.

However, exactly because of the fact that we could not demonstrate the end result in the context of verifiable use cases, we thoroughly understood the importance of completely assessing the true nature of the system we developed. Agent-based models are tricky. They are never completely verified or validated, and their potential is never fully realized, given that we do not give them enough chances to self-organize or emerge. Consequently, their true nature is never fully *understood*.

This leads us to the issue of trust. How do we gain the trust of the end user in the statement that an agent-based model will help them address their issues?

It is obvious that agent-based modelers have not succeeded in this, given the fact that there are no *killer* agent-based applications in the marketplace. This is not to say that we ourselves, the modelers, are not convinced that the system is good and accurate. Of course, we must ensure that the system is, first and foremost, accurate to a sufficient degree in terms of the user specifications and requirements. The issue here is knowing what kind of end user validation will be sufficient for the user to accept the model/system as a new tool in his/her toolbox.

Trust

The use of agent-based models depends on the level of trust that is established between the model and the end user. Proper verification, validation, sensitivity analysis, uncovering self-organization and emergence, and understanding performance of the system under exogenous shocks are critical for securing the trust of the end user. If that does not happen, then the tool will not be used.

However, it is not enough to earn the trust of the end user. Several other issues are often of critical importance to end user's adoption of an agent-based model. What works will depend on the project. In some projects, researchers are expected to figure things out, build a model, test a model, and only then explain everything to the sponsor. Ideally, however, a process with more interactions is strongly desirable and will increase the likelihood of good communication and the sponsor's trust and use of the model. The admonitions that follow have analogues in suggestions by William Rouse in his chapter of this volume, based significantly on experience with large health systems (Rouse 2019).

They include the following:

- Design the model with a clear purpose based on a series of conversations between designers and end users. Of course, the end user has the final say on what the purpose of the system is. However, the model designer must educate the end user on the characteristics of agent-based models, including both what they can do and what they cannot do well. The designer may also want to build in more functionality than the end user originally anticipates wanting.
- Ask the user how he/she will evaluate the system, as well as what needs to happen for the system to be adopted. Consult with all relevant stakeholders, and provide a way to provide feedback on all phases of the project development.
- Assure that end users and all stakeholders participate in evaluation of the intermediate phases and results of the project development. These intermediate phase reviews are critical for uncovering issues early on that, if left unattended, will lead to the failure of the project in the end.

- Define the type, format, and detail of the output of the simulation, and obtain approval from the end user. Have the user interact with mock-ups or prototypes so as better to identify what works and does not work for him or her.
- Similarly, work with the end user to define the nature of interactions between user and system. Present options to the end user, but accommodate to what the end user wants and believes how interactions should *feel* like (no matter how *low tech* they might seem to the designer). As with other aspects of the work, designers may want to build in additional functionality, just in case, but only if it does not interfere with what the user believes is needed.
- Design the utilities necessary for the end user to perform what-if-type analysis in the context of the specified purpose.
- Provide evidence of the verification, validation, and sensitivity analysis, as well as any other testing that has been performed at any stage of the product development. Where possible, include the end user in some of that testing.
- Document all aspects of the system thoroughly, descriptively, operationally, and understandably.

Following the above list will go a long way toward establishing the trust of the end user in the model itself. Of course, each model development is different, and the designer should always expect that the end user would make additional requirements on the model.

Summary

The world is ready for complexity theory. It desperately needs it. Technical connectivity, globalization, scientific and technological advancements, medical advances, social media, and economic growth have made all human-made systems connected. Such systems require complexity-based models to help humans manage and control them.

Agent-based modeling is one of the methods that is particularly well suited for helping end users understand, explain, predict, and prescribe states and conditions of the modern complex adaptive systems that they are dealing with or even in charge of. It is sobering, however, to know that no agent-based model or application has entered the mainstream market as of yet. This seems to indicate that there are serious issues with either the capabilities of agent-based models or users' perceptions of them.

Taking into consideration the issues brought up in this chapter may help. The trust of the end users is the key to their adoption of agent-based models and applications. The trust will not happen until we, the modelers, can demonstrate that the models can be verified, validated, tested, and effectively used by the end user.

The potential for the agent-based modeling is huge. It is up to us, however, to realize that potential. The world is ready and primed for it. Are we?

References

Atkinson, A.B. (1969). The timescale of economic models: how long is the long run? *Review of Economic Studies* 36: 137–152.

Axelrod, R. (1997). Advancing the art of simulation in the social sciences. In: *Simulating Social Phenomena, Berlin* (ed. R. Conte, R. Hegelsmann and P. Terna), 21–40. Heidelberg: Springer.

Bar-Yam, Y. (1997). *Dynamics of Complex Systems*. Reading, MA: Perseus Books.

Chalmers, D.J. (2006). Strong and weak emergence. In: *The Re-Emergence of Emergence* (ed. P. Davies and P. Clayton), 244–254. Oxford University Press.

Chambers, D.J. (2002). Varieties of emergence. http://consc.net/papers/granada .html (accessed 28 September 2018).

Chih-Chun, C., Sylvia, B.N., and Christopher, D.C. (2007). Specifying, detecting and analysing emergent behaviours in multi-level agent-based simulations. In: *Proceedings of the 2007 Summer Computer Simulation Conference (SCSC '07)*, 969–976. San Diego, CA: Society for Computer Simulation International.

Davis, P.K. and O'Mahony, A. (2019, this volume). Priorities for improving social-behavioral modeling. In: *Social-Behavioral Modeling for Complex Systems* (ed. P.K. Davis, A. O'Mahony and J. Pfautz). Hoboken, NJ: Wiley.

Dosi, G., Fagiolo, G., and Roventini, A. (2006). An evolutionary model of endogenous business cycles. *Computational Economics* 27 (1): 3–34.

Durkheim, E. (1982 [1895]). *The Rules of Sociological Method*. New York, NY: Free Press.

Hadzikadic, M., O'Brien, S., and Khouja, M. (eds.) (2013). *Managing Complexity: Practical Considerations in the Development and Application of ABMs to Contemporary Policy Challenges*. Springer.

Holland, J.H. (1998). *Emergence: From Chaos to Order*. Cambridge, MA: Perseus Books.

Kahneman, D. (2003). Maps of Bounded Rationality: A Perspective on Intuitive Judgement and Choice. *American Economic Review* 93: 1449–1475.

Krolikowski, R., Kopys, M., and Jedruch, W. (2016). Self-organization in multi-agent systems based on examples of modeling economic relationships between agents. *Frontiers in Robotics and AI* 3: 41.

McNamara, L., Trucano, T., and Backus, G. (2007). Verification and validation as applied epistemology. *UCLA Lake Arrowhead Conference on Human Complex Systems*, Lake Arrowhead, CA (25–29 April 2007). http://hcs.ucla.edu/ arrowhead.htm.

Ormerod, P. (2004). Information cascades and the distribution of economic recessions in capitalist economies. *Physica A* 341: 556–568.

Ormerod, P. and Rosewell, B. (2004). On the methodology of assessing agent based models in the social sciences. In: *Evolution and Economic Complexity* (ed. J.S. Metcalfe and J. Foster), 24–37. Cheltenham: Edward Elgar.

Ormerod, P. and Rosewell, B. (2009). Validation and verification of agent-based models in the social sciences. In: *Epistemological Aspects of Computer Simulation in the Social Sciences*, Lecture Notes in Computer Science, vol. 5466 (ed. F. Squazzoni), 130–140. Berlin, Heidelberg: Springer.

Pilch, M., Trucano, T., Moya, J. et al. (2000). *Guidelines for Sandia ASCI Verification and Validation Plans – Content and Format: Version 2.0*, 2000–3101. Albuquerque, NM: Sandia National Laboratories, SAND.

Rand, W. and Wilensky, U. (2006). Verification and validation through replication: a case study using Axelrod and Hammond's ethnocentrism model. In: *Proceedings of the NAACSOS*, Notre Dame University, South Bend, IN (21–23 June 2006).

Rouse, W.B. (2019, this volume). Human-centered design of model-based decision support for policy and investment decisions. In: *Social-Behavioral Modeling for Complex Systems* (ed. P.K. Davis, A. O'Mahony and J. Pfautz). Hoboken, NJ: Wiley.

Smith, V.L. (2003). Constructivist and ecological rationality in economics. *American Economic Review* 93: 465–508.

Tesfatsion, L. (2017). Empirical validation and verification of agent-based models. http://www2.econ.iastate.edu/tesfatsi/empvalid.htm (accessed 22 December 2017).

Volkova, S., Hodas, N., and Greaves, M. (2019, this volume). Data, validation considerations. In: *Social-Behavioral Modeling for Complex Systems* (ed. P.K. Davis, A. O'Mahony and J. Pfautz). Hoboken, NJ: Wiley.

Whitmeyer, J. (2013). The ACSES model of Afghanistan: introduction and social theories. In: *Managing Complexity: Practical Considerations in the Development and Application of ABMs to Contemporary Policy Challenges* (ed. M. Hadzikadic, S. O'Brien and M. Khouja), 45–58. Springer.

Windrum, P., Fagiolo, G., and Moneta, A. (2007). Empirical validation of agent-based models: alternatives and prospects. *Journal of Artificial Societies and Social Simulation* 10 (2): 8.

29

Modeling Social and Spatial Behavior in Built Environments: Current Methods and Future Directions

Davide Schaumann and Mubbasir Kapadia

Department of Computer Science, Rutgers University, New Brunswick, NJ, USA

Introduction

Built environments have a major impact on the people who inhabit them: they affect the inhabitants' well-being as well as their ability to partake and excel in various processes that occur within and around buildings (living, working, leisure, etc.). In turn, building inhabitants have a major impact on how buildings are designed and operated, affecting the consumption of various resources over time.

At present, the correlation between a given built environment and its effect on the people who inhabit it becomes known only after the environment has been built and occupied. Post-occupancy evaluations (POE), for example, compare the expected building performance with the actual one to improve an existing situation or gain knowledge that can inform the design of future buildings. Data collected in existing buildings is often analyzed using mathematical methods (such as regression, deterministic, and stochastic) to correlate variables of human behavior and the built environment (e.g. Wang et al. 2005; Mahdavi and Tahmasebi 2015). The assumption that using mathematical models validated in a specific context helps to predict human behavior aspects in a different context, however, may be flawed due to the unique nature of cultural, social, and environmental conditions.

Despite much research into human–environment relationships (Barker 1968; Alexander 1979; Whyte 1980; Gehl 1987; Preiser et al. 2015), no satisfactory method has yet been developed that can anticipate building–user interactions at the time buildings are designed, rather than after they are built and occupied. Failures in doing so may lead to severe consequences, such as underperforming buildings, diminished productivity, general dissatisfaction, and imbalances between expected and actual energy consumptions. Instead,

Social-Behavioral Modeling for Complex Systems, First Edition.
Edited by Paul K. Davis, Angela O'Mahony, and Jonathan Pfautz.
© 2019 John Wiley & Sons, Inc. Published 2019 by John Wiley & Sons, Inc.
Companion website: www.wiley.com/go/Davis_Social-Behavioralmodeling

by effectively modeling and predicting the complex behaviors that are likely to occur when spaces are occupied, architects and engineers will be better able to maximize occupants' physical comfort, social well-being, and job performance while minimizing their collective impact on the natural environment.

To predict the performance of future buildings with respect to efficiency and sustainability considerations, architects and engineers harness the power of computer simulations (Rockcastle and Andersen 2014; Grobman and Elimelech 2016). Such simulations address specific aspects of buildings, such as energy, daylight, plug loads, and structural integrity. In contrast, simulating the impact of buildings on the people who will occupy them has progressed more slowly compared with the aforementioned methods. Even though human behavior simulation is a well-studied topic in the field of computer animations and video games (Kapadia et al. 2015), such a method is yet to be applied to predict and analyze how a building design affects different use patterns occurring in it. Simulations of this kind must include, in addition to information about the building itself, information about the intended users of a building, and the activities they will perform. Modeling users and activities, however, is a challenging task due to the dynamic, stochastic, psychological, contextual, and cultural components involved in describing what people do, where, and when.

Current approaches to investigating human–environment relationships mostly rely on static analyses. These methods expand traditional building models generated with computer-aided design (CAD) or building information modeling (BIM) tools with information about the building's intended users and the activities they perform. A configurational approach has been proposed by Hillier and Hanson to analyze people movement in spaces using a graph-based spatial representation that measures spatial relations and connectivity (Hillier and Hanson 1984). Eastman and Siabiris (1995), Maher et al. (1997), Ekholm (2001), and Wurzer (2010) augmented the information contained in spatial models by providing static descriptions of people activities in space. Kim and Fischer (2014a,b) developed an automated method for mapping specific activities onto appropriate spaces. Dzeng et al. (2015) described a model to optimize functional use allocation based on a set of space usage metrics including people movement and flow. Schultz and Bhatt (2012) and Bhatt et al. (2012, 2013) incorporated into traditional space models aspects of spatial cognition and reasoning into traditional space models to support designers' decision-making. Çekmiş et al. (2014) demonstrated a method to represent patterns of human inhabitation in open-planned spaces. Tashakkori et al. (2015) developed a space model that integrates indoor and outdoor information for emergency response facilitation.

These approaches, however, fail to provide a dynamic, time-based representation of a building in use, which accounts for people decision-making abilities in response to the building function and to the surrounding physical and social environment.

Multi-agent approaches have thus been adopted to model spatiotemporal building–user interactions, such as pedestrian movement (Yan and Kalay 2004), egress situations (Chu et al. 2014), crowd behavior (Thalmann and Musse 2013; Kapadia et al. 2015), and scheduled activities in office (Tabak et al. 2010) and university buildings (Shen et al. 2012). While these approaches help designers test *what-if* scenarios to reveal how a particular design solution may affect the behavior of its inhabitants (Simon 1969), they currently present significant shortcomings when accounting for agents' perceptual, cognitive, and decision-making abilities in a dynamic, physical, and social context (Kapadia et al. 2012a).

A more advanced representation of human behavior in built environments is thus needed to model a variety of behaviors, including emergency evacuations, pedestrian wayfinding, and more holistic use patterns in process-driven facilities, like hospitals. In such settings, behavior is driven by a combination of scheduled procedures (e.g. medical checks) and unscheduled adaptations of such procedures due to dynamic social and spatial conditions (e.g. spatial bottlenecks that affect the flow of human and nonhuman resources).

The remainder of this chapter is organized as follows. First, we review current methods for simulating human behavior. We then discuss the latest advancements in modeling key components of MAS, such as *spaces*, *actors* and *activities*, and review *behavior authoring* frameworks that can be used to coordinate the behavior of multiple agents in a given context. Lastly, we discuss possible avenues of future research, towards practical tools that can help architects, facility managers and other decision makers design future built environments.

Simulating Human Behavior – A Review

Several human behavior simulation methods have been developed to represent different kinds of use scenarios. Various literature reviews (Hamacher and Tjandra 2001; Helbing et al. 2002; Santos and Aguirre 2004; Gilbert and Troitzsch 2005; Shiwakoti et al. 2008; Zheng et al. 2009; Zhou et al. 2010; Duives et al. 2013; Thalmann and Musse 2013; Kapadia et al. 2015; Pelechano et al. 2016) collectively provide a comprehensive overview of existing approaches, mainly focusing on crowd simulations in normal or egress situations. Here, we provide a broader categorization that transcends the domain of application of the model to focus on overall modeling principles. The discussed approaches are the following: *system dynamics*, *process-based*, flow-based, particle-based and *MAS*.

System Dynamics

System dynamics models describe the behavior of nonlinear systems using *stocks*, *flows*, and *feedback loops* (Forrester 1997). *Stocks* are entities that accumulate or deplete over time. *Flows* describe the rate of change in stocks. *Feedback loops* characterize causal relationship among entities of the system (positive or negative impact). Differential equations describe the different states of the variables that depend on each other and vary over time.

Such models are generally used to help decision-makers understand the behavior of industrial and socioeconomic systems, such as the development of land use and population growth (Guhathakurta 2002), and people flow in hospitals (Brailsford 2008). Nevertheless, examples can be found where system dynamics has been used to model emergency egress behaviors in buildings. In the model proposed by Shen (2005), for instance, the rooms of a building are modeled as stocks, while the flow among rooms is regulated by a series of parameters affecting people movement, including the occupants' speed and the density of people in the rooms.

The high level of abstraction imposed by system-level modeling, however, prevents the encoding of information about a building geometry and the individual entities (e.g. the people) that move through a space.

Process-Driven Models

Process-based models represent human behavior in buildings as a structured sequence of activities that require a set of resources (e.g. people, equipment, and spaces) and take a certain (usually stochastic) amount of time. Stochastic values can be associated with the duration of an activity or with the selection process between alternative activities.

In process-based models, space is represented as a network of nodes and edges, where nodes may be rooms and edges indicate the connections between them, with an associated (usually stochastic) traversal time. The flow of people between nodes is described in terms of queuing behavior, whereby a Markov chain process describes the probability of an individual moving from one node of the network to another (Løvås 1994): when the traffic demand exceeds a node's capacity, a queue builds up, causing individuals to wait a certain amount of time.

These kinds of models have been used to analyze resources' flow in buildings where users' behavior is driven by a specific set of procedures, such as hospitals or airports (Sinreich and Marmor 2005; Marmor et al. 2012). Nevertheless, they suffer from several limitations. First, they describe individuals in a homogeneous fashion, averaging away the differences among building occupants. Second, they model spatial features only in terms of the time required to traverse spaces, ignoring the multiplicity of effects that spaces produce on human activities due to people perception and cognition of spatial, social, and environmental aspects. Third, they do not account for activities that emerge due to spatial or social contingencies, which might delay scheduled activities and create spatial congestion. A crowded or uncomfortable waiting area in a hospital ward, for example, might drive visitors to wait in areas that have not been designed for that purpose, causing obstructions to the circulation of hospital staff and equipment and possibly to the interference of medical tasks.

Flow-Based Models

Flow-based models represent human movement and interactions in space by adopting mathematical equations describing physical phenomena. The premise of this approach is that people, under certain conditions (e.g., while trying to evacuate a space or in narrow passages) generate a collective movement pattern that is akin to the behavior of gas molecules, particles, or fluids. The behavior of pedestrian crowds has been described at a macro-scale in such fluid-like terms using Maxwell-Boltzmann theory (Henderson, 1971), the Navier–Stokes equations (Hughes, 2003), and continuum models (Treuille et al., 2006). Such models are mostly used to represent homogenous crowd movement at a coarse level of detail (i.e. without focusing on the movement and behavior of the individual crowd members).

Particle-Based Models

In particle-based models, people are considered as homogenous entities with identical properties, whose movement and interactions are governed by global laws. Helbing and Molnar, (1995), for example, describe crowd behavior as a pattern of attraction and repulsion (social) forces. While such models account for particles' responses to the spatial and social context, they mostly ignore the cognitive effects on human behavior in a space (Torrens, 2016). Accordingly, they can simulate only specific aspects of human movement in a space, failing to represent more complex human behaviors that depend on peoples' autonomous decision-making, and individual profiles.

Multi-Agent Systems

In MAS autonomous agents inhabit dynamic spatial environments and sense, plan, and act autonomously to achieve a specific goal (Wooldridge and Jennings 1995). Agents' behavior is usually triggered by local conditions (both spatial and social) rather than global information. It affects and is affected by the behavior of other agents in a reactive manner: each agent pursues its individual goal (e.g. exit the building, in case of a fire egress simulation) while reacting to the environment it perceives (e.g. fire, smoke), as well as to the actions and behaviors of other agents (e.g. crowding, etc.). Complex behaviors thus emerge from the unfolding of low-level behaviors and interactions among multiple agents (Bonabeau 2002).

Agents' behavior is often driven by a set of rules that account for local information available to the agent. One of the earliest examples is a system developed by Reynolds (1982, 1987), where a finite set of rules (separation, alignment, and cohesion) are used to simulate the flocking behavior of birdlike objects called *boids*. Other examples have been developed to simulate human behavior in small urban settings (Yan and Kalay 2004) and in evacuation scenarios (Pan et al. 2007; Tang and Ren 2012; Chu et al. 2014; Zhao et al. 2014).

While MAS successfully address some of the complexities of simulating human–environment interactions, several challenges must still be addressed to represent human behavior in settings populated by different people that perform different tasks in a shared space. Such challenges broadly involve modeling: (i) the *space* that people inhabit, including its geometrical and semantic properties (e.g. a patient room, a nurse station), which constrains its use to specific types of activities; (ii) the *actors* that inhabit those spaces, including their roles in the building organization as well as their situated perceptual and cognitive abilities; (iii) the *activities* that people engage in, such as navigating a space or participating in work-related or social meetings; and (iv) structured collaborative *behaviors* performed by multiple agents over a succession of spaces that comprise a combination of scheduled activities and unscheduled adaptations to dynamic social and spatial conditions. In the following sections, we describe recent advancements and future directions within MAS for addressing each of these challenges.

Modeling Social and Spatial Behavior with MAS

Human behavior emerges from temporally and physically situated interactions among people and with their surrounding environment. It is inseparable from both its physical and social settings, as well as the activities of other occupants. To address the complexities of representing human behavior in built environments, we propose a comprehensive framework for simulating human social and spatial behavior (Figure 29.1). In the following sections, we detail recent advancements and open challenges to model key components of the proposed framework, namely, (i) the *space*, (ii) the *actors* that inhabit it, (iii) the *activities* they engage in, and (iv) disembodied *behavior-authoring* mechanisms that coordinate complex collaborative behaviors of multiple agents in a given space.

Modeling Spaces

Spaces represent the physical setting where actors perform activities. To support human behavior simulation, space models must store additional information compared with traditional models developed for architectural

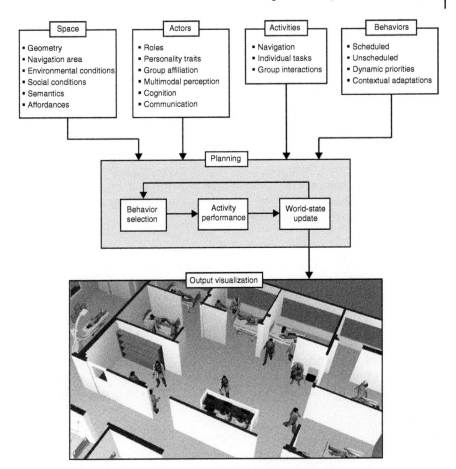

Figure 29.1 A comprehensive human behavior simulation framework.

design and engineering. Such models, in fact, mainly focus on the physical components of the built environments (e.g. beams, slabs, walls, windows, doors) rather than on the space contained within them. Even though BIM tools allow encoding information about the spaces between them (rooms), additional information is required to support a dynamic description of how people move and interact in space.

Graph-Based Approaches

These approaches describe people movement across a network of spaces. In such representations spaces are modeled as nodes and their connection as links (Løvås 1994). While this approach supports various kinds of movement-related analyses (e.g. Hillier and Hanson 1984), it abstracts away key information about the building geometry and space semantics, which indicates how it can be used by the different users and for what activity. Other approaches, instead, divide a space into discrete portions of walkable areas, called *navigation meshes*, to support agents' navigation (Kallmann and Kapadia 2014).

Navigation Meshes

Static navigation meshes use a triangulated representation of free space and only consider static obstacles. Dynamic navigation meshes, instead, coarsely account for dynamic properties of the environment. They store time-varying space properties such as a density field, which affects the cost of choosing a triangle for navigation. Such space subdivision, however, is mostly determined by navigation-related principles, rather than by perceptual, cognitive, environmental, or functional considerations, making this representation limited for supporting other kinds of spatial behaviors.

Grid-Based Approaches

Environments can be discretized into grid cells. This allows for creating a semantically rich collection of cells to define specific kinds of space (e.g. a room, a corridor) and storing information in each cell (or group of cells) about its dynamic changes, including social and environmental conditions (Date et al. 2017a). While this approach enables storing time-varying spatial properties in automatically generated or custom-defined collection of cells, proper fine-tuning of the grid cells is required to yield realistic agent movement. Furthermore, this space representation is challenging when representing more dense crowd situations, since agents can occupy only one cell at each time step.

Semantics and Affordances

To yield realistic social and spatial behavior, space models have been empowered with additional calculation abilities to aid agent's perception in a virtual environment. A list of semantic and affordance attributes can be associated with discrete space portions to describe how a space is used at a given time (Schaumann et al. 2015, 2016a). For instance, in a hospital context, a room can be associated with *patient room* semantics and with a specific number of activity types afforded such as *clinic, social, private,* and *assistance.* The activation of a specific affordance determines how a space is currently used, and it stipulates which other affordances are supported in the same space, therefore allowing or inhibiting other behaviors.

Modeling Actors

Actors represent the inhabitants of the built environment. Different kinds of actor models have been developed to represent people at different levels of abstraction. In system dynamics and process-based models, actors are not explicitly represented. They are considered as resources that flow in a system (much like spaces or equipment). In particle-based models, agents are often represented homogeneously – without variance in their goals, perception, and decision-making abilities. Built environments, however, are populated by different kinds of people, which may have different characteristics, preferences, and work-related roles. More complex agent models are thus required to describe social and spatial behaviors.

Profiles

Actor models have been associated with specific profiles that include personality traits and social group affiliation. Braun et al. (2005) extended homogeneous agent models commonly used is particle-based models (e.g. Helbing and Molnar 1995) by assigning each agent a different personality. Durupınar et al. (2011), Guy et al. (2011), and Kapadia et al. (2012b) incorporated personality traits in crowd simulations. Chu et al. (2014) associated with each individual a specific social group of belonging (e.g. a family), which can affect an agents' decision-making in egress situations. Mascarenhas et al. (2016) integrate culture into virtual agent models. Sopher et al. (2017) described a general framework to model agents' profiles, which comprises a description of static and dynamic features, including their characteristics, status, knowledge, preferences, and abilities. Date et al. (2017b) created different user profiles by assigning to each agent a different threshold of tolerance to an environmental parameter (e.g. noise), which affects actors' navigation.

Perceptual and Cognitive Abilities

Agents' perception, cognition, and communication are key factors to guarantee realistic behavior responses to dynamic social and environmental conditions. So far, particular emphasis has been put on visibility issues. Agents have been equipped with visual perception abilities including line-of-sight ray casts and foveal cone intersections to see the world around them (Kapadia and Badler 2012). Other approaches for vision-based steering have been proposed by Kapadia et al. (2009) and Ondřej et al. (2010). In an effort toward multimodal perception, Huang et al. (2013) and Wang et al. (2014) developed, respectively, a sound propagation and localization methods to equip autonomous agents with auditory perception capabilities. A cognitive-based approach has been proposed by Shao and Terzopoulos (2005), and Kielar and Borrmann (2018). Kullu et al. (2017) proposed an agent-to-agent communication model for simulating interactive crowds.

Modeling Activities

Activities represent the dynamic interactions between the actors and the space, which may involve moving toward a target as well as work-related or social meetings.

Navigation

Navigation is a key activity commonly modeled in MAS due to its dynamic, social, and spatial nature. It consists of a coordinated and goal-directed movement through the environment (Montello 2005). Navigation broadly involves *pathfinding*, *steering*, and *locomotion* tasks (Kapadia and Badler 2012). *Pathfinding* is the process of finding a collision-free path from the current position of an agent to its target location placed beyond sensory experience, usually taking into account static and dynamic aspects of the environment. *Steering* is the act of following a planned path while avoiding static and dynamic obstacles. *Locomotion* animates an actor to a trajectory path output by steering, taking into account the abilities (affordances) of the human actor (e.g. body articulations, speed, etc.). Kapadia and Badler (2012) and Kallmann and Kapadia (2016) provided an extensive review of existing navigation models for autonomous virtual users. Here, we briefly summarize key approaches.

Precomputed Roadmaps

This method regards the environment as static. Examples include the work of Sung et al. (2005), Sud et al. (2007), and Kallmann (2010). Built environments, however, are dynamic due to the movement of people and objects in space and the dynamic environmental conditions. If, for instance, a crowd is temporarily blocking a passage, an agent may be required to devise an alternative route to the destination. Planning approaches are thus required that can quickly return solutions with bounds on suboptimality and iteratively refine the solution while accommodating dynamic changes in the environment (Kallmann and Kapadia 2014).

Reactive Planning Approaches

Such approaches adopt a one-step look ahead to avoid collisions with most imminent threats. Date et al. (2017b), for example, developed a hierarchical navigation system that combines wayfinding and locomotion into a unique framework. Each agent computes at each time step both the global and local paths toward a target while accounting for dynamic social and environmental conditions. Other examples have been developed by Reynolds (1987) and Lamarche and Donikian (2004).

Predictive Planning Approaches

Such approaches approximate the trajectories of neighbor agents in choosing collision-free trajectories. Examples can be found in the work of Paris et al. (2007), Van den Berg et al. (2008), Kapadia et al. (2009), Karamouzas et al. (2009), Karamouzas and Overmars (2010), and Singh et al. (2011).

Decision-Making

In MAS, actors are often equipped with the ability to autonomously determine their navigation target. Such target can either be precomputed or calculated at run time. Actors, for example, can choose a target based on an expected (measurable) utility. Hoogendoorn and Bovy (2004) proposed an activity-based model to analyze travel behavior based on activity scheduling. In Krontiris et al. (2016), the behavior of virtual agents is determined by environmental *attractors* – spatial locations that enable the performance of specific activities. Different types of agent architectures have also been designed to mimic the rational decision-making process of agents to different degrees of complexity. Among others, the belief–desire–intentions (BDI) architecture (Rao and Georgeff 1995) is often used to determine which activity to perform next and where.

Multi-agent approaches, however, can vary in the degree to which individual agents can make autonomous decisions. The assumption that complex behaviors can be modeled by increasing individual agents' complexity, in fact, may be invalid when describing collaborative, goal-oriented activities in complex settings (O'Sullivan and Haklay 2000). Behavior-authoring frameworks have thus been developed to coherently represent the use phenomenon of a building such as a hospital or an airport; vast amounts of knowledge must be stored in each agent before collaborative behaviors based on cultural norms as well as spatial and social circumstances can emerge. A patient check procedure in a hospital, for example, where a doctor and a nurse must meet a patient at a certain place and time to start the activity, depends on resolving such complexities as the presences of the actors at a planned time/space and what to do if one or more of them are delayed.

Behavior Authoring

Overarching controlling mechanisms have been developed to direct multiple agents' behavior in space with the purpose of facilitating an overall coherent representation of building use phenomena, accounting for shared activities among different building inhabitants.

Centralized Scheduling Systems

Such systems assign an activity to different kinds of agents. Goldstein et al. (2010) used data-driven user profiles to determine agents' activities and their locations. Shen et al. (2012) used precomputed schedules to drive agent's movement in space. Tabak et al. (2010) developed a scheduling mechanism that combines into one schedule both work-related and leisure-related activities in office buildings. Schaumann et al. (2017a) used a dynamic scheduler that adopts operation research (OR) techniques to coordinate the performance of individual and collaborative procedures in hospital buildings.

Top-down management systems have also been proposed to coordinate the behavior of large groups of computer-controlled nonplayer characters (NPC) in response to the dynamic behavior of human-controlled avatars (Bates 1992;

Magerko et al. 2004; Shoulson et al. 2011). Such approaches provide dynamic adaptations to the game's unfolding narrative plot to guarantee its coherence and consistency over time.

Event-Centric Approaches

Events are self-contained computational entities that encapsulate a set of procedures for directing the behavior of actors performing goal-oriented activities in some space (Kapadia et al. 2011; Shoulson et al. 2013; Simeone and Kalay 2014; Kapadia et al. 2016; Schaumann et al. 2015, 2016a, 2017b). Event-centric approaches temporarily direct multiple actors to ensure coordinated behaviors that adapt to the dynamic social and spatial context. Actors participating in an event suspend their autonomy and are exclusively controlled by the event to ensure coordination between them. To direct actors' behavior in space, events feature a combination of top-down and bottom-up coordination strategies; on the one hand, they dictate what actors should do, thereby affording the coordination of collaborative behavior patterns. On the other hand, they adapt their execution to bottom-up situations contingent upon the dynamic state of the world, such as the time- and context-dependent status of the spaces and actors involved in the event's execution.

Event Management Systems

An *event manager* dynamically coordinates the performance of scheduled and unscheduled events based on context-dependent priorities. It is responsible for monitoring the state of the world, for determining whether the conditions required to perform specific events are satisfied, and for triggering must urgent/important events at a given time while resolving conflicts among events competing for the same resources (e.g. actors or spaces) (Schaumann et al. 2015, 2017a).

The events triggered can be scheduled or unscheduled. Scheduled events are expected to be performed at specific times of the day (e.g. the doctors' round pattern). Unscheduled events, instead, could manifest themselves at any given time (e.g. the spontaneous interaction between a doctor and a nurse in a hospital context), and they may affect the performance of scheduled events. In a hospital setting, for instance, structured collaborative procedures like a patient check or a medicine administration are often disrupted by unscheduled social interactions whereby visitors may ask information to the medical staff in a hospital corridor causing delays to schedule procedures and even possible mistakes.

Priority values quantify the desirability of the event's performance (or outcome) at a specific time while accounting for the dynamic conditions of the world (both the condition of the entities participating in the event and any other condition that may directly or indirectly affect the event). Such priorities can relate either to the goals of individuals or group of individuals participating in the event (Arentze and Timmermans 2009) or to larger goals of an organization (Soomer and Franx 2008). Event management systems have been proven useful to coordinate the performance of planned and unplanned multiple behaviors in hospital environments, which involve multiple actors and occur in a series of spaces (Schaumann et al. 2017b).

Discussion and Future Directions

In this chapter, we discussed key challenges in simulating the spatial and social behavior of virtual humans. Despite recent advancements in the area, there is much more work to be done towards developing practical tools that can assist architects, urban planners, and other decision-makers in designing settings that better perform from the users' point of view. We believe that the following directions are of particular importance.

Creating Heterogeneous Agents

At present, agents are mostly modeled as homogeneous entities. Different agents (or groups of agents), however, exhibit a different spatial behavior depending on their unique characteristics. We pose the need for hetero-geneous agents, with different goals, needs, abilities (motor, sensor), and behaviors (teenagers, parents with kids, grandparents with toddlers). The behavior of each agent type can affect the behavior of other agents located in the same space, therefore affecting the overall building use process. In hospitals, for example, a wide variety of users (patients, visitors, doctors, nurses, technicians, administrative staff) populate the same space at the same time. The behavior of each user is tightly coupled with the behavior of other users. For example, a visitor that cannot find the location of a patient is likely to interrupt a staff member to ask for information (Schaumann et al, 2016b). Moreover, patients with special needs may require assistance from the hospital staff to perform the most basic day-to-day activities. Modeling various agent characteristics and degrees of sensory-motor abilities can support accessibility analyses as well as more holistic studies on how buildings are used. Such analyses hold potential to lead to meaningful design modifications aimed at better addressing human needs.

Improving Agents' Multi-Modal Perception and Cognition

More complex perceptual and cognitive models are required to account for multi-modal spatial and social stimuli (Hölscher et al., 2004, 2006, 2009, 2011;

Bhatt et al. 2011, 2017). As Information Technologies (IT) make their way into the fabric of the built environment, so-called 'smart' buildings and cities (which are, essentially, technologically-enhanced built environments) are further impacting the social-physical-cultural aspects of people's life. Rather than passive stages, IT-enhanced environments become active participants in the life of a setting because they sense and communicate to the human inhabitants information that lies beyond their motor-sensory perceptive range, thereby affecting their decision-making abilities. Thus, technologically-aided stimuli should also be incorporated in human behavior simulations since they widely affect human movement and interactions (Gath-Morad et al, 2016). Such aspects, combined with advanced behavior authoring tools able to coordinate the behavior of multiple agents in both process-driven (e.g. hospitals, airports, train stations), as well as actor-driven (museums, parks) settings hold potential to improve existing behavior models and yield more realistic results.

Using Human Behavior Simulation as a Decision-Support System in Architectural Design

Architects and building managers are in dire need of analytical tools able to predict how buildings will be used by future inhabitants prior to their construction and occupancy. MAS can be used to simulate what-if scenarios in architectural design options or to test multiple design alternatives against a set of key performance indicators specified by the various design stakeholders (Schaumann, 2018). Based on the simulation results, design stakeholders are able to iteratively evaluate multiple design options (Hong et al, 2016; Hong & Lee, 2018) or optimize architectural layouts against specific criteria (Haworth et al. 2017; Nagy et al. 2017; Berseth et al, 2018). To support optimization, data-driven approaches could potentially be integrated with simulation frameworks to speed up the calculation process and support faster iterations in architectural designs (Qiao et al, 2017; Liu et al., 2017). Efficient ways to communicate simulation results must also be investigated. Traditional agents are often represented as 2D particles that move in space. The high level of abstraction adopted in such simulations prevents domain experts as well as other design stakeholders from easily interpreting the simulation results. Recent advancements in computer graphics support more advanced agent models in terms of graphics and motion. Such advances support a clearer communication of human behavior that helps designers evaluate multiple design solutions (Hong et al, 2016). Communication of heterogeneous agent models can also be greatly enhanced by adding expressivity and variety to the full-body motion of virtual characters (Kapadia et al., 2015).

We hope that the wider perspective we have taken with these suggested research directions will open the way to new topics and accelerated progress in simulating complex building-user interactions.

Acknowledgments

This work has been funded in part by NSF IIS-1703883, NSF S&AS-1723869, DARPA SocialSim-W911NF-17-C-0098, and the Rutgers Murray Fellowship.

References

Alexander, C. (1979). *The Timeless Way of Building*. New York: Oxford University Press.

Arentze, T.A. and Timmermans, H.J. (2009). A need-based model of multi-day, multi-person activity generation. *Transportation Research Part B: Methodological* 43 (2): 251–265.

Barker, R.G. (1968). *Ecological Psychology: Concepts and Methods for Studying the Environment of Human Behavior*. Stanford University Press.

Bates, J. (1992). Virtual reality, art, and entertainment. *Presence: Teleoperators & Virtual Environments* 1 (1): 133–138.

Berseth, G., Khayatkhoei, M., Haworth, B. et al. (2018). Interactive Diversity Optimization of Environments. ArXiv Preprint arXiv:1801.08607.

Bhatt, M. and Schultz, C. (2017). People-centered visuospatial cognition: next-generation architectural design systems and their role in conception, computing, and communication. In: *The Active Image: Architecture and Engineering in the Age of Modeling*, Philosophy of Engineering and Technology (ed. S. Ammon and R. Capdevila-Werning). Springer International Publishing.

Bhatt, M., Holscher, C., and Shipley, T. F. (2011). Proceedings of the Symposium on Spatial Cognition for Architectural Design (SCAD 2011).

Bhatt, M., Schultz, C., and Huang, M. (2012). The shape of empty space: human-centred cognitive foundations in computing for spatial design. In: *IEEE Symposium on Visual Languages and Human-Centric Computing*, 33–40. IEEE.

Bhatt, M., Schultz, C., and Freksa, C. (2013). The 'space' in spatial assistance systems. In: *Representing Space in Cognition: Interrelations of Behaviour, Language, and Formal Models*. Oxford University Press.

Bonabeau, E. (2002). Agent-based modeling: methods and techniques for simulating human systems. *Proceedings of the National Academy of Sciences of the United States of America* 99 (3): 7280–7287.

Brailsford, S.C. (2008). System dynamics: what's in it for healthcare simulation modelers. In: *Simulation Conference, 2008. WSC 2008. Winter*, 1478–1483. IEEE.

Braun, A., Bodmann, B.E., and Musse, S.R. (2005). Simulating virtual crowds in emergency situations. In: *Proceedings of the ACM Symposium on Virtual Reality Software and Technology*, 244–252. ACM.

Çekmiş, A., Hacıhasanoğlu, I., and Ostwald, M.J. (2014). A computational model for accommodating spatial uncertainty: predicting inhabitation patterns in open-planned spaces. *Building and Environment* 73: 115–126.

Chu, M.L., Parigi, P., Law, K., and Latombe, J.-C. (2014). Modeling social behaviors in an evacuation simulator. *Computer Animation and Virtual Worlds* 25 (3–4): 373–382.

Date, K., Schaumann, D., and Kalay, Y.E. (2017a). Modeling space to support use-pattern simulation in buildings. In: *Proceedings of the Symposium on Simulation for Architecture and Urban Design*, 181–188. Toronto.

Date, K., Schaumann, D., and Kalay, Y.E. (2017b). A parametric approach to simulating use-patterns in buildings: the case of movement. In: *Proceedings of the International Conference on Education and research in Computer Aided Architectural Design in Europe*, 503–510. Rome.

Duives, D.C., Daamen, W., and Hoogendoorn, S.P. (2013). State-of-the-art crowd motion simulation models. *Transportation Research Part C: Emerging Technologies* 37: 193–209.

Durupinar, F., Pelechano, N., Allbeck, J.M. et al. (2011). How the ocean personality model affects the perception of crowds. *IEEE Computer Graphics and Applications* 31 (3): 22–31.

Dzeng, R.-J., Wang, W.-C., Hsiao, F.-Y., and Xie, Y.-Q. (2015). An activity-based simulation model for assessing function space assignment for buildings: a service performance perspective. *Computer-Aided Civil and Infrastructure Engineering* 30 (12): 935–950.

Eastman, C.M. and Siabiris, A. (1995). A generic building product model incorporating building type information. *Automation in Construction* 3 (4): 283–304.

Ekholm, A. (2001). Modelling of user activities in building design. In: *Proceedings of the International Conference on Education and Research in Computer Aided Architectural Design in Europe (eCAADe)*, 67–72. Helsinki.

Forrester, J.W. (1997). Industrial dynamics. *Journal of the Operational Research Society* 48 (10): 1037–1041.

Gath-Morad, M., Schaumann, D., Zinger, E. et al. (2016). How smart is the smart city? Assessing the impact of ICT on cities. In: *International Workshop on Agent Based Modelling of Urban Systems*, 189–207. Cham: Springer.

Gehl, J. (1987). *Life Between Buildings: Using Public Space*. New York: Wiley.

Gilbert, N. and Troitzsch, K. (2005). *Simulation for the Social Scientist*. New York: McGraw-Hill Education (UK).

Goldstein, R., Tessier, A., and Khan, A. (2010). Schedule-calibrated occupant behavior simulation. In: *Proceedings of the 2010 Spring Simulation Multiconference*, 180. Society for Computer Simulation International.

Grobman, Y.J. and Elimelech, Y. (2016). Microclimate on building envelopes: testing geometry manipulations as an approach for increasing building envelopes' thermal performance. *Architectural Science Review* 59 (4): 269–278.

Guhathakurta, S. (2002). Urban modeling as storytelling: using simulation models as a narrative. *Environment and Planning B: Planning and Design* 29: 895–911.

Guy, S.J., Kim, S., Lin, M.C., and Manocha, D. (2011). Simulating heterogeneous crowd behaviors using personality trait theory. In: *Proceedings of the 2011 ACM SIGGRAPH/Eurographics Symposium on Computer Animation*, 43–52. ACM.

Hamacher, H.W. and Tjandra, S.A. (2001). *Mathematical Modelling of Evacuation Problems: A State of Art*. Fraunhofer-Institut für Techno-und Wirtschaftsmathematik, Fraunhofer (ITWM).

Haworth, B., Usman, M., Berseth, G. et al. (2017). CODE: crowd-optimized design of environments. *Computer Animation and Virtual Worlds* 28 (6): e1749.

Helbing, D. and Molnar, P. (1995). Social force model for pedestrian dynamics. *Physical Review E* 51 (5): 4282.

Helbing, D., Farkas, I.J., Molnar, P., and Vicsek, T. (2002). Simulation of pedestrian crowds in normal and evacuation situations. *Pedestrian and Evacuation Dynamics* 21 (2): 21–58.

Henderson, L.F. (1971). The statistics of crowd fluids. *Nature* 229: 381–383.

Hillier, B. and Hanson, J. (1984). *The Social Logic of Space*. Cambridge: Cambridge University Press.

Hölscher, C., Meilinger, T., Vrachliotis, G. et al. (2004). Finding the way inside: linking architectural design analysis and cognitive processes. In: *Spatial Cognition IV. Reasoning, Action, Interaction*, 1–23. Berlin, Heidelberg: Springer.

Hölscher, C., Meilinger, T., Vrachliotis, G. et al. (2006). Up the down staircase: wayfinding strategies in multi-level buildings. *Journal of Environmental Psychology* 26: 284–299.

Hölscher, C., Büchner, S.J., Meilinger, T., and Strube, G. (2009). Adaptivity of wayfinding strategies in a multi-building ensemble: the effects of spatial structure, task requirements, and metric information. *Journal of Environmental Psychology* 29 (2): 208–219.

Hölscher, C., Tenbrink, T., and Wiener, J.M. (2011). Would you follow your own route description? Cognitive strategies in urban route planning. *Cognition* 121 (2): 228–247.

Hong, S.W. and Lee, Y.G. (2018). The effects of human behavior simulation on architecture major students' fire egress planning. *Journal of Asian Architecture and Building Engineering* 17 (1): 125–132.

Hong, S.W., Schaumann, D., and Kalay, Y.E. (2016). Human behavior simulation in architectural design projects: an observational study in an academic course. *Computers, Environment and Urban Systems* 60: 1–11.

Hoogendoorn, S.P. and Bovy, P.H. (2004). Pedestrian route-choice and activity scheduling theory and models. *Transportation Research Part B: Methodological* 38 (2): 169–190.

Huang, P., Kapadia, M., and Badler, N.I. (2013). SPREAD: sound propagation and perception for autonomous agents in dynamic environments. In: *Proceedings of the 12th ACM SIGGRAPH/Eurographics Symposium on Computer Animation*, 135–144. ACM.

Hughes, R.L. (2003). The flow of human crowds. *Annual Review of Fluid Mechanics* 35 (1): 169–182.

Kallmann, M. (2010). Shortest paths with arbitrary clearance from navigation meshes. In: *Proceedings of the 2010 ACM SIGGRAPH/Eurographics Symposium on Computer Animation*, 159–168. Eurographics Association.

Kallmann, M. and Kapadia, M. (2014). Navigation meshes and real-time dynamic planning for virtual worlds. In: *ACM SIGGRAPH 2014 Courses*, 3. ACM.

Kallmann, M. and Kapadia, M. (2016). Geometric and discrete path planning for interactive virtual worlds. *Synthesis Lectures on Visual Computing: Computer Graphics, Animation, Computational Photography, and Imaging* 8 (1): 1–201.

Kapadia, M. and Badler, N.I. (2012). *Navigation and Steering for Autonomous Virtual Humans*. Wiley.

Kapadia, M., Singh, S., Hewlett, W., and Faloutsos, P. (2009). Egocentric affordance fields in pedestrian steering. In: *Proceedings of the 2009 Symposium on Interactive 3D Graphics and Games*, 215–223. ACM.

Kapadia, M., Singh, S., Reinman, G., and Faloutsos, P. (2011). A behavior-authoring framework for multiactor simulations. *IEEE Computer Graphics and Applications* 31 (6): 45–55.

Kapadia, M., Shoulson, A., Boatright, C.D. et al. (2012a). What's next? The new era of autonomous virtual humans. In: *Motion in Games. MIG 2012*, Lecture Notes in Computer Science, vol. 7660 (ed. M. Kallmann and K. Bekris). Berlin, Heidelberg: Springer.

Kapadia, M., Shoulson, A., Durupinar, F., and Badler, N.I. (2012b). Authoring multi-actor behaviors in crowds with diverse personalities. In: *Modeling, Simulation and Visual Analysis of Crowds*, 147–180. New York: Springer.

Kapadia, M., Pelechano, N., Allbeck, J., and Badler, N. (2015). Virtual crowds: steps toward behavioral realism. *Synthesis Lectures on Visual Computing* 7 (4): 1–270.

Kapadia, M., Frey, S., Shoulson, A. et al. (2016). CANVAS: computer-assisted narrative animation synthesis. In: *Proceedings of the Symposium on Computer Animation*, 199–209.

Karamouzas, I. and Overmars, M. (2010). A velocity-based approach for simulating human collision avoidance. In: *International Conference on Intelligent Virtual Agents*, 180–186. Springer.

Karamouzas, I., Heil, P., Van Beek, P., and Overmars, M.H. (2009). A predictive collision avoidance model for pedestrian simulation. In: *International Workshop on Motion in Games*, 41–52. Springer.

Kielar, P. M., and Borrmann, A. (2018). Spice: a cognitive agent framework for computational crowd simulations in complex environments. *Autonomous Agents and Multi-Agent Systems* 32 (3): 387–416.

Kim, T.W. and Fischer, M. (2014a). Automated generation of user activity–space pairs in space-use analysis. *Journal of Construction Engineering and Management* 140 (5): 04014007.

Kim, T.W. and Fischer, M. (2014b). Ontology for representing building users' activities in space-use analysis. *Journal of Construction Engineering and Management* 140 (8): 04014035.

Krontiris, A., Bekris, K.E., and Kapadia, M. (2016). ACUMEN: activity-centric crowd authoring using influence maps. In: *Proceedings of the 29th International Conference on Computer Animation and Social Agents*, 61–69. New York: ACM.

Kullu, K., Güdükbay, U., and Manocha, D. (2017). ACMICS: an agent communication model for interacting crowd simulation. *Autonomous Agents and Multi-Agent Systems* 31 (6): 1403–1423.

Lamarche, F. and Donikian, S. (2004). Crowd of virtual humans: a new approach for real time navigation in complex and structured environments. In: *Computer Graphics Forum*, vol. 23, 509–518. Wiley Online Library.

Liu, W., Pavlovic, V., Hu, K. et al. (2017). Characterizing the relationship between environment layout and crowd movement using machine learning. In: *Proceedings of the 10th International Conference on Motion in Games*, 2. ACM.

Løvås, G.G. (1994). Modeling and simulation of pedestrian traffic flow. *Transportation Research Part B: Methodological* 28 (6): 429–443.

Magerko, B., Laird, J., Assanie, M. et al. (2004). AI characters and directors for interactive computer games. *Ann Arbor* 1001 (48): 109–2110.

Mahdavi, A. and Tahmasebi, F. (2015). Predicting people's presence in buildings: an empirically based model performance analysis. *Energy and Buildings* 86: 349–355.

Maher, M.L., Simoff, S.J., and Mitchell, J. (1997). Formalising building requirements using an activity/space model. *Automation in Construction* 6 (2): 77–95.

Marmor, Y.N., Golany, B., Israelit, S., and Mandelbaum, A. (2012). Designing patient flow in emergency departments. *IIE Transactions on Healthcare Systems Engineering* 2 (4): 233–247.

Mascarenhas, S., Degens, N., Paiva, A., Prada, R., Hofstede, G. J., Beulens, A., and Aylett, R. (2016). Modeling culture in intelligent virtual agents: From theory to implementation. *Autonomous Agents and Multi-Agent Systems* 30 (5): 931–962.

Montello, D.R. (2005). Navigation. In: *The Cambridge Handbook of Visuospatial Thinking* (ed. P. Shah and A. Miyake), 257–294. Cambridge: Cambridge University Press.

Morad, M.G. (2017). A virtual city simulation platform to assess the effects of information and communication technologies on pedestrian navigation in urban settings. MSc dissertation. Technion-Israel Institute of Technology.

Moussaïd, M., Kapadia, M., Thrash, T. et al. (2016). Crowd behaviour during high-stress evacuations in an immersive virtual environment. *Journal of the Royal Society Interface* 13 (122): 20160414.

Nagy, D., Lau, D., Locke, J. et al. (2017). Project discover: an application of generative design for architectural space planning. Symposium on Simulation for Architecture and Urban Design.

Ondřej, J., Pettré, J., Olivier, A.-H., and Donikian, S. (2010). A synthetic-vision based steering approach for crowd simulation. *ACM Transactions on Graphics (TOG)* 29 (4): 123.

O'Sullivan, D. and Haklay, M. (2000). Agent-based models and individualism: is the world agent-based? *Environment and Planning A* 32: 1409–1425.

Pan, X., Han, C.S., Dauber, K., and Law, K.H. (2007). A multi-agent based framework for the simulation of human and social behaviors during emergency evacuations. *Ai & Society* 22 (2): 113–132.

Paris, S., Pettré, J., and Donikian, S. (2007). Pedestrian reactive navigation for crowd simulation: a predictive approach. In: *Computer Graphics Forum*, vol. 26, 665–674. Wiley Online Library.

Pelechano, N., Allbeck, J.M., Kapadia, M., and Badler, N.I. (2016). *Simulating Heterogeneous Crowds with Interactive Behaviors*. CRC Press.

Preiser, W.F.E., White, E., and Rabinowitz, H. (2015). *Post-Occupancy Evaluation*, 1e. New York: Routledge.

Qiao, G., Yoon, S., Kapadia, M., and Pavlovic, V. (2017). The Role of Data-driven Priors in Multi-agent Crowd Trajectory Estimation. ArXiv Preprint arXiv:1710.03354.

Rao, A.S. and Georgeff, M.P. (1995). BDI agents-from theory to practice. In: *Proceedings of the 1st International Conference of Multiagent Systems*, 312–319.

Reynolds, C.W. (1982). Computer animation with scripts and actors. In: *ACM SIGGRAPH Computer Graphics*, vol. 16, 289–296. ACM.

Reynolds, C.W. (1987). Flocks, herds and schools: a distributed behavioral model. *ACM SIGGRAPH Computer Graphics* 21 (4): 25–34.

Rockcastle, S. and Andersen, M. (2014). Measuring the dynamics of contrast & daylight variability in architecture: a proof-of-concept methodology. *Building and Environment* 81: 320–333.

Santos, G. and Aguirre, B.E. (2004). A critical review of emergency evacuation simulation models. Presented at the Building Occupant Movement During Fire Emergencies, Gaithersburg, Maryland.

Schaumann, D. (2018). An event-based model – for simulating human behavior patterns in not yet built environments. PhD Thesis. Technion – Israel Institute of Technology.

Schaumann, D., Kalay, Y.E., Hong, S.W., and Simeone, D. (2015). Simulating human behavior in not-yet built environments by means of event-based narratives. In: *Proceedings of the Symposium on Simulation for Architecture & Urban Design*, 1047–1054. Society for Computer Simulation International.

Schaumann, D., Morad, M.G., Zinger, E. et al. (2016a). A computational framework to simulate human spatial behavior in built environments. In: *Proceedings of the Symposium on Simulation for Architecture and Urban Design*, 121–128.

Schaumann, D., Pilosof, N.P., Date, K., and Kalay, Y.E. (2016b). A study of human behavior simulation in architectural design for healthcare facilities. *Annali Dell'Istituto Superiore Di Sanita* 52 (1): 24–32.

Schaumann, D., Breslav, S., Goldstein, R. et al. (2017a). Simulating use scenarios in hospitals using multi-agent narratives. *Journal of Building Performance Simulation* 10 (5–6): 636–652.

Schaumann, D., Date, K., and Kalay, Y.E. (2017b). An event modeling language (EML) to simulate use patterns in built environments. In: *Proceedings of the Symposium on Simulation for Architecture & Urban Design*, 189–196. Toronto.

Schultz, C. and Bhatt, M. (2012). Multimodal spatial data access for architecture design assistance. *AI EDAM* 26 (Special Issue 02): 177–203.

Shao, W. and Terzopoulos, D. (2005). Autonomous pedestrians. In: *Proceedings of the 2005 ACM SIGGRAPH/Eurographics Symposium on Computer Animation*, 19–28. ACM.

Shen, T.-S. (2005). ESM: a building evacuation simulation model. *Building and Environment* 40 (5): 671–680.

Shen, W., Shen, Q., and Sun, Q. (2012). Building information modeling-based user activity simulation and evaluation method for improving designer–user communications. *Automation in Construction* 21: 148–160.

Shiwakoti, N., Sarvi, M., and Rose, G. (2008). Modelling pedestrian behaviour under emergency conditions–state-of-the-art and future directions. In: *Australasian Transport Research Forum (ATRF)*, vol. 31, 457–473. Australia.

Shoulson, A., Garcia, F., Jones, M. et al. (2011). Parameterizing behavior trees. In: *Motion in Games. MIG 2011*, Lecture Notes in Computer Science, vol. 7060 (ed. J.M. Allbeck and P. Faloutsos), 144–155.

Shoulson, A., Marshak, N., Kapadia, M., and Badler, N.I. (2013). ADAPT: the agent development and prototyping testbed. *IEEE Transactions on Visualization and Computer Graphics* 20 (7): 1035–1047.

Simeone, D. and Kalay, Y.E. (2014). An event-based model to simulate human behaviour in built environments. In: *Proceedings of the International Conference on Education and research in Computer Aided Architectural Design in Europe*, 525–531.

Simon, H.A. (1969). *The Sciences of the Artificial*. Cambridge, MA: MIT Press.

Singh, S., Kapadia, M., Hewlett, B. et al. (2011). A modular framework for adaptive agent-based steering. In: *Symposium on Interactive 3D Graphics and Games*, 9. ACM.

Sinreich, D. and Marmor, Y. (2005). Emergency department operations: the basis for developing a simulation tool. *IIE Transactions* 37 (3): 233–245.

Soomer, M.J. and Franx, G.J. (2008). Scheduling aircraft landings using airlines' preferences. *European Journal of Operational Research* 190 (1): 277–291.

Sopher, H., Schaumann, D., and Kalay, Y.E. (2017). Simulating human behavior in (Un) built environments: using an actor profiling method. *World Academy of Science, Engineering and Technology, International Journal of Computer, Electrical, Automation, Control and Information Engineering* 10 (12): 2074–2083.

Sud, A., Gayle, R., Andersen, E. et al. (2007). Real-time navigation of independent agents using adaptive roadmaps. In: *Proceedings of the 2007 ACM Symposium on Virtual Reality Software and Technology*, 99–106. ACM.

Sung, M., Kovar, L., and Gleicher, M. (2005). Fast and accurate goal-directed motion synthesis for crowds. In: *Proceedings of the 2005 ACM SIGGRAPH/Eurographics Symposium on Computer animation*, 291–300. ACM.

Tabak, V., de Vries, B., and Dijkstra, J. (2010). Simulation and validation of human movement in building spaces. *Environment and Planning B: Planning and Design* 37 (4): 592–609.

Tang, F. and Ren, A. (2012). GIS-based 3D evacuation simulation for indoor fire. *Building and Environment* 49: 193–202.

Tashakkori, H., Rajabifard, A., and Kalantari, M. (2015). A new 3D indoor/outdoor spatial model for indoor emergency response facilitation. *Building and Environment* 89: 170–182.

Thalmann, D. and Musse, S.R. (2013). *Crowd Simulation*. London, London: Springer.

Torrens, P.M. (2016). Computational streetscapes. *Computation* 4 (3): 37.

Treuille, A., Cooper, S., and Popović, Z. (2006). Continuum crowds. In: *ACM Transactions on Graphics (TOG)*, vol. 25, 1160–1168. ACM.

Van den Berg, J., Lin, M., and Manocha, D. (2008). Reciprocal velocity obstacles for real-time multi-agent navigation. In: *IEEE International Conference on Robotics and Automation, 2008. ICRA 2008*, 1928–1935. IEEE.

Wang, D., Federspiel, C.C., and Rubinstein, F. (2005). Modeling occupancy in single person offices. *Energy and Buildings* 37 (2): 121–126.

Wang, Y., Kapadia, M., Huang, P. et al. (2014). Sound localization and multi-modal steering for autonomous virtual agents. In: *Proceedings of the 18th Meeting of the ACM SIGGRAPH Symposium on Interactive 3D Graphics and Games*, 23–30. ACM.

Whyte, W. (1980). *The Social Life of Small Urban Spaces*. Washington, DC: The Conservation Foundation.

Wooldridge, M. and Jennings, N.R. (1995). Intelligent agents: theory and practice. *The Knowledge Engineering Review* 10 (02): 115–152.

Wurzer, G. (2010). Schematic systems – constraining functions through processes (and vice versa). *International Journal of Architectural Computing* 8 (2): 201–217.

Yan, W. and Kalay, Y.E. (2004). Simulating the behavior of users in built environments. *Journal of Architectural and Planning Research* 371–384.

Zhao, D., Wang, J., Zhang, X., and Wang, X. (2014). A cellular automata occupant evacuation model considering gathering behavior. *International Journal of Modern Physics C* 26 (08): 1550089.

Zheng, X., Zhong, T., and Liu, M. (2009). Modeling crowd evacuation of a building based on seven methodological approaches. *Building and Environment* 44 (3): 437–445.

Zhou, S., Chen, D., Cai, W. et al. (2010). Crowd modeling and simulation technologies. *ACM Transactions on Modeling and Computer Simulation (TOMACS)* 20 (4): 20.

30

Multi-Scale Resolution of Human Social Systems: A Synergistic Paradigm for Simulating Minds and Society
Mark G. Orr

Biocomplexity Institute & Initiative, University of Virginia, Charlottesville VA 22904, USA

Introduction

Recently, we put forth an initial sketch of what we call the *Resolution Thesis* Orr et al. (2018). The thesis holds that (i) models of cognition will be improved given constraints from the structure and dynamics of the social systems in which they are embedded and (ii) the resolution of social simulations of agents will be improved given constraints from cognitive first principles.[1] This thesis reflects a variety of motivations, the most obvious being the observation that there is little overlap between the cognitive sciences and the generative social science approach, both of which rely heavily on computer simulation to understand aspects of human systems, albeit at different levels of scale. The former focuses almost exclusively on the mind as a scientific object of study for which the lion's share of simulation efforts reflect methods that represent a generalizable conception of the mind, and the latter emphasizes multiple aspects of social systems, the mind being only one of these aspects. Thus, we seem to have some kind of historical trade-off: the details of one level of scale result in the potential oversimplification at the other level of scale. We posit, by the resolution thesis, that an interdependence between cognitive science and generative social science could be leveraged for the purposes of improving our understanding of important phenomena studied by both.

The *Resolution Thesis* can be understood from multiple perspectives. From the generative social science perspective, the *Resolution Thesis* means that the representations of agents in social simulations should be informed closely

---

1 For the purposes of this chapter, cognitive models are theoretical models of human perception, thought, and action that includes, broadly, explanations of emotion, motivation, and affect in addition to more traditional domains of cognitive psychology and cognitive science; one could arguably use the term *psychological first principles* as equivalent to cognitive first principles.

Social-Behavioral Modeling for Complex Systems, First Edition.
Edited by Paul K. Davis, Angela O'Mahony, and Jonathan Pfautz.
© 2019 John Wiley & Sons, Inc. Published 2019 by John Wiley & Sons, Inc.
Companion website: www.wiley.com/go/Davis_Social-Behavioralmodeling

by cognitive science and relevant neurophysiological considerations. This runs somewhat counter to the principled adherence to simplification of the internal processing of simulated agents found in this literature, a simplification that served to show that complex social dynamics can be driven by simple behavioral rules of agents.

More recently, there have been efforts in the generative social sciences that acknowledged that closer ties to the psychological and neurophysiological underpinnings of human behavior may yield benefit in terms of modeling social phenomena. Epstein's neurocognitive approach is a notable effort in this vein (Epstein (2014)); there are other related approaches (e.g. Sakellariou et al. 2008; Malleson et al. 2012; Caillou et al. 2017). However, there still remains a large gap between these recent efforts and the implementation of models from cognitive science and psychology, not necessarily in principle, but in practice. It is worth noting that there are some threads in cognitive science that approach social modeling from the perspective of cognitive science. In particular, Ron Sun's push for multi-agent systems based on cognitive first principles is notable (Sun 2006); other work in this vein exists (Bhattacharyya and Ohlsson (2010), Fu and Pirolli (2007), Gonzalez et al. (2003), Huberman et al. (1998), Lebiere et al. (2000), Reitter and Lebiere (2012), Romero and Lebiere (2014), West and Lebiere (2001), and West et al. (2005). The relatively new field of computational social psychology is clearly relevant (Vallacher et al. 2017) as well as the computational organizational theory approach (Prietula et al. 1998).

From the view of cognitive science, the *Resolution Thesis* means that patterns of organization (e.g. information flow on the Internet, clustering of behaviors in a community) at the social and organizational level should inform cognitive models when appropriate.[2] In other words, these patterns should be included as convergent evidence for a theory or model of cognition. At first considera-tion this notion may seem hard to fathom because the implications of cognition for the structure and dynamics of social systems are little understood from the cognitive science perspective.[3] Without an explanatory scheme that links facets of the cognitive model to aspects of social organization, how do we inter-pret the convergent evidence from social systems? The work mentioned above with respect to the cognitive first principles within the generative social science approach (Bhattacharyya and Ohlsson (2010), Fu and Pirolli (2007), Gonzalez et al. (2003), (Huberman et al. (1998), Lebiere et al. (2000), Reitter and Lebiere (2012), Romero and Lebiere (2014), Sun (2006), West and Lebiere (2001), West et al. (2005)) begins to put in place a better understanding of the implications of cognition on social systems, but does not generally consider simulation out-put as part of the convergent evidence for theory at the cognitive level of scale.

---

2 The appropriateness may not be easily determined; for social cognition it may be obvious, but it may not be as clear for other domains, e.g. categorization.

3 Anderson's Relevance Thesis (Anderson (2002)), a somewhat rare exception, reasons about how cognition may have implications for social organization.

This is a subtle but critical point, so let us put it differently. From the cognitive perspective, simulation of social systems with agents grounded deeply in cognitive first principles does not imply the *Resolution Thesis* unless the simulation results are used to judge the quality of the cognitive first principles that ground the agents.

A third and more general view is that the *Resolution Thesis* is about human social systems. An understanding of any of these levels of scale is dependent, to some degree, on an understanding of the others. In effect, the notion of convergent evidence as originating, in part, from other levels of scale applies to all levels of scale. The implication is that we should leverage information across scale in an iterative and synergistic way.

The *Resolution Thesis*, despite sounding somewhat reasonable at face value, is opposed, to some degree, from both the cognitive science and generative social science perspectives. Simon's notion of nearly decomposable systems – that the temporal dynamics of adjacent levels of scale, in most systems, are little correlated – suggests that we can understand well the dynamics at each level of scale independently of the others (Simon 1962)(see Anderson 1972), for similar arguments in physical systems). The KISS principle (keep it simple, stupid) used heavily in the generative social sciences is clearly akin to Simon's notion and is bolstered by its early wins in understanding the behavior of social systems using simple, noncognitive agents (Axelrod et al. 1995), Epstein (2002), Schelling (1969). In cognitive science, Newell, in considering the time scale of human behavior, suggested that the social band ($> 10^4$ seconds, representing social systems and organizational behavior) is characterized to be weak in strength in the sense that it may not reflect computations in a systematic purposeful way relative to lower temporal bands, e.g. cognitive and neural processes Newell (1990).

These counter arguments notwithstanding, our working assumption is that the degree to which the *Resolution Thesis* is useful is an empirical issue. The state of the art in technology and computing and the tight coupling between them and the current milieu should afford exploration of the Resolution Thesis. To this end, we have developed the *Reciprocal Constraints Paradigm* (henceforth *RCP*), a methodological approach for exploring the *Resolution Thesis*. The value of the *RCP* does not lie in precise formal prescriptions, but in providing a scaffold for growing our understanding of social systems and possibly revealing something new about the interdependencies among levels of scale. We will address this last point in more detail in the "Discussion."

The Reciprocal Constraints Paradigm

Figure 30.1 shows the four primary components of the *RCP*: a cognitive system with potential ties to neurophysiology, a social system, and the constraints

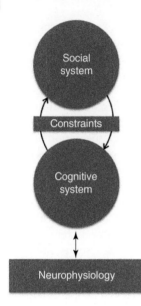

Figure 30.1 The four components of the *RCP* are a cognitive system with potential ties to neurophysiology, a social system, and the constraints on each one from the other. In the *RCP* social systems and cognitive systems are assumed to be derived from and exhibit an abstract set of first principles and properties, called *S* and *C*, respectively. Also captured here is the potential for integrating neurophysiological considerations into the cognitive system when appropriate; these may prove as essential for some social systems (the gray two-headed arrow indicates this potential). The notion of constraint refers to the use of information from *S* and *C* to inform one another in a principled way.

between levels of scale. We assume that social systems and cognitive systems are derived from and exhibit an abstract set of first principles and properties, called *S* and *C*, respectively (e.g. theoretical entities, experimental paradigms, patterns in empirical data in respect to the disciplines that address a particular level of scale). Defining *S* and *C* will depend on the social system or cognitive system of interest. The notion of constraint simply refers to the use of information from *S* and *C* to inform one another in a principled way. An upward constraint refers to information from the cognitive level as informing the social level of scale; downward constraints reverse this relation.

A primary example of entities from the cognitive level of scale is the set of allowable algorithms *A* such that $a \in C$. That is, *A* defines algorithms that are grounded in and thus recognized by work in cognitive science and psychology. Primary examples at the social level of scale are the social structures, channels of information, and dynamic aggregate signals of behavior (e.g. a distribution of degree in a social network over time) that characterize a social system, a large fraction of which are formalized using graph theory/network science. It is important to emphasize that within *S* are notions regarding the behavior of agents, not only social structures.

A central assumption in the *RCP* is that cognitive systems and definitions of agent behaviors in social systems are meant to represent human information processing capacities that can be described as mathematical functions (Van-Rooij 2008).[4] Thus, at the cognitive level of scale *C*, we can define a cognitive system as $\psi_{ct} : I_{ct} \rightarrow \psi_{ct}(i)$ where I_{ct} is the set of allowable inputs and $\psi_{ct}(i)$ is

---

4 This is equivalent to Marr's computational level (Marr (1982).

the output; in S we have a corresponding agent definition as $\phi_{at} : I_{at} \rightarrow \phi_{at}(i)$ where I_{at} is the set of allowable inputs and $\phi_{at}(i)$ is the output for an agent.[5]

Applying the Reciprocal Constraints Paradigm

In practice there are multiple approaches available for application of the *RCP*, but what unites them is the study of human social phenomena, defined either at one level of scale or at multiple levels of scale. Naturally, the first step is to identify social phenomena of interest, a task that is inherently tied to one's perspective. If the perspective is largely cognitive, then the focus would most likely be on understanding the psychological processes, representations, etc. in relation to social systems, but informed in some way by entities at the social level of scale S. Another perspective, at the social scale, would dictate a concern with the social structures and dynamics of the social system (multiple humans interacting) with some degree of constraint from cognitive first principles. These two perspectives are both what we call single-scale approaches to the *RCP*. Of course, one could take a multi-scale perspective that draws from both and likely depends on a simulation approach that captures aspects of C and S in one runnable system.

We will address the obvious issues and difficulties in applying the *RCP* after providing a description of some potential methods for application. The goal in this section is simply to express what it might look like to apply the *RCP*.

Single-Scale Approaches

The single-scale approach of the *RCP* aims to elucidate or refine a model at one level of scale by using some information from another level.

Consider the cognitive scale. The single-scale approach amounts to mapping some properties of social systems to properties of cognitive systems for the purpose of identifying potential implications of social structure and dynamics that should be considered when evaluating a model of cognitive process or representation. For example, suppose one has in hand a cognitive model of how humans acquire attitudes from social interactions (via some learning process) that has not yet been evaluated against empirical data. From the perspective of the *RCP*, proper evaluation of the model (of attitude acquisition) would incorporate or consider aspects of the social level of scale that may have implications for the cognitive system.

In this hypothetical example, we might start with the observation that cognitive learning mechanisms are known to be sensitive to the sequential order in which information is presented to the system.[6] Also, we might observe that

---

5 Social and cognitive systems may define parameters regarding variability among a set of agents; this is not reflected here.
6 We see these phenomena broadly, for example, in attitude formation (Cacioppo et al. 1992), categorization (Heit 1994), and text comprehension (McNamara and Kintsch 1996).

at the social level of scale, the patterning and dynamics of social interactions are dependent upon some properties of S if S contains graph G where $V(G)$ and $E(G)$ are the agents and information channels, respectively. Thus, graph G may have implications for the inputs to the attitude acquisition model in terms of the distribution and sequential ordering of features that the attitude model cares about (e.g. beliefs).

What do these observations in aggregate (one at each level of scale) tell us about how to evaluate the cognitive model of attitudes? The notion, from the perspective of the *RCP*, is that some consideration for how networks affect the distribution of and sequential ordering of the inputs should be incorporated into the design of the empirical observations. Generating random distributions of inputs might not suffice because the hypothesized cognitive model may be representing something in humans that is sensitive to networked information flow. In short, understanding the flow of information on networks might increase the realism of the experimental context used to evaluate the cognitive model.

This is one example of trying to understand the implications of social properties on cognitive systems; what this means precisely would depend on the social phenomena of interest (e.g. early language development may depend on a different $G \in S$ compared with racial stereotypes or large-scale population biases in attitudes).

At the social level of scale, there is a similar approach – mapping the properties of cognitive systems to social systems to reveal potential implications from the former to the latter. An obvious approach is an analysis of the degree to which the definitions of agent behavior ϕ_{at} compares to any cognitive system $\psi_{ct} \in C$ for the purpose of clarifying to what degree a social agent seems to be aligned with cognitive first principles. For the case in which the agent behavior definitions and the cognitive system ϕ_{at} and ψ_{ct} are formally well defined, this might be relatively straightforward,[7] but this is in no way guaranteed.

Certainly, there will be cases for which ψ_{ct} is not formally defined.[8] In fact, a likely scenario is that the closest matching cognitive system ψ_{ct} for the phenomena at hand is only defined in terms of an experimental paradigm, somewhat ill-defined, nonformal theoretical entities and an interpretation of empirical data resulting from application of the experimental paradigm that respects the theoretical entities. So, in this scenario, how does one go about comparing an existing definition of agent behaviors ϕ_{at} defined in a social system to the properties of a cognitive system ψ_{ct} when the nature of the two representations is vastly different? This is not a trivial task, but one approach would be to exploit

---

7 Potential methods for such a comparison would, ideally, focus not only on the comparison of input/output functions but also on the nature of the runnable algorithms.

8 Many theoretical entities in psychology are not formalized in precise mathematical or computational terms but in terms of experimental methods and the relative interpretation of results using statistical methods and reasoning.

the experimental paradigm that is used to define the cognitive system ψ_{ct} in a manner that affords exploring the definitions of agent behavior as defined by ϕ_{at}. In other words, one could mimic the existing experimental paradigm that is the basis for the cognitive system through simulation with agents standing in for humans (assuming ϕ_{at} is defined algorithmically). In essence, this is like running a psychological experiment on artificial agents.[9] The output of the simulations could be compared to the patterning and dynamics of human performance in the original empirical data on humans. This is one suggestion of many potentials, but we hope it illustrates well the potential difficulties.

In summary, although the single-scale approach does not represent the full-blown resolution thesis, it might afford better resolution of a target level of scale by considering some of the implications of other levels of scale. The specific phenomena of interest will drive the precise approaches used.

Multi-scale Approaches

The multi-scale approach is simple in principle: build a simulation platform that simultaneously captures essential aspects of both the social and cognitive systems S and C in respect to a social phenomenon of interest (e.g. an agent-based model of cognitive agents). The upward constraints would mean defining agent definitions ϕ_{at} from cognitive first principles. The downward constraints, generated by some measure of how well the simulation of the social system matches the empirical regularities as defined by the phenomena of interest, would serve as a signal that would suggest modifications to S, C or both. If this scheme seems simplistic, the details of its application are not.

We will illustrate using a hypothetical example. Imagine we are interested in the patterning of obesity by race/ethnicity, a phenomenon of interest with key social and policy aspects (e.g. cultural attractor states, social and shared environmental influence, spatial patterning co-occurring with racial segregation and residential mobility) and, in fact, a preexisting set of social simulations from which to draw (see Nianogo and Arah 2015). Assume we adopt one of the existing simulation approaches, none of which incorporate much in terms of cognitive first principles in C. Given this starting point, one approach would be to substitute a suitable cognitive system ψ_{ct} for the agent behavior definitions ϕ_{at} while keeping the other aspects identical to the original simulation. In other words, we would infuse cognition into the agents.

To do this, because the context is within a simulation environment, there is a minimum requirement that the inputs to and outputs from the cognitive system ψ_{ct} match what the simulation environment has available (for input) and expects as output relative to agents. However, a deeper concern is to

---

9 This approach, comparing human experimental data against an isomorphic simulated experiment where a cognitive model is a stand-in for a human, is common in computational psychology and cognitive science.

find a substitution that is theoretically reasonable (similar to the arguments made above with respect to the single-scale approach to social systems) from a potentially limited but variable set of candidate ψ_{ct} to consider, some or none of which might closely resemble the agent behavior definitions ϕ_{at}. For this hypothetical example, we could draw from several candidates in the health behavior tradition, each of which might be considered as different (e.g. $\{\psi_{ct_i}, \psi_{ct_j}, \psi_{ct_k}, \ldots, \psi_{ct_n}\}$). Choosing among such alternatives may be difficult or may limit the degree to which aspects of the agent behavior defined by ϕ_{at} that can be captured by substitution.

We are now, already, at an interesting point in this hypothetical scenario because it yields potentially difficult decision points. For example, if ϕ_{at} does not match any existing ψ_{ct_i}, what is implied? We might assume that the agent behavior definitions ϕ_{at} might not align with what we know about obesity behavior at the cognitive level of scale. Or it could be argued that the health behavior field, in respect to the processes for which the agent model was developed to study, has yet to develop a cognitive system ψ_{ct_i} that matches ϕ_{at}. One could easily complicate this decision point further or add layers of complication, but we wanted to point out that it gets complicated, fast, with nontrivial solutions, e.g. the development of a cognitive model of a specific health behavior that is grounded empirically takes substantial effort and resources.

However, let us assume that we find a suitable existing cognitive system ψ_{ct} to substitute for the agent behavior definitions ϕ_{at}. Our next task would be to consider the downward constraints. Imagine that along with a preexisting social simulation from which to co-opt, come empirical data, judged of adequate quality, that could be used to compute an objective function with respect to the simulations accuracy given our substitution of ψ_{ct} for ϕ_{at}. This signal, then, would serve as the downward constraint, i.e. a signal that may indicate issues with the cognitive model.[10]

Let us assume that the simulation does poorly in terms of accuracy; what is to be done? One might conduct a set of Monte Carlo simulations to measure the departure from accuracy with respect to the parameters of the cognitive system ψ_{ct} and consider optimizing these parameters to maximize accuracy. But this raises a subtle concern – some would argue that not all parameters in ψ_{ct} should be free to vary on theoretical grounds (Reitter and Lebiere 2010). So, we could take this concern into account for our optimization approach.

It would be reasonable at this point, especially given the potential insights that the Monte Carlo simulations might provide, to revisit the cognitive model and consider it from multiple angles. What is the empirical basis of the model? Is it replicated? What was its purpose originally? Has the model been used across several applied settings? Are the assumed cognitive processes and representations grounded in other similar models of similar phenomena? Given

---

10 One might reasonable use the difference between accuracy given ψ and ϕ instead.

a deep dive into the cognitive model, there might be several options for improving the simulation. Would this spur further experimentation at the cognitive level of scale? What would this look like? Would a new experimental paradigm be generated that respected the structure in the agent-based model? These are all possibilities.

Another concern, aside from issues with the cognitive model, is that unless the properties in S captured in the simulation completely reproduce the empirical data (within a reasonable degree of tolerance), one needs to consider, given the objective function, whether to vary some components in the simulation that represent something in the social system S instead of or in addition to features of the cognitive system C.

The above scenario is but one approach, one that is largely *fixing the social system S and importing a cognitive system C*. But what if we *fix C and generate S* instead? What does this look like? We will stay with the obesity example from above, but change the scenario such that we do not know about any simulation approaches that represent mainly S. Instead, imagine that what we know about the social system S is a set of population-level empirical studies, some of which include information on social networks and the built environment and some theoretical statements about peer influence on social networks. Furthermore, similar to the scenario above, there is a limited set of candidate cognitive systems to consider ($\{\psi_{ct_i}, \psi_{ct_j}, \psi_{ct_k}, \dots, \psi_{ct_n}\}$) in the health behavior tradition, and, likely, other relevant aspects of the cognitive system C may not be represented in them (e.g. categorization, learning, and memory processes). Further, only one of these candidate ψ_{ct_i} is in a computational formalism.[11] We decide that ψ_{ct_i} will serve as our starting point and call it simply ψ_{ct}.

Analysis of ψ_{ct} reveals that it captures the learning and on-the-fly formation of attitudes toward specific health behaviors (considered a precursor to behavior), it is composed in a general manner such that it applies to virtually any health behavior, and it is empirically grounded using traditional health behavior theory measurement techniques (in one particular behavioral domain).

These features are useful, but some key components are missing that are akin to first principles of social system S. In particular, ψ_{ct} is mute with respect to the generation of social structure and related dynamics (e.g. decision-making for initiating/dissolving relationships; social influence mechanisms in terms of how others' behavior or attitudes can potentially serve as the input to the cognitive system ψ_{ct}). Thus, at minimum, some decision points arise in terms of how the simulation implements generation and change in network structure and the mechanism of social influence.

To this end, one could start by implementing a static network topology that captures regularities found in empirical studies of human social networks and

---

11 There do exist a small handful of computationally implemented health behavior theories; see Orr and Chen (2017) for a review.

disallow change in the network ties as the simulation progresses. In terms of social influence, we might assume that what is spreading are attitudes and that exposure to others' attitudes can serve as input to an agent[12] and further there is a knowable stochastic function between attitude and behaviors relating to obesity, e.g. energy balance behaviors relating to caloric intake and use. At this point, we could build out further the social structure and dynamics in relation to the problem of interest using both theoretic and empirical components from S, e.g. racial/ethnic distributions in obesity.

Notice that what is going on here is a process of building from a cognitive kernel and adding layers of assumptions from S where S includes different kinds of information.[13] At some point, we need to run the simulation and determine how to apply the downward constraint.

Once the simulation is runnable, the downward constraint could operate, as described above, by computing an objective function with respect to the accuracy of the simulation compared with extant empirical data at the social level of scale. Here, the issues are mainly the same as in the *fixing S and importing C* case described above. We will not explore this further, as it is described above.

These two examples are largely hypothetical versions of the multi-scale approach to the *RCP*. The value, we hope, is to provide some sketches of what it might look like in practice. Clearly, there are many issues that are raised, even by these sketches, let alone by the general notion of the *RCP* and the *Resolution Thesis*.

Discussion

We have presented the *Resolution Thesis*, its motivation, an approach for acting on its premises, and some examples of what application of the *Reciprocal Constraints Paradigm* might look like. The latter was, largely, an exercise in emphasis – application of the *RCP* is fraught with difficulties on several dimensions, e.g. theories about the individual-level behavior may have little overlap with theories developed in a sociological or demographic context; compute resources may outstrip what is available when integrating cognitive modeling with social modeling; adjudicating over many free parameters across levels of scale given one error signal will not be clear-cut; defining and developing algorithms of behavior from largely ill-defined, nonformal descriptions of individual-level behavior may require conducting clever human subject experiments; there may be issues with the quality and sparsity of data with respect to both observations and theoretical entities. In short, the emphasis

---

12 We have implemented prototypes of this sort Orr et al. (2017).
13 This is very similar to standard practice in social simulation that uses agent-based approaches except that the kernel starts from first principles of computational social psychology.

was really on the difficulties of applying the *RCP*. For the remainder of the chapter, we will shift our focus on developing a deeper understanding of the purpose and goal of the *RCP*.

One way to understand the *RCP* is as a repair for a series of historical accidents – the ones that generated strong divisions between disciplines at or close to the scale of the individual, e.g. neuroscience, psychology, psychiatry and cognitive science on one hand and disciplines at larger scales, e.g. sociology, demography, communications, and economics on the other. The issue, for our thesis and the *RCP*, is that this division is reflected in the various approaches for simulation of human behavior among which the methods for interfacing between them are not well understood. The repair, so to speak, is designed to increase the degree of confidence we have in developing an understanding of social systems from a truly multi-scale perspective, one that provides scientific advancement within and between levels of scale, simultaneously.

But there is more subtlety here, maybe best illustrated by a thought experiment. Imagine an alien is sent to Earth to observe and understand the social behavior of humans. It might observe things like stock markets, automobile traffic dynamics, pedestrian patterns in cities, mating behavior, warfare, the Internet, etc. It could also consider some closer observation of single individuals in controlled contexts or in contexts that are well measured. It might also come to recognize that other living species seem to have similar social organizational principles, at least at some level of abstraction. In short, the alien's approach would be to understand various levels of scale but driven by a unified, holistic perspective; one might call this approach scale-agnostic, to mean that it learns about the parts and interactions as one. Scale is a convenience for understanding parts of the system in the service of understanding the whole.

Some readers might notice parallels between the alien's approach and evolutionary biology and sociobiology (see the classic, Wilson 2000) – so in a strong sense, there is nothing new in the alien's approach. We have designed this thought experiment to emphasize the goal of the alien's approach, a goal that matches that of the *RCP*: a holistic, scale-agnostic way of understanding the varieties of behaviors in a system, such as its components, their interdependencies, and its more macro properties.

The issue we face today is that integrating neurophysiology, cognitive science, game theory, and sociology into an agent-based framework heavily imbued with network science, physics, and computer science perspectives is a patchwork.[14] The goal of the RCP is to transform this patchwork into a unified whole. The key component, or tool if you will, that is offered by the *RCP* is the systematic application of constraints between levels of scale. Over time, through many iterations, the parts and the whole should, we hope, become more aligned in a way that mirrors the product and goal of the alien's approach described above.

---

14 Clearly, there are other disciplines we could add to this list, but these are some of the core.

So, if you forget all but one thing, remember the novelty and strength in the *RCP* lies in its commitment to constraints between levels of scale. We hope this notion will prove useful in moving forward.

Acknowledgments

This research is (partially) based upon work supported by the Defense Advanced Research Projects Agency (DARPA) via the Air Force Research Laboratory (AFRL). The views and conclusions contained herein are those of the authors and should not be interpreted as necessarily representing the official policies or endorsements, either expressed or implied, of DARPA, the AFRL, or the US government. The US government is authorized to reproduce and distribute reprints for governmental purposes notwithstanding any copyright annotation thereon. The National Science Foundation (NSF) supported this research (award #1520359).

The ideas presented in this work are a direct extension of prior work (Orr et al. 2018) and from discussions with the following colleagues: Bill Kennedy, Christian Lebiere, Bianica Pires, Peter Pirolli, and Andrea Stocco.

References

Anderson, P.W. (1972). More is different: broken symmetry and the nature of the hierarchical structure of science. *Science* 177 (4047): 393–396.

Anderson, J.R. (2002). Spanning seven orders of magnitude: a challenge for cognitive modeling. *Cognitive Science* 26 (1): 85–112.

Axelrod, R. (1995). A model of the emergence of new political actors. In: *Artificial Societies: The Computer Simulation of Social Life*, 19–39. Taylor & Francis Group.

Bhattacharyya, S. and Ohlsson, S. (2010). Social creativity as a function of agent cognition and network properties: a computer model. *Social Networks* 32: 263–278.

Cacioppo, J.T., Marshall-Goodell, B.S., Tassinary, L.G., and Petty, R.E. (1992). Rudimentary determinants of attitudes: classical conditioning is more effective when prior knowledge about the attitude stimulus is low than high. *Journal of Experimental Social Psychology* 28 (3): 207–233.

Caillou, P., Gaudou, B., Grignard, A. et al. (2017). A simple-to-use BDI architecture for agent-based modeling and simulation. In: *Advances in Social Simulation 2015, Advances in Intelligent Systems and Computing*, vol. 528 (ed. W. Jager, R. Verbrugge, A. Flache et al.), 15–28. Springer.

Epstein, J.M. (2002). Modeling civil violence: an agent-based computational approach. *Proceedings of the National Academy of Sciences of the United States of America* 99 (Suupl. 3): 7243–7250.

Epstein, J.M. (2014). *Agent_Zero: Toward Neurocognitive Foundations for Generative Social Science*. Princeton, NJ: Princeton University Press.

Fu, W.-T. and Pirolli, P. (2007). SNIF-ACT: a model of user navigation on the world wide web. *Human Computer Interaction* 22 (4): 355–412.

Gonzalez, C., Lerch, F.J., and Lebiere, C. (2003). Instance-based learning in dynamic decision making. *Cognitive Science* 27 (4): 591–635.

Heit, E. (1994). Models of the effects of prior knowledge on category learning. *Journal of Experimental Psychology: Learning, Memory, and Cognition* 20 (6): 1264.

Huberman, B.A., Pirolli, P., Pitkow, J.E., and Lukose, R.M. (1998). Strong regularities in world wide web surfing. *Science* 280 (5360): 95–97.

Lebiere, C., Wallach, D., and West, R. (2000). A memory-based account of the prisoners dilemma and other 2x2 games. In: *Proceedings of International Conference on Cognitive Modeling*, 185–193. Netherlands: Universal Press.

Malleson, N., See, L., Evans, A., and Heppenstall, A. (2012). Implementing comprehensive offender behaviour in a realistic agent-based model of burglary. *Simulation* 88 (1): 50–71.

Marr, D. (1982). *A Computational Investigation into the Human Representation and Processing of Visual Information*. Cambridge, MA: MIT Press.

McNamara, D.S. and Kintsch, W. (1996). Learning from texts: effects of prior knowledge and text coherence. *Discourse Processes* 22 (3): 247–288.

Newell, A. (1990). *Unified Theories of Cognition*. Cambridge, MA: Harvard University Press.

Nianogo, R.A. and Arah, O.A. (2015). Agent-based modeling of noncommunicable diseases: a systematic review. *American Journal of Public Health* 105 (3): e20–e31.

Orr, M.G. and Chen, D. (2017). *Computational Models of Health Behavior*. New York: Psychology Press/Routledge.

Orr, M.G., Ziemer, K., and Chen, D. (2017). *Systems of Behavior and Population Health*. New York: Oxford University Press.

Orr, M.G., Lebiere, C., Stocco, A. et al. (2018). Multi-scale resolution of cognitive architectures: a paradigm for simulating minds and society. *Proceedings of the International Conference SBP-BRiMS*.

Prietula, M., Carley, K., and Gasser, L. (1998). *Simulating Organizations: Computational Models of Institutions and Groups*, vol. 1. Cambridge, MA: The MIT Press.

Reitter, D. and Lebiere, C. (2010). Accountable modeling in ACT-UP, a scalable, rapid-prototyping ACT-R implementation. *Proceedings of the 2010 International Conference on Cognitive Modeling*.

Reitter, D. and Lebiere, C. (2012). Social cognition: memory decay and adaptive information filtering for robust information maintenance. In: *Proceedings of the 26th AAAI Conference on Artificial Intelligence*, 242–248. AAAI.

Romero, O. and Lebiere, C. (2014). Simulating network behavioral dynamics by using a multi-agent approach driven by ACT-R cognitive architecture.

Proceedings of the Behavior Representation in Modeling and Simulation Conference.

Sakellariou, I., Kefalas, P., and Stamatopoulou, I. (2008). Enhancing netlogo to simulate BDI communicating agents. In: *Artificial Intelligence: Theories, Models and Applications. SETN 2008, Lecture Notes in Computer Science*, vol. 5138 (ed. J. Darzentas, G.A. Vouros, S. Vosinakis, and A. Arnellos), 263–275. Springer.

Schelling, T.C. (1969). Models of segregation. *The American Economic Review* 59 (2): 488–493.

Simon, H.A. (1962). The architecture of complexity. *Proceedings of the American Philosophical Society* 106 (6): 467–482.

Sun, R. (2006). *Cognition and Multi-Agent Interaction: From Cognitive Modeling to Social Simulation*. New York, NY: Cambridge University Press.

Vallacher, R.R., Read, S.J., and Nowak, A. (2017). *Computational Social Psychology*. Routledge.

Van Rooij, I. (2008). The tractable cognition thesis. *Cognitive Science* 32 (6): 939–984.

West, R.L. and Lebiere, C. (2001). Simple games as dynamic, coupled systems: randomness and other emergent properties. *Cognitive Systems Research* 1 (4): 221–239.

West, R.L., Stewart, T.C., Lebiere, C., and Chandrasekharan, S. (2005). Stochastic resonance in human cognition: ACT-R vs. Game theory, associative neural networks, recursive neural networks, q-learning, and humans. In: *Proceedings of the 27th Annual Conference of the Cognitive Science Society*, 2353–2358. Mahwah, NJ: Lawrence Erlbaum Associates.

Wilson, E.O. (2000). *Sociobiology*. Cambridge, MA: Harvard University Press.

31

Multi-formalism Modeling of Complex Social-Behavioral Systems

Marco Gribaudo[1], Mauro Iacono[2], and Alexander H. Levis[3]

[1] *Department of Computer Science, Polytechnic University of Milan, Milan, Italy*
[2] *Department of Mathematics and Physics, Università degli Studi della Campania "Luigi Vanvitelli", Caserta, Italy*
[3] *Department of Electrical and Computer Engineering, George Mason University, Fairfax, VA 22030, USA*

Prologue

The value of multi-formalism simulation can be illustrated by considering an example from a past study on methods for assisting military commands in crisis planning. The study developed an illustrative scenario that could be explored with simulation (Levis and Carley 2011). Although many aspects of the scenario were fictitious (decisions, results), the scenario was based on events that occurred in 2002 and used the actual names of countries, regions, and officials at the time. The scenario was as follows:

- The animosity between India and Pakistan has its roots in history and religion and is epitomized by the long-running conflict over the state of Jammu and Kashmir. China administers an area called Aksai Chin at the northeastern corner of Jammu and Kashmir, although India contests Chinese control.
- A sequence of terrorist incidents has occurred in Srinagar, the capital of the Indian-administered state of J&K, and along the line of control (LOC) separating the Northern Areas controlled by Pakistan and Jammu and Kashmir controlled by India. There is growing instability and disaffection with the government in Pakistan, while in India the opposition parties are becoming stronger. The two countries start making a series of escalating moves (e.g. they recall diplomatic staff, move troops toward the LOC, reposition mobile missile batteries, and initiate activities in their nuclear weapon facilities). Events such as bombings are taking place on both sides of the LOC. As usual in such situations, escalations occur in part because each side lacks an understanding of the adversary's intent. It may know facts about, e.g. troop movements, but not about adversary intent and strategy.

Social-Behavioral Modeling for Complex Systems, First Edition.
Edited by Paul K. Davis, Angela O'Mahony, and Jonathan Pfautz.
© 2019 John Wiley & Sons, Inc. Published 2019 by John Wiley & Sons, Inc.
Companion website: www.wiley.com/go/Davis_Social-Behavioralmodeling

- US military commands take notice, particularly US Central Command (CENTCOM) and US Pacific Command (PACOM), which have Pakistan and India within their geographic areas of responsibility (AOR), respectively. China is also concerned because of the danger of nuclear exchanges. The United Nations and Russia contemplate possible involvement. The common objective of the United States and the larger international community is to dissuade or deter the two adversaries from escalating the situation into a nuclear exchange. They want to facilitate a rapid de-escalation of crisis. Because misinformation and misinterpretations (sometimes reflecting domestic pressures) can foster escalation, the US commands are considering an intelligence, surveillance, and reconnaissance (ISR) plan to keep its leaders informed and a complementary information operations campaign in which the United States provides improved information about the state of the conflict to the two adversaries, thus reducing the ambiguities and potential for unintended escalation.

Given this illustrative scenario, the objective in the study was to analyze the situation and to develop and analyze alternative courses of action (COAs) intended to help in de-escalating the crisis. In doing so, the study considered all instruments of national power: diplomatic, information, military, economic (DIME). To do it, a diverse set of social-behavioral models was used.

Two short-duration vignettes were considered for the scenario period from 1 June 2002 to 20 July 2002. The first addressed CENTCOM's and PACOM's observations of the situation and the diplomatic and military maneuvers by India and Pakistan. Each command had a collection of contingency plans (CONPLANs) for their geographic AOR. This situation would cause both commands to more closely observe as well as update their respective CONPLANs and supporting models. Each command would also develop multiple COAs that are focused on broadly defined diplomatic actions and on surveillance and other intelligence activities. The second vignette (B) assumed that the COAs of vignette A had not produced the desired results and that the crisis was escalating. CENTCOM and PACOM assessed the situation and developed COA to de-escalate the crisis (Levis and Carley 2011).

Clearly, no single model could capture the many aspects of the problem. Social Network models drew on documents of the period to represent the key decision-makers in the Pakistani and Indian governments. Colored Petri nets represented organizational structures and processes at CENTCOM and PACOM. Comparisons between the formal organizations, as defined in Petri net models, and the empirical data captured in the Social Networks led to the refinement of both models. An agent-based model then enabled consideration of dynamic interactions. A Timed Influence Net was used for the situational assessment model and to generate potential COA. These COAs were then exercised with the refined models to determine which ones came close to meeting US objectives. Figure 31.1 shows the workflow of analysis, COA development, and evaluation. The modeling and simulation tools are shown on the left. CAESAR III is based on the Petri net formalism, and PYTHIA is Timed Influence Nets, a specialization of Bayesian nets; both were developed at the System Architectures Lab at George Mason University. ORA, a Social Network analysis tool, and Construct, an agent-based modeling tool, were developed at CASOS at Carnegie Mellon University. The figure also indicates the diversity of data sources that were used to build the models.

The approach that was used to address this problem required multi-formalism modeling. But since the models interacted with each other, the question arose: Although technology enables us to connect and interoperate such diverse models, expressed in different mathematical and modeling formalisms, how can we be assured that their interoperation is valid?

Introduction

The richness and variety of modeling formalisms, each of which is tailored on a specific class of problems, not only empowers the modeling process but also results in the proliferation of different models when coping with complex social-behavioral systems. These individual model types may draw from the same or different data sets and can address one or several aspects of the problem. In practice, they might not be aligned to each other and the mutual influences connecting them could be lost, weakening the overall results. Multi-formalism modeling is an approach to modeling that aims to coherently combine different modeling formalisms for different aspects of the same entity while appropriately reflecting mutual influences among models. Multi-modeling, another term often used, includes the interoperation of different models expressed in the same formalism.

"Modeling is the process of producing a model; a model is a representation of the construction and working of some system of interest" (Maria 1997). Figure 31.2 represents the modeling hierarchy where a model is obtained using a modeling tool that applies a modeling formalism to represent a specific system. The model itself should always conform to the modeling formalism used to create it. In this context, the term formalism refers to a formal or formalized syntactical definition for a modeling language (in textual, graphical, or other form) used to describe an abstraction of a system. A formalism employs proper distinguishable syntactical elements to represent atomic or nonatomic components of the system with their properties and syntactical constructs that describe the rules by which elements can be used to form a good representation. It is thus a description tool to capture and abstract some specific aspects of a social-behavioral system and provide a coherent and formal representation of it. Some formalisms are executable (i.e. they generate

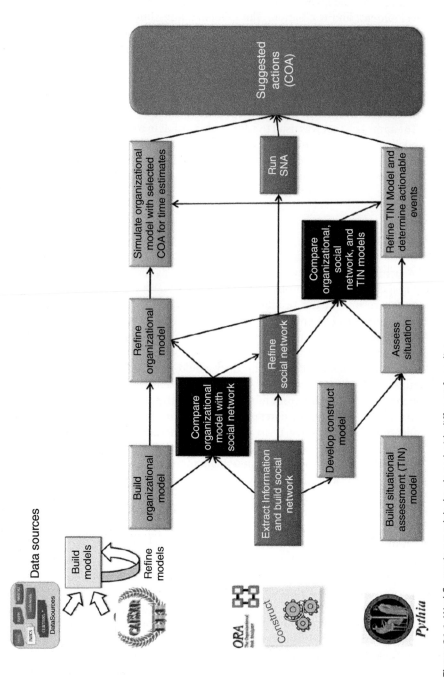

Figure 31.1 Workflow using models developed with different formalisms.

Figure 31.2 Modeling hierarchy.

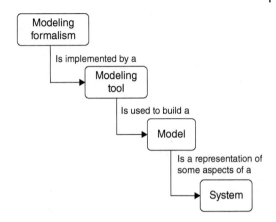

behavior using underlying algorithms or *solvers*; these are often called solvable formalisms), while others are primarily representational (i.e. they represent only the structure but not the behavior).

To address in a structured manner the modeling and simulation issues arising when multiple models must interoperate, four layers are considered. They are analogous to the seven layers of Open Systems Interconnection (OSI) architecture (MacKinnon et al. 1990). The physical layer (hardware and software) is a platform that enables the concurrent execution of multiple models expressed in different formalisms and provides the ability to exchange data and also to schedule the events across the different models. The syntactic layer assures that the right data are exchanged among the models. Once this is achieved, a third problem needs to be addressed at the semantic layer, where the interoperation of different models is examined to ensure that conflicting assumptions in different modeling formalisms are recognized and form constraints to the exchange of data. The use of layers provides a conceptual framework for discussing what needs to be done in order to use multiple interoperating models. Testbeds usually provide the physical and syntactic layers in an integrated way. The implementation of the workflow layer is closely coupled to the syntactic layer. In the workflow layer, valid combinations of interoperating models are considered to address specific issues. Different issues require different workflows. Each of these layers and the research challenges they pose are discussed in subsequent sections in this chapter.

Because the social-behavioral domain is very complex when observed from different perspectives, a classification scheme is needed that can place different modeling approaches and modeling formalisms in context but in a way that is meaningful to empiricists who collect the data, to modelers that need data to develop and test their models, and to theoreticians who use model-generated data to induce theoretical insights. Social scientists have been collecting and analyzing sociocultural data to study social groups, organizations, tribes, urban and rural populations, ethnic groups, and societies. Data may be punctual or expressed by time series, related to an entity or a population, focusing on individuals or their relations: this requires a classification and taxonomy of model types and how data relate to them to bridge the gap between data and models. As a start to the development of a taxonomy for the problem space, three dimensions are considered: social entity or granularity, time, and scope of problem.

Social Entity or Granularity

This is an important dimension because it addresses the issue of improper generalization of the data. Being explicit in specifying the characterization of the social entity that will be modeled and analyzed is essential. The social entity classification can range from a specific individual (e.g. a military leader) to a cell or team (e.g. a terrorist cell), to a clan or tribe, to an ethnic or religious group, and all the way to a multicultural population – the society – of a nation-state. There are serious definitional issues with regard to the social entity that need to be addressed by mathematical and computational modelers such as spatial, temporal, and boundary constraints, sphere of influence, change processes, and adaptation rate. For example, while the boundaries of an individual are impermeable, the boundaries of a tribe are permeable. Marriage may entail crossing clan or tribal boundaries and, interestingly, in both directions. Individuals can be in only one location at a time, whereas groups can have a very complex spatial–temporal presence. As the size of the social entity increases, the sphere of influence size may increase, the physical space covered may increase, and the adaptation rate decreases. Depending on the situation, the time response of individuals, or enterprises, or even nation-states can range from seconds to years; similarly, adaptation to new situations may require minutes, weeks, months, or years to take effect. Finally, the change processes are different with individuals changing as they learn and nation-states changing through processes such as migration, legislation, economic collapse, and war.

Time

The time attribute is complex and cannot be captured by a single variable. First, time spans over periods, requiring variables for at least the starting and ending points. Another possible attribute is the interval between sampling instants. Second, there is the time dimension of the model itself (the epoch that model represents that can be an instant or a period). Choosing an instant leads to static models; the instant of time can be a point in the timeline, a month, a year, or any other epoch. Essentially, a static description of the social-behavioral system is developed that remains unchanged during that epoch. The alternative is considering a dynamic model in which behavior changes. Continuous time, discrete time, and discrete event models are examples of dynamic models. The length of time that the model addresses (its time horizon) is a key temporal variable.

The time dimension is closely related to the scope of the problem and is also related to the size of the social entity being considered – as the time horizon increases, the sphere of influence (i.e. the size of the social entity that will be impacted) is increased. Furthermore, the larger the social entity, the longer the time horizon that may need to be considered to observe effects on the behavior. While exceptions are possible, this consideration restricts meaningful combinations of entity size and time horizon.

In many cases, the mathematical and computational models are created so that they can be used in developing strategies to effect change that will increase the effectiveness of the system. In this case, proper time handling is essential both in the model and the underlying data. An additional challenge with respect to the temporal dimension(s) is considering the persistence of effects on the targeted social entity; it is important to account not only for delays in an effect becoming observable but also for its possible gradual attenuation over time.

Scope of Problem

Military decisions are usually characterized as tactical, operational, and strategic. Often, tactical decisions are concerned with the present and immediate future and with a limited spatial scale, operational decisions with the near-term future and possibly larger spatial scale, and strategic decisions with the long term and again possibly at the geopolitical level. However, information technology has changed this: tactical decisions may have almost immediate strategic implications. The three types have become tightly coupled. For example, a terrorist act, itself a small tactical operation, can have significant strategic impact by changing the long-term behavior of a population (e.g. an unsuccessful terrorist action on an airplane causing significant changes in air travel security worldwide). Similarly, long-term oppression of a population coupled with demographic changes can cause a sudden eruption of violent protest.

These three dimensions (social entity, time, and scope) define some key aspects of the problem space for modeling diverse types of social-behavioral systems (Figure 31.3). Additional dimensions can be used as needed. For example, a fourth dimension can indicate the types of data sources available, while a fifth one may include the modeling formalisms that are appropriate for the issue being investigated.

Given a classification scheme and the appropriate ordering, we can define a set of feasible cells in the three-dimensional space of Figure 31.3. Each cell defines a particular model by specifying (i) the social entity being considered, (ii) the scope of the issue to be addressed, and (iii) the epoch over which the proposed model will apply. Some cells may not correspond to interesting cases.

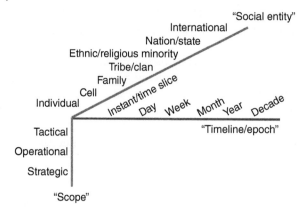

Figure 31.3 A notional representation of the three dimensions for modeling social-behavioral systems.

On Multi-formalism

The classification scheme of Figure 31.3 would place the model of interest in one or more (in this case contiguous) cells. The next step is to identify the formalisms appropriate to the system of interest. The classification scheme of Figure 31.3 acts as the first filter for identifying appropriate formalisms based on social entity, scope, and time dimension and identifying those that are inappropriate because of granularity or temporal considerations. Having identified a set of modeling formalisms, the existence of multiple modeling alternatives leads to two questions: What data are required for each modeling formalism, and how will the various models interoperate?

Also, some mathematical and computational models, by their basic assumptions, may not be able to be associated with specific values of the other axes. For example, applying agent-based models to a single individual is odd, so the corresponding set of cells would be empty. Multiple models may share a cell, depending on the problem scope. The classification scheme selected for each axis should make sense with respect to the problem domain. For example, terrorist organizations exist and operate amidst civilian (noncombatant) populations. Since no single model can capture well both the terrorist cell dynamics and the noncombatant population behavior, a set of interoperating models are needed.

The multi-formalism modeling approach aims to define techniques and methods that allow coordinating the modeling of different aspects of a system by means of interoperating heterogeneous models. This interoperation can be implemented in a variety of ways: from composing a single macro-model by a set of proper sub-models, for each of which the most suited modeling formalism should be applied without losing the coherence of the overall model (Sanders et al. 2007), to a dynamic network of models that are invoked to do their part in accordance with a workflow designed to address an issue.

Multi-formalism is thus a modeling approach that aims to enable modelers to exploit different modeling formalisms for different aspects of the same system while keeping the coherence, coordination, and mutual influences of the different parts of the overall modeling system. Multi-formalism modeling can support the design of complex systems by allowing field experts to use familiar tools through composing them and coordinating their interoperation.

Let *modular multi-formalism* define the case in which it is possible to obtain a model as composition of heterogeneous sub-models or to decompose a model into heterogeneous sub-models (Gribaudo and Iacono 2014). Model composition of this type potentially offers some advantages in managing the complexity of a model and in facilitating sharing and reuse. Let a *composed model* be a model obtained by composition of sub-models (Davis and Anderson 2003, 2004). Let *composition semantics* define a mechanism by which, during a composed model solution process, sub-models can interact. A given composition semantic can be applied in any step of every solution strategy presented. The composition semantics are implemented in the form of a workflow that manages the execution of the composed model.

Note that interactions among the components of the composed model can take a wide variety of forms: two models run in series with the output of one providing an input to the other; one model can be invoked and run inside another; two models run side by side and interoperate; and one model is used to construct another by providing design parameters and constraints or to construct the whole or part of another model. The interoperation can be complementary or supplementary; these are aspects of the need for semantic interoperability.

Issues in Multi-formalism Modeling and Use

The Physical Layer and the Syntactic Layer

This physical layer is a computational platform that enables the concurrent execution of multiple models expressed in different modeling formalisms and provides the ability to exchange data and also, if necessary, to *schedule* the events across the different interoperating models that have different time scales. The second layer is the syntactic layer that assures that the right data are exchanged among the models. Some of the technical and implementation issues regarding the physical layer were resolved more than a decade ago. There are numerous testbeds, based on different principles, which enable the interoperation of models. Three different examples follow.

In SHARPE (Symbolic Hierarchical Automated Reliability and Performance Evaluator) (Trivedi 2002) models are composed of sub-models in a fixed set of different formalisms with different execution algorithms. SHARPE deals with

Markov models, product form queuing networks, and generalized stochastic Petri nets; the sub-models interact by exchanging probability distributions, and the solution process is determined by the user. Modularity is managed at model level by its source code.

SIMTHESys (Structured Infrastructure for Multi-formalism modeling and Testing of Heterogeneous formalisms and Extensions for SYStems) (Barbierato et al. 2011a,b; Iacono et al. 2012) provides a methodology for extensible design and evaluation of multi-formalism models, with model composition and multiple solution methods. It is based on the explicit definition of both syntax and semantics of all atomic components of a formalism and on a set of nonspecialized computational engines that are used to generate (multi)formalism-specific reusable solvers. (A *solver* applies a numerical method to *solve* the set of equations – or formalisms – that represent the *model*.) Model interactions are defined by arc superposition.

The C2 Wind Tunnel (C2WT) is an integrated multi-modeling simulation environment (Karsai et al. 2004; Hemingway et al. 2011). Its framework uses a discrete event model of computation as the common semantic framework for the precise integration of an extensible range of simulation engines, using the run-time infrastructure (RTI) of the high-level architecture (HLA) plat-form. The C2WT offers a solution for a class of multi-formalism simulation by decomposing the problem into model integration and experiment or simulation integration. The key characteristic of the C2WT is that a *federate* is created for each modeling formalism (e.g. a colored Petri net federate, a Simulink federate, etc.). Once the federate for a formalism has been created, then any model that is consistent with that formalism can be inserted and used in the simulation.

These three examples (and there are many more in the literature), while work-ing properly and achieving their goals, also indicate clearly that much more research needs to be done to establish some common standards and proce-dures at the physical and syntactic layers. Each of the existing computational testbeds has been designed to address a specific class of problems; none of them are general enough to accommodate easily new formalisms. For example, the issue of how to deal with the interoperation of time-driven and event-driven models (hybrid systems) has not been solved in general.

Once a suitable testbed has been constructed implementing the physical and syntactic layers, a third kind of problem needs to be addressed at the semantic layer, where the interoperation of different models is examined to ensure that conflicting assumptions in different modeling formalisms are recognized and form constraints to the exchange of data. It is not sufficient to pass geolocation from one model to another; the coordinate system must be the same, or an adaptor must be inserted that converts from one coordinate system to another.

The Semantic Layer

In an effort to formally represent the semantic interoperability of disparate models and modeling languages, Rafi (2010) and Levis et al. (2012) have developed a meta-modeling framework. This approach extends earlier works by Kappel et al. (2006) and Saeki and Kaiya (2006) for a class of modeling formalisms primarily used for social-behavioral modeling problems. This framework leads to a phased approach that uses concept maps, meta-models, and ontologies. An ontology defines a common vocabulary for a particular domain and the relationships between the elements of that vocabulary. The approach is not simple and requires multiple steps. It is based on comparing the ontologies (for each modeling formalism) to help identify the similarities, overlaps, and/or mappings across the model types under consideration.

The fundamental idea is to deconstruct a modeling formalism into its fundamental concepts. If an axiomatic approach was taken to developing a modeling formalism, as the case was with Forrester's (1968) system dynamics, the task is relatively easy. Also, if the formalism is expressed as a formal mathematical model (e.g. differential equations, Bayesian nets, colored Petri nets), then the basic material exists for creating a concept map and from that a formal ontology. The situation is much more challenging when one considers formalisms such as Social Networks, agent-based models, and probabilistic decision models such as Influence and Timed Influence Nets. The relationships in the latter models are inferred from data or postulated by subject matter experts.

The approach starts by specifying a modeling formalism by constructing generalized concept maps (Novak and Cañas 2008) that capture the assumptions, definitions, elements, and their properties and relationships relevant to that formalism. These concept maps address different aspects of the formalism. Although they are a structured representation, they are not formal and are therefore not amenable to machine reasoning. Figure 31.4 is a concept map addressing the following question: What are the concepts underlying Social Networks, and what are the relationships among these concepts? Figure 31.5 addresses a different question: What are the types of analyses that can be carried out with Social Network modeling, and what are the relationships among these analyses?

The concept map representation is then formalized to reveal syntactic considerations and to lay the foundation for ontology. A basic ontology serves as the foundation ontology; it does not contain semantic concepts related to the modeling formalism or modeled domain but rather acts as a skeleton for the ontology. An example for Social Networks is shown in Figure 31.6. In the next step, semantic concepts and relationships are added to obtain the refactored ontology. Refactoring modifies an ontology's structure while preserving its semantics. This refactoring (accomplished with an understanding of the set of formalisms being treated) facilitates the next step of comparing ontologies

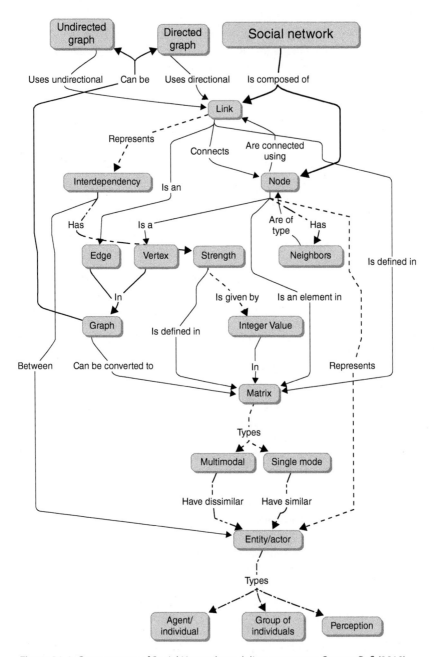

Figure 31.4 Concept map of Social Network modeling constructs. Source: Rafi (2010).

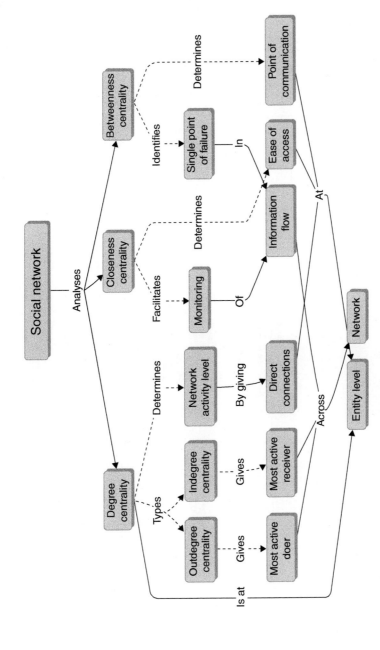

Figure 31.5 Concept map of Social Network analyses (Rafi 2010).

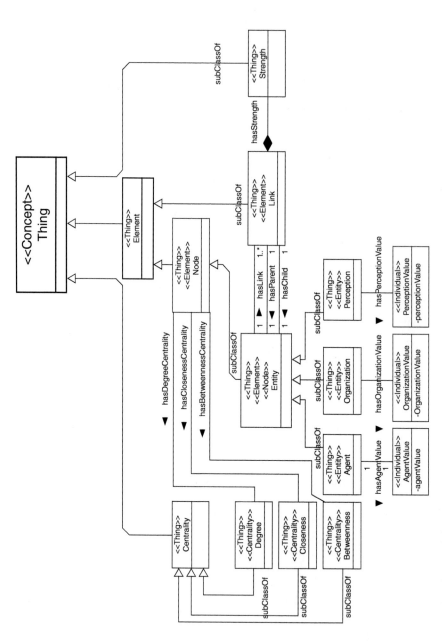

Figure 31.6 Social Network foundational ontology. Source: Rafi (2010).

across formalisms. Each formalism may have its own terms for concepts that appear in both formalisms. Once the individual ontologies are completed, mapping of concepts across the ontologies is begun. The objective is to use the same terminology for the same concept across the refactored ontologies

The refactored ontology for the Social Network formalism is shown in Figure 31.7.

The resulting overall ontology is called an enriched ontology. It should identify mappings among the semantically equivalent concepts of the different formalisms considered so that the exchange of information or analysis results between models constructed using these formalisms can take place. The approach is illustrated for three formalisms in Figure 31.8. This enriched ontology acts as the (template) knowledge container for a specific domain, and it can be reused, since this knowledge is in machine-readable form.

The refactoring of the different ontologies inside the enriched ontology is achieved by defining additional object properties in related classes and is done manually. For instance, the *Agent* and *Organization* classes from the Social Network refactored ontology can be mapped to the *subject* and *object* classes of the Influence Net refactored ontology by adding *hasSubjectValue* and *hasObjectValue* object properties to the existing object property of *Agent* and *Organization* classes (see Table 31.1). Similarly, the *Belief* class in Social Network refactored ontology can be mapped to Influence Net's *Belief* class. Table 31.1 summarizes these mapped concepts between both the refactored ontologies. The ultimate result of this mapping is an *enriched ontology* that is the knowledge container of both Influence Net and Social Network modeling techniques.

The enriched ontology so constructed for the modeling formalisms can be reasoned with using the logical theory supporting the ontological representation (Bechhofer 2003). The mappings suggest possible semantically correct ways to ensure consistency and to exchange information (i.e. parameter values and/or analysis results) between different types of models when they are used in a workflow addressing a specific problem of interest.

The enriched ontology can be represented graphically using an application such as GraphViz (2017), an open-source graph visualization software. The enriched ontology for Timed Influence Nets and Social Networks is shown in Figure 31.9.

Analysis of the enriched ontology indicates what types of interoperation are valid between models expressed in different modeling formalisms. Two models can interoperate (partially) if some concepts appear in both modeling formalisms and have no contradictory concepts that are invoked by the particular application. By refining this approach to partition the concepts into modeling formalism/language input and output concepts and also defining the concepts that are relevant to the questions being asked to address the problem, it becomes possible to determine which sets of models can interoperate to address some or all of the questions of interest and which sets of models use

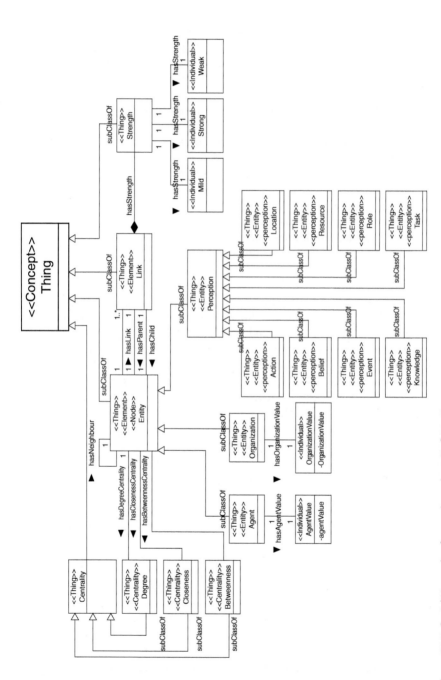

Figure 31.7 Social Network refactored ontology (Rafi 2010).

Table 31.1 Enriched ontology derived from the refactored ontologies for Timed Influence Nets and Social Network formalisms.

Influence Net refactored ontology elements			Social Network refactored ontology elements		
Domain class	Object property	Range class	Domain class	Object property	Range class
Subject	hasSubjectValue	Subject	Agent	hasSubjectValue	Subject
	hasAgentValue	Agent		hasAgentValue	Agent
	hasOrganizationValue	Organization		hasObjectValue	Object
Object	hasObjectValue	Object	Organization	hasObjectValue	Object
	hasAgentValue	Agent		hasAgentValue	Agent
	hasOrganizationValue	Organization		hasSubjectValue	Subject
Verb	hasVerbValue	Verb	Task	hasVerbValue	verb
	hasTaskValue	Task		hasTaskValue	Task
Intent/decision	hasElements some Action and hasElements some subject and hasElements some verb	Action, subject, verb		hasElements some Action and hasElements some subject and hasElements some verb	Action, subject, verb
Action	hasElements some subject and hasElements some verb and hasElements some object	Action, subject, verb	Action	hasElements some subject and hasElements some verb and hasElements some object	Action, subject, verb
	hasActionValue	Action		hasActionValue	Action

(continued)

Table 31.1 (Continued)

Influence Net refactored ontology elements			Social Network refactored ontology elements		
Domain class	Object property	Range class	Domain class	Object property	Range class
Belief	hasElements some subject and hasElements some verb and hasElements some (Ability or Decision or Action or Event) and hasEvidence some Evidence	Subject, object, verb, Ability or Decision or Action or Event, Evidence	Belief	hasElements some subject and hasElements some verb and hasElements some (Ability or Decision or Action or Event) and hasEvidence some Evidence	Subject, object, verb, Ability or Decision or Action or Event, Evidence
	hasBeliefValue some Belief	Belief		hasBeliefValue some Belief	Belief
State	hasStateValue	State	Event	hasStateValue	State
	hasEventValue	Event		hasEventValue	Event
Quality	hasQualityValue	Quality	Knowledge	hasQualityValue	quality
	hasKnowledgeValue	Knowledge		hasKnowledgeValue	Knowledge

Source: Rafi (2010).

Figure 31.8 Enriched ontology construction.

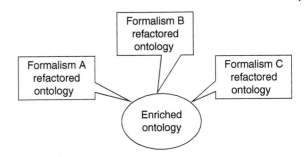

contradictory input and output concepts that are relevant to those questions. If the latter occurs, then the results would not be valid. A well-known example of such a situation is the interoperation of system dynamics models and Leontief input–output models. In the system dynamics formalism, the concept of equilibrium does not exist. Instead it supports the concept of steady state. Conversely, the Leontief model is an equilibrium model. This limits the interoperation of the two models. The key issue is whether the interoperation, as defined by the workflow for a particular application, triggers the contradictory concepts. If the equilibrium solution of a Leontief model is used as the initial state for a system dynamics simulation, then the contradictory concepts are not triggered, and the interoperation is valid.

Finally, for semantic interoperability, models need to be interchanged across tools. This requires model transformations. The transformations are formally specified in terms of the meta-models of the inputs and the outputs of the transformations. From these meta-models and the specification of the semantic mapping, a semantic translator that implements the model transformation is synthesized. While much is being done on model transformation techniques, this has been focused on model-based systems engineering models and much less on social-behavioral models. For example, one early effort to transform one type of model to another (Moon 2007) focused on using the data from a Social Network to generate an Influence Net.

The last two paragraphs indicate two major areas for further research. Formalisms used for social-behavioral modeling need to be deconstructed, their ontologies developed and then refactored so that enriched ontologies can be used to develop sets of rules for valid interoperation of models developed using these formalisms. The desired objective would be a set of rules that will be embedded in domain-specific workflow management languages that are used to construct and manage the simulations.

The Workflow Layer

It has been argued in this chapter that while individual modeling formalisms such as Social Networks, agent-based modeling, system dynamics, colored

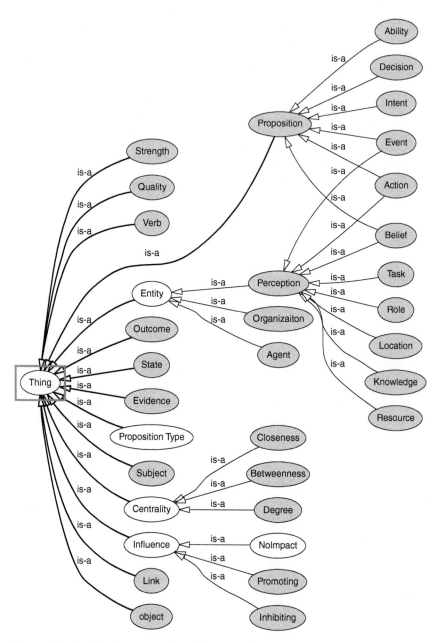

Figure 31.9 Graphical representation of the enriched ontology of Table 31.1. Source: Rafi (2010).

Petri nets, etc. might be capable of addressing specific issues, complex social-behavioral problems may require multiple models interoperating to complement/supplement each other. In the previous sections, the syntactic and semantic challenges of the multi-formalism modeling approach were discussed. But there is one more major challenge in actually implementing multi-formalism modeling and simulation: the design and implementation of the workflows that enable and control the correct interoperation of the various models. At this time, the approach is domain specific, and within a domain, different issues require different workflows.

A multi-formalism modeling workflow is itself a model of a process for generating data expressed as values, tables, or graphs that inform the analysis that is being done to support decision-making. A formal approach to capture such a workflow requires a modeling language with its own rules. Creating workflows using a domain-specific language allows for translating visual views of model interoperation into an executable implementation. There already exist generic techniques for creating and executing workflows such as Business Process Model and Notation (BPMN 2014) and Web Services Business Process Execution Language (WSBPEL 2007). Such a language would be tailored to a problem domain of interest and would offer a high level of expressiveness. This has led to the development of domain-specific multi-formalism modeling workflow languages.

The objective is helping multi-formalism platform users in creating workflows of modeling activities while guaranteeing both syntactic and semantic correctness of the resulting composition of interoperating models. The approach is domain specific because it is driven by the ontologies used at the semantic layer. The first step consists of the identification and characterization of a domain of interest and the modeling formalisms that support it. The domain specification identifies a region containing a number of nonempty cells in the construct of Figure 31.3; it specifies the types of social entities being considered, the scope of the issues being addressed, and the timeline or epoch of the proposed simulation. Domain analysis follows; its aim is to provide formal representations of syntactic and semantic aspects of the domain. A new domain-specific workflow language is then developed to construct workflows that capture multi-formalism activities in the selected domain. A domain ontology resulting from the domain analysis step is utilized to provide semantic guidance that effects valid model interoperation.

A critical challenge concerns the scheduling of the updates in the simulation, i.e. when each interoperating model computes its new state. There are two issues: (i) great disparity in the implicit or explicit time scales (the so called Δt's) and (ii) the merging on the timeline time-driven and event-driven processes. This is usually done in the context of a specific domain through the development of a suitable scheduler that drives the simulation. This scheduler is part of the workflow management scheme. For example, the multi-formalism

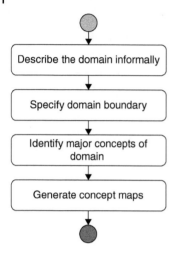

Figure 31.10 Domain identification process. Source: Levis and Abu Jbara (2014).

platforms discussed briefly in the syntactic layer have different schedulers as required by the classes of issues they address and the types of formalisms they support.

The first step in developing such a workflow language deals with characterizing a specific domain of interest in which interoperating models are used to address issues of interest. As shown in Figure 31.10, the design process begins with an informal description of the domain in the form of statements that identify the issues to be addressed, the modeling formalisms to be usually used, data sources and types, and main actors involved including domain experts, modelers, and analysts. Then the domain boundary is identified to provide the basis for excluding unrelated concepts (outside of scope). Several concept maps (Novak and Cañas 2008) are constructed as a semiformal representation of the domain. Generating concept maps is an iterative process until a satisfactory domain representation is reached.

The next step is to carry out domain analysis. Class diagrams derived from the concept maps are produced to capture the structural aspects of the domain. Then a consolidated class diagram is produced that includes interoperations between the various models being used. This consolidated class diagram serves as the basis for the domain-specific multi-formalism modeling workflow language. In addition, ontologies based on the concept maps are constructed to capture the semantic aspects of the multi-formalism modeling activities. These ontologies are then matched to construct a domain ontology.

A development approach similar to the one presented by Mernik et al. (2005) is used for developing an application or subject area domain-specific language. In the first phase, a decision has to be made on whether to use an existing general-purpose language (GPL) or to develop a new domain-specific one. The use of a GPL to create workflows of model interoperations often will not address the specific requirements of each modeling formalism that is being used and the possible interoperations between models. Each application domain has its own characteristics, and interoperations between various models require specific constructs that need to be included in the workflow language. The current research decision has been to develop a new domain-specific multi-formalism modeling workflow language for each domain of interest. It is a challenging open research issue to determine how general a workflow language can be and still be practically useful. An already developed domain-specific workflow language can be reused as the starting point for other similar domains. In the second phase, analysis of the application domain and its constructs takes place. The third phase is where the new language is designed. In general, there are two approaches. One way is to base a new domain-specific workflow language on an existing one (language exploitation). The second approach is the invention of a new language. Developing a completely new language is nontrivial; it requires extensive domain knowledge in addition to expertise in developing modeling languages.

As an example of the first approach, a new domain-specific multi-formalism modeling workflow language was developed for the drug interdiction domain based on the BPMN language (Abu Jbara 2013; Levis and Abu Jbara 2014). BPMN is a graph-based generic workflow language that provides a readily understandable notation. Its core and complete element sets allow expressing a wide variety of workflow constructs. Figure 31.11 shows a basic set of multi-formalism modeling workflow language constructs based on BPMN.

The task concept from BPMN was parsed into two major categories, operations and interoperations. Operations are those activities that a single model

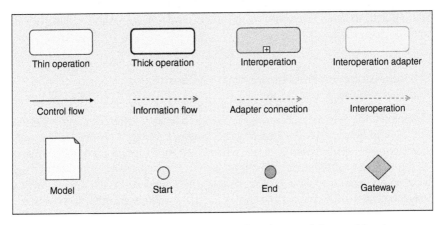

Figure 31.11 Elements of a domain-specific multi-formalism modeling workflow language based on BPMN. Source: Levis and Abu Jbara (2014).

expressed in a single modeling formalism is carrying out internally. Interoperations are those activities that involve interoperation of models expressed in different formalisms. Operations are themselves classified as thin operations for the case when modeling tools have the capability of exposing their functionalities as services and thick operations for the case in which a legacy modeling tool is used as a package. A new element, the interoperation adapter, was introduced to capture the interoperations between models.

In the next phase of the workflow language development process, a formal meta-model of the language has to be created. This is based on the already developed class diagram that represents the domain. This diagram with the addition of elements identified in Figure 31.9 becomes the basis of the language. Once the language is obtained, a platform is required that would allow for both the deployment of the new workflow language and its use in creating workflows.

The research community in this area has addressed this issue from different perspectives, and many approaches have been developed. However, the processes for developing these workflow languages are laborious and not easily generalizable. There is need to streamline such design processes and develop tools that enable users to customize languages to meet the needs of different domains and of the interoperation of different models expressed in a wide variety of formalisms.

Issues in Multi-formalism Modeling and Simulation

In addition to the research challenges in multi-formalism modeling and simulation discussed above in the context of the syntactic, semantic, and workflow layers, there are more research challenges related to multi-formalism. It is useful to partition them into three main topics: model representation, simulation, and results.

The Representation Problem: Information Consistency, Representability, and Sharing

Whenever different formalisms are used together, their different natures should coexist and be represented in a way that allows keeping their original semantics (representability) and keeping information fit for manipulations involving elements from different formalisms (sharing). The representation should offer a common general approach to present all the information in a homogeneous way (consistency). One possible choice for the workflow architecture is a modular organization (hierarchical or networked) achieved by composition of heterogeneous sub-models with syntactical interaction mechanisms as specified by the workflow.

Simulations carried out to obtain some type of results are influenced by the main goals of the specific issue being addressed and depend on the supported multi-formalism approach. Model transformations, typical of model-driven systems engineering, use *transformation description languages* to implement mappings between different formalisms and to convert models. Transformation descriptions operate on formalisms by specifying transformation rules for elements or group of elements between formalisms: this is very suitable for supporting open, vertical multi-formalism (an example is AToM^3 by de Lara and Vangheluwe 2002). However, if the original and target formalisms contain contradictory concepts, the transformation may not be valid: thus, an analysis of the type described in the semantic layer discussion is essential. Developing a set of rules embedded in the domain-specific workflow language that point out (and possibly prevent) inconsistent relations between models would be a major step forward.

The Simulation Level: Process Representation and Enactment

The execution engine of the composed model must support the diverse solution processes (computations or algorithms) of the included models. With fixed formalism sets, the solution process could be predetermined and implicitly described by the workflow; otherwise, it has to be described and implemented on a case-by-case basis using proper computational engines, possibly relying on model transformations.

The nature of a simulation process depends on the nature of the desired results: it may produce qualitative information (e.g. behaviors) or quantitative information (e.g. probability of achieving a desired goal within a specified time interval). The execution of the composed models may be based on individual model *engine orchestration*. In this case, engines are designed to be used as independent services and orchestrated as specified by the workflow or business process (van der Aalst and ter Hofstede 2012). In this approach, simulation engines are independent from the workflow engine and can be provided by different experts or standalone tools. Issues of differences in time scales and the differences between time-driven and event-driven formalisms, as discussed earlier, are of paramount importance.

The Results Level: Local and Global Results Representation, Traceability, Handling, and Reuse of Intermediate Results

The effects of a multi-formalism modeling-based simulation process may generate results that are local to some model elements or global. A correct representation of the results that is easily interpretable by the user requires a formal specification related to the nature of the formalisms, the workflow specification, and the simulation process. As the nature of desired results substantially impacts the organization of a multi-formalism modeling and simulation efforts, the optimal representation of the results depends heavily on the structure of the composed model. The main problems are traceability (that is, the possibility of mapping back to a given component model the data obtained by the simulation process) and management of results that are generated by the composed model and not the individual component models. A general approach to addressing this problem is lacking, and while in each case a unique representation of results has been effected, the development of tools that enable the construction of the visualization of the results (sometimes such tools are called viewers) would be very helpful.

Conclusions

To address in a holistic way social-behavioral problems, a single model expressed in a single formalism is usually not sufficient. Different social theories and different behavioral models have been developed, and each one of them offers some unique insights into the problem of interest but ignores other aspects. Multi-formalism modeling and simulation, when done properly, enables the valid composition and execution of interoperating models expressed in different modeling formalisms. While substantial research has been done in investigating the diverse challenges that multi-formalism poses and some key results have been obtained, more work needs to be done to analyze some of the prevailing formalisms and develop rules that will help researchers expand the range of questions they can ask by composing models and executing them on platforms or testbeds correctly. Specific research challenges have been identified in the description of the syntactic, semantic, and workflow layers, and in the last section, broader challenges have been identified, aimed at making multi-formalism modeling and simulation accessible to researchers working on social-behavioral systems. The current results are limited by being domain specific; whether these can be generalized to become domain independent appears to be a goal too far at this time. The challenges, for example, of using engineering models interoperating with social-behavioral models require much careful thought.

Epilogue

In the study applying these methods to the scenario described in Prologue, investigating potential COA for de-escalating the crisis between India and Pakistan, independent analyses were conducted for CENTCOM and PACOM using the structure of Figure 31.1 but with the appropriate region-specific models. A number of possible actions and their timing were considered. These simulated sequences of actions started with separate interactions between US diplomats and US Regional Command senior staff with their counterparts in India and Pakistan and escalated to interactions between civilian and military leaders in the United States and senior leaders in the two countries. The results were not encouraging. The probability that US actions would help de-escalate the crisis was very small. However, when the CENTCOM and PACOM models for COA development and assessment were merged, thus enabling the development of coordinated COAs by the two US Regional Commands, the situation improved. Both countries seemed to be willing to de-escalate for a while depending on their adversary's actions. But if forces were engaged, then the likelihood that Pakistan would escalate, including preparations for a nuclear option, increased. This was consistent with Pakistan's state policy. Computational experimentation that varied timing enabled the ordering of the multi-state activities for maximum combined effect. The analyses also showed that the timing of the US actions was critical. Once the crisis had started escalating, it was very difficult to de-escalate. Early action was essential with rapid involvement of the senior leaders in the United States interacting with the senior leadership in the two belligerents. Furthermore, through the Social Network analysis and agent-based modeling, it became clear that prior relationships between US leaders and leaders in the two countries were essential for rapid high-level interactions. Although the experiments were based on imperfect simulation, these insights appeared valid.

References

Abu Jbara, A. (2013). On using multi-modeling and meta-modeling to address complex systems. PhD dissertation. George Mason University, Fairfax, VA.

Barbierato, E., Gribaudo, M., and Iacono, M. (2011a). Defining formalisms for performance evaluation with SIMTHESys. *Electronic Notes in Theoretical Computer Science* 275: 37–51.

Barbierato, E., Gribaudo, M., and Iacono, M. (2011b). Exploiting multi-formalism models for testing and performance evaluation in SIMTHESys. *Proceedings of the 5th International ICST Conference on Performance Evaluation Methodologies and Tools*, Paris, France. ICST (Institute for Computer Sciences, Social-Informatics and Telecommunications Engineering).

Bechhofer, S. (2003). OWL reasoning examples. University of Manchester, http:// owl.man.ac.uk/2003/why/latest (accessed 25 July 2017).

BPMN (2014). Business process model and notation. http://www.omg.org/spec/ BPMN (accessed 25 July 2017).

Davis, P.K. and Anderson, R.H. (2003). *Improving the Composability of Department of Defense Models and Simulations*. Santa Monica, CA: RAND Corporation.

Davis, P.K. and Anderson, R.H. (2004). Improving the composability of DoD models and simulations. *Journal of Defense Modeling and Simulation* 1 (1): 24–36.

van der Aalst, W. and ter Hofstede, A. (2012). Workflow patterns put into context. *Software and Systems Modeling* 11: 319–323.

Forrester, J.W. (1968). *Principles of Systems*. Cambridge, MA: Wright-Allen Press.

GraphViz (2017). Graph visualization software. www.graphviz.org (accessed 27 July 2017).

Gribaudo, M. and Iacono, M. (2014). An introduction to multiformalism modeling. In: *Theory and Application of Multi-Formalism Modeling* (ed. M. Gribaudo and M. Iacono), 1–16. Hershey, PA: IGI Global.

Hemingway, G., Neema, H., Nine, H. et al. (2011). Rapid synthesis of high-level architecture-based heterogeneous simulation: a model-based integration approach. *Simulation* 88: 217–232.

Iacono, M., Barbierato, E., and Gribaudo, M. (2012). The SIMTHESys multi-formalism modeling framework. *Computers and Mathematics with Applications* 64: 3828–3839.

Kappel, G., Kapsammer, E., Kargl, H. et al. (2006). Lifting metamodels to ontologies: a step to the semantic integration of modeling languages, International Conference on Model Driven Engineering Languages and Systems MoDELS 2006. In: *Model Driven Engineering Languages and Systems*, 528–542. Berlin: Springer.

Karsai, G., Maroti, M., Lédeczi, A. et al. (2004). Composition and cloning in modeling and meta-modeling. *IEEE Transactions on Control Systems Technology* 12 (2): 263–278.

de Lara, J. and Vangheluwe, H. (2002). AToM3: a tool for multi-formalism and meta-modeling. In: *FASE* (ed. R.-D. Kutsche and H. Weber), 174–188. Heidelberg: Springer.

Levis, A.H. and Abu Jbara, A. (2014). Multi-modeling, meta-modeling and workflow languages. In: *Theory and Application of Multi-Formalism Modeling* (ed. M. Gribaudo and M. Iacono), 56–80. Hershey, PA: IGI Global.

Levis, A.H. and Carley, K.M. (2011). Computational Modeling of Cultural Dimensions in Adversary Organizations. Final technical report, System Architectures Laboratory, George Mason University, Fairfax, VA.

Levis, A.H., Zaidi, A.K., and Rafi, M.F. (2012). Multi-modeling and meta-modeling of human organizations. In: *Advances in Human Factors and Ergonomics. Proceedings of the 4th AHFE Conference, San Francisco, CA* (ed. G. Salvendy and W. Karwowski), 21–25. Boca Raton, FL: CRC Press.

MacKinnon, D., McCrum, W., and Sheppard, D. (1990). *An Introduction to Open Systems Interconnection*. New York, NY: Computer Science Press.

Maria, A. (1997). Introduction to modeling and simulation. *Proceedings of the 29th Winter Simulation Conference*, Atlanta, GA (7–10 December 1997). Washington, DC: IEEE Computer Society.

Mernik, M., Heering, J., and Sloane, A.M. (2005). When and how to develop domain-specific languages. *ACM Computing Surveys (CSUR)* 37: 316–344.

Moon, I.-C. (2007). Destabilization of adversarial organizations with strategic interventions. PhD thesis. School of Computer Science, Carnegie Mellon University, Pittsburgh, PA.

Novak, J.D. and Cañas, A.J. (2008). The Theory Underlying Concept Maps and How to Construct and Use Them. Technical report, IHMC Cmap Tools 2006-01 Rev 01-2008, Pensacola, FL

Rafi, M.F. (2010). Meta-modeling for multi-modeling integration. MS thesis, George Mason University, Fairfax, VA.

Saeki, M. and Kaiya, H. (2006). On relationships among models, meta models and ontologies. *Proceedings of the 6th OOPSLA Workshop on Domain-Specific Modeling*, Portland, OR (22 October 2006).

Sanders, W.H., Courtney, T., Deavours, D. et al. (2007). Multi-formalism and multi-solution-method modeling frameworks: the Möbius approach. In: *Proceedings of Symposium on Performance Evaluation – Stories and Perspectives*, Vienna, Austria (5–6 December 2003), 241–256.

Trivedi, K.S. (2002). *SHARPE 2002: Symbolic Hierarchical Automated Reliability and Performance Evaluator*. DSN IEEE Computer Society. ISBN: 0-7695-1597-5.

WSBPEL (2007). OASIS web services business process execution language [online]. https://www.oasis-open.org/committees/tc_home.php?wg_abbrev=wsbpel (accessed 27 July 2017).

32

Social-Behavioral Simulation: Key Challenges
Kathleen M. Carley

Institute of Software Research, School of Computer Science and Engineering and Public Policy, Carnegie Institute of Technology, Carnegie Mellon University, Pittsburgh, PA 15213, USA

Introduction

Social-behavioral simulation is often turned to as a way of reasoning about the human condition and thinking through what might happen under various interventions. Potential advantages include overcoming human decision bias, thinking through a large number of alternative situations, and reasoning about context and over periods of time that are too complex for the human unaided by the computation to reason about. While these advantages exist, simulation of complex social-behavioral systems is not routine. Nor are the simulation models, necessarily, accurate reflections of reality. There are more creditable models where the logics built into the model reflect established theory or empirical regularities. Yet despite best efforts, the logics in these models may reflect human biases. Further, in many simple models and in those where no theory exists, the model may actually reflect simply the potentially nonsensicle opinions of the creators.

Agent-based (e.g. Bonabeau 2002; Davidsson 2002; van Dam et al. 2012), event history (Box-Steffensmeier and Jones 2004), Petri net (Tabak and Levis 1985; Murata 1989), and system dynamic models (Sterman 2001; Mohaghegh and Mosleh 2009) are generally used as the simulation frameworks when modeling complex systems in the social sciences. No one methodology is proving adequate, so the field is moving toward hybrid modeling and a system of systems approach using interoperable models. These models make explicit the gaps in underlying theories; support theory comparison, integration, and development; and enable users to create a framework in which they can rapidly reason about alternative explanations (useful forensically) or alternative courses of action (useful for planning) (Gilbert and Troitzsch 2005).

Social-Behavioral Modeling for Complex Systems, First Edition.
Edited by Paul K. Davis, Angela O'Mahony, and Jonathan Pfautz.
© 2019 John Wiley & Sons, Inc. Published 2019 by John Wiley & Sons, Inc.
Companion website: www.wiley.com/go/Davis_Social-Behavioralmodeling

Complex systems are generally characterized as systems composed of many interacting heterogeneous components, such that the behavior of the system is nonlinear, is not predictable from the sum of the parts, and often exhibits self-organization and such that actions at one level of granularity lead to emergence at another level of granularity (Bar-Yam 2002; Miller and Page 2009). Complex systems are difficult to understand and explain. It is difficult to predict the effects of actions on them and to predict when new events will occur within them. Social-behavioral systems are classic examples of complex systems. However, they also represent a special case of complexity because humans, who can learn, are key elements of social-behavioral systems.

The field of simulation and the validation of simulations have their roots in shop floor management and operations research. Most of the approaches to model development, the theory of validation, and the expectations for how models operate are based on an operations research conception predicated on models of physical systems. Key assumptions underlying that body of work include stationarity of process and the corollary that components do not learn. These assumptions are violated in social-behavioral systems. Consequently, the traditional science of validation does not apply, and a new science is needed (National Research Council 2006, 2008). Developing such a science will depend on meeting a number of challenges both in terms of how to communicate social-behavioral models and their results and how to develop such models to create reusable and scalable technologies.

Key Communication Challenges

Independent of the scientific challenges, there are a number of communication challenges faced by those interested in developing and using simulation to address complex social-behavioral systems. These challenges take the forms of cognitive biases that humans have when they try to understand the results of models and the models themselves. Two of these will be considered:

Consumer bias: Human consumers of simulations are themselves humans and so have a tendency to assume that they know how humans behave. In contrast, many humans have placed physics on a pedestal, have deep-seated math anxiety, and assume that mechanical and engineered systems are hard to understand. This creates a situation where the consumers of social-behavioral models think they should be able to understand and agree with models of human activities, but do not make the same assumption for physical system models. A consequence is that social-behavioral models are held to a higher standard vis-à-vis communication than other models.

Storytelling bias: Humans have a tendency to think they understand how something works when they can tell a story about it. Storytelling is a cognitive crutch by which we keep track of the order in which things occur and the relations among things. The more complex the story, the harder it is to recall.

Simple models such as the model of racial segregation (Schelling 1969) or the garbage can model (Cohen et al. 1972) or the NK model (Kauffman and Weinberger 1989) lend themselves to stories that are easy to retell even in contexts distinct from that used to justify the original model. Complex models, such as EpiSims (Mniszewski et al. 2008) or BioWar (Carley et al. 2006), do not lend themselves to such simple stories. Humans are also more likely to view as accurate models they think they understand. A result is that consumers of the NK model and the garbage can model think they understand how the models work and are more likely to view them as credible and accurate when, in fact, they really do not understand the scope conditions and the models themselves cannot be validated. In contrast, more complex models, even if they are validated, are harder to tell stories about, are less likely to be viewed as understandable, and so are less likely to be viewed as credible.

Key Scientific Challenges

Independent of the communication challenges, there are a number of scientific challenges faced by those interested in developing and using simulation to address complex social-behavioral systems. Overcoming these challenges is critical for creating scalable, reusable social-behavioral simulation systems and for reducing the high cost of developing credible simulation systems. These key challenges center around speed of development, reuse, and approaches to validation. These challenges are itemized in Table 32.1. They

Table 32.1 Summary of scientific challenges for social-behavioral modeling.

Challenge	Meaning	Example
Reuse across context	Explains/forecasts how phenomena x evolve in sociocultural context y_1 at time n_1 and in sociocultural context y_2 at time n_2	Use model to explain terrorist recruitment for IRA before 2000 in the United Kingdom and then for ISIS recruitment in Syria, 2010–2017
Reuse across level	Explains/forecasts behavior of actor and group of those actors	Use model with data set 1 to explain an individual's anger toward election fraud for a candidate and on data set 2 to explain population-level anger toward election fraud for a national election
Reuse across time	Explains/forecasts behavior of entities at time period n_1 and at time period n_2	Use model with data set 1 to predict state stability in 1910 and on data set 2 to predict state stability in 2010

(Continued)

Table 32.1 (Continued)

Challenge	Meaning	Example
Multi-scale	Explains/forecasts the behavior or change in opinions/beliefs of x, and sets of x in the same simulation run	Use model to forecast behavior of individuals moving between firms, and/or organizations moving between conglomerates or merged firms, using the same set of data and during the same set of simulation runs
Rapid instantiation	Shortens time to prepare and import data to model and adjust model parameters	Construct and use model repeatedly on different data sets, and observe a decrease in the time to prepare and ingest data
Multi-group	Generates model that remains valid as actor granularity is changed	Use model on data set 1 where agent is a human to predict consensus, on data set 2 where agent is an organization to predict consensus, and on data set 3 where agent is a nation state to predict consensus
Multi-temporal	Generates model that remains valid as time steps change	Use model to explain escalation of commitment of individuals to groups' goals using data set 1 where individual commitment is measured each day and data set 2 where individual commitment is measured each year
Multi-spatial	Generates model that remains valid as spatial region is changed	Use model to explain the formation of crime zones using data set 1 where the zones are measured at the city block level and data set 2 where the zones are measured at the city level
Interoperability	Allows model to be used with other models as part of a modeling workflow	Use model in a system of systems, demonstrating the ability to ingest data from and export data to other models
Data fusion	Allows model to use data at different temporal, spatial, and group levels with different levels of fidelity	Use model to ingest and use for instantiation, and then again for validation, with two sets of data that vary in granularity of actor, time period, or spatial region or have other collection inconsistencies
Response surface	Generates boundary conditions and the full response surface	Conduct virtual experiments to estimate the response surface via statistical calculations

overlap with the challenges discussed in overview chapter of this volume (Davis and O'Mahony 2019). Rather than providing a detailed discussion of each challenge, three issues that apply to multiple challenges will be discussed:

Model granularity: Social-behavioral systems are often reasoned about, explained, and measured at different levels of granularity. Among the levels of granularity of interest are time (minute, day, week, year), space (block, city, county, country), agent group (neuron, person, decision-making unit, organization, nation), and knowledge group (word, concept, topic, theme). Typically models that are accurate or designed for one level of granularity do not work at other levels of granularity. For example, models that are accurate at the cognitive and task level do not scale to the community or even large group level, whereas models that are informative about populations are inaccurate at the cognitive level. One reason for this is that the logics of interaction may change fundamentally with scale. Basic research on how logics change with scale is needed. Another reason for this is that the aggregation processes and disaggregation processes needed as one moves up and down the levels of granularity are not known. Two advances in this area are dynamic network modeling and social cognition modeling (Morgan et al. 2017). Dynamic network models provide a mechanism whereby the pattern of interaction among the components can be used to affect the decision logic or behavior of the component whose internal logic can also be represented as a network (networks where the nodes are networks). Social cognition is a set of mechanisms whereby individual humans make sense of the world by reasoning not about every individual but about collectives in the same way they reason about individuals. This includes social cognitive mechanisms such as recognizing and responding to a generalized other, inferring individual traits from perceptions about a group, generalizing group traits from activities of an individual, and recalling average rather than actual behavior. Focusing on these types of logics that support connecting one level of a model with another is critical for improved model accuracy and reuse.

Level of validation: Historically, the science of validation has treated validation as a single process. A model was either valid or not, and there was a single approach to validation. In contrast, for social-behavioral simulations it is generally recognized that there are multiple levels or types of validation and that the level of relevance and the extent of validation depend on the model's purpose (Burton 2003; National Research Council 2008). There are various ways of characterizing the various aspects of validation. For example, Davis and O'Mahony (2019) describe the components as being description, explanation, postdiction, coarse exploration, and prediction. An alternative approach is to think in terms of the empirical precision at which the model is validated and/or matched to the predictions of another model (i.e. docked) – including similarity in the pattern of results, the distribution, the range, or the exact value (Axtell et al. 1996). Still more detailed schemes exist that take into account the number of features on which to validate, the level of empirical match, and the extent of revalidation (Yahja and Carley 2005). To move forward with a science of validation for social-behavioral simulations, a validation framework is needed that accounts both for the various components of validation, the purpose of the model, the level of empirical match, and the nature of the underlying data to support match. See also another paper in this volume discussing validation (Grace et al. 2019).

Identifying the right level and type of validation for a model is further complicated in a reuse model–expansion situation. The crux of the problem has to do with interaction among subtheories. Most social-behavioral simulations are at their core multi-theoretical. Typically the way in which these theories linked together is not known. Model development often means suggesting a theory for these gaps. Validation then becomes a process of creating an empirically grounded mega-theory (e.g. see Schreiber and Carley 2004). This is counter to another assumption of traditional validation science, which assumes that the theory is known and well specified. This is one of the gaps through which human biases creep into these models. In a related vein, typically social-behavioral simulations are built in a building block fashion. That is, the basic model is built, possibly validated, and used to explore an issue. Then the model is reused, generally expanded on, and the process repeats. The issue here is that as models are expanded, the earlier parts of the model may no longer behave as they did previously. This is particularly true in nonlinear complex systems. This means that the results of earlier validation exercises may no longer hold, hence, leading to a situation where the model needs to be revalidated for each reuse/expansion situation.

Validation vs. tuning: Historically, the science of validation for simulation has been based around two assumptions: stationarity of process and measurability. Stationarity of process means that you can collect a data set from time period N or machine y and tune the model until you get a high level of predictive accuracy (e.g. 98%). Tuning is done by changing internal processes in the model. The resultant model is considered validated and can then be used to predict the behavior of the system at other time periods or the behavior of objects from comparable machines. Social-behavioral system violates this assumption. Consequently, tuning the model to a historic case study is actually overfitting. The resultant model is then fragile and unlikely to apply in other situations. Consider, for example, the artificial intelligence/machine learning (AI/ML) models, particularly those that require training sets. Such models typically use data from a single context/time period and split it into two sets. The model is built based on the first part of the data and then applied to the second part. Accuracy of such models can become quite high (e.g. approaching 98%). However, if moved into the wild and used in other contexts, they often will not work, and/or the accuracy can decrease dramatically due to the nonstationarity of process.

Simply tuning a model to two historical cases is not a sufficient solution. Recall that as a model is reused from one situation to the next, new modules are often added. A possible consequence is that if a model is tuned to one historical case and then reused, expanded, and tuned to a second case, the model may no longer be tuned to the first case. In complex social-behavioral simulations, the relationship between model components and validation is so complicated that special artificial intelligence (AI) tools may be needed to track and predict which parts of the model will become invalidated as the model is expanded and retuned (Yahja and Carley 2005).

Another assumption of the traditional science of validation is measurability; i.e. it is possible to delineate all aspects of the physical system and measure its properties. Social-behavioral systems violate this assumption in part because we often cannot know what people are thinking, particularly in situations where people themselves are unable to forecast their own behavior – such as life-threatening situations. For these and other reasons, such as measurement imprecision, and propagating measurement errors, predictive accuracy (i.e. predicting future behavior) of social-behavioral models is often much lower than those for physical systems. While ML/AI models often claim they are doing prediction – as was noted – the high accuracy they achieve is often using data from the same time period and context as they were tuned (i.e. trained) on and rarely have this high a level of prediction for future events.

For such reasons, simulation often plays a different role in the social-behavioral context than in the physical system context. Its value is in forecasting, not prediction. In other words, social-behavioral simulations are generally the most useful for describing the space of future possibilities and the relative likelihood of particular outcomes. Trying to express this difference to policy-makers and those not steeped in simulation is difficult for a variety of reasons (see also Davis and O'Mahony 2019). First is the terminology. As was alluded to the term, prediction means, depending on the user, a priori predicting a specific event in the future (aka point prediction), explaining behavior in one part of the data given the other part of the data, and a priori estimating the relative likelihood of different outcomes (aka probabilistic forecast). Similarly the term prediction also has been used to mean extrapolation (as is often done with regression models), suggesting possibilities (aka qualitative forecast), equivalent to prediction, and a priori estimating the relative likelihood of different outcomes (aka probabilistic forecast or prediction). Using this terminology, the difference can be described as simulation of physical systems as being more capable of point prediction and simulation of social-behavioral systems as being more capable of probabilistic forecasting or qualitative forecasting.

A second difficulty is that the differences are in part a matter of degree. That is, the vast majority of physical system simulations support point prediction. Certainly the science of validation is designed for this type of prediction. In contrast, the vast majority of social-behavioral simulations support either probabilistic forecasting or qualitative forecasting. For those few social-behavioral simulations that can do point prediction, they do so only under specific scope conditions. At this point a science of validation that supports forecasting does not exist in its entirety. Further, the tactics and techniques needed to determine whether the scope conditions and the logics in social-behavioral simulations are such that they will support point prediction are understood more in the abstract than in practice. For example, stationarity of process is needed, but measuring it is often an open challenge.

Toward a New Science of Validation

A consequence of these considerations is that a new science of validation is needed for social-behavioral modeling. At the heart of such a science is the practice of validation in parts and incremental validation, instead of tuning. Validation in parts means that parts of the model (the input, the outputs, and the internal processes) are validated separately, often on different data and sometimes by different teams. Three factors are needed: (i) inputs of the model have been shown to have at least the same distributional properties as real world. In the limit the input data for the model are exactly real-world data streams and/or empirically derived from real-world sensors. Under the philosophy that the model should be able to do the task it seeks to explain, the model must take as input the exact data that the real-world social-behavioral system would. Models that fit these criteria could, in principle, be substitute for the social-behavioral system in other contexts (much as AI personal shoppers could substitute for a human personal shopper). (ii) The distribution of model outputs contains, as special cases, historic examples. However, this does not imply that the mean prediction of the model should agree with the historical data point. Indeed if the model contains discontinuities in some of the variables, the mean prediction of the model might be a situation that logically cannot exist. Logically, one would expect that as the number of historical cases increases, the distribution of historical data points should come to reflect the plausibility distribution from the simulation. (iii) Internal processes in the model have been shown to match at some level to processes observed in the real world. This match might be very qualitative as when confirming the face validity of a model by claiming that each element has an analog in the real world. Or this match might be strongly empirical and precise using exact values in equations that match the equations used to describe the real world.

Incremental validation means that the model is validated in steps, and at each step the number of aspects that is validated and the level at which the model is validated are increased. Incremental validation is an approach meant to overcome the high level of complexity and the extensive number of variables and processes needed to explain social behavior. For incremental validation a set of features to be validated are identified. This set is often modularized by granularity level, or theory being operationalized, or data context. The model is built and it is validated against the base features. Then new modules/theories are added, and the model is revalidated against the old features and also validated against the new set of features. A key advantage of this approach is that it allows theory to be developed to fill in the gaps linking different levels of granularity. Another advantage is that this approach does not require a universal perfect data set, but enables the simulationists and the consumers to make use of large diverse sets of data from multiple sources and contexts that may or may not be fusible.

A science of validation appropriate for social-behavioral simulation may be well served by the development of a model social agent (Carley and Newell 1994). The model social agent is a conceptual framework that defines the elements of a social agent and what behavior should emerge from any model containing agents composed of those elements. A first attempt in this direction was forwarded by Carley and Newell (1994). While not sufficient, the suggested framework shows the power of having such a model in providing guidance for what models should and should not be able to do and so what type of and level of validation might be needed.

Conclusion

The state of the art in simulation of complex social-behavioral systems has advanced in the past several decades. Gone are the days of all models being rational actor models. Gone are the days of model assessment being done by generating a single simulation run, viewing the results, and stating that "it looks right." That being said, the field needs some serious advances to become more mainstream. Overcoming some of the challenges described herein will help.

Nevertheless, it is important to recognize that these are simulations of complex social behavior. It is unreasonable to expect rapid development, high levels of accuracy, and high levels of reuse when the subject being modeled is still so poorly understood. The *laws* governing human interaction, learning, social engagement, and cognition are still being discovered. Each new discovery increases the value of and the ease of building simulations and validating them. In that sense, advancing the field of simulation is codependent with advancing our understanding of human social behavior.

It is also important to recognize that humans and the groups or communities they form are an evolving system. This suggests that the simulations themselves will need to evolve. With complex and evolving systems, ensemble approaches are often of value. For example, major advances in weather forecasting occurred once sets of models were used in an ensemble fashion to provide estimates. Standardization in output formats and levels of granularity for outputs enabled cross-fertilization. Such an approach may do well in the social-behavioral context as well.

A final approach is that social-behavioral modeling is *big science*. Strong, reusable systems require substantial resources to be developed. Large teams are needed to develop and fuse the data needed for model development and validation, for building the infrastructure and tool chains for using models in an interoperable fashion, to run virtual experiments using the model to examine results, and to describe and document the model. The infrastructure around the model needed to ensure its use requires more people and more time than the model design and development itself. If we continue to think of social-behavioral simulation as being comparable with running a statistical analysis of a moderate data set, progress in this area will be slow.

References

Axtell, R., Axelrod, R., Epstein, J.M., and Cohen, M.D. (1996). Aligning simulation models: a case study and results. *Computational and Mathematical Organization Theory* 1 (2): 123–142.

Bar-Yam, Y. (2002). General features of complex systems. In: *Encyclopedia of Life Support Systems*. Oxford: EOLSS UNESCO Publishers. Retrieved 16 September 2014.

Bonabeau, E. (2002). Agent-based modeling: methods and techniques for simulating human systems. *Proceedings of the National Academy of Sciences of the United States of America* 99 (suppl 3): 7280–7287.

Box-Steffensmeier, J.M. and Jones, B.S. (2004). *Event History Modeling: A Guide for Social Scientists*. Cambridge University Press.

Burton, R.M. (2003). Computational laboratories for organization science: questions, validity and docking. *Computational and Mathematical Organization Theory* 9 (2): 91–108.

Carley, K.M. and Newell, A. (1994). The nature of the social agent. *Journal of Mathematical Sociology* 19 (4): 221–262.

Carley, K.M., Fridsma, D.B., Casman, E. et al. (2006). BioWar: scalable agent-based model of bioattacks. *IEEE Transactions on Systems, Man, and Cybernetics – Part A: Systems and Humans* 36 (2): 252–265.

Cohen, M.D., March, J.G., and Olsen, J.P. (1972). A garbage can model of organizational choice. *Administrative Science Quarterly* 17 (1): 1–25.

van Dam, K.H., Nikolic, I., and Lukszo, Z. (eds.) (2012). *Agent-based modelling of socio-technical systems*, vol. 9. Springer Science & Business Media.

Davidsson, P. (2002). Agent based social simulation: a computer science view. *Journal of Artificial Societies and Social Simulation* 5 (1): 1–5.

Davis, P.K. and O'Mahony, A. (2019, this volume). Improving social-behavioral modeling. In: *Social-Behavioral Modeling for Complex Systems* (ed. P.K. Davis, A. O'Mahony and J. Pfautz). Hoboken, NJ: Wiley.

Gilbert, N. and Troitzsch, K. (2005). *Simulation for the Social Scientist*. McGraw-Hill Education (UK).

Grace, E., Blaha, L.M., Sathanur, A.V. et al. (2019, this volume). Evaluation and validation approaches for simulation of social behavior: challenges and opportunities. In: *Social-Behavioral Modeling for Complex Systems* (ed. P.K. Davis, A. O'Mahony and J. Pfautz). Hoboken, NJ: Wiley.

Kauffman, S.A. and Weinberger, E.D. (1989). The NK model of rugged fitness landscapes and its application to maturation of the immune response. *Journal of Theoretical Biology* 141 (2): 211–245.

Miller, J.H. and Page, S.E. (2009). *Complex Adaptive Systems: An Introduction to Computational Models of Social Life: An Introduction to Computational Models of Social Life*. Princeton University press.

Mniszewski, S.M., Del Valle, S.Y., Stroud, P.D. et al. (2008). EpiSimS simulation of a multi-component strategy for pandemic influenza. In: *Proceedings of the 2008 Spring Simulation Multiconference*, 556–563. Society for Computer Simulation International.

Mohaghegh, Z. and Mosleh, A. (2009). Incorporating organizational factors into probabilistic risk assessment of complex socio-technical systems: principles and theoretical foundations. *Safety Science* 47 (8): 1139–1158.

Morgan, G.P., Joseph, K., and Carley, K.M. (2017). The Power of Social Cognition. *Journal of Social Structure* 18 (1): 0_1–0_22.

Murata, T. (1989). Petri nets: properties, analysis and applications. *Proceedings of the IEEE* 77 (4): 541–580.

National Research Council (2006). *Defense Modeling, Simulation and Analysis: Meeting the Challenge, National Research Council*. Washington, DC: The National Academy Press.

National Research Council (2008). *Behavioral Modeling and Simulation, from Individuals to Societies, National Research Council*. Washington, DC: The National Academies Press.

Schelling, T. (1969). Models of segregation. *American Economic Review* 59: 488–493.

Schreiber, C. and Carley, K.M. (2004). Going beyond the data: empirical validation leading to grounded theory. *Computational and Mathematical Organization Theory* 10 (2): 155–164.

Sterman, J.D. (2001). System dynamics modeling: tools for learning in a complex world. *California Management Review* 43 (4): 8–25.

Tabak, D. and Levis, A.H. (1985). Petri net representation of decision models. *IEEE Transactions on Systems, Man, and Cybernetics* 6: 812–818.

Yahja, A. and Carley, K.M. (2005). WIZER: an automated intelligent tool for model improvement of multi-agent social-network systems. In: *Proceedings of the Eighteenth International Florida Artificial Intelligence Research Society Conference (FLAIRS 2005)* (15–17 May 2005), 44–50. Clearwater Beach, FL: AAAI Press.

33

Panel Discussion: Moving Social-Behavioral Modeling Forward

Angela O'Mahony[1,], Paul K. Davis[1,*], Scott Appling[2], Matthew E. Brashears[3], Erica Briscoe[2], Kathleen M. Carley[4], Joshua M. Epstein[5], Luke J. Matthews[1,13], Theodore P. Pavlic[6], William Rand[7], Scott Neal Reilly[8], William B. Rouse[9], Samarth Swarup[10], Andreas Tolk[11], Raffaele Vardavas[1], and Levent Yilmaz[12]*

[1] *Pardee RAND Graduate School, Santa Monica, CA 90407, USA*
[2] *Georgia Tech Research Institute, Atlanta, GA 30318, USA*
[3] *Department of Sociology, University of South Carolina, Columbia, SC 29208, USA*
[4] *Institute of Software Research, School of Computer Science and Engineering and Public Policy, Carnegie Institute of Technology, Carnegie Mellon University, Pittsburgh, PA 15213, USA*
[5] *Department of Epidemiology, Agent-Based Modeling Laboratory, New York University, New York, NY 10003, USA*
[6] *School of Computing, Informatics, and Decision Systems Engineering, School of Sustainability, Arizona State University, Tempe, AZ 85281, USA*
[7] *Department of Marketing, Poole College of Management, North Carolina State University, Raleigh, NC 27695, USA*
[8] *Charles River Analytics, Cambridge, MA 02138, USA*
[9] *School of Systems and Enterprises, Stevens Institute of Technology, Center for Complex Systems and Enterprises, Hoboken, NJ 07030, USA*
[10] *Biocomplexity Institute and Initiative, University of Virginia, Charlottesville, VA 22904, USA*
[11] *Modeling, Simulation, Experimentation, and Analytics, The MITRE Corporation, Norfolk, VA 23529, USA*
[12] *Department of Computer Science, Auburn University, Auburn, AL 36849, USA*
[13] *Behavioral and Policy Sciences, RAND Corporation, Boston, MA 02116, USA*

After working with contributors on their individual chapters, we identified some important questions on which interesting disagreements exist. We posed these questions to contributors and a few others. This chapter is an edited but not iterated recounting of responses to questions that deal with simulation and emergence, how to relate models at different levels of resolution, and how to assure more *humanness* in agents.

*Panel moderators

Social-Behavioral Modeling for Complex Systems, First Edition.
Edited by Paul K. Davis, Angela O'Mahony, and Jonathan Pfautz.
© 2019 John Wiley & Sons, Inc. Published 2019 by John Wiley & Sons, Inc.
Companion website: www.wiley.com/go/Davis_Social-Behavioralmodeling

Simulation and Emergence

Question: Some believe that few if any social-behavioral simulations have truly generated *emergent phenomena*. Rather, it is argued, when simulations have claimed to do so, the emergence was actually baked in. What are your thoughts on this issue, both in assessing the current situation and in looking forward?

Respondents: Kathleen M. Carley, Paul K. Davis, Joshua M. Epstein, Levent Yilmaz, Luke J. Matthews, Theodore P. Pavlic, William Rand, Scott Neal Reilly, Andreas Tolk, Samarth Swarup, and Raffaele Vardavas

Contributors differed on whether simulations can generate *true* emergence but differed also on what *true* means. Andreas Tolk, Kathleen M. Carley, and Joshua M. Epstein argue that, *of course*, the outputs of simulation can only reflect the simulation's structure and inputs. That said, the simulation outputs can be highly important and beyond human capacity to anticipate without the help of models. Luke J. Matthews, Levent Yilmaz, Samarth Swarup, and Raffaele Vardavas – without disputing that a simulation's output is determined by its inputs and structure – note evidence of *distinctly nonobvious* emergence in simulations. Swarup argues that as our ability to build more clever agents improves, we will see increasingly profound emergent phenomena, often the result of agent learning. Levent Yilmaz notes that macro behavior may not even be describable in the vocabulary of the simulation that generates it and posits a need to develop *meta-protocols* defined in terms of the attractor space of the model. Some contributors define emergence as requiring surprise, i.e. unanticipated behavior. That ties emergence to human knowledge. Others reject that definition because many emergent phenomena (in their definition) exist independent of human knowledge and because the phenomenon doesn't change merely because we come to understand it.

Our contributors also explored the *value* of simulating emergence. Paul K. Davis argues that a particularly valuable role for simulation is to reproduce emergence observed empirically, so as to understand it, but sees doing so as often being in the class of *hard problems*. Scott Neal Reilly argues for using simulation to "explore the space of tradeoffs and configurations that are consistent with the phenomenon." Ted Pavlic extends this, arguing that "ultimately, simulations of emergent phenomena should only be employed as part of a scientific process: testing a hypothesis on the mechanisms underlying observed emergent phenomena or generating new plausible hypotheses that may lead to further 'real-world' experimentation."

A number of contributions appeared initially to be fundamentally contradictory. We concluded, however, that the contradictions had origins in (i) the definition of emergence (Do only surprising phenomena count?), (ii) the emotive meaning of *baked-in* (Can something *baked in* be profoundly nonobvious and important?), and (iii) whether some forms of real-world emergence generate something not understandable even in principle by micro-level properties and interactions thereof with the environment.[1] The last of these subdivides. Some phenomena are in principle understandable, but only after accounting for learning and adaptation over time, perhaps including consequences of so-called downward causality. If the reader keeps these issues and distinctions in mind, the relationships among contributions will be clearer. We did not attempt to reconcile the contributions through editing because we wanted to preserve the spirit of the initial language.

Andreas Tolk: Epistemological, Not Ontological Emergence

Natural complex adaptive systems (CASs) may produce something new, like structures, patterns, or properties, that arise from the rules of self-organization. These novelties are emergent if they cannot be understood as any property of the components, but are new properties of the system. One of the leading methods to better understand CAS is the use of their computational representation in simulations and predominantly in agent-based approaches.

Computational science disciplines, such as computational physics, biology, chemistry, and social science, explore the use computer models in their research. The discipline of complex systems research benefitted significantly from these developments, as computers amplified our abilities to model, simulate, and evaluate the computational twins of the natural systems of interest. The principal steps of such a computational study of CASs were captured in detail by Holland (1992) and elaborated by many authors since then.

But can these digital twins produce emergence? To answer this question, it is necessary to understand the difference between epistemological and ontological emergence. In the epistemological view, emergent properties and laws are systemic features of complex systems. Such a system is governed by true, law-like generalizations within a special science that is irreducible to fundamental physical theory for conceptual reasons. As such, this view characterizes the concept of emergence strictly in terms of limits on human knowledge of complex systems. The unpredictability is a matter of human knowledge, not a characteristic of the systems. The ontological view is quite different. Here, the emergent properties are independent of the human knowledge about them. Instead they are novel, fundamental types of properties altogether. Something new emerges that was not there before, and that cannot be explained by the components and their interactions and relations alone.

So, what can we use computer simulations for? The famous artificial intelligence researcher Hubert L. Dreyfus (1972, 1992) is well known for his two books on the limits and constraints of computers. Dreyfus points to known limits that are founded in the nature of computers as captured in the works of Turing, Church, Gödel, and other pioneers of computer

1 In the philosophical literature, the examples sometimes given for such pure emergence include consciousness and quantum mechanical phenomena, such as action at a distance (Clayton and Davies 2006; O'Connor and Wong 2015).

science that were often overlooked by his colleagues. These constraints are still valid, and they are valid for computational simulations even of CASs. Although computers have improved significantly over the recent years, the essential constraint remains: computers can only transform input parameters into output parameters using computable functions to do so. As a result, computers cannot create something new out of nothing, as everything that is produced by a computer must be in the input data or the transforming algorithm. They cannot produce ontological emergence!

However, computational simulations of CAS can reproduce epistemological emergence, and more importantly, they can reduce epistemological emergence by increasing the knowledge about the system: what we understand with the help of the simulation system is no longer emergent! As such, simulations are utmost important in support of gaining scientific knowledge and provide also practical support in the form of tools, but they can't produce ontological emergence.

Kathleen M. Carley: Emergence Does Not Happen Magically in Simulations or the Real World

In practice, two types of emergence have sometimes been confused:

(1) *Emergence due to complexity*: In this case, the emergent behavior is a relation between two or more variables or agents in the model that was not *a priori* predicted by the simulationists nor known to be predicated on the theory being captured by the model. Such behaviors are often the result of a complex chain of interactions, or dependent on a large number of interacting variables, so that most humans, unaided by computational techniques, cannot think through all the interactions and possibilities. This type of emergence is one of the hallmarks of simulation models that are developed by integrating multiple previously unconnected theories.

(2) *Emergence due to level of analysis*: Here the emergent behavior is a pattern of activity at a granularity distinct from (and usually coarser than) the one at which the model is programmed. Micro-level actions then can have unanticipated macro-level results. The classical example is when a flock of birds is simulated with rules that say they can only turn right or left 90°, or go straight. Simulation output shows the flock appearing to move at a 45° angle. In principle, there should be a parallel version of emergence in which macro-level constraints result in micro-level adaptations, but this has been less examined by simulationists.

The notion that there is a kind of *real* emergence where results are not related to the simulation or learning algorithms does not make sense. For human societies, things that *emerge* don't just happen magically but follow from and have their explanations in the way that humans learn and interact, and due to the changes of their physical or sociocultural environment. The same should be so for simulations.

Joshua M. Epstein: Of Course, *Emergent Phenomena* Are *Baked* into Computer Models

I am asked this question frequently, and usually rhetorically, the suggestion being that, if *baked in*, a result can be dismissed as somehow trivial, uninteresting, already known, not useful, or some such.

First, regarding the fraught term *emergent*, I have previously discussed its logical confusions, deistic origins, anti-scientific implications, and a procession of withering philosophical criticisms from Bertrand Russell to Hebert Simon (Epstein 1999). Luckily, the panel question does not actually require a resolution of this hoary debate.

However, to address the panel question coherently, two clarifications are essential. First, whatever may be one's definition of *emergence*, the question itself presupposes that the *emergent phenomena* of (putative) interest here are states or outputs *of a computational simulation*, and *not* states in the biological development of a human embryo, or of phase transitions from liquid to a gas, in which connections the term *emergent* is also encountered.

Second, let us agree that (as most inquisitors would) that *baked in* means "mechanically attained after a finite number of model iterations."

Under these two structures, my answer to the panel question is, "Of course, *emergent phenomena* are baked in, because they are computer outputs, and computer outputs are baked into the code!" Perhaps pedantically,

- *Premise 1*: All (putatively) emergent phenomena are simulation states/ outputs (assumed by the panel question).
- *Premise 2*: All computer states/outputs are baked into the code.
- Therefore, *Conclusion*: All (putatively) emergent phenomena are baked in to the code (Syllogistic).

Only Premise 2 requires discussion. If we are given a simulation's initial conditions, parameters, and updating (inference) rules, then every state is attained by strict deduction (Epstein 1999). That is, every computational model can be written in Turing Machine code, and for every Turing Machine there is a unique equivalent set of equations (in partial recursive functions). Therefore, every attainable state can be strictly (recursively) deduced and is therefore *baked in*. Indeed, all the immortal proofs in mathematics are demonstrations that some proposition, the theorem, is *baked in* to the axioms. That hardly makes the theorem trivial.

Imagine attending Andrew Wiles' Princeton colloquium revealing that – after 300 years of failed attempts by the greatest of mathematicians – he had finally proved Fermat's Last Theorem. Who in their right mind would raise their hand and skeptically quip, "But Professor Wiles, wasn't that just baked into the Axioms?" Of course it was baked in. Therein lies the miracle! All the great proofs demonstrate that the theorem is baked into (is mechanically deducible from) the axioms.

And so it is for all the great computational models. The Schelling model shows (remarkably) that a macroscopic segregated state is *baked into* seemingly remote micro-rules. Conway's Life "gliders" are *baked in* to the seeming remote Cellular Automaton rules. This is the entire point: to discover the seemingly remote micro-rules that generate, or *bake in*, the macro-phenomenon.

One might well ask, "What about randomness?" Staying with food analogies, this is a red herring. Elsewhere, I note that *on computers* we simulate randomness with *deterministic* so-called pseudorandom number generators (Epstein 2006). We do this to ensure that we can replicate runs of particular interest. If we use the same random number *seed*, we will get exactly the same run. So, every particular *stochastic* realization is again a strict deduction and is *baked in*. Even if one uses *naturally occurring* random numbers (from atmospheric phenomena, or radioactive decay), if one wants to *replicate* the run, the random string thus harvested must still be saved. But, if we use the saved string, the result is again baked in. If one uses an unreproducible random string, then the most that can be *probably baked in* is a statistical distribution over a set of individually unreproducible runs (by maximum entropy methods, for example).

Of potential interest could be "Garden of Eden" states. I believe this term was introduced by John Tukey, to denote states that satisfy some criterion, such as being an equilibrium, but are not attainable from any other state of the system under the rules of the game (Epstein and Hammond 2002). Of these configurations, one might say that their *existence* is baked into the axioms, but their *attainment* – and so their *emergence* – is baked out!

In any case, I hope it is clear that the assertion under discussion is itself half-baked at best!

Levent Yilmaz: Emergent Behavior May Have a Higher-Level Ontology

Emergence is a consequence of the interaction of the rules, which are indeed designed. That does not mean that model developers preordain or determine the emergent outcome. *The emergent behavior may even have a higher-level ontology with constructs that are not defined in the vocabulary of the model.* The concept of the formation of a group, which is an emergent regularity, may not even be phrased in terms of the composition of constructs of the model that generates it. Conway's Game of Life is particularly revealing. The *oscillators*, *spaceships*, and *gliders* are not even remotely related to the four basic rules of the game. Similarly, a weather prediction model that simulates the dynamics of gas molecules while using terms such as pressure and temperature may generate a hurricane. However, hurricanes may not have been engineered as a construct into the model. Instead, the interactions of the rules among each other as well as with the continuously evolving environment determines the patterns that result. I suspect that recognition of emergent behavior and using it effectively within the model will require meta-protocols, which are defined in terms of the attractor space of the model. The emergent behavior may even feedback in the form of positive and negative feedback to further constrain the application of the rules. They are not engineered into the model. *Emergence can be both surprising, allowing us to learn from a model, or expected, enabling us to justify the model.* These are not contradictory views of emergence. Both are useful.

Samarth Swarup: The Promise of Clever Agents for True Emergence in Simulations

There are multiple perspectives about emergence. The simplest is that we do simulations to compute the population-level consequences of individual behaviors. This makes sense to do only if the outcomes aren't obvious or known already. If so, outcomes are *emergent*. This can be tied to the notion of computational irreducibility, which refers to computations where the fastest way of determining the outcome is to run the program (i.e. there is or seems to be no simpler way to predict the outcomes from the initial conditions). In this view, a simulation is a way of knowing and, indeed, may be the fastest way of knowing. This is the epistemological perspective. In practice, this applies to any reasonably complex social simulation.

Now let's consider a particular scenario. Imagine a simulation of the aftermath of a nuclear explosion. Suppose that a lot of people rush toward ground zero to find family members. This is risky because of radiation and fallout. Suppose further that there are a number of healthcare locations where people can be treated to some degree for radiation sickness. This might lead to a feedback loop where people search for family members until they get sick, whereupon they get treated and then start looking for family members again, and so on. If the simulation is set up right, we would see emergence of a density dependence on health outcomes. Put simply, if people are falling sick faster than they can recover, most people will end up sick, whereas if they are being treated faster than they are falling sick, most people would end up in good health. If the size of the area and the number of healthcare locations are fixed, this would manifest as a function of the population density. This is not a relationship programmed into the simulation; it emerges from the interactions between the agents and the environment. This is the ontological perspective in which emergence is due to law-like relationships arising from interactions in a system. This can be particularly satisfying if a number of complex micro-level interactions effectively *cancel out*, so that the system can be described simply at a macro level (as when molecular interactions aggregate so that a fluid can be described as a function of just three variables: pressure, volume, and temperature).

In a complex social simulation, we generally do not observe such simple behavior at the population level. However, such a relationship may be embedded in a web of other behaviors, interactions, and environmental effects.

Additional work is then necessary to identify the relationships that emerge and the generating mechanisms. This is an open simulation analytics question, where new methods are needed. Once we have such methods, a third kind of emergence will become possible. Suppose that we design agents that are sophisticated enough to alter their behaviors based on observed population-level relationships that emerge from the simulation. For example, suppose that the agents are allowed to observe the density dependence of health outcomes in the example above. This could be possible by observing the number of people falling sick in different subregions that have different population densities. Particularly clever agents could then choose to shelter if they find themselves in areas of high population density, thus avoiding falling sick. This would be an example of the emergence of *downward causation*, where emergent macro-level phenomena exert a causal influence on micro-level behaviors. This is sometimes referred to as *true emergence*. There are important technical challenges to be solved before we can build a simulation of this type. These include the emergence discovery problem, i.e. how to find emergent relationships and how to design agents that can take advantage of such relationships.

Luke J. Matthews: Examples of True Emergence in Current Agent-Based Models

Yes, clearly some agent-based models have demonstrated true emergence. Definitely there are socio-behavioral simulations in which the emergent phenomenon is not *baked in.*

Two examples would be my work on cultural traditions in capuchin monkeys (Franz and Matthews 2010) and Charlie Nunn's work on simulating the main cognitive features people talk about with regard to culture, including peer pressure and prestige bias (Nunn et al. 2009). Capuchin monkeys can produce group-typical traditions even though they have no peer pressure (a.k.a. conformity bias), no sense of group norms or group identity, and no ability to faithfully imitate. This is because when building a model with both basic reinforcement learning (operant conditioning) and even the simplest social learning mechanism (social enhancement), and letting them interact, produces group-typical traditions! But, group typical-ness is not instantiated in the cognitive capacity either for reinforcement learning or enhancement learning.

Charlie Nunn built an agent-based model with cognitive mechanisms that people like Boyd and Richerson and Joe Henrich have focused on for culture – including conformity bias and prestige bias. What Nunn found is that when natural selection acted on a cultural trait, and that is represented in a spatially explicit social system along with cognitive characteristics, the cognitive characteristics essentially had no effect on the emergent dynamics of the system because migration dynamics and selection were 800 lb gorillas compared to them. So, although people evolved to have these cognitive biases because they are adaptive for them individually, the cognitive biases have virtually no impact on the emergent epidemiological dynamics at least for fitness-relevant traits. That's still not a popular idea because in the evolution of culture literature it is popular to fixate on cognitive mechanisms. This cognitive focus seems especially deeply rooted among Western researchers who approach the work from individualist cultural traditions.

So, in my mind a researcher can definitely create agent-based models with emergence that is not baked in. That said, many people doing agent-based models for humans have fallen into the trap of making extremely complex agent-based models that especially have many cognitive features. It is then very difficult at the end to tell what is causing what. Thus, apparent emergence might be reported, but it was actually baked in as cognitive mechanisms, but the modeler did not realize that initially primarily because the model was just too complex to understand.

Raffaele Vardavas: Importance of Nonlinearity for Emergence

In my view emergence is not just a surprise factor. Emergence occurs when the new environments resulting from the complex interplay of different entities in the system cannot be understood by understanding each entity or groups of entities in isolation. Hence, emergence occurs only in nonlinear systems that include complex feedback loops where new environments cannot be understood by simply summing over each agent's behavior when considered in isolation. Emergence occurs when entities self-organize and adapt to new environments that they, in part, helped to form.

Bill Rand: The Difficulty in Simulating Emergence

Many definitions of emergence contain some notion of surprise, i.e. that emergence is a phenomenon at the macro level that one would not expect from the micro-level rules. That cannot be a good definition, since it means that if scientists and researchers observe the same emergent phenomenon over and over again and are therefore no longer surprised, then emergence no longer occurs. A better definition of emergence would be a definition that somehow states that the emergent phenomenon is not defined by the micro-level entities, but instead is a product of the interaction of multiple entities. However, such a definition might be overbroad by counting any macro-state of the system, such as an average, as an emergent phenomenon. That may or may not be acceptable.

If we accept this second definition of emergence, then – of course – emergence is baked into every simulation. That does not make the results any less valuable, showing that one set of micro-conditions creates a particular pattern of emergence and another does not can be extremely valuable. Moreover, though it is often the case in simple models that the macro-level results seem to be baked into the micro-level rules, this is not the case as the models become more complex. As a quick explanation, we have done work on simulating social media behavior from individual-level rules (Darmon et al. 2013; Harada et al. 2014). When we tried to use these models of individual-level behavior to recreate macro-level, emergent patterns of social media behavior, we failed in our first attempts (Ariyaratne 2016). Even though individual-level rules worked very well, they failed to take into account phenomena outside the system that became much more important at the macro level, such as time of day. These outside patterns, seasonality in this case, are important for synchronizing the micro-level rules, and hence not including them prevents emergent patterns. To create emergent patterns of behavior, it is important to get all the components and their interactions correct, and thus in some ways simulations can be used as tests of minimal and sufficient conditions that are necessary to generate emergent phenomenon.

Paul K. Davis: Reproducing Emergence Through Simulation Is a Valuable Hard Problem to Tackle

My original discipline was theoretical chemistry and physics, so I have always seen emergence as analogous to the phase transitions of those subjects (e.g. precipitation in a liquid or the advent of superfluidity). Anyone observing those for the first time will see them as magical. All examples of emergence that come to my mind *should*, I believe, be reproducible in simulation. Doing so might require sophisticated agents, and variable-structure, self-aware simulation. In this context, the latter means to me not only the capability of identifying which of some *a priori* structures applies at a given point but also the capability to recognize new objects or patterns in itself, giving them names, and proceeding with them as part of structure. Building such a simulation would be no mean feat, but it seems to be a *hard problem*, not something impossible. There may also be types of emergence that cannot even in principle be generated from simulation. Examples sometimes given, all of which are in dispute, include the soul, consciousness, and such spooky aspects of quantum mechanics as action at a distance.

Scott Neal Reilly: Simulations Can Explore How Emergent Behavior Might Occur

Instead of tackling this question head on, let me offer a slightly different perspective. While surprising emergence in simulations might be possible and useful, it isn't typically how I think about these sorts of issues. Instead of creating low-level simulations and seeing what might unexpectedly emerge, I typically start with the higher-level (emergent) phenomenon and try to come up with a lower-level explanation/simulation from which that phenomenon might emerge. That is, I more or less *try* to bake the phenomenon in. Not in the sense that it is explicitly coded in its high-level form, but in the sense that I want it to emerge from the low-level model. If it doesn't emerge, or doesn't emerge in a way that matches reality, that suggests that the underlying explanation isn't correct. Or if it is close to correct, it lets me explore the space of trade-offs and configurations that are consistent with the phenomenon. This does not tell me when a low-level model is correct, but it does let me explore the space of possibly correct models. In some of my earlier work, I was exploring models of human emotion and how they arise from other mental processes. This led me to better understand how the human motivation system must work such that emotions arise as they do. My models rarely surprised me by producing things that I didn't expect, but often surprised me by producing things that I didn't want as they were clearly inconsistent with human emotions, which led me to explore alternate hypotheses (Neal Reilly 1996).

Ted Pavlic: Simulations Can Serve as Existential Witnesses for Emergent Phenomena

What does it mean to say that emergence might be *baked in* to a simulation? As nicely summarized in a complexity primer (Prokopenko et al. 2009), emergence is a property of a *large* (but not too large) *group* of agents participating in an *open system* with *nontrivial interactions* that result in *internal constraints* that lead to *coordinated global behavior* that is *self-organized* in patterns detectable to an outside observer. We call these patterns *emergent*. If the patterns are somehow *encoded* into the agents, then the phenomenon does not fit all of the criteria. For example, if an agent receives information about the state of all other agents and uses that information to place itself so that the group achieves some global task, then the system is *not* self-organized and thus cannot be emergent. This precise definition is in contrast to the fuzziness of whatever is meant by *baked-in*.

Better criticisms of claims about emergence from simulation should focus on: (i) the realism of conditions for self-organization and (ii) whether the emergent patterns are meaningful and adaptive. A simulation can be used as an existential witness for emergent phenomena, but the demonstration that such phenomena are possible does not mean that they are probable. A spectacular coordinated pattern that emerges from nontrivial interactions of simulated fish may be extremely fragile to parameter choices or subtle nuances such as the requirement that all fish behave using identical behavioral programs. Although this is a true example of emergence, the conditions may be contrived and absent in nature. Even if the emergent patterns are robust to parameter changes and heterogeneity, the emergence may or may not be meaningful. The convex hull that surrounds a flock of starlings appears to itself be a macroscopic, amoeboid macrostructure with its own novel patterns of motion that are decoupled from individual starlings. Again, such emergent patterns can be reproduced in a simulation, but in this case the emergence is a mere epiphenomenon of the selfish interactions of birds that are individually minimizing their risk of predation. There is no cost nor benefit of the emergent amoeba. The amoeba is not truly adapting to circumstances.

By contrast if the emergent phenomenon in question appears to be adaptive (as with a CAS), then both sources of criticism vanish. The adaptiveness of an emergent structure that would otherwise require highly contrived parameter choices may increase the possibility of those parameter choices occurring naturally. For example, in the social cycle of the slime mold *Dictyostelium discoideum*, the spontaneous emission and reemission of cyclic AMP leads to spiral waves whose centers attract nearby cells as aggregation sites (Devreoles et al. 1983). These highly stylized patterns can be regenerated with high fidelity in simulation of the hypothesized underlying mechanism and the surrounding physical medium (Höfer et al. 1995). When the cells of these social aggregations come together, they form motile *slugs* that eventually develop into stalks supporting fruiting bodies that effectively disperse spores to remote areas of the habitat.

Whether a natural-system phenomenon is *adaptive* depends, from a behavioral ecologist's point of view, on its *survival value* (Tinbergben 1963) after considering both benefits and costs. A very beneficial macroscopic structure might be selected even if only a narrow set of parameters enables it. An otherwise likely emergent phenomenon may not occur if accompanied by great costs. If the amoeboid envelopes surrounding starlings made them markedly more conspicuous to predators, we would not expect such emergent patterns to persist. Of course, the behavioral ecologist is also concerned with the developmental and evolutionary constraints. Certain emergent phenomena may be unavoidable side effects, whereas other emergent phenomena might be impossible. Consequently, in both natural group phenomena as well as novel group phenomena that emerges out of complex human socio-technical interactions, there may always be latent, emergent endogeneous risks that are impossible to purge from the space of possible outcomes. For example, essentially blind army ants that conduct massive foraging raids navigating only with olfactory bread-crumb-like chemicals deposited on their path will occasionally become stuck in a self-reinforcing circular mill of ants, disconnected from their home nest. They will churn until exhaustion and eventual death (Schneitle 1944). This self-organized pattern is emergent, costly, but always a possibility due to the constraints of how these nearly blind ants navigate. This kind of endogenous risk shares many of the features of a stock market run – where individuals driven by a rise or fall in price make concomitant purchasing or selling decisions that further reinforce the change in price. Long stock market runs are rare, as are circular mills of army ants. However, they are both easy to study in simulation. In fact, simulation may be the best tool for understanding the space of possible albeit improbable deleterious phenomena that cannot be predicted by analyzing individuals alone. Simulation then gains its value not through its ability to reproduce reality in high fidelity, but by allowing discovery of scenarios that are important but too rare to anticipate from data alone.

Ultimately, simulations of emergent phenomena should only be employed as part of a scientific process. They should either be testing a hypothesis on the mechanisms underlying some already observed emergent phenomena or for generating new plausible hypotheses that may lead to further *real-world* experimentation.

Relating Models Across Levels

Question: Consensus exists on the need for social-behavioral models describing phenomena at different levels of detail. A number of chapters in this book mention this, but all are short on details. What are the primary conceptual or methodological impediments? What do you see as desirable and feasible 10-year goals for doing so?

Respondents: Scott Appling, Matthew E. Brashears, Erica Briscoe, Kathleen M. Carley, Paul K. Davis, Corey Lofdahl, Kent Myers, Ted Pavlic, Scott Neal Reilly, and William B. Rouse

Contributors agreed that multi-scale models (or model families, as discussed by Davis) can often provide a better systemic view of phenomena than single-scale models. However, challenges exist both to our ability to build such models and to understand how to use them. Three pervasive challenges emerged from the discussions.

1. There is an insufficient scientific foundation to draw on for understanding how different levels relate to each other (Kathleen M. Carley).
2. Modeling infrastructure currently does not allow for seamless integration of multiple levels or types of models (Levent Yilmaz, Kathleen M. Carley).
3. Mismatches exist between the scale of theories and data – for example, while many theories focus on the individual, data are often aggregated at much higher levels, such as economic theories about why an individual works vs. unemployment data at the national level (Corey Lofdahl, Erica Briscoe, and Scott Appling).

Contributors offered (sometime conflicting) suggestions to improve multi-scale modeling that focused mainly on getting model substance right.

Understand context:

- Recognize that the fundamentally interpretive nature of social reality becomes even more challenging in cross-level modeling (Matthew E. Brashears).

- When relating models at different levels, give mathematical definition to *context* and specify different aggregations and disaggregations for different contexts (Paul K. Davis).

Identify relevant levels of analysis – and their interactions – for addressing the question of interest:

- Differentiate both the different types of models and the different types of modeling techniques that can be used to create those models – and that both top-down and bottom-up techniques are useful regardless of the details of the models themselves (Scott Neal Reilly, Raffaele Vardavas).
- Make models as simple as possible (but not simpler). The level of detail included in a model should be governed by the research question and the hypothesis being tested (Ted Pavlic).
- Focus attention on modeling interactions across levels. As Kent Myers argues, meso-level modeling can reveal connections between micro and macro levels that can provide a valuable alternative to multi-scale modeling.
- Top-down and bottom-up approaches are needed for different problems (e.g. system dynamics models and agent based models, respectively) (William B. Rouse, Paul K. Davis, Erica Briscoe, and Scott Appling).

Matthew E. Brashears: Interpretation Is Crucial in Cross-Level Modeling

A key obstacle to modeling social systems, especially across scales or with high levels of fidelity, is appropriately capturing the fundamentally interpretive nature of social reality. To borrow an example from sociologist Max Weber, when someone picks up an axe, they may be intending to chop wood for their own fireplace, earn a wage, exercise for their own health, vent their rage, defend someone against an enemy, or even commit murder. Predicting the outcome of such a situation requires understanding not simply the physical characteristics of the circumstance (i.e. a person grasping an axe) but also the meanings that are layered over top of it, giving rise to a combination known as the *definition of the situation*. This is key to predicting behavior. For example, contrast the statements "A woman slaps a child." with "A mother spanks a brat;" for many observers the former event seems to be assault of an innocent, while the latter is harsh discipline for misbehavior. But in both cases a female actor (woman/mother) is enacting physical violence (slap/spank) on a weak target (child/brat). Therefore what is different, and relevant for predicting the actions and reactions of others, are the meanings attached to events rather than the events themselves.

Given these challenges, Weber argued for his *Verstehen* method, in which explanations must be sufficient on the level of meaning. Less cryptically, to *explain something*, we must understand what actions mean to each participant in a situation and how those meanings affect their subsequent actions. This is difficult because humans are deft symbol manipulators, as revealed by Rene Magritte's famous painting of a pipe with the caption, "Ceci n'est pas une pipe" ("This is not a pipe"). The painting initially provokes consternation because the observer notes that it *is* a pipe only to realize upon reflection that in fact it is only a painting of a pipe and not a pipe itself. We so readily accept the symbol as what it signifies that being forced to recognize this mental sleight of hand is frustrating – but also enlightening. To predict behavior we need our models to be able to capture not simply the painting, but the pipe as well.

It may seem impossible to account for the meanings individuals attach to events as this appears to require knowing the contents of every mind, but appearances can be deceiving. Meanings are often not individual but rather are socially defined. For example, the large body of work on affect control theory (ACT) shows that the affective (emotional) reactions of individuals to events can be mathematically described and accurately predicted (see www.indiana.edu/~socpsy/ACT). Indeed, this theory provides tools for quantitatively modeling the events described above between a woman and a child, generating predictions of the emotional reactions of onlookers. Likewise, Ann Swidler notes that culture operates as a toolkit, providing sets of symbols, routines, and meanings that are shared by members of a society (Swidler 1986). These don't just passively lie over top of a situation but are deployed by individuals to achieve desired objectives. For example, to show affection on Valentine's Day, a man might select flowers, chocolate, or jewelry for a female partner rather than a gift card to a retailer; these gestures connoting affection are broadly understood and can be deployed actively to influence the ideas and behaviors of others. In short, meanings are not idiosyncratic but are highly patterned and predictable.

In short, modeling across levels and with more detail than the broadest possible overview requires attending to the meanings that human actors attach to circumstances, use to define situations, and ultimately deploy to achieve their ends. These meanings are shared, which is what lends them their power and therefore are measureable and often predictable. With serious effort, it should be possible within 10 years to make considerable progress in adapting well-established research programs, such as ACT, to computational models, as well as linking these affective reactions to behaviors. In that same period, it should be possible to substantially improve our ability to measure, categorize, and model cultural *tools*. However, if researchers simply discard meanings as unimportant, perhaps viewing them as too soft or hazy to deserve attention, then their models will forever be insufficient, focusing only on the painting, while humans respond to the pipe.

Erica Briscoe and Scott Appling: Multi-Scale Modeling Can Exploit Both Data- and Theory-Driven Insights

In simulating human behavior, multi-scale investigations are often necessary because, e.g. not enough empirical data is available to establish the *true* causal relationships at a single level. A multi-scale analysis allows tracing causal relationships between levels (Mayntz 2004; Kittel 2006). This has been termed the Coleman *bathtub* model of social explanation (Coleman 1990). Salthe (1991) contends that *three contiguous levels* are sufficient to represent any real system. This question reflects the *micro-to-macro problem*, the capacity for theory to explain the relationship between the constitutive elements of social systems (individual, micro-level cognitive agents) and the emergent phenomena that result from their interaction on larger scales, such as organizations (at the meso level), and widespread information diffusion (at the macro level). This problem has led to models that concentrate, and thus may restrict, scientific study by scoping investigations to emphasize cognitive, social, or sociological perspectives. To bridge the divide, theoretically grounded models of human cognition should be paired with those that account for the influence of interpersonal interactions, as well as representations of how these individual and group interactions emerge to form phenomena reflected at the society level (Briscoe et al. 2011).

The fact that many behavioral models operate on only a single scale should not preclude efforts toward understanding how multi-scale interactions may be represented in a more comprehensive or *whole system* modeling approach. The field of systems science has grown out of the belief that some systems cannot be studied at only lower levels. Within this field, open systems, i.e. those which interact with their environment, must be represented with a multi-scale approach or *scale-holistic* approach (Klir 1991). What makes a *whole system* depends, of course, on where system boundaries are drawn, but as Goguen and Farela point out, two important properties of capturing a system's *wholeness* is that emergent, or *systems properties*, are able to appear and that the level of chosen complexity precludes the system from any further reduction (Goguen and Farela 1979). This emergence may be a critical component in explaining *black swan* events (Taleb 2007), which occur so rarely as to make it difficult to incorporate them into mathematical models with any precision.

Multi-scale interactions have been traditionally perceived as flowing upward, from micro to macro, where the macro-scale effects are directly created from individual actions. Hedström and Swedberg (1998) refer to this interaction as *transformational*, where individuals' interactions are transformed into a collective outcome. This interaction is readily identified when macro-scale phenomena appear as emergent effects of interdependent agents; however, individual action implicitly eliminates structural features from the relationship directly responsible for a macro-scale phenomenon. For example, in the adoption of new technology (Briscoe et al. 2011), the belief structure of individuals determine whether they will adopt a new behavior, but the shape of the whole process, how quickly it spreads, or whether it diminishes early is entirely dependent upon the social structure in the population and the profile of receptiveness over all individuals. Depicting such matters as a multi-scale problem begs for the use of *both* data-driven (e.g. the characterization of social networks based on social media) and theory-driven approaches (e.g. the application of the technology acceptance model).

Scott Neal Reilly: A Combination of Theory-Driven and Data-Driven Inquiry Is Best

I think there are two separate issues at play here, or at least two different ways of using terminology that I think are important to keep clear. The first issue is about *top-down* vs. *bottom-up*. We often tend to use these terms to relate this to theory-driven (top-down) vs. data-driven (bottom-up) modeling. In my work with colleagues, however, we often start with theories and then look for data to back them up and refine them. Also, when sufficient data is available, we can use data mining and machine learning to help suggest additional hypotheses for a human social scientist to vet and further explore as appropriate. We have proposed that neither is the *right* way and that a hybrid approach that leverages both theory and data is best. I don't see any significant conceptual or methodological impediments to doing this as I see it done all the time. There are plenty of cases where it is done poorly (e.g. sticking with theories that don't fit the data, or using data-driven models that overfit the data and don't work in practice), but those are more issues of practice.

The second issue relates integrating models that work at different levels of detail. For instance, how do you integrate agent-based models of individuals and system dynamics models of society-level phenomena? These models can be built using top-down, bottom-up, or hybrid approaches in the senses just described, so this is largely an orthogonal issue to the first one. And this is where there are more conceptual and methodological challenges. I tend to create single-level models that are customized to the task at hand and so sidestep this problem. Other authors are better able to speak to the multilevel challenges and opportunities.

Corey Lofdahl: Decomposition Is Sometimes Necessary But Creates Issues

When creating, or attempting to create, multi-resolution models, I usually find myself moving from more aggregate, high-level models to lower-level, decomposed models because aggregation is a simplifying assumption that is built into the system dynamics (SD) simulation methodology (Lofdahl 2010). SD stocks are a modeling component that captures integrative processes by calculating changes in a variable from time step to time step, like water flowing into and out of a tub (Forrester 1963; Sterman 2000). That is, the value or level of a stock is determined by the difference of the stock inputs and outputs. This is a significant simplification because stocks are treated as perfectly connected, homogeneous networks, while reality is imperfectly connected and heterogeneous (Rahmandad and Sterman 2008). Decomposition is the process of moving from an abstract and holistic representation to a detailed and fractionated one. There are three primary difficulties associated with this decomposition process – component (i) initialization, (ii) connection, and (iii) evolution.

Decomposing an aggregate geography and analyzing its components provides an example. First, when an aggregate geography, for example, Baghdad, is decomposed into its nine constituent districts, it creates an initialization issue because each new component needs to be initialized separately. For example, if the population of aggregate Baghdad is almost 8 million people, then what is its population by district? One may be tempted to average the total population and use that each district, but this provides no additional analytic insight and is almost certainly incorrect. Aggregate measures can therefore only be decomposed to the resolution at which data can be collected effectively. Second, when regions are decomposed, their spatial relationships can be represented as in a contiguity matrix in which a 1 represents a connection through a shared border and 0 represents spatial disconnection. This then allows for the representation of real-world, spatial diffusion behaviors such as cross-border flows of people, money, and goods (Anselin 2013). Third, these spatial relationships and their associated behaviors vary over time or evolve. For example, as the population of Baghdad grows, new districts are created, or if the population shrinks, districts may unify or coalesce, which requires a change in the representation of spatial relationships. Representing these complex dynamic processes easily and flexibly in multi-resolution frameworks remains an ongoing research and development *opportunity*.

Ted Pavlic: Detailed Models Are Only Sometimes Desirable

Models exist on a spectrum from analogous to metaphorical with the value of a model depending not on how detailed it is but on how insightful it is to a particular research question. Progress will never be made if social scientists fail to adopt modeling because the models are not sufficiently detailed. More detailed models also yield less transferrable and generalizable insight. A highly detailed model should only be used when considering, e.g. "hard operations research problems" of operating a very specific system for a very specific length of time. Insight into general patterns across societies can only be found with *less* detailed models capturing the salient features across those societies. What to include should be governed by the research question and the hypothesis being tested through the use of the model. If the hypothesis is specific (e.g. its scope is the city of Santa Clara, CA), then detail is needed. If the hypothesis is general, e.g. it is to explain a phenomenon observed in Santa Clara, Des Moines, and elsewhere, then such detail would be inappropriate and likely incorrect. Every detail added to a model is another assumption. Models should be parsimonious on their assumptions or else risk providing insights that are fragile. In the next 10 years, sociologists should focus less on improving model fidelity and more on better understanding the proper way to use models within a scientific process. Computational models of complex, nonhuman natural systems have been very successful in guiding the search for knowledge and it was not necessary for many of them to be highly accurate despite the complex behavioral repertoires of the animals being simulated. Similarly, it would be exceptional (and suspicious) to believe that a simulation of the entire population of the United States *must* involve agents with accurate representations of emotional states to be useful. If there is any danger to *baking in* emergent results, it comes from adding more detail into models, not less.

William B. Rouse: Top-Down or Bottom-Up Modeling Serve Different Purposes

Depending on the question of interest, phenomena can be addressed different ways. When we study the impacts of interventions on patients with diabetes, heart disease, or Alzheimer's, we are interested in the impact of patient characteristics. Hence, a bottom-up agent-based model makes sense. A top-down systems dynamics model would aggregate away individual differences that we hypothesize are important.

On the other hand, the costs of these interventions are highly affected by economic inflation. An agent-based approach to predicting inflation would be overwhelming and, at best, marginally useful. A system dynamics economic model would be a good approach in this case.

If we are interested in flows of patients through capacities that provide the interventions of interest, then discrete-event models could predict waiting times, queue lengths, etc. Such a model would be intermediate between bottom-up agents and top-down economies.

A key issue is how these three models play together. Higher-level models tend to operate on longer time scales. For example, climate change models address decades or centuries, while models of consumer choices, which may affect climate, focus on days or weeks.

A related issue is entangled states where the solutions of two models interact, potentially requiring simultaneous solutions. This can sometimes be avoided using *principled heuristics* that enable separability of solutions. Devising such heuristics can be as much an art as a science.

Paul K. Davis: Aggregation and Disaggregation Functions Need To Be Contextual

In my mind's eye, the natural way to view agent-based modeling (ABM) is by analogy to a laboratory in which one conducts microscopic experiments with the intention of discovering or understanding and characterizing relatively more macroscopic regularities, which can then inform corresponding macroscopic theories and related models. Within theoretical physics and chemistry, analogs would be in studying molecular dynamics (either with pure theory or simulation) to better understand equilibrium and some forms of nonequilibrium thermodynamics. Part of the analogy would be to use macroscopic knowledge to inform the design of computational experiments: that is, we already know that interesting macroscopic phenomena exist (social movements and narratives among them). This should help tell us what to look for. Has the ABM community been too enthusiastic about the false idol of understanding phenomena as being generated from agents with simple rules (as in the original Boids)? In practice, *real-world humans* act not only based on their goals and individual characteristics but also because of myriad effects of the environment manifested as constraints and also stimuli causing the humans to change their values and behavior.

As a different matter, I observe that while many authors exhort understanding linkages among models at different levels of detail, not much attention seems to be paid to the mathematical theory required in doing so. As described elsewhere (Davis 2019), the correctness of an aggregation depends on how it will be used and in what context (sometimes, the right aggregation is an average; sometimes, it is the maximum value). This is unfortunate because it implies a great deal of complication. However, it is something that humans do every day: we work with abstractions intended to be roughly right for what we are doing. If the afternoon is different from the morning, we change our abstractions accordingly, without missing a beat. Conceptualizing the analog to this in social-behavioral modeling in, to me, another hard but feasible problem.

Raffaele Vardavas: Bottom-Up Modeling Need Not Be All or Nothing

Simple nonlinear sets of ordinary differential equations (ODEs) can show complex dynamics and even chaos. Thus, equation-based models (EBMs) can provide a lot of insight into the dynamics of these systems. However, they cannot easily include heterogeneities, network effects and more complex behavioral rules of adaptation. Hence, the need for microsimulation and agent-based models. However, these models can easily become highly complicated with many detailed mechanisms and the need to provide parameter estimates for each. This then requires an increased number of ensemble case runs to average over and more and harder data to analyze. Thus, whenever possible different modules of the model should be made as simple as possible. This includes using ODEs for some components of the model. For example, if we have an ABM that models firm behavior and competition, we might want to model the population of citizens who are serviced by the firms, or can invest in them, using EBMs. Hence, my preference is a bottom-up approach where the entities of interest are modeled using agents, but it is not necessary to model all entities in the system this way.

As an additional comment, although ABMs can include many of the agents' heterogeneities to be representative of the starting conditions of a real system, I find that the most revealing ABMs show how these heterogeneities emerge spontaneously from the model dynamics starting from a homogeneous set of agents.

Kent Myers: Meso-Modeling Is a Good Fit for Addressing Concrete Human Problems

A meso-model is intended to reveal connections between micro and macro levels. Typically, detail is reduced at the extremes. This is not everyone's favorite; scholars will tend to see it as compromised for one reason or another. But there are advantages to working with meso-models:

- It is possible to represent significant relationships without laboring over modeling details or big data.
- In particular scenarios, causes and effects will tend to make sense and be traceable.
- The meso-model level covers the same territory as actor mental models. It is the level at which actors in the system typically account for their strategic interventions. If researchers learn something at that level, decision-makers are likely to pay attention and may change their thinking.
- It is possible to add to such a model selectively as events and testing warrants.

I'm not arguing against those who aspire for more richness. I'm only arguing for the benefits of getting things roughly right for systems that are inherently hard to nail down and that are changing as we speak, and for which we (or policy-makers) need understandable guidance now. This is similar to the arguments in favor of *grounded theory*: formulate what is going on in a situation of interest to gain insight that is ready for use in that situation.

We can also use newer framing concepts. For example, Uhl-Bien and Marion describe a meso approach where the micro level is leader decision-making and the macro level is a complex adaptive organization. They point out that many organizational researchers seek to explain what goes on in an organization by favoring one of these levels over another, but they argue in favor of research that employs both simultaneously:

Complexity Leadership Theory ... acknowledges the role of individuals and agency (i.e., what we will call, adaptive leadership) but also recognizes that non-explained social phenomena are part of the explanation (i.e., CAS dynamics). It is this recognition that leads us to propose a model of adaptive functioning based on interdependent (and meso) relationships between agentic actions and emergent forces.
(Uhl-Bien and Marion 2009, p. 637)

In other words, the admonition is to put the attention on the interaction between the CAS and the agents who are able to intervene. The agents can be modeled as biased or as nonrational as one sees fit, and their actions will be filtered through a CAS that won't have linear, predictable responses. Perhaps that approach just doubles the confusion, but it is a realistic approach and keeps the focus on what can be done to make the whole system succeed, fail, or evolve.

Meso-modeling is a promising direction to take over the next 10 years, but not as a stepping stone toward something else, especially not toward the discredited, old-fashioned dreams of complete and accurate representation supporting perfect control. The more sober aim will be to get it roughly right for useful insight. That is, we should use newly available data, surely, but not stop there. The systems we are interested in involve humans and can't help but be constituted in various ways by what people are thinking. To get at this *data*, we need to talk to people and develop expertise in interpreting what they are saying and doing. The findings we develop in this way may even be left out of explicit models, but they can't be left out of the inquiry as a whole. Such thoughts may appear to be sending us backward, toward armchair theorizing, but in fact it is getting us past a major methodological impediment, the assumption that both the inquirer and the actors can be scrubbed out of the picture with reductive controls and that somehow that is more scientific. Arid modeling will surely remain a competitor in the academy, but no one should be surprised if the practical results are meager.

Levent Yilmaz: Improved Development of Hybrid Models Is Possible

Designing hybrid models is indeed a challenge. The Winter Simulation Conference has recently established a track to foster new developments in this area. I wrote on this topic in the early 2000s and suggested the development of simulation model programming features that improve our ability to blend together alternative paradigms in a coherent manner. In the programming language domain, for instance, we observe the emergence of new languages that bring together the object-oriented/imperative and the functional paradigm under a single framework (e.g. Scala, F#). Multi-paradigm modeling (e.g. discrete-event models with system dynamics elements) is still cumbersome due to our lack of understanding of the requirements for seamless interfacing between such paradigms. I am confident that we will be able to generalize from experience and formalize the principles toward methodical development of multi-paradigm models.

Kathleen M. Carley: Distinguishing Challenges of Multilevel and Hybrid Simulation

I believe that the panel question is an ill-formed question as it is conflates two things – multilevel modeling and hybrid models. These have different advantages and different constraints to their development.

Multilevel simulations: In a multilevel simulation, there is a *dimension* along which an activity at different levels of granularity on that dimension follows different logics and where behavior at one level of granularity may constrain or enable behavior at another. An example is the *agent* dimension where neuron-level agents, human-level agents, group-level agents, corporate-level agents, and country-level agents each follow their own logic for how they come into being, how they die or end, how they interact, grow, and so forth. Multilevel simulations are difficult to develop because (i) there may be insufficient theoretical understanding of how the different levels of granularity operate or constrain/enable each other; (ii) there may be insufficient and/or inconsistent data for informing, setting initial values, and validating such models, and hence much of the effort may go into data fusion exercises rather than simulation; (iii) the models themselves are more complex and take longer to develop due to the number of variables; (iv) meaningful virtual experiments that lead to results that are interesting at each level of granularity are difficult to construct; (v) existing simulation engines are not designed to easily facilitate multilevel modeling (while they can be used, there is no built-in support for multilevel activities) and hence the simulationist spends extra time developing the multilevel infrastructure instead of developing the model per se; and (vi) multilevel models are substantially more complex to explain to those wanting to understand how the model works, thus reducing trust in such models. That being said, there are a number of multilevel models that have been developed and used (Lanham et al. 2014; Carley et al. 2017). It is generally thought that taking a multi-modeling approach and linking together a set of independent models into an interoperable suite, or a system of systems, is a promising approach. Unfortunately, existing infrastructures for linking multiple models were developed for models of engineered systems and so do not have the ability to operate simultaneously at diverse levels of granularity for time, space/distance, agent/group size, and specificity or/confidence of results. A better infrastructure would enable more progress. However, the key difference will be developing an understanding of theories regarding and empirical data supporting meso-level modeling.

Hybrid simulations: Hybrid simulations are those in which one part of the overall model is developed using one type of modeling technology, such as system dynamics, and another part is developed using a different type of modeling technology – such as an agent-based model. The co-usage of two or more modeling frameworks is what characterizes a hybrid model (referred to as multi-formalism modeling in Gribaudo et al. 2019). Some hybrid models are used in multilevel modeling. The key challenges to doing hybrid modeling include: (i) most simulationists are trained in a single type of modeling and lack the experience to develop a hybrid model; (ii) hybrid modeling often requires team efforts and correspondingly higher develop-ment and coordination costs; (iii) hybrid models have higher data demands (often in different formats and granularities); (iv) technical infrastructure tools that support linking models from different simulation frameworks don't exist; and last, but not the least, (v) it is often unclear theoretically and empirically under what conditions the hybrid models can be interoperable without violating their scope conditions or making egregious assumptions about how processes link together.

Going Beyond Rational Actors

Question: It is widely agreed that some agents in social-behavioral model-ing need to incorporate much more human-like characteristics in their cognition and behavior, a point made convincingly by Robert Axtell (see Table 33.1) and, in a different stream, Joshua M. Epstein in his Agent_Zero work on cognitively plausible agents. These represent grand challenges for the social-behavioral community. Both establish agendas for future research. What are your reactions and suggestions about, e.g. the framings provided by Axtell and Epstein, the most scientifically difficult aspects of the challenges, the most socially important problem to be solved (i.e. agent features that would be most significant to informing strategies and policies), or technological obstacles?

Respondents: Joshua M. Epstein, Rafaelle Vardavas, Levent Yilmaz, Kathleen M. Carley, Scott Neal Reilly, Ted Pavlic

This question grew out of a discussion spurred by presentations Robert Axtell and Joshua M. Epstein gave at a RAND workshop on social-behavioral modeling in April 2017. Axtell issued a challenge to the social-behavioral modeling community to move from simplistic to realistic social scientific models. Table 33.1 presents the dimensions on which Axtell argued we can make progress. In his presentation, Epstein argued that the first goal going forward for agent-based models is to have cognitively plausible agents, i.e. agents with emotions, deliberative capacity, and social connections, all of

Table 33.1 The Axtell challenge for moving from simplistic to realistic social science models.

Model feature	Simplistic	Realistic
Number of agents	1, 2, ∞	Actual number
Goals, objectives	Static, scalar-valued utility	Evolving, other-regarding
Beliefs, aspirations	Neglected, fixed	Socially determined
Behavior	Rational, random	Empirical
Interactions	Well mixed	Networks
Information	Homogeneous	Heterogeneous
Markets, prices	WMAD, global price vector	Decentralized, local prices
Firms	Unitary actor, rational	Multi-agent, behavioral
Institutions	Neglected	Multilevel
Temporal structure	Static, one-shot, impulse tests	Asynchronous, dynamic
Governance	Benevolent social planner	Self-governance, emergent
Source of dynamism	Exogenous, outside economy	Endogenous, inside economy
Solution concepts	Agent-level equilibrium (Nash, Walras)	Aggregate steady states
Methodology	Deductive, mathematical	Abductive, computational
Ontology	Representative agent, maximum utility	Ecology of interacting agents
Policy stance	Designed from top down	Evolved from bottom up

WMAD refers to a market-focused theory of business cycles associated with Friedrich Hayek.
Source: Adapted slightly from presentation of Robert Axtell at RAND workshop, April 2017. See also Axtell et al. (2016).

which may interact to drive behavior. With this goal in mind, his Agent_Zero work is characterized by agents that are[2]

- Endowed with distinct affective/emotional, cognitive/deliberative, and social modules that are grounded in neuroscience.
- Internal modules that interact to produce individual, often far from rational, behavior.
- Multiple agents interacting to generate a wide variety of collective dynamics related to health, conflict, network dynamics, economics, social psychology, and law.

Epstein's agenda is to "get the synthesis started," with specific components provisional.

In response to this question, panel contributors emphasized that ongoing work and future work agendas are developing more realistic agents. In fact,

---

2 Adapted slightly from presentation at RAND workshop, April 2017. The agenda is elaborated in his book, *Agent_Zero* (Epstein 2013), which includes discussion of classic efforts to include some of the features in agent-based modeling (e.g. SOAR).

as Kathleen M. Carley notes, needing to go beyond rational actor assumptions is *old news*. In particular, contributors highlighted a few promising approaches for developing more realistic agents.

Machine learning: Joshua M. Epstein urges the approach of Inverse Generative Social Science, which draws on machine learning to generate agent rules rather than parameters.

Dynamic network models: Kathleen M. Carley and Raffaele Vardavas discuss how social network models move beyond rationalist assumptions and more closely approximate humans' decision-making processes. Developing more computationally tractable dynamic network models will provide more leverage modeling emergent phenomena.

Quantum cognition models: Levent Yilmaz emphasizes the need to take cognitive biases and limitations into account as we move toward greater agent realism. Quantum cognition models address the order effects and conjunction fallacies observed in actual human behavior.

Decision field theory (DFT) models: Ted Pavlic cautions against the dangers of embedding overly realistic decision rules in agents and proposes building in mechanisms from cognitive psychology meant to characterize decision-making accuracy and latency rather than more complicated agent-specific decision-making processes.

Moving beyond the technological processes of building more adaptive agents, contributors emphasize the importance of answering the following questions when developing agent-based models.

How realistic do agents need to be? Increases in the realism of agents entails costs in terms of data needs, computational requirements, comprehensibility, and the feasibility of validation, as well as other concerns. As Ted Pavlic notes, we gain little when our models become "as inscrutable as the system we are trying to describe." Scott Neal Reilly argues that a model's detail should be that needed to gain robust analytical insights. Greater detail can make it difficult to interpret models, reduce robustness of results, and increase computational burden.

What are the learning and deliberation processes for agents? As agents become more *realistic*, the information sources and learning processes of agents become more important. Vardavas and Carley highlight the need to link internal agent processes to their social networks. Pavlic cautions that more *realistically* delineating the decision rules of agents risks baking in emergent phenomena.

How will agents behave in diverse environments? Models with more *realistic* agents can run the risk of being overfit to a specific context. This may be a particularly acute concern when using evidence-based approaches such as statistical analyses of past behavior or machine learning approaches to identifying agent characteristics. Such approaches can lock in past behaviors and may miss emergent behavior (Vardavas, Carley).

An overarching conclusion from this discussion is that realism for realism's sake is not a good modeling objective. While more realistic agents may be necessary to understand some contexts, building realistic agents entails trade-offs and may result in models that are *less* useful for their purpose.

Joshua M. Epstein: Inverse Generative Social Science – What Machine Learning Can Do for Agent-Based Modeling

While I certainly endorse the broad agenda referred to in the panel question, I am particularly excited by what I call *Inverse Generative Social Science* and by how machine learning (ML) can advance it. As background, artificial intelligence and ML are *displacing* humans, but not *explaining* them. Machines can crush humans at chess, but do not illuminate *how humans play* chess. When asked how he came up with a winning brilliancy against IBM's Deep Blue, Gary Kasparov answered simply, "It smelled right."

While it has not been used in this way, machine learning can help *explain* human behavior, even moving ABM into a new epoch. So far, agents have been iterated forward to generate such *explananda* as settlement patterns, scaling laws, epidemic dynamics,…, all sorts of things. But these are all examples of the forward problem: we design agents and grow the target phenomenon. The *motto* of generative social science is: "If you didn't grow it, you didn't explain it." But there may be many ways to grow it! How do we find *all* the nontrivial generators?

This is Inverse Generative Social Science – agents as model *outputs*, not model *inputs* – and ML can enable it. Compactly, let M be some macro-phenomenon we want to explain. Let m be a candidate micro-world (heterogeneous agents, their *rules*, numerical parameters, environment, initial conditions, etc.). When iterated forward, every micro-m generates some macro-thing, call it $G(m)$. The explanatory fitness of m is the (metric-dependent) proximity of $G(m)$ to M. So, given an explanatory target M (wealth distribution, spatial settlement pattern, conflict dynamic, financial panic, or difficult *conjunctions of these*), use machine learning to (i) *construct* the large search space of admissible heterogeneous agent micro-worlds (m's), (ii) *encode* this space in a searchable way, and (iii) *evolve* (for instance) the fittest explanatory candidates (e.g. mutation and crossover on agent *rules*, not just parameters) under various metrics.

I proposed this earlier (Epstein 1999, 2006), but the components – neurocognitive, computational, and data analytic – were immature then. Independently, these have arrived and the time is ripe to synthesize them and attack inverse generative problems of social science and history.

Notable efforts are underway (Rand 2019). Relatedly, though they don't call it this, the *Inverse Anasazi* modeling of Gunaratne and Garibay (2018) uses evolutionary programming to construct new micro-rules that (slightly) outperform those reported in Axtell et al., though they confirm the overall conclusions. This method thus allows for a kind of *structural* (not merely parametric) sensitivity analysis that will be important in formalizing the robustness of rule-based systems as this promising agenda unfolds.

Raffaele Vardavas: Evidence-Based Models Need to be General Enough to be Realistic Under Alternative Specifications

The way I think about CASs for socioeconomic systems is that individual agents need three components affecting their behavior: (i) their adaptation to personal decisions and experiences, (ii) social network effects reflecting the influence of social contacts, and (iii) the effect of media and social media.

I agree that we want to move away from rational expectations and simplistic utility maximization where agents are thought to make decisions relying completely on deductive thought. Instead, we need to use more realistic models that include inductive thought and social network influences in the decision process.

Models should pass the Lucas critique whereby internal casual mechanisms and behavioral rules should be fundamental and general enough to allow the model to make sensible and realistic predictions under very different alternative environments and policies. This may indeed mean moving away from informing the model entirely using regression models of data when the system was in a given state or at the status quo. Such regressions may not hold true when the system is run far from the status quo conditions. Regression models provide a descriptive account of agents' behaviors. In my opinion, they should be used to help formulate the mechanistic behavioral and causal rules. For example, a regression model may suggest that a person's decision to vaccinate for the flu does not strongly depend on whether he or she caught the flu last year or in past years, but does depends strongly on whether he or she vaccinated in the past year or years. Does that suggest that we should put this directly in our ABM? Suppose that we consider a *what-if* scenario or world with no more seasonal flu. In this scenario, the majority of agents in the model may still be vaccinating for many years after flu was eradicated. This seems to be implausible. Hence, we need to be careful in our effort of having our models highly tied to the data and *evidence based* using these types of descriptive approaches. The fact that the model can be validated to reproduce the status quo conditions does not necessarily mean that it is valid under policies or conditions that are very different from the status quo.

Kathleen M. Carley: Agent-Based Dynamic Network Models Produce More Realistic Agents

This (needing to go beyond rational actors) is old news. For the last 20 years, there have been social science models that are not rational actor models. The only people still building rational actor models are physicists, engineers, and economists. None of the leading computational social scientists do this.

Many modern agent-based models are agent-based dynamic network models. Network-based models were proposed prior to World War II as an alternative to the rational man model. When agent-based models are predicated on evolving social and knowledge networks, they generate behavior that is more realistic. Also, the best agent models at the cognitive level, e.g. Soar and ACT-r, are also not rational actor models. A key challenge is how to scale these models and place them in a social/knowledge network context.

Modern machine learning models, which require training sets, are the new variant of rational actor models. They are predicated on the assumption that tomorrow is like today, that the training set is comprehensive vis the problem space, and that the actor (in this case the ML model) can know and process all information. However, in many cases social-behavioral problems are sufficiently complex and volatile, tomorrow is not like today, and the environmental volatility is such that by the time the training sets are developed they are no longer useful and that it is unlikely that all information can be known and processed – at least in the time available.

Consequently, this class of ML models are, like other rational actor models, (i) too fragile to be used in the wild, (ii) don't reflect human heterogeneity, and (iii) have a very short shelf-life.

Levent Yilmaz: Realistic Models Must Include Cognitive Biases and Limitations

In social and behavioral modeling, realism requires modeling human behavior in such a way that the cognitive biases as well as limitations are accurately captured in accordance with empirical observations. Some may argue that such differences may cancel each other in the aggregate. This may be the case for some applications. Yet, adversarial behavior modeling may require us to account for our opponent's biases and perspectives to develop robust strategies. Similarly, heuristics and biases in human decision-making are widely reported. The limitations of the use of Kolmogorov axioms of probability in classical decision-making has also been reported as error-prone. More recently, quantum cognition models are promoted to address the order effects and conjunction fallacies observed in actual human behavior

(Bruza et al. 2009). More research and development is needed to streamline the programming of agents that can exhibit such quantum like behavior.

Scott Neal Reilly: High Degree Realism Entails Costs That May Not Be Outweighed by Their Benefits

I wholeheartedly agree that we need more realistic models than have often been used in the past. I think the more interesting question is what level of abstraction does the realism need to live at? In other words, I think Axtell's breakdown is great and I find it hard to imagine for people arguing we should live in the *simplistic* column instead of the *realistic* column. That said, I think the two-column presentation suggests more of a dichotomy than a spectrum, and I wonder if Epstein is going too far along that spectrum in his call for *cognitively plausible* agents that are *grounded in neuroscience*. I suspect Prof. Epstein and I would agree that you only need to model down to the right level, as I don't see him calling for our models to go to the level of atoms or quarks. The question is whether cognitive or neurological realism is important for most social science problems. I am sure that he is right that additional levels of realism could provide us with powerful new insights that cannot be gained from a purely behavioral level of realism, but I suspect there is a spectrum from simpler models that provide certain kinds of insights and are useful to certain kinds of problems to something cognitively/neurologically plausible, which would be applicable to other problems. I also think it is important to recognize the complexity, effort, and cost (and, for the record, chance for bugs and other errors) that comes from the kinds of fidelity required for cognitive and neurological plausibility and to determine whether the costs of that kind of fidelity are worth the benefits. Let me give a simple example from a rather different domain: what if you want to model the game of craps to explore the outcomes associated with various strategies? In this case, there is no benefit to modeling the physics of each die falling and bouncing off the table instead of using a random-number generator. If you want to understand the dynamics of tumbling cubes, you need to go down to the physics level, but for understanding the dynamics of the craps game, you just need plausibly realistic behavior on the part of the dice at the level of which face ends up facing up. Similarly, models of human behavior might need to go to the cognitive and neurological levels to answer certain questions, but if we have simpler means to get realistic behavior at the level that matters for interesting social science questions, then I think we need to explicitly reason about the cost–benefit trade-offs of various levels of realism for the problem at hand.

Ted Pavlic: With Additional Realism Comes Additional Liability

It is difficult for me to persuade myself to agree with anything in this prompt, and I do not believe that Table 33.1 is convincing. The term *simplistic* is obviously a straw man meant to be knocked down by (wait for it) *realistic* models.

Again, realism is never the goal in any modeling endeavor. Models need not be *realistic* to be insightful. Moreover, the word *realistic* is a comparable adjective – it is possible for some models to be *more realistic* than others, but there is no such thing as a *most realistic* model. Even when experimenting with real populations of people, any insights gained from those experiments may not generalize to other populations or even other times. With additional realism comes additional liability – every small bit of realism added to a model is another assumption that may not be met in *reality*. Models should be as *realistic* as they need to be to answer a question, but no more. Might it be important that agents can exist in *sad* or *happy* states and have different behavioral rules depending on their state? It's possible. Does it make a model more *realistic* if 1000 (or even 1 million) virtual agents are constrained to a crystalline lattice and update their interactions with their close neighbors based on their current emotional state (*happy* or *sad*)? That still sounds like an *unrealistic* model to me, and it is arguably *less realistic* than a simplistic model with well-mixed agents that have no emotional states. Furthermore, new results from neuroscience show that some aspects of deliberation occur after scans of the brain indicate that a decision has already been made. So, our introspection may create the illusion of deliberation when (at least some of) the underlying decision mechanisms are occurring much more quickly.

Getting back to Tinbergen (1963), it is important not to confound proximate causation (mechanisms) and ultimate causation (adaptive function). In fact, this error poses the greatest risk to *baking in* emergent phenomena. It would be strange to assume that an individual human moves around looking to maximize his or her number of offspring produced, and yet some other behavior (like responding to hunger signals by eating) may tend to increase the success of that individual. Very often when *realism* is added to a model, it provides a back door to taking for granted that which was to be shown. Furthermore, terms like *bounded deliberative capacity* sounds almost as if adaptive function is being introduced at the individual level as opposed to being a product of other mechanisms. Rather than developing some model of *deliberative capacity* and then *bounding* it for an agent, it would be possible to build in mechanisms from cognitive psychology meant to characterize decision-making accuracy and latency – like DFT models (e.g. drift–diffusion modeling for two-alternative force choice [TAFC] tasks). Implementing a DFT mechanism on an agent would be much simpler and would capture the important features of decision-making without having to actually develop a theory of how decisions are made that is almost certainly wrong.

The other downside of making models *too realistic* is that they become as inscrutable as the systems that we are using them to help explain. Once the computational model becomes so difficult to explain that a statistical model is used as a meta-model for the phenomena (i.e. the statistical model describes the computational model that describes the *realistic* phenomena), then it may not be necessary to have the computational model.

In the end, the capabilities of an agent should be tailored to the research question, and there should be some investigation of whether results are even sensitive to special features like *emotion*.

There is no reality. There are only models. Electrons don't exist. Newton's laws are only approximations. The general theory of relativity is just a framework for describing space time. Some enterprising young graduate student may develop new theories of particle physics next year that better explain our observations without having to posit the existence of an electron. Similarly, our understanding of neuroscience is constantly changing. Neural networks themselves, which can be viewed as models, have gone through three major generations of development – from binary to general activation functions to spiking neural networks – and there could be more. The properties of neural networks fundamentally change when moving from second- to third-generation neural networks; however, that does not mean that a good model of human society requires that every agent make decisions that make use of a spiking neural network. Good modeling is both about getting the internal structure right *and* about drawing good boundaries around the things that matter and the things that can be approximated away. "All models are wrong, but some are useful" (Box 1976).

References

Anselin, L. (2013). *Spatial Econometircs: Methods and Models*, vol. 4. Springer Science & Business Media.

Ariyaratne, A. (2016). Large scale agent-based modeling: simulating twitter users. Master of Science. University of Maryland, College Park, MD.

Axtell, R.L., Capella, E., Aasstevens, R. et al. (2016). Agentization: relaxing simplistic assumptions with agent computing.

Box, G.E.P. (1976). All models are wrong, but some are useful. *Journal of the American Statistical Association* 71 (356): 791–799. https://doi.org/10.1080/01621459.1976.10480949.

Briscoe, E., Weiss, L., Whitaker, E., and Trewhitt, E. (2011). Closing the micro–macro divide in modeling technology adoption. *Proceedings from 2nd Annual Conference of the Computational Social Science Society of America*, Santa Fe, NM.

Bruza, P., Busemeyer, J., and Gabora, L. (2009). Introduction to the special issue on quantum cognition. *Journal of Mathematical Psychology* 53: 303–305.

Carley, K.M., Morgan, G.P., and Lanham, M.J. (2017). Deterring the development and use of nuclear weapons: a multi-level modeling approach. *Journal of Defense Modeling and Simulation: Applications, Methodology, Technology* 14 (1): 95–105. https://doi.org/10.1177/1548512916681867.

Clayton, P. and Davies, P.C.W. (eds.) (2006). *The Re-Emergence of Emergence: The Emergentist Hypothesis from Science to Religion*. Oxford: Oxford University Press.

Coleman, J. (1990). *Foundations of Social Theory*. Cambridge, MA: Harvard University Press.

Darmon, D., Sylvester, J., Girvan, M., and Rand, W. M. (2013). Understanding the predictive power of computational mechanics and echo state networks in social media, arXiv:1306.6111v2.

Davis, P.K. (2019, this volume). Lessons on decision aiding for social-behavioral modeling. In: *Social-Behavioral Modeling for Complex Systems* (ed. P.K. Davis, A. O'Mahony and J. Pfautz). Hoboken, NJ: Wiley.

Devreoles, P.N., Potel, M.J., and MacKay, S.A. (1983). Quantitative analysis of cyclic AMP waves mediating aggregation in *Dictyostelium discoideum*. *Developmental Biology* 96 (2): 405–415. https://doi.org/10.1016/0012-1606(83)90178-1.

Dreyfus, H.L. (1972). *What Computers Can't Do: The Limits of Artificial Intelligence*. New York, NY: Harper & Row.

Dreyfus, H.L. (1992). *What Computers Still Can't Do: A Critique of Artificial Reason*. Boston: MIT Press.

Epstein, J.M. (1999). Agent-based computational models and generative social science. *Complexity* 4 (5): 41–60. Retrieved from http://www.uvm.edu/~cmplxsys/legacy/newsevents/pdfs/2013/epstein-complexity-1999.pdf.

Epstein, J.M. (2006). Remarks on the foundations of generative social science. In: *Handbook of Computational Economics* (ed. L. Testfatsion and K.L. Judd), 1585–1604. Amsterdam: Elsevier Science Publishers.

Epstein, J.M. (2013). *Agent_Zero: Toward Neurocognitive Foundations for Generative Social Science*. Princeton, NJ: Princeton University Press.

Epstein, J.M. and Hammond, R.A. (2002). Non-explanatory equilibria: an extremely simple game with (mostly) unattainable fixed points. *Complexity* 7 (4): 18–22.

Forrester, J. (1963). *Industrial Dynamics*. Cambridge, MA: MIT Press (ISBN: 0262560011).

Franz, M. and Matthews, L.J. (2010). Social enhancement can create adaptive, arbitrary and maladaptive cultural traditions. *Proceedings of the Royal Society B: Biological Sciences* https://doi.org/10.1098/rspb.2010.0705.

Goguen, J.A. and Farela, F.J. (1979). Systems and distinctions; duality and complementarity. *International Journal of General Systems* 5 (1): 31–43.

Gribaudo, M., Iacono, M., and Levis, A.H. (2019, this volume). Multi-formalism modeling of complex social-behavioral systems. In: *Social-Behavioral Modeling for Complex Systems* (ed. P.K. Davis, A. O'Mahony and J. Pfautz). Hoboken, NJ: Wiley.

Gunaratne, C. and Garibay, I. (2018). Evolutionary model discovery of factors for fram selection by the artificial anasazi. *Proceedings of the Computational Social Science Society* (November 2017), Santa Fe, NM. arXiv:1802.00435.

Harada, J., Darmon, D., Girvan, M., and Rand, W.I. (2014). Forecasting high tide: predicting times of elevated activity in online social media. *Proceedings from 2015 IEEE/ACM International Conference on Advances in Social Networks Analysis and Mining 2015*, Paris, France.

Hedström, P. and Swedberg, R. (eds.) (1998). *Social Mechanisms: An Analytical Approach to Social Theory*. Cambridge University Press.

Höfer, T., Sherratt, J.A., and Maini, P.K. (1995). Cellular pattern formation during *Dictyostelium* aggregation. *Physica D: Nonlinear Phenomena* 85 (3): 425–444. https://doi.org/10.1016/0167-2789(95)00075-f.

Holland, J.H. (1992). Complex adaptive systems. *Daedalus* 121 (1): 17–30.

Kittel, B. (2006). A crazy methodology? On the limits of macro-quantitative social science research. *International Sociology* 21 (5): 647–677.

Klir, G.J. (1991). Systems profile: the emergence of systems science. In: *Facets of System Science* (ed. G.J. Klir), 337–354. Boston: Springer.

Lanham, M.J., Morgan, G.P., and Carley, K.M. (2014). Social network modeling and agent-based simulation in support of crisis de-escalation. *IEEE Transactions on Systems, Man, and Cybernetics: Systems* 44 (1): 103–110. https://doi.org/10.1109/TSMCC.2012.2230255.

Lofdahl, C. (2010). Governance and society. In: *Estimating Impact* (ed. A. Kott and G. Citrenbaum), 179–204. New York, NY: Springer.

Mayntz, R. (2004). Mechanisms in the analysis of macro-social phenomena. Working paper 03/3 of the Max Planck Institute for the Study of Societies (MPIfG).

Neal Reilly, W.S. (1996). *Believable Social and Emotional Agents*. Pittsburgh, PA: Carnegie Mellon Univesity.

Nunn, C.L., Thrall, P.H., Bartz, K. et al. (2009). Do transmission mechanisms or social systems drive cultural dynamics in socially structured populations? *Animal Behaviour* 77 (6): 1515–1524.

O'Connor, T.I. and Wong, H.Y. (2015). Emergent properties. In: *The Stanford Encyclopedia of Philosophy* (ed. E.N. Zalta), Summer 2015 Edition. Retrieved from https://plato.stanford.edu/archives/sum2015/entries/properties-emergent/.

Prokopenko, M., Boschetti, F., and Ryan, A.J. (2009). An information-theoretic primer on complexity, self-organization, and emergence. *Complexity* 15 (1): 11–28. https://doi.org/10.1002/cplx.20249.

Rahmandad, H. and Sterman, J.D. (2008). Heterogeneity and network structure in the dynamics of diffusion: comparing agent-based and differential equation models. *Management Science* 54 (5): 998–1014.

Rand, W. (2019, this volume). Theory-interpretable, data-driven agent-based modeling. In: *Social-Behavioral Modeling for Complex Systems* (ed. P.K. Davis, A. O'Mahony and J. Pfautz). Hoboken, NJ: Wiley.

Salthe, S.N. (1991). Two forms of hierarchy theory in western discourses. *International Journal of General Systems* 18 (3): 251–264.

Schneitle, T.C. (1944). A unique case of circular milling in ants, considered in relation to trail following and the general problem of orientation. *American Museum Novitates* 1253: 1–26.

Sterman, J.D. (2000). *Business Dynamics: Systems Thinking and Modeling for a Complex World*. Boston: McGraw-Hill (ISBN: 007238915X).

Swidler, A. (1986). Culture in action: symbols and strategies. *American Sociological Review* 51: 273–286.

Taleb, N.N. (2007). *The Black Swan: The Impact of the Highly Improbable*. Allen Lane.

Tinbergben, N. (1963). On aims and methods of ethology. *Zeitschrift für Tierpsychologie* 20: 410–433.

Uhl-Bien, M. and Marion, R. (2009). Complexity leadership in bureaucratic forms of organizing: a meso model. *Leadership Quarterly* 20 (4): 631–650. Retrieved from http://pdfs.semanticscholar.org/749c/c0f16d285b3f7f282eabcd6f312e740c772e.pdf.

Part V

Models for Decision-Makers

34

Human-Centered Design of Model-Based Decision Support for Policy and Investment Decisions

William B. Rouse

Alexander Crombie Humphreys Chair, School of Systems and Enterprises, Stevens Institute of Technology, Center for Complex Systems and Enterprises, Hoboken, NJ 07030, USA

Introduction

We use models to answer questions and solve problems. Often these problems involve designing solutions in terms of physical form, functional capabilities, and policies intended to enable, incentivize, or inhibit particular behaviors. The key point is that models are intended to support problem-solving including broad *problems*, such as "How does this system work? What is going on?"

In other words, models are not ends in themselves. Instead, they are a means for addressing the ends of problem-solving. Thus, not only do we want models to be scientifically and technically valid. We want them to be an acceptable means to the problem solvers of interest. We also want them to be viable in terms of being worth the effort to learn and employ in problem-solving.

It is useful to distinguish between two very different types of overarching questions. Understanding the nature of problems often involves assessing *what is*. This is the realm of, among other things, *big data*. In contrast, solving problems often also involves asking "what if" and "under what conditions?" (Davis 2019). Addressing these questions typically involves exploring possibilities that have never been tried before. Visualization and computational models are used to predict the consequences of such possibilities under various assumptions or, often, to understand the potential consequences while dealing with uncertainty.

This chapter summarizes a long history of personal experiences developing models to address *what-if* questions. Thus, I use the pronoun *I* throughout. These experiences start with submarines in 1967 and end with driverless cars in 2017. I consider the modeler as a user, advisor, facilitator, integrator, and explorer in the contexts of aerospace and defense, electronics and

Social-Behavioral Modeling for Complex Systems, First Edition.
Edited by Paul K. Davis, Angela O'Mahony, and Jonathan Pfautz.
© 2019 John Wiley & Sons, Inc. Published 2019 by John Wiley & Sons, Inc.
Companion website: www.wiley.com/go/Davis_Social-Behavioralmodeling

semiconductors, automobiles, healthcare delivery, and higher education. The modeling lessons learned from these many experiences are summarized. More broadly, I provide observations on model-based problem-solving.

Modeler as User

I started at Raytheon in 1967 after completing my sophomore year as a summer intern. I came back the following summer as well and then worked half time during my senior year as an assistant engineer. I worked at the Submarine Signal Division in Portsmouth, RI.

One major assignment was to determine how many spare parts for the sonar system, for each type of part, should be carried on a submarine. Space, obviously, is at a premium on a submarine, so that was the major constraint. I needed to take into account the failure rates of parts, as well as the consequences of their failing and there being no spares.

The end result was a deck of almost 1000 IBM cards that simulated the failure and repair of the sonar system over the course of a mission. Since failures were randomly generated, multiple runs were needed to generate probability distributions of system availability. This consumed quite a bit of computer time, with the results coming back hours later.

I named this simulator MOSES, which stood for Mission-Oriented Systems Effectiveness Synthesis. I would submit my deck of cards at the computer center and check back in a couple of hours. I got to know the people there pretty well. When I asked if the MOSES results were back, they often responded, "No, he is still wandering in the wilderness."

I was the only person in the Reliability and Maintainability Department that understood MOSES. I was the only user. People posed questions, and I figured out how to get MOSES to answer them, sometimes requiring extensions of the software. Being the only user eventually limited my opportunities on other projects. This not only provided job security but also caused me to feel trapped.

I wanted to build models that other people would use to address questions in their jobs. As the stories that follow illustrate, this turned out to be much more difficult than I anticipated.

Modeler as Advisor

As a faculty member at the University of Illinois at Urbana-Champaign in the mid-1970s, I encountered an interesting problem associated with the Illinois Library and Information Network (ILLINET). They processed millions of requests per year for library and information services, ranging from reference questions to interlibrary loans.

The network had three levels – local libraries, library systems, and major resources libraries. The state compensated each node in the network for each service transaction provided, even those transactions that were unsuccessful. We developed a network model to predict network performance in terms of average time until service success, average cost of service success, and probability of service success.

The goal was to determine request routing strategies that would minimize time and cost and maximize probability of success. An analytic solution of this queuing network problem was derived. The model was parameterized using large data sets available from the state. A forecasting model was developed to enable updating of parameters. Numerous case studies were performed at the system and state level (Rouse and Rouse 1980).

The model served to integrate substantial knowledge about ILLINET. In today's terminology, it was very much evidence based. It was delivered to the state as contractually required. We continued to work with them over several years. Occasional calls asking questions about one thing or another indicated that the model was being used. However, my sense was that usage depended on our being available as advisors.

Modeler as Facilitator[1]

I founded Search Technology in 1980 with Russ Hunt. We focused on simulation-based training and intelligent decision support systems. Our primary customers were electric utilities and the US defense industry. During the 1980s, we developed a methodology termed human-centered design (Rouse 1991, 2007). Human-centered design is a process of considering and balancing the concerns, values, and perceptions of all the stakeholders in a design. By stakeholders, I mean users, customers, developers, maintainers, and competitors. The premise of human-centered design is that the major stakeholders need to perceive methods and tools to be valid, acceptable, and viable.

Valid methods and tools help solve the problems for which they are intended. Acceptable methods and tools solve problems in ways that stakeholders prefer. Viable methods and tools provide benefits that are worth the costs of use. Costs here include the efforts needed to learn and use methods and tools, not just the purchase price.

In the late 1980s, several of our customers asked us to teach them how to do human-centered design. The result was a series of workshops that we delivered quite frequently. Workshop participants suggested that we create a series of

---

1 This section discusses a suite of tools for which I led the development. The central point is not these specific tools, but the lessons learned in creating and employing them to help literally thousands of users.

books that captured the content of the workshops, including the many case studies presented. The result was a series of three books published by John Wiley (Rouse 1991, 1992, 1993).

Once the books were provided during the workshops, participants had another suggestion. One participant put it crisply, "We don't really want to read these books. We would like tools such that in using the tools we would be inherently following the principles in the books." Another participant said, "We don't just want knowledge; we want executable knowledge."

We agreed with the idea but we were very slow getting started. Finally, two customers, independently, offered to buy corporate-wide licenses for the tools even though they – and we – did not know what they were buying. The initial payments toward these licenses provided the resources to get started.

The first tool was the *Product Planning Advisor* (PPA). This tool embodied the principles of human-centered design, built around multi-stakeholder, multi-attribute utility theory (Keeney and Raiffa 1976), and quality function deployment (Hauser and Clausing 1988); see Figure 34.1. This eventually became our best-selling tool (Rouse and Howard 1993), but it did not get there smoothly.

We formed user groups at the two companies that had committed resources and asked everyone in both groups what they wanted the tool to do. We used

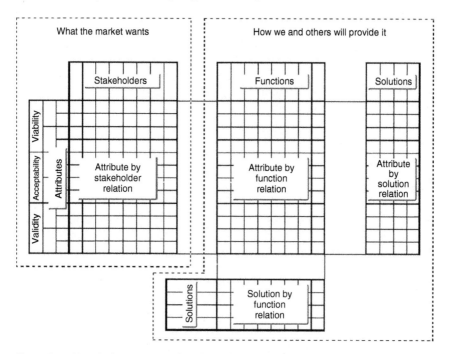

Figure 34.1 Knowledge structure of product planning advisor.

this list of desires to build a tool that provided functions and features that *anybody* had requested, i.e. we provided functions and features covering everything that anyone had asked for. When we demonstrated the prototype, virtually, every user was overwhelmed. We had provided what they wanted, but it had too many options, modes, etc.

We went back to the drawing board and redesigned PPA to only provide the functions and features that *everybody* had requested. This would have also been problematic if the consensus requests were foolish. However, the consensus requests were very sensible.

This version was a success, and we eventually sold many hundreds of copies to over 20 well-known companies. PPA was sold in conjunction with training workshops where participants learned human-centered design and the use of PPA, all in the context of real product planning problems of importance to their companies.

Much to our surprise, the services associated with PPA and our other tools (described below) went well beyond training. We were repeatedly asked to facilitate workshops associated with new product planning endeavors, despite most of the participants in the workshops having been trained earlier. Consequently, we facilitated many workshops for customers that included Digital, Honeywell, Motorola, Rolls-Royce, and Rover.

I asked a senior executive why these services were continually needed. His response was, "I am not at all concerned with the costs of your software and services. I am totally concerned with the overall costs of success. Your involvement lowers those costs." Having facilitated hundreds of product planning workshops across many industries, we could share many lessons learned. These cross-industry perspectives were highly valued.

Our next tool was *Business Planning Advisor* (BPA). This emerged when a senior executive observed, "Your product planning process has helped us tremendously, but it is inconsistent with how we run the overall company. Could you take a look at that for us?" This led to expanding our focus to the overall enterprise. We formed a subsidiary, Enterprise Support Systems, which eventually became an independent company in 1995.

BPA did not sell as well as PPA. There were too many competing software packages in the marketplace. It served a role in that our customers expected us to have such a tool, but it did not have as unique a value proposition as PPA. We did end up facilitating a seemingly countless number of executive off-sites, not because we wanted to have such a service line, but because we did not want customers to retain our competitors.

Next was the *Situation Assessment Advisor* (SSA). This tool was an expert system-based tool where users answered 50+ questions about their company, products and services, and competitors. It would then tell them which situations they were in and where they were headed, all based on an associated book, *Start Where You Are* (Rouse 1996). The most popular feature was the case studies provided of what other companies with their present and projected future situations did and how well it worked.

SSA was modestly priced and was popular with executive teams. However, sales were not impressive. We had created a tool that a small team used perhaps once per year. They would buy one copy. In contrast, many people used PPA frequently. Companies would buy 25–50 copies, motivated in part by a break point in the pricing.

Our next book was *Don't Jump to Solutions: Thirteen Delusions That Undermine Strategic Thinking* (Rouse 1998a). This book came with an expert system tool for assessing the extent to which the reader's company was beset by any of the thirteen delusions. The book was selected by Doubleday as their book of the month and was the main selection, with the software tool included. Many thousands of copies sold, but the price point ($59.95 for book plus tool) needed hundreds of thousands of sales to make it worthwhile. We did get some publicity when the book won an award as one of the top 20 management books published globally that year.

I often asked executives what tools they might want next. One answer was, "How are my current R&D investments likely to affect the financial future of the company, and how should such projections affect what we invest in now?" The result was the *Technology Investment Advisor* (TIA), with Motorola as the lead customer.

The core of TIA was an option pricing model along with product life cycle models and production learning models (Boer 1999; Rouse et al. 2000). TIA was rolled out across Motorola Labs and used extensively by 3M, Coca-Cola, IEEE, Lockheed, Raytheon, and the Singapore Ministry of Defence. It included a substantial service component as the notion of real options required very knowledgeable support (Boer 1999). Over 30 significant engagements resulted (Rouse and Boff 2004).

We marketed PPA, BPA, SAA, and TIA as the *Advisor Series* of planning tools, summarized in Figure 34.2. PPA sold the most copies by far. Both PPA and TIA generated substantial service revenues, the latter because of the conceptual difficulties customers often had buying into the notion of investing in options that might or might not be exercised. People would ask me why one should invest in something that you likely might not use. I would respond, "Does anyone in the room regret that you did not use your life insurance last year?"

All of the tools in the *Advisor Series* involved facilitating groups to build models to address their questions of interest. The tools were primarily just computational frameworks that users populated with their knowledge. Some training was needed, but, as most users were technically educated, this was not difficult. What surprised us was how much ongoing facilitation was needed.

Later in this chapter, I address how these tools were evaluated. At this point, however, it is worth mentioning the primary focus of the facilitation. People

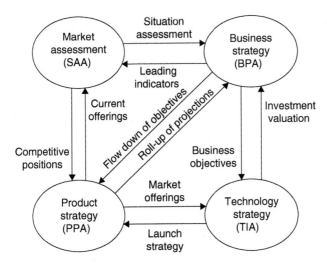

Figure 34.2 The advisor series of planning tools.

needed some help in understanding the underlying mathematics, particularly as it related to the problem at hand. However, much more of the facilitation focused on how to think about markets, stakeholders, and competitors. Often this involved dragging the teams away from technical details before they had fully framed the market and technology opportunities.

Modeler as Integrator

I returned to academia in 2001 as chair of the School of Industrial and Systems Engineering at Georgia Tech. My 13 years of experience running two companies and working with thousands of executives and senior managers as our customers had convinced me that enterprises as complex systems was an area worthy of research (Rouse 2001, 2004). I focused in particular on enterprise transformation in the sense of understanding and enabling fundamental change (Rouse 2006).

I soon became immersed in the healthcare enterprise ecosystem – patients, providers, suppliers, payers, and regulators. I had served in a consulting role to the executive team of the American Cancer Society in the late 1990s. I now became involved in collaborative efforts between the National Academy of Engineering and the Institute of Medicine (now National Academy of Medicine). We soon focused on how systems engineering models could provide value as we strived to create an evidence-based learning health system.

Co-chairing a couple of initiatives with Institute of Medicine (IOM) members, it struck me that we needed to create a real model to show people what it meant and how it could be useful. We decided to create a multilevel model of Emory University's prevention and wellness program for their employees. The question was how to scale up a pilot program for a cohort of 700 employees to 60 000 covered lives. The vice president for human resources (HRs) wanted to be confident in a positive return on investment.

We completed the first version of this model in four months. It quickly became clear that demonstrations of this model at the academies were all just an abstraction if we could not provide committee members hands-on experience varying assumptions and intervention choices and, of course, seeing results. This caused us to pay careful attention to interactive visualizations that made model usage interesting and, hopefully, compelling (Rouse 2015).

At one point, the committee members with medical backgrounds asked, "What do you call this thing?" I said it was a multilevel enterprise model. There were just blank stares. Thinking quickly, I said it is like a flight simulator for evaluating health policies, albeit without the accuracy possible for a purely engineered system like an aircraft. Everyone nodded with understanding. The phrase policy flight simulator was embraced and continues to evolve (Rouse 2014, 2015) (Figure 34.3).

Four projects focused on developing policy flight simulators for Emory Healthcare, Vanderbilt Health, Indiana Health, and Penn Medicine focused on hypertension, diabetes milletus, coronary heart disease, Alzheimer's disease, and transition care for elderly patients, respectively. We also developed a simulation of the New York City healthcare ecosystem that included rich component models for the many (over 60) hospital corporations in the city. This model was used to understand patterns of mergers and acquisitions.

Figure 34.3 Policy flight simulator for New York City health ecosystem.

These model development projects focused on integrating models of clinician and patient decision-making, flow models of delivery processes, microeconomic models of providers and payers investment decisions, and macroeconomic models of economic and health policies. The structures of these models were informed by the medical literature as well as findings from operations research, behavioral economics, and so on. The parameters of these models were estimated from a range of large data sets on diseases, interventions, providers, and payers (Rouse and Cortese 2010; Rouse and Serban 2014).

Teams of modelers and healthcare professionals accomplished all of these projects. These teams identified key questions, relevant data sets, and appropriate component models. Succinctly, the value added was in the integration of the components, rather than the components with which many stakeholders were already familiar.

One particular comment was very cogent. A well-known medical thought leader said, "You have pulled everything together into an integrated, user-friendly interface that includes data from clinical trials, our patient data, medical effectiveness models, and business economics models. I can explore my business issues without having to find all these pieces."

These models provided a means for these teams to explore the issues for which they were responsible. The development process involved coproduction of an evidence-based tool for exploration of these issues. For example, the models were used to understand and redesign processes to enable economic scaling of innovative pilot tests into much broader application.

Modeler as Explorer

Recently, our team at Stevens has been using computational models to explore the dynamics of technology adoption, particularly in the automobile industry. Working with General Motors, we studied the adoption of new powertrain technologies, ranging from internal combustion innovations to hybrid, electric, and hydrogen fuel cell propulsion systems. Working with Accenture, we have also considered likely scenarios for adoption of autonomous vehicle technologies, including the impacts of adoption on the auto insurance and auto finance industries.

These efforts have involved extensive interactions with subject matter experts. For example, the study of the impact of autonomous vehicles on the insurance industry involved over 60 meetings with industry experts from Accenture and their clients. The model served as a means for joint exploration of the problem space. Thus, as elaborated elsewhere in this book (Davis 2019), most of the early discussions focused on the extent to which all relevant phenomena were included and represented appropriately. Discussions also addressed identifying data sets to parameterize representations.

As models matured and provided meaningful predictions, discussions shifted to the validity of predictions and the extent to which unexpected results could be explained. The overall goal was to create a forecast of the impacts of technology adoption with which everyone was comfortable. For example, one particular assessment was discussed in depth, namely, that electric vehicles will not significantly benefit the environment if the electricity for charging the vehicles is produced in coal-fired power plants.

To a great extent these applications were model-based discussions laced with available although often uncertain relevant evidence. The model was a means rather than an end. The intense discussions surrounding the development and use of the models served to raise issues of attributes and structure and eventually facilitated reaching consensus on the scenarios of interest to the companies involved.

Validating Models

As noted in the chapter's Introduction, the models summarized here were developed to address *what-if* questions. The possible futures we were considering inherently did not exist. Further, we were usually predicting years into the future. Hence, empirical validation of our predictions did not make sense.

This was further complicated by our predictions of probability distributions rather than just expected values. In several cases, we were predicting probability distributions for several scenarios. How can predictions of sets of probability distributions be validated?

A limited type of validation can be conducted by predicting the performance of the status quo system. If the models fail this test, something is wrong somewhere. However, full empirical validation is inherently not possible. Thus, the predictive validity of the models will remain in question. How can we really know that our predictions will come to be true?

The simple answer is, "We cannot." Nevertheless, this is not a fatal flaw if the goal is insights rather than predictions. If the goal is to understand why ideas might work or not – might be good ideas or bad ideas – then the models can be used to provide prediction-based insights. We can learn what *might* happen and the conditions under which these possibilities are likely to happen.[2] We might then try to influence these conditions.

Experience has shown that this is a good way to quickly discard bad ideas and to refine good ideas for subsequent empirical evaluation. We can determine

---

2 This experience is precisely what is discussed in Davis (2019) of this volume as exploration with coarse prediction (see also Davis et al. 2018). As in that chapter, we were going beyond what-if questions to ask "under what conditions?" questions.

why seemingly good ideas are actually bad. We can also learn the conditions under which apparently good ideas are likely to succeed. Such insights are the hallmarks of creative and successful problem-solving.

Another approach to validity focuses on the decision-makers. Do they feel the model helps them make better-informed decisions? Is their gut level of comfort higher? As a model builder, I seek feedback on the extent to which the model influenced their decisions. I also look at the decisions made and resources involved. For example, the use of TIA resulted in decisions involving many billions of dollars (Rouse and Boff 2004).

Modeling Lessons Learned

This chapter has summarized my experiences developing models to address a rather broad range of problems over the past 50 years. What have I learned? In this section, lessons learned about modeling are summarized. The next section considers lessons learned about problem-solving.

It is useful to summarize modeling lessons learned in the context of developing and deploying policy flight simulators (Rouse 2014, 2015). There are eight tasks associated with creating and using policy flight simulators:

- Agreeing on objectives – the questions – for which the simulator will be constructed.
- Formulating the multilevel model – the engine for the simulator – including alternative representations and approaches to parameterization.
- Designing a human–computer interface that includes rich visualizations and associated controls for specifying scenarios.
- Iteratively developing, testing, and debugging, including identifying faulty thinking in formulating the model.
- Interactively exploring the impacts of ranges of parameters and consequences of various scenarios.
- Agreeing on rules for eliminating solutions that do not make sense for one or more stakeholders.
- Defining the parameter surfaces of interest and *production* runs to map these surfaces.
- Agreeing on feasible solutions and the relative merits and benefits of each feasible solution.

The discussions associated with performing the above tasks tend to be quite rich. Initial interactions focus on agreeing on objectives, which includes output measures of interest, including units of measure. This often unearths differing perspectives among stakeholders. I have often encountered people who have known each other for years but did not realize they had conflicting perspectives. The modeling process tends to unearth such differences.

Attention then moves to discussions of the phenomena affecting the measures of interest, including relationships among phenomena. Component models are needed for these phenomena, and agreeing on suitable vetted, and hopefully off-the-shelf, models occurs at this time. Also of great importance are uncertainties associated with these phenomena, including both structural and parametric uncertainties.

As computational versions of models are developed and demonstrated, discussions center on the extent to which model responses are aligned with expectations. A typical overall goal is to computationally redesign the enterprise. However, the initial goal is usually to replicate the existing organization to see if the model predicts the results actually being currently achieved.

Once attention shifts to redesign, discussion inevitably shifts to the question of how to validate the model's predictions. As these predictions inherently concern organizational systems that do not yet exist, validation is limited to discussing the believability of the insights emerging from debates about the nature and causes of model outputs. In some cases, deficiencies of the models will be uncovered, but occasionally unexpected higher-order and unintended consequences make complete sense and become issues of serious discussion.

Model-based policy flight simulators are often used to explore a wide range of ideas. It is quite common for one or more stakeholders to have bright ideas that have substantially negative consequences. People typically tee up many alternatives for organizational designs, interactively explore their consequences, and develop criteria for the goodness of an idea. A common criterion is that no major stakeholder can lose in a substantial way. For the Emory simulator, this rule pared the feasible set from hundreds of thousands of configurations to a few hundred.

Quite often, people discover the key variables most affecting the measures of primary interest. They then can use the simulator in a *production mode*, without the graphical user interface, to rapidly simulate ranges of variables to produce surface plots. Discussions of such surface plots, as well as other results, provide the basis for agreeing on pilot tests of apparently good ideas. Such tests are used to empirically confirm the simulator's predictions, much as flight tests are used to confirm that an aircraft's performance is similar to that predicted when the plane was designed *in silico.*

Policy flight simulators serve as boundary spanning mechanisms, across domains and disciplines and beyond initial problem formulations, which are all too often more tightly bounded than warranted. Such boundary spanning results in arguments among stakeholders being externalized. The alternative perspectives are represented by the assumptions underlying and the elements that compose the graphically depicted model projected on the large screen. The debate then focuses on the screen rather than being an argument between two or more people across a table.

These observations are well aligned with my findings from a study involving 100 planning teams and over 2000 participants using one or more of the tools from the *Advisor Series* (Rouse 1998b). Workshop participants were asked what they sought from computer-based tools for planning and design. Here is a summary of their responses:

- Teams want a clear and straightforward process to guide their decisions and discussions, with a clear mandate to depart from this process whenever they choose.
- Teams want capture of information compiled, decisions made, and linkages between these inputs and outputs so that they can communicate and justify their decisions, as well as reconstruct decision processes.
- Teams want computer-aided facilitation of group processes via management of the nominal decision-making process using computer-based tools and large screen displays.
- Teams want tools that digest the information that they input, see patterns or trends, and then provide advice or guidance that the group perceives they would not have thought of without the tools.

Policy flight simulators can provide the *engine* that drives these capabilities, but greater advisory capabilities are needed to fully satisfy all these objectives. A good example comes from the networked version of PPA that enabled teams to work remotely and asynchronously. When asked what they liked best about this version, users did not comment on the modeling capabilities. Instead they expressed great appreciation for a feature that was unique to this version.

The networked PPA kept *minutes* of every user transaction with the tool, including every proposed change and implemented change. Given that the tool knew each user and their issues of interest, we created a function called What's Happened Since I Was Last Here? When users clicked on this option, they were provided with an explanation of how their issues of interest were addressed while they were away. They found this invaluable.

A significant element of this value was due to being able to trace how decisions were made, what assumptions were agreed upon in the process, and why trade-offs were resolved in particular ways. In one instance, PPA was used to plan seven generations of microprocessors, with each new generation beginning by reviewing the PPA analyses of previous generations. The minutes provided the rationale for decisions by people not involved in the new generation and perhaps no longer with the company.

The obvious conclusion from these modeling lessons learned is that people want an environment that helps them address and solve their problems of interest. Computational models coupled to interactive visualizations provide some of this support. However, facilitation – human or otherwise – and capturing of *minutes* are also crucial elements of this support. They really like it when the support system surprises them with suggestions that, upon careful examination, are really good ideas.

Observations on Problem-Solving

Models are intended to support problem-solving. The models developed are means to the ends of solving problems. This chapter has summarized observations of a large number of endeavors in terms of problems addressed and how the problem-solving teams functioned. The observations can be clustered into starting assumptions, framing problems, and implementing solutions.

Starting Assumptions

Who are the stakeholders in the problem of interest and its solution? It is essential that one identify key stakeholders and their interests. All critical stakeholders need to be aligned in the sense that impacts on their interests are understood. The consequences of not understanding these impacts can undermine any chance of consensus.

Look at problems and solutions from the perspectives of stakeholders. How are they likely to be thinking? It is crucial to understand stakeholders' costs of change. One should consider providing stakeholders the necessary services to enable change. These can be seen as elements of the overall solution. Thus, the *solution* may be much larger than originally envisioned.

Articulate and validate assumptions. Significant risks can result when there are unrecognized assumptions. It is important to validate assumptions before deciding to invest in solutions suggested by models. This can sometimes be difficult when key stakeholders *know* what is best (Rouse 1998a).

Understand how other stakeholders may act. The effectiveness of a strategy is strongly affected by competitors. Your solution may cause them to react in ways that decreases the attractiveness of the solution. Using one or more team members to play competitors' roles can often facilitate this.

Framing Problems

Define value carefully. Translating invention to innovation requires clear value propositions, as illustrated by earlier discussions of new product planning and technology strategy. In both cases, value needed to be framed from the perspective of the marketplace, not the inventors. Markets do not see their main role as providing money to keep inventors happy.

Think in terms of both the current business and possible future businesses. Current success provides options for future success but perhaps with different configurations for different markets. The TIA was often used to explore how current products and customers provide options for new products and customers.

Consider possibilities for customizing solutions for different customers and constituencies. The healthcare models discussed earlier required stratification and tailoring of processes to varying health needs. This was critical to the economic viability of these health offerings. Henry Ford, almost 100 years ago, was the last person to believe that everyone wanted exactly the same automobile.

Access and integrate available data sets on customers, competitors, technologies, etc. The healthcare examples showed how data integration increases confidence. Great insights can be gained by mining available data sets, including internal sets, publicly available sets, and purchasable sets.

Plans should include strategies for dealing with legacies. The status quo can be an enormous constraint because it is known, paid for, and in place. It is often necessary to have plans to get legacies *off the books*. Discarding or liquidating assets for which one paid dearly can be painful (Jensen 2000). Nevertheless, it can be exactly the right thing to do, particularly during times of creative destruction (Schumpeter 1976).

Implementing Solutions

It can be great fun to pursue market and/or technology opportunities. Innovators can earn high payoffs, albeit with high risks, as was often the case for PPA and TIA projects. The key is to have the human and financial resources to support and sustain the commitment to innovate.

In stark comparison, crises are not fun. I have often seen the high costs and substantial consequences of delaying change. Typically, the status quo has devoured most available human and financial resources. When change is under-resourced, failure is quite common.

The existing enterprise can hold change back. Several of our engagements involved product companies considering adding service lines of business. The difficulty of changing business models should be carefully considered. New business opportunities may be very attractive, but if success requires substantially new business models, one should assess the enterprise's abilities to make the required changes.

Change should involve stopping as well as starting things. Stopping things will likely disappoint one or more stakeholders. Our model-based efforts to explore future scenarios for research universities (Rouse 2016) illustrated the difficulty of keeping everybody supportive. The consequence is that the status quo dominates, especially when senior management team members were recruited to be stewards of the status quo.

Conclusions

We use models to answer questions and solve problems. Often these problems involve designing solutions in terms of physical form, functional capabilities, and policies intended to enable, incentivize, or inhibit particular behaviors. The key point is that models are intended to support problem-solving.

Initially our models are limited to visualizations, perhaps just sketches. These visualizations may evolve to become more elaborate and perhaps interactive. They enable exploration of connections among entities and how relationships among entities work. Such visualizations are models. They express the problem solver's perceptions of what phenomena matter, how they interact, and key trade-offs.

Sometimes a good visualization is all that is needed. The problem-solving group's discussion and exploration of the visualization can lead to a conclusion on how to proceed. In other situations, deeper explorations are needed. These explorations may involve more formal representations of phenomena and relationships. Deep computation may be warranted but perhaps used sparingly.

The discussions and explorations usually lead to creative suggestions for possible courses of action. All the creativity comes from the group of problem solvers, not computers. In other words, policy flight simulators seldom fly themselves. Instead, computers provide the means to explore the implications of seemingly good ideas. Bad ideas are rejected and good ideas are refined, perhaps for empirical evaluation.

This whole process results in many graphs and perhaps surface plots. Models may be revised and extended. The most powerful impact, however, is that the problem-solving group develops a shared mental model of what effects what, what really matters, and what trade-offs are crucial (Rouse 2007, 2015). Members of the large number of groups with whom I have worked have repeatedly told me that the resulting shared mental model was far more powerful than any of the graphs or plots produced.

It is not unusual for decision-makers to find that agreed upon valid results are not acceptable or viable. In other words, the solution is arguably the right thing to do, but the changes implied are not acceptable, and the investments required are not seen as viable. Such assessments are seldom capricious. A particular solution, while valid, may simply not work in their culture.

Nevertheless, I have found that evidence-based decision-making is not a natural act for many executives and senior managers. They are used to relying on intuition, which makes complete sense for frequent and familiar situations (Klein 2003). However, the infrequent and unfamiliar situation they now face merits a more analytical approach (Rouse 1998a, 2001, 2006, 2007).

There are three notions that can ease them into such an approach. First, it is important that they realize the supporting roles that models play. They do not give you the answer; rather, they provide a means for exploring alternative answers. Second, interactive visualizations are central to such explorations. Finally, hands-on explorations by the decision-makers themselves can make a huge difference. This transforms customers of models into owners of models. Interestingly, the models will have paid their way if the customers then make good decisions, even if the customers explain the decisions in intuitive terms and barely mention the models, much less claim them as a trusted source of insights.

References

Boer, F.P. (1999). *The Valuation of Technology: Business and Financial Issues in R&D*. New York, NY: Wiley.

Davis, P. (2019, this volume). Lessons on decision aiding for social-behavioral modeling. In: *Social-Behavioral Modeling for Complex Systems* (ed. P.K. Davis, A. O'Mahony and J. Pfautz). Hoboken, NJ: John Wiley & Sons.

Davis, P., O'Mahony, A., Gulden, T.R. et al. (2018). *Priority Challenges for Social-Behavioral Research and Its Modeling*. Santa Monica, CA: RAND Corporation.

Hauser, J.R. and Clausing, D. (1988, May–June). The house of quality. *Harvard Business Review* 3: 63–73.

Jensen, M.C. (2000). *A Theory of the Firm: Governance, Residual Claims, and Organizational Forms*. Cambridge, MA: Harvard University Press.

Keeney, R.L. and Raiffa, H. (1976). *Decisions with Multiple Objectives: Preference and Value Tradeoffs*. New York, NY: Wiley.

Klein, G.A. (2003). *Intuition at Work*. New York, NY: Doubleday Currency.

Rouse, W.B. (1991). *Design for Success: A Human-Centered Approach to Designing Successful Products and Systems*. New York, NY: Wiley.

Rouse, W.B. (1992). *Strategies for Innovation: Creating Successful Products, Systems, and Organizations*. New York, NY: Wiley.

Rouse, W.B. (1993). *Catalysts for Change: Concepts and Principles for Enabling Innovation*. New York, NY: Wiley.

Rouse, W.B. (1996). *Start Where You Are: Matching Your Strategy to Your Marketplace*. San Francisco, CA: Jossey-Bass.

Rouse, W.B. (1998a). *Don't Jump to Solutions: Thirteen Delusions That Undermine Strategic Thinking*. San Francisco, CA: Jossey-Bass.

Rouse, W.B. (1998b). Computer support of collaborative planning. *Journal of the American Society for Information Science* 49 (9): 832–839.

Rouse, W.B. (2001). *Essential Challenges of Strategic Management*. New York, NY: Wiley.

Rouse, W.B. (2004). Embracing the enterprise. *Industrial Engineering* (March): 31–35.

Rouse, W.B. (ed.) (2006). *Enterprise Transformation: Understanding and Enabling Fundamental Change*. New York, NY: Wiley.

Rouse, W.B. (2007). *People and Organizations: Explorations of Human-Centered Design*. New York, NY: Wiley.

Rouse, W.B. (2014). Human interaction with policy flight simulators. *Journal of Applied Ergonomics* 45 (1): 72–77.

Rouse, W.B. (2015). *Modeling and Visualization of Complex Systems and Enterprises: Explorations of Physical, Human, Economic, and Social Phenomena*. New York, NY: Wiley.

Rouse, W.B. (2016). *Universities as Complex Enterprises: How Academia Works, Why It Works These Ways, and Where the University Enterprise is Headed*. New York, NY: Wiley.

Rouse, W.B. and Boff, K.R. (2004). Value-centered R&D organizations: ten principles for characterizing, assessing, and managing value. *Systems Engineering* 7 (2): 167–185.

Rouse, W.B. and Cortese, D.A. (eds.) (2010). *Engineering the System of Healthcare Delivery*. Amsterdam: IOS Press.

Rouse, W.B. and Howard, C.W. (1993). Software tools for supporting planning. *Industrial Engineering* 25 (6): 51–53.

Rouse, W.B. and Rouse, S.H. (1980). *Management of Library Networks: Policy Analysis, Implementation, and Control*. New York, NY: Wiley.

Rouse, W.B. and Serban, N. (2014). *Understanding and Managing the Complexity of Healthcare*. Cambridge, MA: MIT Press.

Rouse, W.B., Howard, C.W., Carns, W.E., and Prendergast, E.J. (2000). Technology investment advisor: an options-based approach to technology strategy. *Information Knowledge Systems Management* 2 (1): 63–81.

Schumpeter, J.A. (1976). *Capitalism, Socialism and Democracy*. New York, NY: Harper & Row.

35

A Complex Systems Approach for Understanding the Effect of Policy and Management Interventions on Health System Performance

Jason Thompson[1], Rod McClure[2], and Andrea de Silva[3]

[1] *Transport, Health and Urban Design Research Hub, Melbourne School of Design, University of Melbourne, Parkville, VIC 3010, Australia*
[2] *Faculty of Medicine and Health, School of Rural Health, University of New England, Armidale, NSW 2351, Australia*
[3] *Department of Epidemiology and Preventive Medicine, Alfred Hospital, Monash University, Clayton, VIC 3800, Australia*

Introduction

Traditionally, relationships in medical and health research are understood through the hypothetico-deductive framework. First, hypotheses regarding observed or expected patterns between sets of people differentiated by group features are proposed. Hypotheses are then tested through comparison of features (e.g. gender, age, intervention, or treatment groups) on differences in outcome variables of interest (e.g. depression, well-being, recovery, etc.). Over time, gathered evidence is then used to judge whether repeated observed associations among independent outcome variables are causal.

Hugely successful for understanding simple relationships, health researchers have relied heavily on this structure, of which the patient/problem/population–intervention–comparison–outcome (PICO) framework (U.S. National Library of Medicine 2018) is perhaps the most common example. The widespread adoption of PICO and related frameworks for framing *good* research questions is further facilitated by the widespread availability and training of researchers in statistical and methodological processes that complement such research designs. Increasingly, however, there is a growing understanding that the *real world* of health and rehabilitation systems may not bend as easily to the straightforward research methods that have largely dominated the field (Collie et al. 2018).

In the field of post-injury rehabilitation, researchers readily acknowledge the benefits of longitudinal vs. cross-sectional study designs (e.g. Ponsford et al. 2014). However, limitations associated with linear rather than dynamic modeling processes are less well appreciated (Barton 1994; Kay and Schneider 1994). This is true even within some of the more advanced theoretical and longitudinal literature that attempts to clarify complex associations between predictors and outcomes (O'Donnell et al. 2010, 2013; Thompson et al. 2014b; Murgatroyd et al. 2017). With few exceptions (Hirsh et al. 2007; Iezzi et al. 2007; Liedl et al. 2010), do studies in the field of traumatic injury rehabilitation deal with the potential influence of feedback mechanisms in their designs or discussions.

For instance, in considering the association between levels of depression and return to work among a group of injured people, traditional methodologies conceptualize the relationship between depression and return to work as unidirectional rather than a potentially reinforcing loop (Montgomery et al. 1999; Richmond et al. 2009; Carriere et al. 2015). Despite this approach, which is clearly at odds with the lived experience of patients and rehabilitation providers, policy and practice recommendations are often consequently made from an evidence base built on studies that have investigated this relationship on the basis that their original questions meet criteria for *good* research questions as judged within a PICO framework. As a result, potential feedbacks are ignored, are assumed to not exist, or are mentioned in the limitation section of academic papers only (e.g. Thompson et al. 2014c). In reality, variables or individual factors under study within complex dynamic systems such as health may be at once both independent *and* dependent variables; their isolation and independence contrived more so to benefit internal validity than external.

The process of post-injury physical and mental health rehabilitation is particularly complex, and the trajectory of sometimes lengthy individual recovery is difficult to predict. Rehabilitation occurs amid a dynamic network of relationships among patients, healthcare providers, compensation systems, legal professionals, legislators, rehabilitation coordinators, spouses, workplaces, families, and a host of other stakeholders. Recovery is highly dependent upon initial accident circumstances, individual demographic characteristics, and patients' understanding of their injury in the context of the remainder of their lives (Diller 2005; Thompson et al. 2014a,b). This degree of complexity and the practical difficulties in studying injured populations mean that studies using more linear methodologies to represent individual recovery processes and the interaction of individuals with systems limit their ability to provide effective decision support in *live* settings. System managers acknowledge this, but at present, they have little alternative. They have not been provided with a set of alternative tools that can overcome deficiencies in the traditional well-worn approach.

This may be a counterproductive situation for clinicians and health system managers that rely on such evidence as a basis of decision-making and who must deal with members of populations at hand rather than their more sanitized, trial-ready counterparts (Morin-Ben Abdallah et al. 2016). Whether this is part of the reason for public health research's relatively low levels of influence on public health decision-making (Orton et al. 2011) is unclear, but it is hard to see how it assists. Public health and injury rehabilitation systems do not have the luxury of filtering incoming patient cohorts. They cannot stop treating people who drop out (Thompson et al. 2011), disregard unusual cases or statistical outliers from inclusion or analysis, discount the presence of comorbid disorders, restrict age ranges and injury types, or ignore any other factors often associated with exclusion criteria demanded of typical intervention trial participants (Tuszynski et al. 2007). They must deal with all patients who present, regardless of circumstances and the impact that each may have on overall health system performance.

Understanding Health System Performance

There is argument to suggest that no accepted definition of health systems exists (Bowling 2014); however, this has not restricted effort expended during the past two decades on attempting to create common frameworks for assessing health system performance. Principally, these attempts have been centered on understanding the boundaries of health system responsibilities, the functions they perform, and how these functions translate into achievement of health system goals and outcomes (Murray and Evans 2006; Duckett and Willcox 2015).

Murray and Frenk (2000) describe the functions of health systems as comprising stewardship, resource creation, financing, and service provision. Core goals of the health system are then defined as improved performance across three areas: responsiveness to community expectations, fairness in financial contributions, and, ultimately, overall population health (see Figure 35.1). Adopting this framework, the World Health Organization (WHO) (2000) considers that combined high levels of performance across these three elements is indicative of a well-functioning healthcare system.

Improving health system performance requires an ability to understand the effect of changing external states on system revenue and expenses and to respond appropriately through adjustment of available management and policy levers (Gray et al. 2003). For example, post-crash injury rehabilitation systems must have confidence in predicting levels of road trauma produced by the transport system and the costs of trauma care, rehabilitation, and common law litigation associated with new and existing patients. Against this, expected revenue gathered from taxation, fees, insurance premiums, or investment returns from pooled resources must be set (Duckett and Willcox 2015; Motor Accident Insurance Commission 2016). These efforts to balance competing priorities are not always compatible, and, unless completely independent, each adjustment to system management or policy settings designed to solve current problems sets in motion a new set of issues that may need to be addressed

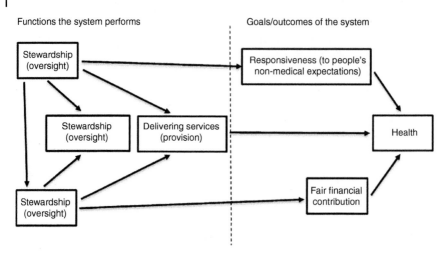

Figure 35.1 Conceptual model of health system performance. Source: As defined by Murray and Frenk (2000) and adapted by the World Health Organization (2000).

in the future (Sterman 2006). Therefore, rather than being studied as entities made up of a series of independent factors and relationships, health systems and their subcomponents may benefit from being investigated as they are – complex systems comprising dynamic, interdependent relationships between factors operating in an environment of fixed resources (Rutter et al. 2017). Health systems lend themselves to study under new methodologies that can account for heterogeneous actors, interdependent factors, and feedback loops not possible using traditional linear techniques. However, while there have been repeated calls for public health scientists to embrace complex systems methods (Pope and Mays 1993; Beresford 2010; Rutter et al. 2017; Tracy et al. 2018), relatively few practical examples exist demonstrating the relative benefits and limitations of such approaches applied to real-world settings.

Agent-based modeling (ABM) is a technique that shows promise in overcoming theoretical and practical difficulties associated with researching and understanding complex socio-technical systems such as health systems (Kanagarajah et al. 2010). To explore the utility of using ABM to trial policy settings within a simulated real-world health system, we conducted an experiment within a state-owned, personal injury motor-vehicle insurance and transport injury rehabilitation provider located in Australia (the health system). In conjunction with management staff, we undertook development of an ABM that reflected the function of the organization at a macro level, incorporating policy and operational mechanisms with the potential to influence overall system performance. The goal of the project was to trial various policy and resource management scenarios to estimate effects on overall health system performance (Nikolic et al. 2013) as conceptualized under the WHO framework (Murray and

Frenk 2000; World Health Organization 2000) and embodied by the system's corporate performance indicators, which included:

1. Patient satisfaction (rated as satisfaction with the system on a scale out of 10).
2. Patient physical and mental health recovery outcomes (rated through self-reported scores on the SF-12 V2 physical and mental health scales (Ware et al. 1996).
3. Duration of time patients had been involved in the health system since their crash.
4. Scheme forward liabilities.
5. Scheme operational costs.

Method

The modeled health system was a stand-alone, state-owned personal injury insurance, compensation, and rehabilitation system that provided support for people following injury sustained in a motor-vehicle crash. The boundaries of the system were considered to lie at the point at which the health system could not influence either the behavior of the scheme or the characteristics of individuals. For this reason, distributions of incoming individual patient characteristics such as gender, injury severity, attributions of responsibility for crashes (Thompson et al. 2014a), social support (Prang et al. 2015), presence of preexisting injuries, illnesses or mental health conditions (Berecki-Gisolf et al. 2015), and a wait-list effect (Wideman et al. 2012) (where patient health improved at a mean rate over time) were held constant over time within the model. The model, itself, was represented by an abstract two-dimensional space within which agents interacted with one another and was built using NetLogo 6.0 (Wilensky 2016). Agents within the model consisted of patients, rehabilitation coordinators, physical health services, mental health services, and plaintiff solicitors. All agents were modeled at the individual level – i.e. there were no *aggregate* services. Patients entered the system at a random location in the environment at a rate of 12 new patients per time step.

Patients

Each injured patient had a number of dynamic states that varied over the duration of their existence in the system, including:

1. Level of mental health.
2. Level of physical health.
3. Level of satisfaction with the health system.
4. Length of time spent in the health system (duration of recovery).

5. Costs incurred through accessing health services.
6. Length of time elapsed since the patient had last visited a physical or mental health service.
7. Current number of pre-approved services available to each client they could use before being required to make a new request.

Without treatment, patients improved their levels of mental health and physical health at an approximate rate consistent with a wait-list effect. They also decreased their satisfaction with the scheme at a rate consistent with previous research demonstrating a reduction in levels of satisfaction with compensation schemes with longer duration recoveries (Thompson et al. 2015).

Rehabilitation Coordinators

Rehabilitation coordinators responded to requests for physical and mental health treatment from patients. When contacted by a patient, the rehabilitation coordinator could pre-approve (with a manipulatable likelihood of between 0% and 100%) a set number of treatment services per request, initialized at 6. Patients could only receive approval of services from rehabilitation coordinators when the rehabilitation coordinator was not occupied dealing with other patients' requests, reflecting capacity constraints within the system. If a rehabilitation coordinator was busy, the patient continued to seek service from other random rehabilitation coordinators until their request had been satisfied.

Physical and Mental Health Treatment Services

Treatment services could only provide treatment to one patient at a time, reflecting capacity constraints within the system. To replicate diminishing returns of treatment services as patients neared full health, treatment increased the physical and/or mental health status of patients by a rate per interaction equal to the square root of the difference between the patient's current physical or mental health score and patients' maximum possible scores within either their physical or mental health domain (70). To recognize the relationship between physical and mental health, physical or mental health scores were then multiplied by the patient's score in the alternative domain divided by the normative population health mean (50). Importantly, both physical and mental health service providers could also refuse to provide treatment to individual patients if they did not want to deal with a *compensable* case, a common phenomenon whereby general practitioners or other medical providers refuse service to compensable patients due to perceived administrative burden and payment delays (Brijnath et al. 2016).

Plaintiff Solicitors

Plaintiff solicitors had goals of attracting injured patients as clients. In order to reflect poorer levels of recovery among patients who initiated litigation, solicitors interacted with patients to reduce patients' satisfaction with the health system, mental health, and physical health by 5% on each interaction (Harris et al. 2008). The overall number of solicitors in the model increased in response to a combination of poor mean patient mental health and low levels of satisfaction with the system. This drove an increased market for services, drawing more plaintiff solicitors into the system (Fitzharris et al. 2013).

Model Narrative

Seeking Treatment Service Approval from the Health System

Injured patients entered the modeled system at an initial rate of 12 per time step. Patients' immediate post-injury physical and mental health was randomly allocated around a normally distributed mean of 35 points with a standard deviation of 5 points. The scoring system was designed to reflect the system's use of the SF-12 for measuring the health of their injured population, a 12-item general health questionnaire routinely used by the system to assess self-reported physical and mental health functioning (Ware et al. 1996; Victorian Transport Accident Commission 2012).

When patients entered the system, they searched within the model space for a rehabilitation coordinator at a random location to approve health services for them if their current levels of mental or physical health were below a treatment approval threshold set by health system policy (initially <45 points). When patients found an available rehabilitation coordinator, the patient could then request treatment services. The rehabilitation coordinator then decided whether to approve the patients' request (initially 90% of treatment requests were approved). Approval of services assumed a satisfaction increase of 1%, while denial of services resulted in a decrease in satisfaction of 1%.

If the patient's treatment request was approved, the patient was provided with six service *credits* to use before needing to return to request further services. If the rehabilitation coordinator was either occupied with other patients or the patient's initial request for services was denied, the patient sought services from another random rehabilitation coordinator in the system, delaying access to treatment.

Seeking Healthcare Services

Once patients had either mental health or physical health services approved by the rehabilitation coordinators, they then searched for a suitable mental or physical healthcare provider and requested treatment. If the health service was not already dealing with requests from other patients, the healthcare provider then either accepted or refused to treat the patient based on a dynamic likelihood of their propensity to accept compensable (i.e. insured by state-run injury compensation system) patients (initialized at 80% acceptance). If the request to treat was accepted, treatment occurred, and, dependent upon whether the treatment was mental or physical in nature, the patient's health improved in that respective domain by the mean effectiveness of physical or mental health treatment. The extent of improvement was rapid in the earlier stages of recovery and slowed toward the final stages as patients reached full health.

Each time a patient used a healthcare service, the allocated number of *pre-approved* services was reduced by one unit until their service credits were exhausted. If they had not yet recovered, patients again sought approval from rehabilitation coordinators for additional treatment services.

In addition to seeking healthcare services through the approval mechanisms of the post-crash response system, the patient's movement within the system exposed them to contact with other patients and plaintiff solicitors, with whom they exchanged information. When patients met other very unsatisfied, *disgruntled* patients in the scheme (with satisfaction scores <5), a small amount of the satisfaction of disgruntled patients was then transferred to other patients, reducing their level of satisfaction by 10%. While the direction of such changes to levels of satisfaction among patients is consistent with both internal organizational and independent research evidences related to the system (Thompson et al. 2015), the magnitude of effect could not realistically be calibrated. Instead, it was tuned during model verification stages to reasonable assumptions of system managers' understanding of effects.

Exiting the Health System

Patients existed within the health system until both their physical and mental health scores reached a threshold level above which they were considered to have recovered and were therefore ineligible to further assistance. Patients' *activity status* was also recorded as whether they had accessed any health services in the prior 100-day period. Patients could become *inactive* if they met thresholds for treatment but either had failed to have services approved by the system or had failed to find a treating healthcare provider in this time. The significance of inactive patients to the scheme is that this group consists of patients who are still eligible to access services, but they have either not done so or are unable to do so. Therefore, they represented an ongoing financial and service liability to the system that could affect future performance.

Policy Scenario Simulation

To test the impact of various policy and management interventions on the performance of the health system, the model was set to operate for a series of 500 steps before being *shocked* by 1 of 9 policy or management interventions. Each policy intervention was designed to reflect realistic options available to the health system to improve performance. Mean effects of each intervention on scheme performance were then monitored over 30 iterations for a further 1000 time steps (roughly equivalent to days) post-implementation to observe average short-, medium-, and longer-term effects. Policy settings tested were:

1. Improving the effectiveness and quality of services available to patients through payment for *premium* healthcare where effectiveness increased alongside investment.
2. Reducing approval rate of services by rehabilitation coordinators (i.e. reducing the approval rate of requested services by rehabilitation coordinators from 90% to 70%).
3. Early intervention (i.e. rehabilitation coordinators actively sought out patients with claim durations of <30 days in order to provide access to services sooner).
4. Improving patient access to health services (i.e. increasing the number of existing services that accepted compensable patients from 80% to 100%).
5. Improving road safety (i.e. reducing incoming patients through investment in increased safety measures resulting in 10% reduced road trauma).
6. Improving availability of rehabilitation coordinators (i.e. increasing numbers of rehabilitation coordination staff by 10% from 300 to 330).
7. Increasing pre-approval rates (i.e. doubling the number of services that could be pre-approved for patients from 6 to 12 before being required to return to rehabilitation coordinators to request further treatment).
8. Reducing eligibility of services (i.e. increasing the threshold of injury severity for patients being eligible to receive services from the insurer).
9. Do nothing (i.e. no intervention).

While the extent of change made within each of these domains was somewhat arbitrary, the nature of changes reflected the real-world experience of system managers participating in the model development.

Results

Results of policy and management interventions on individual health system performance indicators are plotted in Figures 35.2–35.8. Figure 35.9 then shows

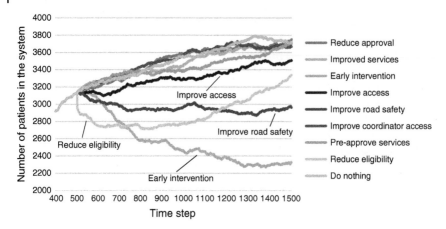

Figure 35.2 Number of patients in the simulated health system from time step 250 to 1500 under each policy scenario.

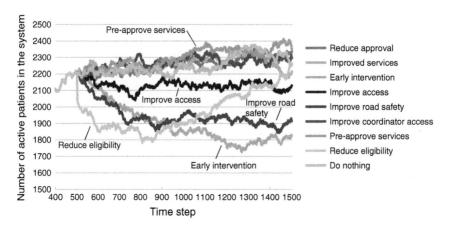

Figure 35.3 Number of *active* patients that had received services in the previous 100 time steps in the simulated health system from time step 250 to 1500 under each policy scenario.

a combined ranking of each policy setting against criteria of overall health system performance (Murray and Frenk 2000). Given that the health system used as the basis for the model was a state-owned monopoly pooled risk personal injury insurer, fair financial contribution was judged as the forward liability estimates of the system as these costs ultimately flowed through to the public in the form of compulsory insurance premiums (NSW Government 2016).

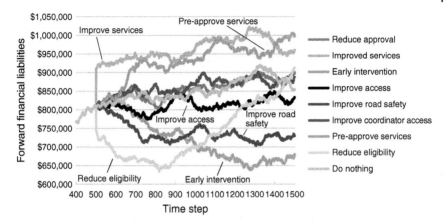

Figure 35.4 Estimated forward liabilities of the simulated health system from time step 250 to 1500 under each policy scenario.

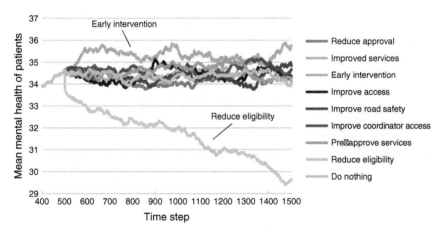

Figure 35.5 Mean mental health scores of patients within the simulated health system from time step 250 to 1500 under each policy scenario.

Figure 35.2 demonstrates that the most successful strategies for reducing overall numbers of patients in the health system were *Early intervention, Improving road safety, Reducing eligibility,* and *Improving access* to existing latent health services in the community. However, the trends for these strategies were not consistent over time. *Reducing eligibility* appeared most successful immediately following implementation (i.e. time step 600) but was then surpassed by *Early intervention* and eventually by *Improving road safety. Early intervention fast-tracked* services to injured patients, whereas

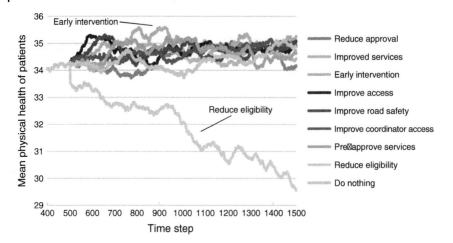

Figure 35.6 Mean physical health scores of patients within the simulated health system from time step 250 to 1500 under each policy scenario.

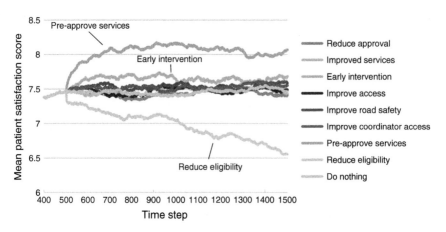

Figure 35.7 Mean satisfaction scores of patients within the simulated health system from time step 250 to 1500 under each policy scenario.

Road safety reduced the numbers of entrants to the system through primary prevention activities. By contrast, *Reducing eligibility* removed the ability of people to access the scheme by increasing thresholds for entry.

While the trends for total patient count and *active* patients (Figure 35.3) were similar, analysis revealed that, in particular, *Pre-approval of services* resulted in a higher *active* patient count. This was due to the greater accessibility of treatment services for patients under this condition, resulting in less delays caused

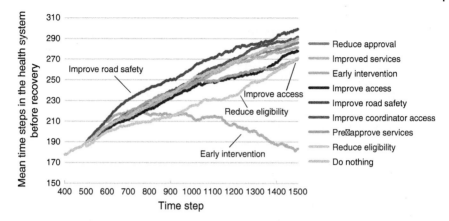

Figure 35.8 Mean recovery duration among patients in the simulated health system from time step 250 to 1500 under each policy scenario.

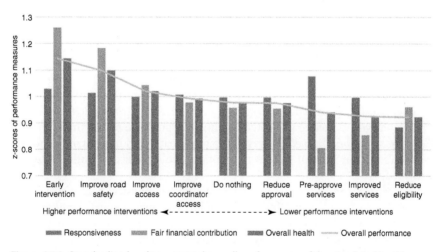

Figure 35.9 Standardized and aggregated overall performance of the simulated health system at time step 1500. Simulation conclusion across elements of responsiveness, fair financial contribution, and overall health under each policy scenario.

by time spent seeking approvals and capacity constraints among rehabilitation coordinators in remaining policy scenarios.

Figure 35.4 shows the estimated forward liabilities of the health system based on calculation of the total number and cost of services used by patients in the previous 100 days. Forward liabilities were estimated by summing the product of the total services used by active clients in the prior 100 days by the cost associated with those services (investment is physical and mental health services increased per unit cost) plus nominal hospital costs that reduced as patents' mental and physical health improved. Inactive patients were given a nominal liability sum on the assumption they had exited the system with little likelihood of return. Following the principle of insight rather than numbers (Nikolic et al. 2013), figures used for this estimation should be considered as monetary *units* only.

Results showed that *Reducing eligibility* to the scheme for patients demonstrated immediate financial benefit for the health system following policy implementation. However, as time elapsed, the advantage of this strategy was then surpassed by all but two other strategies: *Pre-approval of services* and investment in *Improved services*. The potential costliness of investment in improved services appeared to drive the result observed under this scenario. To achieve better outcomes, the health service paid a 25% premium on top of existing rates, which, based on a dollar-for-dollar estimation, produced a 25% increase in treatment effectiveness per interaction. The *Pre-approval* intervention ultimately became the strategy with the highest potential for increasing liabilities because fewer delays and *bottlenecks* existed for patients to be able to access the services they desired. This resulted in more services being accessed by patients per period and therefore a higher projected forward liability estimate.

The most successful strategies for reducing estimated forward liabilities were *Early intervention* and *Improving road safety*. Although the impact of these strategies took longer to be realized by the system than through *Reducing eligibility*, and even increased projected liabilities for an initial period in the case of *Early intervention*, their effect was ultimately greater and more stable, buoyed by the reduction in total patient numbers in the system that both strategies provided.

Figures 35.5 and 35.6 show the mean mental and physical health scores for patients at each time step. Results highlight three main points. Firstly, *Early intervention* was again the most successful strategy in improving overall patient population health. This was likely due to the improvement in health that *Early intervention* made among those at the earliest, most injured stages of recovery.

The second observation is that most other strategies produced similar overall results, also reflecting observed consistency in real-world systems (Victorian Transport Accident Commission 2016). This highlights that in health systems with fixed entry and exit thresholds, attempting to demonstrate health improvements within a patient population using a cross-sectional design may not be sensitive enough to easily distinguish between the performance of interventions because only those patients who remain *active* within the service (or recently inactive) are surveyed (Thompson et al. 2014c). When measured cross-sectionally, each different cohort therefore continues to reflect a general injured population with a similar mean health status containing upper and lower bounds.

However, due to changes in the boundaries of entry and exit thresholds, *Reducing eligibility* was the exception to observed consistency in results across each intervention, having the most detrimental effect on mental and physical health status. This strategy set in train a series of circumstances and feedback mechanisms that reinforced the ill health of the patient population, seeing it continue to decline over the course of the simulation. By increasing the threshold for accessing services in the health system, mean patient mental and physical health was immediately reduced as all patients who were previously well enough to be eligible for services were excluded from the health system and therefore from analysis. The remaining less healthy group also had lower mean levels of satisfaction with the system (see Figure 35.7) (Thompson et al. 2015), leading them to be more likely to engage with a plaintiff solicitor, which further reinforced their dissatisfaction with the health system and reported ill health (Barsky and Borus 1999; Matsuzawa and Dijkers 2014).

Figure 35.7 shows that by contrast with patients' low levels of mean satisfaction in the *Reducing eligibility* intervention, *Pre-approval of services* and *Early intervention* strategies showed the greatest positive impact on satisfaction. While the effect of *Early intervention* is likely tied to the increase in mean physical and mental health scores observed in Figures 35.5 and 35.6, the effect of *Pre-approval of services* can be understood by the fact that patients were half as likely to experience a disapproval of a requested service, which reduced likelihood of incidents leading to dissatisfaction (i.e. service refusals). Consequently, patients were also more likely to be able to access health services faster, further reducing likelihood of becoming dissatisfied (Thompson et al. 2015) and potentially engaging an attorney to sue the health system.

Figure 35.8 shows the mean recovery durations of patients currently existing within the health system. Initially, strategies that focused on improving health service access reduced mean recovery durations more than *Road safety* and *Early intervention*, which produced the highest mean recovery durations of all strategies in the period immediately after implementation. This somewhat counterintuitive result was because *Road safety* reduced the number of new, *young* injuries into the health system, which pushed the mean toward older, existing patients. Similarly, the result for *Early intervention* was due to this intervention assisting newer patients to recover earlier in their claim, again pushing the mean toward older, existing clients. However, despite initial poorer outcomes, both these strategies ultimately proved successful with *Early intervention* producing the lowest recovery durations and *Road safety* on trend to become the second most successful intervention by simulation end.

In contrast, *Reducing eligibility* was the most successful strategy for reducing mean recovery durations immediately following policy implementation. However, by reducing thresholds for system entry, thresholds for exit were also lowered. By simulation end, mean recovery durations for patients in this intervention were trending steeply upward, indicating no long-term benefit.

To assess the overall impact on health system performance (Murray and Frenk 2000; World Health Organization 2000) of each policy intervention at trial completion, three categories were constructed reflecting (i) responsiveness to patients' non-medical needs, (ii) fair financial contribution, and (iii) overall patient health. For ease of comparison, performance indicators across each scenario were standardized through conversion into z-scores. While *Fair financial contribution* and *Responsiveness* were constructed from single variables (i.e. health system liabilities and satisfaction, respectively), *Overall health* was constructed from the mean z-scores of simulated patients' physical health, mental health, and mean recovery durations. Ranked overall results from each strategy can be seen in Figure 35.9.

Of first note in Figure 35.9 are the extreme *high performing and low performing* strategies of *Early intervention* and *Reducing eligibility*, respectively. *Improve road safety* is ranked second most effective strategy, and *Improve access* to existing health services is ranked third. *Improved services* and *Pre-approval* perform reasonably on *Responsiveness* and *Overall health* but are ultimately ranked lower due to estimated higher financial liabilities associated with their implementation. The three mid-ranked strategies were headed by *Improve coordinator access* (fourth) followed by *Do nothing* (fifth) and *Reduce approval* (sixth).

Discussion

This work produced a model of a select real-world health system using a complex systems approach (ABM) that reflected behavior of the system in response to multiple policy interventions. Results showed that interventions focused on patient health including increasing speed and access to care as well as primary prevention activities (e.g. *Road safety*) produced the highest performing healthcare systems as judged by existing international frameworks (Murray and Frenk 2000; World Health Organization 2000). By contrast, strategies focused on short-term financial performance of the system by reducing patient access to services or increased expenditure on services with less than equivalent return on health outcomes per unit of expenditure (Duckett and Willcox 2015) ultimately performed most poorly. In this regard, the difference in health system performance across strategies when judged in the short term vs. long term was stark.

It is important to acknowledge that these results are of a modeled health system built upon a high-level conceptualization and understanding of system actors and relationships that included a combination of empirical data and reasoned estimates from system managers with whom the model was built. It therefore does not represent comprehensive *truth* in either its outputs or the mechanisms by which they are generated. Further, sensitivity analysis to determine which variables were most influential in driving outcomes was not conducted as the magnitude of reasonable ranges to consider within-variable variation is uncertain. However, the model does represent an approach to investigating the effect of health systems that has been widely called for (e.g. Beresford 2010; Rutter et al. 2017). It represents a working, explicit, replicable, and dynamic representation of a select health system built from both inductive and deductive investigations and codeveloped with health system managers through significant stakeholder engagement. At face value, it produces both sensible outcomes and insight into unusual health system behavior that is not easily predictable from thought experiments alone (Epstein 2008). Thus, it demonstrates significant advantage over static representations of health system performance and appears to meet Box's (1976) criteria of being both wrong *and* useful in that it has the potential to advance and accelerate knowledge of health system design through the device of theory–practice iteration.

Therefore, that this model does not contain all potential variants of the experience of patients, the health system, treatment services, staff, or administrators does not render it invalid. As a tool for policy insight rather than *numbers* (Nikolic et al. 2013), it is an advantage. Very quickly, any system manager familiar with parameters contained within the model could use it to test ideas or combinations of interventions to gain better insight into their potential short- and long-term effects.

> "It may be baloney, but at least it's replicable baloney" – Joshua Epstein, Lecture on the 20th Anniversary of "Generative Social Science: Studies in Agent-Based Computational Modeling," George Mason University, 2016.

The potential capacity of this type of low-investment, complex system modeling to reduce uncertainty or amplify intelligence (Vinge and Euchner 2017) in decision-making is therefore compelling (Milliken 1987). Within real-world health systems, management and policy-makers cannot wait for the completion of inevitably imperfect and impractical system-level randomized control trials prior to acting. Decisions will be taken or forced, regardless, and made based on implicit (i.e. ideas, intuition, and conjecture) or explicit models such as that demonstrated here. Neither provides perfect information; however the explicit computational model at least has the advantage of being transparently built, tested, replicated, challenged, recorded, and iterated (Box 1976; Epstein 2008). This can be an extremely beneficial process, enabling managers and participants to visualize an otherwise abstract problem and better understand the mechanics and incentives contained for them and other actors in a system they may otherwise be unaware of (Rouse 2015; Tracy et al. 2018). It is our contention that in the absence of such participatory modeling processes, implemented models are almost destined to be poorly trusted and therefore unlikely to be adopted in any serious manner.

For example, when testing and verifying health system models, techniques used in our research group are to guide system managers through a series of interactive role-playing *games*. Here, we construct either individual- or group-level competitions where participants act as managers of a fictional simulated health system that bears resemblance to the real one they are familiar with and have assisted to build. Reiterating that the toy model is fictional is important as it removes a degree of defensiveness that can arise if managers feel overly protective of strategies they are wedded to in a real system. Workshops range from those where we ask participants to meet or beat predefined performance targets set by a fictional board of directors to those where groups have been pitted against one another to create the highest performing healthcare system. In large-scale efforts involving up to 40 people at a time and immersive multimedia rooms (see Figure 35.10), we have adapted a version of *musical chairs*, asking paired teams to select the likely highest performing systems from a selection of 80 potential examples and explain to the remainder of the group why they chose their given system. All systems are then set to run simultaneously, enabling their performance to be monitored and compared against other teams' selections. We use these methods to both increase engagement with the model and overcome an issue in that the model is in reality both too simple (i.e. there are always potential elements of behavior in a real system not contained in the model that a participant could highlight in an attempt to discredit the exercise) and potentially too complex to comprehensively explain in a one or two hour workshop (literally billions of potential policy combinations are available within even the simplest of representations as described here). When facilitated effectively, participants quickly move into the roles assigned to them. The hands-on nature of the exercise helps to solidify their understanding of the system dynamics in a manner that would be more difficult than through description alone. Though the enhancement of learning associated with this element of the modeling exercise has not been our discrete focus so far, we plan to more formally assess this in the future.

Conclusions

The pursuit of methodological techniques that embrace complexity and reduce reliance upon methodological and statistical reductionism may provide considerable gains in our understanding of how health systems may be best designed and managed into the future. Though many authors have recognized the need to embrace complex systems methods for investigating and understanding health system design (Rutter et al. 2017), few specific applications

Figure 35.10 Interactive workshop. The workshop was held with system managers focused on identifying elements leading to high performance healthcare systems.

exist demonstrating potential limitations, benefits, and outcomes of such work as demonstrated here.

When developed in conjunction with health system managers, we consider that computational social science, including ABM, has the potential to build on the solid foundation of existing research approaches to answer pressing policy questions that can improve the function of health systems and, in turn, public health. The next step-change in improving management, practice, and outcomes in health and injury rehabilitation system design may not be from traditional observational studies, but from *synthetic* evidence and insight gathered via methods of computational social science.

References

Barsky, A.J. and Borus, J.F. (1999). Functional somatic syndromes. *Annals of Internal Medicine* 130 (11): 910–921. https://doi.org/10.7326/0003-4819-130-11-199906010-00016.

Barton, S. (1994). Chaos, self-organization, and psychology. *American Psychologist* 49 (1): 5.

Berecki-Gisolf, J., Collie, A., Hassani-Mahmooei, B., and McClure, R. (2015). Use of antidepressant medication after road traffic injury. *Injury* 46: https://doi.org/10.1016/j.injury.2015.02.023.

Beresford, M.J. (2010). Medical reductionism: lessons from the great philosophers. *QJM* 103 (9): 721–724. https://doi.org/10.1093/qjmed/hcq057.

Bowling, A. (2014, 2014). *Research Methods in Health. [electronic resource]: Investigating Health and Health Services*. Maidenhead: McGraw-Hill Education.

Box, G.E.P. (1976). Science and statistics. *Journal of the American Statistical Association* 71 (356): 791–799.

Brijnath, B., Mazza, D., Kosny, A. et al. (2016). Is clinician refusal to treat an emerging problem in injury compensation systems? *BMJ Open* 6 (1): e009423.

Carriere, J.S., Thibault, P., and Sullivan, M.J. (2015). The mediating role of recovery expectancies on the relation between depression and return-to-work. *Journal of Occupational Rehabilitation* 25 (2): 348–356.

Collie, A., Newnam, S., Keleher, H. et al. (2018). Recovery within injury compensation schemes: a system mapping study. *Journal of Occupational Rehabilitation* https://doi.org/10.1007/s10926-018-9764-z.

Diller, L. (2005). Pushing the frames of reference in traumatic brain injury rehabilitation. *Archives of Physical Medicine and Rehabilitation* 86 (6): 1075–1080. http://dx.doi.org/10.1016/j.apmr.2004.11.009.

Duckett, S. and Willcox, S. (2015, 2015). *The Australian Health Care System*, 5e. South Melbourne, Victoria: Oxford University Press.

Epstein, J.M. (2008). Why model? *Journal of Artificial Societies and Social Simulation* 11 (4): 12.

Fitzharris, M., Liu, S., Shourie, S., and Collie, A. (2013). Factors associated with common law claims lodged to the Transport Accident Commission. Retrieved from Melbourne, Australia.

Gray, B.H., Gusmano, M.K., and Collins, S.R. (2003). AHCPR and the changing politics of health services research. *Health Affairs* W3. (Supplementary Web Exclusives).

Harris, I., Young, J.M., Rae, H. et al. (2008). Predictors of post-traumatic stress disorder following major trauma. *Australian and New Zealand Journal of Surgery* 78 (7): 583–587.

Hirsh, A.T., George, S.Z., Riley, J.L. III, and Robinson, M.E. (2007). An evaluation of the measurement of pain catastrophizing by the coping strategies questionnaire. *European Journal of Pain* 11 (1): 75–81. https://doi.org/10.1016/j.ejpain.2005.12.010.

Iezzi, T., Duckworth, M.P., Mercer, V., and Vuong, L. (2007). Chronic pain and head injury following motor vehicle collisions: a double whammy or different sides of a coin. *Psychology, Health & Medicine* 12 (2): 197–212. https://doi.org/10.1080/09540120500521244.

Kanagarajah, A.K., Lindsay, P., Miller, A., and Parker, D. (2010). An exploration into the uses of agent-based modeling to improve quality of healthcare. In: *Unifying Themes in Complex Systems*, 471–478. Springer.

Kay, J.J. and Schneider, E. (1994). Embracing complexity: the challenge of the ecosystem approach. In: *Alternatives*, July–August 1994, 32+. Retrieved from http://go.galegroup.com/ps/i.do?id=GALE%7CA15660774&v=2.1&u=monash&it=r&p=AONE&sw=w&asid=038e1c5ef6daa4742cf269cc4e5f99a1.

Liedl, A., O'Donnell, M., Creamer, M. et al. (2010). Support for the mutual maintenance of pain and post-traumatic stress disorder symptoms. *Psychological Medicine* 40 (7): 1215–1223. https://doi.org/10.1017/S0033291709991310.

Matsuzawa, Y. and Dijkers, M. (2014). The experience of litigation after TBI. II: coping with litigation after TBI. *Psychological Injury and Law* 8 (1): 88–93. https://doi.org/10.1007/s12207-014-9212-0.

Milliken, F.J. (1987). Three types of perceived uncertainty about the environment: state, effect, and response uncertainty. *Academy of Management Review* 12 (1): 133–143.

Montgomery, S.M., Cook, D.G., Bartley, M.J., and Wadsworth, M. (1999). Unemployment pre-dates symptoms of depression and anxiety resulting in medical consultation in young men. *International Journal of Epidemiology* 28 (1): 95–100.

Morin-Ben Abdallah, S., Dutilleul, A., Nadon, V. et al. (2016). Quantification of the external validity of randomized controlled trials supporting clinical care guidelines: the case of thromboprophylaxis. *The American Journal of Medicine* 129 (7): 740–745. http://dx.doi.org/10.1016/j.amjmed.2016.02.016.

Motor Accident Insurance Commission (2016). 2015/16 Annual Report. Brisbane: Queensland Government.

Murgatroyd, D., Harris, I.A., Chen, J.S. et al. (2017). Predictors of seeking financial compensation following motor vehicle trauma: inception cohort with moderate to severe musculoskeletal injuries. *BMC Musculoskelet Disord* 18 (1): 177. https://doi.org/10.1186/s12891-017-1535-z.

Murray, C.J. and Evans, D. (2006). *Health Systems Performance Assessment*. Office of Health Economics.

Murray, C.J. and Frenk, J. (2000). A framework for assessing the performance of health systems. *Bulletin of the World Health Organization* 78 (6): 717–731.

Nikolic, I., van Dam, K.H., and Kasmire, J. (2013). Practice. In: *Agent-Based Modelling of Socio-Technical Systems* (ed. K. van Dam, I. Nikolic and Z. Lukszo). Dordrecht: Springer.

NSW Government (2016). On the road to a better CTP scheme. In: *Options for Reforming Green Slip Insurance in NSW*. Sydney: NSW Government.

O'Donnell, M., Creamer, M., McFarlane, A. et al. (2010). Does access to compensation have an impact on recovery outcomes after injury? *Medical Journal of Australia* 192 (6): 328–333.

O'Donnell, M., Varker, T., Holmes, A.C. et al. (2013). Disability after injury: the cumulative burden of physical and mental health. *The Journal of Clinical Psychiatry* 74 (2): e137–e143. https://doi.org/10.4088/JCP.12m08011.

Orton, L., Lloyd-Williams, F., Taylor-Robinson, D. et al. (2011). The use of research evidence in public health decision making processes: systematic review. *PLoS ONE* 6 (7): e21704. https://doi.org/10.1371/journal.pone.0021704.

Ponsford, J.L., Downing, M.G., Olver, J. et al. (2014). Longitudinal follow-up of patients with traumatic brain injury: outcome at two, five, and ten years post-injury. *Journal of Neurotrauma* 31 (1): 64–77.

Pope, C. and Mays, N. (1993). Opening the black box: an encounter in the corridors of health services research. *British Medical Journal* 306 (6873): 315–318. https://doi.org/10.1136/bmj.306.6873.315.

Prang, K.-H., Berecki-Gisolf, J., and Newnam, S. (2015). Recovery from musculoskeletal injury: the role of social support following a transport accident. *Health Qual Life Outcomes* 13 (1): 97. https://doi.org/10.1186/s12955-015-0291-8.

Richmond, T.S., Amsterdam, J.D., Guo, W. et al. (2009). The effect of post-injury depression on return to pre-injury function: a prospective cohort study. *Psychological Medicine* 39 (10): 1709–1720.

Rouse, W.B. (2015). *Modeling and Visualization of Complex Systems and Enterprises: Explorations of Physical, Human, Economic, and Social Phenomena*. Wiley.

Rutter, H., Savona, N., Glonti, K. et al. (2017). The need for a complex systems model of evidence for public health. *The Lancet* https://doi.org/10.1016/S0140-6736(17)31267-9.

Sterman, J.D. (2006). Learning from evidence in a complex world. *American Journal of Public Health* 96 (3): 505–514. https://doi.org/10.2105/AJPH.2005.066043.

Thompson, J., Berk, M., Dean, O. et al. (2011). Who's left? Symptoms of schizophrenia that predict clinical trial dropout. *Human Psychopharmacology: Clinical and Experimental* 26 (8): 609–613.

Thompson, J., Berk, M., O'Donnell, M. et al. (2014a). Attributions of responsibility and recovery within a no-fault injury compensation scheme. *Rehabilitation Psychology* 59 (3): 247–255.

Thompson, J., O'Donnell, M., Stafford, L. et al. (2014b). Association between attributions of responsibility for motor vehicle crashes, depressive symptoms, and return to work. *Rehabilitation Psychology* 59 (4): 376–385. https://doi.org/10.1037/rep0000012.

Thompson, J., O'Donnell, M., Stafford, L. et al. (2014c). Attributions of responsibility for motor vehicle crashes, depression and return to work. *Rehabilitation Psychology* 59 (4): 376–385.

Thompson, J., Berk, M., O'Donnell, M. et al. (2015). The association between attributions of responsibility for motor vehicle accidents and patient satisfaction: a study within a no-fault injury compensation system. *Clinical Rehabilitation* 29 (5): 500–508.

Tracy, M., Cerda, M., and Keyes, K.M. (2018). Agent-based modeling in public health: current applications and future directions. *Annual Review of Public Health* 39: 77–94. https://doi.org/10.1146/annurev-publhealth-040617-014317.

Tuszynski, M., Steeves, J., Fawcett, J. et al. (2007). Guidelines for the conduct of clinical trials for spinal cord injury as developed by the ICCP panel: clinical trial inclusion/exclusion criteria and ethics. *Spinal Cord* 45 (3): 222–231.

U.S. National Library of Medicine (2018). PICO Framework. Retrieved from https://www.ncbi.nlm.nih.gov/pubmedhealth/PMHT0029906 (accessed 8 September 2018).

Victorian Transport Accident Commission (2012). TAC Annual Report. Melbourne: Victoria.

Victorian Transport Accident Commission (2016). 2015/16 Annual Report. Melbourne: Victoria.

Vinge, V. and Euchner, J. (2017). *Science Fiction as Foresight: An Interview with Vernor Vinge Vernor Vinge talks with Jim Euchner about his writing and about how companies can use science fiction to see into the future.* Taylor & Francis.

Ware, J.E., Kosinski, M., and Keller, S.D. (1996). A 12-item short-form health survey: construction of scales and preliminary tests of reliability and validity. *Medical Care* 34 (3): 220–233.

Wideman, T.H., Scott, W., Martel, M.O., and Sullivan, M.J. (2012). Recovery from depressive symptoms over the course of physical therapy: a prospective cohort study of individuals with work-related orthopaedic injuries and symptoms of depression. *Journal of Orthopaedic & Sports Physical Therapy* 42 (11): 957–967. https://doi.org/10.2519/jospt.2012.4182.

Wilensky, U. (2016). Netlogo (Version 6.0): Centre for Connected Learning and Computer-Based Modeling. Evanston, IL: Northwestern University. Retrieved from ccl.northwestern.edu/netlogo (accessed 8 September 2018).

World Health Organization (2000). *The World Health Report 2000: Health Systems: Improving Performance.* World Health Organization.

36

Modeling Information and Gray Zone Operations

Corey Lofdahl

Systems & Technology Research, Woburn, MA 01801, USA

Introduction

The twenty-first century has been dominated by technology, which has led to new forms of industry, economics, national wealth, trade, and geopolitics. These advancements have also led to a host of other changes including new forms of international influence and warfare. Initially, the promise of technology in warfare, from the perspective of the nation-state, was bright. Precision targeting and stealth weapon technology helped the United States and its allies achieve victory quickly and easily against Iraq in the first Gulf War after Saddam Hussein invaded Kuwait in August 1990. But even as America celebrated its seemingly easy victory over Iraq a few short months later, technology was changing the nature of geopolitics and warfare in ways that were not obvious at the time and that would play out over decades.

Martin van Creveld (1991) wrote about an emerging style of conflict that would transform the way wars would be fought, which focused more on social and cultural rather than technical factors in the form of five questions. The first question is: by whom will war be fought? In the twentieth-century wars of great power competition, war was fought by expanding nation-states. In the twenty-first century, van Creveld predicted that wars would instead be fought between nation-states and insurgents who seek to challenge the power and legitimacy of the state. This type of warfare, which is referred to variously as *low-intensity conflict* (LIC), *operations other than war* (OOTW), or *counterinsurgency* (COIN), is both venerable and frequent as compared with great power wars but receives comparatively little attention by nation-states.

The second question is: who participates in war? That is, what are the relationships between those who fight wars and the noncombatants? Traditionally, the western way of war was conceived by Carl von Clausewitz (1873), who wrote

Social-Behavioral Modeling for Complex Systems, First Edition.
Edited by Paul K. Davis, Angela O'Mahony, and Jonathan Pfautz.
© 2019 John Wiley & Sons, Inc. Published 2019 by John Wiley & Sons, Inc.
Companion website: www.wiley.com/go/Davis_Social-Behavioralmodeling

during the Napoleonic Wars, as a three-part *trinity* of (i) the state, (ii) the army, and (iii) the population. In the twenty-first century, this Clausewitzian trinity has been expanded to include such other groups as nongovernment organizations (NGOs), multinational corporations (MNCs), terrorist organizations, and cross-border population flows, resulting in a more complex and confusing political environment.

The third question is: how is war fought? Traditionally, wars fought among the great powers focused on similar and symmetric militaries that were controlled by leaders who made quick decisions and engaged in decisive battles. The strategic decisions of the Napoleonic Wars, such as Waterloo, and the blitzkrieg or *lightning war* of the Nazis in World War II are emblematic of this perspective. In the twenty-first century, van Creveld reasoned that war would be fought asymmetrically between armies of the nation-state in uniforms and insurgents without uniforms. Insurgents do not fight decisively but instead seek to wear down nation-states in protracted wars of attrition.

The fourth question is: for what is war fought? War for the nation-state is essentially rational, with Clausewitz calling it "an extension of politics pursued by other means." In the twenty-first century, according to van Creveld, war would become not a logical means to an end but an affective end to itself. That is, war would become increasing less rational and more emotional, with insurgents attempting to draw great powers into their regional conflicts.

The fifth question is: why is war fought? The motivations of individual soldiers, traditionally, have centered on nationalism and patriotism. In the twenty-first century, van Creveld reasoned that the motivations of insurgents would be increasingly driven by culture, religion, and identity politics.

From these five guiding questions, a range of themes can be drawn out that portray a historical trajectory for the nation-state. The wars fought after 1945 have been primarily LICs that the purportedly more powerful nation-states lost. The nation-state armies however consistently train and equip to fight conventional force-on-force wars rather than the LICs they are most likely to face – and lose.

Addressing these challenges requires the nation-state to rethink the training, procurement, and data collection requirements of its militaries who are tasked with engaging and confronting adversaries – both nation-states and insurgents – which is the subject of this chapter. It also requires acknowledging the changing strategies and tactics of adversaries in response to technology, which includes social, narrative, organizational, and doctrinal changes (Arquilla and Ronfeldt 2001). This is done from a complex systems perspective that focuses on the unintended doctrinal and strategic impacts of technology and attendant social changes. Advanced military technologies, while decisive in Gulf War I, led to such unintended consequences as they diffused throughout the international system, and geopolitical competitors and insurgents changed their behavior to avoid their effective use. This analysis is broken up into three parts. First, the writings of two near-peer competitors – China and Russia – are reviewed to understand how they think about these twenty-first-century technology changes and propose to pursue their national interests. Second, a system dynamics (SD) simulation model is developed that explains the impact of information operations on the modern battlefield. Third, the same technologies that impact the strategies of America's geopolitical competitors also drive its *gray zone* response (Mazarr 2015; USSOCOM 2015) to operations that fall short of war but are neither peace. The analytic gaps identified in the SD simulation are used to generate requirements for a more sophisticated modeling capability that better support the analysis of gray zone and information operations. In conclusion, the need for advances in the analysis of complex social systems affected by pervasive technology, information operations, and gray zone operations is discussed.

The Technological Transformation of War: Counterintuitive Consequences

It is generally acknowledged that technology has affected the way war is fought, especially in the twenty-first century, but many of those impacts were not intended nor anticipated. Consequences that are unanticipated, unintended, or counterintuitive occur due to the social systems into which technology is introduced because the resulting system complexity is beyond the ability of human cognition to understand and predict (Forrester 1971). For example, Gulf War I of 1990 and 1991 – with its precision weapons, five-week air campaign, hundred-hour ground campaign, and pervasive television coverage – established the expectation that high-technology weapons would deliver quick and decisive victories with minimal cost and loss of life (Cordesman and Wagner 1994).

China

US television audiences however were not the only people who took note of the effectiveness of America's high-technology weaponry. The Chinese military also observed how quickly and easily the United States and its coalition partners defeated Saddam Hussein's military in Iraq and concluded that the best way to compete with the United States was indirectly, or as the Chinese phrased it in *Unrestricted Warfare*, "the force moves away from the point of the enemy's attack" (Liang and Xiangsui 1999, Chapter 6). That is, the Chinese realized that any battlefield the US military is attacking is inherently dangerous, so it is best to *move away* and avoid the attack. That does not mean however that China planned to avoid competition with the United States. Instead, they recommend competing in alternative dimensions in which they were more likely to prove successful. Liang and Xiangsui (1999, Chapter 2) describe this competitive indirection as "the war god's face has become indistinct." Additional competitive dimensions identified by the Chinese included (i) trade war, (ii) financial war, (iii) the new terror war, and (iv) ecological war, with each different type of *war* presenting an additional separate dimension of geopolitical competition. The *new terror war* merits additional explanation because it encapsulates the crux of China's *Unrestricted Warfare*, which is not *unrestricted* in the traditional sense of extremeness or boundlessness. Instead, it is *unrestricted* in an alternate sense of technology-enabled conflict brought to bear across multiple dimensions against a geopolitical competitor.

These multiple different types of competition are not intended to replace traditional warfare singularly but in combination. Understanding that social systems are complex entails recognizing multiple conflict dimensions – trade, finance, technology-enabled terror, ecology, etc. – and considering that when combined, the target may not even recognize that they are being attacked and will be completely surprised by the result. Liang and Xiangsui (1999, Chapter 8) describe this indirect, *all-encompassing* type of attack in terms of omnidirectionality, synchrony, asymmetry, and multidimensional coordination, but perhaps the clearest example is provided by the South China Sea (Glaser 2015). China does not have a blue-water navy like the United States has but challenges America's traditional right of maritime passage in international waters by building bases on previously uninhabited islands in the South China Sea. These construction activities were accompanied by synchronized diplomatic actions claiming sovereignty over the bases, information operations outlining the historical claims of China to these islands, and military intercepts of planes and ships that come too close in what America maintains are international waters. Even though many of these claims and activities are dubious, the fact that no country was willing to challenge China meant that these combined actions achieved their geostrategic aims and China effectively gained control of the South China Sea without the cost of warfare, though the competition continues (Hong 2018).

Russia

The Russians, like the Chinese, developed their own way to pursue their geopolitical goals, which they call *new generation warfare* (Bērziņš 2014). It too is multidimensional and essentially combines small, undeclared forces and information operations to engage geopolitical competitors in an indirect and subversive manner. The forces used by Russia consist of specially prepared irregular forces that attack in a combined and synchronized manner. These forces may assassinate opposition leaders, terrorize populations, or destroy the economic infrastructure of the target without war being declared. Over time, such actions seem almost *normal* because conflict has been temporally extended to the point that violence short of war becomes a permanent social feature. Information operations supplement and support the limited military operations through deception, disinformation, and misattribution. The goal is to move away from the traditional physical impact of war to a more affective, emotional, and mental impact. The goal is to influence asymmetrically and indirectly the decisions and behavior of the target population without war being declared.

There is no set time for the execution of these operations as the level of conflict is maintained just below that of openly declared war for extended periods. Russia includes the additional factors of nationalist demographics, moral justification, and legal defense in its new generation warfare, though some of these factors have long histories that have been adapted to current exigencies. Russian citizens have long been placed in regions where Russian, Soviet, or imperial Russian leaders have desired increased influence, such as the Baltic nations of Estonia, Latvia, Lithuania, and Belarus (Staliūnas 2007). Violence and instability caused through undeclared military and information operations can then be cited as a reason for further Russian interference on the pretext of protecting the local Russian population. Additionally, Russia seeks legal support and justification for its actions. Even if the arguments reduce to dubious legalism, they support the special and information operations by influencing the mental state and decision-making of the target population.

Russia's *new generation warfare* was used to annex the Crimea in March 2014 and influence the Ukraine afterward. Since the fall of the Soviet Union, Russia has been particularly sensitive to the expansion of Europe's North Atlantic Treaty Organization (NATO) eastward. Since World War II, the Russians have pursued the strategic notion of *depth* – that is, distance between itself and its perceived threats. The three Baltic nations joining NATO in 2004 were difficult for Russia to accept because it affected its *depth* as NATO was now only 160 km from St. Petersburg instead of 1600 km (Bērziņš 2014). When the Ukraine also started to move toward NATO's orbit, this was seen by Russia as a *red line* that could not be crossed because, in Russia's opinion, its security requires a Ukraine that is friendly, subordinate, and Russified. In response, Russia annexed the Crimea through a combination of operations including troops that were already in place, moral reasoning that they were protecting ethnic Russians and the Ukrainians themselves, and a hastily organized referendum in which Crimea voted to join Russia. The result was that Russia annexed the Crimea, its traditional Black Sea base for more than 250 years, in three weeks without a shot being fired (Bērziņš 2014, p. 3). Additionally, Russia sent special forces into the Ukraine under the pretext that they were merely intelligence officers, a new generation military action intended to prevent the Ukraine from being influenced by the west.

This section has concentrated on the way that America's near-peer competitors, China and Russia, have implemented asymmetric strategies to confront America indirectly. However, it can be argued that they have put into geopolitical practice what insurgents have always done – adapted in terms of culture, strategy, and tactics to offset the advantages of more established powers. These adaptations and the reasoning behind them are developed further in the following section.

Modeling Information Operations: Representing Complexity

Describing and understanding the impacts of single military operations is difficult, but analyzing the interactions of multiple operations – like the *kinetic* insurgent and *non-kinetic* information operations of Russia's next generation warfare – is even more difficult. For example, US Army General Stanley McChrystal noted the counterintuitive phenomenon of *insurgent math* in Iraq (Lemieux 2010), which occurs when 10 insurgents are removed from 100 in a region, but instead of 90 remaining insurgents, there are, counterintuitively, 110. How is this possible? Counterintuitive social behaviors are driven by the inherent complexity of social systems (Forrester 1971), which in this case are driven by the combination of (i) kinetic military operations and (ii) non-kinetic information operations described in the previous section. The relationship between these two operational phenomena is specified, quantified, and integrated using the SD computer modeling and simulation methodology, which provides a top-down systems perspective that helps to envision and understand the behavior of complex social systems. SD is a mature methodology (Sterman 2000) that provides a way not just to model complex dynamic systems but also to frame and create such models, which are developed here using the Vensim SD simulation environment (Eberlein 2007).

SD models feedback relationships using stock–flow, accumulative, or integrative relationships as shown in Figure 36.1, which shows an initial, intuitive formulation of insurgent math using Vensim. It shows a stock–flow relationship with the rectangle representing a stock and an outflow on the right that decreases the insurgent population through kinetic operations, which represents the mental model that informed McChrystal's actions. If the number of insurgents is decreased then, he wondered, how can it possibly increase? What follows lays out a story of how to describe and understand some complex phenomena.

Figure 36.2 provides additional system logic that shows how the number of insurgents *can* (depending on detail) increase despite some being removed. The kinetic operations undertaken have unintended consequences, such as civilian casualties, which are used by insurgents to influence the much larger general population using the kind of cultural, religious, and identity-based messages described by van Creveld (1991). Even if a kinetic operation has no civilian casualties, the insurgent may claim such casualties to gain the

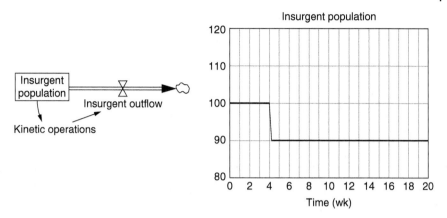

Figure 36.1 Kinetic operations reduce the insurgent population.

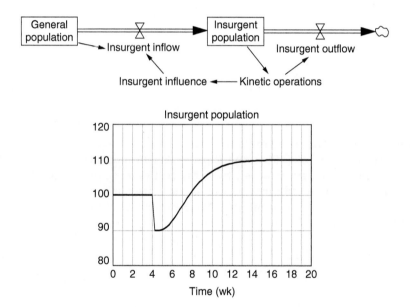

Figure 36.2 *Insurgent math*: insurgent influence operations reverse the gains of kinetic operations.

sympathy of the population, thereby radicalizing a small percentage who quickly replace the insurgents who were removed. In Figure 36.2, as shown on the behavior over time graph, 10 insurgents are removed from the insurgent population at week 4, but insurgent influence causes 20 new people to become insurgents so that at week 16, there are 110 insurgents rather than the initial 100.

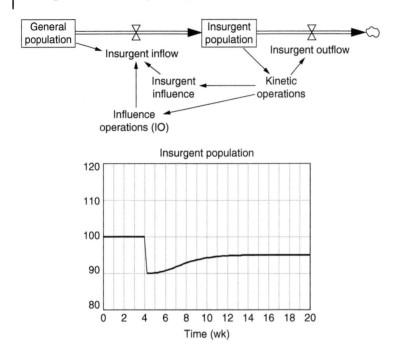

Figure 36.3 Influence operations (IOs) counter insurgent influence.

Figure 36.3 shows that influence operations (IOs) that counter insurgent influence can help to consolidate the gains of kinetic operations. In this example, 10 insurgents are removed from the battlespace through kinetic operations, but IO counters the insurgent messaging that falsely claims civilian casualties, so only 5 new insurgents arrive to replace them, meaning that the original 100 insurgents have been reduced to 100 − 10 + 5 or 95.

Figure 36.4 asserts that using the same IO campaign as Figure 36.3 but with no kinetic operations and therefore no opportunity to claim civilian casualties, the number of insurgents is reduced from 100 to 85. This is a speculative scenario that assumes a degree of success for IO that may not be achievable in the real world, but developing hypotheses through simulation is the kind of scenario-based analysis and *what-if* experimentation that provokes thought, discussion, additional experimentation, creativity, and operational innovation. This use of computer simulation, then, is not *predictive* in the traditional sense, but rather a vehicle for explanation and exploration as discussed in Chapter 2 of this volume (Davis and O'Mahony 2018). If the depiction is used to guide strategy, then it becomes necessary to understand how to make IO effective as postulated. It should be noted that the features of Figures 36.1–36.4 could be discussed in prose without the benefit of diagrams or computer models, but experience shows that these models are often more effective than simple

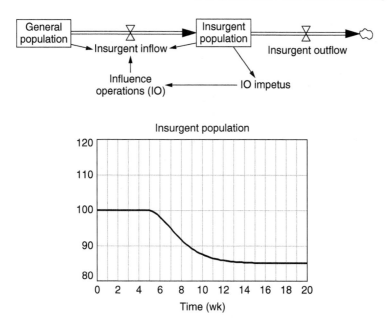

Figure 36.4 Influence operations (IOs) alone without kinetic operations.

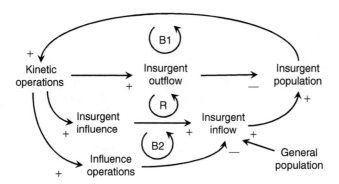

Figure 36.5 Causal loop diagram (CLD) of Figure 36.3 IO model.

prose for sharpening thought, clarifying communication, and stimulating fruitful debate.

The sequence of models developed from Figures 36.1–36.4 shows how seemingly simple relationships can result in confusing complex dynamics. Causal relationships such as these can quickly hinder understanding and overwhelm the human mind's ability to operate productively. Figure 36.5 combines these figures in way that remains comprehensible (with some experience), whereas a full version might be cluttered as the result of showing too many minor contributions. Figure 36.5 shows the systems' feedbacks in the form of a causal loop diagram (CLD) with negative or balancing loops marked with a *B* and positive or reinforcing loops with an *R*. The top loop, B1, is a balancing loop based on the Figure 36.3 model, in which the insurgent population causes more kinetic operations, which in turn increases insurgent outflow, thereby reducing the insurgent population. Note that the positive signs next to the causal arrows denote *change in the same direction,* while negative signs denote *change in the opposite direction.* Continuing with the story, the insurgents, recognizing this, respond with an influence campaign that results in the R feedback loop that works against McChrystal's kinetic operations by increasing the insurgent population. The IO campaign (previously shown in Figure 36.3) creates the B2 feedback loop, in Figure 36.5, which supports kinetic operations and counters insurgent influence. These analytic insights are extended in the following section.

Modeling Gray Zone Operations: Extending Analytic Capability

This study argues that conflict has fundamentally changed in the twenty-first century as America's competitors have employed insurgent-style strategies and tactics to offset and counter US technical innovations. What is less clear is how the United States and its allies should respond to these new capabilities and behaviors. This process starts with the analytic tasks associated with making sense of these challenges.

Figure 36.6 depicts a *gray zone* that exists between peace and war that embodies elements of each. The four elements of national power are the diplomatic, informational, military, and economic, and while the United States was the dominant power of the twentieth century through a com-bination of diplomatic and economic power, its geopolitical competitors

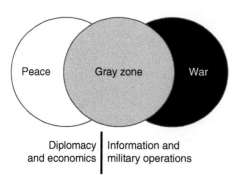

Figure 36.6 The gray zone exists in the tension between the diplomacy and economics view associated with peace and the information and military view associated with war.

have challenged it in the twenty-first century through a combination of asymmetric information and military power (USSOCOM 2015, p. 4). While, traditionally, there has been a desire for speed and decision in the execution of warfare (von Clausewitz 1873), gray zone conflict instead pursues a temporal extension or *gradualism* that achieves the benefits of war without the costs (Mazarr 2015). That is, if a country does not even recognize (or, more precisely, does not appreciate the extent to which) it is being attacked, and no single action is provocative enough to warrant a response, then national interests that traditionally have required war can be pursued short of war. The art of this process comes in planning and executing a campaign that is sufficiently gradual and effective (Bērziņš 2014).

The question then becomes, how can an analyst connect and comprehend the many small, seemingly disparate events that combine to achieve a gray zone operation? The SD models described in the previous section provide a starting point because the separate influences associated in a gray zone campaign can be specified and sometimes quantified and integrated within a coherent computer simulation. The SD models also highlight analytic gaps that point the way to additional information requirements and modeling capabilities. One of the key gaps is how to identify, connect, and contextualize individual events to comprehend how they contribute collectively to gray zone campaigns.

SD simulations capture three key aspects of causal complexity: (i) stock–flow (accumulative and integrative) relationships, (ii) nonlinear relationships, and (iii) feedback relationships, both reinforcing (positive) and balancing (negative). SD achieves this analytic capability through the abstraction of the stock, which represents simply a much more complex underlying reality. As shown in Figure 36.7, an SD stock, as depicted by a rectangle, represents an idealized collection of identical things, while in the real world, a stock is a messy collection of somewhat similar things.

Figure 36.7 System dynamics (SD) stocks and flows can be disaggregated into distributed dynamic networks.

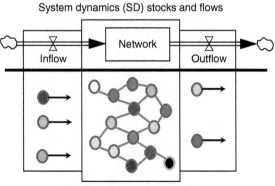

Distributed dynamic networks

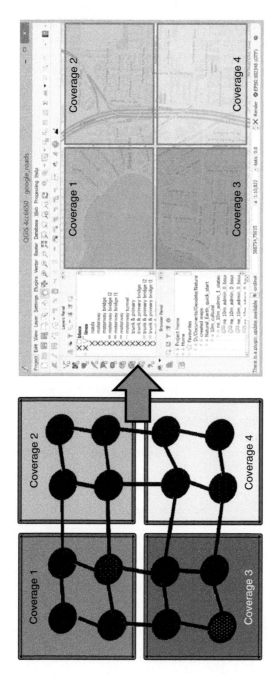

Figure 36.8 Distributed networks can be mapped to regions or *coverages* in a geographic information system (GIS).

More precisely, an SD stock is a representation of a homogeneous and perfectly connected network that, in reality, is a heterogeneous and imperfectly connected network (Rahmandad and Sterman 2008). A rich literature has built up regarding the analysis networks, and much is known about them, but they tend to be unchanging and static, while the networks required to understand how individual events contribute to gray zone campaigns would need to be flexible and dynamic. Therefore, work needs to be done to develop modeling and simulation techniques that further the capabilities of dynamic networks.

Figure 36.7 shows notionally how the stock and flow abstraction of a *top-down* SD simulation can be disaggregated into a *bottom-up* dynamic network. Dynamic complexity is introduced because incoming nodes must be attached somewhere specific within the network, while outgoing nodes must be removed from somewhere, a process that is much simplified by the stock and flow SD abstraction. The networked structure supports the type of opposition adaptation that is characteristic of the gray zone:

> What distinguishes netwar as a form of conflict is the networked organizational structure of its practitioners—with many groups actually being leaderless—and the suppleness in their ability to come together quickly in swarming attacks.
> (Arquilla and Ronfeldt 2001, p. xi)

Combining and reconciling the aggregate dynamics of top-down SD models with the specificity of bottom-up network models remains an ongoing goal of gray zone modeling and analysis research.

Additionally, network representations allow for the visual representation of space, which is vital for military commanders who need to know what happens where. Computer-based geographic information systems (GISs) provide an intuitive way to visualize space through the mapping of network nodes, individually or in groups, to specific GIS regions or *coverages* as shown in Figure 36.8.

Combining maps and models that represent both space and time is an active research area that can potentially benefit gray zone and related analysis (NASEM 2016).

Conclusion

This study has focused on how technology has transformed twenty-first-century warfare in ways that are unexpected, surprising, and counterintuitive. There has long been a belief that war is fought primarily by nation-states contesting rationally for political power. However, as van Creveld (1991) predicted a quarter century ago that conflict would transform in fundamental ways. The first Gulf War seemed to confirm the view of traditional war, which was fought decisively by armies in uniform. China and Russia however took note of how futile it was to fight the United States directly on the battlefield and instead chose to compete with America indirectly, especially with gradual information and military operations that combine to form a gray zone, something between war and peace. As an example of how this indirect, twenty-first-century way of warfare, an SD computer simulation of Stanley McChrystal's *insurgent math* was created and explained showing how removing 10 insurgents from a group of 100 can counterintuitively result in 110 through information operations. Finally, the American response to gray zone operations was examined with an emphasis on analytic innovations that address the gradualism, particularism, and spatiality of information and gray zone operations.

These issues and observations are informed by a singular methodological insight – that social systems, including the international system and its attendant geopolitical conflicts, are complex, which results in their characteristically surprising, unpredictable, and counterintuitive behaviors (Forrester 1971). Computer-based modeling and simulation techniques, such as the SD example developed herein, can be used to account for and address the challenges associated with analyzing complex social systems. While such techniques are well known in the academy (Sterman 2000), they have not transitioned to the operational or policy communities as much as some might have initially anticipated. However, the emerging realities and challenges of pervasive technology have caused the late Stephen Hawking, among others, to predict the twenty-first century "will be the century of complexity" (West 2017, p. 20). As technology causes war to transform and the international system to evolve, its behavior will become ever harder to analyze and comprehend. Computer-based modeling and simulation techniques will prove increasingly necessary to understand these systems and the drivers underlying their dynamics. Several technical innovations have been suggested here to make these tools and techniques more applicable to the analysis of these systems and more potent in the formulation of policies to influence, shape, and maintain them.

This chapter has discussed a military problem that illustrates the complexity of social and behavioral systems that can be addressed through modeling and simulation. It has also illustrated important functions of SD simulation that extend beyond prediction – that is, the function of diagnosing, describing, and explaining complex social and behavioral phenomena as well as exploring the potential consequences of changes in strategy (akin to *interventions* in other examples). Methods for doing so include using visual languages for comprehensibility, using multiple formalisms (in this case both SD and network modeling), using GISs, and addressing issues at different resolutions.

References

Arquilla, J. and Ronfeldt, D. (2001). *Networks and Netwars: The Future of Terror, Crime, and Militancy*. Santa Monica, CA: Rand Corporation.

Bērziņš, J. (2014). Russia's new generation warfare in Ukraine: implications for Latvian Defense Policy. *Policy Paper 2*. Riga: National Defense Academy of Latvia Center for Security and Strategic Research, 2002–2014.

Cordesman, A.H. and Wagner, A.R. (1994). *The Lessons of Modern War: Volume IV. The Gulf War*. Boulder, CO: Westview Press.

Eberlein, R. (2007). *Vensim User's Guide (Version 5)*. Harvard, MA: Ventana Systems.

Forrester, J.W. (1971). Counterintuitive behavior of social systems. *Technology Review* 73 (January): 52–68.

Glaser, B.S. (2015). Conflict in the South China Sea. Contingency Planning Memorandum Update for Center for Preventative Action, Council on Foreign Relations, New York, NY (April).

Hong, N. (2018). *The South China Sea in 2018: What to Expect in the Asia Pacific Maritime Domain*. Washington, DC: Institute for China American Studies (ICAS) Preview.

Lemieux, J. (2010). No, really: is the US military cut out for courageous restraint? *Small Wars Journal* (23 July), p. 1.

Liang, Q. and Xiangsui, W. (1999). *Unrestricted Warfare*. Beijing: PLA Literature and Arts Publishing House.

Mazarr, M.J. (2015). *Mastering the Gray Zone: Understanding a Changing Era of Conflict*. Carlisle, PA: US Army War College.

NASEM (2016). *From Maps to Models: Augmenting the Nation's Geospatial Intelligence Capabilities. A Report of the National Academies of Sciences, Engineering, and Medicine (NASEM)*. Washington, DC: National Academies Press.

Rahmandad, H. and Sterman, J. (2008). Heterogeneity and network structure in the dynamics of diffusion: Comparing agent-based and differential equation models. *Management Science* 54 (5): 998–1014.

Staliūnas, D. (2007). *Making Russians: Meaning and Practice of Russification in Lithuania and Belarus After 1863*. New York, NY: Rudopi.

Sterman, J.D. (2000). *Business Dynamics: Systems Thinking and Modeling for a Complex World*. Boston: McGraw-Hill.

USSOCOM (2015). *The Gray Zone*. Tampa, FL: Headquarters, US Special Operations Command.

Van Creveld, M. (1991). *The Transformation of War*. New York, NY: Free Press.

Von Clausewitz, C. (1873). *On War*, vol. 1. London: N. Trübner & Co. Translated by James John Graham.

West, G. (2017). *Scale: The Universal Laws of Life and Death in Organisms, Cities, and Companies*. London: Weidenfeld & Nicolson.

37

Homo Narratus (The Storytelling Species): The Challenge (and Importance) of Modeling Narrative in Human Understanding

Christopher Paul

RAND Corporation, Pittsburgh, PA 15213, USA

The Challenge

Over the last several years, the Department of Defense (DoD) has been repeatedly excoriated for failing to make sufficient and effective use of narrative in support of military operations. Numerous studies, reports, theses, and papers describe the *battle of narratives* or *battle for the narrative* and decry the fact that the United States has been losing that battle, or, worse, losing by default by failing to contest the narrative battlespace.[1] Enthusiasts point out that narratives are foundational to justification, legitimacy, public opinion, and effective influence (Nissen 2012). Further, legitimacy and public perceptions are essential to success in operations. The strongest statement of this view asserts that "it will not be the military implementation that will determine the success or failure of the present-day campaign. The side with the most compelling narrative will succeed" (Crannel and Sheppard 2010).

This chapter presents lessons learned from a study conducted for the US Marine Corps to understand what operational commands can and should try to accomplish with narratives and identifies implications for representing narratives effectively in modeling. It provides an overview of current thinking on the nature of narratives and how they might be better included in military operations. This operational focus raises issues that should be taken seriously by those concerned with social-behavioral science and its modeling. The chapter highlights several important questions that frame challenges to effective modeling of narrative:

---

1 Portions of this discussion are drawn from C. Paul, K.S. Colley, and L. Steckman, "Fighting Against, With, and Through Narrative," unpublished, RAND Corp.: Santa Monica, CA. A related article is forthcoming in the *Marine Corps Gazette.*

Social-Behavioral Modeling for Complex Systems, First Edition.
Edited by Paul K. Davis, Angela O'Mahony, and Jonathan Pfautz.

- Can models provide sufficient verisimilitude and fidelity to provide valuable simulated experience for end users?
- If narratives are a key to how humans make sense of the world, how should narratives be represented in agents?
- Can modelers anticipate alternative narratives so that agents can, in the course of simulation, shift among those available depending on influences?
- Can modelers go farther and construct simulations in which major narrative adjustments, or even new narratives, emerge in the natural course of simulation?
- Can agents be constructed to recognize conflicts between narratives and accomplishing objectives? Can they then take countermeasures?

Related matters are discussed elsewhere in this volume (Corman et al. 2019; Davis and O'Mahony 2019; Yilmaz 2019), but current social-behavioral modeling is a long way from being able to deal with these challenges.

What Are Narratives?

Despite the constant refrain of the importance of narratives and the repeated demand that we become better at fighting against, with, or through narratives, there is alarmingly little agreement about how exactly to do that or even exactly what a narrative is. This lack of consensus constitutes one of the first challenges for modeling and simulation: what, exactly, should we be trying to model? One prominent scholar notes that while there is a substantial amount of existing theory on narratives, "this body of literature is poorly organized" (Corman 2011). Another notes that the "discussion of stories and narratives is hampered by the fact that there is no widely accepted definition regarding just what a story is" (Casebeer and Russell 2005). Concerningly, this confusion leads the term *narrative* to be conflated with *message* or *theme*, which are indeed related to narratives but lack the depth, complexity, and relation to context of narratives.[2] Modeling and simulation of narrative must go beyond more simplistic efforts to model communication and messaging, though that might be a place to start.

Different views of what narratives are and should be are real and consequential, not a mere matter of semantics. My attempt to synthesize the existing discussion suggests consensus on three matters, which are essential to explaining narratives and understanding why they are important to operations: First, they are about stories and have story properties (settings, characters, plots, resolutions, beginnings, middles, and ends). Second, stories are how human beings

---

2 The Joint Staff J7 has released guidance for CommSync, explicitly stating the differences between narratives, themes, and messages. Whereas a narrative focuses on "context, reason, and desired results [and] … enables understanding for external stakeholders," messages are "narrowly focused communications that support a specific theme … to create a specific effect." Also see Department of Defense (2013).

understand and make sense of the world and their place in it. And third, the stories that get told about the events of military operations and conflicts affect perceptions and understandings of those operations, which in turn affect the perceived legitimacy of those operations and the extent to which one side or the other receives an individual or group's support. Remember one man's rebel is another man's freedom fighter . . . .

Because narratives affect perceptions, and perceptions ultimately affect decisions and actions, there is a role for narrative in live, virtual, or constructive (LVC) simulations of military operations that model or wish to include the attitudes and behaviors of enemy forces and so-called "green" forces and actors, such as civilians, noncombatants, and others whose behavior might affect the progress or outcomes of operations.

What Is Important About Narratives?

What about narratives do commanders, planners, and operators need to understand? Relatedly, what aspects of narrative are important to capture in modeling and simulation for training or exercises? I want to highlight three facts about narratives because they are particularly relevant to thinking about (or simulating) narratives in the defense context:

- People use narratives to make sense of the world and their place in it.
- Compelling narratives have consistency, familiarity, and proof.
- Narratives already exist, and although they can be shaped over time, they cannot always be changed or replaced.

Each of these three points is explained below.

People Use Narratives to Make Sense of the World

Narratives and other mental shortcuts help humans make sense of the things we see and experience in the world. Research shows that people use stories to help structure memory, cue certain approaches to problem-solving, format new information, and define our identities (Narvaez 1998; San Roque et al. 2012). Narratives also often suggest or hint at how we should feel about an event based on the emotional content of the narrative or even imply a value judgment or suggest a course of action, perhaps based on the moral of the story. "Narratives make sense of the world, put things in their place according to our experience, and then tell us what to do" (Laity 2015).

Part of making sense of the world is making sense of our place in it. When exposed to compelling narratives, we subconsciously identify with the actors and struggles explained in them. When we relate to the characters or their struggle, we use the outcomes in the narrative to give us purpose or suggest courses of action. For example, many US military recruiting commercials tell a story about a young man facing difficult personal challenges or defending innocents from "chaos." Potential recruits personally identify with the individual in the commercial and both the struggles he faces and the goals he pursues, making them more likely to join. ISIS similarly offers opportunities for recruits to protect the "persecuted" and for young men to be a part of an organization not afraid to act in the face of "oppression," but in their stories, they use characters, struggles, and goals chosen to resonate with their target audiences. In both cases, potential recruits can identify personally and emotionally and see a path of action to address key events in their worldviews; thus, narratives can be both explanatory and mobilizing.

What happens when we cannot make sense of events witnessed or accounts heard? The human brain wants the information it receives to make sense. When the brain cannot make sense out of incoming information, that information is more likely to be discounted or ignored or recombined with previous information until it does make sense (Seese and Haven 2015). So, if something new happens (say, the arrival of US troops to provide humanitarian aid), it is interpreted based on the existing stories or overarching narratives held by the observing audience or individual. If the dominant existing narrative about American troops is negative (they are villains and only come to hurt, belittle, and occupy us), then the new facts will predominantly be interpreted in a way that is consistent with that narrative, even if this requires the omission of some of the details (the part about the humanitarian aid, perhaps?), leaving the audience with a negative view.

These narrative-based perspectives are referred to in the academic literature as *narrative frames*.[3] Such frames are not necessarily derived from a single story but from an audience's whole collection of stories, created and transmitted within societies over time. These narrative frames (or lenses) shade how we view the world and help with our sensemaking. For example, the Marine Corps has a number of memorable and important specific narratives, such as those relating to Tripoli, Chesty Puller, Guadalcanal, and Iwo Jima. However, there is also a Marine Corps narrative frame, a way of seeing the world and the Marines' role in it, so that it is consistent with all of those stories and is more or less shared by all Marines.

This is true of other peoples as well. Narratives and narrative frames will vary widely in different cultural contexts, because *different groups of people have different collections of stories.* This is important because different groups of people will perceive the same events differently, and make different sense of them, based on their different narrative frames. Just because our narrative frame suggests events be interpreted in a certain way does not mean that frame, perspective, and interpretation are shared by other audiences.

---

3 For more on frames and framing, see Goffman (1974) and Benford and Shaw (2000). A similar concept is called *schemas* in cognitive psychology (McVeee et al. 2005).

Any simulation that implicitly or explicitly models sensemaking (as with agent-based simulation) could benefit from inclusion of narratives and narrative frames. Narrative frames are not strictly and objectively rational. Different groups in a simulation need to be allowed to perceive and interpret events, actions, and messages differently, depending on the mobilized narratives and their narrative frames.

Compelling Narratives Have Consistency, Familiarity, and Proof

Compelling narratives have at least three characteristics: consistency, familiarity, and proof (Case and Mellen 2009). *Consistency* refers both to the internal consistency of a story (e.g. whether the outcome follows logically from the action described and whether the characters' behavior is true to type) and to the story's consistency with other salient narratives or narrative frames. *Familiarity* is about how well known a story or narrative is; more than just awareness of the story, familiarity also implies a level of comfort with the story, which could come from sharing themes in common with stories within a broader narrative frame. *Proof* is about the evidence available in support of the narrative and can vary widely. Proof can hinge on the perceived credibility of what is claimed, perceived credibility of the narrator, eyewitness accounts, or recorded pictures or video. Note that what constitutes proof varies considerably by context and medium. For example, in the United States, the facts in stories presented by television news anchors are accorded high degrees of credibility and generally accepted as strong proof. Elsewhere in the world, however, state-run television news reports are not considered much proof at all, while a story repeated from a friend of a friend might count as strong proof, despite less compelling evidence. Presumably these three attributes (consistency, familiarity, and proof) can be scored and scaled as part of a model, though the attributes of a given narrative will likely differ for different groups (and that will be important to model, too).

Narratives Already Exist and Cannot Always Be Changed or Replaced

As much as different audiences will ascribe different levels of consistency, familiarity, and proof to different narratives, different audiences also have a different collection of stories and narratives available to them and prefer to interpret new events in a way that is consistent with their existing collection of stories. Because of the substantial body of preexisting stories available to any audience, most events they witness or experience immediately fit within, and sensemaking is supported by at least one of those preexisting narratives. "Audiences will without exception always interpret stories in their terms" (Zalman 2010). *This can make it very difficult to present a new or alternative narrative that will have traction.*

In most cases, when US forces act in foreign lands, there will already be one or more narratives in place that are going to be the dominant narratives of those events for relevant foreign audiences, regardless what themes, messages, and images are offered to accompany those actions. So, if preexisting narratives drive the understanding of events in most cases, when and how can US forces oppose, counter, or offer alternatives to those narratives? More briefly, when are there *narrative opportunities*, and what kinds of opportunities are they?

When something happens that people notice and care about, relevant audiences will become aware of it and try to make sense of it. In any given instance, one of three things will happen:

1. The event fits perfectly within one existing narrative, reinforcing that narrative and connecting to all the other content (negative or positive) from that narrative. That narrative becomes the dominant narrative for this event.
2. The event fits reasonably well within more than one available narrative or can be viewed through the lens of more than one relevant narrative frame. The event will be understood through one or more of the available narrative frames, but which one(s) will be dominant is unclear (and perhaps shapeable).
3. The event does not fit well within available narratives or mobilized narrative frames. The event will end up connected to one or more narratives (perhaps new, perhaps old) and viewed through narrative frames, but which ones and how it will be interpreted is an open question.

Each of these three possibilities corresponds to a different level of narrative opportunity. These three levels of opportunity will need to be captured in efforts to model potential friendly force impact on narratives in models:

1. If the event fits perfectly within an existing mobilized narrative, there is very limited narrative opportunity, leaving very few options. These include (i) accepting and embracing all or part of that narrative (if it is positive or has positive or at least tolerable aspects), (ii) adjusting planned actions so that they are not so easily connected to that narrative (if the planned action is going to connect directly to an unfavorable narrative, consider not doing that action, or finding a way to do it that will be perceived and framed differently), or (iii) trying to emphasize aspects of the action that suggest an alternative narrative frame (basically try to make a situation #1 into a situation #2). Just to emphasize, *sometimes the only way to create an opportunity to change the narrative is to change the actions.*
2. If events fit reasonably well within one or more alternative narratives or frames, there is some narrative opportunity. Those trying to fight with, through, and against narratives can pick the available narratives that are most favorable or beneficial to the joint force and try to emphasize aspects of the action that are consistent with those narratives or otherwise try to

frame the event so it is viewed in that way. Provided there is an alternative narrative, there might be an opportunity to emphasize how the event is not like what happens in an unfavorable narrative. Note this is **not** a sufficiently wide-open opportunity to make up a wholly new narrative, just an opportunity to push toward and emphasize favorable available narratives and perhaps push away from unfavorable narratives.

3. If the event is something new or different, people are still going to try to understand it and connect it to existing frames, but there may be greater opportunity to shape which ones are adopted or to introduce new ones. "Since narratives are neither fixed nor infinitely malleable, each side has a window of opportunity in which it may choose to change its narrative in order to address changing circumstances effectively" (Case and Mellen 2009, p. 1). Here, narrative opportunity is greatest, as a much wider range of available narratives or narrative frames can potentially be mobilized to help observers understand the event. It may even be possible to promote a wholly new narrative; however, it would be easier, and would likely have more traction, to try to mobilize some dormant preexisting narrative or lens than to create a wholly new one. A dormant narrative is more likely to be consistent and somewhat familiar, whereas a wholly new narrative, even if there is an opportunity for one, will need to build its consistency, familiarity, and proof from scratch.

When comparing competing narratives or narrative frames, audiences consider and weigh the consistency, familiarity, and proof of each. This subconscious or conscious comparison of competing narratives operates following cognitive processes not unlike those used by a jury during deliberations (Case and Mellen 2009).

Once a given narrative or narrative frame has been associated with an event or series of events, it will be difficult to change that connection. However, there may be opportunities to emphasize different aspects of that narrative, to try to combine it with another salient narrative with more favorable characteristics, or to otherwise shape the narrative. Again, in most situations, a wholly new narrative is unlikely to gain much traction because – as compared with other available narratives – it will lack external consistency and will be unfamiliar, regardless how much proof is associated with it (especially if that proof is more compelling to Western audiences than to relevant audiences).

The degree of narrative opportunity available in a given situation will be critical to capture in efforts to add narrative to modeling and simulation. The success of efforts to shape or change narratives should be heavily influenced by the degree of narrative opportunity identified in the simulation. Simulations should *not* allow success in positively influencing perceptions if the actions naively attempt to propagate a narrative when there is no opening in the (simulated) narrative opportunity space!

What Can Commands Try to Accomplish with Narratives in Support of Operations?

Much of the discussion of narratives surrounds the *strategic narrative* that should accompany US strategy in general, as well as for any region or theater. Such strategic narratives are important but need to be anchored to effective highest-level strategy, an area in which the United States has particularly struggled.[4] My discussion will more modestly focus on *operational* narratives to help relevant audiences make sense of US military operations and related actions and events. These narratives must nest with higher-level narratives in the same way that subordinate objectives and goals must nest with goals, objectives, and end states prescribed at higher levels.[5]

So, what should commands hope to accomplish with narratives in support of military operations? Which of these things can be captured in modeling and simulation? First is *internal coordination*. If humans make sense of events through narrative, then a clear mission narrative will be useful for the troops. Such a narrative needs to fit with existing military and service-specific narrative frames (avoid a narrative that tries to make the Marine Corps feel like the Peace Corps), but done right, a mission narrative makes it easier for everyone to understand and remember mission objectives and to understand their role in the story that will lead to achievement of those objectives. A clear mission narrative can help troops avoid the "say-do gap" that often opens between actions and communications; it can promote unity of effort and diminish the likelihood of information fratricide. A good mission narrative guides follow-on planning, targeting, and execution and enables mission command because subordinates will be better able to judge whether an available course of action is consistent with the narrative and thus preferred. In many simulations, the problems that internal narrative helps reduce are not modeled. If they would be, then it should be trivial to add modeling of the consequences of internal narrative.

Second, commands can use narrative to *offer a positive or alternative explanation to external audiences*. Relevant audiences are going to find narratives and narrative frames to help them make sense of US operations. Left to rely solely on their own histories and experiences, many of these narratives will support views and actions that are contrary to US operations. Countering these existing perceptions is a core challenge of narrative in operations. Commands should seek to promote narratives of their operations that ascribe positive meanings to their actions so that they add up to something that should be supported, or at least patiently tolerated, rather than being viewed

---

4 See, for example, the criticisms discussed in Weitz (2008), Haddick (2012), and Zelleke and Zorn (2014).

5 Ideally, the President or National Security Staff provide the DoD with strategic-level guidance that includes the strategic narrative. In the event the narrative is not provided or unclear, JDN 2-13 (2013) suggests several ways to extract the strategic narrative from other high-level documents.

negatively. The extent to which this is possible will be constrained by the level of narrative opportunity available, as described in the previous section. Done well, promotion of a favorable narrative increases understanding of, tolerance of, and support for US operations. This increased tolerance can increase the likelihood of desired behaviors (noninterference, cooperation, etc.) and can also increase freedom of maneuver because US force actions occur within the confines of a locally accepted narrative of legitimacy.

Third, the command may want to *compete with or undermine narratives at odds with mission objectives*, when there is sufficient narrative opportunity to do so. Many operating environments may contain narratives or narrative frames that do not support US force presence or objectives, or such narratives may be introduced or mobilized by adversarial groups whose interests do not align with those of the United States. In order to reap the benefits of having a broadly accepted legitimating narrative and achieve desired levels of support, US forces may need to find a way to "defeat" these alternative narratives. There is a growing body of literature on ways to attack narratives head on, but that is outside the scope of this chapter.[6]

Defeating hostile narratives must go hand in hand with the promotion of positive narratives. Audiences *will* find a narrative or narrative frame for events, and they will make sense of them, one way or another. It is impossible to defeat a narrative and just leave a narrative vacuum. There must be an alternative narrative that replaces it. "The one thing that replaces [or modifies] a story-based belief ... is a better story" (Seese and Haven 2015).

The second and third uses, offering positive narratives and attempting to compete with or undermine disadvantageous narratives, are at the core of the challenge of including narrative in modeling and simulation for training and exercises. Simulations supporting these levels need to model narrative "play," which is predicated on modeling the structure of available narratives (mobilized and dormant), consequences when actions correspond to an unfavorable narrative, probability of success when there is an effort to connect events to one narrative over another, the slow strengthening or weakening of narratives with certain audiences through contradiction or supporting evidence, and everything else necessary to simulate a realistic and dynamic narrative context.

Moving Forward in Fighting Against, with, and Through Narrative in Support of Operations

Three Kinds of Narrative

With these three objectives – internal coordination, offering a positive narrative, and competing with opposing narratives – in mind, how can US forces go about this in training and in practice? I see three kinds of narratives that

---

6 For instance, see Cobb (2016).

operational commanders and staffs should be thinking about: the command's mission narrative, the command's external narrative, and the narrative desired within the relevant audiences. Making progress toward successful employment of these narratives can follow a crawl, walk, run progression. Beneficial modeling and simulation support required differs slightly at each of these levels.

Thinking at all about narrative is at the *crawl* level, as is preparing and disseminating an internal mission narrative for internal coordination. Modeling and simulation to support *crawl* is pretty minimal: the inclusion of even minimal incentive or opportunity within the context of a simulation to draw users' attention to narrative, space or an input field that accepts an internal mission narrative as an input (even if that input does not actually affect the model), including information on narrative as part of the discoverable information within a simulated scenario, etc.

Identifying how the command's actions are likely to be perceived by external audiences and building an external narrative to project alongside those actions is at the *walk* level, as it requires some knowledge of relevant populations and their narrative frames. Supporting the *walk* level requires a great deal more from modeling and simulation. Models most include actual consequences from various groups/actors/agents' perceptions of the user's forces' actions, as well as being connected to discoverable context – models need to produce (potentially) different perception-driven results for different actions (or different approaches to very similar actions), and must do so in a way that is at least partially discoverable and dependent on characteristics of simulated groups. The requirement for discoverability is also part of the modeling and simulation challenge. Even if a simulation engine is realistically capturing the available narratives, changes to those narratives, and the consequences of those narratives, that information cannot be presented unfiltered to users. The narrative landscape is beset by a *fog of war* problem far out of scale with traditional battlefield uncertainty. Realistically presenting situational awareness of narrative (and of the information environment more broadly) is a huge and important challenge for successful modeling and simulation in this area (McGrath 2016).

Understanding the nuance of available and likely narratives and narrative frames and planning ways to get external audiences to talk (narrate) favorably about the command is more challenging still and on the way to (or part of) the *run* level. Supporting the *run* level will place a significant burden on modeling and simulation, necessitating nuanced cultural models; models for how narratives are mobilized and matched to unfolding events; even more sophisticated modeling of narrative situational awareness; models for how narratives change over time and under pressure; and models for things that are very difficult to reliably predict, such as what message will go *viral* (and when) and which stories or explanations will prove to be *sticky* and which will not.[7]

---

7 See, for example, Heath and Heath (2007).

Developing a Command's Mission Narrative

The command's mission narrative is the simple orienting story the commander will offer to troops to convey the objectives of the mission and their role in accomplishing them. I think that the same process that produces the commander's intent could produce the command's mission narrative with very slight adjustment. Some extra thinking may be required to transform some language to better portray a story in which the troops of the command are among the characters and the essential tasks are their actions. As a staff concludes mission analysis and prepares the commander's guidance and intent, they should also prepare the mission narrative as part of that intent. Ultimately, the mission narrative is just a restatement of the commander's desired end state as the conclusion of a story and the role the commander expects troops to play in bringing that end state about: it captures the essence of the *why* and the *how* of the mission as envisioned by the commander.

An example of an excellent summary phrase for a command's mission narrative is "No better friend, no worse enemy" as used by then MGen James Mattis with First Marine Division beginning with commencement of Operation Iraqi Freedom I in March of 2003 (Mattis 2003). This headline captured two facets of his intent: that his Marines be aggressive and flexible in taking the fight to the enemy, but that civilians and prisoners be treated with chivalry and spared unnecessary harm. It gave clear roles to his Marines – the best of friends, the worst of enemies. It tied into the existing Marine Corps narrative frame, with the same narratives that confirm "every Marine a rifleman" being highly consistent with the "no worse enemy" portion. When he returned to Iraq in 2004 for OIF II, he kept "no better friend, no worse enemy" but also added "first, do no harm" in order to emphasize the relief and reconstruction emphasis of the new mission (Mattis 2004).

Not every command's mission narrative will be as straightforward and short as Mattis' masterpiece, but every such narrative should connect with the narrative frames and identities of the troops addressed, describe the roles those troops will play as the operation unfolds, and state the desired conclusion of the story of the operation. As noted above, the kinds of problems that internal narratives solve or prevent are often not problematized in simulation and are probably not a priority for simulation. Just allowing internal narratives as an input without tying it to any consequences in the simulation engine may be sufficient for most training and exercise contexts. Or, if a simulation includes the possibility of information fratricide or disjoint actions by friendly forces, then internal narratives could be scored and modeled as reducing the likelihood of such unfortunate events.

Developing a Command's External Narrative

Developing a narrative to try to impart to external audiences is much more challenging (and requires much more sophistication to sufficiently simulate).

Ideally, the command's mission narrative and the command's external narrative will be one in the same. It is much more effective to offer the same story of justification, explanation, and purpose to the troops executing the mission and to the audiences witnessing that execution. Unfortunately, a quick and easy narrative that resonates with US troops because it connects to their existing narrative lenses and identities may fall flat with relevant foreign audiences because it is inconsistent with their preexisting narratives, is unfamiliar to them, and lacks proof that they find compelling.

Developing a command's external narrative necessitates work and effort not currently part of operational planning routines and modeling content not currently available in existing LVC training and exercises. Preparing a command's external narrative requires a fairly robust understanding of the relevant audiences (those people whose behavior is instrumental to the success or failure of the campaign); their narrative frames in terms of their history, worldview, and recent events; and the available narratives about the United States, US forces, and their operations and actions. To be able to plan effectively for a command's external narrative, intelligence preparation of the information and operating environment must include attention to these kinds of issues, as must the simulation of this intelligence preparation and simulation of situational awareness more broadly. This preparation may require (or benefit from) media and social media monitoring or available behavioral, cultural, and linguistic subject matter experts who have sufficient knowledge of the operational context to meet the need and the kinds of information gathered as part of target audience analysis as conducted by military information support operations (MISO) personnel. To simulate preparation of a command's internal narrative and model its impact, those same kinds of things will need to be included in modeling and simulation and made part of the scenario context and simulated discoverable information.

Developing and Promoting Desired Narratives Among Relevant Audiences

The story that US forces tell about what they are doing is important, but what is even more important is the stories relevant local groups tell each other and themselves about what US forces are doing and about what US adversaries are doing. The high art of fighting with, through, and against narratives at the *run* level involves getting external narrators to tell and repeat favorable stories about US forces and reducing the prevalence of narratives favorable to adversaries.

If the command's external narrative is perfect, then it will be adopted by relevant audiences as the prevailing narrative. However, while that level of perfection is an aspirational goal, it may be only partially achievable due to numerous variables in the battlespace that are beyond the command's control. More likely, the command's external narrative is heard and becomes part of the local discourse, but is subject to counter-narratives that start with the question, "why are the Americans *really* here?" Selectively promoting or discouraging specific narratives or narrative elements within the broader relevant narrative landscape requires deep understanding and a deft touch. Doing so requires both additional understanding and additional capability. Shaping and fighting narratives at this level requires even more extensive cultural and linguistic inputs; deeper understanding of available myths, memes, and other narrative elements; intelligence about key influencers; capabilities for persuasion and influence (like MISO); and better understanding of the cultural and cognitive aspects of narrative generation and promulgation. To use modeling and simulation to train and exercise at this level will require inclusion of all of these things as meaningful simulation elements. In some cases, that will be particularly difficult, as the necessary social science foundation is still being built. Further, shaping narratives is often a slow process. Realistically matching changes in narratives to a simulation/exercise time scale is yet another challenge to modeling and simulation in this area (McGrath 2016).

Conclusion: Seek Modeling and Simulation Improvements That Will Enable Training and Experience with Narrative

In conclusion, I join the chorus and exhort US military formations to increase their efforts beyond messaging and improve their ability to fight with, through, and against narratives. I also exhort the modeling and simulation community to work to include relevant aspects of narrative (and other aspects of the information environment and information effects) in their efforts.

As described above, in order to begin moving on the crawl, walk, run progression toward being able to fight effectively within narrative terrain, US forces need to be able to do the following five things. In order to train and practice these tasks, modeling and simulation will need to be able to allow and reflect the consequences of users' efforts to:

1. Identify available salient narratives and narrative frames already present in the operating context.
2. Anticipate which of those narratives are likely to be connected to planned US actions and undertakings.
3. Identify which narratives or aspects of narratives are favorable or neutral to US objectives and which are unfavorable.
4. Recognize when altering planned actions can create opportunities for more favorable narratives.
5. Push on and into the information environment to promote more favorable alternatives (when available) or more positive/favorable aspects of unavoidable narrative frames.

Addressing narratives is difficult and including them effectively in simulation is similarly challenging. However, it is only by embracing the complexities of human understanding and the effects of narratives that we will truly be able to fight and win in the information age.

References

Benford, R.D. and Shaw, D.A. (2000). Framing processes and social movements: An overview and assessment. *Annual Review of Sociology* 26: 611–639.

Case, D.J. and Mellen, B.C. (2009). Changing the story: the role of the narrative in the success or failure of terrorist groups. Masters Thesis. Naval Postgraduate School, Monterey, CA.

Casebeer, W.D. and Russell, J.A. (2005). Storytelling and terrorism: towards a comprehensive 'counter-narrative strategy'. *Strategic Insights* IV (3): 4.

Cobb, S. (2016). Narrative assessment of the 'cognitive space' to support influence. *Proceedings from Strategic Multilayer Assessment SOCOM Gray Zone Telecon Series*.

Corman, S.R. (2011). Understanding extremists' use of narrative to influence contested populations. *Proceedings from Workshop on Mapping Ideas: Discovering and Information Landscapes* (29–30 June 2011), San Diego State University.

Corman, S.R.N., Suston, S.W., and Tong, H. (2019, this volume). Toward generative narrative models of the course and resolution of conflict. In: *Social-Behavioral Modeling for Complex Systems* (ed. P.K. Davis, A. O'Mahony and J. Pfautz). Hoboken, NJ: Wiley.

Crannel, M., and Sheppard, B. (2010). Achieving narrative superiority to succeed in Afghanistan (a Light Year-Group Study Paper). Retrieved from http://www.ideasciences.com/library/papers/NarrativeCornerstonetoSuccess.pdf.

Davis, P.K. and O'Mahony, A. (2019, this volume). Improving social-behavioral modeling. In: *Social-Behavioral Modeling for Complex Systems* (ed. P.K. Davis, A. O'Mahony and J. Pfautz). Hoboken, NJ: Wiley.

Department of Defense (2013). Joint doctrine note 2-13 commander's communication synchronization.

Goffman, E. (1974). *Frame Analysis: An Essay on the Organization of Experience*. Cambridge, MA: Harvard University Press.

Haddick, R. (2012). Why is Washington so bad at strategy? *Foreign Policy* (9 March).

Heath, C. and Heath, D. (2007). *Made to Stick: Why Some Ideas Survive and Others Die*. New York, NY: Random House.

Laity, M. (2015). Nato and the power of narrative. *Beyond Propaganda* (22–28 September), p. 23.

Mattis, M.G.J.N. (2003). 1st Marine Division (REIN) Commanding General's Message to all hands. *Memorandum* (March).

Mattis, M.G.J.N. (2004). Letter to all hands. *Memorandum* (March).

McGrath, J.R. (2016). Twenty-first century information warfare and the third offset strategy. *Joint Forces Quarterly* 82 (3d): 16–23.

McVeee, M.B., Dunsmore, K., and Gavelek, J.R. (2005). Schema theory revisited. *Review of Educational Research* 75 (4): 551–566.

Narvaez, D. (1998). The influence of moral schemas on the reconstruction of moral narratives in eighth graders and college students. *Journal of Educational Psychology* 90 (1): 13–24.

Nissen, T.E. (2012). Narrative led operations: put the narrative first. *Small Wars Journal* (17 October). http://smallwarsjournal.com/jrnl/art/narrative-led-operations-put-the-narrative-first.

San Roque, L., Rumsey, A., Gawne, L. et al. (2012). Getting the story straight: language fieldwork using a narrative problem-solving task. *Language Documentation & Conservation* 6: 135–174.

Seese, G.S. and Haven, K. (2015). The neuroscience of influential strategic narratives and storylines. *IO Sphere*, Fall, 33–38.

Weitz, R. (2008). The U.S. Strategy deficit: The dominance of political messaging. Second Line of Defense: Delivering Capabilities to the Warfighter blog.

Yilmaz, L. (2019, this volume). Toward self-aware models as cognitive adaptive instruments for social and behavioral modeling. In: *Social-Behavioral Modeling for Complex Systems* (ed. P.K. Davis, A. O'Mahony and J. Pfautz). Hoboken, NJ: Wiley.

Zalman, A. (2010). Narrative as an influence factor in information operations. *IO Journal* 2 (3): 4–10.

Zelleke, A. and Zorn, J.T. (2014). United States: where's the strategy. *The Diplomat* (5 February).

38

Aligning Behavior with Desired Outcomes: Lessons for Government Policy from the Marketing World

Katharine Sieck

Business Intelligence and Market Analysis, RAND Corporation and Pardee RAND Graduate School, Santa Monica, CA 90401, USA

Marketing and government often face a similar challenge: to understand why people do what they do so that practitioners can shape and influence behavioral patterns toward different ends.[1] In government, an understanding of human behavior helps policy-makers determine the best way to intervene in a situation to improve the health, welfare, and security of the community. Because their work is centrally about facilitating the social good, the government has at its disposal a range of techniques for influencing behavior that includes *harder* (more coercive measures) such as mandates and legislation. Consider the case of how to reduce deaths in car crashes. On the one end, a speed limit sign is a *nudge*, while an expensive ticket is more of a *shove*. Beyond that, the ability to revoke someone's license or imprison them for repeated violations is more coercive a technique for getting people to conform to recommended speed limits.

In marketing, the driving rationale for understanding human behavior is to further such business goals as brand loyalty, profits, engagement, and other indicators of business health and growth. Whether the aim is to attract new customers, increase purchase frequency, shut out competition, or change the way people think about a category of products/services, marketers turn to a range of *soft power* techniques including encouragement, enticement

---

1 In both government and marketing contexts, major issues arise when contemplating *influence* because similar methods can be used for good and bad purposes. Marketing promotes commerce, which is generally considered good, but not when it entices youth to become nicotine addicts. Government influence is good when it promotes participation in elections or awareness of health issues when shopping at the supermarket, but countries such as the United States reject excessive government intrusion and demand integrity rather than propaganda in government communications. In this chapter, I focus on positive forms of influence, although mentioning with some alarm instances where serious issues of ethics and privacy can arise where abuse has occurred.

Social-Behavioral Modeling for Complex Systems, First Edition.
Edited by Paul K. Davis, Angela O'Mahony, and Jonathan Pfautz.

(Roberts 2004), design, and *nudges* (Thaler and Sunstein 2008) to inspire people to do something different. Absent the ability to force conformity toward specific ends, marketers must have a nuanced understanding of what drives behavior, what will inspire changes, where and when to efficiently and effectively communicate these messages, and who has the ability to influence others within a community.

Too often, government officials resort to the *harder* techniques when their understanding of the underlying behaviors is insufficient to compel adherence through *softer* approaches. Consider the situation of childhood immunizations. Unable to create effective policies that address the diverse reasons that parents fail to immunize their children, states are increasingly resorting to more draconian tactics that tie school attendance to mandatory immunizations in all cases, save for a very limited number of extreme medical exemptions. While these policies are effective in boosting coverage rates, they accomplish this through force (e.g. the refusal to admit a child to school), rather than through partnership, increased understanding, or increased access. The consequence is often a greater disdain for the role of government, rather than a sense of partnership and participation with elected officials. To this end, there is much the government leaders can learn from the efforts of those in marketing and business as it considers why people do what they do and how to get them to do something different. While officials have those heavy-handed options, it is often in their best interests to take more subtle approaches to facilitating alignment with the social good.

This chapter presents three efforts that I led within a marketing and design activity in private enterprise, highlighting a different way to think about motivation, behavior, and societal change that may be of particular use for policy practitioners whose job is to shape social welfare. My team and I sought to overcome two common biases in marketing that we believed limited the effectiveness of work. First, in usual studies, people are typically *positioned as individuals*, divorced from the wider contexts that shape their lives. Second, there is a tendency to *dismiss the integrated way in which people live*, by looking at actions solely within one category (e.g. health and wellness) while ignoring how these choices are linked across other categories (e.g. money and finances).Working with my team in marketing, we drew from the very *social* sciences of anthropology and sociology, which posit motivation as being not just "in one's head" but entangled in relationships and shaped through wider structural and cultural dynamics (Shore 1996). Within this framework, we developed a three-step research process to gather ever-nuanced details about why people do what they do. As part of these efforts, we sought out creative ways to integrate qualitative and quantitative approaches to human behavior as a social act. Certainly, these are not the only techniques used in marketing, but they constitute a range of ways to integrate theory, qualitative data, and quantitative data to provide a more comprehensive understanding of behavior and behavior change.

Alongside each marketing example, I discuss ways that these techniques can better inform the practice and work of our government. Just as these are not equally relevant or efficacious for every marketing challenge, they will likely have limits in their policy applications too. However, this general approach of understanding how to better align human motivation and behavior toward desired ends could reduce the number of instances in which policy is heavy handed and thereby inspire greater teamwork and cooperation between those who make laws and those who must live under them.

Technique 1: Identify the Human Problem

Marketing discussions frequently begin with a business challenge: for example, how do we get more people to buy our product? How do we increase revenue over last quarter? How do we overcome a product failure? In government, parallel conversations might include topics such as how do we address the problem of homelessness? How do we slow the opioid epidemic? Or, what is the best way to stop the conversion of young people to a radical terrorist group? Addressing such questions is a critical first stem. We need to understand what matters to people and why.

At the broadest level, we can postulate that there are some shared human motivations. This is the foundational element of things such as Freud's theory of id/ego/superego and Maslow's hierarchy of needs. These grand theories are possible because as diverse as human cultures are, we are all fundamentally human. Across the globe and through time, our shared humanity has resulted in some shared problems – for example, how to raise the young and bury the dead, how to manage conflict and build alliances, how to control and allocate resources, and how to address fears and disappointments. While cultures have often devised particular and unique strategies for such challenges, tailored to the unique circumstances of their geography or resources, the challenges themselves remain constant. This is evident in cross-cultural comparisons of childrearing (Whiting 1963), shame and fear (Lakoff 1987), and resistance (Scott 1985). This is also the foundation of the Human Relations Area File, an expansive database that tracks human behavior cross-culturally and through time (see http://hraf.yale.edu).

If we presume that humans have addressed shared challenges over time and across space, then one way to understand a given human behavior is by studying behaviors that look similar, with attention to the rationale and explanations given for those actions. This technique provides a more limited range of potential motivation for any specific behavior or a range of behaviors for any specific motivation. Two examples from our work in marketing will illustrate each effort. First, in an attempt to understand the motivation for why people wear cosmetics, we pulled examples of *face paint* from theater, war, religious rituals, beauty, and self-defense. The range of reasons that we found provided the client (a cosmetic company) with a new way to think about their product beyond *beauty* or *age defying*. Second, to elucidate the range of potential behavioral response to stigma, we explored the many ways cultures have managed it, including tactics such as humor, distancing, reframing, or subversion. These enabled our client (an investment firm) to consider a broader scope of responses to the anti-Wall Street sentiment that plagued their work in the wake of the 2008 recession.

However, simply understanding the diversity of human behavior and motivations was not sufficient. As noted, these tactics are specific to certain circumstances: they reflect the constellation of beliefs, expectations, and resources available to different cultures in different times. A simple *cut-and-paste* approach to tactics would not necessarily be successful, so we had to engage in a strategic consideration of what would work for any particular client, how the tactics of competitors align against the set of possibilities, and how other contextual factors may shape or constrain variations on these approaches. For example, in the cosmetics case, knowing that face paint was often used in battle to inspire fear and confusion in the opponent is one thing; knowing how to translate that into a viable communications and creative strategy is a different task. But it is an important alternative to simply view cosmetics as an attempt to lure partners or to hide signs of aging, which are tired and rote terrains within the cosmetics industry.

When considering the world of policy, the technique that we used holds possibilities for generating new ideas and approaches, especially in well-trod terrains. Let's consider the case of radicalization. While often portrayed as a relatively new phenomenon, it's actually quite ancient. For example, radicalization has echoes of religious conversion and devotion, and thus there is much to be learned from the historical accounts of Christian crusaders and saints, studies of religious conversion among Pentecostal and Charismatic Christians, and studies of religious cults to explore the ways in which religious communities use language, rituals, and social dynamics to foster a sense of both absence and belonging among new converts. Alternatively, contemporary radicalization has parallels to the literature on charismatic leaders, which offer insights into how these individuals often gain such sway over communities. Yet again, radicalization might be better understood through the lens of stigma and shame, with attention to how individuals are made to feel unworthy but worthy of redemption through specific acts. One could even see radicalization as a rite of passage and study how things like bar mitzvahs and weddings and graduations might shed light on crafting an identity and belonging to a wider community.

These are a few examples of how the behavior in question can be reframed in multiple ways to inspire different ideas about the driving causes and the potential solutions. Using this approach of identifying the human problem, the point is to *understand a behavior through other behaviors that look like it* in order to gain new perspectives on how to intervene. In every case, the next question should be: how have communities themselves addressed these issues? How have they mitigated the negative consequences of these behaviors? This is the kind of detail that can be found in historical, ethnographic, and sociological studies of human behavior. For example, understanding the Crusades – which came to an end only after 200 years of fighting, millions of lives lost, and several failed treaties – may offer insights on the reality of both military and diplomatic options available in this new era of religious uprising. Alternatively, understanding that a *rite of passage* is designed to help a young person secure his or her place as an adult within a community provides an alternative range of interventions for young radicals – one more focused on shoring up the social and economic pathways toward full communal participation that keep them within established (peaceful) folds and out of radicalized terrain.

By situating any specific problem within the wider context of human behavior, government officials and policy analysts are able to consider a much wider range of well-tested approaches for channeling behavior toward socially desirable ends. While this alone will not structure policy, it will provide both new ideas and alternative ways to think about the challenge itself.

Technique 2: Rethinking Quantitative Data

Historically, the core challenge of the social sciences was data sufficiency: specifically, having sufficient information to make broad yet valid pronouncements about human motivations and behaviors. Yet today, social scientists face an almost opposite challenge: we are swimming in behavioral and attitudinal data of all kinds (purchase behavior, exercise data, location information, likes/dislikes, social engagement). In contrast to previous eras, little of this data was collected to elucidate competing theories of human behavior. Little was collected with any theoretical consideration at all, to be blunt. Oddly, this renders it difficult to process, largely because the data is a by-product of other things, not the thing we sought (Madsbjerg 2017). It tends to be a random collection of facts in the absence of a guiding framework.

Faced with so much information, researchers must determine which data sets to explore and how to link them together. We must also be cautious about using interpretative theories to do so. For example, credit agencies in 2007–2008 used a risk model based on the theory that housing, savings, employment, and debt were in independent sectors of the economy. As a result, they greatly underestimated the significance of problems they saw with mortgage bonds. This contributed the massive economic collapse in 2008 (Silver 2012). In failing to see the emergent interconnections between housing, savings, employment, and debt, they built a risk model based on an unrealistic idea of sectoral independence. History is replete with these kinds of missed causal connections resulting from false assumptions about *the way things are* (c.f. Douglas 1992; Bloch 1998; Boholm 2003). In an era of *big data*, our choice of data sets warrants greater consideration when the complementary data are ostensibly there, but we lack the capacity to thoughtfully determine what to include, how to integrate the data, and what the models should be.

This data–theory inversion has been a bit of a crystal ball for behavioral sciences but an admittedly murky one. With so much information available, researchers can often predict certain things without understanding why this is the case. For example, Target can gauge whether its customers are pregnant by what they purchase (Duhigg 2012) and Facebook can predict an impending break-up with relative accuracy (Ferenstein 2014). But neither team has a clear explanation as to *why*: Target does not know what this purchase behavior signals for cultural attitudes about parenting and motherhood, perceptions of dangers and sacrifice, or the influence of experts, social networks, and others on purchase decisions. Similarly, Facebook knows that the key predictor of relationship longevity is an interest in each other, but it cannot say how couples sustain that interest over time in online and offline ways.

The real challenge is finding a way to match theory to data in an organized, thoughtful fashion that enables exploration of competing explanatory theories. It means building a mid-level framework that will sort massive amounts of data into meaningful dimensions for understanding human behavior without constraining interpretive possibilities.

Faced with this challenge, my marketing research colleagues and I took a step back and structured a framework that would sort data in a way that aligns with broad concepts in the social sciences but was one level removed from specific theories. Given that our goal was to identify groups/clusters within massive data sets, we turned to the early work of anthropologists and sociologists to understand how these researchers studied collective behavior (c.f. Evans-Pritchard 1940; Mead 2001 [1928]; Benedict 2005 [1934]; Durkheim 2008 [1912]; Malinowski 2014 [1922]). Historically, anthropologists identified a geographically located community (the Nuer, the Azande, the Kwakiutl) and then detailed how those in the community lived. In reviewing these early ethnographies, we identified 10 *dimensions of community* that appear in different monographs as core axes for defining communities, including:

1. *Landscape*: the physical (and now digital) environments in which people live.
2. *Language*: the medium of communication.
3. *Creation story*: their sense of their history and roots.
4. *Economy*: what and how they exchange things; what has value.
5. *Laws*: the rules that govern behavior.
6. *Threats*: what they fear.
7. *Sacrifice*: what they give up in order to belong to the community.

8. *Rituals*: how they enact their shared commitment.
9. *Symbols*: the things that mean more than what they are and signal identity.
10. *Values*: what matters to them and why.

We then included the work of sociologists to add additional nuance to behaviors within these categories. For example, Mauss' (1990) [1950] and Appadurai's (1988) works on gift exchange and social life were important to a broader understanding of *economy*, whereas Bourdieu (2010) [1984] and Goffman (1959) were significant in helping us understand how people use symbols in a variety of ways. In this step, we did not model in specific theories, but looked across the literature to understand the dynamics. For example, we were not testing Bourdieu's vs. Goffman's approaches to status and presentation of self, but, rather, we noted that symbolic activity and presentation happened in three levels: things on or part of the physical body (e.g. surgical procedures, clothing), mid-range items that signal identity (e.g. employer, frequent flier status), and *big ticket* purchases (e.g. car, neighborhood). Moreover, there was a noted difference among theorists as to people's willingness to publicly embrace and show symbols vs. symbols that were hidden or avoided or unobtainable but desired.

Using these core social sciences, we designed a framework that enabled us to retrospectively identify *community* affiliations through the Survey of the American Consumer.[2] Run by GfK, this is a massive data set that polls 25 000 people yearly on a wide range of topics, including:

- Demographics (age, sex, occupation, income, household composition, home ownership).
- Business responsibilities.
- Personal and business travel.
- Public and leisure activities.
- Online/digital behavior.
- Consumer attitudes and behaviors (brand preferences and purchase behavior).
- Political outlook.
- Psychographic questions.

We reviewed the 120-page survey question by question and sorted them to align with the 10 dimensions pulled from the early ethnographies. This entailed team discussion and review of each question to determine whether and where it fit in the framework. Not every question was included, but every question was reviewed. To give a flavor of what it looked like, these are some of the questions included in our symbols dimension:

---

2 There are other data sets used within marketing, such as Claritas, which is managed by Nielsen. A variety of factors shaped our choice of the GfK data set, but this should not be read as an endorsement of this particular database at the expense of competitor offerings. Decisions of that nature should reflect the research and strategic needs of an organization.

- *It is important to me to be well-groomed.*
- *My cell phone is an extension of my personality.*
- *I'll buy trendy clothes even if they are not the highest quality.*
- *I prefer to buy things my friends or neighbors would approve of.*
- *The vehicle a person owns says a lot about him or her.*
- *Achieving a higher social status is important to me.*
- *My goal is to make it to the top of my profession.*

We followed this process for 9 of the 10 dimensions – we could not find a way to track language in this data set, and so turned to other data sources for that particular dimension.

Within each dimension, we then identified different clusters of people. We tested several clustering techniques for each dimension to see what led to the greatest clarity between the clusters, finally settling on whatever number of clusters was needed to yield distinctive groups that were meaningful. It is worth noting that we continued refining the clusters over time by adding in questions, removing questions, and working toward greater clarity across each dimension. Some continued to be a challenge (e.g. threats), whereas others produced very distinctive groups early on. For example, our symbols dimension split into four unique groups as shown in Figure 38.1.

With this framework, we could run any set of brands against each other to see how their communities compared. We could see how risk averse they were, how much they valued social integration, what role routine played in their lives, and what worried them – and how all of these differed by their preferred brand for everything from toothpaste to insurance to luxury automobiles. Figure 38.2 provides a graphic overview of how these *dimension heat maps* looked, using a client example from the health and fitness industry, but simplifying the figure

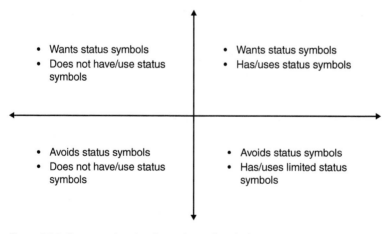

Figure 38.1 Decomposing the dimensions of symbols.

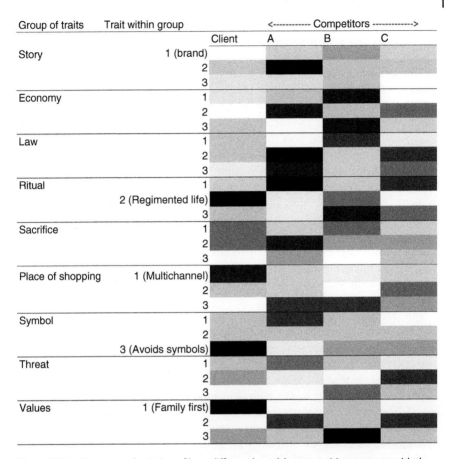

Figure 38.2 A heat map depiction of how different brands' communities resonate with the 10 dimensions and their components. Darkness of cell indicates degree to which company's customers resonate with trait. Black: strong resonance; white: low resonance.

substantially by deleting most of the companies (columns) and most of the sub traits (rows). The shading of a given cell measures the resonance of a company's customers with a given trait. For example, customers of our client (first column in Figure 38.2) were not especially influenced by brand (row 1, part of *Story*). They leaned toward a regimented life (row 2 of *Ritual*), which was sharply in contrast with customers of competitors (especially A and C). Our client's customers were also avid multichannel shoppers (row 1 of *Place*) – i.e. they shopped online, in stores, through their mobile phones, etc. Again, this was a very different behavior than that of the competitors' customers. Similar distinctions showed up in that our client's customers avoided *symbols* and placed family first (under values).

The usefulness of the display was less for a single row than in the pattern across rows and how it revealed something meaningful and understandable. Our client's customers resonated with four traits, one each under rituals, place of shopping, symbol, and values. Noting this helped us to understand a pervasive problem with the structure of our client's program (why people stopped participating) and offered a new solution that drew upon what mattered most to our client's customers: the social relationships that defined their lives. Specifically, our client's customers often put other people's need first – family was their driving consideration (*Values*); they downplayed any tangible symbols of success or wealth for themselves (*Symbols*), while accommodating the needs of everyone else by purchasing what each required by whatever means was convenient (*Places*). In always privileging others, they would forget to take care of themselves, and they only turned to our client's program when their weight started to interfere with their ability to be that central caretaker. They left the program when they felt that participating impinged on others in their family (e.g. by not being home with them). Our solution was to restructure the support component of the program to include a close friend/family member rather than a stranger or professional.

The framework could also be used to identify ways to *lure* competitor's customers into a brand community. Figure 38.3 shows data from an entirely different study, one in the insurance industry (again, simplified for this presentation). By studying it, we concluded that it might be possible to lure customers from that of competitor C because customers of competitor C were similar in several ways to our client's customers (and different in some other ways). They were similar for some traits within sacrifice, symbols, threat, and values but quite different with respect to law and economy. Understanding all this allowed us to suggest a strategy for our client to change its model.

The same methods could also be used to predict potential future customers. For example, on a luxury car project (not shown), we used the framework to identify which mid-range brand community was most like our client's customers: in short, who were the people who buy $40 000 cars and who are most like our people who buy $140 000 cars? And how are they different as well, so that we might understand how product/service offerings should change over time to accommodate their tastes?

Finally, the array approach illustrated by Figures 38.2 and 38.3 could be used to identify brand pairings by starting with a profile and then exploring what brands tended to cluster against it. For example, do people who shop at Whole Foods also tend to shop at certain other stores? If so, what might this reveal about neighborhood dynamics, brand affinities, mental models about what *fits* together in a person's life, and a host of other questions as to why patterns exist across brands and categories?

In every instance, we supplemented this work with other research: primary qualitative work (interviews, observations), secondary research, business

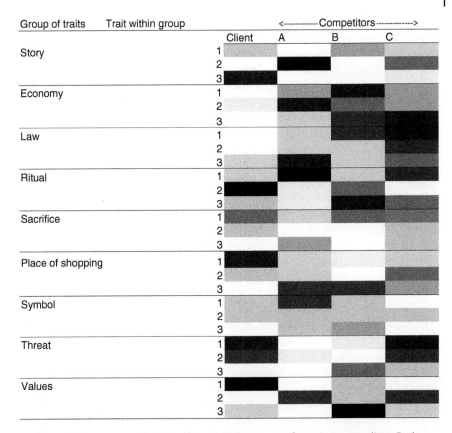

Group of traits	Trait within group		<---------- Competitors ---------->		
		Client	A	B	C
Story	1				
	2				
	3				
Economy	1				
	2				
	3				
Law	1				
	2				
	3				
Ritual	1				
	2				
	3				
Sacrifice	1				
	2				
	3				
Place of shopping	1				
	2				
	3				
Symbol	1				
	2				
	3				
Threat	1				
	2				
	3				
Values	1				
	2				
	3				

Figure 38.3 Heat map used to identify potential customers for an insurance client. Darkness of cell indicates degree to which company's customers resonate with trait. Black: strong resonance; white: low resonance.

analytics, and literature reviews. But over and again, the framework proved robust in giving a unique picture of a community that was more accurate and more actionable than most other segmentation approaches seen in the industry and more rigorous than could be achieved solely with qualitative methods implemented on these time frames (typically measured in weeks).

At this point it is important to illustrate how analogous approaches can be used improperly. Cambridge Analytica, the firm reputed to have influenced the outcomes of the Brexit vote and the 2016 US presidential election (Lapowsky 2017), created just this sort of framework in its *big five* personality profiling assessment (Bisceglio 2017). Like us, they created a mid-level framework grounded in social theory (we used anthropology; they used psychology). In both cases, the work enabled analysts to see both the connective patterns across particular domains and the unique intervention points that could be used to shift behavior.

However, our data gathering techniques differed in important ways. Our team used an existing survey in which participants had been reimbursed for their time and knew the data would be used for marketing purposes. Cambridge Analytica did not. Their wanton disregard for established regulations regarding accessing user data and their disregard for social norms about harvesting data from social networks (see Granville 2018) are important caveats about the power of social science modeling. The Cambridge Analytica work apparently violated contractual requirements and misused personal data in what may have been an illegal effort by a foreign power (Russia) to influence the US election. Russian information warfare efforts were alarming enough to stimulate creation of a special investigation (the Muller inquiry) and multiple congressional hearings interrogating executives and former executives of Cambridge Analytica and Facebook. Thus, unscrupulous use of some social science methods akin to those used properly in the business world has raised major national security issues. More generally, such methods and their efficacy raise major privacy and ethical issues regarding data, issues that have not yet been adequately confronted (Balebako et al. 2019).

Applied to social questions, this technique can be used to provide nuanced insight into entrenched problems. For example, there are likely multiple groups within the *anti-vaxxer* movement, and understanding how they differ and how best to appeal to their specific concerns would enable states to offer tailored communications through pediatricians, rather than simply imposing mandates that all children *must* be vaccinated in order to attend public schools. Moreover, some of these groups may have parallel challenges in other policy terrains. For example, if some are *anti-vaxxers* because of logistics around getting children to public health clinics, these same individuals may struggle with regular medical care and would benefit from a school-based or home-visiting program. The broader point is that employing a framework that assesses behavior and motivation across policy terrains may provide more nuanced insights into how to encourage greater alignment with social goods than mandates by fiat. This kind of mid-level structure that sorts massive data set while maintaining open theoretical interpretation could prove a powerful ally in designing effective policies, when done with respect for existing regulations governing consent and access to personal data.

Technique 3: Rethinking Qualitative Research

If one of the end goals of both policy and marketing is to change behavior, then it helps to understand why, when, and how people do – or do not – change their behavior. This is the terrain of qualitative methods: they shine in their ability to provide richness and contextual perspective on a particular topic. For example, we may know that more than 1/3 of American adults are overweight, but if we ignore the varying narratives as to *why* people become overweight and why they prefer some interventions over others, subsequent policies will only work for a subset of the intended audience, if at all. Designing programs to provide greater access to healthy foods will do little if the driving reason lies in ideals of beauty or resource constraints for food preparation.

Qualitative methods, however, will not reveal motivation in some magical way. In fact, our default reliance on interviews and focus groups can obscure motivation when researchers are not attuned to the broader dynamics at play in people's lives. To this end, we knew we needed to go beyond verbal accounts of past events because, as Herman (2004) noted, the cognitive models that shape our understanding of a story change with new events and new experiences, so that our assessment of past choices will alter based on the subsequent outcome of those decisions. Similarly, prospective accounts of imagined futures can be equally unreliable because they require people to hypothesize about too many contextual variables. For qualitative methods to be useful, they must be organized in a coherent and strategic way so as to capture the unique dynamics regarding behavior as it unfolds in daily life.

While some of the social sciences specialize in qualitative methods that allow us to see behavior as lived in the moment (e.g. anthropology, qualitative sociology, social psychology), researchers in marketing and policy do not have the luxury of time typically afforded to such projects. Instead, we often needed a way to *simulate* these critical moments in order to trigger richer reflections on motivation and decision-making. Moreover, because we were interested in behavior change, we wanted specific tactics to elicit desired changes and learn from those moments.

We turned again to the social sciences and, specifically, to anthropology to review how cultures have organized behavior change. While there is no shortage of long-term strategies including behavior modeling and enculturation, we needed something that could be more efficient in instigating change. We found a solution in rituals. Turner (1970) describes rituals as "culturally salient experiences designed to transform a person, and thus their relationships to others and to material goods." They accomplish this through two tactics. First, they elevate sensory features, employing touch, sound, taste, smell, and images to reinforce lessons through an *embodied experience*. As Basso (1992) notes, these sensory components become triggers or reminders, in future moments: a wedding ring is a daily reminder of promises made and social roles embraced by the couple. Second, they create a shared template within a culture that connects people together through the experience. This builds bonds across generations, across geography, and links people to a shared identity: Catholics can attend mass in almost any corner of the world and know roughly what is happening because of the ritual's structure.

Cultures use ritual to structure or instantiate behavior change. Rituals teach new social roles and expectations through sensory-rich, emotionally complicated moments of change and guide people on an interpretive journey to an alternative understanding of their world and their place in it. But they do so in different ways. We identified four categories of rituals frequently used by cultures to address different types of behavior change:

- *Remembrance*: rituals that focus attention on important things/people/events from the past as a way to connect them to the present moment – Memorial Day, Easter, Hanukkah, Mawlid an-Nabi.
- *Deprivation*: rituals that require a deliberate avoidance of everyday things as a way to refocus attention on what matters and to purify/clarify one's relationship with the world and with deities – Lent, Yom Kippur, Ramadan.
- *Inversion*: rituals that subvert the standard social order as a way to channel transgressive impulses, as in Halloween, Holi, Carnival, Mardi Gras.
- *Transition*: rituals that explicitly change the social status of the participants – weddings, funerals, bat mitzvahs, confirmations, graduations.

Our guiding question became: can rituals help us understand the types of behavior change we would want to see in business? And if so, *can we craft research experiences that mimic rituals as a way to get better insights on the barriers to and opportunities for behavior change*? In short, rather than rely on yet another round of interviews or focus groups, could we shift our perspective on qualitative research to have it mimic what we'd want to see anyway – behavior change – and have people document this experience in visual, narrative, and sensory-rich ways that allowed us to better understand motivation?

Our first task was aligning business challenges with the behavioral changes created through each type of ritual and then exploring how to craft a research experience that addresses these challenges. The discussions below were a fruitful starting point, but were not comprehensive: other likely business and communications challenges were not covered, just as additional types of rituals might have been included.

Remembrance

- Rituals of remembrance help address challenges around customer loyalty by reminding people why this relationship is important and valuable for their lives. It is most effective for products that are central to a person's sense of self – e.g. running shoes for a runner, cooking gear for a chef, and tools for a carpenter. In market research on these products, when people are asked why they select a particular brand or engage in a particular activity, it is not uncommon for them to respond, "It's just what I do," or "I can't really explain it. It's just personal."
- In the same way that couples will recount their courtship on their anniversary, we asked people to talk about key events in their lives as an athlete/chef/carpenter. To the extent that relationship dynamics change over time, with relatively less attention to the good and more attention to

the points of frustration or disappointment (see Aron et al. 2003), such rituals serve to anchor people in historical narratives that may predate them or in commitments made that may no longer seem as compelling as when they began. For example, when working with elite runners, we had them review their collections of race bibs and medals to talk about different marathons, how they trained, what happened during those specific events, and what it taught them about being a runner over time. These stories revealed challenges and frustrations in ways that were difficult to elicit through standard conversation and provided key insights into how brands could make it better to keep being a runner.

- In such ways organizations translate the dynamics of interpersonal relationships into communication strategies: as the things we love to do become challenging, companies can step in to remind us why these things matter and to transform difficulties into opportunities for renewed commitment and engagement. Translating this into policy, this could help address situations where commitment wanes over time – for example, tenure in the public schools or among social service case workers, military service, police/fire/emergency responders, and even *insider threat* challenges to cybersecurity or national defense.

Deprivation

- Rituals of deprivation address the challenges of product/service distinction by helping people gain a new appreciation for the role of a *taken for granted* item in their daily/weekly routine. Unlike the products that fall into the remembrance terrain, deprivation focuses on the everyday things in life – toilet paper, laundry detergent, and garbage bags – items where people have a preference and tend to purchase by habit, but may also swap if something is cheaper or more accessible. They tend not to have the same strong loyalty ties as those we see in remembrance terrain, but they do fulfill much of that *background* space as well.
- As a research tool, deprivation projects involved asking people to give up certain items for a period of time – typically a few days or more. We ran one on behalf of a company that made kitchen products (garbage bags, containers, plastic wrap). In this category, people will typically say that they purchase a brand as *habit* or because it's *affordable*. There is little thought as to what role these products play. However, when they no longer had access to them, they struggled to make daily life work without these items, and they gained a deep appreciation for how these products enabled their everyday lives to flow as expected.
- When forced to give up items or to find substitutes for them, people quickly see how things are integrated into their wider lives. This realization can foster either a greater appreciation for something or a desire to restrict its

influence. In the policy world, this has tremendous potential to shed light on the challenges people face when trying to give up a negative behavior (e.g. drugs, alcohol, smoking, sweets, risky sex, etc.). By situating these behaviors within an integrated view of a person's life, researchers can see what else is lost in that transition and how these losses may inspire people to continue with negative behaviors. Moreover, they provide insight into positive experiences that could be expanded by that individual.

- Finally, as a policy teaching tool, deprivation can also be a powerful way to facilitate empathy for the challenges people face trying to feed a family on welfare payments, or find a doctor who takes Medicaid, or manage finances while living on minimum wage. By taking away the privileges afforded those who do not rely on government support systems, policy-makers themselves can gain a new appreciation for the challenges of navigating the very systems they are charged with managing.

Inversion

- Rituals of inversion help organizations explore opportunities for innovation by loosening the rules and expectations around how people use goods or services and encouraging them to think differently. Inversion rituals suspend category standards and deliberately challenge conventions as a way to document and explore what happens when *anything goes*. In the colloquial language of business, inversion helps organizations identify *white spaces* of possibility.
- In our work, we typically used this approach when a category was dominated by a particular model or narrative. For example, when working with a Caribbean nation on its tourism account, we conducted a semiotic analysis of the language and imagery used by competitor nations in their tourism work. It was, as we noted, a *sea of sameness*: beaches, drinks, escape, and *tune out*. All downplayed the challenges of navigating their country; they also provided only limited engagement with local communities – often because such engagements can be complicated or uncomfortable. We recruited travelers who recently visited our client country to understand what made it interesting and unique and why they selected this destination over the many other options. By focusing on the differences and designing a visual and communication language that was the inverse of other competitors, we transformed the conversation around travel itself and grew the nation's GDP by 4% as a result of increased tourism.
- Inversion works well when conversations essentially stall. In the policy world, this happens all too frequently when political opponents entrench in opposing positions and essentially refuse to consider alternatives. Or, when people have come to accept a subpar system as *just the way things are*. Inversion

activities create possibilities for considering alternative futures, alternative perspectives, and alternative solutions. For example, 30% of parents who have a child with special needs ultimately leave the workforce entirely in order to manage the challenges and demands of raising their kids (USDHHS 2004). Rather than just accept that "this is really hard," inversion activities would turn that experience around to consider a multitude of ways to change the situation: perhaps through extending parental leave, facilitating flexible work schedules, allowing less than full-time positions, or providing specialty care services during nontraditional work times (evenings, weekends, etc.).

Transition

- Rituals of transition help companies when they need to grow their customer base by welcoming new members. Transitions shift people's sense of themselves from one type of person to another. Whether one prefers butter or margarine, Coke or Pepsi, or any other brand over a competitor, very often these preferences become emotional and entangled in the kind of person one wants to be. To the extent that one's identity is linked to products and services, attracting new customers entails more than just be accessible and affordable. It means being desirable.

- Transition projects work especially well when people have an aversion to a specific brand or category of products. For example, we worked with a luxury sports car brand that was either worshiped or reviled by people, with few having only a neutral attitude about the company. Hence our challenge was to get people who swore they were *not that kind of person* to become exactly that kind of person, if only for a brief while. We did it by mimicking a key part of transition rituals: pairing our *initiates* (those who hated the brand) with *elders* (people who owned and loved this brand). We had them talk and spend time together focusing on the cars. We also had our *initiates* drive the cars through their typical daily routines for part of a day to experience their life as a person they swore they were not. In the end, most *initiates* walked away with a greater appreciation for the car – two even purchased them – and a changed understanding of why people love them.

- Transition projects help counter stigma, lack of familiarity, and interpersonal barriers. In policy, they can be potentially fruitful for a host of issues in which people struggle to understand the lives of other people. For example, to bring to light the challenges facing teachers, policy-makers can try that job for a few days (run the classroom, grade the work, deal with students' personal lives, etc.). To get people with an opioid and alcohol addiction to seek treatment when they did not perceive a problem in their behavior, one RAND program found success in encouraging them to *just try it* for a few weeks, resulting in 30% of participants staying in treatment (Watkins et al. 2017).

Summary

Getting adults to change behavior is a potentially fraught process, even under the best of circumstances, as most marketing executives will attest. While policy-makers can resort to mandates and draconian measures, these should always be a last resort. The approaches outlined in this piece offer a few ways to gain insights into how to encourage, or *nudge*, people into better options. In an ideal situation, projects include all three components: start with asking the right question, explore community dynamics through quantitative data sets, then create qualitative projects that pilot the desired changes. Given the political, economic, and legal challenges that arise from poorly designed policies, these methods offer a way to better understand how to bring people along with proposed changes and how to design better options from the start.

References

Appadurai, A. (1988). *The Social Life of Things*. Cambridge: Cambridge University Press.

Aron, A., Aron, E., and Norman, C. (2003). Self-expansion model of motivation and cognition in close relationships and beyond. In: *Blackwell Handbook of Social Psychology: Interpersonal Processes*, Chapter 19 (ed. G.J.O. Fletcher and M.S. Clark), 478–501. London: Blackwell Press.

Balebako, R., O'Mahony, A., Davis, P.K., and Osoba, O. (2019, this volume). Lessons from a workshop on ethical and privacy issues in social-behavioral research. In: *Social-Behavioral Modeling for Complex Systems* (ed. P.K. Davis, A. O'Mahony and J. Pfautz). Hoboken, NJ: Wiley.

Benedict, R. (2005 [1934]). *Patterns of Culture*. New York, NY: First Mariner Books.

Basso, K. (1992). *Western Apache Language & Culture: Essays in Linguistic Anthropology*. Tucson: University of Arizona Press.

Bisceglio, P. (2017). The dark side of that personality quiz you just took. *The Atlantic* (13 July 2017). https://www.theatlantic.com/technology/archive/2017/07/the-internet-is-one-big-personality-test/531861 (accessed 5 October 2017).

Bloch, M. (1998). *How we Think They Think: Anthropological Approaches to Cognition, Memory & Literacy*. London: Taylor & Francis.

Boholm, A. (2003). The cultural nature of risk: can there be an anthropology of uncertainty. *Ethnos* 68 (2): 159–178.

Bourdieu, P. (2010 [1984]). *Distinction: A Social Critique of the Judgment of Taste*. Oxford: Routledge Press.

Douglas, M. (1992). *Risk & Blame: Essays in Cultural Theory*. London: Routledge Press.

Duhigg, C. (2012). How companies learn your secrets. *The New York Times*. http://www.nytimes.com/2012/02/19/magazine/shopping-habits.html?pagewanted=all&_r=0 (accessed 5 October 2017).

Durkheim, E. (2008 [1912]). *The Elementary Forms of Religious Life*. Oxford: Oxford University Press.

Evans-Pritchard, E.E. (1940). *The Nuer: A Description of the Modes of Livelihood and Political Institutions of a Nilotic People*. Oxford: Oxford University Press.

Ferenstein, G. (2014). Predicting love and breakups with Facebook data. https://techcrunch.com/2014/02/14/facebook-love-data (accessed 5 October 2017).

Goffman, E. (1959). *The Presentation of Self in Everyday Life*. New York, NY: Doubleday Press.

Granville, K. (2018). How Cambridge Analytica harvested Facebook data, triggering a new outcry. *The New York Times* (19 March 2018). https://www.nytimes.com/2018/03/19/technology/facebook-cambridge-analytica-explained.html (accessed 21 March 2018).

Herman, D. (2004). *Story Logic: The Problems and Possibilities of Narrative*. Lincoln: University of Nebraska Press.

Lakoff, G. (1987). *Women, Fire and Dangerous Things: What Categories Reveal About the Mind*. Chicago: University of Chicago Press.

Lapowsky, I. (2017). What did Cambridge Analytica really do for Trump's campaign. *Wired* (26 October 2017). https://www.wired.com/story/what-did-cambridge-analytica-really-do-for-trumps-campaign (accessed 18 October 2017).

Madsbjerg, C. (2017). *Sensemaking: The Power of the Humanities in the Age of the Algorithm*. New York, NY: Hachette Books.

Malinowski, B. (2014 [1922]). *Argonauts of the Western Pacific*. New York, NY: Routledge Classics.

Mauss, M. (1990 [1950]). *The Gift: The Form and Reason for Exchange in Archaic Societies*. New York, NY: W.W. Norton & Company.

Mead, M. (2001 [1928]). *Coming of Age in Samoa: A Psychological Study of Primitive Youth for Western Civilization*. New York, NY: First Perennial Classics.

Roberts, K. (2004). *Lovemarks: The Future Beyond Brands*. New York: A. G. Lafley.

Scott, J. (1985). *Weapons of the Weak: Everyday Forms of Peasant Resistance*. New Haven, CT: Yale University Press.

Shore, B. (1996). *Culture in Mind: Cognition, Culture and the Problem of Meaning*. Oxford: Oxford University Press.

Silver, N. (2012). *The Signal and the Noise: Why So Many Predictions Fail – And Some Don't*. New York, NY: Penguin Books.

Thaler, R.H. and Sunstein, C.R. (2008). *Nudge: Improving Decisions About Health, Wealth & Happiness*. New York, NY: Penguin Books.

Turner, V. (1970). *Forest of Symbols: Aspects of Ndembu Ritual*. New York, NY: Cornell University Press.

U.S. Department of Health and Human Services, Health Resources and Services Administration, Maternal and Child Health Bureau (USDHHS) (2004). *National Survey of Children with Special Health Care Needs Chartbook 2001.* Rockville, MD: U.S. Department of Health and Human Services.

Watkins, K.E., Ober, A.J., Lamp, K. et al. (2017). Collaborative care for opioid and alcohol use disorders in primary care the SUMMIT randomized clinical trial. *JAMA Internal Medicine* 177 (10): 1480.

Whiting, B. (ed.) (1963). *Six Cultures: Studies in Child Rearing.* New York, NY: Wiley.

39

Future Social Science That Matters for Statecraft

Kent C. Myers

Net Assessments, Office of the Director of National Intelligence, Washington, DC 20511, USA

Perspective

This chapter is less about my own research than about the perspective of a strategist in the Office of the Director of National Intelligence as he looks across many social science activities with an eye toward their value for intelligence. It not only is a personal account but also reflects conversations with academic scholars and government personnel who are seeking a more fruitful linkage between researchers and decision-makers. My intent is more one of noting shortcomings and opportunities in the research endeavor than extolling the virtues of what is already solid.

Recent Observations

As of 2018, a study committee at the National Academies of Science (NAS) is identifying lines of social science inquiry that promise to be beneficial to national security over the next 10 years (National Academies of Science 2018). Efforts to survey the next decade have worked well in some fields, such as astrophysics, where a tight-knit community of scientists needs to set priorities on the design and use of very expensive, shared devices. In social and behavior science, however, such a survey is fiendishly difficult. I see three underlying conditions that lead to this difficulty:

- The interests of social scientists are quite varied and dispersed. They work alone or within relatively isolated groups that maintain their own references, standards, methods, and funding sources.

Social-Behavioral Modeling for Complex Systems, First Edition.
Edited by Paul K. Davis, Angela O'Mahony, and Jonathan Pfautz.
© 2019 John Wiley & Sons, Inc. Published 2019 by John Wiley & Sons, Inc.
Companion website: www.wiley.com/go/Davis_Social-Behavioralmodeling

- Social scientists pursue understanding but often without a clear beneficiary. Some claim that their work is cumulative and will achieve something grand and useful, but quite a lot of the work is a matter of tearing down rival concepts and restarting with a new twist or *turn*. In addition, of concern to those of us in the government who seek help, some scholars eschew any contact with national security organizations on moral grounds, though this sentiment has moderated in recent years.
- Social scientists are not professionally required to or rewarded for looking far ahead. Groups certainly organize around a set of methods or concerns, but there are rarely any large, shared investments to manage. Those who do look too far ahead may be criticized for being speculative or even grandiose.

Several scholars who have contributed to the NAS decadal survey have had difficulty shifting their focus. While wonderfully articulating about their work, they can be struck dumb when asked by a committee member what benefits to national security they can imagine over the next decade. It's not a question they are ever expected to answer. Nevertheless, the question exposes a gulf between the research that is performed and the need to govern wisely and securely. If, as many believe, we are at a major point of transition toward complex social modeling, we might consider how the question of usefulness appears in this new context and of what a fresh answer might consist.

The NAS committee's search for lines of research that will be useful to national security rests on an understanding of evolving security problems. Members of the intelligence community (IC) have provided insights into what policy-makers need and what intelligence advisors are offering. While the IC isn't the only conduit through which leaders learn about social reality that can inform and improve their actions, it is an important conduit that is constantly assessed for its truth and effectiveness in fostering understanding. Through these discussions, both social researchers and intelligence analysts agree that policy-makers will not be consuming social science findings directly. In several senses, scientific findings need to be *translated*:

- Scientists typically seek generalizations about multiple cases under standardized conditions, while policy-makers face unique situations with a specific set of circumstances. If the generalizations are to apply, one has to reconsider how well they apply to particular situations.
- Scientific findings are often expressed in terms that are difficult for decision-makers to interpret. Scientists recognize the difficulty when educating undergraduates. They will often revert to stories and colloquial language in lectures, and students are assigned readings from classic works that would never pass muster if offered as a journal article or dissertation.
- However well the scientists communicate with undergraduates, they may fail to "speak the language" of policy-makers or anticipate what they want to know. A familiarity with statecraft makes the communication of scientific findings to such consumers more effective.

Interactions with the Intelligence Community

Enter the IC analysts and briefers, whose advisory products and briefings are tuned to what decision-makers will be able to grasp and apply. While social science findings have not traditionally had a deep and direct effect on intelligence products, there is a growing need and willingness to use such sources. As situations have become more complex and confusing and where stock answers have become less convincing, some IC briefers have said that policy-makers are now asking for deeper understanding. They are more willing to sit for *deep dives* where complexities, uncertainties, and different ways of framing the problem are elaborated. This is an opening for social science.

In what is called *anticipatory intelligence*, analysts venture beyond a factual update to present a more systemic understanding, replete with dynamics that are expected to shape events. A separate project at the National Academies is meant to support this burgeoning specialty in the IC. The theories that describe various social and behavioral phenomena, such as deterrence, are being gathered and synthesized for the purpose of reminding the analyst of the many factors and interactions proven in the scientific literature to be either relevant or irrelevant.

Translation of this kind is certainly needed, and there is much that can be done at every stage of intelligence production to gather the value of social research and make it available to support wise statecraft. Yet this process of translation rests on some assumptions that deserve further examination. Rather than assuming a wide and permanent gap between social science researchers and the policy-maker and placing the intelligence analyst as the necessary conduit between them, would it be possible to reduce the gap, even to the point of not needing the intelligence intermediary at all?

Senator Patrick Moynihan, a social scientist turned politician, ably advocated for policy that had a strong grounding in research. This is not to say that his policy positions always proved to be superior, but he forced a deeper understanding of family and cultural dynamics among his peers, and social policies that he participated in shaping were not drawn in ignorance or by prejudice. We may also point to influential advisors who drew on academic research, such as Henry Kissinger and Zbigniew Brzezinski. (Kissinger, interestingly, barred intelligence advisors from policy discussions.) There is no chance that the United States would ever, as a rule, appoint what we might disparagingly call Mandarins or philosopher kings, but a bit more education and research experience among our leaders and advisors might reduce the gap. To some extent, this is achieved through the professionalization of Congressional and National Security Council staff, who often perform translation (of a somewhat different kind from intelligence reporting) by bringing both reasonableness and social science knowledge to policy deliberations.

A recent instance of gap shortening from both sides is instructive. DoD's Strategic Multilayer Assessment (SMA) program brings together a large number of military personnel with a large number of social researchers (Strategic Multilayer Assessment 2018). The military personnel are not standing back from the discussion, nor are they requiring advice directly applicable to situations preconceived as military threats. They are eager to gain understanding of patterns that underlie current challenges, especially for situations where conventional military practices are no longer convincing. For example, when Russians attempt to subvert countries with information operations, are these attacks effective, and if they are, can or should these actions be countered or deterred? There are many new features to these situations that require new thinking backed by research.

Like their military counterparts, the researchers who are drawn to the SMA program often break frame from the typical academic role and speak about specific applications of their findings. Those in the academic mainstream may look down upon this work and not reward it as they would publications in prestigious scientific journals. But there are different rewards, such as government-funded studies, consulting opportunities, and service to the nation.

What is especially interesting is that very few intelligence personnel attended the last SMA conference. One would think that IC translators would find this event to be an excellent opportunity to develop relationships on both sides of the gap. But let's put that issue aside and dwell further on how social sciences can better serve statecraft over the next 10 years. I believe that this will require some breaks with conventional standards and practices. In the next sections, I review relatively recent shifts in the research community that appear promising, beginning with phronetic social science.

Phronetic Social Science

Duncan Watts has sounded a warning: "Ten years into the era of what is now called computational social science, it seems to me that more data, and even better data, is not enough. Nor has the influx of physicists and computer scientists into the social sciences over the past two decades clearly ameliorated the coherency problem. Far from the social sciences acquiring a coherent physics-inspired core of empirically validated theoretical knowledge, they have instead acquired a whole new batch of physics-inspired models that have, if anything, added to the confusion" (Watts 2017).

Watts argues that social science can make headway on the problem of incoherence through *use-inspired research*. Instead of developing many incommensurable theories for the same facts, organizing around practical problems may result in findings that build upon each other. Perhaps so, but let's pause at the word *use*. That is a recommendation to take on a perspective, inevitably an interest in society.

Flyvbjerg reaches back to Aristotle for guidance at this juncture (Flyvbjerg 2001). Aristotle taught that the social and political realm requires *phronesis*, a type of knowledge different from that of the physical realm. A leader with *phronesis* has good judgment and act with wisdom. We moderns don't intuitively grasp that the ability to negotiate political situations with good judgment is a type of knowledge, though this is what we want leaders to do. It is possible for social scientists to contribute to this knowledge if they address the leader's problem directly. This doesn't mean that all the existing methods need to be thrown out nor that we dispense with objectivity or theory, but it does mean that the work of social science becomes directed toward use within political reality, as Watts recommends. Flyvbjerg summarizes: "The primary purpose of phronetic social science is not to develop theory, but to contribute to society's practical rationality in elucidating where we are, where we want to go, and what is desirable according to diverse sets of values and interests: The goal of the phronetic approach is to add to society's capacity for value-rational deliberation and action" (Flyvbjerg 2016).

One implication of use-inspired phronetic research is that we embrace the case study as a means toward knowledge. The unique characteristics of a case are exactly what a decision-maker faces, and scientists should not be deterred from dealing with the case as is or believe that case inquiry is somehow inferior. In fact, there are good arguments that it can be superior (Flyvbjerg 2007).

While we take on a perspective, that does not make us blind to other perspectives. In fact, perspectives become thematic, aimed at truly understanding the different views and motivations and how they clash. This is particularly important with the international and intercultural issues that the IC deals with. For those, different actors can frame the situation very differently. It becomes crucial to understand the just aims of all the actors as they see them, even though American understanding differs.

In this approach, systems are not being isolated from their environments in order to make them analytically tractable. Similarly, phenomena favored by one discipline (such as economics) are not isolated from phenomena favored by other disciplines (such as politics). Influences from environments are ever present in reality, and studies that omit them from consideration render misleading results, creating a sense of surety that is not justified. An awareness of environmental factors and interpenetrating systems helps a decision-maker gain an accurate understanding of contingency and uncertainty. This requires that the researcher take at least an interdisciplinary approach (but it is better not to isolate systems and disciplines in the first place).

The resulting studies may seem vague and inconclusive but appropriately so. Results offered in the form of an optimal strategy within an intentionally simplified and isolated representation of the problem are misleading. We are often led to believe that, while a true representation of the problem isn't delivered today, it is just around the corner, typically with more data, more epicycles, and more funding. The wise decision-maker requires something else. As operations researchers once said, he or she needs a decision aid, not a decision made. The decision aid will be a richer understanding of the complex thicket of influences and the potentials for shaping and developing the situation. This kind of science is not entirely unfamiliar. Some scientists will claim that it is exactly the *muddled thinking* that we are trying to get past. But observe what has been happening in the field of economics. After a lengthy detour into a mathematized standard model, several strains of open, systemic inquiry are returning, and decision-makers are finding it enlightening and useful. This includes institutional economics, behavioral economics, ecological economics, and even economic history, all supported by big data, but not at the same time dominated by simplifying assumptions and formalisms.

The rear guard will continue to fight back, spouting *rigor* and other shibboleths to keep science pure in a particular way. Yet these methodological commitments, enforced by academic incentives, can limit the contribution of social science to sensemaking in complex environments. It is interesting to see how this tension has played out over the years in the American Political Science Association. In the 1960s, some members wanted to speak more freely about the political controversies of the day. A compromise was reached, whereby members who wanted to organize nonsanctioned events were given space to meet after the conclusion of the conference. They were eventually allowed to list these events in the back pages of the conference book, apart from the officially sanctioned agenda. Over the years, the official sessions become staid. These sessions were necessary for careerists but were of middling interest to those who wanted to understand politics. In recent years, the "back of the book" sessions have commanded the largest spaces and attracted the largest audiences. Some of the most popular sessions are run by political philosophers whose level of scholarship is unquestionably high though clearly out of step with the remnants of positivism and behaviorism.

Closer to home, this tension was visible in a prior National Academies project sponsored by the IC, where social scientists were tasked to examine the craft of intelligence (National Research Council 2011). In one of the public meetings, an exasperated scholar asserted that the intelligence process was unscientific and needed the standards that he was using. Members of the IC didn't take kindly to this suggestion; they viewed the bulk of social science as a game that is irrelevant to the needs of statecraft (Myers 2010).

Quite recently, another NAS committee of social scientists was asked to assess the value of social science in terms of addressing national priorities (National Academies of Science 2017). In other words, is social science relevant, and in particular, is it relevant to national security? The panelists cautiously concluded that some social science does have significant value, but which part? A questioner at the presentation of this study wanted to know whether the panelists, all of whom made their careers as quantitative scholars holding central positions in their fields, were speaking only of the brand of social science that they pursue. Did they think that qualitative research had any value or only minor value? Years ago, the response might have been a polite acceptance of qualitative research, perhaps as an unavoidable preliminary to the conduct of *real* science. But the answer this time was quite different. All panelists indicated surprise at the question. They saw no essential distinction between qualitative and quantitative research and asserted that both types of research could be rigorous and highly valuable.

Based on convergent methodological trajectories, the gap between case-oriented intelligence and social science may appear to be collapsing, but the communities continue to thrive in parallel worlds with very limited contact. Eventually, one would hope to see that the IC is both drawing from and contributing to the phronetic portion of the scientific corpus on a regular basis. The evidence so far is scant, though the potential for greater two-way collaboration abounds (Argrell and Treverton 2015).

Cognitive Domain

A recent *turn* in social science is toward the cognitive domain. While this movement takes many directions, a major starting point was to simply drop the behavioristic rule that behavior, because it is measurable, is all that matters. What people are thinking has always mattered for statecraft, and social science has found a way to rejoin the conversation about political thought, bringing along big data computation with new kinds of measurement (McCubbins and Turner 2012).

What is an actor? Interpretive social science answers this question differently from most reductive methods. Jerome Bruner, an early proponent of a practical American strain, spoke of the "story structure of experience" (Bruner 1990), meaning that "action is understood in the same way as a plot in a narrative. Action is always symbolically mediated, symbols acting as a quasi-text that allows conduct to be interpreted" (Ezzy 1998). While many scholars launch into flights of hermeneutical elaboration at this point, others address some very workaday concerns for the IC. Al-Qaeda and ISIS have little more to offer than stories, after all, and look at how durable their stories are and the dramatic actions that result.

The NAS decadal committee ran a full-day workshop on narrative, sensing that there is much more to be learned about it that will benefit national security. We now have recorded linguistic data as never before, full of symbols that mediate action, posing a major interpretive challenge. While computation can't do the job alone, it can do quite a lot to identify patterns of meaning and communication that will harm or help US national interest. The military recognizes that *nonkinetic* conflict, waged with words and symbols, has become serious business. Russian influence operations, in particular, are an acute concern.

The analysis of such attacks involves much more than counting and categorizing key words. Advanced sentiment analysis software can do a better job of characterizing the barrage of messages, but it remains difficult to tell what the actual effects are, plus what an appropriate response will entail. One has to consider, for example, social identity, a vague but certainly powerful way in which people frame words and events and are moved to act.

Some motivations defy easy characterization. LTG Michael Nagata pointed out in an SMA session that, while we can describe the ISIS ideology of death, it is not obvious why many people find this ideology sensible and appealing. Without that understanding and related understanding of motivations, our strategies may be ineffective (Davis and Cragin 2009). One of the white papers submitted to the NAS committee argued that conventional psychology has failed to fully illuminate the appeal of ISIS ideology and that it is more fruitfully framed as a spiritual disorder, of a piece with the millenarian outbreaks during the Middle Ages (Franz 2017). A decadal committee member recognized that this was a neglected direction for cognitive research, violating certain rules about what psychology may legitimately address yet clearly of practical significance to national security.

In more mundane affairs of state, we have a lot to learn about the emotional component of cognition. Studies show that a certain ratio of positive emotions to negative emotions (between roughly 3 to 1 and 6 to 1) is correlated with flourishing, understood as a state of positive functioning and positive feeling most of the time. People who have a lower ratio or higher ratio are less likely to be flourishing. That might sound like a measurement nightmare, but the concepts align with human experience, and reliable ways have been found to measure them at the individual level (Fredrickson 2013).[1] What is unexpected is that this finding scales up from individuals to whole countries. The 20 top flourishing countries have populations in the 3–6 positivity ratio range, the same range that would predict high flourishing at the individual level. See Figure 39.1 below (Myers 2018).

It is not obvious why this relationship should repeat at the national level. Future research may hold some surprises. If the ratio in a country were to rise, would this facilitate development, rather than merely reflect it? If leaders are truly able to influence the emotions of a population, might emotional *jawboning* be more consequential than we suppose? This is all highly speculative, of course. I am merely illustrating that cognition has something to do with statecraft and that use-inspired inquiry into cognitive phenomena has potential. Novel data collection will be required, along with novel theory, measurement, and modeling.

---

1 Fredrickson (2013) withdrew her prior claim for a specific, optimal positivity ratio, but she stands by a more relaxed claim used here that a "higher positivity ratio is better, within bounds."

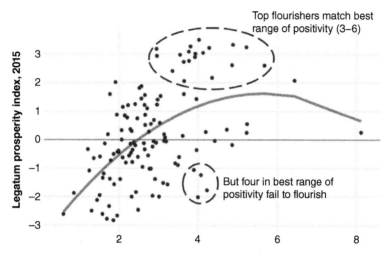

Figure 39.1 Positivity ratio and level of flourishing for 123 national populations.

Reflexive Processes

We want our models to represent social processes accurately, but we often use limiting assumptions consciously or not, in order to make the models work. It is possible to relax these assumptions. Behavioral economics, for example, recognizes that people use biases that create results that deviate from the assumptions of a rational actor. Complexity theory relaxes the assumptions of conventional economic theory concerning equilibrium and diminishing returns. Another troublesome assumption remains, one that cuts much deeper.

This common assumption is that a social system doesn't change in relation to what it is observing about its own actions and perceptions and particularly about its expectations. This act of self-observation is referred to as reflexivity, a major dynamic that generates uncertainty and perhaps increasing uncertainty, as social systems become increasingly informed and more densely linked.

The effects of reflexivity are most easily viewed in the stock market. Fallible investors perpetually misprice stocks according to their biases. When a price begins to rise, investors will follow, sensing that others perceive increasing value. At a certain point, there is a large gap between what many perceive as the value vs. the *actual* value, and at that point, the belief in the perception collapses. Thus, a run-up of a stock price often begins when many view a stock to be underpriced in relation to its future earnings. The price will rise gradually, which by itself promise higher future earnings, generating more investment, but then, the price collapses quickly after it reaches a peak that is no longer believable. It drops even faster due to the use of leverage when purchasing stock. To try to analyze this situation simply with data trends or supply and demand misses why and how the market works. Mimicry among market players observing each other is itself an influence on what the market contains (Harmon et al. 2015).

George Soros analyzed financial market in this manner and says that this is how he made his billions, by assessing contagious perceptions, typically in situations far from equilibrium (Soros 2014). He explains: "Central to my world-view is the idea that human affairs—events with thinking participants—have a fundamentally different structure from natural phenomena. The latter unfold without any interference from human thought; one set of facts follows another in the causal chain. Not so in human affairs. The causal chain does not lead from one set of facts to the next, but connects the situation and the participants' thinking in a two way, reflexive feedback loop... since there is always a divergence between the participants' view and the actual state of affairs, reflexivity introduces an element of uncertainty into the course of events that is absent in natural phenomena" (Soros 2009, p. 284). He goes on to argue that fallibility and reflexivity are endemic to all social systems and are particularly important for actors to be aware of who must formulate expectations to guide decisions with broad consequences. This mode of analysis does not guarantee the elimination of errors, in that all social systems are indeterminate and unknown biases remain, but it can create a useful edge. It also renders suspect other kinds of modeling that purport to offer something more reliable, based on what are imagined to be constants in society.

This is not to say that existing analytics are worthless, only that they have a range of application, typically within what Dave Snowden (developer of the Cynefin framework for aiding decision-making) calls the *complicated* environment, where professional rules of thumb work and where relationships persist over time. In the *complex* environment, where there is greater volatility and stresses that change relationships in unpredictable ways, analytics that encompass reflexivity are called for. Snowden recounts the time when he presented this distinction between complex and complicated to Admiral John Poindexter, then of DARPA, explaining that it is a mistake to apply the methods that work in the complicated regime to the complex regime where they don't work. Admiral Poindexter quickly grasped the implications and burst out, "That explains 50 years of failure in American foreign policy! We treated complex as complicated" (Snowden 2018). This remark was surely an overstatement, but Poindexter might have been thinking of Robert McNamara's inappropriate application of *best practices* from the complicated auto industry to measure and manage progress in the complex Vietnam war.

An account of complex social reality, one that embraces human agency, cognition, and subjectivity, will necessarily be path-dependent and uncertain. Models that aid the interpretation of sentiments will certainly be useful. What might be helpful as well is a combination of scenario writing with system dynamics modeling, where self-modifying relationships are more easily drawn. Snowden has additional techniques that seem strange because they involve a great deal of analyst creativity and interpretation rather than automated procedures. We may need to think less about setting parameters in deductive, deterministic machines and resume efforts to develop sensemaking skills akin to the *phronesis* of wise decision-makers.

Developments ensue not only from the interaction of focal variables but also from the field itself. The *ground* is in motion, and there will be trends in the environment that will have a shaping effect on reflexive situations. They constitute an envelope of plausibility through which reflexive scenarios can be drawn. The National Intelligence Council's (NIC's) publication, *Global Trends*, has admirably formulated the environmental background of statecraft (National Intelligence Council 2016). The integration of this material with situational analysis is unfinished business. It was revealing that James Clapper, the former director of National Intelligence, said that he never had the time to read this publication when he was on the job, but after he retired, he was able to read it at his leisure and found it very helpful. Somewhere along the sensemaking process, somebody will need to take the time to lash long-term environmental factors with the challenging situations of the day.

Conclusion

Some of the social science movements I have mentioned have been controversial, not least because they seem to sacrifice the ideals of simplicity and clarity, yet they aspire to comprehend society as we find it, not as we prefer to find it under simplifying assumptions.

Social science can help policy, but that will not be an automatic result of having more data, bigger models, and more findings. For the purpose of wise statecraft, the chronic gap between social science and policy-making needs to be reduced, and this requires adjustments on both sides of the gap. Researchers need to conduct use-inspired phronetic social science. Policy-makers, for their part, need to develop as sensemakers who are conversant with the dynamics of reflexive social systems and new methods of inquiry appropriate to the complex domain.

Where does that leave the intelligence function? There is enough institutional inertia to keep intelligence analysts in business, but secret factual briefings are a thin leg to stand on, especially when so much can be learned from open sources these days, often with faster turnaround than the IC can muster. Could intelligence inquiry, as it might develop under the new directions and standards I have sampled, come to resemble a new phronetic social science? Indeed, new developments in anticipatory intelligence move in this direction by:

- Playing out indeterminate simulations as a series of contingencies and inter-actions.
- Illuminating mental models and subjecting them to examination.
- Gaining a sense of how background environments are shifting and affecting focal events.

In this context, big data modeling efforts risk aimlessness and excess, especially to the extent they resemble prior efforts that misrepresented systems and failed to usefully inform policy. But there are promising new areas on which to train the computers. One is the relatively neglected cognitive domain, where agreements and disagreements continually construct the unpredictable reality that decision-makers must negotiate.

And then, there are the beleaguered humans, both the expert analysts/advisors and the decision-makers. We have mentioned that they need to become familiar with new methods and informed by their products – indeed that is the thrust and value of this volume – but they should also not succumb to either of two abnegations: that their human expertise is dispensable or that their gut impulses are just fine and in no need of development or refreshment. Gary Klein has pointed out that there is an attack on expertise from research in five areas where the weaknesses of human decision-making have been exposed: decision research, heuristics and biases, sociology, evidence-based performance, and information technology, including especially artificial intelligence (Klein 2018). Klein makes a stout argument that "none of these [expert-denying] communities poses a legitimate threat to expertise. Left unchallenged, the overstatements and confusions that lie behind these claims can lead to a downward spiral in which experts are dismissed." Rather, the findings should be used to strengthen the work of expertise, the illimitable element of wise governance.

References

Argrell, W. and Treverton, G. (2015). *National Intelligence and Science: Beyond the Great Divide in Analysis and Policy*. Oxford: Oxford University Press.

Bruner, J. (1990). *Acts of Meaning*. Cambridge: Harvard University Press.

Davis, P. and Cragin, K. (eds.) (2009). *Social Science for Terrorism: Putting the Pieces Together*. Santa Monica, CA: RAND.

Ezzy, D. (1998). Theorizing narrative identity: symbolic interactionism and hermeneutics. *The Sociological Quarterly* 39 (2).

Flyvbjerg, B. (2001). *Making Social Science Matter*. Cambridge: Cambridge University Press.

Flyvbjerg, B. (2007). Five misunderstandings about case-study research. In: *Qualitative Research Practice: Concise Paperback Edition* (ed. C. Seale, G. Gobo, J. Gubrium and D. Silverman), 390–404. London and Thousand Oaks, CA: Sage.

Flyvbjerg, B. (2016). What is phronesis and phronetic social science? (25 May 2016). https://www.linkedin.com/pulse/what-phronesis-phronetic-social-science-bent-flyvbjerg-%E5%82%85%E4%BB%A5%E6%96%8C- (accessed 5 June 2018).

Franz, M. (2017). Analyzing the mindset of religiously inspired terrorists. White Paper Submitted to National Academies Decadal Study of Social and Behavioral Sciences for National Security, 2017.

Fredrickson, B. (2013). Updated thinking on the positivity ratio. *American Psychologist* 68: 814–822.

Harmon, D., Lagi, M., de Aguiar, M.A.M. et al. (2015). Anticipating economic market crises using measures of collective panic. *PLoS One* 10 (7): e0131871.

Klein, G. (2018), The war on experts: five professional communities are trying to discredit expertise. *Psychology Today Blog*. https://www.psychologytoday.com/us/blog/seeing-what-others-dont/201709/the-war-experts (posted 6 September 2017, retrieved 5 June 2018).

McCubbins, M. and Turner, M. (2012). Going cognitive: tools for rebuilding the social sciences. In: *Grounding Social Sciences in Cognitive Sciences* (ed. R. Sun), 387–414. Cambridge, MA: MIT Press.

Myers, K. (2010). *Reflective Practice: Professional Thinking for a Turbulent World*, 67. Palgrave Macmillan.

Myers, K. (2018). National flourishing and shared positive emotions. *Second International Conference on Complexity and Policy Studies*, George Mason University, Fairfax, Virginia (April 2018).

National Academies of Science (2017). *The Value of Social, Behavioral, and Economic Sciences to National Priorities: A Report for the National Science Foundation*. The National Academies Press https://www.nap.edu/catalog/24790/the-value-of-social-behavioral-and-economic-sciences-to-national-priorities.

National Academies of Science (2018). Social and behavioral sciences for national security: a decadal survey. http://sites.nationalacademies.org/DBASSE/BBCSS/SBS_for_National_Security-Decadal_Survey/index.htm (accessed 5 June 2018).

National Intelligence Council (2016). Global Trends: Paradox of Progress. https://www.dni.gov/index.php/global-trends-home.

National Research Council (2011). *Intelligence Analysis for Tomorrow: Advances from the Behavioral and Social Sciences*. Washington, DC: The National Academies Press.

Snowden, D. (2018). Phronesis in Cynefin. Blog Entry (19 January 2018). http://cognitive-edge.com/blog/phonesis-in-cynefin.

Soros, G. (2009). *The Crash of 2008 and What It Means*, 2e. New York: Public Affairs.

Soros, G. (2014). Fallibility, reflexivity, and the human uncertainty principle. *Journal of Economic Methodology* 20 (4): 309–329.

Strategic Multilayer Assessment (2018). http://nsiteam.com/sma-description.

Watts, D. (2017). Should social science be more solution-oriented? *Nature Human Behaviour* 1 (1): 0015.

40

Lessons on Decision Aiding for Social-Behavioral Modeling

Paul K. Davis

RAND Corporation and Pardee RAND Graduate School, Santa Monica, CA 90407, USA

One important function of social-behavioral modeling (SBM) should be decision aiding (Davis et al. 2018; Davis and O'Mahony 2019), but for that function social-behavioral (SB) models will need to have characteristics that are currently unusual. In this chapter, I discuss those characteristics by drawing on experience in other domains in which model-based decision aiding has been important. The focus is primarily on informing strategic-level planning.

Strategic Planning Is Not About Simply Predicting and Acting

A common image of decision aiding is that leaders have a well-posed problem to which they need a solution. In this image, various model-based analyses predict which option would provide the best results, i.e. which option would be optimal. Classic methods include decision analysis and linear programming. The models employed are assumed to be mathematically sound and the data accurate, although perhaps expressed in statistical terms. The methods may be so good that the solution found can be accepted with confidence. If uncertainty analysis is necessary, it can be simple sensitivity analysis. For example, models may predict heavy snow and ice tomorrow with 20% probability, thereby aiding a city's decision-makers as they contemplate whether to have emergency storm groups on duty.

Strategic decision-making is different. Strategic-level decision-makers are commonly faced with ill-defined problems and may have only a vague and tentative sense of what the problem actually is. They may not recognize some of the factors that will affect the consequences of choice, much less how factors interact and create, e.g. feedbacks, time delays, and network effects

Social-Behavioral Modeling for Complex Systems, First Edition.
Edited by Paul K. Davis, Angela O'Mahony, and Jonathan Pfautz.
© 2019 John Wiley & Sons, Inc. Published 2019 by John Wiley & Sons, Inc.
Companion website: www.wiley.com/go/Davis_Social-Behavioralmodeling

(Forrester 1963; Dörner 1997; Sterman 2000). Uncertainties abound, especially regarding predictions about the future (Morgan and Henrion 1992). Profound disagreements may exist about facts, likelihoods, and consequences of taking a given option.

Three examples of strategic planning may be useful. First, consider defense planning to establish the size, capabilities, and postures of future military forces. Planners do not know the circumstances of future war: how it will start, what strategies will be employed, and so on. Strategic and operational surprises have been common (Bracken et al. 2008). A second example involves climate change. How should planners proceed amid uncertainties regarding the magnitude of such effects as sea-level rise? How high should seawalls be? When, if ever, should the effort to protect coastal urban areas be abandoned? What should be done to mitigate damage from the next big tropical hurricane? How should such mitigation efforts proceed given the varied interests of the tourist, fishing, farming, urban development, and industrial sectors? As a final example, consider a healthcare system. It faces major uncertainties about legislative developments and related economics, the pace of medical developments and their costs, what epidemics will arise, and the rise of antibiotic-resistant bacteria. No model exists that can provide a correct prediction of *the* future. Think of the unexpected turbulence associated with the Affordable Healthcare Act, efforts to repeal it after the 2016 election, and inconsistent efforts to prop it up selectively when repeal efforts failed. On the disease front, think of such surprises as the opioid epidemic of this century or the earlier HIV/AIDS epidemic.

A related problem is that, despite myths to the contrary, decision-makers often do not even know their ultimate objectives. They may express objectives, but those often change (March 1996; Morgan 2003; National Research Council 2014). For example, war objectives often end up different to those initially expressed. Also, when multiple stakeholders negotiate, compromise solutions reflect socialization processes in which the various stakeholders come to recognize (perhaps grudgingly) additional values. This is the realm of so-called wicked problems (Churchman 1967; Rittel and Webber 1973; Rosenhead and Mingers 2002). Yet another reality is that many issues will be revisited over time, with many facts and opinions also changing. Decisions at a given time may be more like establishing an initial rough direction than choosing a final destination. Sometimes, success will amount to *muddling through* (Lindblom 1965, 1979) or *logical incrementalism* (Quinn 1978).

Aiding strategic decision-making, then, is not just a matter of predicting the future, calculating the best option, and taking action accordingly. Finally, even the best decision aiding can only go so far because decision-makers must consider factors outside the realm of what analysts can help with (e.g. personal commitments and the give-and-take of politics). Some of this complexity is suggested by Figure 40.1, which depicts a picture of what is intended to

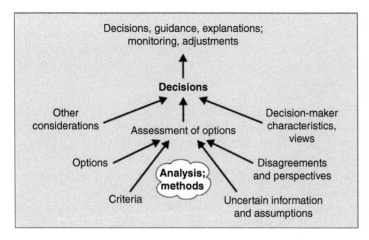

Figure 40.1 An image of analysis aiding decision-making. Source: Adapted from Davis (2014).

be rational decision-making followed by implementation, monitoring, and adjustments over time.

Characteristics Needed for Good Decision Aiding

Sizable academic literatures exist on decision-making and strategic planning, but in this chapter I draw most heavily on particular threads of work with which I am personally familiar. These have included theory, methods, and working directly with senior officials.

In the mid-2000s, an Under Secretary of Defense asked for an approach to decision aiding that would help him conduct *capability area reviews* to decide, in each of various military problem areas, what new military systems to develop and acquire. He saw current reviews as disorganized, shallow, and insufficiently strategic. He asked for extension and generalization of some prior work that I had led for the director of another agency. The subsequent study accomplished that and, in the process, noted principles much more general than the problem of defense acquisition (Davis et al. 2008a; Davis 2014). The principles emphasized that analysis should help decision-makers with the following:

- See and reason about the whole (top-down thinking).
- Recognize the multiple criteria for assessing options.
- See options as alternative portfolios of investments or actions applying diverse instruments to address the multiple criteria in a balanced way.
- Consider an appropriate range of options that anticipate intelligent reactions (e.g. by a military or corporate adversary).

- Evaluate the options in appropriate depth while confronting uncertainty and disagreement.
- Assist in finding budget-informed strategies that would be as flexible to changing missions, adaptive to circumstances, and robust to events as feasible (i.e., the strategy should seek *FARness*).

Although the F, A, and R in FARness measure significantly different things, for brevity in what follows I refer in shorthand to *planning for adaptiveness* while meaning to include all components of FARness. Terminological confusion is rampant in the literature with authors using words like robustness, agility, robust adaptiveness, flexibility, and even resilience to mean much the same thing (Davis 2014).

This chapter is not the place to discuss all of these matters, particularly option construction, but the next few sections discuss gaining a strategic perspective by taking a systems approach, studying issues both broadly and sometimes in detail, confronting uncertainty systematically, and finding adaptive strategies. The final section of the chapter draws implications for SBM if it is to be used for analysis to aid policy-makers.

Systems Thinking for a Strategic View

Concepts

The version of *systems thinking* that I write about here was pioneered in part by the introduction of systems analysis in the 1950s and 1960s (Quade and Boucher 1968), first at RAND and then in the Department of Defense (DoD) (Enthoven and Smith 2005). The original context was the interplay of budgeting, strategy, and resource allocation in DoD planning. Systems analysis broadened to cover social issues and also morphed into a more comprehensive approach called policy analysis with applications to many domains of social policy (Quade and Carter 1989). More or less in parallel, but with notable differences, today's systems thinking was also pioneered by MIT's Jay Forrester in what became the method of system dynamics (Forrester 1963; Sterman 2000). Later systems thinkers sometimes have taken more qualitative approaches for strategic planning in the corporate world and have emphasized learning and adaptation (Senge 2006). Although these strands are the ones I emphasize here, systems thinking has a much longer history associated with figures like Ludwig von Bertalanffy in biology, Kenneth Boulding in economics, and Peter Checkland in operations research (Boulding 1956; Checkland 1999; von Bertalanffy 1951, 1969). Some, like Checkland, have drawn sharp distinctions between *hard* and *soft* versions of systems thinking. I am less comfortable doing so because it is useful to slide from one to the other as befits the context.

The systems approach is different from the common social science approach associated with discrete hypothesis testing and competitive but fragmentary theories. Not accidentally, it is consistent with the challenge posed elsewhere that social science needs more unifying theories for use in SBM (Davis and O'Mahony 2019). Such work emphasizes a broad and comprehensive, causal top-down perspective (but without the negative connotation of simplistic top-down thinking). This implies the need to incorporate *all* the variables needed to understand system behavior in causal terms. Paraphrasing a point that Forrester often noted, to omit a variable is equivalent to assuming that its value is 0 or 1 (depending on whether it would be in an additive term or a multiplier). Systems thinkers most certainly believe in parsimony, but not the version of parsimony common in social science. Systems thinkers retain variables that are significant in the future whether or not their significance has been observed to date; also, they also retain *soft* variables, some variables that are difficult to measure (e.g. mental stress), perhaps treating them as uncertain parameters.

Examples

A first example illustrates the differences between systems thinking and discrete hypothesis thinking. In 2008, DoD requested that RAND conduct a comprehensive review of the social science on terrorism. Our interdisciplinary team was quickly struck by the myriad alleged explanations of terrorism in the literature: terrorism was due to mental illness; no, it was due to poverty and relative deprivation; no, it was due to a particular religion; no, it was due to any intolerant religion; no, it was a response to tyranny; no, it reflected nationalism or an extended version thereof (to mention only some of the explanations). Although researchers commonly pursue such separate threads, policy-makers need a more integrated picture. This implies that models to aid decision-making should also be more integrated. The subtitle of our book became *Putting the Pieces Together* (Davis and Cragin 2009).

No unified model existed in the realm of quantitative social science. And even if it had, it would have been in the form of regression models with limited value as aids to strategic decision-making. Fortunately, qualitative social science yielded rich insights allowing us to construct a more nearly comprehensive model, albeit one that was intentionally both qualitative and static. Reportedly, the model proved helpful to DoD officials and commanders. The review also led to follow-up research, including empirical case studies to validate and enrich the sub-model describing public support for terrorism and insurgency (Davis et al. 2012). That was later turned into an uncertainty-sensitive computational model (Davis and O'Mahony 2013, 2017).

In roughly the same time period, the DoD's Joint Staff sought to turn a new field manual for counterinsurgency into something more analytical. The manual was the product of a joint effort of the Army and Marines led by

Generals David Petraeus and James Mattis, respectively (Nagl et al. 2007). It had been informed by historical materials and fresh contributions by social scientists. It was, however, a field manual, not something easily used by analysts or modelers. The Joint Staff's Brett Pierson worked with PA Consulting to develop a system dynamics model to represent the concepts of that manual (Pierson et al. 2008). Doing so proved useful because the qualitative version of the model (the causal loop diagrams, rather than model outputs) allowed discussion even by senior officers, of complicated interactions and feedbacks with implications for military planning. Even simple modeling of this sort can be quite informative, as illustrated by another chapter in this volume (Lofdahl 2019).

Going Broad and Deep

How Much Detail Is Needed?

A common point of controversy in modeling and analysis is how much detail to include. Scientific inquiry often calls for going deep, and the urge often exists to go deep so as to exploit modern computers and low-level data. Analysis in support of policy-makers, however, needs to be different. Even with infinitely fast computers, policy-makers need to reason about, discuss, and sometimes reach consensus regarding options to be taken. This implies the need for low-resolution causal models (mental or computerized). That is, the process of abstraction is not just an analytic convenience, but something essential for decision-making in all walks of life. None of us can have coherent and convergent discussion when talking about tens or hundreds of variables. It follows that we also need models that depend only on abstract (i.e. aggregate) variables. Again, this is general across domains. For example, a physician will ordinarily discuss issues at the level of organs and tissues, rather than cells or molecules.

The feasibility of well-grounded reasoning at an aggregate level depends on nature cooperating. Fortunately, the real world is often amenable to approximate description at different levels of aggregation because of nature's nearly decomposable systems (Simon 1996). So also, the phenomena of social science are describable at different levels of detail (Zacharias et al. 2008), although the neatness of separation is not yet well understood, as noted by another chapter in this volume (Orr 2019).

This said, the devil is often in the details. We all know that aggregate descriptions can be insidiously misleading as with "Sure, you can swim in the pool safely: on average, it's only two feet deep" or "On average, students do better in classrooms with fewer students, so reducing class size is a priority." Many policy errors are made with analogous simplifications.

A Dilemma?

It may appear that a dilemma exists: on the one hand, we need the high-level (low-resolution) synoptic view, but on the other hand, we must worry about higher-resolution details. To proceed, we need to understand the dilemma better. Upon reflection, a number of distinct problems arise:

- *Improper abstraction* (incorrect theory): Averages are often a poor basis for aggregation. The correct aggregation depends on the context in which it will be used. To use the first example above, it may be better to represent the pool's depth by its greatest depth or by the depth at the stream's clearly marked crossing point. To use the second example, a local school board might care about the average value of reducing class size across only those schools comparable to itself in teaching quality, socioeconomic status of students, and physical security. This context dependence of aggregation is seldom noted when authors show diagrams postulating that higher-resolution models should somehow be mapped into lower-resolution models.

- *Low-level failures*: The validity of an abstraction's value depends on estimates of lower-level variables. For example, the safety of a nuclear reactor estimated by engineers will be seriously wrong if *any* of the reactor's critical components is built by corrupt contractors with substandard materials. Similarly, failure of an international intervention operation to enforce peace may be doomed to failure in *any* one of several critical components of the challenge that are sufficiently troubled (security, economics, governance, or sociology). Such components interact nonlinearly (Davis 2011).

- *System instability*: Sometimes, an aggregate-level model has been accurate for a long time, but the complex adaptive system (CAS) that it represents finds itself in a state in which it is extremely sensitive to perturbations. The model may still be correct on average, but with huge variations is possible with systems on the *edge of chaos*. As an example, a crowd in a social gathering may seemingly be normal and the security force attending to it may be adequate by historical standards, but the crowd may be ready to erupt because of recent emotional events. If it erupts, no one knows how it will behave or what will happen.

Resolution of Dilemma

These and other reasons exist for worrying about high-level (aggregate- or abstract-level) reasoning. Nonetheless, the comprehensive high-level (low-resolution) reasoning is crucial. The admonition, then, is that

- Analysis for higher-level decision aiding should provide a high-level comprehensive view and provide selective detail *where needed*.

That is, decision aiding should provide the high-level view and allow *selective* zooming into detail. The detailed views will only sometimes be necessary. This also reflects common experience. For example, a health system might rule out use of a drug knowing only that it is expensive and has common bad side effects and that less costly alternatives exist. Occasionally, however, for a particular patient who reacts poorly to the alternatives, assessing the drug in question in more detail would be appropriate.

When being provided analyses, decision-makers appreciate zoom capability, even though they seldom have the time or inclination to go into depth. They can expect their staffs and advisors to do so. Further, even a policy-maker or top staffers will sometimes want more depth to test the mettle of those who prepared the analysis, to understand the underlying reasoning better, or to check on special issues on which they have received backdoor warnings. And to be sure, a policy-maker may demand depth when things have gone very wrong and it is necessary to understand how and why.

This need for selective zooming applies in a number of dimensions. *Resolution* can be higher or lower in, for example, the objects treated (e.g. individuals vs. society), the richness of attributes ascribed to the objects (e.g. a simple economic utility function or behavior driven by values, psychology, and perceptions), spatial and temporal scales, or the richness with which relationships and interactions are described (e.g. interaction only with adjacent objects or with objects anywhere within the relevant social network) (Davis and Hillestad 1993). This is worth noting because so much discussion of resolution and scale overlooks the different dimensions. An agent-based model with a million individual-level agents may be rich in some respects, but simplistic in others – if, for example, the agents use the same simple decision rule based on maximizing economic utility.

How much selective zooming is necessary for decision aiding? Realistically, the answer is that it all depends. A classroom rule of thumb is that the analyst needs to work at the nominal level of detail, one level deeper to know where pitfalls lie and one level higher to appreciate what simplifications are useful for the policy-maker. This need for multiresolution view has important implications for modeling and analysis.

Confronting Uncertainty and Disagreement

Uncertainty is a core reality for higher-level planning. So is disagreement, which logically may be considered part of uncertainty but manifests itself differently. Confronting these issues is more difficult than recognizing them. Intuitively, people often assume that going with a best estimate is good enough. Or they have in mind only one or two discrete uncertainties, such as how fast the economy will grow next year. Uncertainties, however, exist in many dimensions, and best estimates may be counterproductive. In the past, attempting to confront uncertainty was often paralytic, which is why uncertainty analysis has so often been absent. In the late 1980s, for example, doing better on such matters was highlighted as one of the truly hard and important problems of policy analysis (Walker et al. 1987). At that time, the concept of multiscenario analysis was just appearing (Davis 1988). Much has happened in the last 30 years and we can now do much better.

Normal and Deep Uncertainty

An admirable textbook discusses uncertainty analysis in depth for variables with uncertainties that can be characterized by parameter ranges or probability distributions (Morgan and Henrion 1992). Elegant analysis is possible, to include analysis informed by Bayesian models accounting for both objective and subjective uncertainties.

More recent developments address *deep uncertainty*, a term that traces to Nobelist Kenneth Arrow (Bankes 2002). Aspects of deep uncertainty have been referred to over the decades as scenario uncertainty, future uncertainty, model uncertainty, or *real* uncertainty. More precisely, one definition is:

> Deep Uncertainty: the condition in which analysts do not know or the parties to a decision cannot agree upon (1) the appropriate models to describe interactions among a system's variables, (2) the probability distributions to represent uncertainty about key parameters in the models, and/or (3) how to value the desirability of alternative outcomes. (Lempert et al. 2003)

As a simple example, how the ball will fall in roulette is uncertain, but we know the odds (normal uncertainty). In contrast, how some war or plague will start in 2030 is not something for which we have useful statistics or reliably predictive models (deep uncertainty).

A crucial first step in addressing deep uncertainty is systematically characterizing the dimensions of uncertainty so that, in a given study, analysts can decide which to address. Some methods for doing so have proven useful, and I describe two of them here.

Exploring Uncertainty in Scenario Space

An early example was in 1990s defense planning, which included conceptualizing the *scenario space* that planning should consider. The left column of Table 40.1 shows suggested dimensions (Davis 1994, p. 82). First, what is the war or crisis all about? What are interests? Allies and adversaries? From what did the crisis or war develop? Has there been a long preparation period for war, or did something pop up quickly? Second, what are the various participants' objectives and strategies? Third, what are the strengths of the participants: the size and character of their armies, navies, and air forces? Fourth, what are the

Table 40.1 Dimensions of scenario space.

Defense planning	Generic
1. Political–military setting (e.g. origin of crisis; alliances; broad interests; and timing of warning, alerts, mobilization, deployment, etc.)	Context
2. Operational objectives and strategies (for the US, opponents, allies, and Third countries)	Objectives and strategies
3. Forces and other instruments of power (e.g. orders of battle, structure of units)	Power and resources
4. Weapon system and individual-force capabilities (e.g. accuracy of precision munitions, the movement rate of armored units, efficiency of command-control systems, and the qualitative effectiveness of officers and men resulting from training, morale, and other factors)	Effectiveness of resources
5. Geographic and other aspects of environment (e.g. weather, terrain, transportation networks, and port facilities)	Geographic location and its generalizations
6. The processes that govern military operations, including combat (e.g. the equations describing the phenomena of combat and movement)	Relationships and processes governing systemic changes

Source: Davis (1994, p. 82).

capabilities of the participants' weapon system and force units? Fifth, what are they physical circumstances of geography, terrain, weather, and infrastructure? And, last but definitely not least, what do we know and not know about all the interactions and processes? To illustrate the last item, a conceit of combat modeling had long been that the models could predict the results of combat. Realistically, the combat models were (and are) unreliable except when results are overdetermined, as with massive force superiorities. Ultimately, the models had a base of sand (Davis and Blumenthal 1991).

The right column of Table 40.1 is generic, applying to many social-policy problems and defense studies. Different terminology may be appropriate, but the breakdown is a good starting point. For example, in a given study, one might replace *Objectives and strategies* by *Risk-reduction areas* because the sponsor of analysis likes to think in terms of risk reduction. As another example, row 6 (relationships and processes) might correspond to different assumptions about economic elasticities, the exponents of a Cobb–Douglas equation, or whether tax cuts can stimulate the economy enough to pay for themselves.

With this construct, a case or scenario can be seen as a point in scenario space as indicated in Figure 40.2. This is merely a cartoon, but viewing the analysis challenge in this way helps participants to internalize the need to

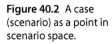

Figure 40.2 A case (scenario) as a point in scenario space.

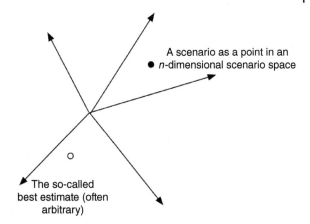

A scenario as a point in an *n*-dimensional scenario space

The so-called best estimate (often arbitrary)

address not just best estimate points and a few convenient sensitivities, but the larger problem.

As complex as the scenario space already is with the six dimensions of Table 40.1, each dimension has many components and levels of detail. Nonetheless, it has proven valuable at the beginning of an analysis project to review uncertainties prodded by Table 40.1. Analysts quickly add specifics appropriate to the project. Two half-day sessions can be sufficient to open minds ("Oh, drat, yes we actually need to vary that"), to explicitly narrow scope ("That's important, but not relevant to our study for the particular sponsor office"), to decide on how to represent the many abstractions ("we'll represent this by warning time and consider only two values"), and to set the stage for establishing requirements for modeling and data collection (to include representing uncertainties). At least some public documentation illustrates how this was turned into down-to-earth practicalities in early post-Cold War analysis (Fox 2003).

Exploration Guided by the XLRM Framework

A different framework has been used in numerous social-policy studies dealing with, e.g. climate change, allocating water resources, disease, and epidemics. This is the XLRM framework shown in Table 40.2, as constructed from a study on robust decision-making (RDM) (Lempert et al. 2003) and various informal materials used by the study's authors. L refers to policy levers, X refers to *exogenous factors* outside control of the decision-makers, M are performance standards that decision-makers and their constituents may use to rank scenario desirability, and relationship R describes the ways in which factors relate to each other and govern evolution of the future based on the decision-makers' choices of levers and the various X factors. The XLRM approach is similar to the scenario-space approach in many ways but gives greater visibility to a few policy levers and a few measures of effectiveness. It is the framework

Table 40.2 Illustrative use of the XLRM framework.

Uncertainty factors (X)	Policy levers (L)
Uncertain factors not controlled by planning organization (i.e. exogenous factors)	Strategies available to planning organization

Models (R)	Performance metrics (M)
Models to estimate metrics of performance (M) for each strategy (L) across ensembles of assumptions about uncertain factors (X)	Metrics used to evaluate performance of alternative strategies

that I would recommend for many studies using models of social and behavioral phenomena.

The Nuts and Bolts of Coping with Dimensional Explosion

Given the dimensionality of the problem, how can one do analysis? A variety of tactics make the analysis feasible. They can be used separately or in unison in a given study:

1. Use low-resolution models as abstractions suitable for exploration.
2. Decide how the abstract dimensions of uncertainty will or will not be addressed (concrete variables and parameter-value sets).
3. Use appropriate test cases or use sampling methods rather than attempting to cover all cases.
4. Look for patterns that make it unnecessary to look at huge portions of the scenario space (e.g. find contours such that results are unacceptable beyond boundary lines or, conversely, are fully acceptable).
5. Hold some independent variables constant because they are known accurately or believed to be relatively insignificant (e.g. the gravitational constant or the average number of numbers in a string that can be remembered by a human).

These tactics are all subtle in practice and analysis continues to be a demanding activity. For example, the model used for exploration must be structurally valid for that purpose (Davis et al. 2018; Davis and O'Mahony 2019). Decisions on which uncertainties to include are inherently judgmental. Finding appropriate test cases may require analytical sophistication, as in identifying *spanning sets* of test cases (Davis 2014). Sampling methods (e.g. built-in tools for Monte Carlo analysis) usually assume independence of variables even though many of the variables of exploratory analysis are correlated. For example, in assessing a military defense, it would be folly to imagine that being surprised and being distracted at the time of attack were independent low probability events: a smart adversary would use tactics to achieve both at the same time. Because of such correlations, typical exploratory analysis at RAND does not assign probability distributions. Instead, judgments about likelihood are deferred to the end of analysis where one or a few judgments are understandable and meaningful rather than buried (Lempert et al. 2003). Finding boundaries beyond which exploration is unnecessary is nontrivial in an n-dimensional scenario space (although data-mining methods address this). As for holding many variables constant, choices on this are not uncommonly a source of error, e.g. when one accepts uncritically the validity of an official database or assumes that a system will perform according to its original specifications.

Despite such problems of detail, this method of organizing exploratory analysis has served well in numerous studies.

Finding Strategy to Cope with Uncertainty

Although exploratory analysis can be very interesting for the analyst, it is the divergent-thinking part of a study and does not in itself yield anything. Success in decision aiding must include a convergent process of finding strategies. A number of approaches can be taken, each with its own name, process diagrams, terminology, and gimmicks. The literature abounds with such approaches. In what follows I touch upon only three while providing pointers to more general literature.

Planning for Adaptiveness with a Portfolio of Capabilities

The approach taken in a number of studies has been to find, working iteratively with policy-makers, strategies that are flexible, adaptive, and robust (achieving FARness) or, again as shorthand, finding ways to plan for adaptiveness. The process has been as follows (Davis et al. 2008a; Davis 2014):

1. *Frame the problem as portfolio analysis*: Commonly, policy-makers have numerous objectives and should therefore have numerous criteria for evaluating options. They also have numerous instruments to help in reaching their objectives. The decision, then, is about what mix of instruments should be used to address the mix of objectives in an acceptable way and at an acceptable cost. An early example was the 1950s concept of investing in a mix of stocks and bonds to achieve both long-term capital gain and reduce risk of losing unacceptably large amounts of capital (Markowitz 1952).
2. *Identify criteria*: Identify the multiple criteria needed to address separate objectives and separate major uncertainties. This requires grouping to avoid dimensional explosion.

3. *Use policy scorecards* to discuss the options by the multiple criteria. Some of the criteria can be unabashedly subjective (e.g. quality of life). Do not attempt to *optimize* with some a priori objective function. Because policy-makers often do not know their objective function (or even have one), they should review the various options with multi-criteria scorecards as illustrated below.

4. *Construct diverse and creative options*: Decision-makers frequently lament the poverty of the options that come up from their organizations. Finding better ones can be nontrivial, involving broad knowledge, organizational sophistication, and analytic talents. Sometimes, computational methods can help, but it is unwise to fall into the trap of ignoring uncertainty when doing so. Classic methods for seeking Pareto-optimal options can be generalized to deal with uncertainty (Davis et al. 2008b), something not discussed further in this chapter.

5. *Compare options with uncertainty-sensitive displays* at different levels of detail. Scorecards provide such a comparison at a high level, zooming down into more detailed scorecards can elaborate, as can separate charts and tables communicating dependence on the primary uncertain variables. When doing so, the approach favors *region charts, trade-off charts*, and other devices that preemptively address *beyond-what-if* questions. Figure 40.3, as explained below, is an example. It might be the result of running a great many simulations for points within the region plot shown and summarizing the results as shown.

6. *Later, build parameterized objective functions* to represent alternative perspectives. After decision-makers have compared options at the scorecard level and directed adjustment of options that balance their concerns appropriately, construct objective functions (frequently nonlinear) to capture their preferences and use those to *neaten up* the analytical work by employing optimization methods. The neatening may be important for translating high-level decisions into more detailed ones.

7. *Show option comparisons with alternative strategic perspectives*: Do not obfuscate important differences by, e.g. using a single-objective function as a linear weighted sum of very different ones. As an example, some relevant policy-makers may be short-term oriented, while others are long-term oriented; some policy-makers might have faith that a particular initiative would work roughly as proposed by its proponents, while others might be convinced that it would prove to be not only ineffective but also counter-productive due to side effects. Such disagreements should be highlighted, not buried.

Several tables and charts can help elaborate. Table 40.3 illustrates a fictitious *policy scorecard*, the concept for which traces back to the 1970s (Goeller et al. 1983), but the analogue of which is familiar today in, say, *Consumers*

Table 40.3 An illustrative policy scorecard.

Option	Effectiveness (% immune)	Side effects 0, very low; 10, very high (discomfort, not danger)	Acceptance by public (low, medium, high)	Cost ($B)
Do nothing	5	0	High	0
Option 1	50	2	High	3
Option 2	80	3	Medium	5
Option 3	98	5	Low	6

Report. Table 40.3 compares four options for a disease-prevention program by the effectiveness of treatment, level of side effects, public acceptance (e.g. willingness to go to the trouble and expense, endure possible side effects, etc.), and cost. Although an enthusiast for quantitative work might monetize the three criteria (the second to fourth columns) and add them up with some subjective weighting factors to get a composite index number, the decision-maker would be ill served: he or she needs to see the analysis results at this level. Option 1 might be favored because of low political risks, significant value, and low cost.

Table 40.4 shows another illustrative scorecard, this one using effectiveness in each of two scenarios as criteria and, within each, distinguishing between expected results in a nominal version of the scenario or a more stressful one (e.g. a smarter adversary or a higher-intensity storm). Another criterion (Technical risks) is whether the option can actually be implemented – an issue because some involve new technology and may encounter human obstacles (reluctance to change procedures, conflicting interests, etc.).

Table 40.4 Illustrative policy scorecard with criteria being effectiveness in alternative scenarios.

Option	Effectiveness Scenario 1 (nominal)	Scenario 1 (worse case)	Scenario 2 (nominal)	Scenario 2 (worse case)	Technical risks (0, very low; 10, very high)	Cost ($B)
Baseline	Failure	Failure	Failure	Failure	0	0
1	Success	Failure	Failure	Failure	2	5
2	Success	Failure	Success	Failure	2	10
3	Success	Uncertain	Success	Failure	3	12
4	Success	Success	Success	Success	5	10

Figure 40.3 illustrates with a simplified example going beyond-what-if analysis to show the performance of options across a state space. It is a *region plot*. Suppose the policy issue is vulnerability to a potential shock, e.g. a hurricane or epidemic. Huge uncertainties exist about how big the shock might be or how fast it might develop. Figure 40.3a characterizes how well the shock could be dealt with today, depending on its size and speed. This provides much more information than just knowing that it would possible to deal with a particular planning case (the point shown). Suppose, however, that the decision-maker is worried about events that have made bigger and faster-developing shocks possible and even likely. How should he or she think about options? The intent would be improvement of capabilities that would expand the region of good results and decrease the region of failure. An improvement option (actually, a package of actions) might achieve this as shown in Figure 40.3b. Even if the most likely planning cases were deemed to be not too bad, the improvement option would hedge against bigger-than-expected shocks moving faster than expected.

The process of assessing capabilities across the relevant space is exploratory analysis. The DoD introduced the related concept of capabilities-based planning in 2001. The intent was and is to have military forces able to address whatever challenges arise, rather than preparing only for some very specific scenarios (Rumsfeld 2002). How much can be accomplished in this regard depends on the budget available, the feasibility of meeting the challenges, and the clever use of hedging. Some of the subtleties and controversies about the approach are discussed elsewhere (Davis 2014, Appendix B).

Where do charts such as Figure 40.3 come from if one is using models, such as simulation models, to evaluate developments in a particular scenario? For many problems, such charts can be generated directly from analytical models by analogy with what economists, physicists, and chemists have done for many years. For more complex problems, such as those involving social and behavioral phenomena that might be described by simulations, a simulation run generates just a single point in scenario space. Further, it may not be clear how to display results because the model may depend on dozens or hundreds of variables. Even if one projects results onto a well-chosen set of two axes, the results may not have the clarity of Figure 40.3. Figure 40.4 illustrates what results might look like, drawing schematically from a notional analysis in which religiously inspired terrorism was modeled by analogy to an epidemic with the population having a mix of people who are naturally immune to the disease, vulnerable, or immune after recovery (Davis et al. 2007). The course of the disease (terrorism) varies with other scenario assumptions. In the figure, the results after some length of time are shown as open circles for good (the disease is dying out) to dark dots for very bad (the disease is still growing). Note that the good points and bad points can be found intermingled throughout the space.

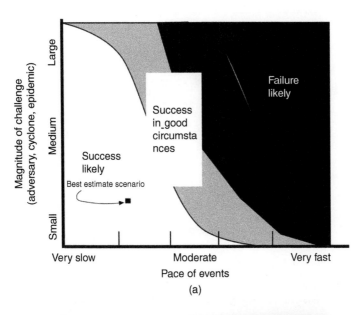

(a)

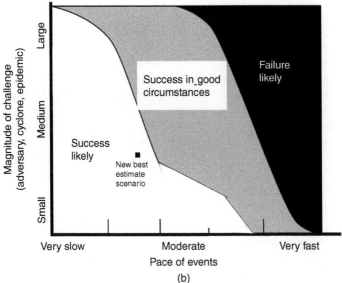

(b)

Figure 40.3 Illustrative region plot: capability to deal with potential events. (a) Current capability and (b) capability after improvement measures. A given scenario would appear as a point in this display. The scenario's outcome would be favorable, ambiguous, or negative (white, gray, or black) depending on the magnitude of shock and pace of events.

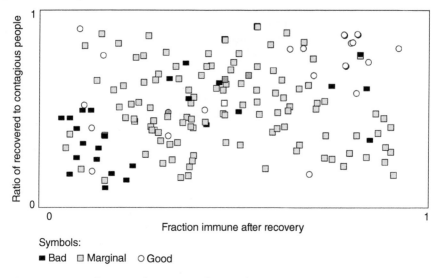

Figure 40.4 Two-dimensional projection of raw exploration outcomes. Fictitious results to illustrate point. For actual analysis results with colored points, see Davis et al. (2007).

This is because the result is a function of many variables, but the results are being projected onto only two dimensions (Davis et al. 2007).

Figure 40.5 shows the result of processing data akin to that from Figure 40.4 but from actual model runs (Davis et al. 2007) using average results in the neighborhood of a cell across the cell. A pattern is now evident, one looking more like that of simple analyses such as in Figure 40.3. The point is simply that in making sense of exploratory analysis, it is crucial to make use of computational methods akin to filtering. Fortunately, some excellent methods for doing so exist.

One more example is apt. Figure 40.6 illustrates a notional display indicating the circumstances under which a system might be suitable for policy intervention. It imagines a phenomenon, the results of which depend on the magnitude of change introduced by the intervention and the pace of that change. The

Ratio of recovered to contagious	Fraction immune after recovery									
	0	0.05	0.1	0.15	0.2	0.25	0.3	0.35	0.4	0.45
1.0	1.1	1.3	1.6	1.8	2.1	2.4	2.7	3.3	3.2	3.9
0.9	1.1	1.3	1.5	1.8	2.1	2.5	2.5	2.9	3.5	3.3
0.8	1.0	1.3	1.5	1.7	2.1	2.4	2.6	2.8	3.1	3.8
0.7	1.0	1.2	1.5	1.8	2.1	2.2	2.6	2.8	3.2	3.1
0.6	1.0	1.2	1.6	1.6	1.9	2.1	2.4	2.8	3.2	3.4
0.5	1.0	1.1	1.3	1.6	1.7	2.2	2.3	2.2	3.0	2.8
0.4	0.9	1.0	1.3	1.4	1.5	1.5	1.7	1.8	2.1	1.8
0.3	1.0	0.9	1.2	1.5	1.5	1.5	1.7	1.8	2.1	1.8
0.2	0.8	0.8	1.0	1.0	1.4	1.2	1.4	1.6	1.5	1.3
0.1	0.6	0.7	0.7	1.0	1.0	0.8	1.1	1.6	1.5	1.2

Figure 40.5 Processed results revealing pattern. Source: Davis et al. (2007).

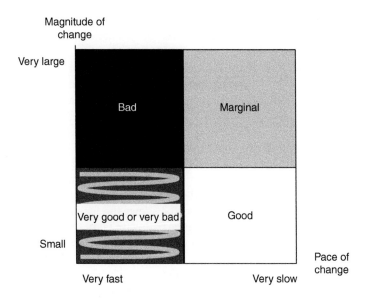

Figure 40.6 Flagging instability regions when considering interventions. Color of region indicates model-estimated consequence of intervention. Results are very uncertain where system is unstable (bottom left).

claim is that results will be good for modest changes introduced slowly (bottom right), bad for major changes introduced rapidly (top left), and unpredictable for even small changes introduced rapidly (bottom left). This type of stability chart might be much more useful for decision aiding than showing results for one or a few detailed scenarios, coupled with verbal caveats. This, then, is a tangible example of how even CAS with their notorious sensitivity to conditions can be characterized usefully in policy analysis. For the notional case, the admonition is: "Don't even think about an intervention that will implemented quickly."

Finding Adaptive Strategies by Minimizing Regret

In another overlapping strand of research at RAND called RDM, colleagues have developed what they refer to as scenario discovery methods (Lempert et al. 2006; Groves and Lempert 2007). The general approach they take is similar to the one described above in many ways, but they make greater use of computer search rather than visual inspection. This includes using the PRIM algorithm from data mining to find regions of the n-dimensional input space that lead (usually) to good outcomes and other regions that lead (usually) to bad outcomes. Their methods are part of the larger RDM toolkit on which substantial documentation exists, as well as a many completed applied studies. Links can be found at an associated website: https://www.rand.org/topics/robust-decision-making.html. Let it suffice here for me to point to a particular option in the toolkit, which is to use computer search to find decision choices that will minimize a regret function evaluated over an ensemble of potential scenarios (Lempert et al. 2006).

Other authors have employed scenario discovery methods in a variety of ways (see, e.g. (Kwakkel et al. 2013)). Somewhat related methods are used to summarize simulation studies (Parikh et al. 2016) with large numbers of agents.

Planning Adaptive Pathways

A third example of decision aiding analysis draws on work at Delft University in the Netherlands. The basic idea is to go about strategic planning in a way that allows decisions to be revisited and major changes of strategy made. The idea was discussed nearly three decades ago (Lempert and Schlesinger 2000), but the ideas have come far since then. Figure 40.7 shows the essence of the concept in a *Metro Chart* that identifies alternative pathways (akin to different rail lines in a subway system). The city or country at issue might proceed for now along Pathway B (e.g. limited broadening of a river going through a city), but – after gathering information about actual changes in sea level and updated projections – switch to another path that would accelerate mitigation programs (e.g. greatly broaden the river or add a new channel) or shift to a branch that would actually slow development. The analysts in such work emphasize that they have no illusions about politicians adhering rigorously to the particular strategy illustrated. Rather, it is quite significant if political leaders merely act upon the concept of preparing for possible shifts of pathway – creating options for the future that would otherwise not exist. Doing so has costs (R&D, contingency planning, contingent contracts), but the potential benefits are large.

For other methods, interested readers may want to examine the website of the Society for Decision Making Under Deep Uncertainty for a sense of recent work and discussion. The URL is www.deepuncertainty.org.

Implications for Social-Behavioral Modeling

Ultimately, the purpose in this chapter has been less to discuss analysis than implications for SBM if it is to support higher-level analysis to aid decision-makers. Table 40.5 summarizes admonitions that mostly follow from the discussion of decision aiding in earlier sections. The first admonition is a bit different. It urges having an explicit campaign plan for analysis and decision aiding from the start of a project. Instead of just building a model and planning to use it, such a campaign may involve bringing to bear a wide range of models, methods, tools, and information sources at different phases of the research, analysis, and interaction with decision-makers. Doing so will not happen without conscious planning and a shared understanding among contributors of how their efforts support the project (something akin

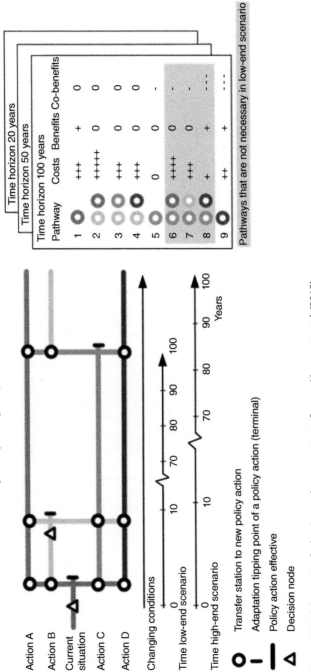

Figure 40.7 Schematic of adaptive pathways strategy. Source: Haasnoot et al. (2013).

Table 40.5 Implications for social-behavioral modeling.

Admonition for social-behavioral modelers and analysts	Comment
Begin with campaign plan for analysis and decision aiding	Plan to exploit diverse models, methods, tools, and information sources. Include interactive methods for discovery, testing, and education
Use systems approach to comprehend the whole and to frame the problems within it	See the whole and interactions of its parts. Include all relevant factors, variables, and processes
Use causal models	Understand, reason, and communicate
Recognize different classes of system state; use different models or variable-structure models	Account for changes in character of system, including emergent phenomena
Use data to test, inform, and calibrate causal models	Use theory-informed approach to data analysis, but also encourage competition to challenge existing theories and allow data-driven approaches to show their virtues
Construct multiresolution models or model families	View issues at deeper level to avoid blunders and at higher level to see simplifications
Design *from outset* for exploratory analysis of uncertainties and disagreements	Understand where intervention is and is not feasible, controllable, and valuable; understand vulnerabilities; anticipate possible problems; plan for adaptation (requires ability to explore at low resolution and to zoom as appropriate)
Generate uncertainty-sensitive outputs	Go beyond what-if questions with region plots, trade-off curves, phase diagrams, and other devices
See decision aiding to help understand *the system*, *game board*, and possible moves, not about predicting future	Move from predict-and-act paradigm to paradigm of exploring possibilities and planning for flexibility, adaptiveness, and robustness (FARness) (planning for adaptiveness for short)

to what another chapter in this volume discusses in terms of establishing a *lingua franca*) (Garibay et al. 2019). Table 40.5 illustrates how thinking about different methods and tools reminds us of their different strengths and weaknesses. The example comes from defense planning, but the points being made are generic. For example, simple models (perhaps as simple as a one-line formula) are valuable for quick and agile analysis and high-level summary discussion; they are, however, very limited in other ways. Simulation models can be quite rich and informative (but only if sufficiently intelligent, as with adaptive agents representing human behavior). They are, however, more difficult to comprehend and less suitable for illuminating strategic-level truths analytically. Human games, field experiments, and historical analysis all have their strengths.

The ratings shown in Figure 40.8 are merely illustrative with particular context in mind. In a particular study, the ratings might be quite different.

In what way are the admonitions of Table 40.5 conventional wisdom, and in what way are they provocative or otherwise nontrivial for those engaged in SBM? The answer varies because the range of current practices is so large. By and large, however (Davis and O'Mahony 2019), SBM needs to move farther toward causal system modeling and farther away from the usual correlation statistics methods of mainstream social science. Dealing with CAS implies the need to anticipate the CAS equivalent of phase transitions, either with a set of model types to use in the different domains or models that can operate across the transitions (e.g. variable-structure simulations) (Yilmaz 2019). Relating to empirical data is different when the intent is decision aiding because the goal is to test, improve, and calibrate a causal model used for reasoning and deeper inquiry, rather than to improve accuracy in the sense of fitting empirical data precisely with an algorithm from machine learning. Further, new work needs to take on the challenge of achieving multiresolution models or model families, rather than being satisfied sometimes with linking models together across scales or perhaps comparing results of models in different formalisms. Designing models for exploratory analysis and generation of beyond-what-if displays is crucial because, without designing the models suitably, such work will not easily be possible late in the game without substantial problems. As a final elaboration on Table 40.5, I note that one of the more robust conclusions from numerous examples of higher-level decision aiding has been to see model-supported decision aiding as being a process of educating and preparing decision-makers to understand their problem domain so that they can artfully construct and choose among options and later monitor progress and adjust (Rouse 2019; Thompson et al. 2019).

Acknowledgments

Several people were kind enough to read and comment on a draft version, including William Rouse, Richard Given, and Cory Lofdahl. I also benefited from earlier discussions with Steven Popper.

		Analytical					Phenomenology	
	Resolution	Agility	Crea-tivity	Trans-parency	"Rigor"	Scope	Physical	Human
Simple models (force ratios, formula models)	Low	5	2	5	5	1	1	1
Capability models (simple computer models)	Low	5	4	5	5	1	3	2
Seminar war gaming	Low	5	5	5	3	5	1	3
Strategic red teaming	Low	5	5	5	3	5	1	4
Tactical red teaming (conceptual only)	Low	5	5	5	3	1	1	3
Tactical red teaming with mockups	High	3	5	5	3	1	5	3
Ideal campaign simulation[a]	Medium	5	5	3	5	5	3	3
Actual campaign simulations	Medium	1	1	1	5	3	3	1
Large human gaming (scores of people)	Medium	1	3	3	1	5	3	3
Detailed models (mission level)	Medium	3	3	3	3	3	3	3
Detailed models (weapons, forces)	High	1	1	1	5	1	5	1
Historical aggregate quantitative analysis	Low	1	1	3	5	1	1	1
Historical quantitative/case-study analysis	Medium	5	5	3	5	1	5	5
Field experiments (weapons systems)	High	1	3	3	5	1	5	1
Historical research (cases, biographical, etc.)	High	1	1	3	3	5	5	5
Field experiments (command control)	High	1	3	3	5	3	5	5

[a]Aspirational but rare with political and military adaptive agents, multiresolution mode, exploratory analysis, and optional human play at key positions.

Figure 40.8 Different tools for different strengths. Source: Davis (2014).

References

Bankes, S.C. (2002). Tools and techniques for developing policies for complex and uncertain systems. *Proceedings of the National Academy of Sciences, Colloquium* 99 (Suppl. 3): 7263–7266.

Boulding, K. (1956). General systems theory – the skeleton of science. *Management Science* 2 (3): 197–208.

Bracken, P., Bremer, I., and Gordon, D. (eds.) (2008). *Managing Strategic Surprise*. New York: Cambridge University Press.

Checkland, P. (1999). *Systems Thinking, Systems Practice: Includes a 30-Year Retrospective*. Chichester: Wiley.

Churchman, C. (1967). Wicked problems. *Management Science* 14 (4): B-141.

Davis, P.K. (1988). The Role of Uncertainty in Assessing the NATO/Pact Central-Region Balance. *N-2839-RC*.

Davis, P.K. (1994). Institutionalizing planning for adaptiveness. In: *New Challenges in Defense Planning: Rethinking How Much Is Enough* (ed. P.K. Davis), 73–100. Santa Monica, CA: RAND Corporation.

Davis, P.K. (ed.) (2011). *Dilemmas of Intervention: Social Science for Stabilization and Reconstruction*. Santa Monica, CA: RAND Corporation.

Davis, P.K. (2014). *Analysis to Inform Defense Planning Despite Austerity*. Santa Monica, CA: RAND Corporation.

Davis, P.K. and Blumenthal, D. (1991). The Base of Sand: A White Paper on the State of Military Combat Modeling. *N-3148-OSD/DARPA*.

Davis, P.K. and Cragin, K. (eds.) (2009). *Social Science for Counterterrorism: Putting the Pieces Together*. Santa Monica, CA: RAND Corporation.

Davis, P.K. and Hillestad, R. (1993). Families of models that cross levels of resolution: issues for design, calibration, and management. Proceedings from Proceedings of the 1993 Winter Simulation Conference, San Diego, CA.

Davis, P.K. and O'Mahony, A. (2013). A Computational Model of Public Support for Insurgency and Terrorism: A Prototype for More General Social-Science Modeling. *TR-1220*.

Davis, P.K. and O'Mahony, A. (2017). Representing qualitative social science in computational models to aid reasoning under uncertainty: national security examples. *Journal of Defense Modeling and Simulation* 14 (1): 1–22.

Davis, P.K. and O'Mahony, A. (2019, this volume). Improving social-behavioral modeling. In: *Social-Behavioral Modeling for Complex Systems* (ed. P.K. Davis, A. O'Mahony and J. Pfautz). Hoboken, NJ: Wiley.

Davis, P.K., Bankes, S.C., and Egner, M. (2007). Enhancing Strategic Planning with Massive Scenario Generation: Theory and Experiments. *TR-392-OSD*.

Davis, P.K., Shaver, R.D., and Beck, J. (2008a). *Portfolio-Analysis Methods for Assessing Capability Options* (August 11, Trans. MG-662-OSD). Santa Monica, CA: RAND Corporation.

Davis, P.K., Shaver, R.D., Gvineria, G., and Beck, J. (2008b). Finding Candidate Options for Investment Analysis: A Tool for Moving from Building Blocks to Composite Options (BCOT). *TR-501-OSD.*

Davis, P.K., Larson, E., Haldeman, Z. et al. (2012). *Understanding and Influencing Public Support for Insurgency and Terrorism* Santa Monica, CA: RAND Corporation.

Davis, P.K., O'Mahony, A., Gulden, T. et al. (2018). Priority Challenges for Social-Behavioral Research and its Modeling. Santa Monica, CA: RAND Corporation.

Dörner, D. (1997). *The Logic of Failure: Recognizing and Avoiding Errors in Complex Situations* (ed. trans.: R. Kimber and R. Kimber), Trans. paperback ed.). Cambridge, MA: Perseus Books.

Enthoven, A. and Smith, K.W. (2005). *How Much is Enough: Shaping the Defense Program, 1961–1969*, 2e (with a new introduction by K.J. Krieg and D.S.C. Chu). Santa Monica, CA: RAND Corporation.

Forrester, J. (1963). *Industrial Dynamics.* Cambridge, MA: MIT Press. ISBN: 0262560011.

Fox, D. (2003). Using exploratory modeling. In: *New Challenges, New Tools for Defense Decisionmaking* (ed. S. Johnson, M. Libicki and G.F. Treverton), 258–298. Santa Monica, CA: RAND Corporation.

Garibay, I., Gunaratne, C., Yousefi, N., and Schinert, S. (2019, this volume). The agent-based modeling canvas: a modeling lingua franca for computational social science. In: *Social-Behavioral Modeling for Complex Systems* (ed. P.K. Davis, A. O'Mahony and J. Pfautz). Hoboken, NJ: Wiley.

Goeller, B.F., Abraham, S.C., Abrahamse, A., et al. (1983). Policy Analysis of Water Management for the Netherlands. Summary Report. Santa Monica, CA: RAND Corporation.

Groves, D.G. and Lempert, R.J. (2007). A new analytic method for finding policy-relevant scenarios. *Global Environmental Change* 17 (1): 78–85.

Haasnoot, M., Kwakkel, J.H., Walker, W.E., and Maat, J.T. (2013). Dynamic adaptive policy pathways: a method for crafting robust decisions for a deeply uncertain world. *Global Environmental Change* 23 (2): 485–498.

Kwakkel, J.H., Auping, W.L., and Pruyt, E. (2013). Dynamic scenario discovery under deep uncertainty: the future of copper. *Technological Forecasting and Social Change* 80 (4): 789–800.

Lempert, R.J. and Schlesinger, M.E. (2000). Robust strategies for abating climate change. *Climatic Change* 45 (3–4): 387–401.

Lempert, R.J., Popper, S.W., and Bankes, S.C. (2003). *Shaping the Next One Hundred Years: New Methods for Quantitative Long-Term Policy Analysis* (August 8, Trans.). Santa Monica, CA: RAND Corporation.

Lempert, R.J., Groves, D.G., Popper, S.W., and Bankes, S.C. (2006). A general analytic method for generating robust strategies and narrative scenarios. *Management Science* 4: 514–528.

Lindblom, C.E. (1965). *The Intelligence of Democracy: Decision Making Through Mutual Adjustment*. New York: Free Press.

Lindblom, C.E. (1979). Still muddling, not yet through. *Public Administration Review* 39 (6): 222–233.

Lofdahl, C. (2019, this volume). Modeling information and gray zone operations. In: *Social-Behavioral Modeling for Complex Systems* (ed. P.K. Davis, A. O'Mahony and J. Pfautz). Hoboken, NJ: Wiley.

March, J.G. (1996). Continuity and change in theories of organizational action. *Administrative Science Quarterly* 41 (2): 278.

Markowitz, H.M. (1952). Portfolio selection. *Journal of Finance* 7: 77–91.

Morgan, P.M. (2003). *Deterrence Now*. Cambridge NY: University Press.

Morgan, M.G. and Henrion, M. (1992). *Uncertainty: A Guide to Dealing With Uncertainty in Quantitative Risk and Policy Analysis*. New York: Cambridge University Press.

Nagl, J.A., Petraeus, D., Amos, D., and Seall, S. (2007). *U.S. Army/Marine Corps Counterinsurgency Field Manual*. Chicago, IL: University of Chicago Press.

National Research Council (2014). *U.S. Air Force Strategic Deterrence Analytic Capabilities: An Assessment of Methods, Tools, and Approaches for the 21st Century Security Environment* Washington, DC: National Academies Press.

Orr, M.G. (2019, this volume). Multi-scale resolution of human social systems: a synergistic paradigm for simulating minds and society. In: *Social-Behavioral Modeling for Complex Systems* (ed. P.K. Davis, A. O'Mahony and J. Pfautz). Hoboken, NJ: Wiley.

Parikh, N., Marathe, M., and Swarup, S. (2016). Summarizing simulation results using causally relevant states. Proceedings from International Conference on Autonomous Agents and Multiagent Systems (AAMAS 2016): Autonomous Agents and Multiagent Systems.

Pierson, B., Barge, W., and Crane, C. (2008). The hairball that stabilized Iraq: modeling FM 3-24. Proceedings from The Human Social Cultural Modeling Workshop, Boca Raton, FL, Washington, DC.

Quade, E.S. and Boucher, W.I. (eds.) (1968). *Systems Analysis and Policy Planning: Applications for Defense*. New York: Elsevier Science Publishers.

Quade, E.S. and Carter, G.M. (eds.) (1989). *Analysis for Public Decisions*, 3e. New York: North Holland Publishing Company.

Quinn, J.B. (1978). Strategic change: logical incrementalism. *Sloan Management Review* 20: 7–21.

Rittel, H.W.J. and Webber, M.M. (1973). Dilemmas in a general theory of planning. *Policy Sciences*.

Rosenhead, J. and Mingers, J. (eds.) (2002). *Rational Analysis for a Problematic World Revisited: Problem Structuring Methods for Complexity, Uncertainty and Conflict*, 2e. New York: Wiley.

Rouse, W.B. (2019, this volume). Human-centered design of model-based decision support for policy and investment decisions. In: *Social-Behavioral Modeling for*

Complex Systems (ed. P.K. Davis, A. O'Mahony and J. Pfautz). Hoboken, NJ: Wiley.

Rumsfeld, D. (2002). Transforming the military. *Foreign Affairs* 81 (3): 20–32, 23.

Senge, P.M. (2006). *The Fifth Discipline: The Art & Practice of the Learning Organization*. New York: Penguin Random House.

Simon, H.A. (1996). *The Sciences of the Artificial*, 3e. Cambridge, MA: The MIT Press.

Sterman, J.D. (2000). *Business Dynamics: Systems Thinking and Modeling for a Complex World*. Boston, MA: McGraw-Hill. ISBN: 007238915X.

Thompson, J., McClure, R., and DeSilva, A. (2019, this volume). A complex systems approach for understanding the effect of policy and management interventions on health system performance. In: *Social-Behavioral Modeling for Complex Systems* (ed. P.K. Davis, A. O'Mahony and J. Pfautz). Hoboken, NJ: Wiley.

von Bertalanffy, L. (1951). Problems of general system theory. *Human biology* 23 (4): 302.

von Bertalanffy, L. (1969). *General System Theory: Foundations, Development, Applications (Revised Edition) (Penguin University Books) (Revised ed.)*. George Braziller Inc.

Walker, W.E., Builder, C.H., Draper, D. et al. (1987). Important Hard Problems in Public Policy Analysis. *P-7282*. Santa Monica, CA: RAND Corporation.

Yilmaz, L. (2019, this volume). Toward self-aware models as cognitive adaptive instruments for social and behavioral modeling. In: *Social-Behavioral Modeling for Complex Systems* (ed. P.K. Davis, A. O'Mahony and J. Pfautz). Hoboken, NJ: Wiley.

Zacharias, G.L., MacMillan, J., and Van Hemel, S.B. (eds.) (2008). *Behavioral Modeling and Simulation: From Individuals to Societies*. Washington, DC: National Academies Press.

Index

Social-Behavioral Modeling for Complex Systems, First Edition.
Edited by Paul K. Davis, Angela O'Mahony, and Jonathan Pfautz.
© 2019 John Wiley & Sons, Inc. Published 2019 by John Wiley & Sons, Inc.
Companion website: www.wiley.com/go/Davis_Social-Behavioralmodeling